Encyclopedia of
Plant Physiology

New Series Volume 12 B

AF792933

Editors

A. Pirson, Göttingen
M. H. Zimmermann, Harvard

Physiological Plant Ecology II

Water Relations and Carbon Assimilation

Edited by

O.L. Lange P.S. Nobel C.B. Osmond H. Ziegler

Contributors

J.D. Bewley K.J. Bradford S. von Caemmerer I.R. Cowan
R.M.M. Crawford H.A. Douthit Jr. G.D. Farquhar A. Hadas
A.E. Hall T.C. Hsiao P.G. Jarvis D. Koller J. Krochko
J.A. Milburn C.B. Osmond J.B. Passioura P.W. Rundel
J. Schönherr E.-D. Schulze A.S. Sussman W. Tranquillini
M.T. Tyree P.E. Weatherley K. Winter H. Ziegler
M.H. Zimmermann

With 153 Figures

Springer-Verlag Berlin Heidelberg New York 1982

Professor Dr. O.L. LANGE
Lehrstuhl für Botanik II der Universität Würzburg
Mittlerer Dallenbergweg 64, 8700 Würzburg/FRG

Professor P.S. NOBEL
Department of Biology
Division of Environmental Biology
of the Laboratory of Biomedical and Environmental Sciences
University of California
Los Angeles, California 90024/USA

Professor C.B. OSMOND
Department of Environmental Biology
Research School of Biological Sciences
Australian National University
Box 475, Canberra City 2601/Australia

Professor Dr. H. ZIEGLER
Institut für Botanik und Mikrobiologie
der Technischen Universität München
Arcisstraße 21, 8000 München 2/FRG

ISBN-13: 978-3-642-68152-3 e-ISBN-13: 978-3-642-68150-9
DOI: 10.1007/978-3-642-68150-9

Library of Congress Cataloging in Publication Data. Main entry under title: Physiological plant ecology II. (Encyclopedia of plant physiology; new ser., v. 12B) Bibliography: p. Includes index. 1. Plant-water relationships. 2. Plants, Effect of carbon dioxide on. 3. Plants–Assimilation. 4. Botany–Ecology. I. Lange, O.L. (Otto Ludwig) II. Title: Carbon assimilation. III. Series. QK711.2.E5 new ser., vol. 12B 581.1s 82-10342 [QK870] [581.1].

Softcover reprint of the hardcover 1st edition 1982

List of Contributors

J.D. BEWLEY
Department of Biology
University of Calgary
Calgary, Alberta T2N 1N4/Canada

K.J. BRADFORD
Department of Vegetable Crops
University of California
Davis, California 95616/USA

S. VON CAEMMERER
Department of Environmental Biology
Research School of Biological Sciences
Australian National University
Box 475, Canberra City A.C.T. 2601/Australia
Present address:
Plant Biology Laboratory
Carnegie Institution of Washington
290 Panama Street
Stanford, California 94305/USA

I.R. COWAN
Department of Environmental Biology
Research School of Biological Sciences
Australian National University
Box 475, Canberra City A.C.T. 2601/Australia

R.M.M. CRAWFORD
Department of Botany
The University
St. Andrews
Fife KY16 9AL/United Kingdom

H.A. DOUTHIT, Jr.
Division of Biology
University of Michigan
Ann Arbor, Michigan 48109/USA

G.D. FARQUHAR
Department of Environmental Biology
Research School of Biological Sciences
Australian National University
Box 475, Canberra City A.C.T. 2601/Australia

A. HADAS
Institute of Soils and Water
Agricultural Research Organization
Volcani Center
P.O. Box 6
Bet Dagan 50250/Israel

A.E. HALL
Department of Botany and Plant Sciences
University of California
Riverside, California 92521/USA

T.C. HSIAO
Department of Land, Air and Water Resources
University of California
Davis, California 95616/USA

P.G. JARVIS
Department of Forestry and Natural Resources
University of Edinburgh
King's Buildings
Edinburgh EH9 3JU/United Kingdom

D. KOLLER
The Hebrew University of Jerusalem
Department of Agricultural Botany
Rehovot/Israel

J. KROCHKO
Department of Biology
University of Calgary
Calgary, Alta T2N 1N4/Canada

J.A. MILBURN
Department of Botany
University of New England
Armidale, N.S.W. 2351/Australia

C.B. Osmond
Department of Environmental Biology
Research School of Biological Sciences
Australian National University
Box 475, Canberra City 2601/Australia

J.B. Passioura
CSIRO Division of Plant Industry
P.O. Box 1600
Canberra City, A.C.T. 2601/Australia

P.W. Rundel
Department of Ecology and Evolutionary Biology
University of California
Irvine, California 92717/USA

J. Schönherr
Institut für Botanik und Mikrobiologie
der Technischen Universität München
Arcisstraße 21
D-8000 München 2/FRG

E.-D. Schulze
Lehrstuhl Pflanzenökologie
der Universität Bayreuth
Universitätsstraße 30
D-8580 Bayreuth/FRG

A.S. Sussman
Horace H. Rackham School of Graduate Studies and Division of Biology
University of Michigan
Ann Arbor, Michigan 48109/USA

W. Tranquillini
Forstliche Bundesversuchsanstalt
Außenstelle für subalpine Waldforschung
Rennweg 1
A-6020 Innsbruck/Austria

M.T. Tyree
Department of Botany
University of Toronto
Toronto M5S 1A1/Canada

P.E. Weatherley
Department of Botany
University of Aberdeen
St. Machar Drive
Aberdeen AB9 2UD/United Kingdom

K. Winter
Lehrstuhl für Botanik II
der Universität Würzburg
Mittlerer Dallenbergweg 64
D-8700 Würzburg/FRG

H. Ziegler
Institut für Botanik und Mikrobiologie
der Technischen Universität München
Arcisstraße 21
D-8000 München 2/FRG

M.H. Zimmermann
Harvard University
Harvard Forest
Petersham, MA 01366/USA

Contents Part B

Contents Part A, C and D

Part A: Physiological Plant Ecology I
Responses to the Physical Environment

Part C: Physiological Plant Ecology III
Responses to the Chemical and Biological Environment

Part D: Physiological Plant Ecology IV
Ecosystem Processes: Mineral Cycling, Productivity and Man's Influence

Introduction

O.L. LANGE, P.S. NOBEL, C.B. OSMOND, and H. ZIEGLER

In the original series of the Encyclopedia of Plant Physiology, plant water relations and photosynthesis were treated separately, and the connection between phenomena was only considered in special chapters. O. STOCKER edited Volume III, *Pflanze und Wasser/Water Relations of Plants* in 1956, and 4 years later, Volume V, Parts 1 and 2, *Die CO_2-Assimilation/The Assimilation of Carbon Dioxide* appeared, edited by A. PIRSON. Until recently, there has also been a tendency to cover these aspects of plant physiology separately in most textbooks. Without doubt, this separation is justifiable. If one is specifically interested, for example in photosynthetic electron transport, in details of photophosphorylation, or in carbon metabolism in the Calvin cycle, it is not necessary to ask how these processes relate to the water relations of the plant. Accordingly, this separate coverage has been maintained in the New Series of the Encyclopedia of Plant Physiology. The two volumes devoted exclusively to photosynthesis are Volume 5, *Photosynthesis I,* edited by A. TREBST and M. AVRON, and Volume 6, *Photosynthesis II,* edited by M. GIBBS and E. LATZKO. When considering carbon assimilation and plant water relations from an ecological point of view, however, we have to recognize that this separation is arbitrary. Since growth is essential to plant survival, the most crucial problem facing terrestrial plants is the balancing of the need to minimize plant water loss against the need to assimilate sufficient carbon for growth. In response to this dilemma, plants have evolved a close link between water status and the CO_2 assimilation processes. This linkage has to be taken into consideration in any ecological interpretation of these phenomena. Because a precise understanding of the manner in which the link is accomplished might lead to improvement in plant productivity, an integrative approach to photosynthesis and plant water relations is also of particular relevance in agronomy.

Until fairly recently, experimental approaches to water relations and photosynthetic processes in plants, respectively, were very different, making a combined analysis of the total water-photosynthesis syndrome difficult. Only during the last three decades have methods become available which allow adequate ecophysiological studies of both aspects of plant performance to be carried out. Whereas the discovery of the radioactive isotope carbon-14 permitted enormous advances in studies of carbon assimilation, the availability of tritium at about the same time had scarcely any impact on plant water relation investigations. Indeed, the combined tritium/carbon-14 porometer is a relatively recent invention; and it was only a few years earlier that the use of infrared gas analyzers in parallel with dewpoint hygrometers permitted rapid advances in the simultaneous measurement of CO_2 and H_2O exchange. The recognition among ecophysiologists of the need for a more unified approach to the study of photosynthesis and plant water relations has increasingly led to the use of the newer techniques for measuring gas exchange in conjunction with methods for investigating plant water status, i.e., pressure chamber and psychrometric methods. Such investigations have contributed immensely to our understanding of plant photosynthetic response in relation to plant water economy, and have made possible the concept of the present volume, namely, the integrated treatment of photosynthesis and water relations within an ecophysiological framework.

With the exception of submerged freshwater plants which comprise a trivial portion of plant biomass and production, photosynthetic activity shows marked responses to the water potential of the plant environment. In terrestrial plants, which comprise about 70% of plant biomass, water relations processes are integrated to maintain leaf water potential within acceptable limits. Generally speaking these limits are set by the requirements that living roots usually have access to water in the soil at water potentials in the range of about 0 to −30 bar, while the living leaf is exposed to the atmosphere with an equivalent water potential of −500 to −2000 bar. The dynamics of plant water relations (Chap. 1, this Vol.) are such as to ensure water potentials of functional leaves within the range of about −2 to −50 bar. However, photosynthesis and other metabolic functions over this range of water potentials are only possible if the osmotic potential of vacuolar sap is sufficient to maintain a positive turgor pressure in the cytoplasm of about 1 to 5 bar (Chap. 2). Although direct methods for the measurement of this important parameter in cell water relations have been devised, its application and integration in complex tissues remains a problem.

The principles of water uptake by roots and of water transport through the plant are well established (Chaps. 3 and 5) but the study of key elements, such as water transfer from soil to root, and across the root, remains difficult. The complexity of the soil environment and the difficulty of differentiating symplastic and apoplastic pathways in the root continue to try our ingenuity. Maintenance of leaf water potential and turgor within acceptable limits is largely due to the formation of cuticular layers at the plant-atmosphere interface (Chap. 6) and consequently due to stomatal control of water fluxes across the large water potential gradient between leaf mesophyll cell walls and the atmosphere. Since the physiological bases of stomatal function have been reviewed in Volume 7 of this series, our treatment is confined to ecologically significant stomatal responses (Chap. 8) and their function in photosynthesis and growth.

Although stomatal closure to preserve leaf water status often occurs, it is inevitably associated with a decrease in intercellular CO_2 partial pressure, which often results in a reduction in photosynthesis and growth (Chap. 7). Furthermore, if stomatal closure and cuticular resistance are insufficient to prevent excess water loss, then photosynthetic and other processes are reduced as a result of water stress, or even damaged due to serious dehydration (Chaps. 9 and 11). The significance of the compromise that stomata effect between the hunger of photosynthesis and the thirst of transpiration has been highlighted by comparative studies of plants with different pathways of CO_2 fixation (Chap. 15). Among other things, the differences in the biochemistry of C_3

and C_4 plants allow C_4 plants to achieve equal or greater rates of CO_2 fixation than C_3 plants with more tightly closed stomata, at much lower intercellular CO_2 partial pressures, with consequent lower water costs of CO_2 fixation. In spite of this, both photosynthetic types display midday depression of photosynthesis due to stomatal closure in response to atmospheric and soil water stress. The possible ecological significance of these differences in photosynthetic pathway and water use pattern has been appreciated for some time, but only recently have we seen comparative evaluations of these processes in terms of plant performance (productivity) and survival (distribution).

In succulent plants with CAM the above compromise is extreme; extraordinarily low stomatal and cuticular conductance can sometimes prevent CO_2 exchange altogether, and confer drought avoidance at the expense of photosynthesis and growth. These plants endure drought as functional photosynthetic systems at high water potential. By contrast, some poikilohydric thallophytes and cormophytes endure drought as nonfunctional photosynthetic systems at low water potential, but these systems are readily reconstituted on rehydration (Chap. 10). Aside from their intrinsic ecological interest, it is likely that these systems will provide fundamental insights into the integration of CO_2 fixation and photochemistry in photosynthesis. In ecological terms they are fascinating mechanisms to preserve the advantages of autotrophy, and to avoid the hazards of seed or spore germination (Chaps. 12 and 13) and seedling establishment. There are interesting analogies between germination and the reconstitution of metabolism in poikilohydric plants, and there are virtues in establishing some principles of water transport in these nonvacuolated quiescent systems. Similar processes are presumably involved in water uptake by plant organs other than roots (Chap. 4). On the other hand, in some cases excessive water (e.g., flooding) has also led to functional adaptations of certain organs (roots) of terrestrial plants. Through special metabolic pathways they are able to avoid generation of poisonous products which might occur due to the anaerobic situation (Chap. 14).

An increasing body of evidence relating daily and seasonal patterns of transpiration and carbon uptake has stimulated the development of the theory that stomata function so as to minimize the water cost of carbon gain (Chap. 17). These notions lead to provocative interpretations of transpiration patterns and the biochemistry of photosynthetic carbon uptake (Chaps. 8 and 16). Extension of the concept of optimal stomatal behavior to embrace water use patterns during growth and development, will undoubtedly provide a framework within which long-held notions of plant form and function (Chap. 18) can be tested. Progress in these areas is likely to be limited more by our ability to digest and distil data than by the acquisition of data.

Although the infrared gas analyzer and the pressure chamber facilitated this explosion in ecological research, new technologies already promise integration. For example, the serendipitous observation that C_3 and C_4 plants differ in their discrimination against carbon-13 has made the ratio mass-spectrometer indispensable to studies of photosynthetic metabolism on a geographical scale. Measurements of $\delta^{13}C$ values also provide integrated historical accounts of composition and competition between C_3 and C_4 plants in communities, as well as CO_2 fixation processes and water use patterns in succulent plants. Further refinements in our understanding of these processes now suggest that the $\delta^{13}C$ value can indicate the average intercellular CO_2 partial pressure in leaves of C_3 and C_4 plants and thereby permit inferences of stomatal behavior and water status in response to environmental water stress.

We are confident that the contributions to this volume, and the continued development of reductionist, as well as integrative technologies, will have a significant impact on physiological plant ecology.

1 Water in the Soil-Plant-Atmosphere Continuum

J.B. Passioura

"Assimilation of carbon dioxide is the *sine qua non* of competitive success in green plants. All other aspects of physiological functioning are ancillary to it. Even the production of seed may be thought of as a strategy to establish new centres of carbon fixation and, in that general sense, the associated metabolic cost is no different in kind from that expended, say, by a tree in maintaining its foliage in the sun" (Cowan 1978).

CONTENTS

1.1 Introduction

When higher plants assimilate carbon dioxide they inevitably lose water. Evolution has provided these plants with facilities both to control and to make good this loss, so as to maintain adequate hydration of their tissues. The development of roots and a vascular system enabled plants to invade the land by ensuring a supply of water to their shoots; the development of a cuticle punctured with stomata enabled them to conserve water when the supply from their roots was outstripped by the evaporative demand.

VAN DEN HONERT'S (1948) classic analysis of the resistances in the transpiration stream showed the pivotal importance of the stomata in controlling water loss. Stomatal behaviour has consequently dominated the study of plant water relations. But a resistance analogue such as VAN DEN HONERT'S (1948) gives us an essentially instantaneous view of the behaviour of a plant. More sophisticated treatments add capacitors (e.g., COWAN 1972a) and thus allow us to study time-dependent behaviour, but even these rely on an instantaneous view of the main structure of a plant, and the phenomena dealt with have time scales no greater than one day. These short-term phenomena tell us a great deal about a plant, but if our interest is in photosynthetic productivity and its relationship to ecological processes, we must ultimately concern ourselves with phenomena occurring at an ontogenetic or even demographic time-scale. The behaviour of a plant in the short term may tell us little about its prospects for long-term productivity. For example, a plant that photosynthesizes rapidly and hence appears to be productive, will die if its concomitant prodigal transpiration exhausts a limited supply of water before the next rain; a conservative plant, in these circumstances, may be the more productive.

Water deficiency influences the behaviour of a plant at all levels of organization: metabolism, physiology and gross morphology are all affected. Of these three it is probably the physiology of the droughted plant that has received the most attention, particularly where it has been concerned with the behaviour of stomata. Yet it is the control of leaf area and morphology that is often the most powerful means a plant has of influencing its water economy, and ultimately its productivity, if it is subjected to drought in its natural environment: To control transpiration is to control the amount of incoming energy that is dissipated as latent heat. A plant that maintains a large leaf area index can control this amount by varying its stomatal conductance, and can conserve water during a drought by closing its stomata. But this may be an unsatisfactory solution if it means that the plant, though still respiring, spends a long time with a large area of hot leaves slowly losing water through their cuticles while

Table 1.1. Physiological processes affecting plant water relations

Process	Time scale
Root growth	Days, weeks
Leaf growth (area, thickness)	Days, weeks
Leaf shedding or senescing	Days, weeks
Changes in hairiness	Days
Conditioning (adaptation at cellular level)	Days
Changes in albedo	Hours, days
Leaf movement (photonasty, rolling)	Hours
Wilting	Hours
Changes in hydraulic resistance	Minutes, hours
Stomatal movement	Minutes, hours

Table 1.2. Environmental processes affecting plant water relations

Process	Time scale
Seasonal evaporative demand	Weeks, months
Run-down in soil water	Days, weeks
Diurnal evaporative demand	Hours
Rain, irrigation	Seconds, hours
Movement of clouds and other shadows	Seconds, hours

barely photosynthesizing. Such a plant is likely to be inefficient in its use of water. A better solution may be to control leaf area.

The point of this preamble is that, when considering photosynthetic productivity in relation to water use, it is necessary to consider processes occurring at a wide range of time scales, not only within the plant, but also within the plant's environment. Tables 1.1 and 1.2, adapted from PASSIOURA (1976), list processes occurring both in the plant and in its environment that affect plant water relations. This introductory chapter will deal with some of the interactions between these processes insofar as they affect photosynthetic productivity. It will firstly cover what is meant by the water status of a plant, then how the water status is influenced by short-term processes, i.e., those we normally think of in relation to the resistance analogue of transpiration, and finally how the long-term processes influence the water relations and photosynthetic productivity.

1.2 Water Status

1.2.1 Definitions

"Water status" is a vague though nevertheless useful term. It has no units, for it can refer to any measure of the general state of a plant in relation to water, and is therefore used only in a relative sense. It may refer to the plant's water potential, its relative water content, or its turgor. In general, it is perhaps best used to qualitatively describe turgor or relative water content rather than water potential, for a given plant could be wilting badly at a moderate leaf water potential while another could be turgid and growing well at a substantially lower water potential. One would normally think of the latter as having the higher water status.

In an attempt to avoid vagueness in describing the energy status of plant water, SLATYER and TAYLOR (1960) introduced thermodynamics to plant physiology, and provided a valuable service in removing many unsatisfactory features of the existing terminology. They showed how terms like "diffusion pressure deficit" were related to chemical potential, and they showed (TAYLOR and SLATYER 1962) how force fields, such as gravity, could be coherently taken into account. Their treatment, and especially the subsequent development of it (NOY-

MEIR and GINZBURG 1967; SLATYER 1967) is now very widely accepted. Unfortunately, however, it has led to two areas of confusion in the literature.

The first of these is due to the plant physiologists' preference to think of the energy status of water in "mechanical" rather than in "thermodynamic" terms, i.e., to think in terms of pressure rather than thermodynamic potential. TAYLOR and SLATYER (1962) originally defined water potential, Ψ, as $\Delta\mu_w$, i.e., the difference in chemical potential of water between the test system and the standard state, which is usually pure, free water at atmospheric pressure. This definition evidently proved to be unpopular, for it was subsequently changed by dividing $\Delta\mu_w$, which has units of J/mol, by the partial molar volume of water $\bar{V}_w$, which has units of m^3/mol, to give $\Psi = \Delta\mu_w/\bar{V}_w$, which has units of pressure, i.e., J/m^3. This concept of water potential is unsatisfactory on theoretical grounds for the reasons outlined by OERTLI (1969), namely that, while the definition of equilibrium in an isothermal system uninfluenced by external force fields is that μ_w is constant throughout, $\bar{V}_w$ is not necessarily constant, so that Ψ may vary. In practice the variation in $\bar{V}_w$ is small, provided we concern ourselves only with the liquid phase, and is much less than the errors associated with the measurement of Ψ. As our experiments become more precise, the use of Ψ must eventually cause some embarrassment; indeed ZIMMERMANN and STEUDLE (1979) think that it does so already. Nevertheless, there is much to be said for retaining the mechanical view of the energy status of plant water: the most widespread instrument for measuring this status is the pressure chamber; the concept of osmotic pressure is clear, familiar, and firmly established; and the concept of turgor pressure is much more vivid to those untrained in thermodynamics than is "pressure potential". In normal experimental work the only difficulty arises when we are measuring the energy status with a psychrometer; it is then that we have to adapt the slightly shaky definition of Ψ as $\Delta\mu_w/\bar{V}_w$. But if we are to retain the mechanical view it seems to me essential that we drop the use of the term "potential" unless we actually mean thermodynamic potential. The recent fashion for changing the sign of osmotic pressure and calling it "osmotic potential" and for calling pressure "pressure potential" results from erroneous pedantry and should be abandoned. The use of pressure potential in this way is particularly grotesque, for the true (i.e., thermodynamic) pressure potential, Ψ_p, has a definite relation to pressure, P, namely, $\Psi_p = \int_0^P \bar{V}_w \, dP$ (BOLT et al. 1976). Having said this, I now have to retract slightly, for the use of Ψ to represent plant water "potential" is so firmly entrenched that there seems to be no alternative for the present but to continue to use it, as DAINTY (1976) recommends. For the components of Ψ, however, it is easy to avoid the use of the term "potential", as in the following discussion.

The second problem with the standard thermodynamic treatment is in its splitting up of Ψ into components to produce the familiar equation:

$$\Psi = P - \Pi - \tau \tag{1.1}$$

where P is hydrostatic pressure, Π is osmotic pressure, and τ is matric suction. Now the splitting of Ψ into components is essential if we are to understand the water relations of a plant, whose performance is more likely to be related to its turgor or the osmotic pressure of its cell sap than to its water potential (HSIAO et al. 1976). The problem with Eq. (1.1) is that τ, which is supposed to account for the influence of solids on Ψ, arises because of an inconsistent definition of pressure. It is zero if P is defined, consistently, as the hydrostatic pressure in the liquid phase (no matter whether this be in the vacuole, the cytoplasm, or the apoplast) and if the constrained counterions that accompany any charged solid be accounted for under Π. τ is zero in the equation not because solids have no influence on Ψ, but because any such influence is implicitly accounted for by P and Π. If we wish to account explicitly for the influence of a solid, we must define τ (as it is in soil physics, where it was first used) as the difference in Ψ between the test system and its equilibrium dialysate when both are at the same height and temperature and are subject to the same ambient pressure.

Thus, we must replace Eq. (1.1) with:

$$\Psi = P - \Pi = P - \Pi_\tau - \Pi_D = -\tau - \Pi_D \qquad (1.2)$$

where Π_D is the osmotic pressure of the equilibrium dialysate of our system, and Π_τ is the effective osmotic pressure due to the constrained counterions that accompany any charged solid. These arguments are discussed in detail by PASSIOURA (1980b).

An alternative scheme for splitting Ψ into components has been proposed by SPANNER (1973), who divides Ψ into an "energetic" term, which combines the influences of pressure and internal energy, and an "entropic" term, which primarily covers osmotic influences. There are two major difficulties with this proposal. The first is that the terms are virtually meaningless to almost all plant physiologists, and unnecessarily obscure the extremely useful concept of pressure. The second is that while these terms may have meaning for a single phase, they have no useful meaning for a multi-phase system such as plant tissue. For example, the entropic term, which is $T(\partial\Psi/\partial T)_P$, and thus describes the dependence of Ψ on temperature, will in general be different in, say, the vacuole from what it is in the cell wall; $(\partial\Psi/\partial T)_P$ measured on a tissue will therefore be some sort of weighted average of the values in the different compartments. As WEATHERLEY (1970) has pointed out, such weighted averages are devoid of physiological meaning.

Many authors include gravitational potential, Ψ_z, as a component of Ψ. However, as Ψ is defined in terms of μ_w (which ignores external force fields), it is better to use the term *total* water potential, Φ $(=\Psi+\Psi_z)$, if we wish to consider the effects of gravity (SLATYER 1967). Except in tall trees, it is usually safe to ignore Ψ_z.

1.2.2 Importance of Water Status

The short-term productivity of a plant depends on its maintaining its photosynthetic tissue at a high water status while its stomata remain open. Normally this means that it maintains positive turgor, i.e., that P in the symplast be everywhere positive. Depending on the values of Π in the symplast, it is clear from Eq. (1.2) that positive turgor could be associated with a very wide range in Ψ. In practice, there is an upper limit to symplastic Π in natural conditions that is generally in the range of 1.0 to 2.5 MPa in mesophytes but may be more than double that in some xerophytes (LARCHER 1975). Positive turgor may not be essential to continued photosynthesis (RAWSON 1979) but is certainly associated with rapid photosynthesis; it is strongly associated with large stomatal conductance, at least in mesophytic crop plants (TURNER 1974), and also with leaf growth (HSIAO et al. 1976; BUNCE 1977; see also Chaps. 7 and 9, this Vol.).

Measurements of Ψ and of symplastic P and Π are therefore crucial to the understanding of the performance of a plant under water stress. How these may be made are discussed in the following chapter of this volume. Bulk measurements, such as those made on a leaf, provide insights into the control of short-term productivity and into how the tissue may have adapted to water stress, for example by increasing symplastic Π at full turgor so that turgor remains positive down to lower values of Ψ (JONES and TURNER 1978).

Of equal importance, but rather more difficult to make, are measurements of symplastic P and Π in small regions of tissue such as meristems, or groups of expanding cells, or pulvini. These tissues profoundly influence both the short- and long-term responses of a plant to water stress. Pulvini, where present, can alter the orientation of leaves so as to improve their water economy (BEGG 1980; SHACKEL and HALL 1979); the growth of meristems, and the expanding tissues they give rise to, affect the leaf area and hence the evaporative demand on a plant. These systems are particularly important because falling water status typically affects growth well before it affects photosynthesis (HSIAO et al. 1976; see also Chap. 9, this Vol.).

The most difficult measurements of all to make are those on single cells (HÜSKEN et al. 1978) especially those on stomatal cells (e.g., EDWARDS and MEIDNER 1975), which are particularly important because of the influence of turgor on stomatal conductance.

When the water status of a plant at or near full turgor does start to fall, which it will if the evaporative loss from the leaves exceeds the influx of water from the roots, the rate of fall of Ψ, and, in the short term, of symplastic P, depends not only on the net rate of water loss, but also on the bulk elastic modulus of the cells, ε, which is given by $V\,dP/dV$, where P is the turgor pressure, and V the symplastic volume. If ε is large, a small change in relative water content will bring about a large change in Ψ. If ε is small, Ψ is much more strongly buffered against changes in relative water content, and this may be an advantage if it means that positive turgor is maintained for longer, although not necessarily so if conservative behaviour is called for (see Sect. 1.4.2.2).

So far I have been largely concerned with "static" aspects of water status. In a transpiring plant, there are gradients in Ψ. It is these gradients that are the driving forces for the transport of water across perfect semi-permeable membranes. For transport across leaky membranes however, i.e., those having a reflection coefficient less than 1, the osmotic component of Ψ does not fully manifest itself, and it is necessary to consider the pressure-induced and osmotically induced flows separately, as discussed in detail by DAINTY (1976) and ZIMMERMANN and STEUDLE (1979). Where there are no membranes, as in soil, or within the apoplast, or perhaps even within the symplast, the driving force for the flow of water is the gradient in τ; since Π_D in the apoplast is usually small (TYREE 1976), gradients in τ within it are virtually identical with those in Ψ; the same is not necessarily true for the symplast.

The current water status of a plant depends largely on the recent history of the evaporative losses from it and the fluxes of water into and through it. These fluxes are discussed in the next section.

1.3 Transport of Water in the Soil-Plant-Atmosphere Continuum

The electrical analogue of VAN DEN HONERT (1948) and its theoretically sounder successors (e.g., COWAN 1965, 1972b; RICHTER 1973) have been of great help in providing a conceptual framework for thinking about the transpiration stream in general terms. When dealing with particular parts of the pathway, however, the faithful application of the electrical analogue is often more confusing than enlightening, particularly where there is a change of phase, or where the transport of water is non-linear, or where active transport of solutes is involved. It is often expedient to abandon the idea that the flux of water between two points is given by $\Delta\Psi/r$, where r is a resistance, and to use a formalism that is more appropriate.

The following discussion explores the factors influencing the movement of water through soil and plant to the atmosphere, assuming quasi-steady conditions, i.e., that the main structure of the plant remains essentially unchanged, and that changes in the plant's environment are slow compared with the rate at which disturbances in Ψ are propagated through the plant.

1.3.1 Through Soil to Root

The old controversy about whether or not soil water was equally "available" to plants all the way down to the "permanent wilting point" was defused during the 1950's and early 1960's when it was realised that availability had to be considered in terms of how fast water could move through the soil to a plant's roots rather than simply in terms of how firmly the water was held by the soil (see Chap. 3, this Vol.). This controversy has subsequently reemerged in a more sophisticated form; some have argued that the rate of flow of water through the soil may limit its uptake by roots even when the soil is quite moist (GARDNER 1960; COWAN 1965; CARBON 1973; GREACEN 1977); others have argued that in real situations, the rooting density is so high, and the rate of uptake of water per unit length of root is so low, that it is only when the soil is quite dry and τ approaches 1.5 MPa that the rate of movement becomes limiting (NEWMAN 1969; LAWLOR 1972; HANSEN 1974; TAYLOR and

KLEPPER 1975). The issues involved have been recently and comprehensively reviewed by TINKER (1976), but for the sake of completeness the main arguments are outlined below.

The flux density of water in soil, F, obeys Darcy's Law, which, for one-dimensional flow is $F = K(d\tau/dx)$, where x is distance and K is the hydraulic conductivity. K is a rapidly increasing function of θ, the volumetric soil water content, and may vary by many orders of magnitude. In unsaturated soil it is often more convenient to describe F in terms of gradients in θ rather than τ, so that $F = -D(d\theta/dx)$, where D is the diffusivity of the soil water, and is related to K by $D = -K(d\tau/d\theta)$; D is usually an exponential function of θ when $\tau < 1.5$ MPa, and typically ranges from $10^{-9}\ m^2\ s^{-1}$ in fairly dry soil, with τ, say, 1.0 MPa, through $10^{-8}\ m^2\ s^{-1}$ with $\tau \simeq 0.1$ MPa, to $10^{-7}\ m^2\ s^{-1}$ in moist soil, with τ, say, 10 kPa (e.g., ROSE 1968). The characteristic time for transport through a distance l is of the order of l^2/D, so we can see that if $D = 10^{-9}\ m^2\ s^{-1}$ ($\simeq 1\ cm^2\ day^{-1}$), many days are required to transport much water out of a slab of soil a few centimetres thick. This highlights the importance of a plant having a dense root system if it is to extract water rapidly from soils having $\tau > 0.1$ MPa.

The geometry of a root system is complicated but it is useful to idealize it by assuming (a) that water flows radially through the soil to a root, (b) that the rate of uptake per unit length of active root, I, is uniform, and (c) that each root has sole access to the water within a hollow cylinder of soil whose inner radius, a, is that of the root, and whose outer radius, b, is $(\pi L)^{-\frac{1}{2}}$, where L is the rooting density, i.e., the length of root per unit volume of soil. Given these assumptions, and some others to describe the boundary conditions, i.e., the flows or τ or θ at a and b, one may derive a variety of equations for describing the flow of water through the soil to a root. One of the most accurate of these is the following (PASSIOURA 1980a):

$$\int_{\theta_a}^{\theta_b} D(\theta)\, d\theta \simeq \tfrac{1}{2} Q b^2 (\tfrac{1}{2} - \ln b/a) = I(\tfrac{1}{2} - \ln b/a)/2\pi \qquad (1.3)$$

where $Q(=IL)$ is $d\bar{\theta}/dt$, the quasi-steady rate of change of mean water content, $\bar{\theta}$, with time, t. If D is assumed to be constant, the integral simplifies to $D(\theta_b - \theta_a) \simeq D(\bar{\theta} - \theta_a)$.

Disagreements about whether or not equations such as (1.3) imply that there is a major difference in τ between a and b have arisen for two main reasons. The first is that few experimenters have taken pains to measure carefully the hydraulic properties of their soils, and since K, in particular, varies so rapidly with θ, it is easy to be in error by an order of magnitude. Furthermore, both $D(\theta)$ and $K(\theta)$ depend strongly on the texture of the soil; for example, a sand (e.g., CARBON 1973) behaves very differently from a loam (e.g., GARDNER 1960; NEWMAN 1969), so that some disagreements may be more apparent than real, being due to different authors having used different soils.

The second reason is that opinions have differed widely about what proportion of the total root length is effective in taking up water. Those who assume that most of the roots are active (e.g., NEWMAN 1969) predict minor gradients of τ near a root; those who assume that few roots are active predict major ones (e.g., CALDWELL 1976). Such differences will remain unresolved until we find ways of measuring effective root length. Indeed, we may never find them and may have to rely on indirect evidence such as that provided by substituting $(\pi L')^{-\frac{1}{2}}$ for b in Eq. (1.3), where L′ is the effective rooting density, and solving the equation for L′ after accurately measuring all the other unknowns. Very few such measurements have been made, but for young wheat plants growing in a loam it seems that L′ is about 0.3 of the total rooting density, L (PASSIOURA 1980a). For wheat grown in sand, however, HERKELRATH et al. (1977a) found that L′ appeared to be only 0.01 L, which prompted them to suggest that there was a major hydraulic resistance at the interface between root and soil in their system (see below).

There is little doubt that Eq. (1.3) is useful for fairly dry soil, and is irrelevant for wet soil, but before we get a clear idea of its relevance in between we must wait for many more experiments to be done in which accurate measurements of the main variables and parameters are made, in particular, $D(\theta)$ or $K(\theta)$, $\tau(\theta)$, L, and τ_a.

The above discussion has implicitly assumed that the roots are uniformly distributed through the soil. While this may be true of pot-grown plants it is certainly not true in the field, where we may often have to consider not only the movement of water to roots from within the rooting zone, but also the movement of water from outside the rooting zone. This uptake of water by a non-uniform root system has been the subject of many reviews, of which good recent ones are by GREACEN (1977) and TAYLOR and KLEPPER (1978).

1.3.2 Across the Root-Soil Interface

While there is some doubt about whether or not roots induce large local gradients of τ in soil, there is little doubt that plants often find it much more difficult to take up water from soil than from solution even when the soil is quite moist. Two recent papers provide good illustrations of this.

1. HERKELRATH et al. (1977a) showed that the extraction of water by wheat roots growing in a sandy soil decreased rapidly once θ fell below 0.1, even though τ was small and D was large at $\theta = 0.1$ being approximately 0.02 MPa and $3 \times 10^{-8}\ m^2\ s^{-1}$ respectively; in order to get even approximate agreement between their data and (a relative of) Eq. (1.3), they had to assume that only 1% of the measured root length was taking up water.

2. FAIZ and WEATHERLEY (1978) calculated from experiments with sunflower growing in sand or soil, that there appeared to be a major drop in Ψ between soil and root that could not be explained in terms of an equation similar to Eq. (1.3).

Both sets of authors concluded from their results that there must have been a major hydraulic resistance at the interface between root and soil, and that the most likely cause of this resistance was poor contact. HERKELRATH et al. (1977b), following COWAN and MILTHORPE (1968), suggested that the hydraulic conductance of the root might depend on the proportion of its epidermis that was in contact with the soil water, which to a first approximation would be proportional to θ, so that the hydraulic conductance of the root: soil system might depend more on θ than on D. This is an interesting suggestion, and implies that the major radial resistance within the root is at the epidermis, rather than at the endodermis as is usually assumed, but it does not explain why the uptake of water by roots can sometimes be satisfactorily explained in terms of Eq. (1.3) without the need to invoke a major interfacial resistance (PASSIOURA 1980a). If the conductance does depend on θ in some cases, why not in all?

Another, and perhaps more likely explanation is that the roots shrink as the soil starts to dry, and in so doing lose some of their hydraulic contact with the soil. If the major hydraulic resistance within the root is in the endodermis rather than in the bulk of the cortex or in the epidermis, the water potential in the cortex, Ψ^{cortex}, will be close to that of the soil, Ψ^{soil}, rather than to that of the xylem, Ψ^{xylem}. Since Ψ^{xylem} is typically much less than Ψ^{soil} (e.g., HERKELRATH et al. 1977a), any loss of contact due to shrinkage of the root will, as pointed out by FAIZ and WEATHERLEY (1978), be self-amplifying, for it will result in Ψ^{cortex} moving from Ψ^{soil} to the substantially lower Ψ^{xylem}, which will lead to further shrinkage. If hydraulic contact is important, and if it is influenced by shrinkage, then the rate of change of Ψ^{soil} may have a major influence on the hydraulic properties of the system; if Ψ^{soil} changes slowly enough for the roots to maintain turgor by osmoregulation, there may be no shrinkage; but if Ψ^{soil} changes rapidly, the roots may not be able to maintain turgor and may then shrink. The need to invoke an interfacial resistance in some cases (HERKELRATH et al. 1977a; FAIZ and WEATHERLEY 1978) but not in others (PASSIOURA 1980a), may have been due to the fairly large differences in $d\Psi^{soil}/dt$ that existed between the experiments.

Another possible explanation for a large apparent interfacial resistance is that there may be a major pile-up of solutes at the surface of the root (or perhaps at the endodermis; NULSEN and THURTELL 1978), owing to the solutes being carried to the root by convection at a faster rate than they are being taken up (TINKER 1976). Such a pile-up would manifest itself osmotically and would lead to a large difference between Ψ^{xylem} and Ψ^{soil} that would appear to be due to a large hydraulic resistance. This effect would be small in a moist soil, for the solutes can quickly diffuse away against the transpiration stream, but if the hydraulic connections between root and soil become scarce, they may remain adequate for the flow of solution into the cortex, but inadequate for the diffusion of the rejected solutes back into the soil. These and other aspects of behaviour at the root:soil interface are discussed in more detail by TINKER (1976) and PASSIOURA (1980a, 1981) and, in Chapter 3, this Volume.

1.3.3 Through the Root

The largest drop in Ψ within a plant usually occurs somewhere between the surface of the root and the xylem, and probably at the endodermis. There is considerable doubt about what is the main pathway for water through the cortex, although NEWMAN (1976) concludes that it is through the symplast rather than the cell walls, for measured permeabilities of walls seem to be too low. If he is right it would imply that the major radial resistance is at the epidermis rather than at the endodermis, which would be compatible with the hypothesis of HERKELRATH et al. (1977a) discussed earlier. However, NEWMAN's (1976) calculations of permeabilities refer to transverse flow across the walls; the permeabilities for tangential flow, which are much more important, may be much larger, especially if there is flow along irregularities in the surfaces of the walls and if fillets of water occupy the intercellular spaces, as seems likely (PASSIOURA 1981).

The nature of the radial hydraulic resistance in the root is even more mysterious than the pathway for the water. Some have found the resistance to be highly variable, and, within limits, inversely related to transpiration rate, so that $\Delta\Psi$ across the root is independent of transpiration rate (WEATHERLEY 1975). Others have found the resistance to be constant (HAILEY et al. 1973; NEUMANN et al. 1974; PASSIOURA 1980a). WEATHERLEY (Chap. 3, this Vol.) discusses this puzzle in detail. One possible explanation, which will be touched on again later, is that much of the variable hydraulic resistance within the plant which is usually assumed to be in the roots may well be in the leaves.

Axial flow in the roots usually suffers little hindrance, particularly in dicots, whose facility for secondary growth endows them with abundant xylem unless vascular disease or a physically constricting soil causes some disruption (TAYLOR and KLEPPER 1978). There are however several reports of large axial resistances in the roots of the Gramineae, whose lack of secondary growth sometimes results in a meagre vascular system. WILSON et al. (1976) for example showed that blue grama (*Bouteloua gracilis*) seedlings grew poorly if drought prevented the development of nodal roots. The large hydraulic resistance of the small xylem vessels of the seminal roots and the sub-coleoptile internode apparently prevented the leaves from getting an adequate supply of water. A similar problem can occur in droughted wheat plants, which may have to rely on a few seminal roots containing only one substantial xylem vessel each to extract water from a moist subsoil, although, in this case, the large resistance may be an advantage if it results in the plants conserving water when young for later use during the critical periods of flowering and grain filling (PASSIOURA 1977).

Axial resistance in the xylem is usually assumed to follow Poiseuille's equation, and, for smooth-walled vessels such as occur in grasses, this is a fair assumption (GREACEN et al. 1976). Where the walls are rough, however, which they typically are in dicots, Poiseuille's equation may grossly underestimate the resistance (GIORDANO et al. 1978; JEJE and ZIMMERMANN 1979), although there is little evidence that the actual resistance has a major influence on the water economy of the plant.

1.3.4 Through the Shoot

Resistance to the longitudinal flow of water in the shoot is generally small (JARVIS 1975), although there have been reports of large resistances in the stems and branches of Sitka spruce (HELLKVIST et al. 1974) and of hemlock (TYREE et al. 1975) and in the stems of tobacco (BEGG and TURNER 1970) and of wheat (DEMMEAD and MILLER 1976), and even in the leaves of wheat (MEIRI et al. 1975). Little is known about what factors influence these resistances, or what their possible functional or adaptive significance (if any) might be. They may enable a plant to buffer Ψ^{leaf} against rapid change if the soil starts to dry, for the loss of Ψ in the soil will be partly counterbalanced by a reduced loss in the shoot when the transpiration rate falls; they may simply be the inevitable concomitant of a protective system for preventing the catastrophic spread of any embolisms (or vascular disease) throughout the plant; or they may provide a means of regulating the way in which water stored in tree trunks is used as a buffer during the course of a day. Although the longitudinal flow of water occurs almost exclusively in the xylem the resistances may well arise from flow outside the xylem, for example across nodes, where the water probably has to traverse living cells.

The hydraulic resistance of a whole plant has often been found to be variable, with a tendency to keep Ψ^{leaf} constant over a wide range of transpiration rates (HAILEY et al. 1973). It was thought, for some time, that this variable resistance resided solely in the roots (WEATHERLEY 1976), but there is now compelling evidence of major variable resistances within leaves (BOYER 1974, 1977; BLACK 1979). These resistances are generally calculated as the ratio between ($\Psi^{stem} - \Psi^{leaf}$) and transpiration rate; it is worthwhile digressing briefly to examine the meanings of Ψ^{leaf} and Ψ^{stem}.

Measurements of Ψ^{leaf} are generally made using either thermocouple psychrometry or a pressure chamber (see Chap. 2, this Vol.). Agreement between the two techniques is usually reasonable, there being no consistent discrepancies, except in certain trees. It is usually assumed that the psychrometer measures Ψ in the leaf tissue generally (which is dominated by the mesophyll), while the pressure chamber measures Ψ in the xylem of the leaf or perhaps even of the petiole. Yet if there is no consistent descrepancy between the two they must both be measuring the same thing. Ψ^{stem}, on the other hand, is either zero (by definition) if a detached leaf is being observed with its petiole in water, or is given by Ψ^{leaf} when the leaf is prevented from transpiring by being enveloped in sheet plastic or Al-foil – such covered leaves are thought to act as though they were tensiometers plugged into the stem. A large difference between Ψ^{leaf} and Ψ^{stem} implies that there is a large hydraulic resistance within the leaf. Once transpiration is stopped, this resistance would no longer be evident, and Ψ^{leaf} and Ψ^{stem} would converge. Unless the resistance were clearly localized, Ψ would vary throughout the transpiring leaf, and the measured Ψ^{leaf} would be some volume-weighted (or, rather, capacitance-weighted) average of the variation in Ψ. The contribution of Ψ^{xylem} to this average would, however, be very small because the capacitance (i.e., $dV/d\Psi$) of the xylem is very small. If the resistance is localized to the junction between xylem and mesophyll, the measured Ψ^{leaf} would be virtually identical to $\Psi^{mesophyll}$ in the transpiring leaf, and given the small capacitance of the xylem (probably $\leqslant 1\%$ of the leaf volume/MPa), the time constant (i.e., capacitance x resistance) for the convergence of $\Psi^{mesophyll}$ and Ψ^{xylem} would be small and much less than the 150 s found by BOYER (1977) for the efflux of water from a leaf which had been subjected to a step-change in Ψ. The general agreement between Ψ^{leaf} as measured by psychrometry or by pressure chamber is therefore not surprising, even though there may be a large hydraulic resistance within the leaf, for as we have seen, the pressure chamber does not necessarily measure Ψ^{xylem} in the petiole. This is perhaps most strikingly illustrated by the experiments of JANES and GEE (1973) in which Ψ^{leaf} as measured by pressure chamber was $\simeq -0.3$ MPa even though Ψ^{xylem} in the petiole was actually positive owing to the roots having been pressurized; when the leaf was cut off, water exuded from the stump of the petiole.

The location of the resistance in the leaf is unknown, although the major pathway of the transpiration stream is probably through the cell walls after it has left the vascular tissue (COWAN 1977). In an expanding leaf, in which a small but important fraction of the water entering the leaf does not evaporate but instead enters the expanding cells, a large resistance to flow across the cell membranes could be responsible for a large difference between Ψ^{leaf} and Ψ^{stem} as BOYER (1974, 1977) has cogently argued. Furthermore, he has pointed out, this difference may *diminish* as transpiration rate increases, if the generally lower Ψ^{leaf} results in a slower expansion of the cells, and may partly counteract the increased gradient in Ψ in the transpiration stream. The result would be that, within limits, Ψ^{leaf} is insensitive to changes in transpiration rate, just as is often found in practice.

Unfortunately, these interesting ideas are convincing only for an expanding leaf, and are difficult to adapt to the fully expanded leaves transpiring at a steady rate which also show a marked insensitivity of Ψ^{leaf} to transpiration rate (BLACK 1979). This behaviour remains mysterious, although it is worth noting here that, because the bulk elastic modulus of turgid leaf cells may be very large ($\varepsilon \gg 10$ MPa) (JONES and TURNER 1978), small amounts of water lost by the leaf or some of its tissue while on the way to the pressure chamber or psychrometer could lower Ψ^{leaf} substantially (TURNER and LONG 1980). Where this happens, the true hydraulic resistance of the leaf may be substantially less than that which we calculate.

It is the control of stomatal conductance that provides a plant with its most powerful means of controlling transpiration rate (at least in the short term), and our interest in the resistances experienced by the transpiration stream is largely, though usually implicitly, directed towards understanding their influence on the water relations of the leaf and hence on the conductance of the stomata. Yet except in the most intricate laboratory experiments (e.g., EDWARDS and MEIDNER 1975) the only readily accessible, seemingly relevant, data are bulk measurements of Ψ^{leaf} and its components, and these we would expect to be dominated by the mesophyll because of its large capacitance. The epidermis, and particularly the guard cells, may have a water potential that is substantially different from Ψ^{leaf}, particularly if much of the evaporation from a leaf takes place from the inner walls of the guard and subsidiary cells, as it seems to at low light intensity, although not at high (SHERIFF 1979). The influences of plant water relations on stomatal behaviour are discussed with a wealth of detail by COWAN (1977) and RASCHKE (1979) and in Chapters 7, 8 and 17, this Volume. For our present purposes, we may note that despite the tenuous causal link between Ψ^{leaf} and stomatal conductance, the correlation between the two is often robust enough for critical values of Ψ^{leaf} to exist (for a given species in a given situation) below which stomatal conductance decreases rapidly. This correlation may be less tenuous than it at first appears if loss of turgor in the mesophyll triggers the production of ABA which in turn causes the stomates to close (PIERCE and RASCHKE 1980).

1.3.5 Evaporation from the Shoot

The driving force for the flow of water from the soil through the plant derives from the evaporation of water (that is, the dissipation of energy as latent heat) from the shoot (see Chap. 1, Vol. 12 A). This evaporation is usually sustained by energy from solar radiation but in dry environments advection can be a major source. The rate of evaporation from a well-developed canopy is usefully described, in the absence of advection, and if stomatal resistance is low, by the Penman equation (MONTEITH 1965), which relates the evaporation to the available radiant energy and the humidity deficit, and, less directly, to air temperature and wind speed. If the stomatal resistance is high, this equation is not applicable, but MONTEITH (1965) has tried to make it so by incorporating into it a variable which he calls surface resistance, which is some function of stomatal resistance. MONTEITH's approach has been strongly criticized by PHILIP (1966) on theoretical grounds, but has nevertheless survived on pragmatic ones (e.g., DUNIN et al. 1978).

When the leaf area index (LAI) of a community is large (say, >3), stomatal conductance is the only property of the plants that has a major effect on transpiration rate, although there are many morphological changes (BEGG 1980) that can have minor ones. When the LAI is low, however, which it usually is in arid environments or during the establishment of a crop, it too can have a large effect (MONTEITH 1965; RITCHIE 1974). Indeed, its integral effect during the life of a crop may be very much greater than that of stomatal conductance (LEGG et al. 1979). The nature of the effect depends on many things, including the arrangement of the plants and the orientation of their leaves, but as a rough guide, the transpiration rate is linearly related to LAI up to LAI $\simeq 2.5$, beyond which point (if stomatal conductance is high) it depends entirely on the environmental conditions (see Sect. 1.4.2.2).

1.3.6 Through the Plant as a Whole

In many experiments, particularly those carried out in the field, it is not feasible to explore the flow of water through soil and plant in great detail, and we have to be content with determining the "hydraulic resistance", r, of the system as a whole, so that we have

$$r = (\Psi^{soil} - \Psi^{leaf})/q \qquad (1.4)$$

where q is the flux of water through the plant. Many attempts have been made to partition r between plant (r^p) and soil (r^s), by estimating r^s using the theory of GARDNER (1960) (e.g., LAWLOR 1972; TAYLOR and KLEPPER 1975; REICOSKY and RITCHIE 1976; BURCH 1979). These attempts have all shown that $r^p \gg r^s$ over most of the range of available water, so that we would not expect the plants to have difficulty in extracting water from the soil until there is little available water left; this does, in fact, seem to be so in many cases (RITCHIE 1973; BURCH et al. 1978). Yet in other cases the transpiration rate is sensitive to θ over a wide range and may even be linearly related to it (BURCH 1979; EAVIS and TAYLOR 1979). This is puzzling behaviour and could imply that r^p increases with decreasing water content, or that Ψ^{leaf} does, assuming, that is, that the critical Ψ^{leaf} for stomatal closure remains unchanged. Since r^p includes any interfacial resistance between root and soil, it is feasible that it could be related to θ (HERKELRATH et al. 1977b), although invoking an interfacial resistance does not seem to help explain the results of EAVIS and TAYLOR (1979), for they found that while θ influenced transpiration rate, root length did not, yet we would expect it to if the interface were a problem. Perhaps it is significant that both BURCH (1979) and EAVIS and TAYLOR (1979) worked with pots, i.e., using disturbed soil, while RITCHIE (1973) and BURCH et al. (1978) worked in the field; the generally higher bulk density and shear strength of field soils might favour good contact between roots and soil.

1.3.7 Summary and Preview

The preceding pages have discussed the flow of water from the soil to the atmosphere through a static plant. The rate of transpiration depends on the atmospheric conditions and on several properties of the plant including the area and orientation of its leaves, and its stomatal and cuticular conductance to water vapour. The stomatal conductance depends on the water status of the plant, which in turn depends on the water status of the soil, on the hydraulic resistances experienced by the transpiration stream, on the transpiration rate, and on the osmotic pressure of the symplast. Our interest in water status usually stems from its influence on photosynthesis, or, more generally, on photosynthetic productivity. This influence is mediated partly through the stomates and through changes in leaf morphology, and partly through more direct effects on the behaviour of chloroplasts (BOYER 1976).

Both photosynthesis and transpiration depend on the leaf area and other morphological properties of the shoot and these in turn depend on the *past* water status of the plant, as also do some metabolic and physiological features, such as the osmotic pressure of the symplast in those plants that osmoregulate in response to water stress. In exploring the interactions between water relations and photosynthetic productivity, therefore, the behaviour of the static plant is merely a starting point. The following sections briefly explore the dynamic interactions between the two with the aim of integrating the behaviour of the plant over longer and longer periods of time, subject to various constraints on water supply in relation to evaporative demand, until we reach a demographic time scale.

1.4 Water Relations and Photosynthetic Productivity

We normally think of photosynthetic productivity as being the net increase in biomass of a plant, or a community of plants, during a growing season, or perhaps a year. This section is concerned with the connection between productivity, in this sense, and the quasi-steady or instantaneous behaviour of plants that was discussed in the previous section, given the constraint of a limited water supply. On a seasonal scale, it is useful to think of the productivity (B) of a plant as being the product of the amount of water transpired (W) and the efficiency with which that water is used to produce biomass, that is, $B = W \times WUE$, where WUE is the dry matter produced per unit water transpired, and typically refers to aboveground dry matter only. These two factors are to a first approximation independent of each other, and are sufficiently robust functions of major environmental and physiological variables, such as precipitation, pan evaporation, and photosynthetic pathway, to give us a crude but nevertheless useful understanding of variation in B due to drought (FISCHER and TURNER 1978). It is obvious that both the plant and its environment influence both W and WUE, but it is worth emphasizing that a plant's influence on W can be very large, even after allowing for the area of ground that it occupies; the vigour and extent of its root system, its ability to extract water from the topsoil before it evaporates directly, or its ability to channel rain down its stems so as to make it more accessible, all influence W and hence all affect B. The following discussion explores the ways in which a plant can influence both W and WUE during given periods of time, so as to maximize B in the face of a fluctuating environment. I will assume that this behaviour is evolutionarily advantageous (for detailed discussion of water use and optimization of carbon assimilation see Chap. 17, this Vol.).

1.4.1 Daily Production

I take as a starting point a single leaf. The connection between its evaporation rate per unit area, E, and its assimilation rate per unit area, A, has been discussed in detail by COWAN and FARQUHAR (1977) and by COWAN (1978). In general, when E and A change in response to stomatal conductance, g, E(A) is positively curved as shown in Fig. 1.1. Furthermore, since the environmental conditions experienced by a leaf change during the course of a day, E(A), though still remaining positively curved, will change its shape during the day.

COWAN and FARQUHAR (1977) have argued that the optimal stomatal behaviour of a leaf faced with a limited supply of water is that which allows maximal assimilation during a day of given evaporation, or its converse, that which allows minimal transpiration during a day of given assimilation. They point out that this behaviour ensures that any small deviations in stomatal conductance from its optimal function of time g(t) are such that the concomitant deviations, δE, in transpiration rate per unit leaf area satisfy the inequality $\int \delta E dt \geq 0$,

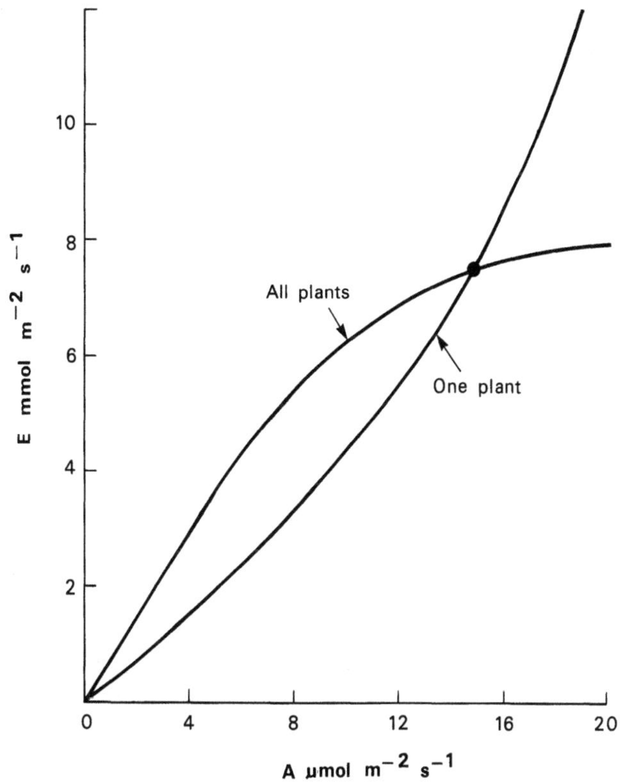

Fig. 1.1. Hypothetical relationships between rate of transpiration, E, and rate of assimilation, A, for an individual plant in an extensive community of plants. The plant is assumed to function normally at the point of intersection of the two lines, one of which represents the variation which would occur if stomatal aperture in the individual, only, were to vary and the other represents the variation which would occur if stomatal apertures in all the plants in the community were to vary in unison. (Reproduced from COWAN 1978)

given that the mean net assimilation rate, $\bar{A}$, is fixed for the day, i.e., that $\int \delta A dt = 0$. This inequality implies that behaviour is optimal if $\delta E/\delta A$ is a constant (λ) throughout the day, providing that $(\partial^2 E/\partial A^2)$ is positive. COWAN and FARQUHAR (1977) chose to concentrate their treatment on stomatal behaviour, but COWAN (1978) has pointed out that the criterion, $\partial E/\partial A = \lambda$, applies for any short-term reversible behaviour of the leaf that influences E, notably those listed here in Table 1.1 with time scales of hours or less; the influence of nastic leaf movement on improving leaf microclimate can be very large (SHACKEL and HALL 1979; BEGG 1980), as can the passive movement associated with wilting (RAWSON 1979). The criterion that $\partial E/\partial A$ be constant for a single leaf also applies to a whole plant and to every leaf on the plant.

Constancy in $\partial E/\partial A$ has some interesting consequences, the most important of which is that when evaporative demand is high, g may be reduced so much that E actually decreases, a phenomenon that has now been observed several times in natural environments and that is discussed in detail in Chap. 7, this Volume. FARQUHAR (1979) has pointed out that a decrease in E with increasing evaporative demand cannot be explained in terms of feedback based on some sensing mechanism within the leaf, for the water status of the leaf is actually improved. The explanation must be in terms of feedforward, which implies that the sensing mechanism (which is perhaps related to peristomatal transpiration) is on the outside of the leaf.

COWAN and FARQUHAR's (1977) hypothesis is an extremely interesting and ingenious one, and FARQUHAR et al. (1980) have recently provided evidence verifying it for the leaves of two species exposed to a range of humidities in a controlled environment. But it is based on an implicit assumption that the water supply of the plant is not only limiting, but is also fixed. This assumption seems appropriate for isolated plants, such as commonly occur in arid environments, but where the plants are competing for water, optimal behaviour may be to maintain g as large as possible whenever A is positive, for water that is saved in the interests of improving $\bar{A}/\bar{E}$ (where the bars denote mean values) may become water that is lost to a competitor. In the context of the hypothesis that $\partial E/\partial A = \lambda$, this means making λ as large as possible, which brings us to a question that cannot be answered by studying diurnal behaviour alone, namely, what is the optimum size of λ? To answer this question requires us to extend our time scale and to adopt a criterion of optimal behaviour that includes the amount of water transpired during the period of time in which we are interested. This leads to considerable complications, for while it is worthwhile assuming, as we implicitly have done, that a plant does not grow during the course of our interest in it, this assumption is untenable when we consider behaviour during several days or weeks. The plant does grow, and its growth influences not only $\bar{A}$ and $\bar{E}$ but also the size of its potential water supply, through, for example, the growth of the root system. Furthermore, the way in which the growth is partitioned, especially between water-harvesting and water-using structures, profoundly influences both W and WUE. These issues are discussed in the next subsection.

1.4.2 Weekly or Monthly Production

The net rate of assimilation of a whole plant, A*, is substantially less than the net rate of assimilation of its leaves. Respiratory losses from heterotrophic tissues such as roots, stems, and possibly fruits, explain the difference. Let us assume for the moment that plants behave optimally if they maximize $\bar{A}^*$ over a given period of say several weeks, that is that they maximize the product of W (or $\bar{E}^*$, the evaporation rate from the whole plant) and WUE (or $\bar{A}^*/\bar{E}^*$). What are the compromises that they must make? An obvious one is in the partitioning of assimilate. To keep $\bar{A}^*/\bar{E}^*$ large requires that the proportion of heterotrophic tissue in the plant be kept small, especially that in the roots, for there is evidence that they respire (or at least consume assimilate) much more rapidly per unit weight than does the shoot (SAUERBECK and JOHNEN 1976). But sacrificing roots to improve $\bar{A}^*/\bar{E}^*$ may result in a substantial drop in $\bar{E}^*$ and hence in $\bar{A}^*$, owing to the roots being unable to extract as much water from the soil. Another obvious compromise is in the value of λ: a large λ may increase $\bar{E}^*$; a small one will increase $\bar{A}^*/\bar{E}^*$.

There are many such compromises, the nature of which depends on a plethora of climatic and edaphic factors and on competition from neighbours. Their complexity suggests that elegant hypothetical generalizations, such as that achieved by COWAN and FARQUHAR (1977) for diurnal behaviour, may be unattainable. Nevertheless, there is much to be said of a qualitative nature, and the rest of the discussion will concentrate on this, exploring in turn (a) behaviour which primarily influences the water supply of a plant, and (b) behaviour which primarily influences the effectiveness with which that water is used.

1.4.2.1 Water Supply

There are several ways in which a plant can influence the amount of water accessible to its roots. The most obvious of these concerns the size of the root system, but other, less obvious, ways include (1) the channelling of rainfall down stems, so that it tends to concentrate close to the centre of the plant, (2) the improvement of the permeability of the soil so that there is less run-off during heavy rains, and (3) the shading of the soil surface (by leaves or by litter), which protects any surface water from rapid evaporation by the sun. The extent to which the roots can ultimately lower Ψ^{soil} will also influence the amount of available water, although the variation may not be great. Most plants that grow in drought-prone environments are capable of lowering Ψ^{soil} to less than -2.0 MPa and some can even lower it to -10 MPa (NOY-MEIR 1973). But the amount of water held by soil between Ψ^{soil} of -2.0 and -10 MPa is usually small unless the soil is very heavy, and its advantage to those plants that can extract it is probably in aiding their survival during severe drought rather than in improving their medium-term productivity.

Variation in the size of root systems is enormous, ranging from those of desert ephemerals and cacti that might tap a soil volume of only a few litres, to those of some trees that may penetrate tens of metres to deep water-tables. I am concerned here not so much with variation between species as with environmental influences on the growth of roots into previously untapped volumes of soil.

The major influences of temperature, aeration, mechanical resistance, and nutrition, have been recently and comprehensively discussed by RUSSELL (1977). Insofar as these adversely affect root growth and thereby limit a plant's water supply, the largest effects are usually in the subsoil.

Apart from riparian and deep-rooted perennial plants which can often tap a permanent supply of water from the subsoil, it is those plants that grow in climates having alternating long dry and long wet spells that particularly rely on subsoil water during much of their lives. Established perennial plants usually have their roots ramified through the subsoil in such circumstances, but annual plants, and establishing perennial ones, may not. There are several examples, particularly amongst cereal crops, in which the roots failed to extract substantial amounts of water from moist subsoils even though the plants were suffering from drought (e.g., SCHULTZ 1971; BLUM 1974; HURD 1974; WALTER and BARLEY 1974). These failures appear often to be due not so much to the roots failing to penetrate the subsoil, as to their failure to develop adequate densities there (JORDAN and MILLER 1980).

Subsoils typically are dense and have a shear strength large enough to slow down root growth. Furthermore, the effect of a high shear strength is exacerbated at the low temperatures that often prevail in the subsoil in cold or temperate climates (GREACEN 1977). In Mediterranean and steppe climates, for example, the temperature of the topsoil rises rapidly during the spring, thus encouraging root growth, but the temperature of the subsoil does not, at least at depths below about 50 cm, where the annual variation in temperature is small, and in any case lags well behind the mean air temperature. Nutritional problems are also rife in subsoils, which are generally poor in macronutrients such as N, P, and K, and are often poor in micronutrients as well; it is particularly important for Ca and B to be available locally because they travel so poorly in the phloem and are essential for root growth (KRAMER 1969).

Surprisingly, soil water potential per se does not seem to have a major effect on root growth, except in quite dry soil or in soil with a high bulk density (TAYLOR and GARDNER 1963). NEWMAN (1966), TAYLOR and RATLIFF (1969) and PORTAS and TAYLOR (1976) have all shown substantial root growth in soil having $\Psi < -1.0$ MPa, and PORTAS and TAYLOR (1976) even found that roots of maize and tomato grew with $\Psi^{soil} < -4$ MPa. This is remarkable behaviour and presumably means either that the roots are able to generate very large osmotic pressures, or that they are hydraulically isolated from dry soil and that there is a substantial flow of water to them from other roots in wetter parts of the soil. The amount of available water contained in such dry soil is so small that it may seem pointless for a plant to expend assimilate in growing its roots in this way, but the growth may be more a reflection of the continuing vigour of these roots, which seem to be able to respond rapidly if their dry soil is rewatered (PORTAS and TAYLOR 1976). Rapid responses of this sort may not be particularly important in the subsoil, but they may be crucial in the topsoil if the roots are to harvest a substantial proportion of a light fall of rain before it evaporates directly from the soil surface.

Since roots are heterotrophic, their growth ultimately depends on the supply of assimilate from the shoot. This supply does not seem to be diminished by mild water stress, and often seems to be even improved by it. It has often been observed that drought increases the ratio of root to shoot (MOONEY 1972; LARCHER 1975), and there are several reports of root growth actually increasing during drought (BENNETT and DOSS 1960; HSIAO and ACEVEDO 1974; SCHULTZ 1974; CUTLER and RAINS 1977). The mechanism for this behaviour is unknown, but it may be due to water stress affecting shoot growth more than it does photosynthesis, so that increased assimilate becomes available to the roots (HSIAO and ACEVEDO 1974). An increase in the root-shoot ratio of a plant during drought has obvious advantages in helping the plant match its water supply to the evaporative demand on its leaves, but it is worth reemphasizing that such a response has a respiratory cost that may greatly reduce WUE.

While the growth of the roots has a large influence on the uptake of water already contained in the soil, the architecture of the shoot can have a large influence on the gain of usable water by the soil, through its effects both on the interception of light falls of rain and on the partial channelling down the stem of heavy falls of rain. The interception capacity of foliage varies considerably and may exceed 2 mm (SLATYER 1967), so that a substantial proportion of light falls of rain may evaporate directly from the leaves and be of little use to the plant. On the other hand, the influence of the shoot on directing intercepted rain to the main stem can be very large. Most of the interest in "stem flow" has concerned that in trees, where it may account for up to 40% of heavy falls of rain. But stem flow can also be great in non-woody plants, for example, potato, for which SAFFIGNA et al. (1976) showed that stem flow could also be up to 40% of rainfall.

The direction of rainfall to the base of the main stem may be of little use in improving the accessibility of the water to the plant unless the permeability of the soil there is adequate to cope with the increased water flow. In general, this permeability is largely a property of the soil, but there are two effects that a plant can have on improving infiltration rate. The first is that it often provides litter under its canopy which slows down any run-off of water and allows more time for the water to sink in. The second is that it may actually improve the permeability of the soil in its vicinity. SLATYER (1962) found that the permeability of the soil near the base of a mulga tree was much greater than that of the soil some distance away, and what is more, the ponded infiltration rates near the tree became virtually constant after about 20 min, which suggests that the water was flowing predominantly down large pores and was not moving as a well-defined wetting front. The distribution of large pores is presumably influenced by the growth of roots, particularly those of perennial plants. The ability of a plant to harvest and sequester water in this way undoubtedly has a major influence on its performance during drought.

Channelling the rain to the base of the stem not only keeps the water within the rooting zone of the plant; it also ensures that it penetrates much more deeply into the soil than would be the case if the infiltration were uniform. The water is therefore much less likely to be lost by direct evaporation from the soil. Such evaporative losses can be a major part of light to medium falls of rain (FISCHER and TURNER 1978) particularly on heavy soils, where >10 mm of rain may be needed to wet the top 50 mm of soil from air-dry to a water potential >-1.5 MPa. As mentioned earlier, the ability of the roots in dry soil to grow rapidly once the soil is re-wet is very important in such circumstances.

1.4.2.2 Water Use and Efficiency of Water Use

I have so far been discussing the amount of water potentially accessible to a plant's root system over a period of some weeks. The way in which the plant actually uses that water depends on both the size of its shoot and its stomatal conductance. RITCHIE (1974) has discussed the effect of the size of the shoot, expressed in terms of leaf area index (LAI), on the transpiration rate per unit ground area, E′, of a community of like plants. Figure 1.2 shows, in schematic form, the behaviour of unstressed plants, i.e., ones having a large stomatal conductance. Little information is available about the behaviour of

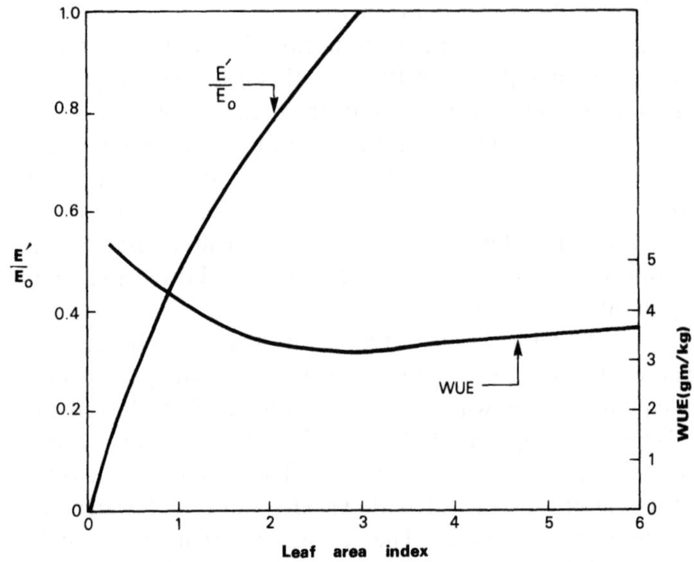

Fig. 1.2. The ratio of actual to potential transpiration rate per unit ground area (E'/E_0), and water use efficiency (WUE), as functions of leaf area index. (After RITCHIE 1974)

stressed plants, but we can probably assume that E' is scaled down by a factor that is approximately uniform over a wide range in LAI.

It is useful to think in terms of two contrasting strategies that plants may have evolved for dealing with a limited water supply (see also Chap. 18, this Vol.).

a) Conservative: This would be appropriate for plants with access to a supply of water that is unlikely to be diminished by competitive neighbours, i.e., plants growing essentially in isolation. Stomatal behaviour would be that discussed in Section 1.4.1, i.e., $\partial E/\partial A$ would be kept at a constant value λ, on a daily basis, and would presumably decline progressively as the water supply diminished. It is difficult to make precise statements about the time course of λ, but its optimal value at any given time will no doubt depend on the water status of the soil, and on the expected duration of the drought, the "expectation" having been built into the genome of the plant through evolution. This issue is dealt with at length in Chapter 17, this Volume. The optimal LAI of a conservative plant would tend to be small, and both the growth and the shedding of leaves would be sensitive to water stress.

b) Prodigal: This would be appropriate for plants that are competing for a limited water supply or that are subject to mild droughts of short duration. Stomatal conductance and growth rate are high, and the water supply is used rapidly until it is almost gone, at which point the plants eke out the remainder so as to survive until the next rain. This behaviour seems to be common in natural ecosystems (FISCHER and TURNER 1978). Many crop plants also appear to behave in this way (RITCHIE 1973, 1974) even though with a crop, it is the behaviour of the community rather than of the individual that is important, so that one might expect any selection for drought tolerance to have produced conservative plants. Actually, depending on the severity of the drought, there are two distinct, and indeed, contradictory strategies that may result in drought tolerance. If a drought is likely to be long, conservative behaviour is appropriate, for it will improve the WUE of the crop without prejudicing W, the amount of water ultimately transpired. But if early relief of drought is likely, or if there is a series of short droughts, prodigal behaviour will probably produce higher yields, for, although it may decrease WUE, it does not necessarily do so (as we shall see below) and it may substantially increase W, through, for example, a greater infiltration of rain into the drier soil, and a smaller evaporative loss of water directly from the soil surface.

Physiologically, the most important distinction between prodigal and conservative plants is that the former probably have a greater ability to maintain turgor when faced with a poor supply of soil water. Variation in this ability is discussed in detail in Chapter 9, this Volume, particularly the ability that some plants have to osmoregulate, i.e., to increase the number of osmoles in their cells in response to falling water status. Morphologically, the most important distinctions are in LAI and in rooting density, both of which must be large if soil water is to be extracted quickly.

So far this section has been primarily concerned with how the plant's behaviour may influence its cumulative transpiration, W. The following discussion concerns the efficiency with which the plant uses water in producing dry matter, and in particular returns to the issue of the influence of $\partial E/\partial A$ on $\bar{A}/\bar{E}$.

COWAN and FARQUHAR's (1977) hypothesis that $\partial E/\partial A$ should be constant if $\bar{A}/\bar{E}$ is to be maximized applies only if E(A) is positively curved when stomatal conductance varies. As discussed in Section 1.4.1, positive curvature would be expected for a single plant whose stomatal conductance varies independently from that of the other plants in its community. Where the plant is part of a large community of similar plants, however, such as in a crop, the stomatal conductance of all plants may vary in unison, whereupon E(A) may become negatively curved as shown in Fig. 1.1 (COWAN 1978). The reason for this is that there is an upper limit to the rate of evaporation from a community, so that when the actual evaporation rate is near this limit, as it is for a crop having a well-developed canopy and a high stomatal conductance, further opening of the stomata will have little influence on E but will still result in an increased A. In these circumstances, a maximal $\bar{A}/\bar{E}$ is achieved by having the stomata wide open whenever there is enough light for photosynthesis. This does not mean that COWAN and FARQUHAR's (1977) hypothesis is generally inapplicable to crop plants, for during a drought, when both g and LAI are small, the transpiration rate of the crop will be much less than the potential rate and E(A) should be positively curved.

$\bar{A}/\bar{E}$ (or WUE) also increases with increasing LAI in a crop that is transpiring at or near potential rate (Fig. 1.2). When LAI > 3, further increase in LAI has a negligible influence on E′, yet may still harvest more light for use in photosynthesis (RITCHIE 1974). Thus the prodigal plant will be the more efficient under well-watered conditions. Where E'/E_0 is small, however, the conservative plant will be the more efficient, not only by keeping $\partial E/\partial A$ constant, but also by keeping LAI small (see Fig. 1,2), if we accept RITCHIE'S (1974) argument that WUE increases with decreasing LAI when LAI < 1.5.

Finally, I reiterate the point touched on in previous sections that heterotrophic tissues, such as roots, necessarily lower WUE, especially if this is defined, as it usually is, in terms of the aboveground production of dry matter. Optimum behaviour would require that the extra water obtained by investing a parcel of assimilate in the roots must enable the plant to at least replace that parcel of assimilate. The compromises reached by wild plants in their allocation of assimilate have been discussed by MOONEY (1972) and in Chapter 18, this Volume. With crop plants, where it is the performance of the community rather than of the individual that is important, competition for water loses its meaning, and we might therefore expect that domestication has reduced the root/shoot ratio. Whether or not this has happened is hard to say, for no data appear to be available. But given the tremendous concentration of roots that some crops have in the topsoil (BARLEY 1970), with rooting densities perhaps ten times larger than that needed to extract all the available water at a reasonable rate, it would seem that there is much scope for reducing the root/shoot ratio without prejudicing the supply of water to the shoot.

1.4.3 Seasonal Productivity

The previous section is largely concerned with physiological and morphological behaviour that may maximize the photosynthetic productivity of a plant over a period of some weeks when it is faced with a limiting water supply. When one looks at the longer time scale of a complete growing season, it is necessary to consider additional factors, particularly where they influence the productivity over several generations.

With long-lived perennials such as shrubs and trees, persistence rather than short-term procreation is usually the key to the productivity of a species, and while different strategies have evolved for coping with the great variation in climate through the year that is common in drought-prone environments [e.g., see the discussion by FISCHER and TURNER (1978) on aridoactive and aridopassive plants], optimal behaviour can usually be discussed along the lines of the previous section. With annual plants, however, whose long-term productivity depends on the ability to produce an ample supply of seeds each year, it is appropriate to ask what behaviour maximizes seed production, rather than simply what maximizes the production of general dry matter. A thorough discussion of this question has been recently provided by FISCHER and TURNER (1978), and it is sufficient to point out here that the production of seed is often very sensitive to conditions at particular times during the life cycle, for example at around flowering or during grain filling, especially with plants of determinate habit. Phenology in relation to climate is therefore of great importance, as also is the way in which a plant meters out any large store of water that it may have access to in the soil; conservative behaviour may have a much profounder effect on seed production than on dry-matter production if it improves the supply of water to the plant during critical periods.

1.5 Conclusions

The theme of water in the soil-plant-atmosphere continuum is usually discussed at time scales short enough for any growth of the plant to be ignored. Photosynthetic productivity on the other hand is generally discussed in relation to a growing season. The object of this chapter has been to explore, in general terms, the connection between the two when the plant is subjected to a limiting water supply. The connection depends largely on the medium-term behaviour of the main physiological and morphological variables, namely stomatal conductance and leaf area, respectively, and these in turn depend on the pattern of allocation of assimilate within the plant, particularly as it affects the compromise that must be made between water-harvesting and water-using structures. The emphasis has been on adaptive responses that a plant of given genetic constitution may make during its life-span. The genetic responses to drought that have resulted in the evolution of a wide range of life-forms and different photosynthetic pathways have been largely ignored, but are discussed in detail later in this Volume (see for instance Chaps. 15 and 18).

1.6 Symbols and Abbreviations

Symbol	Description	Unit
a	radius of root	m
b	$(\pi L)^{-\frac{1}{2}}$, i.e., half the distance between adjacent roots	m
x	distance	m
$\underline{A}$	assimilation rate	mol m^{-2} s^{-1}
$\bar{A}$	mean assimilation rate during a given period	mol m^{-2} s^{-1}
$\bar{A}$*	mean assimilation rate per plant	mol s^{-1}
B	biomass produced	kg
$\underline{E}$	transpiration rate	mol m^{-2} s^{-1}
$\bar{E}$	mean transpiration rate during a given period	mol m^{-2} s^{-1}
$\bar{E}$*	mean transpiration rate per plant	mol s^{-1}
E′	transpiration rate per unit ground area	mol m^{-2} s^{-1}
E_o	potential transpiration rate per unit ground area	mol m^{-2} s^{-1}
F	flux density of water	ms^{-1}
I	rate of water uptake per unit length of root	m^3 m^{-1} s^{-1}
K	hydraulic conductivity	m^2 s^{-1} Pa^{-1}
L	rooting density, i.e., the length of root per unit volume of soil	m m^{-3}
LAI	leaf area index	m^2 m^{-2}
P	pressure	Pa
$\underline{Q}$	rate of change of soil water content	m^3 m^{-3} s^{-1}
$\bar{V}_w$	partial molar volume of water	m^3 mol^{-1}
W	water transpired	kg
WUE	efficiency of water use (=B/W)	kg kg^{-1}
ε	volumetric elastic modulus	Pa
θ	soil water content	m^3 m^{-3}
λ	$\partial E/\partial A$	
μ_w	chemical potential of water	J mol^{-1}
τ	matric suction	Pa
Ψ	water potential ($\Delta\mu_w$) or water "potential" ($\Delta\mu_w/\bar{V}_w$)	J mol^{-1} Pa
Π	osmotic pressure	Pa

References

Barley KP (1970) The configuration of the root system in relation to nutrient uptake. Adv Agron 22:159–201

Begg JE (1980) Morphological adaptations of leaves to water stress. In: Turner NC, Kramer PJ (eds) Adaptation of plants to water and high temperature stress. Wiley-Interscience, New York, pp 33–42

Begg JE, Turner NC (1970) Water potential gradients in field tobacco. Plant Physiol 46:343–346

Bennett OL, Doss BD (1960) Effect of soil moisture level on root distribution of cool-season forage species. Agron J 52:204–207

Black CR (1979) The relationship between transpiration rate, water potential, and resistances to water movement in sunflower (*Helianthus annuus* L.). J Exp Bot 30:235–243

Blum A (1974) Genotypic responses in sorghum to drought stress I. Response to soil moisture stress. Crop Sci 14:361–364

Bolt GH, Iwata S, Peck AJ, Raats PAC, Rode AA, Vachaud G, Voronin AD (1976) Soil physics terminology. Bull Int Soc Soil Sci 49:26–35

Boyer JS (1974) Water transport in plants: mechanism of apparent changes in resistance during absorption. Planta 117:187–207

Boyer JS (1976) Water deficity and photosynthesis. In: Kozlowski TT (ed) Water deficits and plant growth, vol IV. Academic Press, London New York, pp 153–190

Boyer JS (1977) Regulation of water movement in whole plants. In: Jennings DH (ed) Integration of activity in the higher plant. Univ Press, Cambridge, pp 455–470

Bunce JA (1977) Leaf elongation in relation to leaf water potential in soybean. J Exp Bot 28:156–161

Burch GJ (1979) Soil and plant resistances to water absorption by plant root systems. Aust J Agric Res 30:279–292

Burch GJ, Smith RCG, Mason WK (1978) Agronomic and physiological responses of soybean and sorghum crops to water deficits. II. Crop evaporation, soil water depletion and root distribution. Aust J Plant Physiol 5:169–177

Caldwell MM (1976) Root extension and water absorption. In: Lange OL, Kappen L, Schulze E-D (eds) Water and plant life. Ecol Stud Vol 19. Springer, Berlin Heidelberg New York, pp 63–85

Carbon BA (1973) Diurnal water stress in plants grown on a coarse soil. Aust J Soil Res 11:33–42

Cowan IR (1965) Transport of water in the soil-plant-atmosphere system. J Appl Ecol 2:221–239

Cowan IR (1972a) An electrical analogue of evaporation from, and flow of water in plants. Planta 106:221–226

Cowan IR (1972b) Oscillations in stomatal conductance and plant functioning associated with stomatal conductance: observations and a model. Planta 106:185–219

Cowan IR (1977) Stomatal behaviour and environment. Adv Bot Res 4:117–228

Cowan IR (1978) Water use in higher plants. In: McIntyre AK (ed) Water: planets, plants and people. Aust Acad Sci, Canberra, pp 71–107

Cowan IR, Farquhar GD (1977) Stomatal function in relation to leaf metabolism and environment. In: Jennings DH (ed) Integration of activity in the higher plant. Univ Press, Cambridge, pp 471–505

Cowan IR, Milthorpe FL (1968) Plant factors influencing the water status of plant tissues. In: Kozlowski TT (ed) Water deficits and plant growth, vol I. Academic Press, London New York, pp 137–193

Cutler JM, Rains DW (1977) Effects of irrigation history on responses of cotton to subsequent water stress. Crop Sci 17:329–335

Dainty J (1976) Water relations of plant cells. In: Lüttge U, Pitman MG (eds) Transport in plants II Encyclopedia of plant physiology New Ser vol II, Part A. Springer, Berlin Heidelberg New York, pp 12–35

Denmead OT, Miller BD (1976) Water transport in wheat plants in the field. Agron J 68:297–303

Dunin FX, Aston AR, Reyenga W (1978) Evaporation from a *Themeda* grassland. II. Resistance model of plant evaporation. J Appl Ecol 15:847–858

Eavis BW, Taylor HM (1979) Transpiration of soybeans as related to leaf area, root length, and soil water content. Agron J 71:441–445
Edwards M, Meidner H (1975) Micromanipulation of stomatal guard cells. Nature 253:114–115
Faiz SMA, Weatherley PE (1978) Further investigations into the location and magnitude of the hydraulic resistances in the soil:plant system. New Phytol 81:19–28
Farquhar GD (1979) Feedforward responses of stomata to humidity. Aust J Plant Physiol 5:787–800
Farquhar GD, Schulze E-D, Küppers M (1980) Responses to humidity by stomata of *Nicotiana glauca* L. and *Corylus avellana* L. are consistent with the optimisation of carbon dioxide uptake with respect to water loss. Aust J Plant Physiol 7:315–327
Fischer RA, Turner NC (1978) Plant productivity in the arid and semiarid zones. Annu Rev Plant Physiol 29:277–317
Gardner WR (1960) Water availability to plants. Soil Sci 89:63–73
Giordano R, Salleo A, Salleo S, Wanderlingh F (1978) Flow in xylem vessels and Poiseuille's law. Can J Bot 56:333–338
Greacen EL (1977) Mechanisms and models of water transfer. In: Russell JS, Greacen EL (eds) Soil factors in crop production in a semi-arid environment. Univ Queensland Press, St Lucia, pp 163–196
Greacen EL, Ponsana P, Barley KP (1976) Resistance to water flow in the roots of cereals. In: Lange OL, Kappen L, Schulze E-D (eds) Water and plant life. Ecol Stud Vol 19. Springer, Berlin Heidelberg New York, pp 86–100
Hailey JL, Hiler EA, Jordan WR, Van Bavel CHM (1973) Resistance to water flow in *Vigna sinensis* L. (Endl.) at high rates of transpiration. Crop Sci 13:264–267
Hansen GK (1974) Resistance to water transport in soil and young wheat plants. Acta Agric Scand 24:37–48
Hellkvist J, Richards GP, Jarvis PG (1974) Vertical gradients of water potential and tissue water relations in sitka spruce trees measured with the pressure chamber. J Appl Ecol 11:637–668
Herkelrath WN, Miller EE, Gardner WR (1977a) Water uptake by plants: I. The divided root experiments. Soil Sci Soc Am J 41:1033–1038
Herkelrath WN, Miller EE, Gardner WR (1977b) Water uptake by plants: II. The root contact model. Soil Sci Soc Am J 41:1039–1043
Honert van der TH, (1948) Water transport in plants as a catenary process. Discuss Faraday Soc 3:146–153
Hsiao TC, Acevedo E (1974) Plant responses to water deficits, water-use efficiency, and drought resistance. Agric Meteorol 14:59–84
Hsiao TC, Fereres E, Acevedo E, Henderson DW (1976) Water stress and dynamics of growth and yield of crop plants. In: Lange OL, Kappen L, Schulze E-D (eds) Water and plant life. Ecol Stud Vol 19. Springer, Berlin Heidelberg New York, pp 281–305
Hurd EA (1974) Phenotype and drought tolerance in wheat. Agric Meteorol 14:39–55
Hüsken D, Steudle E, Zimmermann U (1978) Pressure probe technique for measuring water relations of cells in higher plants. Plant Physiol 61:158–63
Janes BE, Gee GW (1973) Changes in transpiration, net carbon dioxide assimilation and leaf water potential resulting from application of hydrostatic pressure to roots of intact pepper plants. Physiol Plant 28:201–208
Jarvis PG (1975) Water transfer in plants. In: Vries de Da, Afgan NH (eds) Heat and mass transfer in the biosphere I. Transfer processes in the plant environment. Halsted, Washington DC, pp 369–394
Jeje AA, Zimmermann MH (1979) Resistance to water flow in xylem vessels. J Exp Bot 30:817–827
Jones MM, Turner NC (1978) Osmotic adjustment in leaves of sorghum in response to water deficits. Plant Physiol 61:122–126
Jordan WR, Miller FR (1980) Genetic variability in sorghum root systems: implications for drought tolerance. In: Turner NC, Kramer PJ (eds) Adaptation of plants to water and high temperature stress. Wiley-Interscience, New York, pp 383–399
Kramer PJ (1969) Plant and soil water relationships: a modern synthesis. McGraw-Hill, New York
Larcher W (1975) Physiological plant ecology. Springer, Berlin Heidelberg New York

Lawlor DW (1972) Growth and water use of *Lolium perenne*. I. Water transport. J Appl Ecol 9:79–98
Legg BJ, Day W, Lawlor DW, Parkinson KJ (1979) The effect of drought on barley growth: models and measurements showing the relative importance of leaf area and photosynthetic rate. J Agric Sci 92:703–716
Meiri A, Plaut Z, Shimshi D (1975) The use of the pressure chamber technique for measurement of the water potential of transpiring plant organs. Physiol Plant 35:72–76
Monteith JL (1965) Evaporation and environment. In: Fogg GE (ed) The state and movement of water in living organisms. Univ Press, Cambridge, pp 205–234
Mooney HA (1972) The carbon balance of plants. Annu Rev Ecol Syst 3:315–346
Neumann HH, Thurtell GW, Stevenson KR (1974) In situ measurements of leaf water potential and resistance to water flow. Can J Plant Sci 54:175–184
Newman EI (1966) Relationship between root growth of flax (*Linum usitatissimum*) and soil water potential. New Phytol 65:273–283
Newman EI (1969) Resistance to water flow in soil and plant. I. Soil resistance in relation to amounts of root: theoretical estimates. J Appl Ecol 6:1–12
Newman EI (1976) Water movement through root systems. Philos Trans R Soc London Ser B 273:463–478
Noy-Meir I (1973) Desert ecosystems: environment and producers. Annu Rev Ecol Syst 4:25–51
Noy-Meir I, Ginzburg BZ (1967) An analysis of the water potential isotherm in plant tissue. I. The theory. Aust J Biol Sci 20:695–721
Nulsen RA, Thurtell GW (1978) Osmotically induced changes in the pressureflow relationship of maize root systems. Aust J Plant Physiol 5:469–476
Oertli JJ (1969) Terminology of plant-water energy relations. Z Pflanzenphysiol 61:264–265
Passioura JB (1976) The control of water movement through plants. In: Wardlaw IF, Passioura JB (eds) Transport and transfer processes in plants. Academic Press, London New York, pp 373–380
Passioura JB (1977) Grain yield, harvest index, and water use of wheat. J Aust Inst Agric Sci 43:117–121
Passioura JB (1980a) Transport of water from soil to shoot in wheat seedlings. J Exp Bot 31:333–345
Passioura JB (1980b) The meaning of matric potential. J Exp Bot 31:1161–1169
Passioura JB (1981) Water collection by roots. In: Paleg L, Aspinall D (eds) Physiology and biochemistry of drought resistance in plants. Academic Press, London New York, pp 39–53
Philip JR (1966) Plant water relations: some physical aspects. Annu Rev Plant Physiol 17:245–268
Pierce M, Raschke K (1980) Correlation between loss of turgor and accumulation of abscisic acid in detached leaves. Planta 148:174–182
Portas CAM, Taylor HM (1976) Growth and survival of young plant roots in dry soil. Soil Sci 121:170–175
Raschke K (1979) Movements of stomata. In: Haupt W, Feinleib ME (eds) Physiology of movements. Encyclopedia of plant physiology New Ser vol. VII. Springer, Berlin Heidelberg New York, pp 383–441
Rawson HM (1979) Vertical wilting and photosynthesis, transpiration, and water use efficiency of sunflower leaves. Aust J Plant Physiol 6:109–120
Reicosky DC, Ritchie JT (1976) Relative importance of soil resistance and plant resistance in root water absorption. Soil Sci Soc Am Proc 40:293–297
Richter H (1973) Frictional potential losses and total water potential in plants: a reevaluation. J Exp Bot 24:983–994
Ritchie JT (1973) Influence of soil water status and meteorological conditions on evaporation from a corn canopy. Agron J 65:893–897
Ritchie JT (1974) Atmospheric and soil water influences on the plant water balance. Agric Meteorol 14:183–198
Rose DA (1968) Water movement in porous materials. III. Evaporation of water from soil. Br J Appl Phys Ser 2, 1:1779–1791

Russell RS (1977) Plant root systems: their function and interaction with the soil. McGraw-Hill, London
Saffigna PG, Tanner CB, Keeney DR (1976) Non-uniform infiltration under potato canopies caused by interception, stemflow, and hilling. Agron J 68:337–342
Sauerbeck D, Johnen B (1976) Der Umsatz von Pflanzenwurzeln im Laufe der Vegetationsperiode und dessen Beitrag zur "Bodenatmung". Z Pflanzenernaehr Bodenkde 3:315–328
Schultz JE (1971) Soil water changes under fallow-crop treatments in relation to soil type, rainfall and yield of wheat. Aust J Exp Agric Anim Husb 11:236–242
Schultz JE (1974) Root development of wheat at the flowering stage under different cultural practices. Agric Rec 1:12–17
Shackel KA, Hall AE (1979) Reversible leaflet movements in relation to drought adaptation of cowpeas *Vigna unguiculata* (L.) Walp Aust J Plant Physiol 6:265–276
Sheriff DW (1979) Water vapour and heat transfer in leaves. Ann Bot (London) 43:157–171
Slatyer RO (1962) Methodology of a water balance study conducted on a desert woodland (*Acacia aneura* F. Muell.) community in central Australia. Arid Zone Res 16:15–26
Slatyer RO (1967) Plant-water relationships. Academic Press, London New York
Slatyer RO, Taylor SA (1960) Terminology in plant-soil-water relations. Nature London 187:922–924
Spanner DC (1973) The components of the water potential in plants and soils. J Exp Bot 24:816–819
Taylor HM, Gardner HR (1963) Penetration of cotton seedling taproots as influenced by bulk density, moisture content, and strength of soil. Soil Sci. 96:153–156
Taylor HM, Klepper B (1975) Water uptake by cotton root systems: an examination of assumptions in the single root model. Soil Sci 120:57–67
Taylor HM, Klepper B (1978) The role of rooting characteristics in the supply of water to plants. Adv Agron 30:99–128
Taylor HM, Ratcliff LF (1969) Root elongation rates of cotton and peanuts as a function of soil strength and soil water content. Soil Sci 108:113–119
Taylor SA, Slatyer RO (1962) Proposals for a unified terminology in studies of plant-soil-water relations. Arid Zone Res 16:339–349
Tinker PB (1976) Transport of water to plant roots in soil. Philos Trans R Soc London Ser B 273:445–461
Turner NC (1974) Stomatal response to light and water under field conditions. In: Bieleski RL, Ferguson AR, Creswell MM (eds) Mechanisms of regulation of plant growth. Roy Soc NZ Bull 12:423–432
Turner NC, Long MJ (1980) Errors arising from rapid water loss in the measurement of leaf water potential by the pressure chamber technique. Aust J Plant Physiol 7:527–537
Tyree MT (1976) Negative turgor pressure in plant cells: fact or fallacy? Can J Bot 54:2738–2746
Tyree MT, Caldwell C, Dainty J (1975) The water relations of hemlock (*Tsuga canadensis*). V. The localization of resistances to bulk water flow. Can J Bot 53:1078–1084
Walter CJ, Barley KP (1974) The depletion of soil water by wheat at low, intermediate and high rates of seeding. Proc 10th Int Congr Soil Sci Moscow 1:150–158
Weatherley PE (1970) Some aspects in water relations. Adv Bot Res 3:171–206
Weatherley PE (1975) Water relations of the root system. In: Torrey JG, Clarkson DT (eds) The development and function of roots. Academic Press, London New York, pp 397–413
Weatherley PE (1976) Introduction: water movement through plants. Philos Trans R Soc London Ser B 273:435–444
Wilson AM, Hyder DN, Briske DD (1976) Drought resistance characteristics of blue grama seedlings. Agron J 68:479–484
Zimmermann U, Steudle E (1979) Physical aspects of water relations of plant cells. Adv Bot Res 6:46–119

2 Water in Tissues and Cells

M.T. TYREE and P.G. JARVIS

CONTENTS

2.1 Introduction

As already explained in the previous chapter, generally the water potential, Ψ, of a plant cell is expressed as the sum of three components, as follows (Dainty 1976):

$$\Psi = P + \pi + \tau \tag{2.1}$$

where P, π and τ are the pressure, osmotic and matric potentials, respectively. In a tissue or organ, Eq. (2.1) applies to each cell individually but if different cells have different solute concentrations, moduli of elasticity or volumes of apoplasmic water, then even at equilibrium the cellular values of P, π and τ may differ from cell to cell. Although measurements of π have frequently been made on individual cells using plasmolytic methods and measurements of P have recently been made on individual cells of higher plants using the pressure probe (Steudle et al. 1975; Hüsken et al. 1978, 1980), it is in general not practical to make a large number of such measurements in the tissues and organs of higher plants. In spite of this problem Tyree and Hammel (1972) have argued that it is meaningful to define bulk values of the water relations parameters to characterize the water relations properties of tissues and organs. They derived *volume-averaged,* or more precisely *weight-averaged*, values of pressure potential, $\bar{P}$, and osmotic potential $\bar{\pi}$, from analysis of the Ψ isotherms of individual cells, with the following specific meanings:

$$\bar{\pi} = \sum^{n} \frac{W_i}{W} \pi_i \tag{2.2}$$

and

$$\bar{P} = \sum^{n} \frac{W_i}{W} P_i \tag{2.3}$$

where W is the weight of water in the tissue symplasm, and W_i the weight of water in the ith cell of osmotic-potential π_i and pressure potential P_i. The bar represents weight-averaged values, the summation being made over all n cells in the tissue. Thus for a tissue at equilibrium, Eq. (2.1) should be replaced by the appropriate weight-averaged parameters, i.e.,

$$\Psi = \bar{P} + \bar{\pi} + \bar{\tau}, \tag{2.4}$$

recognizing that the components of Ψ are averaged over each of the heterogeneous phases in the tissue, i.e., vacuole, cytoplasm *or* cell wall.

In principle, weight-averaging over heterogeneous phases is unsatisfactory because, strictly speaking, Ψ and its components are intensive variables which vary from point to point in a cell and tissue (Weatherley 1970). This always obtains when water flow occurs but even at equilibrium when Ψ is the same at all points the components can and do change with position as long as their
recognizing that the components of Ψ are averaged over each of the heterogeneous phases in the tissue, i.e., vacuole, cytoplasm *or* cell wall.

sum, $(\bar{P}+\pi+\tau)$, is constant. For example, at equilibrium, values of $\bar{P}+\pi+\tau$ might be (10)+(−20)+(−0) bar in the vacuole, (12)+(−15)+(−7) bar at a particular point near a charged surface in the cell wall. However, since at the present time we are unable to measure such point potentials, weight-averaged values provide a practical means of characterization of average tissue properties. Nonetheless, their use should not be allowed to obscure the fundament definition of the component potentials. There is general agreement on the meaning of Ψ and how it should be measured for a tissue. However, much less agreement has been reached on the meaning and measurement of the components of Ψ, particularly of τ (e.g., PASSIOURA 1980).

2.2 The Components of Tissue Water Potential

2.2.1 Matric Potential, τ

Matric potentials in tissues were originally conceived as a component of the total potential resulting from interactions between the water and the matrix by analogy with the identification of a matric component in soils (TAYLOR and SLATYER 1961). Matric potentials in soils were considered to be largely the result of capillarity between the pores with "surface forces", especially in clays, important at low potentials (BOLT and FRISSEL 1960). It was assumed that similar potentials would exist in cell walls and possibly also in the cytoplasm where swelling phenomenon such as imbibition were well known. Since the original postulation, the physical meaning of $\bar{\tau}$ has been the subject of unresolved controversy. Some have defined it in terms of a measurement procedure without much regard to thermodynamic principles (e.g., WIEBE 1966; WARREN WILSON 1967a; ACOCK 1975): others have taken more cognizance of the thermodynamic restrictions imposed on definitions based on experimental procedures (e.g., NOY-MEIR and GINZBURG 1967). All such definitions are unsatisfactory because they are derived from tissue properties obtained by volume or weight-averaging over the heterogeneous phases of vacuole, cytoplasm, and cell wall. A satisfactory definition of matric potential must be based upon consideration of it as an intensive property acting at a point. To pursue this concept and consequences quantitatively, we must stray a little from purely thermodynamic theory into a mixture of thermodynamics and statistical mechanics. This approach has been taken with some success by TYREE and KARAMANOS (1980) for plant tissues. Some of the basic concepts were originally put forward for soil by BOLT and MILLER (1958), but unfortunately the merits of this earlier paper appear to have been largely unheeded by plant physiologists.

2.2.1.1 The Physical Origin of τ

Matric forces are interfacial forces, which occur between liquid and solid phases. The forces are short-range and therefore influence only a very small fraction of the total water even in systems with large surface to volume ratios, e.g., clays and plant cell walls. At uncharged surfaces the force interactions are largely London-van der Waals forces or hydrogen bonds, and probably extend for only one or two diameters of the water molecule, 0.3 to 0.6 nm. At charged surfaces in contact with electrolytic solutions the longer range electrostatic forces extend a few nm. The surface charges are balanced by swarms of counterions that arrange themselves in a diffuse layer near the surface. Very high, localized concentrations of counterions can develop and contribute to low localized values of π. The impact of these counterions on π has been calculated according to the Gouy-Chapman theory by TYREE and KARAMANOS (1980) who showed that localized concentrations can exceed 2 molar resulting in values of π below -50 bar. However, it does not seem appropriate to identify τ with such osmotic effects of swarms of counterions because there is no sharp distinction between π within the diffuse layer and outside it; π declines gradually from the microregion where the electric effects are negligible to the sites where they are important.

Within the electric fields of the surface charge there is another effect that reduces the energy of water molecules. This is the interaction of the water dipole with the electric field. Plant cell wall surfaces and clay surfaces have a net negative charge. This causes the water dipole to orient itself with the positive (hydrogen) end a little nearer the surface than the negative (oxygen) end because like charges are repelled and opposite charges are attracted. The net effect is a lowering of the free energy of the water molecules within the electric field. In terms of water potential the magnitude of the effect is given by

$$\tau = -\left(\frac{N_0^2 P_0^2}{3V_w RT}\right) F^2, \tag{2.5}$$

where N_0 is Avogadro's number, R is the gas constant, T is the Kelvin temperature, V_w is the volume of a mole of water, P_0 is the dipole moment of water, and F is the electric field.

Figure 2.1 shows the magnitudes of the components of Ψ near a charged surface according to the Gouy-Chapman theory. In this case bulk water containing 10 mM NaCl at $\Psi = -15$ bar is at equilibrium with water near a surface with a net charge of -4×10^{-5} C cm^{-2}. The large positive pressure that develops near the charged surface can be identified as the swelling pressure of the gels. The pressure arises because of the force with which water molecules are drawn towards the charged surface. The molecules piling on top of each other produce a pressure in much the same way that gravity produces pressure under deep water. Given that π and τ are both very negative near charged surfaces, large positive values of P are necessary to balance Eq. (2.1) since at equilibrium Ψ is uniform over space.

2.2.1.2 The Effect of τ on Water Potential Isotherms

Since the forces contributing to τ are short-range (Fig. 2.1), they influence only a small fraction of the total water volume even in cell walls. Figure 2.2

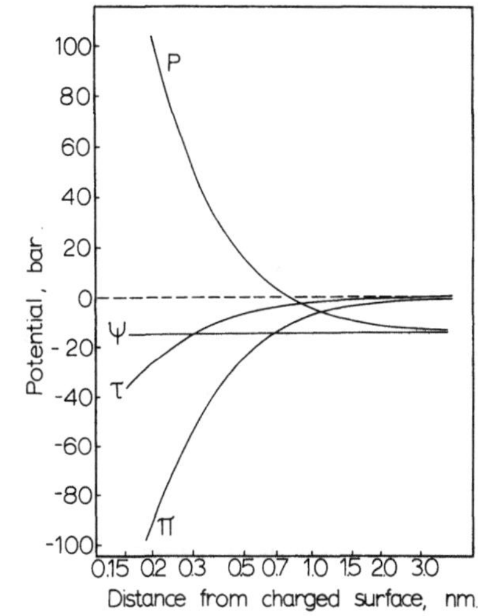

Fig. 2.1. The components of water potential near a charged surface according to the Gouy-Chapman double layer theory with a surface charge of -4×10^{-5} C cm^{-2}. The univalent ion concentration outside the double layer is taken as 10 mM. π and τ are calculated at distances from the charged surface from the calculated electrical potential and electric field, respectively; $P=\Psi+\pi+\tau$. (TYREE and KARAMANOS 1980)

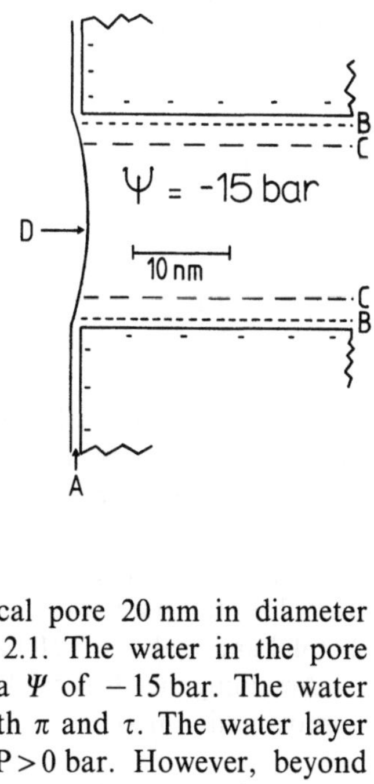

Fig. 2.2. A scale diagram of the air–water interface near a water-filled, charged pore. The pore is 20 nm in diameter and is filled with water at $\Psi=-15$ bar. The other parameters are as in Fig. 2.1. *A* is a thin layer of water on the charged surface held there by π and τ; *B* is the distance from the charged surface at which $\Psi_p=0$; *C* is the distance at which Ψ is more than 90% attributable to negative P; *D* is the air–water interface with a radius of curvature of 96 nm. (TYREE and KARAMANOS 1980)

shows a scale diagram of the water in a cylindrical pore 20 nm in diameter in a plant cell wall based on the profiles in Fig. 2.1. The water in the pore and the water vapour above it are taken to have a Ψ of -15 bar. The water layer *A* close to the charged surface is held by both π and τ. The water layer between the surface and the boundary *B* is at $P>0$ bar. However, beyond boundary *C*, P accounts for more than 90% of Ψ, i.e., $P<-13.5$ bar. Thus most of the water is outside the force effects of the charged surface. The negative pressure in the pore in Fig. 2.2 arises from the surface tension of the air–water interface which has a radius of curvature of 96 nm at -15 bar in this figure. In the xylem negative pressures are included in P and for consistency and clarity we feel that such surface tension effects in pores should also be included in P, rather than in τ, as suggested by WARREN WILSON (1967a) and others.

Only if the air–water interface intrudes into the pore as a result of evaporation of water from the cell walls, for example, will the thin layer of water remaining under the influence of τ become a substantial proportion of the water present.

However, for a pore diameter of 20 nm, Ψ would have to fall below -140 bar for the air–water interface to intrude, and in virtually all but the resurrection plants, the cells would be dead. Consequently we feel that for all practical purposes and certainly within the range of values of Ψ found in nearly all living plants, the effects of τ and of charged surfaces are unimportant and can be excluded from consideration in spite of the prominence given to them in some earlier analyses of Ψ isotherms (e.g., WARREN WILSON 1967a; ACOCK 1975). In our view, matric potential is not an important component of water potential: surface tension and cell turgor can be accommodated by the pressure potential component and all solute effects can be included in the osmotic potential.

2.2.2 The Osmotic Potential, π

For a discussion of the thermodynamic meaning of osmotic potential readers should consult SLATYER (1967) and ROBINSON and STOKES (1959). Readers are advised to consult AIKMAN (1979) for an introduction to the molecular interpretation of π.

2.2.2.1 Ideal Solutions

The osmotic potential of an ideal osmotic solution is

$$\pi_{\text{ideal}}=\frac{RT}{\bar{V}_w}\ln x_w, \tag{2.6}$$

where $\bar{V}_w$ is the partial molar volume of water ($\simeq 18\ \text{cm}^3\ \text{mol}^{-1}$) and x_w is the mole fraction of the water in the solution, i.e., $n_w/(n_w+N)$, where n_w is the number of moles of water in a solution containing N moles of solutes. If the solution contains several solutes, N is the sum of all solutes with the ionic species calculated separately. In the biological range, x_w normally varies from 1.00 for pure water and $\pi=0$, to 0.98, for a solution of approximately 2 molal and π about -50 bar.

Assuming the ideal gas law and Raoult's law, and making the mathematical approximation that

$$\frac{1}{\bar{V}_w}\ln x_w \simeq -\frac{1}{\bar{V}_w}\frac{N}{n_w}, \tag{2.7}$$

it can be shown that for ideal solutions of up to 2 molal Eq. (2.6) can be approximated with an error of less than 2% by

$$\pi_{\text{ideal}}=-RTN/V=-RT\rho_w C^*, \tag{2.8}$$

where V ($=\bar{V}_w n_w$) is the volume of water and C* ($=$N/W) is the solute concentration defined as the total number of moles of solute, N, ions counted separately, per unit mass of *water*, W. ρ_w is the density of water in the solution and is equal to $M_w/\bar{V}_w$, where M_w is the molar mass of water ($=18$ g mol^{-1}). Since $\bar{V}_w$ changes slightly with solution composition and temperature, ρ_w deviates slightly with conditions, but this deviation is usually ignored and ρ_w is taken as 1000 kg m^{-3} or 1.0 g cm^{-3}. (In the latter case ρ_w is often omitted, but this makes the equation dimensionally incorrect.) Thus the definition of π for an ideal solution as proportional to 1/W depends on two thermodynamic idealizations and a mathematical approximation in which x_w is approximated by the first term of a power series.

2.2.2.2 Non-Ideal Solutions

A real solution can have an osmotic potential differing from the ideal value. The size and direction of disagreement depends on the concentration and species of molecules or ions. In the concentration range of 0 to 2 molal, simple salts have values of π that are up to 20% less negative than the ideal value; π of sucrose is up to 20% more negative than the ideal value; π of polyethylene glycol 6000 MW is up to 100% more negative. At low concentrations values the π of inorganic acids and organic salts are less negative than predicted by Eq. (2.8), and more negative at high concentrations. To take account of this deviation, SLATYER (1967) introduced an activity coefficient, such that $\gamma x_w = a_w$, where a_w is the activity of water. ROBINSON and STOKES (1959) used both a rational coefficient, g, and an osmotic coefficient, ϕ, so that

$$\pi = g\frac{RT}{\bar{V}_w}\ln x_w, \quad \text{or} \quad \pi = -\phi RTN/V.$$

For ideal solutions γ and g are equal to 1.0 and $\phi = \frac{n_w}{N}\ln(1-N/n_w)$. For practical purposes, ϕ is also equal to 1.0 for an ideal solution if Eq. (2.8) is accepted as an approximation for an ideal solution. For non-ideal solutions ϕ deviates considerably from unity (Fig. 2.3).

Non-ideality is not a problem in practice because π of free solutions is usually measured by a vapor equilibration or a freezing point method which has been calibrated against known standards based on direct methods of physical chemistry (see ROBINSON and STOKES 1959, Chap. 8 for the methods and LANG 1967 for an example of their application). Commercial freezing point devices usually measure the osmolal concentration, C, of the solution. In terms of the osmotic coefficient, $C = \phi C^*$, hence the osmolal concentration is the concentration of an ideal solution which has the same value of π as the real solution, so C is sometimes less than C* and sometimes larger. The value of π is then calculated from

$$\pi = -RT\rho_w C = -RTN_s/V, \tag{2.9}$$

where N_s is the number of osmoles of solute in the cell sap of the sample and C is N_s/W. In some solutions ϕ is temperature-dependent even though

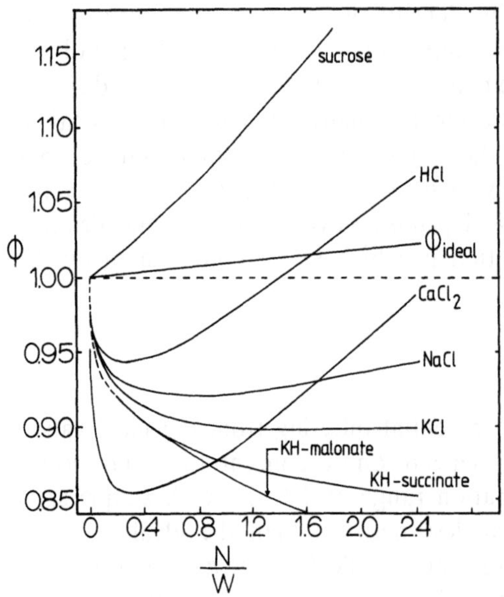

Fig. 2.3. The relation between the osmotic coefficient ϕ and N/W for a number of solutions of single compounds or salts using data from ROBINSON and STOKES (1959). The curve marked ϕ ideal shows the ideal values of ϕ in the thermodynamic sense of ideality as defined by Eq. (2.6) and with $\phi_{\text{ideal}} = -\frac{n_w}{N}\ln(1-N/n_w)$. (TYREE and RICHTER 1981)

C*, the chemical concentration, is constant. Consequently an error of up to 10% can arise when π is measured at the freezing point, T_f (K), and converted to a higher temperature, T (K), i.e., π (at T) is not always equal to $\frac{T}{T_f}\pi$(at T_f). This can best be avoided by measuring π at temperature T by a psychrometric technique.

WARREN WILSON (1967a) proposed that all deviations of π of a solution from ideality should be added on to the matric potential. His rationale appears to be that some water binds to solute molecules with the consequence that this water is unable to participate in osmotic exchange in the solution. The implication is that this bound water becomes part of the solute and decreases the mole fraction of water in the solution, x_w in Eq. (2.6), and hence π. This hypothesis is not consistent with all the facts. It might explain non-ideal deviation for organic molecules but not for salts, the behaviour of which deviates in the wrong direction in spite of water being bound in the hydration shells of ions. The concept of bound water is nebulous since it is defined in terms of the method used to measure it and the values obtained depend on the method used (KRAMER 1955). It is best to avoid the concept of bound water and to treat the non-ideal behaviour of solutions by the use of osmotic coefficients or "rational" coefficients, as discussed earlier (ROBINSON and STOKES 1959).

2.2.2.3 Solute Compartmentation

Solutes lower the osmotic potential in all three of the main heterogeneous phases of the cell – the vacuole, the cytoplasm and organelles and the cell walls. The pressure potential in vacuole, cytoplasm and cytoplasmic organelles must be approximately equal since it is unlikely that the bounding membranes can restrain any appreciable pressure. Consequently the osmotic potential must be uniform throughout the cell, other than close to surfaces as discussed earlier. However, the solutes involved in the different phases are generally totally different. The vacuole contains inorganic ions, such as Na^+ and Cl^-, organic acids like malic acid and also sugars such as sucrose, whereas the cytoplasm contains a range of "compatible" solutes which can accumulate in high concentrations without detriment to the cell biochemistry (HELLEBUST 1976; CRAM 1976; WYN JONES 1981). The compatible solutes which accumulate in the cytoplasm are much less well known; they include K^+ and a number of amino acids or related species, such as proline, glycinebetaine, and polyols such as sorbitol, mannitol and glycerol as well as sugars. Because of these differences in solute species, substantial differences in the degree of ideality of behaviour during loss of water from the two phases are to be expected. This has rarely been considered, the osmotic properties of the cell usually being regarded as those of the solution in the vacuole. The apoplasmic water in the cell wall, in contrast, consists of a dilute salt solution, $\pi = -0.1$ to -0.5 bar, that changes little in concentration during loss of water from the tissue and contributes little to the local value of Ψ which is largely made up of a negative P.

2.2.2.4 Measurement of π by Sap Extraction

π can be measured on extracts of sap obtained from crushed and/or frozen tissue (e.g., NOY-MEIR and GINZBURG 1969). Alternatively, Ψ can be measured on tissue killed by freezing or other means and the measured values of Ψ equated to $\pi + \tau$; π is then computed from a somewhat dubious evaluation of τ (e.g., WARREN WILSON 1967a, b, c; ACOCK 1975). These methods are not to be recommended because the cell sap is diluted with apoplasmic water after crushing or freezing and π is overestimated. The magnitude of this error can be typically 20% and is higher if some of the cells pass through the crushing or freezing procedure unharmed (BOYER and POTTER 1973; TYREE 1976a; KYRIAKOPOULOS and RICHTER 1977). If W_c is the weight of water released from the cell sap by crushing or freezing and W_a is the weight of apoplasmic water, then the ratio of measured π:actual π of the cell sap is $W_c:(W_c + W_a)$. This ratio applies only if no substantial ion exchange occurs between the symplasmic ions released by killing and the ion exchange sites of the cell walls. Ion exchange interactions could raise or lower the ratio depending on the ionic species involved in the exchange.

2.2.2.5 Measurement of π by Pressure–Volume Curves

This method involves the measurement and analysis of a tissue water potential isotherm, and is to be preferred because it yields a weight-averaged value of $\bar{\pi}$ of the undiluted cell sap. In this method the tissue is progressively dehydrated well past the turgor loss point, Ψ of the tissue being measured as a function of the weight of water lost, W_e. Several water relations parameters, including the osmotic potential as a function of water content, can then be determined from the analysis of the Ψ isotherm discussed below.

A disadvantage of this method is that an hour or more of experimental time is required to evaluate $\bar{\pi}$, whereas the sap extraction and tissue killing methods can give results in minutes. In some cases speed may be important. In soybean (*Glycine max*) π is light-dependent and can change rapidly when irradiance changes (WENKERT 1980). WENKERT used the sap–extraction method and corrected for dilution effects by use of the pressure–volume method in darkness, when π was stable.

2.2.3 Analysis of Ψ Isotherms

As water is lost from the tissue, either by successive increments of pressure in a pressure bomb or by evaporation from the organ, $\bar{P}$ falls to zero at the turgor loss point. If negative turgor does not occur, then at lower potentials $\Psi = \bar{\pi}$ and, from Eq. (2.9)

$$\frac{1}{\Psi} = \frac{-W}{RTN_s\rho_w} = -\frac{(W_o - W_e)}{RTN_s\rho_w} = \frac{W_e}{RTN_s\rho_w} - \frac{W_o}{RTN_s\rho_w}, \tag{2.10}$$

where W_o is the weight of symplasmic water at full hydration and W_e is the weight of water lost from the whole sample. This equation is valid only if the amount of apoplasmic water, W_a, does not change during dehydration so that all the water lost comes from the symplasm. This assumption is discussed in more detail in Section 2.2.6. A plot of $1/\Psi$ versus W_e will yield a "straight" line with an intercept equal to $-W_o/RTN_s\rho_w = 1/\pi_o$ where π_o is the osmotic potential at full hydration, and a slope of $1/RTN_o\rho_w$. At any other state of hydration π can be calculated from

$$\bar{\pi} = \bar{\pi}_o\, W_o/W. \tag{2.11}$$

Thus this analysis yields π as a function of W over the whole range of water contents.

It is also possible to calculate $\bar{\pi}_o$ by plotting $1/\Psi$ against relative water content, R* (HELLKVIST et al. 1974; TYREE 1976b; RICHTER 1978). This is so because

$$R^* = \frac{W + W_a}{W_o + W_a} \tag{2.12}$$

and hence

$$W = (W_o + W_a)R^* - W_a. \tag{2.13}$$

Substituting for W in Eq. (2.10) gives

$$\frac{1}{\Psi} = \frac{-W}{RTN_s\rho_w} = \frac{W_a}{RTN_s\rho_w} - \frac{R^*(W_o + W_a)}{RTN_s\rho_w}. \tag{2.14}$$

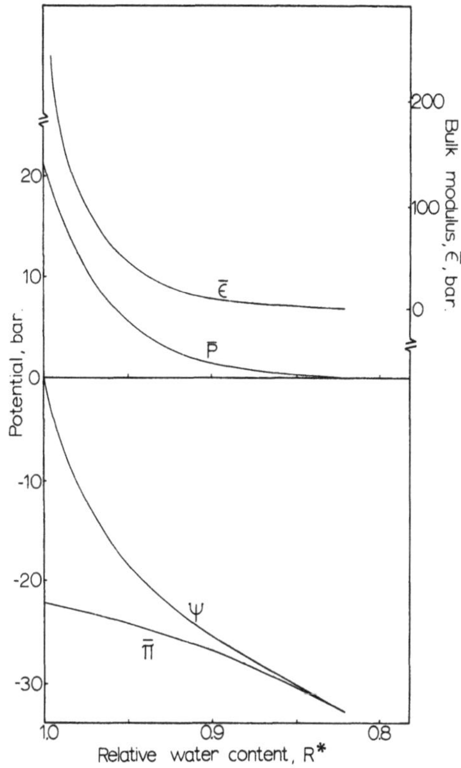

Fig. 2.4. A contemporary Höfler diagram derived from analysis of the Ψ isotherm. The relationships between relative water content (R*) and the total potential (Ψ), the osmotic potential ($\bar{\pi}$), the pressure potential ($\bar{P}$) and the bulk modulus of elasticity ($\bar{\varepsilon}$) of a shoot of Sitka spruce (*Picea sitchensis*) taken from a well-exposed position in forest canopy in mid-winter. $\bar{P}$ was calculated from Eq. (2.17) and $\bar{\varepsilon}$ from Eq. (2.21). (From data given by HELLKVIST et al. 1974)

Thus a plot of $1/\Psi$ against R* also yields a straight line when $\bar{P}$ is zero provided that W_a is constant. π_o is found from extrapolation of the line to $R^*=1$ when $1/\Psi = -W_o/RTN_s\rho_w = 1/\pi_o$.

From Eq. (2.14) it is also clear that extrapolation of the relation between $1/\Psi$ and R* to $1/\Psi=0$ yields an estimate of the relative water content of apoplasmic water,

$$R_a^* = W_a/(W_o + W_a), \tag{2.15}$$

as the intercept on the R* axis.

Thus a plot of $1/\Psi$ against R* provides a description of tissue water relations in terms of the relationships between Ψ and R*, and $\bar{\pi}$ and R* as well as an estimate of R_a^*. From these estimates of $\bar{\pi}$ and Ψ, $\bar{P}$ can be calculated as a function of R* and a Höfler diagram constructed showing how Ψ, $\bar{\pi}$ and $\bar{P}$ change with R* as shown in Fig. 2.4 (e.g., HELLKVIST et al. 1974; RICHTER 1978). Subsequently a quick measurement of Ψ on a *comparable* sample can immediately yield information on the probable value of $\bar{\pi}$ and $\bar{P}$ and R*. Seasonal and environmentally induced changes in these parameters can also be deduced by measuring the change in the Ψ-isotherms (see TYREE et al. 1978).

Recently, RICHTER (1978) proposed that there were certain advantages in the derivation of parameters from the inverse plot of Ψ against $1/R^*$ and this approach has been developed by RICHTER et al. (1980) and HINCKLEY et al.

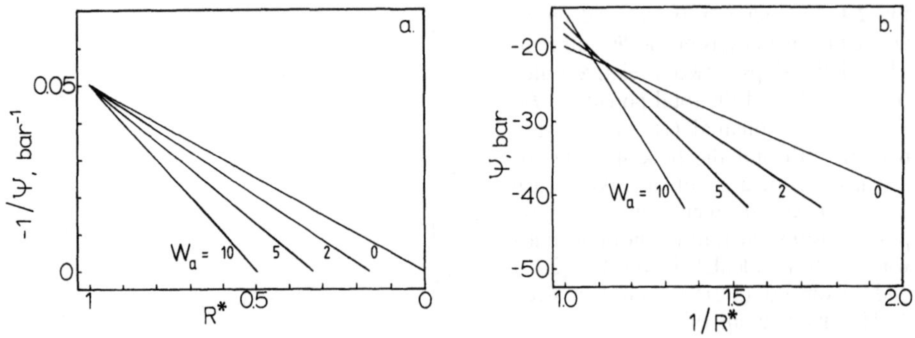

Fig. 2.5 a, b. An illustration of the effect of apoplasmic water on the shape of water potential isotherms. Ψ=water potential in bar, R*=relative water content [$=(W+W_a)/(W_0+W_a)$ where W is the symplasmic water content, W_0 is W at full hydration (when tissue $\Psi=0$), and W_a is the constant apoplasmic water content]. For $W_0=10$ g, W_a was assigned values of 0, 2, 5, and 10 g. **a** is a plot of the form of Eq. (2.14). The intercept on the abscissa is the value of the relative apoplasmic water content R* ($=W_a/(W_0+W_a)$). **b** is a Richter-type plot of Ψ versus 1/R* (RICHTER 1978). Linear regressions were made for values of $1/R^*>1.26$ for $W_a=2$, >1.25 for $W_a=5$, and >1.17 for $W_a=10$ (TYREE and RICHTER 1981)

(1980). However, Eq. (2.14) clearly shows that in the presence of apoplasmic water, a plot of Ψ against 1/R* is not equivalent to a plot of $1/\Psi$ against R*. The inverse of Eq. (2.14) is

$$\Psi=\frac{RTN_s\rho_w}{(W_o+W_a)R^*-W_a}. \tag{2.16}$$

This relation between Ψ and 1/R* is non-linear because of the presence of W_a. Consequently, estimation of $\bar{\pi}_o$ by linear extrapolation introduces errors in the values of $\bar{\pi}_o$. The relation given by Eq. (2.16) is a gently curved line and is linear only if W_a is zero.

TYREE and RICHTER (1981) have made a systematic comparison of the errors in π_o arising from the use of Eq. (2.16) in comparison with Eq. (2.14) for different values of W_a. The data were obtained by computing a range of values of π versus W from Eq. (2.11) and then calculating R* and 1/R* for the same values of W and for different values of W_a using Eq. (2.12). Figure 2.5 A shows that plots of $1/\Psi$ against R* all have the same intercept at $R^*=1$, irrespective of the value of W_a. In contrast, Fig. 2.5 B shows a range of intercepts at $R^*=1$. The correct value for $\bar{\pi}_o$ of -20 bar is only obtained when $W_a=0$; π_o is substantially underestimated at higher values of W_a.

Non-ideality also mitigates against obtaining a linear relationship with Eq. (2.10). Non-ideality, in this context, has the effect of making N_s differ from the actual solute content in the symplasm and as W changes N_s changes slightly. The evaluation of π by Eq. (2.10) is normally done by making W_o change by about 50%. Data given by ROBINSON and STOKES (1959) show that this corresponds to a change in N_s because of non-ideality of 5% or less for most solutes at the concentrations normally found in plant tissue. TYREE and RICHTER

(1981) show that the resulting non-linearity is too small to detect and that the possible errors in $\bar{\pi}_0$ are small, even for a strongly non-ideal solute such as sucrose: if Eq. (2.14) is used the errors resulting from non-ideality are smaller (<0.2 bar) than with Eq. (2.16) (<0.8 bar).

Systematic error in the measurement of Ψ at each value of W also leads to non-linearity in the relations between Ψ and R* in Eqs. (2.14) and (2.16). TYREE and RICHTER (1981) show that neither of the extrapolated estimates of $\bar{\pi}_0$ are in error by exactly the amount of the systematic error, but that Eq. (2.16) gives a slightly better result for solutions in which ϕ increases with increasing concentration, whereas Eq. (2.14) is better when ϕ is constant or decreases. They also demonstrate that non-ideality, together with random error in the measurement of Ψ in relation to W, leads to underestimation of $\bar{\pi}_0$ by both, but that Eq. (2.14) is somewhat superior to Eq. (2.16), giving an estimate of $\bar{\pi}_0$ closer to the true value and with smaller standard error.

Thus from the point of view of non-ideality and error in the measurements, Eq. (2.14) has only a marginal advantage over Eq. (2.16), but in the presence of substantial amounts of apoplasmic water the traditional transformation of Eq. (2.14) is strongly to be preferred.

2.2.4 The Pressure Potential, P

Until recently pressure potentials of cells and tissues were generally inferred as the difference between the total water potential and the osmotic potential. However, it is now possible to measure the pressure potentials of individual cells directly as well as inferring a weight-averaged value of $\bar{P}$ for tissues and organs from an analysis of Ψ isotherms.

At the cellular level great strides have been made towards the direct measurement of turgor potential by use of the pressure probe technique (e.g., ZIMMERMANN 1977; HÜSKEN et al. 1978). The pressure probe consists of a microcapillary tube filled with silicon oil. The capillary is tightly connected to a small pressure chamber which also contains silicon oil; the chamber also contains a small silicon pressure transducer transforming pressure in the chamber into a proportional voltage. The tip of the capillary is inserted into a single cell and the pressure potential is transmitted directly to the pressure transducer via the silicon oil. Although the silicon oil is nearly incompressible, the volume changes of the pressure probe caused by internal pressure changes are still a serious problem because the amount of sap that must flow from a small cell to pressurize the probe is enough to reduce the P of the cell. This problem is overcome in a more advanced model of the pressure probe by fixing the oil/cell sap boundary in the very tip of the microcapillary. The position of the boundary is adjusted and regulated by an electrical feedback mechanism (HÜSKEN et al. 1978).

The pressure probe has been used in large marine and freshwater algae and in multicellular plant cells down to 2×10^{-2} mm^3 in volume or about 35 µm in diameter. Pressure potentials can be measured with an accuracy of ± 0.03 to 0.05 bar. Physical limitations on the apparatus make it impossible to measure P in cells below 20 μm in diameter. Although many vascular plant cells are smaller than that, the technique has led to great advances in our understanding of the physiological role of P at the cellular level in osmoregulation, turgor-dependent ion fluxes (ZIMMERMANN 1978) and the movements of guard cells (EDWARDS and MEIDNER 1979), and will permit a useful re-evaluation of our understanding of the physics of cell enlargement (GREEN et al. 1971).

At the tissue and organ levels, $\bar{P}$ can be deduced from the analysis of water potential isotherms. Once the relationship between water loss, W_e, or relative water content, R*, and $\bar{\pi}$ has been established (Sect. 2.2.3.3) $\bar{P}$ is simply evaluated at any given value of R* or W_e by

$$\bar{P} \simeq \Psi - \bar{\pi}. \tag{2.17}$$

The $\bar{P}$ calculated in this way for an organ is a weight-averaged figure for all the cells because individual cells will have different values of P wherever their values of π differ. For example, values of π in the phloem in sugar beet are twice as negative as in the surrounding parenchyma (GEIGER et al. 1973) and similarly large differences have been reported between the values of π in guard cells, subsidiary cells and epidermal cells of *Tradescantia virginiana* (EDWARDS and MEIDNER 1979). When such cells are at equilibrium with each other, so that each cell has the same value of Ψ, P will vary from cell to cell.

The theory for the analysis of Ψ isotherms (Sect. 2.2.3) assumes that negative pressure potentials (i.e., negative turgor) do not develop in living cells. Much of the contemporary evidence for negative turgor is based on a negative difference being obtained when Eq. (2.17) is applied to measurements of Ψ and π made on the same or parallel samples of a tissue, π being measured after destruction of the selective permeability of the cell membranes by heating or freezing the tissue sample (e.g., WARREN WILSON 1967a; BEADLE et al. 1978). At the present time, we have no independent means of knowing if negative turgor really exists. TYREE (1976a) has argued strongly that negative values are obtained from Eq. (2.17) because destruction of the cell membranes results in dilution of the cell sap by the apoplasmic water.

Negative turgor would produce curvature in the Ψ isotherm. The question, therefore, arises as to how much negative turgor could occur and escape notice, particularly in view of the other contributions to non-linearity discussed in Section 2.2.3. To answer this question TYREE and RICHTER (1982) have computed the theoretical isotherm of a hypothetical shoot having eight different cell types all capable of developing negative turgor. They conclude that curvature resulting from negative turgor will be detectable when it reaches about -2 bar at $\Psi = -20$ bar. Thus negative turgor could contribute up to 10% of the measured water potential without being detected. Whilst this would have little effect on the estimate of $\bar{\pi}_0$ it may affect the estimate of R* by over 50%.

Recently there has been some discussion over the relative merits of the pressure probe and pressure bomb or psychrometer for the evaluation of water relations parameters of cells and tissues (e.g., ZIMMERMANN and STEUDLE 1978). Although the pressure probe is of incomparable value for studying the related physiology of single cells it is unlikely that it can easily give a representative picture of a complex tissue because the P-volume relationships of cells inside a tissue are probably strongly influenced by the "tissue-pressure" of surrounding cells, i.e., if the volume and P of one cell increases it will influence the P of surrounding cells in mechanical contact with it. Since a pressure probe cannot be inserted deeply into a tissue without considerable mechanical damage to surrounding cells or to the tip of the probe, it is difficult to assess the probable importance of "tissue-pressure". However, even if "tissue-pressure" is not important it is clearly impractical to build up a composite picture of the behaviour of a complex tissue by measuring and integrating the effects of the large number of individual cells in the tissue.

It is difficult to reach a useful comparison of the errors in measuring P of single cells and $\bar{P}$ of tissues and organs. Superficially, the error in determining $\bar{P}$ ought to be about the same as the error in measuring the balance pressure ($=-\Psi$) with the pressure gauge. The mechanical pressure gauges commonly used can resolve the balance pressure to ± 0.07 bar (Tyree and Hammel 1972). With a better gauge or pressure transducer it is possible to resolve *a* balance pressure with an even greater accuracy (e.g., ± 0.02 bar) that compares favourably with the accuracy of the pressure probe of ± 0.03 to 0.05 bar.

Nevertheless, the assessment of the accuracy of $\bar{P}$ is complicated by the time scale of the measurement. In the time required to measure the complete Ψ-isotherm of a tissue by the pressure bomb or psychrometer the isotherm may become distorted because of: (1) plastic deformation of cell walls or growth, (2) gradual changes in π, and (3) small internal disequilibria causing small shifts in the apparent value of the balance pressure. The pressure probe, therefore, *appears* to be more accurate because of the short time scale required for the measurement of P of a cell. However, weight-averaged values of $\bar{P}$ for a tissue can be obtained much more rapidly from the measurement and analysis of Ψ-isotherms than would be possible by the integration of measurements made with the pressure probe on individual cells. Over a lengthy period of measurement pressure probe values are also likely to change for the reasons given above and, in addition, as a result of solute exchange with the bathing medium in which it is necessary to maintain the tissue during measurement. The presence of the bathing medium does, of course, make it difficult to reproduce tissue water contents similar to those which occur in the field.

For these reasons, and since the financial investment and technical expertise needed to make measurements of Ψ-isotherms with pressure bombs and psychrometers are less than for the pressure probe, bulk values of $\bar{\pi}$ and $\bar{P}$ made with these methods will continue to be preferred by many ecologists.

2.2.5 Cell Wall Elasticity and Plasticity

When a stress (=a stretching force) is applied to material a strain (=a deformation of shape) results. The strain that results from a stress can be elastic, plastic or both. A totally plastic strain or deformation is said to occur if the material retains completely its deformed state after the stress is released. A totally elastic strain occurs when the material completely returns to its prestressed shape after the stress is released.

Many materials, including plant cell walls, have both plastic and elastic properties (HAUGHTON et al. 1968). A growing cell has a more plastic wall than a mature cell, but even mature cells retain measurable plasticity. Generally speaking elastic deformations arise sooner after the onset of strain than plastic deformations. Plastic strain is a slow "creep", taking minutes or hours to reach completion; the plastic "creep" stops when the material has "hardened" sufficiently so that the applied stress is completely taken up by a new arrangement of molecular bonds. In a weakly cross-linked polymer sometimes even elastic deformations can take a long time to reach completion because time is required for polymer coils to change shape governed by the time for cross-linking bonds to break and reform. This tends to blur the distinction between elastic and plastic strains, since the distinction is based primarily on their relative permanence after the stress is released or on their speed of formation when stress is applied.

In a living plant cell, it is the elastic property of the strained cell wall that applies the restoring force on the cell sap and thus produces the positive internal pressure, P. As the wall stretches to enclose a greater volume of cell sap, the internal pressure increases in much the same way as the pressure of air inside a rubber balloon must be increased to stretch it to enclose a larger volume. The plastic properties, on the other hand, allow a permanent increase of volume (=growth) without a permanent increase of P, although a high P is needed to produce plastic stretch. Since even the cell walls of mature plants retain some plasticity, the relationship between P and cell volume changes. If the cell volume increases by Δv_i the increase in P may be more than the decrease in P when the cell volume is decreased some hours later. Such hysteresis in the P-volume relationships will also cause hysteresis in the water potential isotherm (WENKERT et al. 1978b).

The bulk modulus of elasticity, ε, describes the amount by which a small change in volume brings about a change in P (e.g., ZIMMERMANN 1978):

$$\Delta P = \varepsilon \frac{\Delta v}{v} \tag{2.18}$$

or

$$\varepsilon = \frac{dP}{dv} \cdot v \tag{2.19}$$

The bulk modulus, ε, is thus the proportionality parameter between the observed change in P, ΔP, and the change in relative cell volume, $\Delta v/v$, where v is the cell volume. Since ε is not a constant, but varies with pressure, the formulation of Eq. (2.19) is more appropriate. The more elastic the cell wall, the smaller the value of ε.

Using the pressure probe, ε of a single cell can be measured directly by injecting a known small volume of Δv of oil rapidly into the cell and measuring the resulting instantaneous increase in P, ΔP. So far only a few estimates of ε have been made with the pressure probe on single cells of higher plants. Such values of ε range from zero at $P \simeq 0$ to ca. 100 bar at $P \simeq 5$ bar in a selection of eight kinds of cell (ZIMMERMANN and STEUDLE 1980). The measurement is sufficiently sensitive to discriminate between the ranges of values of ε in the subsidiary, epidermal and mesophyll cells of *Tradescantia virginiana*. In the mesophyll cells of the CAM plant *Kalanchoë daigremontiana*, ε varied between 13 and 128 bar with a mean value of 42 bar in the range of P from 0 to 3.4 bar (STEUDLE et al. 1980). This relatively low value of ε allows substantial changes in cell volume and water storage as N_s increases as a result of malic acid synthesis during the dark period. The resulting changes in P may regulate the entire CAM syndrome including malic acid synthesis in the mesophyll and stomatal movements in the epidermis.

Pressure probe measurements of ε of single cells of algae and higher plants all reveal that ε is pressure dependent, i.e., ε is smaller at low P than at high P (Fig. 2.6). In addition ε is "volume-dependent", i.e., small growing cells have lower values of ε than large mature cells (ZIMMERMANN 1978). The pressure dependence may simply reflect non-linearity of the stress–strain relationship of cell walls; the volume dependence may indicate a correlation between elasticity and plasticity. Small growing cells must be plastic and it is possible that the structural peculiarities that make cell walls plastic may also make them more elastic. In the case of cells of *Nitella* and *Chara* species, pressure and volume dependence is clearly the result of large differences in ε of different regions of the internodal cells (ZIMMERMANN and STEUDLE 1975).

The walls of cells are subject to stresses and strains in several directions at once because of their geometry and attachment to other cells. In addition the cell wall is anisotropic, i.e., it is more "stretchable" in one direction than another and a stretch in one direction may cause a contraction in the perpendicular direction. If a perfectly spherical cell with isotropic walls did exist, then ε would depend in a simple way on the wall thickness, δ, cell radius, r, and Young's modulus of elasticity ε^*. Young's modulus relates stress to strain in one direction. In this hypothetical case

$$\varepsilon = \frac{2\delta\varepsilon^*}{3r} \tag{2.20}$$

but in real plant cells ε depends on geometrical properties in a much more complicated way (see, for example, VINTERS et al. 1977). Nonetheless, it is probably safe to say qualitatively that small thick-walled cells have larger values of ε than large thin-walled cells, since for an isotropic spherical cell this is the case.

At the tissue level it is also possible to define a weight-averaged bulk elastic modulus, $\bar{\varepsilon}$.

$$\bar{\varepsilon} = \frac{d\bar{P}}{dW} \cdot W \tag{2.21}$$

where $\bar{P}$ is as defined by Eq. (2.3) and W is the weight of water in the symplasm.

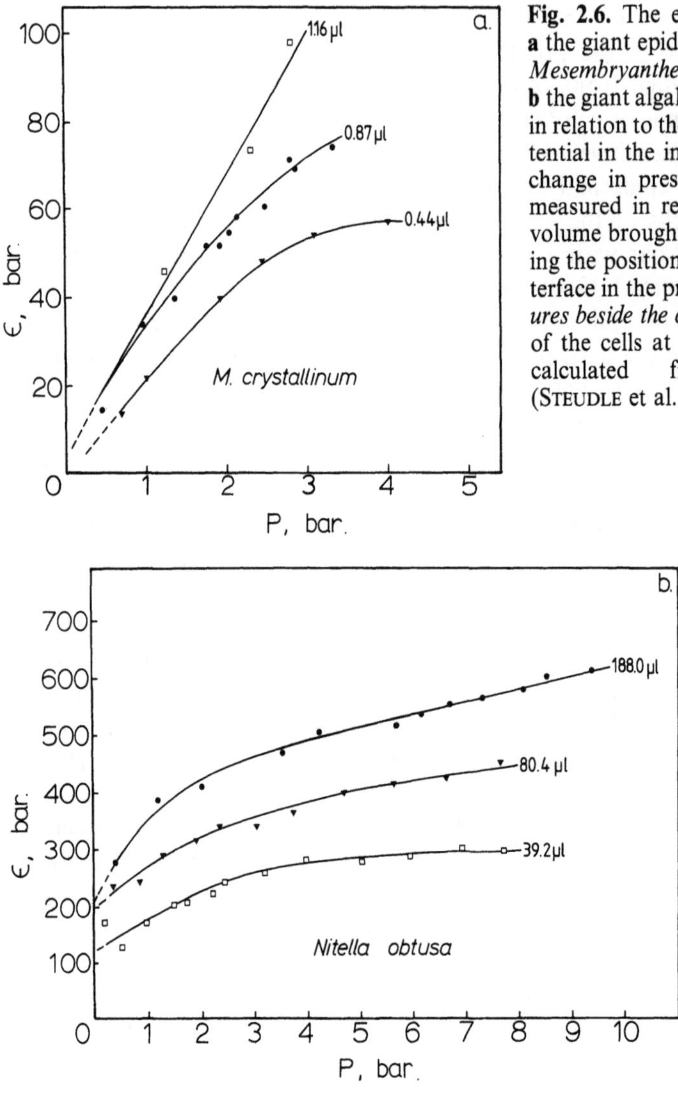

Fig. 2.6. The elastic modulus, ε, of **a** the giant epidermal bladder cells of *Mesembryanthemum crystallinum* and **b** the giant algal cells of *Nitella obtusa* in relation to the pressure (turgor) potential in the individual cells, P. The change in pressure in the cells was measured in response to changes in volume brought about by manipulating the position of the cell sap/oil interface in the pressure probe. The *figures beside the curves* are the volumes of the cells at full turgor, v_0. ε was calculated from $\varepsilon = V_0 \Delta P / \Delta V$. (STEUDLE et al. 1977)

To estimate $\bar{\varepsilon}$, $\bar{P}$ is measured as described in Section 2.2.4 and plotted against W ($= W_0 - W_e$). $\bar{\varepsilon}$ is then calculated as the slope of the line, $d\bar{P}/dW$, at particular values of W. Figure 2.7 shows that, as in single cells, $\bar{\varepsilon}$ for a tissue is also pressure-dependent (HELLKVIST et al. 1974; CHEUNG et al. 1976), but reaches much lower values at low $\bar{P}$ than does ε at low P because of the effect of weight-averaging. In a population of cells at low $\bar{P}$ many of the cells may have no turgor at all, whereas at high $\bar{P}$ all the cells have some turgor. When Ψ decreases at already low $\bar{P}$ the cells with turgor will experience a decrease in both W and $\bar{P}$; the cells at zero turgor will also lose water and thus contribute to dW, but they will not contribute to $d\bar{P}$ because their pressure potential is already zero. This leads to low values of $d\bar{P}/dW$ and $\bar{\varepsilon}$ is correspondingly small. As with ε, $\bar{\varepsilon}$ is also volume-dependent, being less in growing leaves than in mature leaves (TYREE et al. 1978) (see Fig. 2.7).

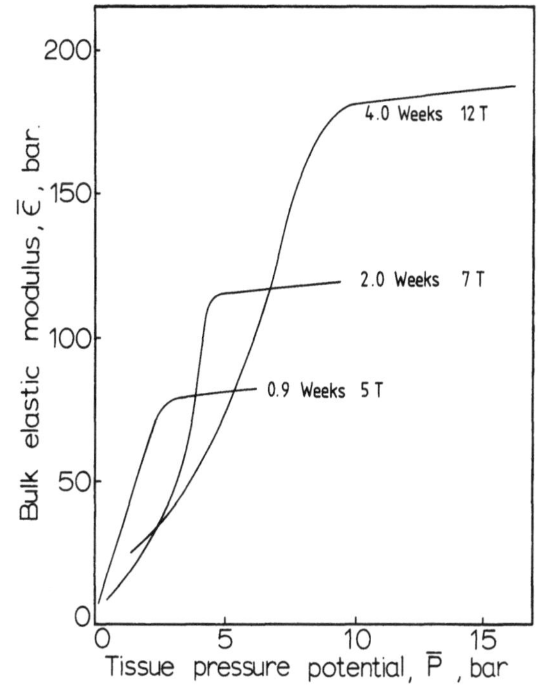

Fig. 2.7. The bulk elastic modulus, $\bar{\varepsilon}$, of leaves of *Populus* in relation to pressure (turgor) potential in the tissue, $\bar{P}$. The *figures beside the curves* are the age of the leaves and the leaf number from the top of the plant. $\bar{\varepsilon}$ was calculated from Ψ-isotherms measured with the pressure bomb using Eq. (2.21). (Unpublished data of TYREE and MCGREGOR)

However, ε and $\bar{\varepsilon}$ are not exactly comparable quantities, so the comparison should not be pushed too far. In particular ε is defined in terms of cell volume [Eq. (2.19)] and $\bar{\varepsilon}$ in terms of the weight of symplasmic water [Eq. (2.21)]. Treating both in terms of the weight of water, TYREE (1981) has shown that

$$\bar{\varepsilon}=\sum^{n}\left(\frac{dw_i}{dW}\varepsilon_i\right)+\sum^{n}\left(\frac{dw_i}{dW}P_i\right)-\bar{P} \tag{2.22}$$

where w_i is the amount of water in the ith cell of a population of cells and the summation is carried out over all n cells in the tissue or organ which contains an amount of water W $(=W_0-W_e)$. Thus $\bar{\varepsilon}$ is not identical with ε but in practice is numerically very close to it. Using a model of a tissue containing nine cell types varying in π_0, W_o and ε, TYREE (1981) has shown that generally ε lies within 2% of the first term on the right of Eq. (2.22).

Whilst such quantitative similarity can be demonstrated by simulation, qualitative comparison between ε and $\bar{\varepsilon}$ is rendered rather obscure because plastic hysteresis effects must affect their quantitative values as well as their heuristic meaning. Furthermore, when comparing values of ε or $\bar{\varepsilon}$ it is necessary to bear in mind that minor differences in the way in which they are measured can affect their values. For example, if the time scale over which ε is measured is long, then a much smaller value of ε may be computed than if the time scale of the measurement is short (ZIMMERMANN and HÜSKEN 1980; VINTERS et al. 1977).

The accuracy with which $\bar{\varepsilon}$ can be determined with the pressure bomb is about the same or better than the accuracy with which ε can be obtained using the pressure probe (TYREE 1981). With the pressure bomb $\Delta\bar{P}$ can be measured with an error of about ± 0.07 bar, although this can be improved with the use of better pressure gauges, which can be compared with an error of ± 0.04 bar for the pressure probe. The larger error in measuring ΔP with the pressure bomb is offset by a higher accuracy in measuring W or W_0. With the pressure bomb W_0 is evaluated from the isotherm and probably has an error of about 5% of the fresh weight. In the pressure probe technique, however, v must be measured optically on cells which may not be of regular geometric shape; the resulting error can be 10% to 30% of the true value depending on the cell.

2.2.6 Apoplasmic Water

Apoplasmic water consists of all water in the plant not in the symplasm. In a tissue, such as the mesophyll, this is almost wholly water of hydration in the cell walls, but in an organ such as a leaf or shoot, apoplasmic water includes water in the lumens of the xylem vessels, tracheids and fibres, as well as water in the cell walls.

The Ψ-isotherm theory so far developed has assumed that the amount of apoplasmic water, W_a, remains constant as the tissue or organ is dehydrated and that it can be estimated as the intercept on the R^* axis by extrapolation of the straight line described by Eq. (2.14) to $1/\Psi = 0$ [see Eq. (2.15)]. TYREE and RICHTER (1982) point out that this procedure should be viewed with caution since constancy of W_a with change in Ψ is by no means certain. Whilst the pores in the cell walls are unlikely to drain at the negative pressure potentials obtaining ($\Psi > -30$ bar), in the much larger lumens of the xylem vessels and tracheids the water may cavitate or be withdrawn at these values of P. In addition, both cell walls and xylem vessels and tracheids will shrink as a result of the tensions which develop within them, the degree of shrinkage depending on their bulk compressive modulus.

Changes in W_a during dehydration result in curvature of the linear relationship expressed by Eq. (2.14). Using a model of a tissue containing eight different cell types, TYREE and RICHTER (1982) showed that the curvature induced has only a small effect on the estimation of $\bar{\pi}_0$ (an underestimation of 0.4 bar in 10 bar in an extreme case) but can result in a 50% underestimation of R_a^*.

TYREE and RICHTER (1982) also demonstrated that non-ideality of the cell sap solution can lead to curvature in the Ψ-isotherm which may result in error in the estimate of R_a^* of up to ± 0.05. Such errors could be significant when R_a^* is small (e.g., 0.1), and can also lead to apparent negative values of R_a^* when ϕ decreases with increasing solute concentration.

Cavitation of xylem vessels and tracheids (see Chap. 5, this Vol.) is a regular occurrence as a consequence of the relatively moderate, diurnal changes in xylem pressure potential (MILBURN and MCLAUGHLIN 1974), and large seasonal changes in the water content of stem sapwood occur as a result (e.g., WARING and RUNNING 1978). Loss of apoplasmic water as a result of cavitation would be most likely to occur during determination of Ψ-isotherms by vapor pressure methods. In the pressure bomb, on the other hand, the water potential at the balance point is zero and cavitation is unlikely. However, water may be driven out from the lumens of the vessels and tracheids if the size of the pit pores is sufficient to allow the passage of the water–air interface. Ψ-isotherms for conifer sapwood measured with the pressure bomb show that about 15% of the water is removed by an applied pressure of 10 bar (EDWARDS and JARVIS 1982) in contrast to previously published Ψ-isotherms measured by vapor pressure equilibration which show a fall in R* of approximately 80% for a reduction in Ψ of 10 bar (WARING and RUNNING 1978; WARING et al. 1979). In sapwood most of the water is apoplasmic occurring in the dead conducting cells, the ray parenchyma occupying only a very small proportion of the volume (5%), whereas in a leaf or leafy shoot, the xylem contains a very much smaller proportion of the total water content. Nonetheless, some careful measurements of changes in xylem water content during the determination of Ψ-isotherms with the pressure bomb are desirable to establish the magnitude of the change in relation to the total organ water content.

The relationship between the tensions which develop in the apoplasmic water in association with changes in W_a can be expressed in terms of the bulk compressive modulus, $\bar{\varepsilon}_c$ as follows:

$$\bar{\varepsilon}_c = \frac{d\bar{P}_a}{dW_a} \cdot W_a$$

where $\bar{P}_a$ is the pressure potential in thc apoplasmic structures and has a negative value almost equal to Ψ. In an incompressible structure $\bar{\varepsilon}$ is very large. Since cell walls are made up of closely cross-linked, dense polymers, we would expect to find the largest values of $\bar{\varepsilon}_c$ there, with smaller values, in decreasing order, in the lumens of sclerenchyma fibres, tracheids and vessels. However, since the walls of vessels are structurally modified to withstand substantial tensions, their values of $\bar{\varepsilon}_c$ are likely to be substantially larger than the values of the bulk elastic modulus of mesophyll cells experiencing positive turgor (see Sect. 2.2.5). Less than 5% of the diurnal swelling and shrinkage which occurs in the stem of a conifer is associated with swelling and shrinkage of the sapwood, the remainder occurring in the living cells of phloem and cambium. The values of $\bar{\varepsilon}_c$ for wood are of the order of 3,000 bar (SKAAR 1972) and probably largely reflect changes in lumen dimensions, values for cell walls being much larger. Consequently compressibility of cell walls can probably be regarded as a negligible source of change in W_a in leaves.

2.3 Ecological Implications of Tissue-Water-Potential Characteristics

2.3.1 Osmotic Potential

2.3.1.1 Origin of Changes in π

Decreases in π in response to water and salt stress have been widely reported over many years (e.g., WALTER 1955; KREEB 1963). In some cases the fall in π has been seen as a passive response to loss of water but more recently considerable emphasis has been put on the concept of osmo-regulation or osmotic adjustment (see Chaps. 1 and 2, Vol. 12 C). The ability to adjust π in response to stress is considered to be an adaptation conferring particular advantages to plants which are tolerant of water and salt stress (TURNER 1979; WYN JONES 1981), the most significant consequence being the maintenance of a positive pressure potential, i.e., turgor, to lower tissue water contents and potentials.

It can be seen from Eq. (2.9) that changes in π can be brought about by changes in N_s, the number of osmoles of solute in the symplasm, or in W, the weight of water in the symplasm. This can be expressed quantitatively by writing the full differential of $\bar{\pi}$ as follows:

$$d\bar{\pi} = -\frac{RT\rho_w}{W} dN_s + \frac{RTN_s\rho_w}{W^2} dW$$

A decrease in π ($=d\pi<0$) can be brought about by an increase in N_s ($=dN_s>0$) or by a decrease in W ($=dW<0$).

An increase in N_s may result from the uptake of additional inorganic salts or from the synthesis of organic solutes. In several investigations, attempts, some more successful than others, have been made to account for the decrease in π by the increase in N_s of various solutes (e.g., TURNER et al. 1978; DAVIES and LAKSO 1979; MUNNS et al. 1979; JONES et al. 1980; FORD and WILSON 1981). For example, CUTLER and RAINS (1978) were able to account for only about 60% of π in cotton leaves by the summed contributions of K^+, NO_3^-, reducing sugars, sucrose, citrate and malate, although they were able to account for a rather larger proportion of the decrease in π resulting from falling R*. By accounting for rather more solutes in other cases, the concentration of ions and organic acids has seemed sufficient to account for π (CRAM 1976; MUNNS et al. 1979; JONES et al. 1980; FORD and WILSON 1981) but in yet other cases the osmotic equivalent of the solutes present exceeded measured π by several bar (FORD and WILSON 1981).

However, on the whole, such attempts have not adequately considered compartmentation of solutes within the cell or the possible range of solutes, particularly compatible solutes within the cytoplasm.

Values of $\bar{\pi}$ would be expected to be similar throughout the cell within the bounds of the plasmalemma. However, there is evidence to suggest that of the inorganic ions K^+ is preferentially located in the cytoplasm and Na^+ and Cl^- in the vacuole (JESCHKE 1979; WYN JONES et al. 1979). High concentrations of electrolytes, including organic anions such as malic, aconitic, oxalic, citric and pectic acids, are probably confined to the vacuole and are not tolerated in the cytoplasm. In the CAM plant *Kalanchoë daigremontiana,* for example, diurnal changes in π are strongly correlated with the diurnal changes in amount of malic acid present in the vacuole (LÜTTGE and BALL 1974, 1977).

The PEP-carboxylase in the cytoplasm would be substantially feedback inhibited by a malic acid concentration in the cytoplasm of only 1% of the total amount synthesized in the dark period (KLUGE and OSMOND 1972).

Whilst a number of enzymes are stimulated by low concentrations of electrolytes (e.g., 50 mol m^{-3}), most seem to be inhibited substantially by salt concentrations over about 100 mol m^{-3} (FLOWERS et al. 1976, 1977). Thus to maintain π in the cytoplasm there is a requirement for "compatible", non-toxic, water-soluble osmotica, carrying no net charge, which can be tolerated at high concentrations by the organelles, membrane-bound processes and other enzymes (WYN JONES et al. 1979). A number of solutes may carry out this role in different species, particularly free amino acids like proline (e.g., BARNETT and NAYLOR 1966; JÄGER and MEYER 1977; HANSON et al. 1977; MUNNS et al. 1979) and glycinebetaine (WYN JONES and STOREY 1978). Other probable cytoplasmic osmotica for which there is evidence in particular species include: sorbitol, prolinebetaine, β-alaninebetaine, choline, dimethylsulphonio propionate, asparagine, glutamine, aspartic acid, glutamic acid and a number of other amino acids (MUNNS et al. 1979; WYN JONES et al. 1979; JONES et al. 1980; WYN JONES 1981). However, whether sugars also fall into this category is at present doubtful. The very large amounts of glucose, fructose and sucrose which are found in some Monocotyledones, including some of the Gramineae, and in xerophytic genera such as *Eucalyptus,* suggest that free sugars mainly occur in the vacuole (WYN JONES et al. 1979). Sucrose may be a compatible solute in some species but glucose and fructose are most probably not. Amino acids such as proline, and betaines such as glycinebetaine are probably synthesized from glycine and serine, intermediates in the photorespiratory carbon oxidation (PCO) cycle (HANSON and NELSON 1978; WYN JONES and STOREY 1981; STEWART 1981; STEWART and HANSON 1980). The reduction in carbon dioxide concentration within photosynthesizing leaves as a result of stomatal closure, shifts the balance from carboxylation to oxygenation of ribulose-1,5-bisphosphate with consequent increase in the proportion of carbon passing through the PCO cycle as compared with the photosynthetic carbon reduction (PCR) cycle, so that there is an increase in the availability of substrate for the synthesis of such osmotic compounds (LAWLOR and FOCK 1977a, b, 1978; LAWLOR 1979). Thus increase in the concentration of some compatible solutes is likely to be an almost inevitable consequence of the effects of stress in causing stomatal closure.

Since the osmotic solutes are compartmentalized within the cell, and the cytosol and organelles occupy only a small proportion of the cell volume, separate measurements of N_s of particular solutes, in relation to the volume of the compartment in which they occur, are necessary if the contributions of different solutes to the overall osmotic potential are to be assessed.

A decrease in W can result from a net water loss from the leaf which will lower R*, *or* from a redistribution of water between the symplasm and apoplasm *without* any change in R*. For example, HELLKVIST et al. (1974) and TYREE et al. (1978) found large seasonal changes in the proportion of symplasmic water content, $W_0/(W_0+W_a)$, in leaves of Sitka spruce and maple from 0.80 in June to 0.60 in September. Thus a fall in π at constant R* does not necessarily mean that there has been an increase in N_s *per cell* in a tissue. This additional mechanism has not been generally recognized, but changes in π of this kind probably involve growth of cell walls and take place over weeks rather than days. A decrease in π as a result of loss of water, i.e., a straightforward increase in solute concentration as a result of dehydration according to Eq. (2.9), cannot be regarded as an osmotic adjustment to maintain a positive pressure potential in the cell: such a fall in π is accompanied by a decrease in $\bar{P}$. On the other hand, either an increase in N_s or a change in the distribution of water between symplasm and apoplasm leading to a decrease in π may represent adjustments to maintain turgor as W decreases.

2.3.1.2 Osmotic Adjustment

Recently considerable emphasis has been placed on demonstrating changes in π of plant tissues that can be ascribed to changes in N_s (see review by TURNER and JONES 1980), and changes in π of this kind are summarized in the following paragraphs.

Diurnal changes in π with amplitudes ranging from 4 to 16 bar have been reported in higher plants. The amount of change resulting from diurnal fluctuation of N_s per cell is from 2 to 8 bar (TURNER 1975; HSIAO et al. 1976; POWELL and BLANCHARD 1976; CUTLER et al. 1977; DAVIES and LAKSO 1978, 1979; WENKERT et al. 1978b). It is not known how diurnal shifts in N_s per cell are induced. Diurnal fluctuations in water stress might affect metabolic pathways or ion pumps, change the balance between the PCO and PCR cycles or cause an imbalance between photosynthesis and translocation. In giant algal cells a reduction in P leads to stimulation of K^+ transport into the cell (see reviews by ZIMMERMANN 1977, 1978) and a similar feedback system seems to exist in the guard cells of *Tradescantia virginiana* (EDWARDS and MEIDNER 1979).

Medium-term reductions in π have been induced in the leaves of stressed plants in comparison with unstressed controls by one or more cycles of soil drying or the application of polyethylene glycol or salt solutions to the rooting medium. Reductions in $\bar{\pi}$ of from 1 to 10 bar as a result of an increase in N_s have been found, the changes being completed in 3 days to 3 weeks (JARVIS and JARVIS 1963a; LAWLOR 1969; SIMMELSGAARD 1976; CUTLER and RAINS 1978; MORGAN 1977; HANSON et al. 1977; HANSON and NELSON 1978; JONES and TURNER 1978, 1980; TURNER et al. 1978; STOREY and WYN JONES 1978; WYN JONES and STOREY 1978; MUNNS et al. 1979; CUTLER et al.; 1980a, b; WILSON et al. 1980). Similar osmotic adjustments have been observed in the floral apex of wheat (MUNNS et al. 1979) and in the root tips of *Zea mays* (SHARP and DAVIES 1979).

The decreases in π in these experiments are largely drought-induced but in some cases they may have ontogenetic (e.g., TYREE et al. 1978) or microenvironmental origins (e.g., TURNER 1974). For example, osmotic adjustment of 2 to 5 bar has been found in the upper levels of canopies as compared with the lower levels (TURNER 1974; HELLKVIST et al. 1974; HSIAO et al. 1976; KIRKHAM and SMITH 1978) although WILSON (1977) found no differences with respect to insertion level in leaves of a *Panicum maximum* cultivar.

Long-term or seasonal changes in π range from 2 to 18 bar. These changes occur over several weeks or months and it is difficult to know how much of the change results from increases in N_s per cell or from changes in water distribution between the symplasm and apoplasm. In some cases the decline in π is a response to drought and is larger on unirrigated than irrigated plants (GOODE and HIGGS 1973; HSIAO et al. 1976; ELSTON et al. 1976; ROBERTS and KNOERR 1977; TURNER et al. 1978; FERERES et al. 1978; HINCKLEY et al. 1980). In other cases the decline in π may be ontogenetic in origin (CLINE and CAMPBELL 1976; TYREE et al. 1978; HINCKLEY et al. 1980), or in evergreens, a response to the onset of winter. In sugar maple (*Acer saccharum*), for example, π_0 declined during leaf development in May and June and then remained more or less constant until leaf fall in October. In *Tsuga canadensis,* in contrast $\bar{\pi}_0$ cycled seasonally in phase with air temperature reaching the lowest values (−27 bar) in mid winter and the highest (−15 bar) in late summer (TYREE et al. 1978).

The degree of diurnal, medium-term and seasonal osmotic adjustment apparently varies widely between species. There are some species which have on investigation shown little or no adjustment and these have not been referred to here. There have been too few studies that clearly demonstrate the presence or absence of osmotic adjustment to allow generalizations to be made. On the whole there is considerable evidence for osmotic adjustment in crop plants and halophytes, perhaps because they have been more intensively investigated, and some evidence that osmotic adjustment is less well developed in woody shrubs and trees (see TURNER 1979; WYN JONES 1981; HINCKLEY et al. 1980; TURNER and JONES 1980).

2.3.1.3 Advantages of N_s-Induced Osmotic Adjustment

Low values of $\bar{\pi}$ throughout the plant should enhance the ability of plants to take up soil water under dry or saline conditions (TYREE 1976b). This advantage is probably marginal in sandy soils because the available water reserves at soil water potentials below −4 bar are very small and the plant's ability to grow deep roots is probably of greater advantage. In clay soils, however, there are considerable soil water reserves at water potentials below −4 bar, so that low leaf or root π may be as important as root growth in assisting water uptake. In salt marsh conditions and saline media reduction in $\bar{\pi}$ within the plant also assists water uptake (e.g., SLATYER 1961; WYN JONES 1981).

Low $\bar{\pi}$ in the leaf also enables turgor to be maintained above zero at higher values of $\bar{\mathrm{P}}$ than would otherwise be the case as Ψ falls. This allows the maintenance of open stomata with larger apertures and higher stomatal conductances and rates of net photosynthesis down to lower values of Ψ than would be the case if π did not adjust (e.g., MCCREE 1974; BROWN et al. 1976; FERERES et al. 1978; BEADLE et al. 1978; JONES and TURNER 1978; TURNER et al. 1978; DAVIES and LAKSO 1979; JONES and RAWSON 1979; HINCKLEY et al. 1980; LUDLOW 1980).

Finally N_s-induced lowering of π also enables maintenance of high turgor pressure for growth. This last possibility deserves special consideration and is discussed in Section (2.3.4).

There are apparently limits to the accumulation of osmotic solutes and the decline of $\bar{\pi}$ in tissues. Further osmotic adjustment does not occur once a certain level of $\bar{\pi}$ is reached in particular species and further environmental drought or osmotic stress may lead to loss of solutes and reversal of adjustment

(LAWLOR 1969; MORGAN 1977; JONES and RAWSON 1979; TURNER and JONES 1980; CUTLER et al. 1980a). In rice (*Oryza sativa*) for example, CUTLER et al. (1980a) found that $\bar{\pi}_0$ fell to ca. −18 bar with both rapid (4–6 days) or slow (20–22 days) drought treatment and would not decline further. The following three possible reasons for a limit to osmotic adjustment can be advanced.

At high concentrations even the so-called compatible solutes may become inhibitory to certain enzymes and processes. Glycinebetaine, for example, has been shown to be relatively nontoxic to mitochondria, chloroplasts, protein synthesis and a range of other enzymes up to concentrations of about 500 osmol m^{-3} but starts to inhibit significantly at higher concentrations (FLOWERS et al. 1977; POLLARD and WYN JONES 1978; LARKUM and WYN JONES 1979). Whilst K^+ is the main electrolyte which accumulates in the cytoplasm in higher plants (CRAM 1976) it too is probably not tolerated at concentrations above ca. 200 mol m^{-3}. Secondly, the production of organic solutes also requires a considerable expenditure of energy. According to WYN JONES (1981), the energy cost of producing one osmole of K_2^+ malate^{2-} from gaseous CO_2, via the PCR cycle and PEP-carboxylase, is 13 mol of ATP, and the cost of producing one osmole of glucose from gaseous CO_2, via the PCR cycle, is 54 mol of ATP, whereas the cost of accumulation of one osmole of NaCl is only 0.54 mol of ATP. Whilst accumulation of Na^+ and Cl^- in the vacuole is inexpensive with respect to both energy and carbon, accumulation of organic acids and sugars as osmotica is much more demanding of both energy and carbon. The energy and carbon costs of compatible solutes are likely to be of the same order. In addition, there is a substantial nitrogen cost in the accumulation of nitrogen dipoles such as proline and glycinebetaine which may contain up to 20% of the nitrogen in a leaf (STOREY et al. 1977; STEWART and HANSON 1980). Clearly, if growth is nitrogen-limited, the capability to adjust osmotically may also be restricted.

Thirdly, at high vacuolar and cytoplasmic concentrations of osmotica, there will be a strong tendency for leakage of both solutes and electrolytes across the cell membranes. To maintain high concentrations of osmotica within the cell and compartmentalization of electrolytes and compatible solutes between vacuole and cytoplasm, and possibly also between cytosol and organelles, requires the continuous expenditure of energy which could therefore add up to substantial amounts (RAVEN and SMITH 1977; JESCHKE 1979).

2.3.2 Bulk Modulus of Elasticity, $\bar{\varepsilon}$

The bulk modulus of elasticity, $\bar{\varepsilon}$, depends on structural properties of the tissue and walls of the individual cell and is also pressure and volume-dependent. Consequently, $\bar{\varepsilon}$ would be expected to vary widely between species and to change as a result of ontogeny and environmental stress.

Differences in $\bar{\varepsilon}$ are partly responsible for the extent to which turgor changes as a result of a change in R*; the slope of the relation between $\bar{P}$ and Ψ is largely dependent upon $\bar{\varepsilon}$. In an elastic tissue, i.e., one with small $\bar{\varepsilon}$, the cells and tissue shrink following loss of water. This results in a smaller decrease in turgor for the loss of a given volume of water than in a more rigid tissue. Thus for a particular value of Ψ, a more elastic tissue has lower R*, lower π and higher $\bar{P}$ than a more rigid tissue (WEATHERLEY 1970). A small ε thus contributes to turgor maintenance in much the same way as osmotic adjustment (TURNER 1979). However, a proper appreciation of the value of R* or Ψ at which turgor is zero for a particular tissue depends upon full knowledge of the relationship between $\bar{\varepsilon}$ and $\bar{P}$ and cannot be derived from knowledge of a single value of $\bar{\varepsilon}$, since both the maximum value of $\bar{\varepsilon}$ and $d\bar{\varepsilon}/d\bar{P}$ vary from tissue to tissue (CHEUNG et al. 1976; TYREE et al. 1978).

Since the definition of a coefficient of enlargement by BROYER (1952), inferences regarding changes in ε have been made from variations in the relationships between Ψ versus R* and $\bar{P}$ versus R* in different species and treatments. These relationships differ widely amongst leaves of different species (e.g., JARVIS and JARVIS 1963b; GARDNER and EHLIG 1965; MAXWELL and REDMANN 1978) and show ontogenetic, seasonal and water stress-induced changes (e.g., JARVIS and JARVIS 1964; KNIPLING 1967; NOY-MEIR and GINZBURG 1969; KASSAM and ELSTON 1974; SHEPHERD 1976; FERERES et al. 1978). The coefficient of enlargement, e (variously defined — see GARDNER and EHLIG 1965; WARREN WILSON 1967c, for example), has shown substantial differences amongst species (e.g., JOHNSON and CALDWELL 1976; MAXWELL and REDMANN 1978) and it is likely that similar differences would emerge from calculation of $\bar{\varepsilon}$ as a function of $\bar{P}$. However, since the values of e available are not defined in relation to $\bar{P}$, we feel that not too much should be read into them. Nonetheless, the often determined relationship between Ψ and R* is informative and useful. We may define the capacitance of a tissue for water storage as $dW/d\Psi = W_0\,dR^*/d\Psi$ and this can be shown to be equal to $W/(\bar{\varepsilon}+\bar{\pi})$ (see ZIMMERMANN and STEUDLE 1978). Thus the slope of the Ψ/R* relationship, which describes the amount of water exchanged per unit change in water potential, depends on both the amount of water in the tissue and on its bulk modulus.

Comparative measurements of $\bar{\varepsilon}$ are few and of the $\bar{\varepsilon}/\bar{P}$ relationship even fewer. CUTLER et al. (1980b) and JONES and TURNER (1980) found no changes in the $\bar{\varepsilon}/\bar{P}$ relation in rice and sunflower, respectively, as a result of drought pretreatment, whereas the pretreatment caused osmotic adjustment of 2 to 6 bar.

Indication of probable adjustment in $\bar{\varepsilon}/\bar{P}$ can be found in the data given by ELSTON et al. (1976) for beans (*Vicia faba*), JONES and TURNER (1978) for *Sorghum bicolor*, TYREE et al. (1978) for maple (*Acer saccharum*), DAVIES and LAKSO (1979) for apples (*Malus domestica*) and WILSON et al. (1980) for four tropical forage species. In the beans and apples, $\bar{\varepsilon}$ decreased during the spring and early summer before increasing again in late summer and autumn, the range of variation being ca. 20 to 40 bar, whereas in the sorghum $\bar{\varepsilon}$ increased as a result of drought preconditioning by 11 to 16 bar. In these cases $\bar{\varepsilon}$ varied not only as a result of season or treatment but also because of ontogenetic changes in cell size and seasonal stress-induced changes in $\bar{P}$. Thus, as in the case of changes in π in the presence of dehydration, it is not possible to evaluate the influence of changes in $\bar{\varepsilon}$ as an adjustment mechanism in the presence of changes in $\bar{P}$. However, TYREE et al. (1978) showed that in sugar maple the asymptotic, maximum value of $\bar{\varepsilon}$ changed from 50 bar in May to 180 bar in July, declining again to about 70 bar in October. In this case adjustment in $\bar{\varepsilon}$ clearly contributed to shifts in the relationship between Ψ and R* and $\bar{P}$ and Ψ. Evaluation of the combined effects of adjustment of π and $\bar{\varepsilon}$ on these relationships must take into account their volume and pressure dependence and therefore requires a model such as used by TYREE (1981).

2.3.3 Turgor Maintenance

The maintenance of turgor as Ψ or R* falls can result from equivalent reduction in either π or in $\bar{\varepsilon}$. In the absence of information as to the response of $\bar{\varepsilon}$ in most studies, turgor maintenance has generally been explained in terms of the reductions in $\bar{\pi}$ as a result of increase in the cell content of N_s (see review by TURNER and JONES 1980).

Partial diurnal turgor maintenance was first described by HSIAO et al. (1976) who found variation in $\bar{P}$ restricted to 2 to 6 bar as Ψ varied between -3 and -12 bar during the course of a day. Similar observations of diurnal change in $\bar{\pi}$ leading to partial stabilization of $\bar{P}$ have been reported in one or two other cases (e.g., DAVIES and LAKSO 1979) but observations of diurnal turgor maintenance are few.

Partial, medium-term turgor maintenance has been observed in several cases (e.g., BEADLE et al. 1978; JONES and TURNER 1978). For example, TURNER et al. (1978) found that Ψ for stomatal closure fell from -17 to -27 bar in *Helianthus annuus* during a dry period in the growth of the crop, but as a result of reduction in $\bar{\pi}$ of 0.6 bar per bar reduction in Ψ the mean value of $\bar{P}$ at stomatal closure remained at 5 bar.

HSIAO et al. (1976) draw attention to full and medium-term turgor maintenance resulting from osmotic adjustment (i.e., $\Delta\pi/\Delta\Psi=1$) in *Sorghum bicolor* (see also FERERES et al. 1978). Since then a number of other cases have been identified (see for examples JONES and TURNER 1980; WILSON et al. 1980) and TURNER and JONES (1980) list 15 other examples of full turgor adjustment dating back to 1972.

Medium term turgor maintenance has been evaluated quantitatively in a number of cases as the relation between $\bar{P}$ and Ψ (TURNER 1974; JOHNSON and CALDWELL 1976; JOHNSON and BROWN 1977; JONES and TURNER 1978; JOHNSON 1978; CUTLER et al. 1980a, b). JONES and TURNER (1978) tried to separate the effects of π and ε in bringing about turgor maintenance in sorghum. Since drought pretreatment resulted in an increase in ε rather than a decrease, the substantial turgor maintenance observed was attributed solely to osmotic adjustment. In the experiments by JOHNSON and co-workers (see above), a range of values of leaf Ψ were obtained by withholding water for 6–10 days from potted plants of a number of species growing in controlled environments. The species able to maintain turgor to low values of Ψ were also the species assessed by more subjective criteria to be the most drought-resistant. *Zea mays*, for example, maintained $\bar{P}$ over a much smaller range of Ψ than a number of species of range grasses and *Agropyron* hybrids. The ability to maintain positive $\bar{P}$ to low Ψ was shown to depend on both the value of $\bar{P}$ at $\Psi=0$, i.e., on

$\bar{\pi}_0$, and on the slope of the straight line, $d\bar{P}/d\Psi$. The value of $\bar{\pi}_0$ depends on the degree of osmotic adjustment which has taken place whereas $d\bar{P}/d\Psi$ depends on several properties of the tissue, including the non-ideality of the osmotic solutes, but probably most strongly on $\bar{\varepsilon}$. Reductions in the slope $d\bar{P}/d\Psi$ at low turgor have been cited as evidence of reduction in $\bar{\varepsilon}$ (CUTLER et al. 1980a, b). Thus by inference from the variation in $d\bar{P}/d\Psi$ substantial variations in $\bar{\varepsilon}$ probably also exist amongst species and have considerable influence on the value of Ψ at zero turgor. Lower growth temperatures also change the values of both parameters but generally have a larger effect on $\bar{\pi}_0$, which was reduced in some species by up to 7 bar by a change in temperature regime to lower day and night temperatures (TURNER and JARVIS 1975; JOHNSON 1978).

Substantial long-term turgor maintenance has also been found over almost the whole growing season. FERERES et al. (1978) found that the decline in leaf $\bar{\pi}$ in *Sorghum bicolor* was sufficient to maintain $\bar{P}$ at an average value of 5 bar for 100 days of the growing season. In *Zea mays*, on the other hand, osmotic adjustment maintained $\bar{P}$ only partially in unirrigated plants. Similar partial turgor maintenance was found by GOODE and HIGGS (1973) in apple. The complete turgor maintenance in sorghum allowed stomata to remain open at values of leaf Ψ down to -20 bar.

2.3.4 Turgor Maintenance and Growth

2.3.4.1 The Growth Equations

The physics of growth of plant cells is best understood in *Nitella* (GREEN et al. 1971) but there is circumstantial evidence that similar relations hold in higher plant cells (HSIAO et al. 1976). In its simplest formulation, growth consists of two separable phenomena. The first is the plastic deformation of the cell walls to encompass a larger cell volume. With a slight modification of Green's original formulation this relation can be written:

$$r=\frac{dV}{dt}=m(P-Y), \tag{2.23}$$

where r, the growth rate, is equated to the rate of volume increase of the cells, dV/dt. The driving force for growth is the amount by which P exceeds the yielding point of the wall, Y. The rate at which the plastic deformation occurs is governed by m which incorporates the wall fluidity and some shape factors.

The second phenomenon is the permeation of water across the cell membrane which can be described by a water transport equation,

$$\frac{dV}{dt}=L\,\Delta\Psi=L(\Psi_e-P-\pi), \tag{2.24}$$

where L is the membrane hydraulic permeability times surface area and $\Delta\Psi$ is the driving force for water flow, equal to the external water potential, Ψ_e, minus the internal water potential Ψ_i ($=\pi+P$). In considering growth at the whole plant level, $\Delta\Psi$ can be considered as the water potential difference between the soil and growing cell and L the corresponding conductivity of the hydraulic pathway. It is important to realize that the driving force for plastic deformation $(P-Y)$ is much smaller than Y and that Y is under metabolic control. Hence P, although it does partly control growth, is a very insensitive indicator of long-term growth, because metabolism can change the value of Y and can thus dominate the real driving force $(P-Y)$.

The value of Y can change with time at a rate given by

$$\frac{dY}{dt}=rh-s, \tag{2.25}$$

where s is the rate of wall softening, and is probably a complex function of temperature, light, water stress history, metabolism and plant hormones, and h is a mechanical hardening coefficient, which depends on the physical properties of the cell wall. Equation (2.25) shows that in the absence of metabolic softening r will cause Y to increase $\left(\frac{dY}{dt}>0\right)$ so that growth will eventually stop when $P-Y=0$ (Eq. 2.23). Growth depends on the long-term, integrated effects of a wide range of factors on the time integral of Eq. (2.25). Even in the steady state when Y does not change $\left(\frac{dY}{dt}=0\right)$, it can be argued that growth is totally controlled by s because then

$$r=s/h. \tag{2.26}$$

Thus we have three equations (2.23, 2.24 and 2.26) that describe growth in terms of three separate driving forces, i.e., $(P-Y)$, $\Delta\Psi$, and s or metabolism.

2.3.4.2 What Controls the Rate of Short-Term Growth?

Better insight into the overall rate control of short-term growth can be obtained by combining the above equations. Equating Eq. (2.23) to Eq. (2.24), solving for P and substituting into Eq. (2.23) gives

$$r=\frac{mL}{m+L}(\Psi_e-\pi-Y), \tag{2.27}$$

or

$$r=\frac{mL}{m+L}(\Delta\Psi+P-Y). \tag{2.28}$$

Equation (2.27) demonstrates the role of osmotic adjustment in controlling the ability of plants to grow at low external water potentials, Ψ_e: decreases in π lead to increases in r. Equation (2.28) emphasizes the control of $(P-Y)$ over plastic wall deformation and the influence of $\Delta\Psi$ on the flow of water

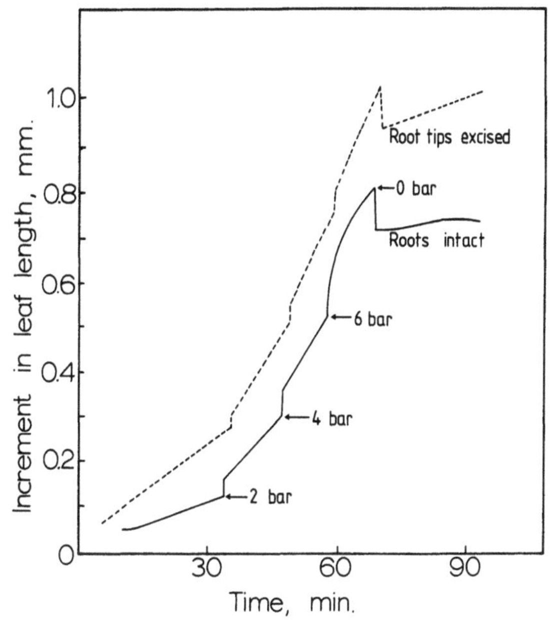

Fig. 2.8. The response of leaf elongation in *Oryza sativa* cv. Kinandang Patong to stepwise increase in pressure applied to the roots of plants in dilute nutrient solution. The *figures beside the curves* indicate the time at which the pressure was raised to the value indicated. The length of an expanding leaf on a seedling in the seven-leaf stage was measured with a linear displacement transducer. (CUTLER et al. 1980c)

from the source, for example, the soil to the cell. The fraction $mL/(m+L)$ can be regarded as a growth coefficient. If $m \gg L$, then $mL/(m+L) = L$, L dominates the magnitude of the growth coefficient and $\Delta\Psi$ is larger and more important than $(P-Y)$. At the other extreme if $L \gg m$ then $mL/(m+L) \simeq m$, m dominates the growth coefficient and $(P-Y)$ is larger and more important than $\Delta\Psi$.

In *Nitella* it is clear that m and $(P-Y)$ dominate short-term growth because $L > m$ (GREEN et al. 1971), but in this case L is determined only by the membrane hydraulic conductivity. In a whole plant L may be less than m because the water pathway includes the series-linked conductivities of the soil + root + xylem + cell wall–membrane complex. In soybean hypocotyl, on the other hand, BOYER and WU (1978) found m and L to be approximately equal. Thus in this case wall extensibility and membrane hydraulic conductivity control cell enlargement about equally. Because of the comparatively low values of L, values of Ψ in the enlarging tissue were -1.7 to -2.5 bar during rapid growth. Similar values of Ψ were predicted with an appropriate water transport model (MOLZ and BOYER 1978).

CUTLER et al. (1980c) have argued that in whole rice plants m is larger than L so that growth is rate-limited by L, and $\Delta\Psi$ in Eq. (2.28) is larger than $(P-Y)$. In their experiments, the rooting medium of water-cultured rice plants was pressurized. This caused a rapid "elastic elongation" of the leaves followed by an increased growth rate (see Fig. 2.8). CUTLER et al. (1980c) suggest that the observed increase in growth rate after the "elastic elongation" is the result of increase in $\Delta\Psi$ as a consequence of the root pressurization. This seems unlikely, however. The increase in $\Delta\Psi$ will be transitory if transpiration does not also increase because of rehydration of the leaf cells. During the transitory increase in $\Delta\Psi$ the leaf will elongate more rapidly either because of growth or because of rehydration of the cells. It is unlikely that the stepwise increase or decrease in leaf length after an increase or decrease in root pressure is the result of cell hydration or dehydration alone because the speed of the changes in length is too rapid. The kinetics of rehydration or dehydration of leaves in a pressure bomb indicate that several minutes are required to complete the process in *Fagus grandifolia* (TYREE and CHEUNG 1977), in *Helianthus annuus* (CRUIZIAT et al. 1980) and in *Populus* spp. (TYREE et al. 1978). Consequently, the increases in elongation rate after the stepwise changes in pressure are probably the result of both growth and rehydration of the leaf tissue. It is thus unclear exactly how much growth has been increased. If, on the other hand, it is argued that rice leaves do rehydrate in a matter of seconds as a result of root pressurization, then P will have increased and $\Delta\Psi$ will have been reduced almost to the prepressurization value. In this case the increased growth rate could have been the result of an increase in (P−Y).

We think it is possible that L and $\Delta\Psi$ do limit short-term growth in whole plants. The problems involved in the interpretation of Fig. 2.8 serve to illustrate the difficulties in establishing the true limitation on short-term growth. More work will need to be done to answer the question and measurements of P during growth may be the only way to provide the answer.

2.3.4.3 What Controls the Rate of Long-Term Growth?

In some recent studies attempts have been made to find a simple correlation between maintenance of high $\bar{P}$ and long-term growth rate. JOHNSON and BROWN (1977) and JOHNSON (1978) compared the response of $\bar{P}$ to Ψ in the leaves of several grass species. They tested the hypothesis that the grasses in which zero $\bar{P}$ is reached at the lowest values of Ψ would grow better at lower Ψ than the others. When several grass species were grown at night/day temperatures of 10°/15 °C, this was the case. At other night/day temperatures, however, the ranking of the species changed, the suggestion being that temperature affected Y. BUNCE (1977) found linear relations between leaf elongation rate and $\bar{P}$ in soybean in three different environments. However, the relationship changed with environment and in drier conditions lower $\bar{P}$ was required to maintain the elongation rate, suggesting a lower value of Y. In contrast, CUTLER et al. (1980d) found that rice plants that had been conditioned by three previous drought cycles required higher values of $\bar{P}$ to maintain daily leaf elongation rates than did well-watered controls, although the maximum rate of elongation in drought-conditioned plants did not differ from that of controls. Thus maintenance of $\bar{P}$ by osmotic adjustment in the conditioned plants was not sufficient to ensure the same rates of leaf elongation as in the controls at the same values of Ψ; other factors were involved. They attributed this result to control of the growth rate by $\Delta\Psi$ rather than by $\bar{P}$ or by ($\bar{P}$−Y) [see Eq. (2.28)], but proper account was not taken of possible shifts in Y between droughted and control plants so the interpretation remains unclear.

In grasses considerable problems arise in measuring growth-controlling variables at the right place – the meristem. WATTS (1972, 1974) showed that leaf growth in young maize plants was much more sensitive to variations in soil temperature than in air temperature because of the location of the meristem close to the soil surface. CUTLER et al. (1980c) point out that Ψ of enclosed non-transpiring rice leaves was 8 to 10 bar higher than in the transpiring leaves, the implication being that Ψ at the meristem would be similarly different. Thus calculation of $\bar{P}$ from measurements of Ψ and π on transpiring leaves may lead to wholly irrelevant estimates for use in growth equations.

It is not surprising that little correlation has been found between $\bar{P}$ and long-term growth even though correlations can be found in the short term (e.g., WENKERT et al. 1978a). This is probably because of long-term shifts in Y which is under the influence of many different factors. Since the net driving force for plastic wall deformation ($\bar{P}-Y$) may be 0.3 bar or less (GREEN et al. 1971), long-term shifts in Y of several bars would destroy a simple correlation between $\bar{P}$ and growth or drought hardiness. Thus failure to find simple correlations between parameters derived from Ψ-isotherms and drought hardiness (e.g., CUTLER and RAINS 1978; MAXWELL and REDMANN 1978) may be based on long-term changes in Y. Given our present state of knowledge, all we can say is that long-term growth patterns are controlled by wall softening. In terms of Eq. (2.26) this means that net growth is the time integral of r which in turn depends on s, i.e.,

$$\text{net growth} = \int r\,dt = \int \frac{s}{h}\,dt.$$

There is little doubt that water stress and the history of water stress a plant has experienced influence growth, but the mechanisms by which some plants either tolerate or avoid water stress are probably very important and influence growth via complex and unknown feedback mechanisms. That is not to say that turgor maintenance is unimportant, but that turgor maintenance is only one of several interdependent factors determining growth and drought hardiness.

2.3.5 Apoplasmic Water

CARR and GAFF (1961) and GAFF and CARR (1961) were the first to suggest in recent times that water in cell walls – apoplasmic water – could act as a reservoir protecting cells from water loss by acting as a buffer for the symplasmic water. Recently this suggestion has been taken up by CUTLER et al. (1977) who have suggested that a high apoplasmic water content is a common feature of drought-adapted plants.

A number of estimates of the apoplasmic water fraction, R_a^* ($=W_a/(W_0+W_a)$) exist, but some (e.g., WARREN WILSON 1967b; ACOCK 1975) are based on dubious methods of analysis of water potential isotherms (see Sect. 2.2.4). Selected values of R_a^* ranging from 0.05 to 0.4 are shown in Table 2.1, but it is difficult from this table alone to draw any correlations between R_a^* and drought adaptation. It is very unlikely, however, that apoplasmic water *in the leaf* can act as much of a buffer against the loss of symplasmic water. In the middle of the day the amount of water transpired hourly can be a large fraction of

Table 2.1. Some relative apoplasmic water volumes

Species	Method of analysis	$R_a^* = (W_a/W_0 + W_a)$	Comments and references
Acer saccharum	P-B	0.2 (June) 0.4 (Sept.)	Single leaves (1)
Picea abies	P-B	0.4 (mature leaves) 0.3 (during leaf flush)	Shoots (1)
Picea sitchensis	P-B	0.2 (Sept.) 0.4 (Jan.)	Shoots (8)
Helianthus annuus	P-B		Single leaves (unpublished)
Populus spp	P-B	0.2 to 0.3	Single leaves (2) and unpublished
Fraxinus pennsylvanica	P-B	0.15	Single leaves (2)
Clematis ligustifolia	P-B	0.05	Single leaves (2)
Cornus stolonifera	P-B	0.05	Single leaves (2)
Acacia aneura	V-E	0.18	Phyllodes (3, 4)
Ceratonia siliqua	V-E	0.08	Single leaves (3, 5)
Pennisetum typhoides	T-S	0.05	Single leaves (3, 6)
Eucalyptus globulus	G	0.4	Leaf tissue (7)

(1) TYREE et al. (1978); (2) CHEUNG et al. (1975); (3) TYREE (1976a); (4) SLATYER (1960); (5) NOY-MEIR and GINZBURG (1967); (6) BEGG et al. (1964); (7) GAFF and CARR (1961); (8) HELLKVIST et al. (1974)

P-B Pressure bomb isotherm analysis (xylem water included)

V-E Vapor equilibrium isotherm analysis (note in this method the xylem conduits may be empty so xylem water does not contribute to W_a)

G Removal of cell wall material and gravimetric analysis

the total leaf water content. For example, in *Eucalyptus* the hourly water loss amounts to 50% to 120% of the total leaf water content (GAFF and CARR 1961), and in cotton less than 5% of the total amount of water transpired comes from net water loss *from the leaf* during the day (WEATHERLEY 1970). JARVIS (1975) calculated for several different kinds of vegetation canopies and transpiration rates that the normal diurnal reduction in R* of leaves could wholly supply transpiration for only 15 to 70 min. Furthermore, since most of the apoplasmic water in the cell wall requires large tensions to remove it (see the earlier discussion in Sects. 2.2.1 and 2.2.4), this apoplasmic water must be regarded as unavailable, so that the daily reductions in leaf R* must be largely at the expense of symplasmic water: the apoplasmic water in the cell walls would be retained and only become available in extreme drought conditions (TEOH et al. 1967). In trees, however, the vastly larger water reservoirs in the stem and branches can act as a significant buffer against net water loss from the leaves (JARVIS 1975; LANDSBERG et al. 1976; POWELL and THORPE 1977; WARING and RUNNING 1978; WARING et al. 1979; WHITEHEAD and JARVIS 1981; Chap. 5, this Vol.). In wheat grown in a temperate climate, water stored in the shoot may also act as a significant buffer to net water loss from leaves (JONES 1978). In these examples there is no evidence that apoplasmic water in the cell walls constitutes a significant fraction of the available water, and only in trees is there evidence of the continued importance of apoplasmic stored water under conditions of high evaporative demand. The significance of cell wall apoplasmic water must therefore lie elsewhere.

Table 1 shows large variations amongst species in R_a^* and more extensive lists are given by CHEUNG et al. (1975) and TYREE et al. (1978). At the present time, there seems to be little apparent discrimination amongst the species with 5% or 40% apoplasmic water with respect to habitat or environment. The only consistent difference between species is the larger values of R_a^* when Ψ-isotherms are determined for shoots rather than for individual leaves, as is usually the case with conifers.

In addition, large seasonal variations in R_a^* and the symplasmic water fraction, $= W_0/(W_0 + W_a)$, have been found, particularly in conifers (Table 2.1) (HELLKVIST et al. 1974; TYREE et al. 1978). Some of this variation is ontogenetic in origin, higher values of R_a^* being found in mature leaves than in young developing leaves (WENKERT et al. 1978b; TYREE et al. 1978), but, in the conifers in particular, the substantial increase in R_a^* in the autumn seems to be associated with cold hardening (HELLKVIST et al. 1974; TYREE et al. 1978). How this relates to changes in the symplasm or in cell wall structure is still completely unknown. However, there is no doubt that certain features of the Ψ-isotherms, including the proportion of apoplasmic water, can be related to the relative amount of dry matter or cell wall material present. Tissues with Ψ-isotherms indicating a large ε also contain a large proportion of dry matter, W_d, per unit saturated water content $(W_s - W_d)$, i.e., large $W_d/(W_s - W_d)$, as well as a large apoplasmic water fraction (P.G. JARVIS and M.S. JARVIS, unpublished, FERERES et al. 1978; RICHTER et al. 1980). The seasonal increase in R_a^* may therefore be the result of growth in cell wall thickness or changes in cell wall structure in "mature" leaves. Such shifts in water distribution could account for some of the observed decrease in $\bar{\pi}$ over the medium and long term, even if N_s/cell remains constant. However, the significance of secondary wall growth is unlikely to lie solely in its effect of decreasing $\bar{\pi}$, because this is an energetically wasteful way to lower $\bar{\pi}$.

2.3.6 Cell Size

Reduction in cell size has often been noted in plants which have developed under conditions of water stress (see HENCKEL 1964). Recently CUTLER et al. (1977) have restated the case for the importance of small cell size to drought adaptation. They measured N_s/cell, cell size and cell wall thickness in leaves of cotton plants (*Gossypium hirsutum*) grown under water stress and found statistically significant differences in cell size (32 versus 37 μm), in π (an increase of about 3 bar) and in cell wall thickness (0.76 versus 0.68 μm) between the stressed plants and the controls. Rather similar observations were made by WILSON et al. (1980) on *Panicum maximum, Heteropogon contortus* and the tropical legume *Macroptilium atropurpureum*.

Having observed that water stress caused a reduction in cell size in cotton leaves, CUTLER et al. (1977) examined the consequences of *diminishing* cell size on $\bar{\pi}$ under the following three theoretical conditions (1) N_s/cell increases linearly with decreasing cell size, (2) N_s/cell is constant and (3) N_s/cell decreases linearly with decreasing cell size. In all three cases they calculated that N_s/cell will increase and hence that π will decrease with decreasing cell size and so lead to more effective turgor maintenance. However, their assumption of linear changes in N_s/cell with changing cell size leads to misleading results. If, during growth, N_s/cell, and hence $\bar{\pi}$, are to remain constant N_s/cell must change in proportion to the *cube* of the cell dimension rather than in linear proportion to the dimension, since to maintain a constant concentration N_s must be proportional to the volume of the cell. It is, therefore, not surprising that their analysis yields a remarkable but, in our view, wholly unconvincing result. We do not doubt the older literature they quote which indicates that drought-adapted plants tend to have smaller cells with thicker walls than mesic plants, or their observations on changes in cell size in cotton. However, we remain unconvinced that the adaptive advantage of small cells results from a decrease in $\bar{\pi}$ which is wholly consequent upon the small size of the cells.

The case relating cell size or structure to function is an interesting one but is still open to speculation, observation and experiment. More careful anatomical and physiological studies of the changes occurring in tissues during acclimation to water stress are required, especially in circumstances when Ψ and R_a^* are changing, coupled with more realistic models of cell water relations such as that of CHEUNG et al. (1976).

2.3.7 Resurrection Plants

A wide variety of plants appear to be able to survive extremely low water potentials for prolonged periods (see Chap. 10, this Vol.). Besides thallophytes such as many algae, lichens, and mosses about 50 species of angiosperms and 23 species of ferns can be dried to equilibrium with water vapor at 10% relative humidity or less ($\Psi < -3{,}000$ bar) for many months and are still viable and resume metabolism within a few hours or days after rehydration (GAFF and HALLAM 1974; GAFF 1980). There appears to be no consistent pattern between structural features and desiccation tolerance. Although resurrection plants are usually pioneer xerophytes, they often lack the classical xeromorphic characteristics. Most Dicotyledons and ferns retain chlorophyll throughout the dehydration and rehydration cycle but chlorophyll is lost during dehydration in many Monocotyledones. About two days are required for the reappearance of chlorophyll after rehydration. Rapid drying of these angiosperms will preserve chlorophyll but not viability. The preservation of cellular fine structure after dehydration also varies from almost complete preservation to complete destruction in resurrection plants (GAFF and HALLAM 1974; GAFF et al. 1976; GAFF 1977). Almost complete destruction of the fine structure also occurs in intolerant plants after desiccation (FELLOWS and BOYER 1978). One can only presume that tolerant plants undergo an orderly dismantlement. There must be very subtle relationships between structure and survival of resurrection plants. Despite the considerable interest shown in these plants by GAFF and his colleagues (see above) in recent years, we are still far from understanding how the plants do it. In view of the extreme desiccation tolerance shown by resurrection plants, more detailed study of the cellular changes occurring during dehydration and rehydration could provide important leads to metabolic water stress parameters in more ordinary plants (GAFF 1980).

References

Acock B (1975) An equilibrium model of leaf water potentials which separates intra- and extracellular potentials. Aust J Plant Physiol 2:253–263

Aikman DP (1979) Movement of water trough membranes. In: Wierzchowski KL, Przestalski S (eds) Biophysics of Membrane Transport. Wroclaw, Poland pp 19–28

Barnett NM, Naylor AW (1966) Amino acid and protein metabolism in Bermuda grass during water stress. Plant Physiol 41:1222–1230

Beadle CL, Turner NC, Jarvis PG (1978) Critical water potential for stomatal closure in Sitka spruce. Physiol Plant 43:160–165

Begg JE, Bierhuizen JF, Lemon ER, Misra DK, Slatyer RO, Stern WR (1964) Diurnal water and energy exchanges in bullrush millet in an area of high solar radiation. Agric Meteorol 1:294–312

Bolt GH, Frissel MJ (1960) Thermodynamics of soil moisture. Neth J Agric Sci 8:57–78

Bolt GH, Miller RD (1958) Calculation of total and component potentials of water in soils. Trans Am Geophys Union 39:917–928

Boyer JS, Potter JR (1973) Chloroplast response to low leaf water potentials, I. Role of turgor. Plant Physiol 51:989–992

Boyer JS, Wu G (1978) Auxin increases the hydraulic conductivity of auxin-sensitive hypocotyl tissue. Planta 139:227–237

Brown KW, Jordan WR, Thomas JC (1976) Water stress induced alterations of the stomatal response to decreases in leaf water potential. Physiol Plant 37:1–5

Broyer TC (1952) On volume enlargement and work expenditure by an osmotic system in plants. Physiol Plant 5:459–469

Bunce JA (1977) Leaf elongation in relation to leaf water potential in soybean. J Exp Bot 28:156–161

Carr DJ, Gaff DF (1961) The role of the cell-wall in the water relations of leaves. Proceedings of the Madrid symposium on plant-water relationships in arid and semi-arid conditions. Arid Zone Res 16:117–125

Cheung YNS, Tyree MT, Dainty J (1975) Water relations parameters on single leaves obtained in a pressure bomb and some ecological interpretations. Can J Bot 3:1342–1346

Cheung YNS, Tyree MT, Dainty J (1976) Some possible sources of error in determining bulk elastic moduli and other parameters from pressure-volume curves of shoots and leaves. Can J Bot 54:758–765

Cline RG, Campbell GS (1976) Seasonal and diurnal water relations of selected forest species. Ecology 57:367–373

Cram WJ (1976) Negative feedback regulation of transport in cells. The maintenance of turgor, volume and nutrient supply. In: Lüttge U, Pitman MG (eds) Transport in plants IIA. Encyclopedia of plant physiology new ser Vol 2. Springer, Berlin Heidelberg New York, pp 284–316

Cruziat P, Tyree MT, Bodet C, LoGullo MA (1980) The kinetics of rehydration of detached sunflower leaves following substantial waterloss. New Phytol 84:293–306

Cutler JM, Rains DW (1978) Effects of water stress and hardening on the internal water relations and osmotic constituents of cotton leaves. Physiol Plant 42:261–268

Cutler JM Rains DW, Loomis RS (1977) The importance of cell size in the water relations of plants. Physiol Plant 40:255–260

Cutler JM, Shahan KW, Steponkus PL (1980a) Dynamics of osmotic adjustment in rice. Crop Sci 20:310–314

Cutler JM Shahan KW, Steponkus PL (1980b) Alteration of the internal water relations of rice in response to drought hardening. Crop Sci 20:307–310

Cutler JM, Steponkus PL, Wach MJ, Shahan KW (1980c) Dynamic aspects and enhancement of leaf elongation in rice. Plant Physiol 66:147–152

Cutler JM, Shahan KW, Steponkus PL (1980d) Influence of water deficits and osmotic adjustment on leaf elongation in rice. Crop Sci 20:314–318

Dainty J (1976) Water relations of plant cells. In: Lüttge U, Pitman MG (eds) Transport in plants II. Encyclopedia of plant physiology new ser Vol II A. Springer, Berlin Heidelberg New York, pp 12–35

Davies FS, Lakso AN (1978) Water relation in apple seedlings. Changes in water potential components, abscisic acid levels and stomatal conductances under irrigated and non-irrigated conditions. J Am Soc Hortic Sci 103:310–313

Davies FS, Lakso AN (1979) Diurnal and seasonal changes in leaf water potential components and elastic properties in response to water stress in apple trees. Physiol Plant 46:109–114

Edwards M, Meidner H (1979) Direct measurements of turgor pressure potentials, IV. Naturally occurring pressures in guard cells and their relation to solute and matric potentials in the epidermis. J Exp Bot 30:829–837

Edwards WRN, Jarvis PG (1982) Relations between water content, potential and permeability in stems of conifers. Plant Cell Environ (in press)

Elston J, Karamanos AJ, Kassam AH, Wadsworth RM (1976) The water relations of the field bean crop. Philos Trans R Soc London Ser B 273:581–591

Fellows RJ, Boyer JS (1978) Altered ultrastructure of cells of sunflower leaves having low water potentials. Protoplasma 93:381–395

Fereres E, Acevedo E, Henderson DW, Hsiao TC (1978) Seasonal changes in water potential and turgor maintenance in sorghum and maize under water stress. Physiol Plant 44:261–267

Flowers TJ, Ward ME, Hall JL (1976) Salt tolerance in the halophyte *Suaeda maritima:* some properties of malate dehydrogenase. Philos Trans R Soc London Ser B 273:523–540

Flowers TJ, Troke PF, Yeo AR (1977) The mechanism of salt tolerance in halophytes. Annu Rev Plant Physiol 28:89–121

Ford CW, Wilson JR (1981) Changes in levels of solutes during osmotic adjustment to water stress in leaves of four tropical pasture species. Aust J Plant Physiol 8:79–91

Gaff DF (1977) Desiccation tolerant vascular plants of Southern Africa. Oecologia 31:95–109

Gaff DF (1980) Protoplasmic tolerance of extreme water stress. In: Turner NC, Kramer PJ (eds) Adaptation of plants to water and high temperature stress. Wiley and Sons, New York, pp 207–230

Gaff DF, Carr DJ (1961) The quantity of water in the cell wall and its significance. Aust J Biol Sci 14:299–311

Gaff DF, Hallam ND (1974) Resurrecting desiccated plants. In: Bieleski RL, Ferguson AR, Creswell MM (eds) Mechanisms of regulation of plant growth. Roy Soc NZ Bull 12:389–393

Gaff DF, Zee S-Y, O'Brien TP (1976) The fine structure of dehydrated and reviving leaves of *Borya nitida* Labill. – a desiccation-tolerant plant. Aust J Bot 24:225–236

Gardner WR, Ehlig CF (1965) Physical aspects of the internal water relations of plant leaves. Plant Physiol 40:705–710

Geiger DR, Giaquinta RT, Sovonick SA, Fellows RJ (1973) Solute distribution in sugar beet leaves in relation to phloem loading and translocation. Plant Physiol 52:585–589

Goode JE, Higgs KH (1973) Water, osmotic and pressure potential relationships in apple leaves. J Hortic Sci 48:203–215

Green PB, Erickson RO, Buggy J (1971) Metabolic and physical control of cell elongation rate – in vivo studies in *Nitella*. Plant Physiol 47:423–430

Hanson AH, Nelson CE (1978) Betaine accumulation and ^{14}C formate metabolism in water stressed barley leaves. Plant Physiol 62:305–312

Hanson AD, Nelson CE, Everson EH (1977) Evaluation of free proline accumulation as an index of drought resistance using two contrasting cultivars. Crop Sci 17:720–726

Haughton PM, Sellen DB, Preston RD (1968) Dynamic mechanical properties of the cell wall of *Nitella opaca*. J Exp Bot 19:1–12

Hellebust JA (1976) Osmoregulation. Annu Rev Plant Physiol 27:485–505

Hellkvist J, Richards GP, Jarvis PG (1974) Vertical gradients of water potential and tissue water relations in Sitka spruce trees measured with the pressure chamber. J Appl Ecol 11:637–668

Henckel RA (1964) Physiology of plants under drought. Annu Rev Plant Physiol 15:386–408

Hinckley TM, Duhme F, Hinckley AR, Richter H (1980) Water relations of drought hardy shrubs: osmotic potential and stomatal reactivity. Plant Cell Environ 3:131–140

Hsiao TC, Acevedo E, Fereres E, Henderson DW (1976) Water stress, growth and osmotic adjustment. Philos Trans R Soc London Ser B 273:479–500

Hüsken D, Steudle E, Zimmermann U (1978) Pressure probe technique for measuring water relations of cells in higher plants. Plant Physiol 61:158–163

Hüsken D, Zimmermann U, Schulze E-D (1980) Water relations of leaves of *Tradescantia virginiana*: direct turgor pressure measurement. In: Spanswick RM, Lucas WJ, Dainty J (eds) Plant membrane transport: current conceptual issues. Elsevier/North-Holland Biomedical Press, Amsterdam New York, pp 469–470

Jäger H-J, Meyer HR (1977) Effect of water stress on growth and proline metabolism of *Phaseolus vulgaris* L. Oecologia 30:83–86

Jarvis PG (1975) Water transfer in plants. In: Vries de DA, Afgan NH (eds) Heat and mass transfer in the biosphere I. Transfer processes in plant environment. Scripta, Washington DC, pp 369–394

Jarvis PG, Jarvis MS (1963a) Effects of several osmotic substrates on the growth of *Lupinus albus* L. seedlings. Physiol Plant 16:485–500

Jarvis PG, Jarvis MS (1963b) The water relations of tree seedlings IV. Some aspects of the tissue water relations and dought resistance. Physiol Plant 16:501–516

Jarvis PG, Jarvis MS (1964) Presowing hardening of plants to drought. Phyton (Buenos Aires) 21:113–117

Jeschke WD (1979) Univalent cation selectivity and compartmentation in cereals. In: Laidman DL, Wyn Jones RG (eds) Recent advances in the biochemistry of cereals. Academic Press, London New York, pp 37–61

Johnson DA (1978) Environmental effects on turgor pressure response in range grasses. Crop Sci 18:945–948

Johnson DA, Brown RW (1977) Psychrometric analysis of turgor pressure response: a possible technique for evaluating plant water stress resistance. Crop Sci 17:507–510

Johnson DA, Caldwell MM (1976) Water potential components, stomatal function and liquid phase water transport resistances of four arctic and alpine species in relation to water stress. Physiol Plant 36:271–278

Jones HG (1978) Modelling diurnal trends of leaf water potential in transpiring wheat. J Appl Ecol 15:613–626

Jones MM, Rawson MM (1979) Influence of rate of development of leaf water deficits upon photosynthesis, leaf conductance, water use efficiency, osmotic potential in sorghum. Physiol Plant 45:103–111

Jones MM, Turner NC (1978) Osmotic adjustment in leaves of sorghum in response to water deficits. Plant Physiol 61:122–126

Jones MM, Turner NC (1980) Osmotic adjustment in expanding and fully expanded leaves of sunflower in response to water deficits. Aust J Plant Physiol 7:181–192

Jones MM, Osmond CB, Turner NC (1980) Accumulation of solutes in leaves of sorghum and sunflower in response to water deficits. Aust J Plant Physiol 7:193–205

Kassam AH, Elston JF (1974) Seasonal changes in the status of water and tissue characteristics of leaves of *Vicia faba*. L. Ann Bot (London) 38:419–429

Kirkham MM, Smith EL (1978) Water relations of tall and short cultivars of winter wheat. Crop Sci 18:227–230

Kluge M, Osmond CB (1972) Studies on phosphoenolpyruvate carboxylase and other enzymes of crassulacean acid metabolism of *Bryophyllum tubiflorum* and *Sedum praealtum*. Z Pflanzenphysiol 66:97–105
Knipling EB (1967) Effect of leaf aging on water deficit-water potential relationships of dogwood leaves growing in two environments. Physiol Plant 20:65–72
Kramer PJ (1955) Bound water. In: Ruhland W (ed) Handbuch der Pflanzenphysiologie Vol I. Springer, Berlin Göttingen Heidelberg, pp 223–242
Kreeb K (1963) Hydrature and plant production. In: Whitehead FW, Rütter AJ (eds) The water relations of plants. Blackwell, Oxford, pp 272–288
Kyriakopoulos E, Richter H (1977) A comparison of methods for the determination of water status in *Quercus ilex* L. Z Pflanzenphysiol 82:14–27
Landsberg JJ, Blanchard TW, Warrit B (1976) Studies on the movement of water through apple trees. J Exp Bot 27:579–596
Lang ARG (1967) Osmotic coefficients and water potentials of sodium chloride solutions from 0 to 40° C. Aust J Chem 20:2017–2023
Larkum AWD, Wyn Jones RG (1979) Carbon dioxide fixation by chloroplasts isolated in glycinebetaine: A putative cytoplasmic osmoticum. Planta 145:393–394
Lawlor DW (1969) Plant growth in polyethylene glycol solutions in relation to the osmotic potential of the root medium and the leaf water balance. J Exp Bot 20:895–911
Lawlor DW (1979) Effects of water and heat stress on carbon metabolism of plants with C_3 and C_4 photosynthesis. In: Mussell H, Staples RC (eds) Stress physiology in crop plants. Wiley and Sons, New York, pp 303–326
Lawlor DW, Fock H (1977a) Photosynthetic assimilation of $^{14}CO_2$ by water-stressed sunflower leaves at two O_2 concentrations and the specific activity of the products. J Exp Bot 28:320–328
Lawlor DW, Fock H (1977b) Water stress induced changes in the amounts of some photosynthetic assimilation products and respiratory metabolites of sunflower leaves. J Exp Bot 28:329–337
Lawlor DW, Fock H (1978) Photosynthesis, respiration and carbon assimilation in water-stressed maize at two oxygen concentrations. J Exp Bot 29:579–593
Ludlow MM (1980) Adaptive significance of stomatal response to water stress. In: Turner NC, Kramer PJ (eds) Adaptation of plants to water and high temperature stress. Wiley and Sons, New York, pp 123–138
Lüttge U, Ball E (1974) Proton and malate fluxes in cells of *Bryophyllum daigremontianum* leaf slices in relation to potential osmotic pressure of the medium. Z Pflanzenphysiol 73:326–338
Lüttge U, Ball E (1977) Water relation parameters of the CAM plant *Kalanchoe daigremontiana* in relation to diurnal malate oscillations. Oecologia 31:85–94
Maxwell JO, Redmann RE (1978) Leaf water potential, component potentials and relative water content in a xeric grass, *Agropyron dasystachyum* (Hook) Scribn. Oecologia 35:277–284
McCree KJ (1974) Changes in the stomatal response characteristics of grain sorghum produced by water stress during growth. Crop Sci 14:273–278
Milburn JA, McLaughlin ME (1974) Studies of cavitation in isolated vascular bundles and whole leaves of *Plantago major* L. New Phytol 73:861–871
Molz FJ, Boyer JS (1978) Growth induced water potentials in plant cells and tissues. Plant Physiol 62:423–429
Morgan JM (1977) Differences in osmoregulation between wheat genotypes. Nature (London) 270:234–235
Munns R, Brady CI, Barlow EWR (1979) Solute accumulation in the apex and leaves of wheat during water stress. Aust J Plant Physiol 6:379–389
Noy-Meir I, Ginzburg BZ (1967) An analysis of the water potential isotherm in plant tissue. I. The theory. Aust J Biol Sci 20:695–721
Noy-Meir I, Ginzburg BZ (1969) An analysis of the water potential isotherm in plant tissue. II. Comparative studies on leaves of different types. Aust J Biol Sci 22:35–52
Passioura JB (1980) The meaning of matric potential. J Exp Bot 31:1161–1169

Pollard A, Wyn Jones RG (1978) Enzyme activities in concentrated solutions of glycinebetaine and other solutes. Planta 144:291–298

Powell DBB, Blanchard TW (1976) The quantitative water analysis of water potential/symplastic volume curves. J Exp Bot 27:597–607

Powell DBB, Thorpe MR (1977) Dynamic aspects of plant-water relations. In: Landsberg JJ, Cutting CV (eds) Environmental effects on crop physiology. Academic Press, London New York, pp 259–285

Raven J, Smith FA (1977) Characteristics, functions and regulation of active proton extrusion. In: Marrè E, Ciferri O (eds) Regulation of cell membrane activities in plants. Elsevier/North-Holland Biomedical Press, Amsterdam New York, pp 25–40

Richter H (1978) A diagram for the description of water relations in plant cells and organs. J Exp Bot 29:1197–1203

Richter H, Duhme F, Glatzel G, Hinckley TM, Karlic H (1980) Some limitations and applications of the pressure-volume curve technique in ecophysiological research. In: Grace J, Ford ED, Jarvis PG (eds) Plants and their atmospheric environment. Blackwell, Oxford, pp 263–272

Roberts SW, Knoerr KR (1977) Components of water potential estimated from xylem pressure measurements in five tree species. Oecologia 28:191–202

Robinson RA, Stokes RH (1959) Electrolyte solutions, 2nd edn. Butterworth, London

Sharp RE, Davies RE (1979) Solute regulation and growth by roots and shoots of water-stressed maize plants. Planta 147:43–49

Shepherd W (1976) Indices of plant moisture stress. Comparisons within and between crop species. J Appl Ecol 13:205–210

Simmelsgaard SE (1976) Adaptation to water stress in wheat. Physiol Plant 37:167–174

Skaar C (1972) Water in wood. Univ Press, Syracuse, New York

Slatyer RO (1960) Aspects of the tissue water relationships of an important arid zone species (*Acacia aneura* F. Muell) in comparison with two mesophytes. Bull Res Counc Isr Sect D 8:159–168

Slatyer RO (1961) Effects of several osmotic substrates on the water relationships of tomato. Aust J Biol Sci 14:519–540

Slatyer RO (1967) Plant-water relationships. Academic Press, London New York

Steudle E, Lüttge U, Zimmermann U (1975) Water relations of the epidermal bladder cells of the halophytic species *Mesembryanthemum crystallinum*: direct measurements of hydrostatic pressure and hydraulic conductivity. Planta 126:229–246

Steudle E, Zimmermann U, Lüttge U (1977) Effect of turgor pressure and cell size on the wall elasticity of plant cells. Plant Physiol 59:285–289

Steudle E, Smith JAC, Lüttge U (1980) Water-relation parameters of individual mesophyll cells of the crassulacean acid metabolism plant *Kalanchoë daigremontiana*. Plant Physiol 66:1155–1163

Stewart CR (1981) Proline metabolism. In: Paleg LG, Aspinall D (eds) Physiology and biochemistry of drought resistance in plants. Academic Press, London New York

Stewart CR, Hanson AD (1980) Proline accumulation as a metabolic response to moisture stress. In: Turner NC, Kramer PG (eds) Adaptation of plants to water and high temperature stress. Wiley and Sons, New York, pp 173–189

Storey R, Wyn Jones RG (1978) Salt stress and comparative physiology in the Gramineae. I. Ion relations in two salt and water-stressed barley cultivars, California Mariout and Arimar. Aust J Plant Physiol 5:801–816

Storey R, Ahmad N, Wyn Jones RG (1977) Taxonomic and ecological aspects of the distribution of glycinebetaine and related compounds in plants. Oecologia 27:319–332

Taylor SA, Slatyer RO (1961) Proposals for a unified terminology in studies of plant-soil water relationships. Proceedings of the Madrid symposium on plant-water relationships in arid and semi-arid conditions. Arid Zone Res 16:339–349

Teoh TS, Aylmore LAG, Quirk JP (1967) Retention of water by plant cell walls and implications for drought resistance. Aust J Biol Sci 20:41–50

Turner NC (1974) Stomatal behaviour and water status of maize, sorghum and tobacco under field conditions. II At low water potential. Plant Physiol 53:360–365

Turner NC (1975) Concurrent comparisons of stomatal behaviour, water status and evaporation of maize in soil at high or low water potential. Plant Physiol 55:932–936

Turner NC (1979) Drought resistance and adaptation to water deficits in crop plants. In: Mussell H, Staples RC (eds) Stress physiology in crop plants. Wiley and Sons, New York, pp 344–372

Turner NC, Jarvis PG (1975) Photosynthesis in Sitka spruce (*Picea sitchensis* (Bong.) Carr) IV. Response to soil temperature. J Appl Ecol 12:561–576

Turner NC, Jones MM (1980) Turgor maintenance by osmotic adjustment: a review and evaluation. In: Turner NC, Kramer PJ (eds) Adaptation of plants to water and high temperature stress. Wiley and Sons, New York, pp 87–104

Turner NC, Begg JE, Tonnet ML (1978) Osmotic adjustment of sorghum and sunflower crops in response to water deficits and its influence on the water potential at which stomata close. Aust J Plant Physiol 5:597–608

Tyree MT (1976a) Negative turgor pressure in plant cells: fact or fallacy? Can J Bot 54:2738–2746

Tyree MT (1976b) Physical parameters of the soil-plant-atmosphere system: breeding for drought resistance characteristics that might improve wood yield. In: Cannell MGR, Last FT (eds) Tree physiology and yield improvement. Academic Press, London New York, pp 329–348

Tyree MT (1981) The relationship between the bulk modules of elasticity of a complex tissue and the mean modulus of its cells. Ann Bot (London) 47:547–559

Tyree MT, Cheung YNS (1977) Resistance to water flow in *Fagus grandifolia* leaves. Can J Bot 55:2591–2599

Tyree MT, Hammel HT (1972) The measurement of the turgor pressure and the water relations of plants by the pressure-bomb technique. J Exp Bot 23:267–282

Tyree MT, Karamanos AJ (1980) Water stress as an ecological factor. In: Grace J, Ford ED, Jarvis PG (eds) Plants and their atmospheric environment. Blackwell, Oxford, pp 237–261

Tyree MT, Richter H (1981) Alternative methods of analysing water potential isotherms: some cautions and clarifications. I. The impact of nonideality and of some experimental errors. J Exp Bot 32:643–653

Tyree MT, Richter H (1982) Alternative methods of analysing water potential isotherms: some cautions and clarifications. II. Apoplastic water and negative turgor. Can J Bot (in press)

Tyree MT, Cheung YNS, Macgregor ME, Talbot AJB (1978) The characteristics of seasonal and ontogenetic changes in the tissue-water relations of *Acer, Populus, Tsuga* and *Picea.* Can J Bot 56:635–647

Vinters H, Dainty J, Tyree MT (1977) Cell wall elastic properties of *Chara corallina.* Can J Bot 55:1933–1939

Walter H (1955) Water economy and the hydrature of plants. Annu Rev Plant Physiol 6:239–252

Waring RH, Running SW (1978) Sapwood water storage: its contribution to transpiration and effect upon water conductance through the stems of old-growth Douglas fir. Plant Cell Environ 1:131–140

Waring RH, Whitehead D, Jarvis PG (1979) The contribution of stored water to transpiration in Scots pine. Plant Cell Environ 2:309–318

Warren Wilson J (1967a) The components of leaf water potential. I. Osmotic and matric potentials. Aust J Biol Sci 20:329–347

Warren Wilson J (1967b) The components of leaf water potential II. Pressure potential and water potential. Aus J Biol Sci 20:349–357

Warren Wilson J (1967c) The components of leaf water potential. III. Effects of tissue characteristics and relative water content on water potential. Aust J Biol Sci 20:359–367

Watts WR (1972) Leaf extension in *Zea mays.* II. Leaf extension in response to independent variation of the temperature of the apical meristem, of the air around the leaves and of the root-zone. J Exp Bot 76:713–721

Watts WR (1974) Leaf extension in *Zea mays.* III. Field measurements of leaf extension in response to temperature and leaf water potential. J Exp Bot 25:1085–1096

Weatherley PE (1970) Some aspects of water relations. Adv Bot Res 3:171–206
Wenkert W (1980) Measurement of tissue osmotic pressure. Plant Physiol 65:614–617
Wenkert W, Lemon ER, Sinclair TR (1978a) Leaf elongation and turgor pressure in field-grown soybean. Agron J 70:761–764
Wenkert W, Lemon ER, Sinclair TR (1978b) Water content-potential relationship in soya bean: changes in component potentials for mature and immature leaves under field conditions. Ann Bot (London) 42:295–307
Whitehead D, Jarvis PG (1981) Coniferous forests and plantations. In: Kozlowski TT (ed) Water deficits and plant growth vol VI. Academic Press, London New York, pp 49–152
Wiebe HW (1966) Matric potential of several plant tissues and biocolloids. Plant Physiol 41:1439–1442
Wilson JR (1977) Variation of leaf characteristics with level of insertion on a grass tiller. III. Tissue water relations. Aust J Plant Physiol 4:733–743
Wilson JR, Ludlow MM, Fisher MJ, Schulze E-D (1980) Adaptation to water stress of the leaf water relations of four tropical forage species. Aust J Plant Physiol 7:207–220
Wyn Jones RG (1981) Salt tolerance. In: Johnson CB (ed) Physiological processes limiting plant productivity. Butterworth, London, pp 271–292
Wyn Jones RG, Storey R (1978) Salt stress and comparative physiology in the Gramineae. II. Glycinebetaine and proline accumulation in two salt and water-stressed barley cultivars. Aust J Plant Physiol 5:817
Wyn Jones RG, Storey R (1981) Betaines. In: Paleg LG, Aspinall D (eds) Physiology and biochemistry of drought resistance in plants. Academic Press, London New York
Wyn Jones RG, Brady CJ, Speirs J (1979) Ionic and osmotic relations in plant cells. In: Laidman DL, Wyn Jones RG (eds) Recent advances in the biochemistry of cereals. Academic Press, London New York, pp 63–118
Zimmermann U (1977) Cell turgor pressure regulation and turgor pressure mediated transport processes. In: Jennings DH (ed) Integration of activity in the higher plant. Univ Press, Cambridge, pp 117–154
Zimmermann U (1978) Physics of turgor- and osmoregulation. Annu Rev Plant Physiol 29:121–148
Zimmermann U, Hüsken D (1980) Elastic properties of the cell wall of *Halicystis parvula*. In: Spanswick R, Lucas WT, Dainty J (eds) Plant membrane transport: current conceptual issues. Elsevier/North-Holland Biomedical Press, Amsterdam New York, pp 471–472
Zimmermann U, Steudle E (1975) The hydraulic conductivity and volumetric elastic modulus of cells and isolated cell walls of *Nitella* and *Chara* spp: pressure and volume effects. Aust J Plant Physiol 2:1–12
Zimmermann U, Steudle E (1978) Physical aspects of water relations of plant cells. Adv Bot Res 6:45–117
Zimmermann U, Steudle E (1980) Fundamental water relations parameters. In: Spanswick RM, Lucas WJ, Dainty J (eds) Plant membrane transport: current conceptual issues. Elsevier/North-Holland Biomedical Press, Amsterdam New York, pp 113–130

3 Water Uptake and Flow in Roots

P.E. WEATHERLEY

CONTENTS

3.1 Introduction

The loss of water by evaporation from the leaves of plants (transpiration) is an inevitable accompaniment of the photosynthetic absorption of carbon dioxide from the atmosphere. Water is absorbed from the soil by the roots to make good this transpirational loss. There is thus a flow of water through the plant from the soil to the atmosphere – the so-called transpiration stream – and it is the purpose of this chapter to focus attention on one part of this stream, namely the flow of water from the soil into, and through the roots. However, to abstract this partial process from the whole root-plant-atmosphere system can be misleading and a note of caution is necessary.

In 1948 VAN DEN HONERT resuscitated and developed Gradmann's analysis of the transpiration stream in which he described the process as a catena of resistances which could be treated as an analogue of Ohm's Law (cf. Chap. 1, this Vol.). VAN DEN HONERT brought out the cardinal fact that in the soil-plant-atmosphere continuum, the resistance of the gas phase (the diffusion of water vapour from the leaf cell surface to the surrounding atmosphere) is many times greater than that of other parts of the system. It follows that the transpirational loss of water from the leaves is the controlling process in the soil-plant-atmosphere system.

This dominance of transpiration makes a study of the resistance of the root difficult, since in the intact plant changes in root resistance will have only an indirect effect via changes in stomatal aperture. Simplifying the system, by cutting off the stem a few centimetres above the roots makes some aspects of our investigations easier. Sap exudes from the stump of such a decapitated plant and the rate of flow and composition of this root exudate can be measured. That this exudation could exert a considerable pressure was demonstrated long ago by attaching a manometer to the stump. This is the so-called root pressure. A study of exudation has revealed that the root behaves like an osmometer, water moving from the external medium across the cortex to the tracheae of the stele in response to a difference in osmotic potential. There does not appear to be any active transfer of *water,* the active element being a continuous secretion of salts into the tracheae to maintain this difference in osmotic potential.

With the transpiring plant there is the additional motive force of a hydrostatic tension in the tracheae arising from the pull of transpiration. These two forces do not seem to be simply additive. An increase of hydrostatic pressure difference across the cortex can cause a disproportionate increase in flow which may be related to interactions with the solute flux and/or pressure-induced changes in the hydraulic conductance. This effect is particularly striking in certain cases where, over a wide range of transpiration rates, there is very little change in the drop of water potential through the plant. This implies that the conductance of the root rises in proportion to the increase in transpirational flux.

It is important for an understanding of the nature of water movement across the root cortex to know the pathway of flow. Three parallel pathways are recognizable: the apoplast (cell walls), the symplast (cytoplasm with plasmodesmata interconnecting the cells) and the vacuolar system (water passing through the cell membranes on entering and leaving each cell). In response to a difference of osmotic potential and hydrostatic pressure between the outside of the root and the conducting elements in the stele, there would be a flow in each of these three pathways. The question is: what are their relative resistances, or in other words what are the relative rates of flow in these pathways?

It is implicit in the above discussion that the cellular pathway through the cortex and stele imposes the major resistance to flow, the resistance of the xylem tracheae usually being relatively small. Of the resistances within the plant: roots, stems and leaves, the root is regarded as having the greatest resistance, and being upstream of the major organs of the plant, changes in its magnitude control the degree of water stress in these organs with its attendant effects on metabolism, growth etc.

So far consideration has been restricted to the internal water relations of the root, which has been regarded as immersed in culture solution or saturated soil. The water stress developing in the leaves under these conditions is a function of the rate of transpiration and the resistance of the plant, the resistance of the soil being negligible. However, if the roots are immersed in unsaturated soil in which water is held by surface tension forces in the soil matrix, a greater depression of water potential in the plant is necessary to abstract water from the soil and for water to move up to the surface of the root, there must be a gradient of Ψ_s in the perirhizal soil (i.e., soil surrounding each individual root). Whether this gradient is so small as to be negligible or so large as to play an important part in the system depends on the velocity of movement of the water up to the root surface and the magnitude of the hydraulic conductance of the soil at the root surface. Neither of these parameters is easy to measure, and it is still a matter of controversy whether perirhizal gradients arise under normal conditions. A further complication is that the interface between the soil and the root surface may present a resistance to flow. The nature of such an interfacial resistance is a matter of speculation. It might arise from a limitation of the area of water transfer to points of contact between soil particles and root surface. If so the interfacial resistance would increase with increasing water stress in the plant, for it has been demonstrated that the tissues of roots as well as shoots contract in volume with increasing stress. Such a root contraction would tend to cause gaps to form between the root surface and the surrounding soil, reducing the points of contact and rendering the pathway for flow more tenuous. If this picture is correct it would clearly form a positive feed-back system – increasing stress in the plant leading to an increased resistance at the soil-root interface and this in turn causing an increased stress. Such a rising stress would be moderated by stomatal closure.

3.2 The Root as an Osmometer

If the stem of a plant is severed a little below the lowest leaf, exudation of sap from the cut surface of the basal part, i.e., the root system, often occurs and may continue for many hours. If a manometer is attached to the stump, considerable pressure may be generated, the so-called root pressure. This process of root exudation has been extensively studied (for a review of the earlier work see KRAMER 1949). The exuding sap was found to contain largely inorganic ions. The osmotic pressure of the sap could be measured, as could that of the surrounding medium in which the plant was growing and if at the same time the rate of exudation was determined, it was found (SABININ 1925) that this rate of exudation (J_v) was directly related to the difference in osmotic pressure between medium and sap ($\Delta\pi$). This could be expressed as:

$$J_v = L\Delta\pi \tag{3.1}$$

where L is a constant, the hydraulic conductance coefficient. EATON (1943) found for cotton plants that a plot of J_v against $\Delta\pi$ was almost linear and passed through the origin. A linear relationship is what would be expected if the cells between the root surface and the xylem tracheae behaved like a single

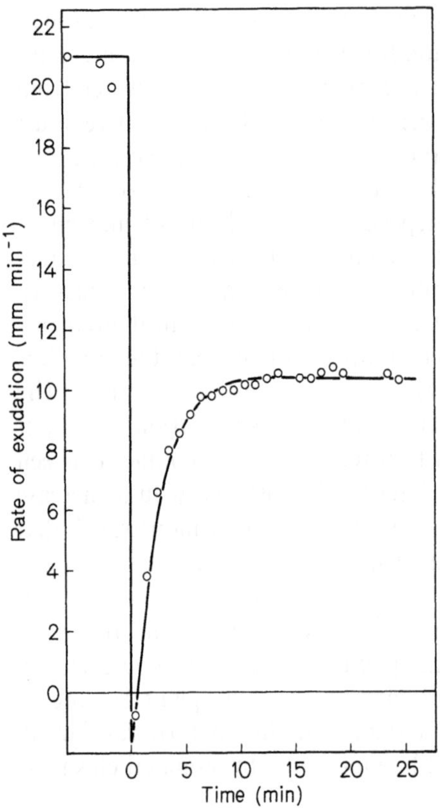

Fig. 3.1. Course of change in rate of exudation from a detopped tomato plant following a sudden increase in the osmotic value of the external medium. (ARISZ et al. 1951)

semipermeable membrane. Of course the root differs from an osmometer in that a steady rate of exudation may be maintained for long periods with little dilution of the xylem sap occurring. This is because there is a steady uptake of solutes by the root which are stored by the cells of the cortex and stele and steadily metered out to the xylem.

In their classic researches in this field, ARISZ et al. (1951) measured the rate of exudation from a tomato root system at 30-s intervals whilst the external medium was rapidly changed for another to which mannitol had been added and which had therefore a higher osmotic value (measured as depression of freezing point). An example of their results is shown in Fig. 3.1. On changing the external medium there is a very rapid response in rate, a large increase in the concentration of the medium such as shown in Fig. 3.1 causing the exudation to become temporarily negative, i.e., sap was drawn back into the root system. After this sudden fall in rate there follows a rise until after 10 min a new steady rate, lower than the original one, is attained. These data are explicable in terms of Eq. (3.1). The initial rate of exudation (J_{v_1}) is given by:

$$J_{v_1} = L(\pi_{x_1} - \pi_{m_1}) \tag{3.2}$$

where π_{m_1} and π_{x_1} are the osmotic pressures of the medium and exudate respectively. The minimum rate (J_{v_2}) following the change of medium to π_{m_2} is reached

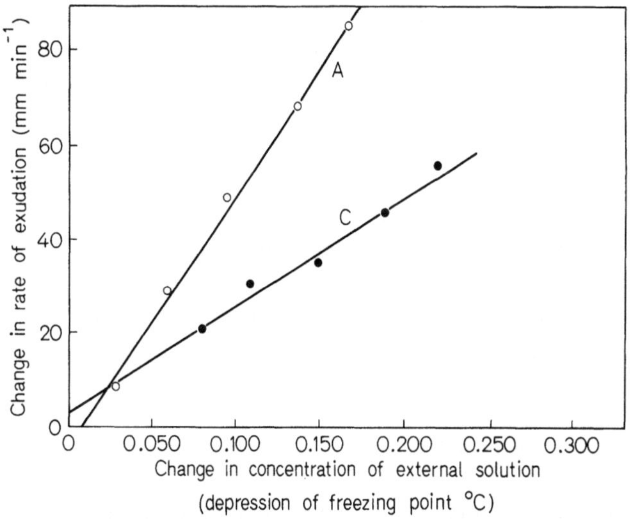

Fig. 3.2. Relationship between the change in osmotic value of the external medium (in terms of depression of freezing point) and the resulting change in rate of exudation (in arbitrary units). *A* experiment in water, *C* in Hoagland solution. (After ARISZ et al. 1951)

in a few seconds and during this time it is reasonable to suppose that the osmotic pressure of the exudate (π_{m_1}) suffered little change. Thus:

$$J_{v_2} = L(\pi_{x_1} - \pi_{m_2}). \tag{3.3}$$

The value of J_{v_2} may be obtained by extrapolation as shown in Fig. 3.1. The subsequent rise to a new steady rate of exudation results from the continued secretion of salts into the xylem tracheae, the exudation rising to a new steady J_{v_3} and the osmotic pressure of the sap rising to π_{x_2} when:

$$J_{v_3} = L(\pi_{x_2} - \pi_{m_2}). \tag{3.4}$$

The osmotic model of the root system can be tested in the following way. The immediate change in rate of exudation ($J_{v_1} - J_{v_2}$) can be obtained by subtracting Eq. (3.3) from (3.2):

$$J_{v_1} - J_{v_2} = L(\pi_{m_2} - \pi_{m_1}). \tag{3.5}$$

Thus when the drop in exudation corresponding to the change in osmotic pressure of the external medium was measured for a number of cycles a plot of ΔJ_v against $\Delta\pi$ could be made and should be linear according to Eq. (3.5). That it was so is shown in Fig. 3.2. Also as expected the lines pass, within the limits of experimental error, through the origin.

More recently, the possibility that the root cells may be to some extent permeable to the solutes in the medium and the sap has been taken into account by inserting the reflection coefficient (σ), into the above equations. The reflection coefficient is the ratio of the apparent osmotic pressure to the osmotic pressure that would occur if the solutes were truly non-penetrating (DAINTY 1963). Thus $\sigma=0$ when the solutes can freely pass through the membrane and $\sigma=1$ when the membrane allows no solute to pass. Equation (3.1) now becomes:

$$J_v = L\sigma\Delta\pi. \quad (3.6)$$

It is implicit in the osmotic model that the flux of water is passive, i.e., entirely accounted for by the difference in osmotic pressure between the external medium and the xylem tracheae. But the question arises whether a fraction of the flow is in fact *active*, i.e., impelled against the thermo-dynamic gradient with the expenditure of metabolic energy. Any such fraction (Φ) would be added to Eq. (3.6):

$$J_v = L\sigma\Delta\pi + \Phi. \quad (3.7)$$

Φ would be manifest as a residual flow when $\Delta\pi=0$, i.e., when the external medium and the sap are isotonic and would be detectable as an intercept on the J_v axis when J_v is plotted against $\Delta\pi$.

The data of EATON (1943) do not permit ready extrapolation to $\Delta\pi=0$ and the instantaneous drop method of ARISZ et al. (1951) cannot detect the existence of Φ. However, in some experiments ARISZ et al. measured the osmotic pressure of the exudate as well as the medium and found that there was a small iso-osmotic flow. Similar evidence for an iso-osmotic flow has been obtained by HOUSE and FINDLAY (1966); KLEPPER (1967); GINSBURG (1971); ANDERSON and COLLINS (1969); ANDERSON and REILLY (1968). However, other explanations for iso-osmotic flow are available without having recourse to an active flux of water. It is assumed that the osmotic pressure in the tracheae of the absorbing region of the root is identical with the osmotic pressure of the sap emerging from the cut end of the stem or root. But as these authors have recognized, it is possible that solutes are abstracted from the tracheal sap as it moves upwards, and so the emergent sap is more dilute than that in the operative region. Such a dilution has been demonstrated in barley plants by OERTLI (1966). Clearly if the roots were surrounded by a solution isotonic with the emergent sap, it would be hypotonic to the tracheal sap in the absorbing region and flow would occur. A similar hypothesis is that the areas for salt secretion and water absorption within the root are not coextensive, the salt-secreting zone being distal and the zone for water uptake extending beyond the salt-secreting zone towards the base of the root. Thus as the sap flows in the tracheae, osmotic influx of water will occur and the sap will become progressively more dilute. This model is referred to as the Standing-Gradient Osmotic Flow (DIAMOND and BOSSERT 1967; ANDERSON et al. (1970).

In this connexion the experiments of GINSBURG and GINZBURG (1970) are interesting since they used root "sleeves", i.e., lengths of root from which the steles had been pulled out, the rupture taking place at the endodermis and a cylinder of cortex remaining. The sleeves were irrigated internally and externally with solutions of known osmotic pressure. With this arrangement the above adjustments of the internal solution could not take place. Nevertheless when J_v was plotted against $\Delta\pi$, straight lines were obtained which, when extrapolated, showed a significant J_v when $\Delta\pi=0$. On the face of it this appeared to be incontrovertible evidence for the existence of an active water transport, but here again an alternative explanation is possible in terms of water moving in the symplasm and an iso-osmotic flow resulting from a difference in the values for σ at the outer and inner surfaces of the sleeves (GINSBURG 1971).

The present position seems to be that iso-osmotic flow can occur across the root cortex and does not necessarily require an intact endodermis. It comprises only a small fraction of the total flow and always seems to be associated with some sort of solute flux (ANDERSON et al. 1970; TYREE 1970). In those cases which have been investigated it does not appear to be a manifestation of an active flux of water, the existence of which has not so far been unequivocally demonstrated.

It has long been known that when the root system is subject to metabolic inhibition, the shoot suffers an increased water stress. Thus lowering the temperature of the root system to near 0 °C can result in leaf wilting. Anaerobiosis and excessive concentration of carbon dioxide round the roots also induce stress. While these factors are clearly of considerable ecological significance, in the context of the above discussion, their effects are perhaps of less importance. They are interpreted as reducing the hydraulic conductance of the root cells and are not regarded as providing evidence for an active component of water flux.

3.3 The Transpiring Plant

In the transpiring plant, water is drawn up the xylem tracheae and because of the resistance to flow through the root cortex and stele a fall in hydrostatic pressure occurs in the tracheae of the root in response to which water moves from the external medium to the stele. Thus:

$$J_v = L\Delta P \tag{3.8}$$

where J_v is the rate of flow through the root which, at a steady state, is equal to the rate of transpiration, ΔP is the drop in pressure between the external medium and the tracheae and L is, as before, the hydraulic conductance coefficient. As in the exuding root system so in the transpiring plant, salt secretion takes place and the force driving water across the cortex is osmotic as well as hydrostatic. Thus combining (3.6) and (3.8):

$$J_v = L(\sigma\Delta\pi + \Delta P). \tag{3.9}$$

The relative magnitudes of $\Delta\pi + \Delta P$ depend on the rate of salt secretion and the rate of water movement. Thus if the rate of transpiration is low, ΔP will be low and $\Delta\pi$ relatively high. But as the rate of transpiration rises, ΔP will

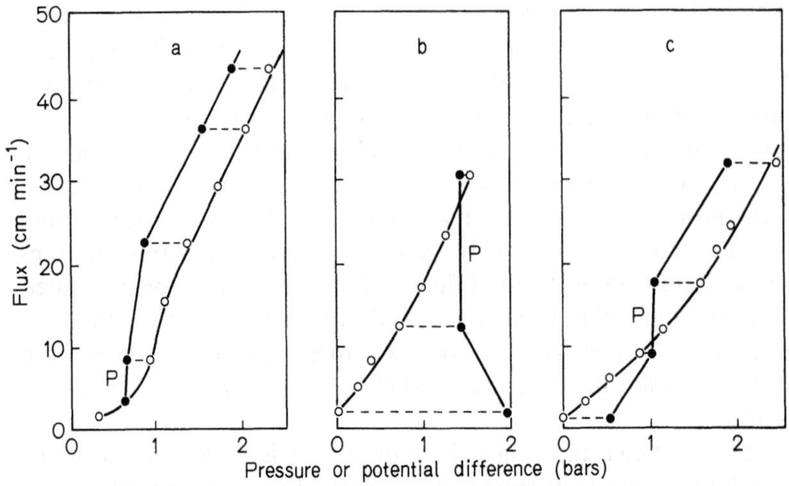

Fig. 3.3a–c. Graphs showing the relationship between flux of water and: *1* hydrostatic pressure difference across the root cortex (o—o) and *2* the total water potential difference (●—●). Corresponding values are connected by *broken lines*. Flux units are equivalent to J_v. (MEES and WEATHERLEY 1957)

rise [Eq. (3.8)] and $\Delta\pi$ will fall since the sap will be diluted by the increased influx of water. At high rates of transpiration the osmotic contribution becomes negligible.

This model cannot be tested with the intact plant since the sap for measuring the osmotic value cannot be extracted without destroying the plant. Experiments have therefore been carried out with detopped root systems with which the transpirational pull is simulated by attaching a suction pump to the stump or by placing the root system in a pressure chamber and forcing water through the roots under pressure. Using the latter technique it was evident that the plot of J_v against ΔP was not linear (MEES and WEATHERLEY 1957; LOPUSHINSKY 1964) but rose more steeply for each rise of pressure, suggesting an increase in hydraulic conductance with increasing pressure (Fig. 3.3). However, this is not necessarily the explanation, for as pointed out in connexion with Eq. (3.9), $\Delta\pi$ will fall with increasing ΔP owing to dilution of the sap and the rise in J_v is less than expected from Eq. (3.9). But as ΔP increases and $\Delta\pi$ declines the increment of J_v for a given increment of ΔP rises, i.e., the curve in Fig. 3.3 steepens and gradually becomes linear and coincides with the plot of Eq. (3.8) when the osmotic contribution becomes negligible. FISCUS (1975) and DALTON et al. (1975) independently put forward mathematical models based on the above suppositions and concluded that it was not necessary to invoke a change in hydraulic conductance to explain the curvilinear rise in J_v plotted against ΔP.

MEES and WEATHERLEY (1957) measured the osmotic pressure of the exudate at each steady pressure and could therefore obtain the total water potential in the xylem and hence the drop in water potential across the root cortex. The plot of J_v against difference in total water potential showed (Fig. 3.3) the seemingly anomalous result that at one particular pressure increment marked

P on each graph there was such a large fall in osmotic pressure of the sap that there was little change in difference of total water potential although there was a large rise in J_v. MEES and WEATHERLEY took this to mean that there was a considerable increase in hydraulic conductance. But FISCUS attributed this effect to there being a low reflection coefficient and his model certainly predicted such an effect if σ were 0.5 or lower. DALTON et al. (1975) came to similar conclusions. NEWMAN (1976), however, devised a test for the FISCUS model which indicated that some of the data of LOPUSHINSKY (1964), and MEES and WEATHERLEY (1957) could not be explained without recourse to a change in hydraulic conductance. In view of this FISCUS (1977) has applied his model to soybean data and contends that no change in hydraulic conductance is necessary. MICHEL (1977) working with intact and decapitated soybean plants came to the conclusion that the hydraulic conductance was linearly related to the flow of water through the plant and that his own results and those of MEES and WEATHERLEY (1957); LOPUSHINSKY (1964) and BROUWER (1954) could only be explained in terms of a change in hydraulic conductance. Thus the issue remains controversial.

Turning from the detopped to the intact transpiring plant we are faced with an even more puzzling situation. The tension in the xylem of the root cannot be measured directly, but if the xylem conduits constitute a low resistance to flow as they demonstrably do, then the water potential of the leaves will not be very different from that of the root xylem. A number of workers have measured the leaf water potentials corresponding to various steady rates of transpiration. Leaf water potentials were usually measured psychrometrically and the transpiration rates controlled by maintaining different levels of atmospheric humidity or light intensity. It would be expected that a plot of leaf water potential against rate of transpiration would be similar to that obtained using pressures applied to decapitated plants i.e. curvilinear at low rates of transpiration giving way to virtually a linear relationship at higher rates of transpiration. In fact experiments of this kind have yielded widely different results for different species and for the same species at the hands of different experimenters (MACKLON and WEATHERLEY 1965; TINKLIN and WEATHERLEY 1966; STOKER and WEATHERLEY 1971; BARRS 1970; NEUMANN et al. 1974; BLACK 1974; ASTON and LAWLOR 1979).

An example of the most contradictory results is shown in Fig. 3.4. Both sets of data are for sunflower plants grown in culture solution. NEUMANN and his colleagues obtained linear relationships indicating a constant hydraulic conductance. Those of BLACK show, at low rates of transpiration, a steep fall in water potential with increasing transpiration followed by a sudden levelling out, the leaf water potential becoming constant over a wide range of transpiration rates. This constancy indicates an increase in hydraulic conductance linearly related to the transpirational flow. Furthermore it has been found (STOKER and WEATHERLEY 1971) that at very high rates of transpiration there was a further linear fall in leaf water potential (Fig. 3.5) and if the roots were killed by a short immersion in hot water there was no adjustment of hydraulic conductance with rate of transpiration and a linear relationship similar to the data of NEUMANN et al. was obtained. STOKER found that the adjustment in conduc-

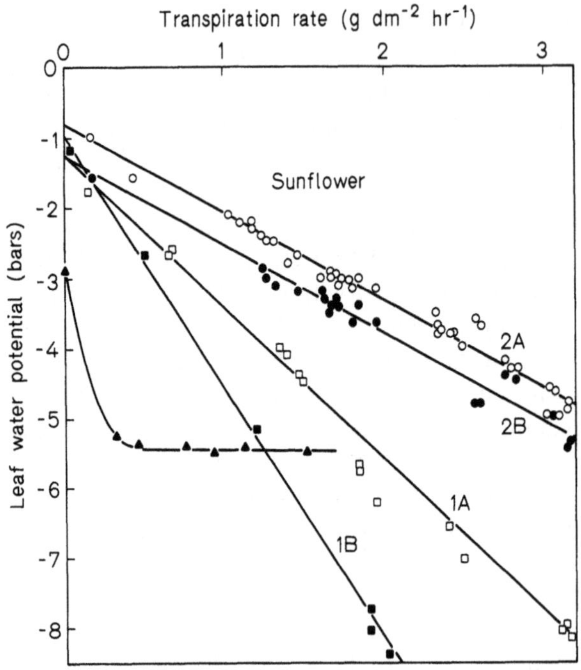

Fig. 3.4. Relation between leaf water potential and rate of transpiration for sunflower plants. Data of NEUMANN et al. (1974): ○, ●, □, ■; data of BLACK (1974); ▲ Figure after NEUMANN et al. (1974) with Black's data added. (WEATHERLEY 1976)

tance of living roots was only partially reduced by cooling the root system to 0 °C (see WEATHERLEY 1971). However, the absence of any adjustment in the killed roots indicates that it is a function of the living cells and therefore metabolic in nature. If the adjustment was in some way metabolic it would be expected that the response to a sudden change in transpiration rate might take some time. Such a delay would be manifest as a transient fall in the water potential of the leaf. Such a transient fall might be missed if the leaf water potential was being followed by periodic sampling for psychrometry. Some kind of continuous record was desirable and this was provided by a β gauge fitted to a single leaf. This instrument indicates changes in leaf thickness only, but these reflect changes in water content and hence water potential. It is very sensitive and rapid in response. STOKER (1968) using cotton plants found that with a threefold increase in rate of transpiration maintained for several hours, the β gauge reading remained constant throughout.

It has been suggested that this constancy of the leaf water potential over a wide range of transpiration rates might be a property of the leaf rather than the root. Whilst it has indeed been shown (POSPISILOVA 1972; BLACK 1979a) that such a constancy is found in detached leaves, the latter author (BLACK 1979b) has demonstrated a similar constancy in the tension in the xylem of the stem. This could only have arisen through a change in root resistance.

Another aspect of this adjustment in conductance of the roots has been recognized for a long time. It is that the root is differentiated longitudinally in rate of water uptake. This was demonstrated by BREWIG (1936) and later by BROUWER (1953 and 1954). They fitted a number of potometers to a single

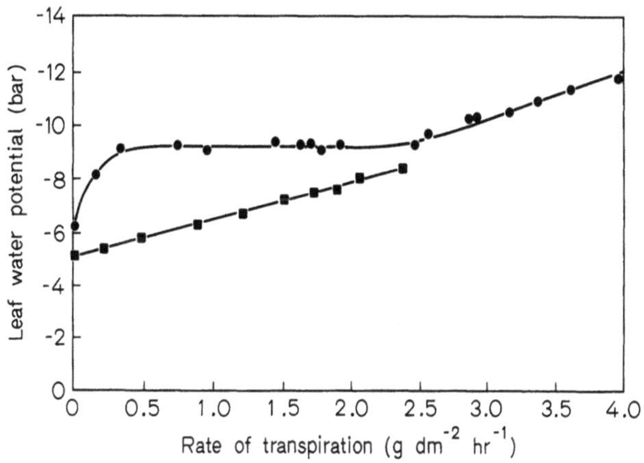

Fig. 3.5. Relationship between rate of transpiration and leaf water potential for sunflower plants in water culture. Living root system ●—●. Heat-killed root system ■—■. (After STOKER and WEATHERLEY 1971)

root of *Vicia faba* to measure simultaneously the uptake by segments along the length of the root from apex to base. It was found that whilst at low rates of flow through the root most of the water was absorbed by the apical zone and little by the basal zone, at high rates of flow most of the water was absorbed by the basal zones.

At first sight this seems to offer some explanation of the rise in conductance of the whole root system: as the transpirational flow increases so does the surface of the absorbing root. However, this is not the whole story. BROUWER was able to estimate the resistances of the various root zones separately and found that whereas that of the apical zone increased slightly with increased flow of water, the resistance of the basal zones showed a considerable decline. It would appear then, that the proportionate increases in hydraulic conductance with increase in transpiration rate arises from an adjustment in the basal regions of the root, but of course this gives us no information on how the adjustment came about. BROUWER suggested that it was a question of the turgescence of the root cells, the lower the turgor pressure the greater the conductance of the pathway through the tissue. It is difficult, though, to see how this offers an explanation for the data in Fig. 3.5. The level values represent a constant drop in water potential across the root cortex and if the osmotic potential of the cortical cells remains constant, no change in their turgor can accompany the changes in transpirational flux. HYLMO (1955) has suggested an explanation based on ERBE'S (1933) method for measuring the distribution of pore size in ultra filters. If the pores are initially filled with an immiscible fluid and water forced through them at progressively increasing pressures, the fluid will be ejected from the larger pores first where it is held by low surface tension forces and from progressively narrower pores with higher pressures. Thus with increasing pressures an increasing number of pores become available for flow of water and the conductance of the membrane increases. Apart from the difficulty of envisaging this in physiological terms, like BROUWER'S hypothesis it offers no explanation of data such as shown in Fig. 3.5 where there is no increase in pressure gradient. A simple model which does behave like this is the rotameter flow meter (TINKLIN and WEATHERLEY 1966). In this the fluid passes upwards round a float within a tapered tube which widens upwards. As the rate of flow increases the float is lifted and the annular space between it and the tube increases in size to let more fluid pass and the float takes up a steady position. At all rates of flow the drop of pressure does not vary as it is a function of the weight of the float. Here again it is difficult to see this in physiological terms and as a hypothesis it does not suggest any critical experiments.

3.4 The Pathway of Water Movement

The cortex, endodermis and stele impose a hydraulic resistance some five orders of magnitude greater than the tracheae of the xylem and they constitute a major resistance to flow through the plant. There are three possible pathways through these tissues (Fig. 3.6a, b, and c). Water might move from vacuole to vacuole (a) at each cell boundary crossing the tangential cell walls, plasmalemma, cytoplasm and tonoplast. The second possibility (b) is via the symplast. Here water enters through the plasmalemma of the epidermal cells and then traverses the cortex and stele within the symplast, moving from cell to cell through the plasmodesmata and leaving via the plasmalemma of the parenchyma cells surrounding the tracheae. This pathway thus entails crossing only two cell membranes. The third pathway (c) is provided by the cell walls (apoplast) wherein water would move in response to a gradient of hydrostatic pressure. This pathway is however interrupted at the endodermis by the Casparian strips which, being suberized, are regarded as being impervious to water. Here then the water must be diverted through the endodermal cells to emerge again into the apoplast of the stele. In this diversion the water must pass through the plasmalemma at entry and exit. Thus the symplastic and apoplastic pathways each involve crossing the cell membrane twice but differ as to the position of the crossing and the nature of the radial pathway: the one being the cellulose cell walls, the other the cytoplasm and plasmodesmata. No doubt water will move to some degree along all three pathways, but what we wish to know is the relative rates of flow through each, i.e., their relative resistances to flow. Valid comparisons are not easy to make. For any one case all the necessary parameters have not been measured and estimates are often difficult to make or are unreliable (for discussion of water transport see also Chap. 5, this Vol.).

3.4.1 The Vacuolar Pathway

There is a strong evidence that movement from vacuole to vacuole can only constitute a small fraction of the total flow. Following the argument of BRIGGS (1967) if J_v ($cm\ s^{-1}$) is the rate of swelling or contraction of the cortical cells when immersed in an osmoticum and the difference in water potential between the osmoticum and the vacuole is $\Delta\Psi$ (bars) then:

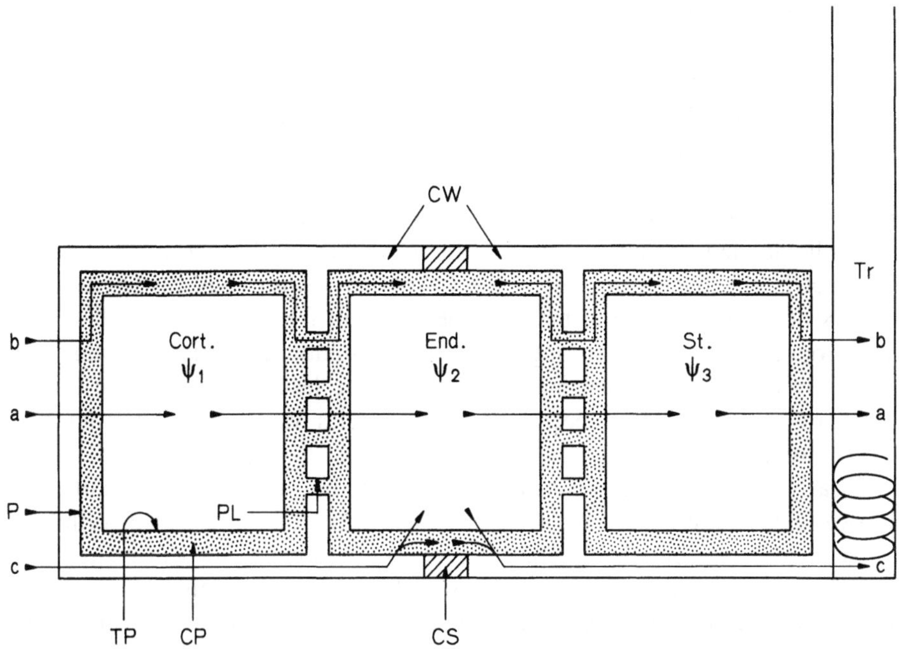

Fig. 3.6. Diagram showing three pathways for water movement across the root cortex. *a* vacuole to vacuole; *b* movement in the symplast; *c* movement in the cell walls. *Cort* cortical cell; *End* Endodermal cell; *St* Stelar cell; *Tr* Trachea; *CP* Cytoplasm; *CS* Casparian strip; *CW* Cell wall; *P* Plasmalemma; *PL* Plasmodesm; *TP* Tonoplast. (WEATHERLEY 1975)

$$J_v = Lp\,\Delta\Psi \quad \text{or} \quad \Delta\Psi = \frac{J_v}{Lp} \tag{3.10}$$

where Lp is the permeability coefficient of the cell. Now when water moves from vacuole to vacuole across the cortex it will encounter two resistances at each cell, one on entering, the other on leaving, and if there are n cells across the cortex, the total drop of water potential will be:

$$\Delta\Psi = \frac{2nJ_v}{Lp}. \tag{3.11}$$

Unfortunately it does not appear that n, J_v and Lp have ever been measured on the same species let alone on the same root system. For the cells of higher plants values of Lp around 10^{-6} cm s^{-1} have been found by STADELMANN (1963) whilst a value as low as 10^{-7} is quoted by BRIGGS (1967). Values as high as 10^{-5} cm s^{-1} have been obtained with giant algal cells (DAINTY and HOPE 1959). The number of cells across the cortex can be as few as five or more than ten. Application of these data in Eq. (3.11) gives a drop of water potential of between 2 and 20 bar per cell, i.e., a drop across the cortex of between 10 to 200 bar. The upper end of this range is out of the question, and even the lower end is larger than usually found (e.g., see Fig. 3.5). It would appear that unless root cells are exceptionally permeable to water, the vacuolar pathway is unlikely to be important.

3.4.2 The Cell Wall Pathway

It has long been known that the cell walls of the root cortex are readily permeable to many solutes (RUFZ DE LAVISON 1910; STRUGGER 1938–9, TANTON and CROWDY 1972). Soluble dyes and readily detectable ions which cannot cross the plasmalemma are nevertheless taken up by the cell walls as far as the endodermis, whilst it has been demonstrated (CLARKSON and ROBARDS 1975) that colloidal lanthanum hydroxide can move across the cortex in the cell walls but is effectively stopped by the Casparian strips. Thus BERNSTEIN and NIEMANN (1960) were able to measure the apparent free space (cell wall space) of the root cortex by equilibrating the roots with standard solutions of mannitol or sodium chloride and then measuring the solute leached out in pure water. For barley, wheat, kidney bean and other species values ranging from 5% to 30% were obtained.

Clearly apoplastic space is available for flow, but its resistance is not easy to ascertain. If the spaces between the microfibrils of the cell wall cellulose are analogous to a bundle of capillary tubes, the Poiseuille law can be applied. RUSSELL and WOOLLEY (1961) in this way found the calculated conductance of an artificial cellulose cell wall was some 20 times that of the vacuolar pathway, whilst BRIGGS (1967) states that the diameter of the interfibrillar spaces in a cellulose wall ranges from 10^{-7} to 10^{-6} cm and with a pressure gradient of 1 bar cm^{-1} the rate of flow comes out as 1.56×10^{-6} cm s^{-1} which is at least ten times that expected through the vacuoles.

Of course the cell wall is not like a bundle of capillaries and the microfibrils in the primary wall do not run parallel, but form a criss-cross pattern. Perhaps the most realistic calculation is that of PRESTON (1974). This is based on the lamella hypothesis of cell wall structure in which water would move in effectively flattened tubes. The equivalent of the Poiseuille equation for such a lamella is:

$$Q = \Delta p\,(4b)\,a^3/12\,l\eta, \tag{3.12}$$

where Q = rate of flow ($\mathrm{cm}^3\,\mathrm{s}^{-1}$); b and l are the radial and tangential dimension of the cell (cm); a is the diameter of the lamella tube (cm); Δp the drop of pressure through the cell (dynes cm^{-2}); η = viscosity of water (10^{-2} poise).

Inserting as typical values $b = 20 \times 10^{-4}$ cm, $l = 20 \times 10^{-4}$ cm, $a = 4 \times 10^{-7}$ cm then for $\Delta p = 1$ bar, $Q = 2 \times 10^{-12}\,\mathrm{cm}^3\,\mathrm{s}^{-1}$. The equivalent flow through a plasmalemma of $Lp = 10^{-7}$ comes out as $4 \times 10^{-13}\,\mathrm{cm}^3\,\mathrm{s}^{-1}$. Thus the conductance of the cell wall lamella is five times that of the plasma membrane. But there are about 50 lamellae per µm thickness of wall thus the conductance of such a wall could be 250 times greater than that of the membrane. On the other hand for a more permeable membrane with $Lp = 10^{-6}$ the wall would be no more than 25 times more conductive. But the conductance increases in proportion to the thickness of the wall, thus if the cortical cell walls were 2 µm thick and there were five cells across the cortex, the conductance of the whole cortex would be $10^{-5}\,\mathrm{cm\,s}^{-1}\,\mathrm{bar}^{-1}$. $10^{-5}\,\mathrm{cm\,s}^{-1}$ represents a high rate of flow through the root but this would evidently only need a pressure drop of 1 bar across the cortex. A. PETTY (personal communication) has recently made a direct measurement of the conductance of the walls of spruce tracheids. Plugs of wood were completely dried out and the lumens of the tracheids filled with paraffin wax. The molten wax does not penetrate the cell walls. When the wax had solidified the walls were allowed to imbibe water and the flow of water through them measured on the application of a known pressure. A figure of 40×10^{-12} cm^3 s^{-1} b^{-1} was obtained which is of the same order of magnitude as that resulting from PRESTON's (1974) calculation. It would thus seem that the cell wall can provide a pathway for the transpiration stream of sufficiently high conductance to require only a modest drop of pressure between the outside of the root and the endodermis. It is the plasmalemmae of the endodermis which will present the greatest resistance in the apoplastic pathway.

3.4.3 The Symplast Pathway

In addition to the resistance of the plasmalemmae of the epidermal cells and the stelar cells surrounding the xylem tracheae, the main resistance to flow in the symplast will probably be in the plasmodesmata where the pathway is much restricted. Here, as with the apoplast, the only way of attempting to arrive at the conductance of the plasmodesmata is by the application of the Poiseuille equation. But lack of knowledge of the details of plasmodesmatal structure makes it difficult to assign reliable figures to some of the parameters. Two possible structures are shown in Fig. 3.7. Clearly flow is more restricted in A than B and the radius of the tube, to which the calculation is very sensitive (4th power), is not easy to measure precisely. Again the viscosity of the moving fluid is not known. The viscosity of pure water is 0.01 poise, but that of cytoplasm can be more than an order of magnitude greater than this. However, it is not suggested that it is the cytoplasm that is flowing, if it were, it would accumulate in the stele and would need to travel back by some form of cyclosis — an unlikely procedure. Somehow the water must flow through the cytoplasm, i.e., between more or less fixed structures which one would think would exert a considerable drag on the flow. Electron micrographs of plasmodesmata reveal them as containing stain and this suggests that they may be occluded and are thus rather poor candidates for bulk flow transport. Of course the occluding material may be an artefact resulting from fixation etc. Details of their structure cannot be seen in the living state. For a fundamental discussion of flow through plasmodesmata the reader is referred to TYREE (1970) and GUNNING (1976).

The rates of water flow through plasmodesmata have been measured by GRAHAM et al. (1973). Working with barley and marrow they fitted micropotometers along a single root so that water uptake could be measured for consecutive 3.5 mm segments. They found (Fig. 3.8) that the water uptake rose sharply behind the tip but at 8–10 cm it fell steeply to a steady rate which was maintained along the rest of the axis. Examination of the endodermis revealed that the steep fall coincided with the change from State 1 to State 2, i.e., to the suberization of the entire wall. Thus the slow steady rate represented uptake where the flow into and out of the endodermis was restricted to the plasmodesmata,

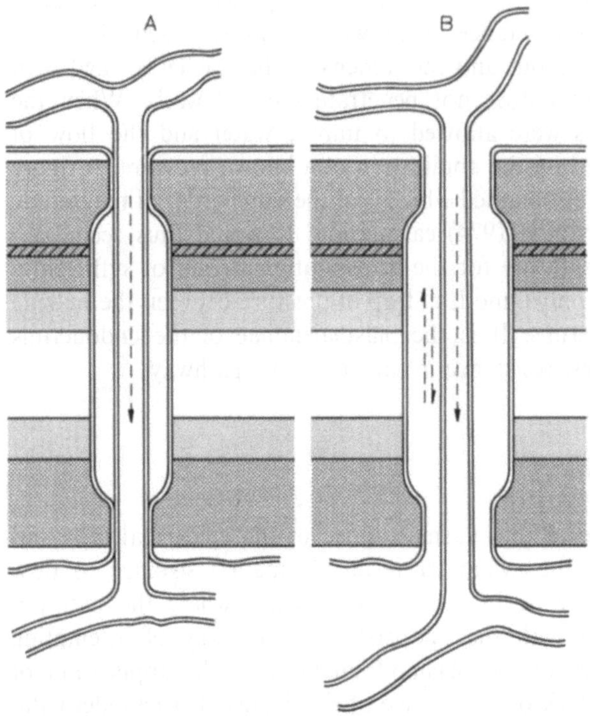

Fig. 3.7 A, B. Possible pathways through the wall cavity created by a plasmodesmos. **A** Situation envisaged by ROBARDS (1971) and seen frequently in electron micrographs of fixed tissue. **B** Hypothetical situation in vivo. (CLARKSON and ROBARDS 1975)

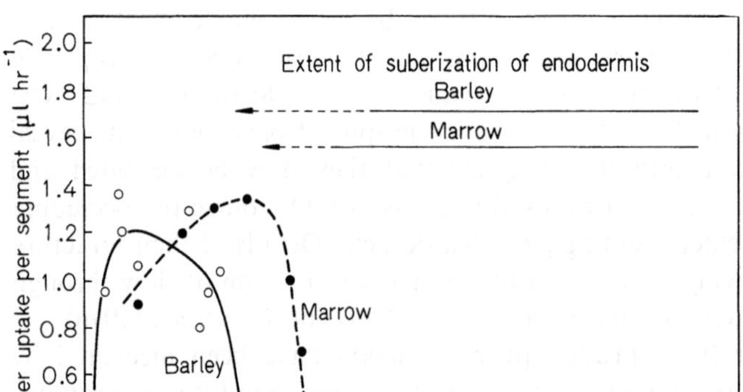

Fig. 3.8. A comparison of water uptake by 3.5 mm segments of intact root and suberization of the endodermis. Length of axis where the endodermis is only partly suberized is shown as *broken lines*. (GRAHAM et al. 1974)

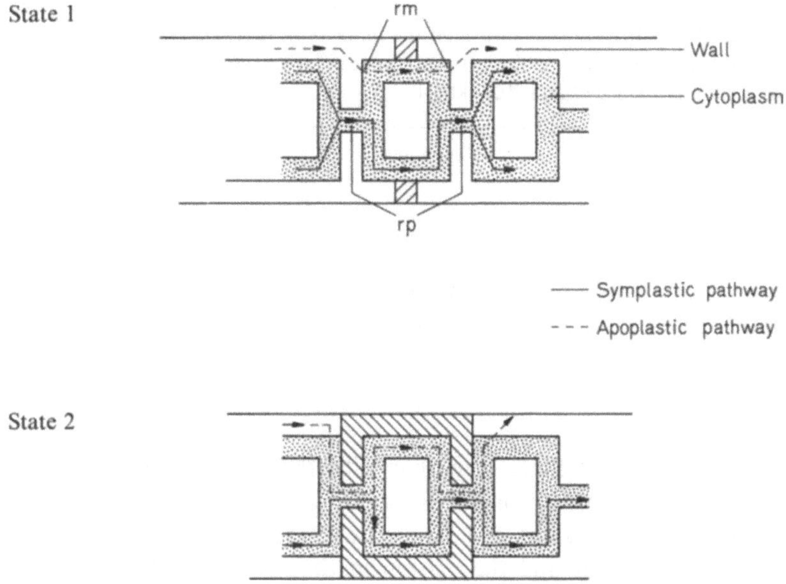

Fig. 3.9. Diagrammatic representation of apoplastic and symplastic pathways for water movement through the endodermis in State 1 and 2. In State 1 the suberization is confined to the Casparian Strip. In State 2 the walls of the endodermal cells are completely suberized

see Fig. 3.9. The frequency, length and radius of the plasmodesmata of the inner tangential wall were measured (CLARKSON et al. 1971; CLARKSON and ROBARDS 1975). Two possible radii were measured, that of the desmotubule (5×10^{-6} mm) and that enclosed by the plasmalemma (10^{-5} mm). Also two values for the viscosity of the moving solution were selected (2 and 50 × the value for pure water). The rate of flow (J_v) was 2.5×10^{-6} cm s^{-1}. Application of the Poiseuille equation using these values gives the pressure differences (ΔP) through the plasmodesmata (see Table 3.1 column 4) necessary to produce this rate of flow. The two values for the radius and the two for viscosity are combined to give 4 possible pressure differences. Corresponding values for the hydraulic conductance of the inner tangential wall ($=J_v/P$) are given in the last column. It will be seen that apart from the last figure (lower radius and higher viscosity) the values are within the range found for the plasmalemmae of plant cells. These figures suggest that the plasmodesmata are unlikely to provide an adequate pathway for water flow across the cortex of the root. From Table 3.1 it appears that the pressure drop would need to be about 1 bar at each cell layer. If the cortex consisted of 10 cell layers, the total drop of pressure would be about 10 bar – certainly more than would be expected. Also it must be borne in mind that in the application of the Poiseuille equation it is implicit that the tubes are free of obstruction – which they hardly appear to be judging from electron micrographs.

There is a further argument for the plasmodesmata not providing the major pathway across the cortex. In passing along the root from the region with

Table 3.1. Pressure differences and hydraulic conductance calculated for a measured rate of flow (2.53×10^{-6} cm s^{-1}) through the inner tangential wall of the suberized endodermis where flow is restricted to the plasmodesmata. Two values for pore diameter and viscosity were used. (CLARKSON and ROBARDS 1975)

Case	r(nm)	η (poise)	ΔP (bar)	Hydraulic Conductivity cm s^{-1} bar^{-1}
I	10	2×10^{-2}	6.06×10^{-2}	4.18×10^{-5}
II	10	5×10^{-1}	15.10×10^{-1}	1.68×10^{-6}
III	5	2×10^{-2}	9.69×10^{-1}	2.61×10^{-6}
IV	5	5×10^{-1}	2.42×10	1.05×10^{-7}

the endodermis at State 1 to that with the endodermis at State 2 the only change is the suberization of the walls. The frequency of the plasmodesmata is the same (CLARKSON and ROBARDS 1975), the cortical plasmodesmata are presumably similar and the tension in the xylem will be little different since the xylem has a high longitudinal conductance. Thus the plasmodesmatal flow should be the same in the State 1 and State 2 segments and if the plasmodesmatal pathway were the major one, there should be little change in rate along the root. Fig. 3.8 shows that in fact there was a 6:1 fall with barley and 7:1 fall with marrow. Clearly there was a six or sevenfold increase in hydraulic resistance concommitant with the suberization of the walls of the endodermis and this suggests that the major radial resistance resides in the endodermis. Reference to Fig. 3.9 shows that in State 2 as a result of suberization both the apoplastic and vacuolar pathways are blocked and flow is canalized through the plasmodesmata of the tangential walls of the endodermis.

Let us consider flow through one face of the endodermis. The ratio of the conductances of the plasmamembrane and the plasmodesmatal pathways can be found in the following way:
For State 1

$$J_m = \frac{\Delta\Psi}{r_m}, \quad J_p = \frac{\Delta P}{r_p} \quad \text{(see Fig. 3.9)}$$

where J_m is the flow through the plasmalemma in response to $\Delta\Psi$, the difference in water potential between the cytoplasm of the endodermis and the cell wall of the adjacent stelar cell. ΔP is the difference in turgor pressure between the endodermal cells and the cells of the outermost layer of the stele, i.e., the pressure drop through the plasmodesmata. It can be shown that if the osmotic potentials in the cytoplasm of the endodermal and adjacent stelar cells are the same, then $\Delta\Psi = \Delta P$. r_m and r_p are the resistances of the plasmalemma and plasmodesmatal pathway respectively.

Thus total flow in State 1:

$$J_{v_1} = J_m + J_p = \Delta P\left(\frac{1}{r_m} + \frac{1}{r_p}\right). \tag{3.13}$$

For State 2:

$$J_m = \frac{\Delta P}{r_m + r_p} \quad \text{and} \quad J_P = \frac{\Delta P}{r_p}.$$

Thus total flow in State 2:

$$J_{v_2} = \Delta P \left(\frac{1}{r_m + r_p} + \frac{1}{r_p} \right). \tag{3.14}$$

Now $J_{v_1}/J_{v_2} = (13) \div (14) = 6/1$ for barley. Solving the resulting quadratic equation it is found that $r_p : r_m = 10.1$.

Thus to account for the 80% drop in rate of flow on passing from State 1 to State 2 manifest in Fig. 3.8, the hydraulic resistance of the plasmodesmata of the tangential walls must be ten times the resistance of the plasmalemmae encountered by the flow on leaving and entering the apoplast as it crosses the endodermis.

The above analysis assumes that the suberization of the endodermis causes a complete stoppage of the flow of water through the cell wall material. There is some evidence that this may not be so (CLARKSON et al. 1978). However, the argument is not weakened – for if there was a partial flow through the suberized wall which had been attributed to the plasmodesmata, then the flow through the latter would have been even less and their resistance correspondingly greater. A more questionable assumption is that the conductances of the plasmodesmata in State 1 and State 2 are the same. The 6:1 fall in rate could have been due to a fall in plasmodesmatal conductance related to the laying down of the suberin lamellae (see refs in GUNNING and ROBARDS 1976).

If the above conclusion is correct and the conductance of the cortical cell walls is high, the site of the main resistance might well reside in the endodermis. The work of GINSBURG and GINZBURG (1970), however, provides evidence to the contrary. They used root segments from which the steles had been pulled out leaving cortical sleeves in which the endodermis was ruptured and therefore played no part in the water movement. They measured the flux of water into or out of the sleeves in response to differences of osmotic pressure and obtained values of L of approximately 10^{-6} cm s^{-1} not dissimilar to an intact root. This suggests that the endodermis is not the site of the main resistance. Further in contrast to the intact root the cell walls in the sleeves could not have provided a pathway since there was no difference in hydrostatic pressure across the sleeves.

GINSBURG and GINZBURG (1970) concluded that the movement of water was in the symplast and regarded flow through the cell wall as inconsistent with their finding that the reflexion coefficient was unity (no leakage through the cell walls) and flow was sensitive to inhibition by DNP, KCN, and CCCP. Such inhibition they attributed to an effect on movement within the symplast, since the permeability of membranes to water is insensitive to inhibitors. A study of the flux of water across the sleeves in response to a difference of hydrostatic pressure would be interesting since this would induce a flow through the cell walls if such could take place. However, the sleeves are too delicate to withstand the necessary pressure difference.

Meantime we are faced with a conflict of evidence both theoretical and experimental. The ranges of values attributable to the various parameters involved is so great that unequivocal discrimination between the possible pathways is not possible. In my view the balance of evidence is in favour of the cell wall. Its conductance seems to be greater than that of the alternative pathways and teleologically it fits in with the structure of the endodermis with its Casparian strips. Indeed, with the apoplastic pathway the endodermis assumes a central role. Here the flow of water must cross the plasmalemma twice and within the endodermis it must briefly travel in the symplast. The function of the symplastic pathway via the plasmodesmata is seen as the transport of solutes probably by diffusion (TYREE 1970). For a contrasting view on the pathway for water movement see NEWMAN (1976).

3.4.4 Additional Points Relevant to the Pathway

With this pathway before us we can return to certain topics considered earlier in this Chapter.

3.4.4.1 The Mechanism of the Variable Root Conductance

That this can lead to a constant leaf water potential over a wide range of transpiration rates was left unexplained. Recently POWELL (1978) has put forward a convincing hypothesis for this in terms of the endodermis of the root. It is suggested that the solute content of the endodermal cells is maintained by an active transfer process, but the transpirational flow of water, canalized through the endodermis, flushes out these solutes so that the greater the rate of transpiration the lower the osmotic value of the endodermal cells. This leads to a fall in turgor pressure in these cells. Now ZIMMERMANN and STEUDLE (1975) have found that the hydraulic conductance of the membranes of the internodal cells of spp. of *Nitella* and *Chara* rises with a fall in turgor pressure, thus if this applied to the root cells of higher plants there would be a rise in conductance with a rise in rate of transpiration. Combining the fundamental equations for these processes and applying realistic values for the various parameters involved, POWELL (1978) demonstrated that in terms of the hypothesis a constant leaf water potential over a wide range of rates of transpiration could be accounted for. Presumably any factor which leads to a fall in turgor of the endodermis should bring about an increased conductance. Thus lowering the water potential round the roots, e.g., by soil drying or flooding with a non-penetrating osmoticum, should lead to an increased conductance. It is interesting to note here that MACKLON and WEATHERLEY (1965) found that when the osmotic pressure of the root medium of transpiring *Ricinus* plants was raised by 3.5 bar by the addition of polyethylene glycol, there was no response in the leaf water potential which remained near to 5 bar. Could this have been due to the reduced difference in water potential through the plant being offset by an increased conductance? However, a completely opposite effect

— a reduction of conductance — was found by LAWLOR (1973) applying polyethylene glycol to wheat plants.

3.4.4.2 The Effect of Rapid Changes in the External Solution

It is implicit in the concept of flow in the apoplast that solutes in the medium round the roots can readily enter the cell wall system and indeed must be drawn in with the transpiration stream as far as the endodermis. If the cell membranes are permeable to a particular solute, it will be taken up by the cortical cells or pass through the endodermis to the stele. But if the plasmalemma is less permeable to the particular solute than it is to water, then the solute will accumulate at the endodermis and will diffuse back against the flow of water and a steady reverse gradient will arise. In view of this, how rapid will be the response in uptake of water on changing the osmotic pressure of the outside solution? If mannitol is added to the solution it might be expected that this will have little effect until it reaches the endodermis. However, it is evident from the experiments of ARISZ et al. described earlier in this chapter, that the response occurs in a matter of a few seconds (Fig. 3.1). The question arises how rapidly does water move through the cortex of the root? A rate of 10^{-5} cm^3 cm^{-2} s^{-1} (BROUWER 1953) can be taken for the root of a transpiring plant. If the distance across the cortex is 0.03 cm and the water were moving across the whole surface of the root, then allowing for the acceleration due to the diminishing surface with inward movement, it would take about 30 min to cross the cortex. If the water movement was confined to the cell walls and these occupied a tenth of the tangential surface, the time taken would be 3 min and if half the volume of the cell wall was occupied by the pathway the time would be 1.5 min. A lag of this duration might be expected before the sudden drop manifest in the experiments of ARISZ et al. (1951). No lag is apparent. Of course although the vacuolar pathway through the epidermal and cortical cells may be of relative high resistance, a sudden rise in osmotic pressure of the medium will draw water out of the epidermal cells and this reverse movement will spread through the cortex, endodermis and stele to the xylem tracheae. But this spread, involving a change in water content of all the cells might be expected to take a noticeable time before *exudation* was reversed. But the reversal (Fig. 3.1) appears to be virtually immediate. Perhaps this fact can be reconciled with apoplastic flow if the roots of the tomato plants as used by ARISZ were very fine and so presented a very short distance across the cortex. Alternatively perhaps the cell wall is differentially permeable with respect to water and mannitol, i.e., the reflection coefficient is greater than zero and so the surface of the epidermis acts osmotically and there would be no lag in change of rate of uptake in response to a sudden change in the medium.

3.4.4.3 The Significance of Endodermal Suberization

Reverting to the change in water uptake along the root, the data presented in Fig. 3.8 whilst showing a sharp reduction in rate of water uptake with the suberization of the endodermis, do show as pointed out by GRAHAM et al. (1974) that the measured uptake by the suberized roots of barley and marrow was equal to about a sixth of that by the younger roots and in the whole root system there is probably a far greater amount of suberized relative to unsuberized root. However, the suberization does lead to a considerable increase in hydraulic resistance and one is led to speculate what advantage the suberization confers. One suggestion (D.T. CLARKSON personal communication) is that the older roots may often occupy the drier regions of the soil and this could lead to the partial desiccation of the cortex leaving the endodermis as a barrier to further drying out but possibly allowing uptake of water via the plasmodesmata on rewetting. Another suggestion is in relation to the possible interfacial resistance between the soil and root surface to be discussed in the next section. It is possible that the interfacial resistance is a function of the rate of water movement across the interface. Now a high endodermal resistance will reduce the rate of uptake and so the interfacial resistance will be less or may not arise at all. Also the cortex being on the upstream side of the endodermal resistance will not suffer an additional water stress and will therefore not contract in volume and no interfacial gap will form (see next section) to exacerbate the interfacial resistance. On this hypothesis the development of an endodermal resistance is seen as a positive advantage.

3.5 The Soil-Rooted Plant

So far consideration has been restricted to the internal water relations of the root which has been regarded as immersed in free water. The water stress developing in the leaves (Ψ_l) under these circumstances is a function of the rate of transpiration (J) and the resistance of the plant (r_p). As pointed out above r_p is often found not to be a constant, but decreases with an increase in rate of flow through the plant. An additional factor arises if the roots are immersed in unsaturated soil. Now water is held by surface tension forces (τ) in the soil matrix and a lower water potential must be developed in the plant if water is to be abstracted from the soil. Furthermore movement of water through the soil to the surface of the root must be down gradients of τ, the magnitude of which is related to the rate of flow of the water (J) and the hydraulic conductance of the soil (K) in accordance with Darcy's Law:

$$J = -\frac{K\,d\tau}{dx} \tag{3.15}$$

x being the distance between the source of water supply within the soil and the surface of the root. The question as to whether these gradients are significant or negligible is a matter of discussion and experiment and is dealt with in some detail by PASSIOURA in Chapter 1, this Volume. Here I shall merely illustrate the situation with reference to the data from a specific experiment (FAIZ and WEATHERLEY 1978) represented in Fig. 3.10.

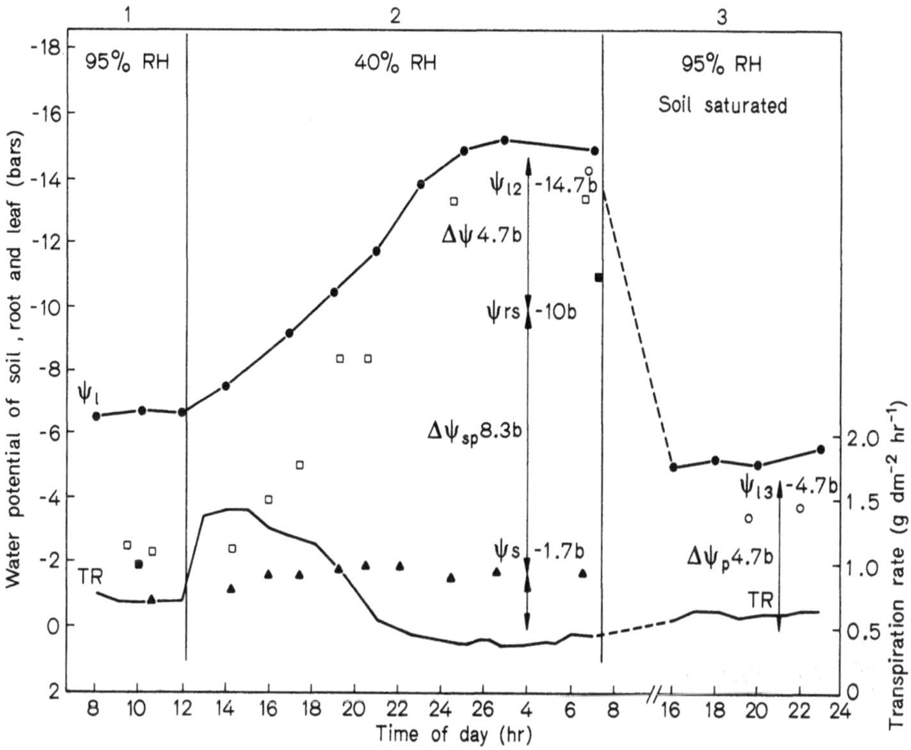

Fig. 3.10. Sunflower plants were grown in soil in pots with upper and lower compartments separated by fine nylon gauze which confined the roots to the upper compartment, but allowed upward capillary rise of water from a water-table maintained at 27 cm below the gauze. The *curves* show the effect of changing the humidity of the airstream on the rate of transpiration (*TR*) and the leaf water potential Ψ_l measured with the psychrometer (●) and with the pressure bomb (○). The water potential of the soil was measured from time to time at several locations within the rooting zone. Means of such values are shown (▲). In addition, the potential of the roots (Ψ_r) was measured: in situ (□) and from samples (■). For further information see text. (After FAIZ and WEATHERLEY 1978)

As was pointed out earlier it has been demonstrated that the leaves of plants in water culture manifest a water potential of -5 to -8 bar which remains constant over a wide range of rates of transpiration. Plants rooted in water-saturated soils behave similarly. In unsaturated soil Ψ_l remains at the steady value at low rates of transpiration (Fig. 3.10 period 1) but at high rates (period 2) there is a steep fall in Ψ_l to a value approaching -15 bar. That such a fall in Ψ_l is dynamic in origin can be demonstrated by the fact that there is complete "recovery" on returning to a slow rate of transpiration (not shown in Fig. 3.10).

This fall in Ψ_l (period 2) is accompanied by a rise in the hydraulic resistance (r) of the soil:plant system as can be seen in Table 3.2 in which the data refer to the end of periods 1, 2 and 3. There is clearly a three- to four-fold increase in resistance when there is high evaporative demand. The resistance of the soil:plant system (r_t) consists therefore of two resistance in series: the resistance of the soil (r_s) and the resistance within the plant (r_p).

Table 3.2. Showing the change in the resistance of the soil: plant system (r) in conditions of lesser (periods 1 and 3) and greater (period 2) stress (Data from Fig. 3.10)

Period	1	2	3
Ψ_l	−6.5	−15.0	−5.0
Ψ_s	−0.9	−1.7	0
Transp	0.55	0.4	0.6
$\Psi_s - \psi_l$	5.6	13.3	5.0
$r = \Psi_s - \Psi_l/\text{transp}$	10.2	33.3	8.3

Water potentials in bar. Transpiration: gm dm^{-2} leaf h^{-1}

$$r_t = r_s + r_p \tag{3.16}$$

Thus the question is: what is the relative magnitude of these resistances and how do they change? The difficulty is that there is no method of measuring the water potential at the root surface – answers can therefore only be inferred and certainty is not attainable. We will consider the two resistances in turn.

3.5.1 The Plant Resistance

The leaf water potential in Fig. 3.10, period 3 (Ψ_{l_3}) is a measure of the drop in water potential ($\Delta\Psi_p$) through the plant, since the water potential at the root surface is zero, the soil being saturated. The mean value of $\Delta\Psi_p$ is 4.7 bar. Does this figure also apply to the end of period 2 in Fig. 3.10? Certainly the rate of transpiration is nearly the same but Ψ_l is much lower at the end of period 2. There is no clear evidence whether this would alter the resistance, but the measurements of water potentials of the roots strongly suggest that the plant resistance did not change much during the experiment. The water potential of the root tissues was measured psychrometrically either in situ using capsules buried in the soil (FISCUS 1972) or from samples of roots obtained from soil borings. It will be seen in Fig. 3.10 that during period 1 the drop of water potential between the roots and the leaves was about 4.5 bar; whilst at the end of period 2 it was about 3 bar. Thus there was not a large change in potential drop through the plant between periods 1 and 2. The water potential of the roots (Ψ_r) was near to that of the soil in period 1 but a large gap (11 bar) between soil and roots opened up in period 2. This perirhizal drop in water potential, i.e., difference in water potential between the bulk soil and the root surface, can be calculated from the data in Fig. 3.10, the potential drop through the plant being taken to be equal to Ψ_{l_3} in period 3 when the roots were in saturated soil. The water potential of the leaves Ψ_{l_2} in period 2 = −14.7 bar. Therefore the potential at the root surface, Ψ_{rs} is 4.7 − 14.7 = −10 bar and this, less the water potential in the soil (Ψ_s) gives the perirhizal drop ($\Delta\Psi_{sp}$), i.e., 10 − 1.7 = 8.3 bar. This is rather less than twice the drop in water potential through the plant. In a similar experiment with sand the perirhizal drop was rather greater than twice that of the plant.

It appears then with a fair degree of certainty that a low level of water potential can develop at the surface of the root. The question now arises how far is this a manifestation of a resistance in the soil?

3.5.2 The Soil Resistance

A model based on Darcy's Law has been developed by GARDNER (1960) which gives the drop in water potential $\Delta\Psi$ through the perirhizal soil as:

$$\Delta\Psi = \frac{Jv}{K \cdot 2\pi L_t} \ln \frac{b}{r}. \quad (3.17)$$

where Jv is the rate of flow, K the hydraulic conductivity coefficient, L_t the total length of the root, r the mean radius of the root and b the distance between the surface of the root and the soil supplying the water, is taken as half the distance between adjacent roots and is obtained from the identity $b = 1/(L_v)^{\frac{1}{2}}$, where L_v is the root density, i.e., the length of root/volume of soil (ANDREWS and NEWMAN 1969).

All the parameters required in Eq. (3.17) were measured for the plants used in Fig. 3.10. K was measured using the pressure plate technique (see CHOWDHURY 1979). Unfortunately the sample of soil used for this was not that actually used in the experiment, but it was garden soil taken from the same location.

$\Delta\Psi$ is found to be only 13 mm of water, a negligible value compared with the drop in Ψ found in Fig. 3.10, i.e., 8.3 bar (8,466 cm of water). Clearly under the conditions of this experiment detectable perirhizal gradients of water potential could not arise.

Before discussing the implication of this, a note of caution must be sounded. Firstly the measurement of the total length of the roots is subject to considerable error. The technique involves washing the roots before length measurement and inevitably there is a loss of very fine roots. Secondly uptake of water is not uniform over the total area of the roots. Uptake may not occur with the older roots so that the effective L_t may be smaller. However, it is unlikely that $\Delta\Psi$ is more than an order of magnitude out whilst it is several orders of magnitude too small to constitute a significant perirhizal gradient. This is in line with Newman's conclusion (NEWMAN 1969a, b) that under normal conditions appreciable perirhizal gradients will not arise.

If the perirhizal gradient is negligible, how can the drop in water potential of 7.8 bar (Table 3.3) be accounted for? It is suggested that a resistance arises at the interface between soil and root. There is no a priori reason for choosing this location, but it is difficult to see where else it could be and possible ways in which a resistance might arise at the interface can be visualized. It might for example simply be a matter of reduction of the area of contact between soil and root. Also it would be expected that the relative size of the roots and the soil particles would be important. The testing of such hypothesis is a matter of future work. If it is confirmed that the soil-root interface is a major resistance in the soil-plant system, and therefore an important cause of water stress in plants, such work assumes a major importance. A further related occurrence is the contraction of the roots with increased stress in the plant. This would lead to gaps forming between the root and soil which would increase the interfacial resistance. Such gaps have been envisaged by PHILIP (1957, 1958); BONNER (1959); DE ROO (1969) and HUCK et al. (1970), the last authors finding that on a dry day the roots of cotton plants shrank by as much as 40%. Such a contraction would lead to a positive feedback increasing the stress in the plant which would be moderated by stomatal closure. This presumably was the cause of the fall in transpiration during period 2 of our experiments. For a further discussion of this subject see TINKER (1976) and NEWMAN (1974).

3.6 Root Hairs

It is perhaps appropriate to discuss root hairs last, not as a measure of their unimportance, but because their possible functions may be related to several of the aspects of root water relations dealt with in previous sections. It has been assumed in the past that root hairs have an absorption function and that their importance lies in presenting a greater surface area than the epidermis alone. There is, however, little information about their function (NEWMAN 1974).

ROSENE (1943) showed, by fixing micro-potometers round individual root hairs, that they are capable of rapid uptake of water. But the root system was in a humid atmosphere and it is not certain how relevant this uptake is to normal submerged conditions. CAILLOUX (1972) used micro-potometers of special designs to measure the water uptake by localized areas of a single hair. He showed that uptake seemed to be restricted to regions where there was underlying cytoplasm. Here again it is difficult to relate the results to behaviour under natural conditions.

The occurrence of root hairs is hardly suggestive of their role. They are limited to a region just behind the elongating region of the root and are often short-lived but not always so. In water culture they are usually sparse or non-existent, whilst they are most prolific where the root passes through larger air filled gaps in the soil – but their function in such a situation would seem to be limited.

Where they occur, root hairs could increase the surface area for absorption. But the water they absorb still has to pass through the inner tangential wall of the epidermal cell. So there would be no advantage unless entry into the cell presents a greater resistance than passage from cell to cell within the root. This could be the case with the symplastic pathway if the plasmalemma were of higher resistance than the plasmodesmata. However, as shown above (Sect. 3.4.3), this did not appear to be true of the endodermis. Another point which argues against the root hair facilitating uptake of water is that the conductance of the soil surrounding the root is 10^5 times higher than the conductance of the pathway offered by the root hairs (NEWMAN 1974). Thus water would move more rapidly up to the epidermal surface through the soil than through the root hairs – the root hairs would indeed be a disadvantage!

Root hairs might, however, have an advantage over the epidermal surface in that their small dimensions permit them to penetrate and mould themselves into the irregular pores between the soil particles, thus making a much closer contact with the soil – a matter perhaps of importance if the soil-root interface has a high resistance. Also they might have an important mechanical function in connexion with the root contraction discussed in the previous section. When the root contracts the root hairs would presumably be dragged inwards pulling with them the soil particles to which they adhere. Thus a zone of small gaps would develop round the roots rather than large cylindrical gaps. Here the mucigel, the layer of mucilage with which the younger roots are invested, might play an important part. (For a further discussion on the role of the mucigel see RUSSEL 1977).

Lastly, the root hairs might have an anchorage function – not in the crude way of preventing the plant from being uprooted, but in forming a solid base for the tips of the growing root to thrust forward through the soil. All these possible functions of the root hair are speculative and will remain so until appropriate experiments can be done. Such experiments are however not easy to carry out.

3.7 Conclusions

In the above discussion of the movement of water through the soil-root system attention has been focussed on three features: (a) The hydraulic resistance of the root, (b) the pathway through the root, (c) the hydraulic resistance of the soil up to the root surface. Each of these present problems the answers to which are still by no means clear.

The exuding root system seems to behave like an osmometer. The rate of flow is a function of the difference in osmotic potential between the surrounding medium and the xylem tracheae. When, however, a difference of hydrostatic pressure is imposed either artificially with a pump or by increasing the rate of transpiration, the hydraulic resistance falls. Several hypotheses for this have been put forward, but as yet no convincing experimental confirmation for any of them has been achieved. The position is further obscured by the fact that even working with the same species contradictory results have been obtained by different workers. Some find a constant resistance, others a changing resistance with changing rate of transpiration.

Of the three possible pathways followed by the transpiration stream across the root cortex, endodermis and stele, there seems to be general agreement that the resistance of the root is not very different from that of a single layer of cells. Therefore flow through the membranes at each cell layer seems unlikely. There remain the apoplastic and symplastic pathways. These are not easy to differentiate since both pathways involve entering the symplast through one membrane and leaving it through another and this constitutes the major resistance. In addition the apoplastic pathway involves a resistance to flow through the cortical cell walls, the symplastic pathway a resistance through the symplast of these same cells, the major resistance in the former being in the interfibrillar spaces of the cell wall and in the latter the flow through the plasmodesmata. As to which of these is the greater is not yet certain.

There is good evidence that there can be a marked drop of water potential between the soil and the root. This could arise in the perirhizal soil if the rate of flow of water were large enough and the hydraulic conductance of the soil low enough. Measurements show that at soil water potentials down to −4 bar or so, no gradient of water potential would be detectable in the perirhizal soil. One is driven to the conclusion therefore that the resistance lies in the soil-root interface. No explanatory model of this interface has been put forward, but it is evident that the demonstrable contraction of the roots resulting from hydrostatic tensions in the plant would increase the interfacial resistance. Thus a positive feedback system would be formed mediated by stomatal control of the rate of transpiration.

References

Anderson WP, Collins JC (1969) The exudation from excised maize roots bathed in sulphate media. J Exp Bot 20:72–80

Anderson WP, Reilly EJ (1968) A study of the exudation of excised maize root after removal of the epidermis and outer cortex. J Exp Bot 19:19–30

Anderson WP, Aikman DP, Meiri A (1970) Excised root exudation – a standing – gradient osmotic flow. Proc R Soc London Ser B 174:445–458

Andrews RE, Newman EJ (1969) Resistance to water flow in soil and plant III Evidence from experiments with wheat. New Phytol 68:1051–1058

Arisz WH, Helder RJ, Nie van R (1951) Analysis of the exudation process in tomato plants. J Exp Bot 11:257–297

Aston MJ, Lawlor DW (1979) The relationship between transpiration, root water uptake and leaf water potential. J Exp Bot 114:169–181

Barrs HD (1970) Controlled environment studies of the effect of variable atmosphere water stress in photosynthesis, transpiration and water status of *Zea mays* L. and other species. In: Proc Symp Plant Response to Climatic Factors. UNESCO, Uppsala, pp 249–258

Bernstein L, Niemann RH (1960) Apparent free space of plant roots. Plant Physiol 5:589–598

Black CR (1974) The sites and relative magnitudes of the resistances to transpirational water flow in sunflower plants. PhD Thesis, Univ Aberdeen

Black CR (1979a) The relationship between transpiration rate, water potential and resistances to water movement in sunflower (*Helianthus annuus* L.) J Exp Bot 30:235–243

Black CR (1979b) The relative magnitude of the partial resistances to transpirational water movement in sunflower (*Helianthus annuus* L.) J Exp Bot 30:245–253

Bonner J (1959) Water transport. Science 129:447–450

Brewig A (1936) Die Regulationserscheinung bei der Wasseraufnahme und die Wasserleitgeschwindigkeit in *Vicia faba* Wurzeln. Jahrb Wiss Bot 82:803–828

Briggs GE (1967) Movement of water in plants. Botanical monographs. Blackwell, Oxford

Brouwer R (1953) Water absorption by the roots of *Vicia faba* at various transpiration strengths. II. Causal relation between suction tension, resistance and uptake. Proc K Ned Akad Wet C56:129–136

Brouwer R (1954) Water absorption by the roots of *Vicia faba* at various transpiration strengths. III. Changes in water conductivity artificially obtained. Proc K Ned Akad Wet C57:68–80

Cailloux M (1972) Metabolism and the absorption of water by root hairs. Can J Bot 50:557–573

Chowdhury MAM (1979) Studies in soil: plant water relations. PhD Thesis, Univ Aberdeen

Clarkson DT, Robards AW (1975) The endodermis, its structural development and physiological role. In: Torrey JB, Clarkson DT (eds) The development and function of roots. Academic Press, London New York, pp 415–436

Clarkson DT, Robards AW, Sanderson J (1971) The tertiary endodermis in barley roots: Fine structure in relation to radial transport of ions and water. Planta 96:292–305

Clarkson DT, Robards AW, Sanderson J, Peterson CA (1978) Permeability studies on epidermal-hypodermal sleeves isolated from roots of *Allium cepa* (onion). Can J Bot 56:1526–1532

Dainty J (1963) Water relations of plant cells. Adv Bot Res 1:279–326

Dainty J, Hope AB (1959) The water permeability of cells of *Chara australis* R. Br. Aust J Biol Sci 12:136–145

Dalton FN, Raats PAC, Gardner WR (1975) Simultaneous uptake of water and solutes by plant roots. Agron J 67:334–339

De Roo HC (1969) Water stress gradients in plants and soil-root systems. Agron J 61:511–515

Diamond JM, Bossert WH (1967) Standing gradient osmotic flow: a mechanism for coupling of water and solute transport in epithelia. J Gen Physiol 50:2061–2083

Eaton FM (1943) The osmotic and vitalistic interpretation of exudation. Am J Bot 30:663–674

Erbe F (1933) Die Bestimmung der Porenverteilung nach ihrer Größe in Filtern und Ultrafiltern. Kolloid Beih 63:277–285

Faiz SMA, Weatherley PE (1978) Further investigation into the location and magnitude of the hydraulic resistances in the soil: plant system. New Phytol 81:19–28

Fiscus EL (1972) In situ measurement of root water potential. Plant Physiol 50–191

Fiscus EL (1975) The interaction between osmotic- and pressure-induced water flow in plant roots. Plant Physiol 55:917–922

Fiscus EL (1977) Determination of hydraulic and osmotic properties of soybean root systems. Plant Physiol 59:1013–1020

Gardner WR (1960) Dynamic aspects of water availability to plants. Soil Sci 89:63–73

Ginsburg H (1971) Model for iso-osmotic flow in plant roots. J Theor Biol 32:147–158

Ginsburg H, Ginzburg BZ (1970) Radial water and solute flows in roots of *Zea mays*. I. Water flow. J Exp Bot 21:580–592

Graham J, Clarkson DT, Sanderson J (1974) Water uptake by the roots of marrow and barley plants. Annu Rep ARC Letcombe Lab

Gunning BES (1976) Introduction to plasmodesmata. In: Gunning BES, Robards AW (eds) Intercellular communication in plants: studies on plasmodesmata. Springer, Berlin Heidelberg New York, pp 1–13

Gunning BES, Robards AW (eds) (1976) Intercellular communication in plants: studies on plasmodesmata. Springer, Berlin Heidelberg New York

House CR, Findlay N (1966) Water transport in isolated maize roots. J Exp Bot 17:344–354

Huck MG, Klepper B, Taylor HM (1970) Diurnal variation in root diameter. Plant Physiol 45:529–530

Hylmo B (1955) Passive components in the ion absorption of the plant. I. The zonal ion and water absorption in Brouwer's experiments. Physiol Plant 8:433–449

Klepper B (1967) Effects of osmotic pressure on exudation from corn roots. Aust J Biol Sci 20:723–735

Kramer PJ (1949) Plant and soil water relationships. McGraw-Hill, New York

Lawlor DW (1973) Growth water absorption of wheat with parts of the roots at different water potentials. New Phytol 72:297–305

Lopushinsky W (1964) Effect of water movement on ion movement into the xylem of tomato plants. Plant Physiol 39:494–501

Macklon AES, Weatherley PE (1965) Controlled environment studies of the nature and origins of water deficits in plants. New Phytol 64:414–427
Mees GC, Weatherley PE (1957) The mechanism of water absorption by roots. I. Preliminary studies on the effects of hydrostatic pressure gradients. Proc R Soc London Ser B 147:367–380
Michel BE (1977) A model relating root permeability to flux and potentials. Application to existing data from soybean and other plants. Plant Physiol 60:259–264
Neumann HH, Thurtell GW, Stevenson KR (1974) In situ measurements of leaf water potential and resistance to water flow. Can J Plant Sci 54:175–184
Newman EI (1969a) Resistance to water flow in soil and plant. I. Soil resistance in relation to amounts of root: theoretical estimates. J Appl Ecol 6:1–12
Newman EI (1969b) Resistance to water flow in soil and plant. II. A review of experimental evidence on the rhizosphere resistance. J Appl Ecol 6:261–272
Newman EI (1974) Root and soil water relations. In: Carson EW (ed) The plant root and its environment. Univ Press, Charlottesville Virginia, pp 363–440
Newman EI (1976) Interaction between osmotic- and pressure-induced water flow in plant roots. Plant Physiol 57:738–739
Oertli JJ (1966) Active water transport in plants. Physiol Plant 19:809–817
Philip JR (1957) The physical principles of soil water movement during the irrigation cycle. Proc Int Congr Irrig Drain 8:125–154
Philip JR (1958) The osmotic cell, solute diffusibility and the plant water economy. Plant Physiol 33:264–271
Pospisilova J (1972) Variable resistance to water transport in leaf tissue of kale. Biol Plant 14:293–296
Powell DBB (1978) Regulation of plant water potential by membranes of the endodermis in young roots. Plant Cell Environ 1:69–76
Preston RD (1974) The physical biology of plant cell walls. Chapman and Hall, London
Robards AW (1971) The ultrastructure of plasmodesmata. Protoplasma 72:315–323
Rosene HF (1943) Quantitative measurement of the velocity of water absorption in individual root hairs by a microtechnique. Plant Physiol 18:588–607
Rufz de Lavison J de (1910) Du mode de pénétration de quelques sels dans la plante vivante. Rev Gen Bot 22:225–241
Russell MB, Woolley JT (1961) Transport processes in the soil-plant system. In: Zarrow MX (ed) Growth in living systems. Basic Books, New York, pp 695–721
Russell RS (1977) Plant root systems their function and interaction with the soil. McGraw-Hill, New York
Sabinin DA (1925) On the root system as an osmotic apparatus. Bull Inst Rech Biol Univ Perm 4, Suppl 2:129–136
Stadelmann E (1963) Vergleich und Umrechnung von Permeabilitäts-Konstanten für Wasser. Protoplasma 57:660–678
Stoker R (1968) Studies in the water relations of the transpiring plant with emphasis on the role of the roots. PhD Thesis, Univ Aberdeen
Stoker R, Weatherley PE (1971) The influence of the root system on the relationship between the rate of transpiration and depression of leaf water potential. New Phytol 70:547–554
Strugger S (1938–9) Die lumineszenzmikroskopische Analyse des Transpirationsstromes in Parenchymen. Flora 133:56–68
Tanton TW, Crowdy SH (1972) Water pathways in higher plants. II. Water pathways in roots. J Exp Bot 23:600–618
Tinker PB (1976) Transport of water to the plant roots in soil. Philos Trans R Soc London Ser B 273:445–461
Tinklin R, Weatherley PE (1966) On the relationship between transpiration rate and leaf water potential. New Phytol 65:509–517
Tyree MT (1970) The symplast concept. A general theory of symplast transport according to the thermodynamics of irreversible processes. J Theor Biol 26:181–214
van den Honert TH (1948) Water transport in plants as a catenary process. Discuss Faraday Soc 3:146–153

Weatherley PE (1970) Some aspects of water relations. Adv Bot Res 3:171–206
Weatherley PE (1975) Water relations of the root system. In: Torrey JG, Clarkson DT (eds) The development and function of roots. Academic Press, London New York, pp 397–413
Weatherley PE (1976) Introduction: water movement through plants. Philos Trans R Soc London Ser B 435–444
Wilson JD, Livingston BE (1937) Lag in water absorption by plants in water culture with respect to changes in wind. Plant Physiol 12:135–150
Zimmermann U, Steudle E (1975) The hydraulic conductivity and volumetric elastic modulus of cells and isolated cell walls of *Nitella* and *Chara* spp: Pressure and volume effects. Aust J Plant Physiol 2:1–12

4 Water Uptake by Organs Other Than Roots

P.W. Rundel

CONTENTS

4.1 Introduction

The extensive early literature on foliar uptake of moisture, spanning nearly three centuries, has been reviewed by Stone (1957a, 1970) and Gessner (1956a). Early field studies by a number of workers demonstrated water uptake by leaves or stems of intact plants (Lloyd 1905; Wetzel 1924; Krause 1935). These and other laboratory experiments (see Stone 1957a, 1970) led to a unanimity of opinion on the ability of leaves to absorb water, but a strong divergence of views on the ecological and physiological significance of this uptake. Remarkably, however, the flow of papers on the subject over the past 50 years has done little to resolve the questions of significance, and foliar water uptake remains as controversial a subject as ever.

There is no question of what the theoretical basis of foliar uptake of water should be. The physical requirements are a gradient of decreasing water potential from the atmosphere to a leaf and through a plant to the soil. Such conditions can be expected to occur in arid and semi-arid habitats where atmospheric vapor pressure deficits are low at night and soil water potentials may be -2 to -3 MPa or lower. In theory, neither saturated atmospheric conditions nor dew point temperatures on leaf surfaces are required for water uptake and transfer to occur. Only a source of water and a favorable water potential gradient are necessary.

Given this background, it is important to set up a series of ecophysiological criteria by which reported examples of foliar uptake of water should be judged. Such uptake must occur in sufficient amounts to increase plant water potentials significantly through direct means rather than indirectly through reduction of transpiration. Thus the following conditions should be met for a critical acceptance of a study:

a) a gradient of decreasing water potential from atmosphere to plant (or even soil) must be present;
b) morphological or anatomical structures must be present to speed water uptake;
c) significant amounts of water must be absorbed and redistributed within the plant; and
d) plant water potentials must be significantly increased by direct effects.

A great many studies have demonstrated that individual criterion are met in specific circumstances. Transport of water from air to root systems and even into soil has been demonstrated in a number of studies (Breazeale et al. 1950, 1959a, b; Breazeale and McGeorge 1953a, b; Haines 1952, 1953; Slatyer 1956). Some conflicting results to these studies have been reported, however (Höhn 1954). Slatyer (1960) reviewed this subject in some detail and discussed the potential significance of leaf and root resistance to water transport, as well as the importance of the magnitude of water potential gradients.

Observations from early agricultural studies that natural or artificial dew could increase the growth of certain plants (see Stone 1970) have long been cited as evidence for foliar uptake of water. More recently similar observations have been reported for a number of species of conifers. Drought-stressed seedlings of *Pinus* and other conifers in California have increased longevity when artificial dew is applied to their foliage (Stone et al. 1950; Stone 1957b, 1964; Stone and Fowells 1955; Stone et al. 1956). *Pinus mugo* (Eisenzopf 1952) and *P. sylvestris* (Härtel and Eisenzopf 1953) in Europe show similar responses. Waisel (1958) found increased water potential of *P. halepensis* following overnight exposure to dew, and later demonstrated direct water uptake into needles using tritiated water (Vaadia and Waisel 1963). Dew absorption through the leaves of drought-stressed *Ceratonia siliqua* (a broad-leaved evergreen) overnight increases their water content by 10% (Waisel 1958). Johnston (1964) reported that plantations of *Pinus radiata* were able to maintain subcritical needle water contents under drought conditions when frequent light rains and fogs were present. An anatomical basis for foliar uptake of water by *P. radiata* has been suggested (Leyton and Armitage 1968).

While these studies would collectively appear to provide strong evidence for accepting the foliar uptake of water as an important physiological mechanism, the individual studies still fail to meet all of the criteria for critical acceptance which were described above. The tritiated water studies (Vaadia and Waisel 1963) lack a demonstration of the magnitude of absorption and the degree of redistribution of this water through the plant. Other studies where increased water contents or increased water potentials were observed have all failed to separate the indirect effects of transpiration reduction on these factors from the direct effects of foliar uptake of water.

On a purely physical basis, MONTEITH (1963) has suggested that evaporation of dew would be unlikely to decrease transpiration significantly in mesophytic plants. He estimated the ratio of potential condensation to potential evaporation as being roughly 1:7 in humid environments and 1:14 in arid environments. He felt then that dewfall would evaporate rapidly after sunrise and would not influence midday gradients of vapor pressure between leaf and air. Nevertheless, it should be obvious that decreased early morning transpiration due to evaporation of dew would be very important to drought-stressed xerophytes with good stomatal control. Limited experimental data has suggested that dew can significantly reduce transpiration (PISEK and CARTELLIERI 1939; JONES 1957). More experimental data is badly needed, however. There is no question of the indirect importance of dew to plant water relations through concentration and drip or stem flow to the soil beneath plant canopies. The process is much the same as the well-known phenomenon of fog drip (AZEVEDO and MORGAN 1974). Many desert fog zone plants have morphological structures and canopy architecture which promote this type of dew condensation and drip. These include divaricate branching patterns, rosette growth forms with upright leaves, and specialized spines (RUNDEL unpublished data). Dew condensation may also occur from less than saturated water vapor through the presence of salt films on leaf surfaces (MOONEY et al. 1980a).

Entry of water into leaves depends on several factors, most notably the wettability of the cuticle. Since the ability of water or any liquid to wet a surface is a function of its contact angle on that surface, the nature of a leaf surface as well as surface tension influences wettability. Dense pubescence or hydrophobic layers of wax or other chemicals on leaf surfaces therefore reduce wettability.

Specialized morphological structures on leaf surfaces have been shown to speed foliar water uptake. These may include specialized trichomes (STUART 1968; BENZING 1976), "hydathodes" (absorptive trichomes) (GRUNDELL 1933; MEIDNER 1954), and specialized cuticle structure (LEYTON and ARMITAGE 1968). It is doubtful that physiologically significant uptake of dew occurs in plant species which lack these structures to facilitate direct foliar uptake.

Cuticular absorption of water is the remaining potential pathway of foliar water uptake. For such a process to occur in xerophytic plants (which typically show low conductance), SLATYER (1960) has pointed out that a rapid increase in cuticular conductance must occur on wetting of the leaf surface. Data to demonstrate such an effect are totally lacking, however. Although the relative importance of stomata as sites of water absorption is uncertain, available data indicate that stomatal uptake of water is negligible (CRAFTS 1933; TURRELL 1947).

In the following sections, reports of foliar uptake in five groups of ecologically and/or physiologically distinct groups of vascular plants are critically reviewed in some detail.

4.2 Foliar Uptake of Water in Vascular Plants

4.2.1 Foliar Uptake of Water in Arid Zone Flowering Plants

A large part of the literature on water absorption through the leaves and stems of vascular plants has focused on the possible significance of this pathway in arid environments. The suggestion that foliar water uptake might be important in desert plants was first made by VOLKENS (1887) in his studies of the physiological anatomy of the flora of Egyptian areas within the Sahara Desert. This subject was further investigated at the Desert Botanical Laboratory at Tucson in the early part of this century (LLOYD 1905; CANNON 1906; SPAULDING 1906a, b). These studies, utilizing a variety of simple experimental techniques to determine if water uptake occurred in moistened leaves or stems, concluded that species with mesophytic leaves readily absorbed moisture, but the results with xerophytes and succulents were variable. In summarizing the most extensive experimental work, SPAULDING (1906b) concluded that foliar absorption was so slow that it was insignificant in comparison to root uptake.

Studies on South African succulents (*Crassula, Mesembryanthemum* (=*Trichodiadema*) and *Anacampseros*) by SCHONLAND (1909) also indicated that water absorption by leaves and stems was so low that it was ecologically unimportant. MARLOTH (1907), however, came to opposite conclusions in his investigations of two species of *Mesembryanthemum* (=*Trichodiadema*) with apical hairs on their succulent leaves. He suggested that these hairs readily absorbed moisture. In later work, MARLOTH (1926) reported water uptake by hairs in species of Liliaceae and Amaryllidaceae. BARTHLOTT and CAPESIUS (1974) quantified cuticular uptake of tritiated water by the same succulents studied by SCHONLAND (1909) and MARLOTH (1907, 1926), and suggested that absorption through paper-like stipules of *Anacampseros,* the epidermis of *Crassula,* and the trichomes of *Trichodiadema* were significant enough to be an adaptation for dew uptake. The small uptake of tritiated water through the spines, epidermis, and areole trichomes of Cactaceae in probably too small to be of ecological significance (SCHILL and BARTHLOTT 1973; BARTHLOTT and CAPESIUS 1974). None of these studies on succulents, however, has met all of the criteria for critical demonstration of foliar or stem uptake.

Australian studies of *Atriplex vesicaria* by WOOD (1925) reported that significant absorption of water into the foliage took place under conditions of nearly saturated atmosphere. Later work on this species has suggested that foliar uptake of moisture is not significant (CARRODUS and SPECHT 1965). Since water on the leaves would dissolve surface salt crusts, it is unlikely that such moisture would be available to leaf cells.

AKHATAR and SHAUKAT (1976, 1979) suggested dew utilization in three species of plants in Pakistan on the basis of survival of exposed plants in soil moisture regimes below the permanent wilting point (−1.5 MPa). No critical demonstration of foliar uptake was included, however, nor was there any suggestion of the physiological tolerance to water stress of non-exposed plants.

WAISEL (1958) carried out a series of experiments on dew absorption by desert and mediterranean plants in Israel. He found dew absorption was variable in different species, but in no case did significant amounts of uptake and redistribution occur. He also noted that the rate of dew absorption increased as soil humidity decreases. Other studies in Israel with *Tamarix* have suggested that presence of dew and water vapor around the foliage is very important to the success of plantations of these trees (GINDEL 1964). Further east in the Asiatic desert belt, SVESHNIKOVA (1972) has described water vapor absorption by leaves and stems of a variety of plants from the Kara-Kum desert region of Turkestan. Again, however, unambiguous foliar uptake has not been demonstrated.

Only two modern studies in North American deserts have reported ecologically significant levels of water uptake through vascular plant leaves, and neither of these is supported by strong data. STARK and LOVE (1969) reported that *Peucephyllum schottii, Atriplex hymenelytra,* and *Larrea divaricata* in Death Valley, California, can all absorb water into the parenchyma of their leaves after light rain or exposure to high humidity. These results were not consistent in their experimental trials, and it is not at all clear if the observed increases in leaf moisture content in their study were actually taken up into living leaf cells or instead absorbed or bound in hygroscopic surface materials. BABU and WENT (1978) described growth chamber experiments for *Tidestromia oblongifolia* and *Atriplex hymenelytra* from Death Valley in which significant foliar uptake of dew occurred. Dew, however, is extremely rare in Death Valley.

The direct leaf uptake of fog moisture has been described for two species of plants in the coastal Namib Desert. Experiments with *Trianthema hereroensis* (Aizoaceae), one of the two dominant species of dune plants from the coast to 60 km inland, has been shown transfer to foliar-applied tritiated water to root tissues within two hours (SEELY et al. 1977). The ecological significance of such uptake to overall water balance in *Trianthema* is difficult to interpret, however. A long tap root extends several meters into the sand, allowing stored ground water to be tapped. Following extended drought of approximately 10 years, it has been observed that only plants low on the dune slopes survived, suggesting a critical importance of soil moisture (SEELY in press). It has also been shown that mature plants will not survive after their roots are severed, even when liquid water is present on the leaves (SEELY in press). These observations suggest that foliar absorption of water is not significant in *Trianthema*. Nevertheless, the distribution of this species is strongly correlated with the occurrence of fog. Apparently suitable dune habitats to the east of the fog zone have greater amounts of precipitation but lack *Trianthema* (SEELY et al. 1977). The ability of *Trianthema* to survive and even maintain reproductive activity throughout the year during long drought periods may well be due more to concentration of fog drip under the plants than to actual uptake of water through the leaves. This type of utilization of fog condensation occurs in *Stipa agrostis sabulicola* in the same habitats (SEELY in press).

Another Namib species for which foliar uptake of moisture has been suggested is *Welwitschia mirabilis*. BORNMAN et al. (1973) demonstrated water uptake through leaf surfaces of *Welwitschia* on the basis of experimental field studies with labeled isotopes. WALTER (1971), however, describes the leaves of *Welwitschia* as unwettable. If foliar uptake does indeed occur, it would seem that it must be a very slow process of limited ecological significance.

One of the most celebrated reports of foliar uptake among desert plants is the example of *Prosopis tamarugo,* a tree mesquite from the virtually rainless Atacama Desert of northern Chile. Studies of this species by SUDZUKI (1969), SUDZUKI et al. (1973) and WENT (1975) have all indicated that the major supply of moisture for this species came from atmospheric water vapor. Recent studies by MOONEY et al. (1980b), however, provide evidence to suggest ground water is the predominant and likely sole source of moisture for these trees.

4.2.2 Foliar Uptake of Water by Bromeliads

While the significance of foliar uptake of water by vascular plants may be controversial in many groups, there is no question of the critical importance of such absorption for many epiphytic Bromeliaceae. The mechanism of water uptake through the leaves of tillandsias and certain other bromeliads has been studied in considerable detail (HABERLANDT 1914; MEZ 1904; BENZING 1976; BENZING and BURT 1970; BENZING and DAHLE 1971; BENZING et al. 1976, 1978). The absorbing qualities of trichomes on the leaves of these bromeliads was first described nearly 100 years ago by SCHIMPER (1888). While most members of the Bromeliaceae possess foliar trichomes, tillandsioid trichome structures are the most highly evolved (BENZING 1976). These peltate trichomes are frequently referred to as scales or shields. The shield is composed of an orderly arrangement of flat cells arrayed only a single cell deep (Fig. 4.1). These arrangements of cells have been described in detail (MEZ 1904; STEINBRINCK 1905; BIEBL 1964; DOLZMANN 1964, 1965; BENZING 1976). At the center of the shield form equal-sized and thick-walled empty cells connect to a column of living stalk cells, usually three to five in number. While the morphology of the shield may vary considerably between species, it is thought to be quite constant within an individual species.

Both shield and stalk cells are cutinized on their exterior surfaces, producing a waterproof tube and valve system. As liquid water strikes the leaf surface of a *Tillandsia,* the wettable surface spreads out droplets to form a thin uniform film of water. The shield cells rapidly fill with water at this time. At this stage the central disc cells become turgid, forcing their upper walls outward, while simultaneously the wings of the shield are flexed downward against the leaf surface (Fig. 4.2). This combined action produces a small suction which draws liquid water in under the shield and down into the stalk cells through osmosis. From these cells water can diffuse throughout the leaf. DOLZMANN (1964, 1965) has suggested on the basis of electron micrographs that water uptake in *Tillandsia* involves a type of pinocytosis.

With subsequent drying and dehydration of the shield cells, the central disc cells shrink and lower to their former position, and the wings flex upward. In this configuration the shield is tightly sealed to prevent water loss from the thin-walled stalk cells. In effect the shields operate as efficient one-way valves.

Since the stomata of *Tillandsia* species are characteristically located flush on the leaf surface among the trichome shields, the presence of even a thin

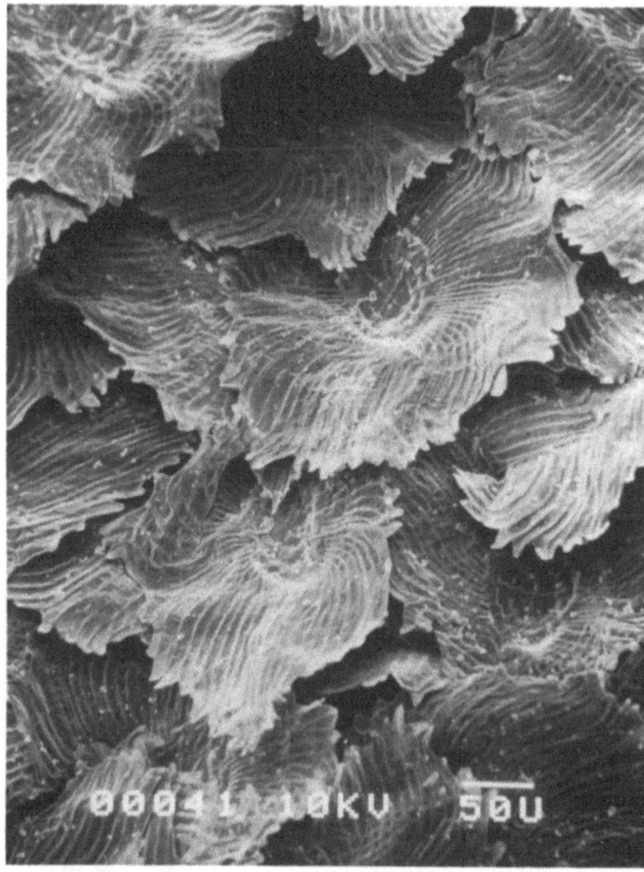

Fig. 4.1. Surface view of absorbing trichome of *Tillandsia geissei* (Paposo, Chile)

film of water on the leaf surface significantly reduces CO_2 exchange and thus restricts photosynthesis. For this reason regular wetting and drying cycles are necessary for *Tillandsia* survival. It is not surprising, therefore, to find that the stands of such species are generally lacking in regions subject to long moist periods.

While BENZING (1976) suggests that the genus *Tillandsia* is only able to absorb liquid water, a number of studies have found significant absorption of water vapor (PENFOUND and DEILER 1967; VIRZO DE SANTO et al. 1976; MARTIN 1980). The exact mechanism of water vapor uptake in *Tillandsia* is unknown, although it appears to be a physical process since dead plants behave similarly to live plants. MARTIN (1980) has hypothesized that water absorption by guard cells may enhance stomatal opening. Whatever the mechanism, nocturnal CO_2 uptake is decreased at low nighttime relative humidities (MARTIN 1980; LANGE and MEDINA 1979).

The water balance of one terrestrial bromeliad, *Tillandsia straminea* has been studied in the field in Peru (ALVIM and UZEDA: in WALTER 1971). The

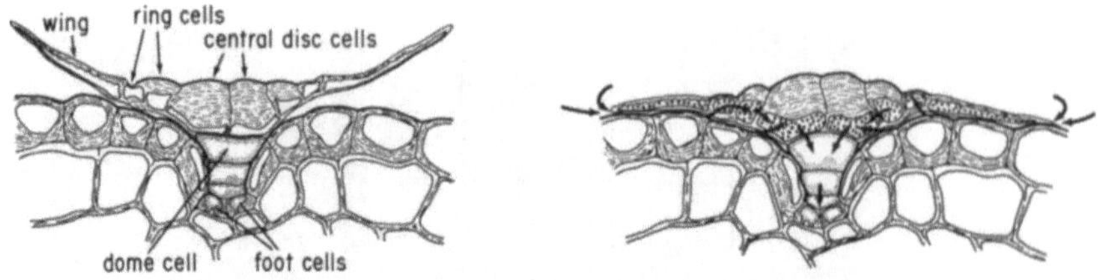

Fig. 4.2. Cross-section of a typical tillandsioid trichome in a dry condition (*left*) and in the process of water absorption (*right*). (Redrawn from BENZING et al. 1976)

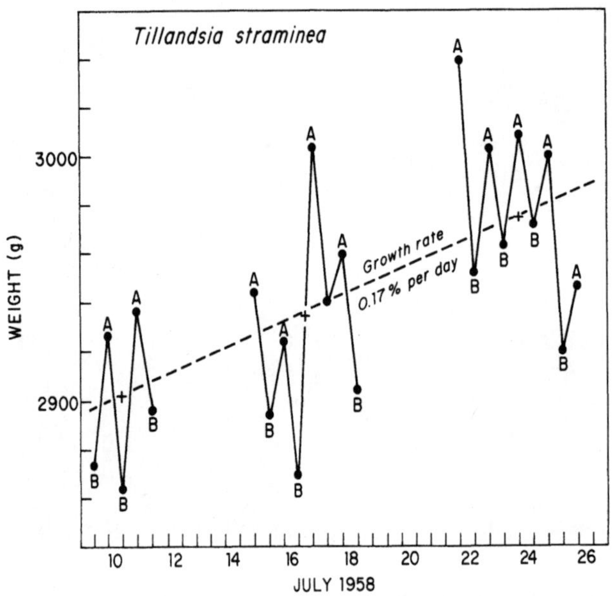

Fig. 4.3. Daily cycle of water relations of *Tillandsia straminea* in the coastal fog desert of Peru. Plant fresh weights were measured at 9 h (*A*) and 17 h (*B*) daily. (Redrawn from ALVIM and UZEDA in WALTER 1971)

water content of several plants was followed with morning and afternoon weighings over a two-week period during July 1958. Despite the lack of precipitation their results (Fig. 4.3) show a clear cyclical water uptake at night and loss through transpiration during the day. They calculated a mean growth increase of 0.17% per day. If sufficient fog was present to maintain this net growth rate for four months of the year, an annual growth in excess of 20% would be possible, without any water input from precipitation.

4.2.3 Foliar Uptake of Water by Orchids

Although epiphytic Orchidaceae are clearly able to utilize liquid water from fogs or mists, there has only recently been data to suggest that significant foliar uptake of moisture takes place. Most orchid leaves have very heavy cuticles that would prevent any uptake.

The presence of absorbing trichomes, however, has now been established in the subtribe Pleurothallidinae (PRIDGEON 1981). These pleurothalloid trichomes are constructed in much the same way as the more elaborate tillandsioid trichomes. Stalk cells with heavily cutinized outer walls and noncutinized transverse walls through which water flows are present in both groups. Both draw water in using much the same mechanisms (PRIDGEON 1981). Absorbing trichomes similar to those of the Pleurothallidinae have been described for at least three other subtribes of the Orchidaceae (SOLEREDER and MEYER 1930). There appears to be a strong correlation between presence of absorbing trichomes and possession of some type of water-storage organ such as a pseudobulb or hypodermis (PRIDGEON 1981).

Some small amount of foliar moisture absorption is also possible in thin-leaved epiphytic orchids from very humid tropical environments (see WENT and SHEPS 1969), but the roots of such species are still thought to be the major route of water absorption (SANFORD 1974). Some controversy is present in the literature on the importance of water absorption by specialized aerial roots in orchids. DYCUS and KNUDSON (1957) found that the velamen (spongy outer tissue layer) of aerial roots was impermeable to water except in the apical region. They ascribed a mechanical function to the velamen to prevent loss of stored moisture, and suggested that water uptake occurs through root surfaces adpressed to a substrate. More recently, however, workers have questioned these results (BARTHLOTT 1976).

4.2.4 Foliar Uptake of Water by Poikilohydric Angiosperms

Poikilohydric angiosperms with leaves which are able to rehydrate from air-dry condition have been described from many arid and semi-arid habitats (cf. Chap. 10, this Vol.). These include both dicots and monocots, representing such families as the Scrophulariaceae, Gesneriaceae, Myrothamnaceae, Velloziaceae, Liliaceae, Cyperaceae, and Poaceae. A large number of poikilohydric angiosperms, many of which are thought to show significant foliar uptake of water, have been described from southern Africa (GAFF 1971, 1977; GAFF and ELLIS 1974). The water relations of poikilohydric angiosperms have been broadly reviewed by GESSNER (1956a, b) and ZIEGLER and VIEWEG (1970).

The most carefully studied of these species is *Myrothamnus flabellifolia*. CHILD (1960) concluded that rehydration of *Myrothamnus* leaves occurred through root uptake of water. GAFF (1977), however, disputed this contention and provided experimental data to indicate that foliar uptake of water was equally as important as uptake through the xylem system. Hydathodes have been described for *Myrothamnus* (GRUNDELL 1933) and these may in reality function as absorbing trichomes.

Air-dry leaves of *Craterostigma plantagineum* (Saxifragaceae) rehydrate more rapidly in water than air-dry roots. Leaf surfaces take up three times the water that cut petioles do following 8 h of rehydration (GAFF 1977). The drought response of this species has also been studied (HORNBY and HORNBY 1964; GALLACE 1974).

Although recovery of full turgor in detached air-dry leaves is commonly a relatively slow process requiring 12 h, some species rehydrate very rapidly. *Chamaegigas intrepidus* (Saxifragaceae), a desiccation-tolerant hydrophyte, rehydrates much more quickly than other poikilohydric species (HEIL 1924; HICKEL 1967; GAFF 1977). There are no available data, nevertheless, to indicate that significant rehydration from water vapor occurs in any poikilohydric angiosperm.

4.2.5 Foliar Uptake of Water by Poikilohydric Ferns and Fern Allies

A number of species of poikilohydric ferns and fern allies have been shown to have leaf trichomes similar in structure to those of tillandsias. Early studies with *Polypodium polypodioides,* an epiphytic fern from the southeastern United States, demonstrated that such trichomes or scales on the lower epidermis of the leaves function in distributing and absorbing water (PESSIN 1924). The mechanism and physiological significance of these structures was described in detail by STUART (1968). A drop of water placed on the dorsal surface of the leaf spreads rapidly from trichome to trichome by capillarity. Rehydration of dried *Polypodium* leaves does not occur when the petiole or rhizomes of dried plants are placed in water. The leaves only regain turgidity when their lower surfaces are in direct contact with liquid water. Partial turgidity with rehydration to 50% of turgid levels may occur after 2–3 days exposure to saturated water vapor. This level of hydration is sufficient to support a level of 40% maximum photosynthesis (STUART 1968). Water uptake of intact dry leaves is slow until the leaves reach about 20% of their turgid water content and then increases in rate considerably. This latter phase of increased uptake is inhibited by anaerobic conditions, indicating a metabolic influence on water uptake. Although the cause of this inhibition has not been investigated, it may be related to changes in membrane permeability (STUART 1968).

The relationship of leaf trichomes to water uptake has also been shown in *Ceterach officinarum,* a poikilohydric fern from the Mediterranean region and the Middle East (ROUSCHAL 1938a; OPPENHEIMER and HALEVY 1962) and appears to be very similar to that described above for *Polypodium.* Detailed studies of the water relations of other poikilohydric ferns and fern allies have not been made, but it would appear that foliar uptake of precipitation is important in many such species. *Selaginella lepidophylla,* the so-called resurrection plant of the southwestern United States, is well known for its rapid foliar uptake of liquid water. Although the morphological basis of water uptake in this species has not been described in ecological terms, the physiological nature of photosynthetic recovery after wetting has been investigated (EICKMEIER 1979). Foliar uptake of water does not necessarily occur in poikilohydric species, however. NOBEL (1978) does not describe any such uptake in his studies of the water relations of *Notholaena parryi* in the Sonoran Desert. GAFF (1971, 1977) reports a large number of poikilohydric ferns and fern relatives from southern Africa. Many of these do show foliar uptake of water, but there is no evidence of any appreciable rehydration of dried leaves from dew (GAFF 1977).

4.3 Water Uptake by Bryophytes

4.3.1 Background

As with lichens, the metabolic activities of bryophytes are directly related to the water content of the thallus. For most species photosynthetic activity increases with hydration to a maximal level at intermediate to high levels of water content (OECHEL and COLLINS 1976; TOBIESSEN et al. 1977, 1979; RUNDEL and LANGE 1980). At higher levels of water content, much of the moisture may be held as surface films on the leaves and as droplets in the leaf axils. Under these conditions, increased resistance to CO_2 transfer may lower photosynthetic rates (FRAYMOUTH 1928; TOBIESSEN et al. 1977, 1979; RUNDEL and LANGE 1980). Respiration rates steadily increase to a maximum level with increasing water content (TOBIESSEN et al. 1977).

Since most bryophytes lack any type of absorptive organs penetrating into the soil, their thalli are largely dependent on atmospheric inputs of moisture. This input may be direct through precipitation of dew condensation, or indirect through growth on substrates where runoff water collects. In considering the water uptake of bryophytes, therefore, two mechanisms of water transport must be considered – external conduction along the outer surface of the thallus and internal conduction in specialized cells.

Although the literature on the water relations of bryophytes is large, there is still a poor understanding of the basic mechanisms of water uptake and conduction. Few studies using modern physiological techniques have been carried out. A detailed synthesis of the historical development of concepts of the water relations of bryophytes beginning with the earliest observations made in the 18th century has been made by HÉBANT (1977). Because of the thoroughness of this work, no attempt will made here to review this older literature.

4.3.2 External Conduction of Water

External conduction of water in bryophytes is facilitated by a variety of specialized surface characteristics on gametophytic tissue, which act as capillary channels. Rhizoids and scales on the ventral surface of many thalloid liverworts form very efficient capillary channels. It has been shown in *Conocephalum* (Marchantiales) that water travels very rapidly between the rhizoids and ventral scales along the base of the thallus to the apical tips where this moisture is readily absorbed (CLEE 1939; MCCONAHA 1939, 1941). This pathway of external conduction has been confirmed in labeled carbon studies (RIFOT and BARRIERE 1974). In *Fimbriaria* (Marchantiales) external conduction is channelized into a ventral groove along the stalk between scales where tuberculate rhizoids are concentrated (BOWER 1935). This channelization supplies the majority of moisture to the reproductive areas on the thallus. Young sporophytes of *Pellia* (Metzgeriales) have also been shown to receive most of their water from external conduction (CLEE 1939).

In mosses and leafy liverworts the pattern of branching and leaf arrangement may strongly influence external water transport. Many mosses, for example, have closely imbricated leaves whose strongly decurrent bases form long narrow channels (HÉBANT 1977). Dry leaves of many arid zone species may be closely appressed but spread rapidly with wetting (RUNDEL and LANGE 1980). This movement strongly influences the level and efficiency of external conduction (MÄGDEFRAU 1935; BAYFIELD 1973). The morphological arrangement of leaves in leafy liverworts has also been shown to affect rates of external conduction of water (CLEE 1937). Morphological structures such as lamellae and filaments on bryophyte leaves may be important in this respect as well.

Although experimental data are very limited, a variety of stem structures in mosses may influence external water transport. These include specialized foliose or filamentous expansions of the stem and the frequent occurrence of rhizoids on the lower stem surface (HÉBANT 1977). Colony growth form is also important in determining rates of water uptake (GIMINGHAM and SMITH 1971; OECHEL and SVEINBJORNSSON 1979).

Since the major importance of external conduction of water in bryophytes is to serve to redistribute moisture, efficient absorption related to this conduction should be expected. While this phenomenon has been shown in many cases, there are many studies that have shown efficient water uptake by bryophyte leaves (ISAAC 1941; ANDERSON and BOURDEAU 1955; ZACHERL 1956). Uptake is particularly efficient in apical areas in comparison to thicker-walled tissues (BOWER 1931, 1933). It must be remembered, however, that bryophytes have motile sperm and require external films of water for fertilization.

4.3.3 Internal Conduction of Water

While internal water conduction may occur in tissues on the basis of simple diffusion between cells, many mosses have specialized water-conducting cells which are dead and hollow at maturity. These cells are most frequently termed hydroids, but have also been referred to as Begleiter, hadrom cells, or stenocysts. In mosses these water-conducting hydroids form central strands in the leafy stem or seta. Similar structures forming portions of the midribs of leaves do not typically interconnect with the axial strands of the stem. Numerous studies over the past 80 years have documented the preferential movement of internal water through these pathways (see HÉBANT 1977). Rates of water conduction through hydroids forming large central strands of bryophyte stems are comparable to rates of xylem flow in vascular plants (ZACHERL 1956). Despite their structural and functional similarity with vascular plant tracheids, a major difference of hydroids is their lack of lignified secondary thickenings on the conducting cell walls. Although a considerable body of literature exists on this subject, no unambiguous reports of the presence of true lignin exist for bryophytes (HÉBANT 1977).

In mosses with both leafy aerial stems and creeping branches, the conducting cells of the upright stems are usually much more highly developed (LORCH 1931). In the Polytrichales, however, a high degree of structural complexity is present in all tissues. The subterranean branches of species of this group have an organization comparable to that of a vascular plant root in many ways (HÉBANT 1977). Large cells of the internal cortex of these have even been interpreted as an endodermis (TANSLEY and CHICK 1901; SCHEINER 1974), but HÉBANT (1977) questions this view. Despite these structures, little convincing evidence exists for uptake of water from these underground branches to the rest of the thallus.

The role of rhizoids in the uptake of water through internal conduction appears to be minor, but there is some disagreement in the literature (see CLEE 1943; BOPP and STEHLE 1957; RIFOT and BARRIERE 1974). Low rates of absorption may be explained in part by the presence of a cuticle on the rhizoids (TRACHTENBERG and ZAMSKI 1979). On the other hand, studies of rhizoids of young fern gametophytes have shown efficient uptake of water (SMITH 1972). In *Radula flaccida,* an epiphyllous leafy liverwort, rhizoids penetrate through the cuticle and epidermal cells of their leaf substrate, and take up water and nutrients from the host substrate (BERRIE and EZE 1975).

In addition to specialized water-conducting cells, living support cells (stereids) adjacent to the hydroids may also function in water conduction (TRACHTENBERG and ZAMSKI 1979). It has long been noted that many xerophytic mosses have unusually large masses of such supporting cells (GOEBEL 1905). These cells are far more than those necessary for mechanical support (LORCH 1931). Heavy masses of these supporting cells in the leaves and leaf bases of many mosses may serve important roles in water uptake and transfer to internal leaf cells and water-conducting cells of the stem (ZAMSKI and TRACHTENBERG 1976; TRACHTENBERG and ZAMSKI 1979).

It has been hypothesized that water held in the thick cell walls of xerophytic bryophytes acts as a buffering system for internal water content during periods of water stress (HÉBANT 1977). PATTERSON (1964), however, strongly doubts that there are any significant structural adaptations for drought in mosses. The significance of such apoplastic water is discussed in more detail in Chapter 2 of this Volume.

The rates of water uptake in bryophytes exposed to liquid water are rapid. Detailed measurements of rates of water uptake for stem and leaf tissue of tundra mosses have been made by GIMINGHAM and SMITH (1971) and OECHEL and SVEINBJORNSSON (1979).

A variety of morphological features of mosses may facilitate condensation and absorption of water under conditions of very light rain or dew (WATSON 1914). Architectural considerations of leaf arrangement may also influence rates of water uptake and loss (WATSON 1914; TOBIESSEN et al. 1977) but PATTERSON (1964) questions whether bryophytes possess special structures which influence rates of water uptake and loss. New experimental work in the significance of leaf arrangement on water relations is needed to resolve this question clearly.

A fundamental difference between mechanisms of water uptake by bryophytes and lichens is the apparent inability of most bryophytes to absorb sufficient water vapor under humid atmospheric conditions to allow significant levels of positive net photosynthesis to occur (RUNDEL and LANGE 1980). A number of studies have documented the low levels of water absorption which do occur in equilibrium experiments in humid atmospheres (MÜLLER 1909; RENNER 1933; CLAUSEN 1952; PATTERSON 1943; ANDERSON and BOURDEAU 1955; TALLIS 1959; WILLIS 1964). Contrary to this general pattern, however, LANGE (1969) was able to show significant levels of photosynthesis in *Hypnum cupressiforme* following equilibration at 96% ralative humidity.

4.4 Water Uptake by Lichens

The water relations of lichens have been an active area for research as evidenced in recent reviews (BLUM 1973; FARRAR 1973; HARRIS 1976). The majority of data on this subject, however, has concentrated on aspects of ecological and physiological responses to changes in thallus water content. The biophysical and biochemical consequences of water uptake in lichens is very poorly understood, particularly with respect to the absorption of water vapor.

Water uptake, as well as loss, is a rapid and purely physical process and there are no strong data to suggest any active metabolic control of the uptake process (cf. BLUM 1973). Nevertheless, morphological and anatomical structure of lichen thalli may be very significant in determining the rates. In addition to uptake from precipitation, runoff, and dew, most lichens are able to hydrate beyond their moisture compensation point in humid atmospheres and this property sets them aside from most vascular plants and most bryophytes.

Uptake of liquid water by lichens is a rapid process which takes place over the entire thallus surface. For most species, full saturation is reached in 1–4 min or less (STOCKER 1927; SMITH 1962; BLUM 1973). Comparative data for the thallus water content of several groups of lichens following immersion of air-dry thalli for 30 s is shown in Table 4.1. These contents are typically 70%–100% of saturation levels. The process of this rapid uptake is thought to occur by both absorption of water by the thick-walled fungal hyphae and capillary movement external to the hyphae. At this stage of saturation, the water content of most foliose and fructicose lichens is 100%–300% of the dry weight of the thallus. Limited data for crustose lichens suggest similar values (RIED 1960; BLUM 1973). No significant differences occur in water contents of fructicose and foliose taxa. However, gelatinous lichens in the Collemataceae have structural characteristics which allow far greater saturation water contents.

Water storage within the lichen thallus appears to be very uneven. The majority of storage is certainly extracellular, with thick hyphal walls and algal sheaths the most important storage sites. BACHMANN (1923) reported that water accumulation in *Umbilicaria* (*Gyrophora*) is concentrated in the medulla. SMITH

Table 4.1. Comparative thallus water content (fresh wt/dry wt) of groups of air-dry lichens after a 30-s immersion in liquid water. (Adapted from BLUM 1973)

Fructicose thalli		Foliose thalli	
Cladonia rangiformis	118	*Anaptychia ciliaris*	120
C. crispata	133	*Umbilicaria grisea*	152
C. gracilis	125	*U. hirsuta*	157
C. furcata	126	*U. pustulata*	197
C. uncialis	129	*Hypogymnia physodes*	128
C. mitis	140	*H. tubulosa*	202
C. rangiferina	145	*Parmelia ryssolea*	122
Cetraria islandica	109	*P. sulcata*	151
Ramalina fastigiata	139	*P. vagans*	124
R. farinacea	141	*P. prolixa*	160
Bryoria implexus	162	*Dermatocarpon miniatum*	142
Bryoria jubatus	145	*D. vellereum*	149
Usnea hirta	154	*Lobaria pulmonaria*	150
Pseudevernia furfuracea	185	*Evernia prunastri*	149
Mean	139	Mean	152
Gelatinous lichens		Semi-gelatinous lichens	
Collema cristatum	394	*Peltigera polydactyla*	247
		P. rufescens	254
C. flaccidum	566	*P. canina*	275
Mean	480	Mean	259

(1960) found 25% more water per unit of dry weight in the medulla of *Peltigera* than in the upper cortex and algal layer. TOBLER (1925), however, described just the opposite pattern in saxicolous lichens with the cortex and algal layer more important than the medulla. Finally, SHOWMAN and RUDOLPH (1971) found water storage concentrated in the algal layer of *Umbilicaria*. The water storage significance of central cords in the medulla of *Usnea* and certain members of the Ramalinaceae has not been investigated.

Differential water storage within lichen thalli is clearly induced by wetting and drying cycles. As fungal hyphae in the upper cortex swell with water uptake, a stage is reached where gas exchange to the algal layer is restricted. Rates of photosynthetic assimilation for most lichens increase directly with thallus moisture content up to this stage, but commonly decline as moisture content approaches saturation. During a drying cycle water loss occurs rapidly from the outer cortex, thereby decreasing resistance to gas exchange. The result is a hysteresis effect with significantly different moisture contents for maximal rates of photosynthetic assimilation in wetting and drying cycles (Fig. 4.4).

Moisture uptake from water vapor and dew condensation are extremely important to many lichens. The ecological and physiological significance of these sources of water have been documented in considerable detail for lichens in the Negev Desert. LANGE et al. (1975) summarize these studies. Dew fall

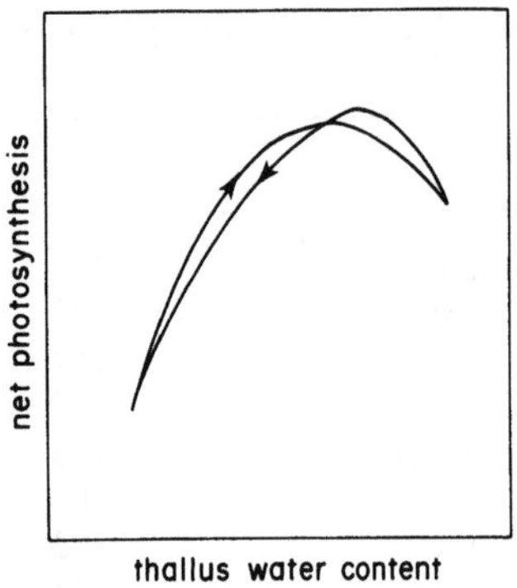

Fig. 4.4. Relationship of thallus moisture content to photosynthetic rate in typical lichens. Note the hysteresis effect differentiating wetting and drying cycles

in the Negev is frequently enough to cause sufficient hydration for several hours of active photosynthesis in the early morning. Over the course of a year, moisture input principally from dew can account for net production rates of 8%–15% of the original thallus biomass (LANGE et al. 1975; KAPPEN et al. 1975, 1979). This is a level equal to or greater than that of vascular plants in the same region.

Under conditions of high relative humidity, lichen thalli will slowly absorb moisture until they reach a level with an internal vapor pressure in equilibrium with the atmosphere. Most lichens will reach a water content of 40%–50% in a saturated atmosphere, but a number of days are required to reach this stage. Exposure for several hours to relative humidities above 90% is usually sufficient to bring thallus moisture contents up above the photosynthetic compensation point. It is not surprising, therefore, to observe that areas of the world with consistent patterns of high humidity but low precipitation support rich lichen floras. This diversity and associated ecological importance of lichens is particularly notable in desert fog zones (RUNDEL 1978a).

Despite the general suggestion in many reviews of lichen-water relations that water uptake and loss is very similar in most lichens, there is good evidence that morphological factors are very important in determining rates of these processes. LARSON and KERSHAW (1976) have shown how branch surface-to-weight ratio, surface water-holding capacity, and growth habit may all influence leaf resistance to water vapor transport in lichens. Patterns of very similar characteristics have been shown to be correlated with specific habitat types for several species of *Ramalina* (RUNDEL 1974, 1978b). The observation by BLUM (1973) that sorediate and isidiate forms of lichens characteristically have more rapid rates of uptake than forms without these structures may well be interpreted on the basis of increased surface area for absorption.

In addition to morphological characteristics, both organic and inorganic compounds deposited on the surface of lichen thalli may influence water uptake. GOEBEL (1926) suggested that the presence of nonwettable secondary lichen compounds on the outer surface of medullary hyphae may help maintain air chambers in an otherwise saturated thallus. Although there is no experimental data to support this suggestion, it would be interesting to look for such compounds in correlation with the presence of cyphellae, pseudocyphellae, and other structures which hypothetically serve as breathing pores (RUNDEL 1978c).

FOLLMANN (1967a, b) described heavy salt encrustations on coastal lichens in central Chile and speculated that hydrophilic salt crusts might provide a favorable water potential gradient for uptake of atmospheric water vapor. Although his published data are difficult to interpret because of unrealistic time axes, RUNDEL (1978a) has shown that such crusts do increase water vapor uptake considerably and that many coastal lichens are able to maintain high rates of photosynthesis under the accompanying levels of low osmotic potential.

4.5 Conclusions

Despite the large existing body of literature suggesting foliar uptake of water in vascular plants, many more studies are needed to assess the physiological significance and ecological importance of direct foliar uptake of dew, precipitation and/or water vapor. Existing data on water uptake by tillandsioid bromeliads are the best-documented examples of direct foliar uptake. All four of the criteria for critical acceptance of foliar absorption which were described in the introduction to this paper have clearly been met. Gradients of decreasing water potential from atmosphere to plant frequently exist, morphological specializations for water absorption are present. Significant amounts of water are absorbed and redistributed and plant water potentials are reduced. Although these criteria are not so clearly met for many epiphytic Orchidaceae and poikilohydric vascular plants, the existing evidence is very strong. For other vascular plants, however, the data are far less convincing. *Prosopis tamarugo,* frequently cited as the best-documented example of foliar uptake of water, almost certainly does not use this mechanism (MOONEY et al. 1980b). In virtually every case, reports of direct foliar uptake of water for other vascular plants lack sufficient data to meet the necessary criteria for critical acceptance unambiguously. While these comments are in no way meant to suggest that all such reports are in error, more rigorous experimental studies are badly needed to validate earlier work. Several areas of study are particularly important for future research. First, what is the potential significance of small amounts of foliar uptake in reducing transpiration loss? Consideration must be given not only to total transpirational loss but also to water-use efficiencies and maximizing total rates of photosynthetic production in water-limited ecosystems. Secondly, the direct effects of foliar uptake of water and the secondary effects of concentration and utilization of fog-drip by roots need to be more clearly differentiated. These effects are frequently blended. Finally, much more consideration must be given to anatomical and morphological adaptations in species for which foliar uptake has been suggested. Only when these new data are assembled will it be possible to present a clear overview of the occurrence of foliar uptake.

For bryophytes and lichens the situation is much less controversial but new research initiatives are nevertheless badly needed. A large part of the descriptive studies of thallus water relations in these groups is now more than half of century old, yet many current studies utilize fundamentally the same approach.

Fuller utilization of modern techniques of physiological ecology and cell physiology would do a great deal toward providing a better elucidation of mechanisms of water uptake. An evolutionary perspective of adaptive mechanisms is far too frequently ignored in current investigations.

References

Akhatar P, Shaukat SS (1976) Role of dew in the survival and phenology of *Gossypium hirsutum* L. cv. Qalandri. Pak J Bot 8:151–155

Akhatar P, Shaukat SS (1979) Drought endurance and dew utilization capacity of *Sorghum bicolor* (L.) Moench and *Ipomoea pes-caprae* (L.) Sweet. Pak J Bot 11:85–91

Anderson LE, Bourdeau PF (1955) Water relations in two species of terrestrial mosses. Ecology 36:206–212

Azevedo J, Morgan DL (1974) Fog precipitation in coastal California forests. Ecology 55:1135–1141

Babu VR, Went FW (1978) The effect of dew on plant water balance in desert shrubs. Annu Arid Zone 17:1–11

Bachmann E (1923) Untersuchungen über den Wasserhaushalt einiger Felsenflechten. Jahrb Wiss Bot 62:20–64

Barthlott W (1976) Struktur und Funktion des Velamen radicum der Orchideen. Dtsch Orchideen Ges Proc Eighth World Orchid Conf, Frankfurt, pp 438–443

Barthlott W, Capesius I (1974) Wasseråbsorption durch Blatt- und Sprossorgane einiger Xerophyten. Z Pflanzenphysiol 72:443–455

Bayfield NG (1973) Notes on water relations of *Polytrichum commune* Hedw. J Bryol 7:607–617

Benzing DH (1976) Bromeliad trichomes: structure, function and ecological significance. Selbyana 1:330–348

Benzing DH, Burt KM (1970) Foliar permeability among twenty species of the Bromeliaceae. Bull Torrey Bot Club 97:269–279

Benzing DH, Dahle C (1971) The vegetative morphology, habitat preference and water balance mechanisms of the bromeliad *Tillandsia ionantha* Planch. Am Midl Nat 85:11–21

Benzing DH, Henderson K, Kessel B, Sulak J (1976) The absorptive capacities of bromeliad trichomes. Am J Bot 63:1009–1014

Benzing DH, Seemann J, Renfrow A (1978) The foliar epidermis in the Tillandsioideae (Bromeliaceae) and its role in habitat selection. Am J Bot 65:359–365

Bernick W (1938) Untersuchungen über den Taufall auf der Insel Hiddensee und seine Bedeutung als Pflanzenfaktor. Mitt Naturwiss Ver Neuvorpommern Rügen 65/66:67–127

Berrie GK, Eze JMO (1975) The relationship between an epiphyllous liverwort and host leaves. Ann Bot (London) 39:955–963

Biebl R (1964) Zum Wasserhaushalt von *Tillandsia recurvata* L. und *Tillandsia usneoides* L. auf Puerto Rico. Protoplasma 5:345–367

Blum OB (1973) Water relations. In: Ahmadjian V, Hale ME (eds) The lichens. Academic Press, London New York, pp 381–400

Bopp M, Stehle E (1957) Zur Frage der Wasserleitung in Gametophyten und Sporophyten der Laubmoose. Z Bot 45:161–174

Bornman CG, Botha CEJ, Nash CJ (1973) *Welwitschia mirabilis*: observations on movement of water and assimilates under foehn and fog conditions. Madoqua 2:25–31

Bower EJ (1931) Water conduction in *Polytrichum commune*. Ann Bot (London) 45:175–200

Bower EJ (1933) The mechanism of water conduction in the Musci considered in relation to habitat. Ann Bot (London) 47:401–422, 635–661

Bower EJ (1935) A note on the conduction of water in *Fimbriaria blumeana*. Ann Bot (London) 49:844–848

Breazeale EL, McGeorge WT (1953a) Exudation pressure in roots of tomato plants under humid conditions. Soil Sci 75:293–298
Breazeale EL, McGeorge WT (1953b) Influence of atmospheric humidity on root growth. Soil Sci 76:361–365
Breazeale EL, McGeorge WT, Breazeale JF (1950) Moisture absorption by plants growing in an atmosphere of high humidity. Plant Physiol 25:413–419
Breazeale EL, McGeorge WT, Breazeale JF (1959a) Movement of water vapor in soils. Soil Sci 71:181–185
Breazeale EL, McGeorge WT, Breazeale JF (1959b) Water absorption and transpiration by leaves. Soil Sci 72:239–244
Cannon WA (1906) The effects of high relative humidity on plants. Torreya 2:21–25
Carrodus BB, Specht RL (1965) Factors affecting the relative distribution of *Atriplex vesicaria* and *Kochia sedifolia* (Chenopodiaceae) in the arid zone of South Australia. Aust J Bot 13:413–434
Child GT (1960) Brief notes on the ecology of the resurrection plant (*Myrothamnus flabellifolia*). JS Afr Bot 26:1–8
Clausen E (1952) Hepatics and humidity. Dan Bot Ark 15:5–80
Clee DA (1937) Leaf arrangement in relation to water conduction in the foliose Hepaticae. Ann Bot (London) 1:325–328
Clee DA (1939) The morphology and anatomy of *Pellia epiphylla* considered in relation to the mechanism of absorption and conduction of water. Ann Bot (London) 3:105–111
Clee DA (1943) The morphology and anatomy of *Fegatella conica* in relation to the mechanism of absorption and conduction of water. Ann Bot (London) 7:185–193
Crafts AS (1933) Sulfuric acid as a penetrating agent in arsenical sprays for weed control. Hilgardia 8:125–142
Dolzmann P (1964) Elektronenmikroskopische Untersuchungen an den Saughaaren von *Tillandsia usneoides* (Bromeliaceae). I. Feinstruktur der Kuppelzelle. Planta 60:461–472
Dolzmann P (1965) Elektronenmikroskopische Untersuchungen an den Saughaaren von *Tillandsia usneoides* (Bromeliaceae). II. Eine Beobachtung zur Feinstruktur der Plasmodesmen. Planta 64:76–80
Drable E, Drable H (1907) The relation between the osmotic strength of cell-sap in plants and their physical environment. Biochem J 2:117–132
Duvdevani S (1957) Dew research in arid agriculture. Discovery 18:330–334
Dycus AM, Knudson L (1957) The role of the velamen of the aerial roots of orchids. Botan Gaz (Chicago) 119:78–87
Eickmeier WG (1979) Photosynthetic recovery in the resurrection plant *Selaginella lepidophylla* after wetting. Oecologia 39:93–106
Eisenzopf R (1952) Ionenwirkung auf die kutikulare Wasseraufnahme von Koniferen. Phyton (Horn Austria) 4:149–159
Farrar JF (1973) Lichen physiology: progress and pitfalls. In: Ferry BW, Braddeley MS, Hawksworth DL (eds) Lichens and air pollution. Univ Press, Toronto, pp 238–282
Follmann G (1967a) Zur Bedeutung der Salzbestäubung für die Krustenflechten. Ber Dtsch Bot Ges 80:206–208
Follmann G (1967b) Fördern Salzkrusten die Wasseraufnahme von Krustenflechten? Umsch Wiss Tech 13:420
Fraymouth J (1928) The moisture relations of terrestrial algae. III. The respiration of certain lower plants including terrestrial algae with special reference to the influence of drought. Ann Bot (London) 42:75–100
Gaff DF (1971) The desiccation tolerant higher plants of Southern Africa. Science 174:1033–1034
Gaff DF (1977) Desiccation tolerant vascular plants of Southern Africa. Oecologia 31:95–109
Gaff DF, Ellis RP (1974) Southern African grasses with foliage that revives after dehydration. Bothalia 11:305–308
Gallace A (1974) The fine structure and physiology of drought tolerant plants. B Sc Honors Thesis, Monash Univ Melbourne, Aust
Gessner F (1956a) Die Wasseraufnahme durch Blätter und Samen. In: Ruhland W (ed)

Encyclopedia of plant physiology Vol III. Springer, Berlin Göttingen Heidelberg, pp 215–246
Gessner F (1956b) Wasserspeicherung und Wasserverschiebung. In: Ruhland W (ed) Encyclopedia of plant physiology Vol III. Springer, Berlin Göttingen Heidelberg, pp 247–256
Gimingham C, Smith R (1971) Growth form and water relations of mosses in the maritime Antarctic. Br Antarct Surv Bull 25:1–21
Gindel I (1964) Attraction of atmospheric moisture by woody xerophytes in arid climates. Commonw For Rev 45:297–321
Goebel K (1900) Organography of plants. Hafner, New York (reprint 1969)
Goebel K (1926) Ein Beitrag zur Biologie der Flechten. Ann Jard Bot Buitenz 36:1
Grundell R (1933) Zur Anatomie von *Myrothamnus flabellifolia* Welw. Symb Bot Ups 2:1–17
Haberlandt GFJ (1914) Physiological Plant Anatomy. MacMillan, London
Haines FM (1952) The absorption of water by leaves in an atmosphere of high humidity. J Exp Bot 3:95–98
Haines FM (1953) The absorption of water by leaves in fogged air. J Exp Bot 4:106–107
Halket AC (1911) Some experiments on absorption by the aerial parts of certain salt-marsh plants. New Phytol 10:121–139
Harris GP (1976) Water content and productivity of lichens. In: Lange OL, Kappen L, Schulze E-D (eds) Water and Plant Life. Ecol Stud Vol. 19 Springer, Berlin Heidelberg New York, pp 452–468
Härtel O, Eisenzopf R (1953) Zur Physiologie und Ökologie der kutikularen Wasseraufnahme durch Koniferennadeln. Zentralbl Gesamte Forstwes 72:47–64
Hébant C (1977) The conducting tissues of bryophytes. Kramer, Vaduz
Heil H (1924) *Chamaegigas intrepidus* Dtr., eine neue Auferstehungspflanze. Beih Bot Zentralbl 41:41–50
Hickel B (1967) Zur Kenntnis einer xerophilen Wasserpflanze: *Chamaegigas intrepidus* Dtr. aus Südwestafrika. Int Rev gesamten Hydrobiol Hydrogr 52:361–400
Hiltner E (1930) Der Tau und seine Bedeutung für den Pflanzenbau. Wiss Arch Landwirtsch Abt A 3:1–70
Hiltner E (1932) Der Tau, ein vernachlässigter Lebensfaktor der Pflanzen. Mitt Dtsch Landwirtsch Ges 47:825–827
Höhn K (1954) Untersuchungen über das Wasserdampfaufnahme- und Wasserdampfabgabevermögen höherer Landpflanzen. Beitr Biol Pflanz 30:159–178
Hornby HE, Hornby RM (1964) The reaction of *Craterostigma plantagineum* Hochst. to desiccation. Kirkia 4:217–220
Isaac I (1941) The structure of *Anthoceros laevis* in relation to its water supply. Ann Bot (London) 5:339–351
Janes BE (1954) Absorption and loss of water by tomato leaves in a saturated atmosphere. Soil Sci 78:187–197
Johnston RD (1964) Water relations of *Pinus radiata* under plantation conditions. Aust J Bot 12:111–124
Jones RL (1957) The effect of surface wetting on the transpiration of leaves. Physiol Plant 10:281–288
Kappen L, Lange OL, Schulze E-D, Evenari M, Buschbom U (1975) Primary production of lower plants (lichens) in the desert and its physiological basis. In: Cooper PJ (ed) Photosynthesis and productivity in different environments. Univ Press, Cambridge, pp 133–143
Kappen L, Lange OL, Schulze E-D, Evenari M, Buschbom U (1979) Ecophysiological investigations on lichens of the Negev Desert. VI. Annual course of the photosynthetic production of *Ramalina maciformis* (Del) Bory. Flora (Jena) 168:85–108
Krause H (1935) Beiträge zur Kenntnis der Wasseraufnahme durch oberirdische Pflanzenorgane. Oesterr Bot Z 84:241–270
Lange OL (1969) CO_2-Gaswechsel von Moosen nach Wasserdampfaufnahme aus dem Luftraum. Planta 89:90–94
Lange OL, Medina E (1979) Stomata of the CAM plant *Tillandsia recurvata* respond directly to humidity. Oecologia 40:357–363

Lange OL, Schulze E-D, Kappen L, Buschbom U, Evenari M (1975) Adaptations of desert lichens to drought and extreme temperatures. In: Hadley NF (ed) Environmental physiology of desert organisms. Dowden, Hutchinson and Ross, Stroudsberg, pp 20–37

Larson DW, Kershaw KA (1976) Studies on lichen-dominated systems, XVIII. Morphological control of evaporation in lichens. Can J Bot 54:2061–2073

Lehman P, Schanderl H (1942) Tau und Reif, pflanzenwetterkundliche Untersuchungen. Springer, Berlin

Leyton L, Armitage IP (1968) Cuticle structure and water relations of the needles of *Pinus radiata* (D. Don) New Phytol 67:31–38

Lloyd FE (1905) The artifical induction of leaf formation in the ocotillo. Torreya 5:175–179

Lorch W (1931) Anatomie der Laubmoose. In: Linsbauer K (ed) Handbuch der Pflanzenanatomie, vol VII. Borntraeger, Berlin

Mägdefrau K (1935) Untersuchungen über die Wasserversorgung der Gametophyten und Sporophyten der Laubmoose. Z Bot 29:327–375

Marloth R (1907) Notes on the absorption of water by aerial organs of plants. Trans R Soc S Afr 1:429–433

Marloth R (1926) Weitere Beobachtungen über die Wasseraufnahme der Pflanzen durch oberirdische Organe. Ber Dtsch Bot Ges 44:448–455

Martin CE (1980) Field and laboratory studies of crassulacean acid metabolism in the epiphyte *Tillandsia usneoides* L. (Spanish moss). Ph D Dissertation, Duke Univ, Durham, NC

McConaha FA (1941) Ventral structures effecting capillarity in the Marchantiales. Am J Bot 28:301–306

McConaha M (1939) Ventral surface specializations of *Conocephalum conicum*. Am J Bot 28:301–306

Meidner H (1954) Measurements of water intake from the atmosphere by leaves. New Phytol 53:423–426

Mez C (1904) Physiologische Bromeliaceen-Studien. I. Die Wasser-Ökonomie der extrem atmosphärischen Tillandsien. Jahrb Wiss Bot 40:157–229

Monteith JL (1963) Dew: facts and fallacies. In: Rutter AJ, Whitehead FH (eds) The water relations of plants. Blackwell, London, pp 37–56

Mooney HA, Gulmon SL, Ehleringer JR, Rundel PW (1980a) Atmospheric water uptake by an Atacama Desert shrub. Science 209:693–694

Mooney HA, Gulmon SL, Rundel PW, Ehleringer JA (1980b) Further observations on the water relations of *Prosopis tamarugo* of the northern Atacama Desert. Oecologia 44:177–180

Müller K (1909) Untersuchungen über die Wasseraufnahme durch Moose und verschiedene andere Pflanzen und Pflanzenteile. Jahrb Wiss Botan 46:587–598

Nehler N (1977) Bromelienstudien. II. Neue Untersuchungen zur Entwicklung, Struktur und Funktion der Bromelien-Trichome. Akad Wiss Lit Mainz Trop Subtrop Pflanzenwelt 20:1–40

Nobel PS (1978) Microhabitat, water relations, and photosynthesis of a desert fern, *Notholaena parryi*. Oecologia 31:293–309

Oechel WC, Collins NJ (1976) Comparative CO_2 exchange patterns in mosses from two tundra habitats at Barrow, Alaska. Can J Bot 54:1355–1369

Oechel WC, Sveinbjornsson B (1979) Primary production processes in arctic bryophytes at Barrow, Alaska. In: Tieszen LL (ed) Vegetation and production ecology of an Alaskan arctic tundra. Ecol Stud Vol 29. Springer, Berlin Heidelberg New York pp 269–298

Oppenheimer HR, Halevy AH (1962) Anabiosis of *Ceterach officinarum* Lam. et DC. Bull Res Counc Isr Sect D 2:127–147

Patterson PM (1943) Some ecological observations on bryophytes. Bryologist 46:1–13

Patterson PM (1964) Problems presented by bryophytic xerophytism. Bryologist 67:390–396

Penfound WT, Deiler FG (1967) On the ecology of Spanish moss. Ecology 28:445–458

Pessin LJ (1924) A physiological and anatomical study of the leaves of *Polypodium polypodioides*. Am J Bot 11:370–381

Pisek A, Cartellieri E (1939) Zur Kenntnis des Wasserhaushaltes der Pflanzen. IV. Bäume und Sträucher. Jahrb Wiss Bot 88:22–68
Pridgeon AM (1981) Absorbing trichomes in the Pleurothallidinae (Orchidaceae). Am J Bot 68:64–71
Potts R, Penfound WT (1948) Water relations of the polypody fern *Polypodium polypodioides*. Ecology 29:43–53
Renner O (1933) Zur Kenntnis des Wasserhaushalts javanischer Kleinepiphyten. Mit einem Anhang zu den osmotischen Zustandgrößen. Planta 18:215–287
Ried A (1960) Stoffwechsel und Verbreitungsgrenzen von Flechten. II. Wasser- und Assimilationshaushalt, Entquellungs- und Submersionsresistenz von Krustenflechten benachbarter Standorte. Flora 149:345–385
Rifót M, Barriere G (1974) La conduction dans le thalle de l'hepatique *Conocephalum conicum* (L) Dum. I. Etude du transit de l'eau a l'aide d'une solution d'acetate de sodium ^{14}C. Rev Bryol Lichenol 40:45–52
Rouschal E (1938a) Eine physiologische Studie an *Ceterach officinarum* Willd. Flora 132:305–318
Rouschal E (1938b) Zur Ökologie der Macchien. I. Jahrb Wiss Bot 87:436–523
Rouschal E (1939) Beiträge zum winterlichen Waserhaushalt von *Cheiranthus cheiri* und anderer wintergrünen Gartenpflanzen. Oesterr Bot Z 88:148–154
Rundel PW (1974) Water relations and morphological variation in *Ramalina menziesii* Tayl. Bryologist 77:23–32
Rundel PW (1978a) Ecological relationships of desert fog zone lichens. Bryologist 81:277–293
Rundel PW (1978b) Evolutionary relationships in the *Ramalina usnea* complex. Lichenologist 10:141–156
Rundel PW (1978c) The ecological role of secondary lichen substances. Biochem Syst Ecol 6:157–170
Rundel PW, Lange OL (1980) Water relations and photosynthetic response of a desert moss. Flora 169:329–335
Sanford WW (1974) The ecology of orchids. In: Withner CL (ed) The orchids: scientific studies. Wiley and Sons, New York, pp 1–100
Scheiner DC (1974) The anatomy, histochemistry and ultrastructure of the gametophore of *Dendrolignotrichum dendroides* (Hedw) Broth. (Bryopsida: Polytrichaceae). PhD Diss Penn State Univ
Schill R, Barthlott W (1973) Kakteendornen als wasserabsorbierende Organe. Naturwissenschaften 60:202–203
Schimper AFW (1888) Die epiphytische Vegetation Amerikas. Bot Mitt Tropen 2:1–162
Schonland S (1909) On the absorption of water by the aerial organs of some succulents. Trans R Soc Afr 1:395–401
Seely MK (1979) Irregular fog as a water source for desert dune beetles. Oecologia 42:213–227
Seely MK Ecological importance of fog to Namib vegetation. In: Juvik JO (ed) The role of fog in ecosystem. Univ Press, Cambridge in press
Seely MK, Vos de MP, Louw GN (1977) Fog imbibition, satellite fauna and unusual leaf structure in a Namib Desert dune plant *Trianthema hereroensis*. S Afr J Sci 73:169–172
Showman RE, Rudolph EE (1971) Water relations in living, dead and cellulose models on the lichen *Umbilicaria papulosa*. Bryologist 74:444–450
Slatyer RO (1956) Absorption of water from atmospheres of different humidity and its transport through plants. Aust J Bot 9:552–558
Slatyer RO (1960) Absorption of water by plants. Bot Rev 26:331–392
Smith DC (1960) Studies in the physiology of lichens. Ann Bot (London) 24:186–199
Smith DC (1962) The biology of lichen thalli. Biol Rev (Cambridge) 37:537–570
Smith DL (1972) Localization of phosphatases in young gametophytes of *Polypodium vulgare* L. Protoplasma 74:133–148
Solereder H, Meyer FJ (1930) Systematische Anatomie der Monokotyledonen. VI. Microspermae. Borntraeger, Berlin

Spaulding VM (1906a) Biological relationships of desert shrubs. II. Absorption of water by leaves. Bot Gaz (Chicago) 41:262–282
Spaulding VM (1906b) Absorption on atmospheric moisture by desert shrubs. Bull Torrey Bot Club 33:367–375
Stark N, Love LD (1969) Water relations of three warm desert species. Isr J Bot 18:175–190
Steinbrinck D (1905) Einführende Versuche zur Cohäsionsmechanik von Pflanzenzellen nebst Bemerkungen über den Saugmechanismus der wasserabsorbierenden Haare von Bromeliaceen. Flora 94:464–477
Stocker O (1927) Physiologische und ökologische Untersuchungen an Laub- und Strauchflechten. Flora 121:334–415
Stocker O (1956) Wasseraufnahme und Wasserspeicherung bei Thallophyten. In: Ruhland W (ed) Handbuch der Pflanzenphysiologie Vol III, Springer, Berlin Göttingen Heidelnerg, pp 160–172
Stone EC (1957a) Dew as an ecological factor. I. A review of the literature. Ecology 38:407–413
Stone EC (1957b) Dew as an ecological factor. II. The effect of artificial dew on the survival of *Pinus ponderosa* and associated species. Ecology 38:414–422
Stone EC (1964) The role of dew in pine survival in soils below the wilting point and its measurement. In: Methodology in plant eco-physiology. UNESCO, Paris, pp 421–427
Stone EC (1970) The ecological importance of dew. Quart Rev Biol 38:328–341
Stone EC, Went FW, Young CL (1950) Water absorption from the atmosphere by plants growing in dry soil. Science 111:546–548
Stone EC, Fowells HA (1955) The survival value of dew as determined under laboratory conditions with *Pinus ponderosa*. For Sci 1:183–188
Stone EC, Shachori AY, Stanley RG (1956) Water absorption by needles of ponderosa pine seedlings and its internal redistribution. Plant Physiol 31:120–126
Stuart TS (1968) Revival of respiration and photosynthesis in dried leaves of *Polypodium polypodioides*. Planta 83:185–206
Sudzuki F (1969) Absorcíon foliar de humedad atmosférica en tamarugo, *Prosopis tamarugo* Phil. Bol Tech Fac Agronomía Univ Chile 30:1–23
Sudzuki F, Bitti C, Acevedo E (1973) Relaciones hidricas del tamarugo (*Prosopis tamarugo* Phil.) en la localidad de Canchones. Bol Tech Fac Agronomía Univ Chile 37:1–23
Sveshnikova VM (1972) On the absorption of water vapor by the overground parts of the Kara-Kum Desert plants (Russian). Bot Zh (Leningrad) 57:880–886
Tallis JH (1959) Studies in the biology and ecology of *Rhacomitrium lanuginosum* Brid. II. Growth, reproduction and physiology. J Ecol 47:325–350
Tansley AG, Chick E (1901) Notes on the conducting tissue-system in Bryophyta. Ann Bot (London) 15:1–38
Tobiessen PL, Mott KA, Slack NG (1977) a comparative study of photosynthesis, respiration and water relation to their vertical distribution. Bryophytorum Bibliotheca 13:253–277
Tobiessen PL. Slack NG, Mott KA (1979) Carbon balance in relation to drying in four epiphytic mosses growing in different vertical ranges. Can J Bot 57:1994–1998
Tobler F (1925) Biologie der Flechten. Borntraeger Berlin
Trachtenberg S, Zamski E (1979) The apoplastic conduction of water in *Polytrichum juniperinum* Willd. gametophytes. New Phytol 83:49–52
Turrell FM (1947) Citrus leaf stomata structure, composition and pore size in relation to penetration of liquids. Bot Gaz (Chicago) 108:476–483
Vaadia Y, Waisel Y (1963) Water absorption by the aerial organs of plants. Physiol Plant 16:44–51
Virzo De Santo A, Alfani A, Luca De P (1976) Water vapor uptake from the atmosphere by some *Tillandsia* species. Ann Bot (London) 40:391–394
Volkens G (1887) Die Flora der ägyptisch-arabischen Wüste auf Grundlagen anatomisch-physiologischer Forschungen. Borntraeger, Berlin
Wahlin BJO (1943) Daggen dess betydelse for vaxternas vattenhushallning. Sven Vall Mosskulturforen Medd 5:348–367

Waisel Y (1958) Dew absorption by plants of arid zones. Bull Res Counc Isr Ser D 6:180–186

Walter H (1971) Ecology of tropical and subtropical vegetation. Van Nostrand, New York

Watson W (1914) Xerophytic adaptations of bryophytes in relation to habitat. New Phytol 13:149–169, 181–190

Went FW (1975) Water vapor absorption in *Prosopis*. In: Vernberg FJ (ed) Physiological adaptations to the environment. Intext Educational Publ, New York, pp 67–75

Went FW, Sheps LO (1969) Environmental factors in regulation and development: ecological factors. In: Plant physiology: a treatise, vol Va:299–406

Wetzel K (1924) Die Wasseraufnahme der höheren Pflanzen gemäßigter Klimate durch oberirdische Organe. Flora 117:221–269

Willis AJ (1964) Investigations on the physiological ecology of *Tortula ruraliformis*. Trans Br Bryol Soc 4:439–445

Wood JG (1925) The selective absorption of chlorine ions and the absorption of water by the leaves in the genus *Atriplex*. Aust J Exp Biol Med Sci 2:45–56

Zacherl H (1956) Physiologische und ökologische Untersuchungen über die innere Wasserleitung bei Laubmoosen. Z Bot 44:409–436

Zamifirescu N (1931) Cercetari asupra absorptiunii apei prin organele aeriene ale plantelor. Supl Bul Minst Agric Si Romeniilor 3(5–6), 1–105

Zamski E, Trachtenberg S (1976) Water movement through hydroids of a moss gametophyte. Isr J Bot 25:168–173

Zattler F (1932) Agrarmeteorologische Beiträge zum Tauproblem auf Grund von Messungen im Hopfgarten. Wiss Arch Landwirtsch Abt A Archiv Pflanzenbau 8:371–404

Ziegler H, Vieweg GH (1970) Poikilohydre Kormophyten. In: Walter H, Kreeb K (eds) Die Hydratation und Hydratur des Protoplasmas der Pflanzen und ihre ökophysiologische Bedeutung. Protoplasmatologia Vol II. Springer, Wien New York, pp 88–108

5 Transport and Storage of Water

M.H. Zimmermann and J.A. Milburn

CONTENTS

5.1 Introduction

This chapter consists of two parts, the first part (by M.H.Z.) describes the structural basis of water conduction from roots to leaves. Xylem anatomy has been known for a long time. However, those aspects which concern water movement through the plant as a whole have remained rather elusive and therefore neglected. The problem was partly that we are dealing here with microscopic structures spread throughout the whole plant body. We need the microscope to look at the structure, but it is too myopic to give us the information we seek.

Whole-plant-xylem anatomy is an essential factor in plant competition and survival. For example, a tree can afford to lose a branch by cavitation of the xylem water, but it could not survive extensive cavitation in the trunk. How does it control this? The answer to this question seems to be "clever" construction.

The second part (by J.A.M.) concerns water storage in the plant body, a property that varies widely among species. Some plants do not store much water but have other means of surviving adverse conditions. An important question is how much of the water in the plant body is available as storage water. Small herbaceous plants can probably afford to transpire water made available by cavitation, if they refill the xylem by root pressure during the following night. This might be a mechanism for the regulation of flow (cavitation under stress reduces conductivity). But to what extent this can happen in trees, if at all, has not yet been resolved.

5.2 The Structural Basis of Water Conduction

DIXON'S (1914) cohesion theory postulates that xylem water of most plants exists under conditions of negative pressure most of the time, and in many tall trees probably all the time. It is difficult to imagine water in a perennial metastable state, particularly in the tops of trees whose branches are tossed about in the wind. As a result, many plant physiologists doubted that the cohesion theory is the correct explanation for sap ascent, until SCHOLANDER et al. (1965) introduced the pressure bomb which made measurements of xylem tensions possible. The reason why water can exist in xylem in a condition of negative pressure continuously, in some cases for many years, is due to the fact that it is confined to small compartments (see also Chaps. 1 and 2, this Vol.). The longer one studies the three-dimensional structure of the xylem, the more one is amazed how well the water-conducting system is protected by built-in safety features.

5.2.1 Conducting Units: Tracheids and Vessels

The functional significance of the conducting unit is containment of embolism. If vessels were continuous capillaries from roots to leaves, a single injury at any point in a capillary would permit entry of air and the whole track would be air-blocked. Our definition of the conducting unit is therefore the volume enclosed entirely by cell wall. Water flows into this space, and out of it, across pit membranes. Water-conducting pits (i.e., tracheid-to-tracheid, tracheid-to-vessel, or vessel-to-vessel) are always bordered to facilitate flow without loss of mechanical strength. In a bordered pit the secondary wall (the mechanical structure) arches over the primary wall (the pit membrane), thus exposing a maximum area of pit membrane between vessels (ZIMMERMANN 1978a). Vessel

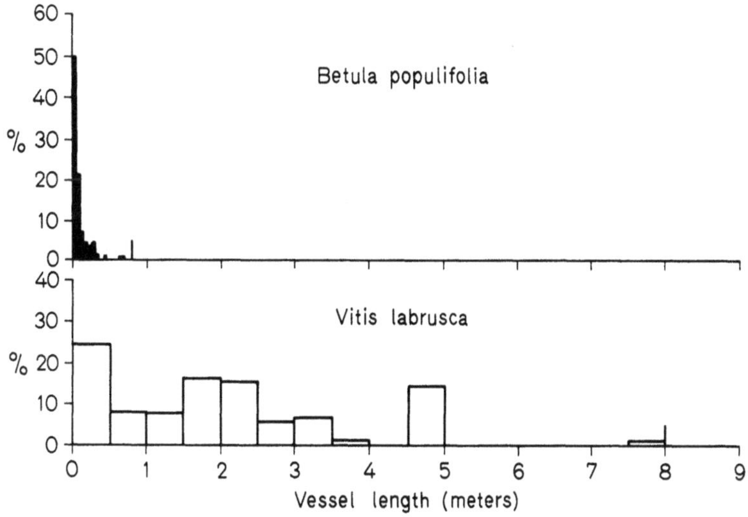

Fig. 5.1. Vessel-length distribution in grey birch (*Betula populifolia*), a tree species with vessels, ca. 60 µm wide, and in the grapevine (*Vitis labrusca*), a species with vessels ca. 300 µm wide. Maximum vessel lengths (80 cm and 8 m respectively) are indicated by a *small vertical line*. (After ZIMMERMANN and JEJE 1981)

ends almost invariably overlap for considerable distances, which increases the area of bordered-pit contact between vessels (ZIMMERMANN 1971). Occasionally one sees, in the light or electron microscope, incompletely broken cell walls at the perforation plates. These structures are emphatically *not* vessel endings.

5.2.2 Vessel Length

In the past, only very sketchy information on vessel length has been available. There was information on the longest vessels, and certain measurement of average length (see ZIMMERMANN and BROWN 1971 for literature citations). The first attempt to measure the distribution of vessel lengths has been published by SKENE and BALODIS (1968) for a *Eucalyptus* species. They found that although a few (less than 1%) of the vessels were about 3 m long, most of them were shorter (e.g., ca. 17% were around 10 cm long). More recently, vessel-length distribution has been measured in a number of North American woody species (ZIMMERMANN and JEJE 1981). As in *Eucalyptus*, all species had far more short than long vessels. Figure 5.1 shows vessel-length distributions of grapevine (*Vitis labrusca*), a wood with vessel diameters around 300 µm and lengths ranging up to 8 m, and grey birch (*Betula populifolia*), a wood with vessel diameters of the order of 60 µm and lengths ranging to 80 cm. In wide-vessel trees the longest vessels may be as long as the whole stem, at the same time the latewood vessels are narrow and short. In other words, Fig. 5.1 may represent a ring-porous tree like oak (*Quercus* sp.): the length distribution of the wide earlywood vessels is like that of *Vitis* and the length distribution of the narrow latewood vessels is like that of *Betula*.

5.2.3 The Vessel Network

The network formed by vessels within the "matrix" of a xylem growth ring was first described by BRAUN (1959), but it was not until the method of cinematographic analysis was available that the three-dimensional layout of vessels in wood became better known (ZIMMERMANN and TOMLINSON 1966; ZIMMERMANN 1971). The cambium produces vessels at successively different angles to the stem axis, therefore, the direct vessel path from any one point in a woody dicotyledonous stem fans out tangentially. This is clearly shown by dye ascents (e.g., KOZLOWSKI and WINGET 1963).

5.2.4 Safety Versus Efficiency

According to the equation of Hagen-Poiseuille, hydraulic conductance of capillaries is proportional to the fourth power of the radius (e.g., ZIMMERMANN 1978b). Vessel diameter is therefore the most significant contributor to the efficiency of water conduction. Evolution of wide vessels increased efficiency enormously, but at the same time it reduced safety. Safety is more difficult to describe mathematically. All we can say at the present is that smaller vessels are safer water conductors. A numerical description of vessel safety would have to be a statistical expression. Under given stress condition, so many vessels of such and such dimensions, representing such and such a percentage of the total functional wood, go out of function within a certain time. It is obvious that if vessels of a species are large, there are relatively few of them, therefore every failure blocks a larger percentage of the total path. The different water-conducting strategies of north temperate trees, namely a long-lasting but relatively inefficient vs. a "disposable" highly efficient system, have been discussed elsewhere (ZIMMERMANN and BROWN 1971). The maximum "practical" vessel diameter in nature appears to be of the order of 0.5 mm. This limit has been reached in many different plant families, larger diameters seem to be simply too unsafe. It is interesting to note, for example, that the greatest tensile strength of water (ca. 350 bar) has been measured in fern annuli where water is confined to a very small volume (RENNER 1915). Nevertheless, it is quite difficult to interfere with sap ascent mechanically, for example by making multiple sawcuts into a tree stem. Consideration of three-dimensional vessel anatomy explains this: most vessels are short, the vessel path spreads tangentially, and spiral grain changes from growth ring to growth ring, sometimes even within a single ring.

5.2.5 Failure of the Water-Conducting System

Failure of conduction could take place for the following reasons:

1. Failure of uptake
2. Injury to the wall of the conducting units (air blockage)
3. Embolism in the absence of injury (cavitation)

4. Plugging of the xylem path
5. Interference with stomatal regulation.

Failure of water uptake can be due to root damage or to drought. The result is an excessive drop of xylem pressure and fracture of water columns (i.e., failure No. 3 above). If stomata open unusually widely, or if vapor pressure difference is unusually large, transpiration may become so excessive that water columns are ruptured (i.e., also failure No. 3). The question of vessel plugging is still insufficiently resolved. In some cases at least air embolism must precede plugging. This is certainly true for bacterial wilt diseases: no bacterium can enter a vessel without puncturing the wall severely enough to admit air. In reality then, there are only three direct causes for failure of water conduction: injury of the vessel walls (air blockage), embolism in the absence of injury (cavitation) and plugging. Injury to walls and plugging are permanent injuries, while embolism can be reversible if the pressure is temporarily raised above ambient. There is evidence that this may happen frequently in certain herbs which develop positive root pressure during the night as indicated by guttation in the morning. Indeed, it is suspected that this mechanism can regulate water flow (MILBURN 1973a). If a water vapor embolus is not eliminated by high pressure within reasonable time (perhaps 24 h or so), diffusion of gas from the wet vessel walls into the vessel lumen could make the blockage a permanent one.

5.2.6 Functional Adaptation of Xylem

Efforts have been made during recent years to interpret xylem structure in terms of adaptive value. One way of doing this is by surveying wood types in different climatic areas of the world, and trying to find meaningful correlations. These efforts have yielded useful summaries of information, but their value is limited, because knowledge about the functional significance of structural features is still restricted. What we really need is rigorous experimental work and measurement of parameters under given environmental conditions. For example, CARLQUIST (1975) puts a great deal of emphasis on mechanical strength of vessel elements. But collapse of functional wood in nature is extremely rare (an example has been reported by LUTZ 1952). Xylem water cavitates long before vessels are in danger of collapse. Drought kills plants by rupturing water columns, not by collapsing xylem vessels.

5.2.6.1 Vessel Size

In Section 5.2.1 of this chapter it has been pointed out that vessels (not vessel elements) are the operating conducting units. Wood anatomists spent a great deal of time measuring element length in different parts of trees and in different climatic regions of the world. However, at this point we know of no functional significance of vessel element length, it appears to be merely the accidental result of other variables of tree growth (e.g., the mechanical requirement for fiber length). It would probably be more meaningful to measure vessel length, which, together with diameter, gives us a measure of vessel volume, a likely measure of safety (in terms of cavitation). But vessel-length measurements are very time-consuming. We also need a precise expression of safety, as discussed in the preceding Section 5.2.4. At this point we can merely state that, in general, vessels are smaller in plants that are subject to greater water stress simply because smaller vessels are safer vessels. This is obviously a very crude statement, but it can nevertheless be illustrated very nicely by the genus *Quercus* whose temperate species have very large vessels, but evergreen species of dry areas have very small vessels.

5.2.6.2 Vessel-to-Vessel Pits

The greatest mechanical stress is not imposed upon vessel walls when vessels are functioning, but when one or a few are air- or vapor-blocked next to intact conducting vessels. Vessel-to-vessel pit membranes permit passage of water, but at the same time prevent the passage of a gas-water interface. When a vessel is embolized, the pressure in its lumen is anywhere between zero (vacuum) and +1 bar (atmospheric pressure) while the neighboring intact vessel is still under negative pressure. A considerable pressure gradient thus exists directly across the pit membranes. The smaller the pores in the pit membranes, the larger the pressure gradient necessary to cause failure to hold the gas-water interface confined to the damaged vessel. One would thus have to anticipate conspicuously small pores in the vessel-to-vessel pit membranes of desert plants. Pit-pore sizes have unfortunately never been measured. Pit membranes are primary walls. Pore sizes in primary walls have only been measured in leaves and found to be of the order of 20 nm. Pore sizes in vessel-to-vessel pit membranes are obviously of importance to plant pathology. Virus particles, for example, can only be distributed via xylem if they are smaller than the pores in the vessel-to-vessel pit membranes.

Vessel-to-vessel pit membranes must offer large surfaces to facilitate water flow from one vessel to the next. At the same time they must be mechanically strong in order not to collapse when one of the vessel embolizes. The bordered pits (secondary wall arching over the primary wall) are unique structures serving both purposes. Some dicotyledons have so-called "vestured" vessel-to-vessel pits (BAILEY 1933; OHTANI and ISHIDA 1976). Here, the tertiary wall of the bordered area develops projections (vestures) into the pit cavity. In the case of embolism, when the pit membrane is forced toward the water-filled element, the vestures support the membrane and prevent it from being torn (ZWEYPFENNIG 1978).

5.2.6.3 Quantity of Xylem

Some early German authors (see Table IV-I, p. 181 in ZIMMERMANN and BROWN 1971) noted that the transverse-sectional area of xylem, when expressed per supplied leaf quantity, increases as one goes from wet to dry habitats (excluding the special case of succulents). This may seem strange at first sight. However, if one considers that at the same time vessel lumen must decrease for reasons of safety, the situation becomes understandable. Volume flow is proportional to the fourth power of capillary diameter. In order to achieve the same conductance the xylem with smaller vessels must have more transverse-sectional area (a second-power relationship). To compensate for this, small-vessel plants must have more xylem area (ZIMMERMANN 1978b).

5.2.6.4 Scalariform Perforation Plates and Wall Sculptures

Features of wall structure such as scalariform perforation plates and spiral thickening may be of adaptive significance. Although scalariform perforation plates are evolutionary relics, the question arises whether they are, under certain circumstances, functionally useful. Remnants of perforation plates and other wall sculptures may be effective in keeping bubbles separated during the thawing of frozen xylem water, thus facilitating their redissolution (ZIMMERMANN 1978b). It has been reported that the incidence of scalariform perforation plates and spiral wall thickenings increases as one moves from tropical lowlands to high altitudes, and from tropical and subtropical to temperate and cold latitudes (VAN DER GRAAF and BAAS 1974; BAAS 1976). Another interesting phenomenon is the fact that a capillary with smooth walls has more resistance to flow than one with spiral thickenings with equal inside diameters (JEJE and ZIMMERMANN 1979). However, we have to be cautious when considering relatively small differences in resistance to flow, because so many other factors (transpiration rate, soil water availability, etc.) are more significant.

5.2.6.5 Adaptation in Aquatic Angiosperms

Although there are aquatic plants that lack roots, there is increasing evidence that hydrophyte roots are functionally similar to those of terrestrial plants: they take up mineral nutrients and synthesize cytokinins and other growth substances which are then transported via xylem to stem and leaves (BRISTOW 1975). In other words, roots are not mere holdfasts, but sources of essential growth factors (KURSANOV 1957). If the plant is entirely submerged, the structural requirements of the xylem are totally different from those of terrestrial plants, because xylem pressures are always positive. Such conduits do not need thick and lignified walls, but might resemble the resin ducts of conifers. While the large cavities in stems and petioles of aquatic plants are gas ducts, the smaller ones within vascular bundles seem to be xylem ducts. This is an interesting area for future investigations.

5.2.7 Patterns of Water Flow Through the Stem

Water rarely flows from a specific root to a specific branch. The flow path in the stem is usually such that an individual root can supply many parts of the crown. In herbaceous plants and arborescent monocotyledons this distribution pattern is based upon the three-dimensional primary vascular anatomy (for palms see p. 179 in ZIMMERMANN and BROWN 1971). In woody plants with secondary xylem, the cambium must produce additional initials as the stem increases in diameter. This is usually accomplished by pseudotransverse divisions of the S or Z type, and the prevalence of one or the other causes the "grain" of the wood to follow a helical rather than an "axis-parallel" path. Spiral grain and its changing angle from year to year has been investigated in conifers with dye ascents by VITÉ and coworkers (e.g., VITÉ and RUDINSKY 1959). In dicotyledons, a spread of dye solution within each growth ring is prevalent, due to the vessel network described in Section 5.2.3 of this chapter. These complex flow patterns make water conduction in the stem amazingly safe (for a more detailed discussion see ZIMMERMANN and BROWN 1971).

5.2.8 The Hydraulic Architecture of Trees

It is of considerable interest to look at sap ascent in plants as a whole. We still know very little about this. There are few attempts to quantify the vascular tracks to specific points in the crown, although there are numerous literature reports on comparative xylem conductivities (e.g., TYREE et al. 1975) in different plant parts, and quantitative information on pressure gradients etc. in different parts of trees which reflect conductivities (e.g., HELLKVIST et al. 1974; RICHTER 1976).

An early attempt at studying what we like to refer to as the hydraulic architecture of trees was made by HUBER (1928), who measured xylem transverse-sectional areas in stems and branches and expressed them per fresh weight of supplied leaves. Huber restricted his measurements to reasonably small individuals of *Abies*, thus avoiding the problem of nonfunctional xylem (heartwood). This, as well as the fact that conductivity is enormously sensitive to conduit diameter, can be overcome by measuring conductance and expressing it per supplied leaf weight or leaf surface area. This has been called leaf-specific conductivity (ZIMMERMANN 1978c). Only a few such measurements are available thus far, concerning some diffuse-porous tree species. Stems have the highest, branches have lower, and twigs have still lower leaf-specific conductivities. One of the most interesting observations was the presence of hydraulic bottlenecks at branch and twig junctions (ZIMMERMANN 1978c). An even sharper constriction seems to be represented by the petiole insertion (LARSON and ISEBRANDS 1978). Flow causes a sharp pressure drop from twigs to leaves. Under conditions of stress the pressure in the petiole xylem may drop so much that cavitation takes place and the leaves are lost. During periods of continued stress lateral branches are often lost (even if the tree is standing in the open), because the resistance to flow from roots to lower lateral leaves may be as great or greater than the resistance to flow to the topmost leaves in the tree. This is an obvious adaptive advantage, because the stem and the top of the crown are the most important parts of the tree. Loss of leaves, twigs, and lower branches is comparable to the loss of the lizard's tail: by losing it the lizard can save its body and grow another tail later. By losing peripheral organs, the tree can save its stem and grow new leaves and branches later. This is particularly important for palms which cannot replace lost stem xylem because of the lack of a cambium.

Much work remains to be done in the area of hydraulic architecture of whole plants.

5.3 Storage and Adversity

Water storage is most evident in succulent plants. Over 5,000 kg of water can be stored by the giant saguaro cactus in its concertina-like stems. Though the physiology has been studied less thoroughly than crop plants and mesophytes, it appears that they can utilize their water reserves extremely frugally when necessary. Three mechanisms facilitate water conservation during adversity. Their stomatal control is highly effective. The metabolism can be reduced and it can also shift from C_3 to CAM. The latter enables gas exchange to occur at night or even cease because it is recycled internally. These physiological adaptations allow succulents to persist at relatively high water potentials for long periods (see also Chap. 15, this Vol.). Water conservation mechanisms are triggered below −10 bar in succulents such as *Portulacaria* and *Peperomia* (HANSCOM and TING 1978).

In contrast, xerophytic plants such as the creosote bush (*Larrea* spp.) have no obvious stores of water and are adapted to tap external stores of water even at very low water potentials. *Larrea* can maintain high rates of transpiration, photosynthesis and cell division at internal water potentials from −8 to −26 bar but these functions are reduced below −45 bar (see BARBOUR et al. 1977). Similarly, mangroves (such as *Avicennia*) which normally desalinate seawater (water potential ca. −25 bar) have large reserves of water available but must be capable of normal metabolism with internal water potentials ranging from −35 to −55 bar (SCHOLANDER 1968). PASSIOURA (1972) suggested that crop plants might be induced to leave water effectively stored in the soil profile at high water potentials through limited rates of utilization. This notion is interesting from the viewpoint of agricultural monoculture but provides no advantage to plants under normal competitive conditions. (For detailed discussion of water use and optimization of carbon assimilation see Chaps. 17 and 18, this Vol.).

Some mesophytic plants, when abundantly supplied with water and subjected to strong sunshine, can lose their own fresh weight as water daily or even hourly. Most mesophytes cannot reduce their diffusive resistance much below 10% of that of normal transpiration rates, so it is unlikely that the 10% to 20% of water available before severe wilting occurs provides a significant store during drought. Nevertheless, certain plants which are nominally mesophytes such as *Tradescantia* or *Pelargonium* (p. 156, MILBURN 1979) have the capacity to survive for long periods on water reserves in their fleshy stems especially when protected from strong light.

Tree trunks can store considerable quantities of water. A 60-m *Sequoia* tree was estimated by RUNNING et al. (1975) to contain 4,000 kg of water. However, the quantity of water in trunks seems to vary on a seasonal basis (WARING and RUNNING 1976, 1978). JARVIS (1975) estimated that water stored in sapwood might support normal transpiration for 0.3 to 14 h. However, since the cells and conduits in wood are strongly lignified and relatively inelastic, it seems that this water can only become available at the expense of embolization following cavitation. The benefit of such a store, which if used must impair the conduction system itself, may legitimately be questioned.

5.3.1 Water Potentials of Stored Water in Plant Tissues

Many techniques can be used to measure the quantities of water held in plant tissues at a range of water potentials (see Chap. 2, this Vol., also RITCHIE and HINCKLEY 1975). All techniques suffer from some defects but the errors and discrepancies result from the different methods and are generally not serious. In addition to quantitative methods there are qualitative measurements such as dendrometry (see KOZLOWSKI 1972) and electrical probes (e.g., SPOMER 1968) which indicate changes in water content of bulky tissues.

5.3.2 Cavitation and Gas Emboli

During the period 1898 to 1914, DIXON (1914) steadily accumulated evidence that water must cohere more strongly than hitherto thought possible. Yet evidence indicating that sap columns did not always withstand severe tensions accumulated gradually from microscopic observation. It was known from gravimetric studies on wood that the moisture content of wood decreased progressively with age. The question then arises if water loss is always irreversible.

The technique of acoustic detection of cavitation by MILBURN and JOHNSON (1966) has been used with increasing effect to analyze the extent and occurrence of cavitation in plants. During its subjection to negative pressure the xylem conduit tends to become distorted mechanically. If the pressure drops low enough, a gas phase develops, either by drawing a gas bubble through the conduit wall or by expanding a microscopic bubble within the conduit. The moment this happens the walls of the conduit release their distortion energy in the form of a vibration which sounds like a "click". Evidence that "clicks" do in fact correspond to cavitation have built up gradually (MILBURN 1973a, b; MILBURN and MCLAUGHLIN 1974; MILBURN and MCLELLAN 1978; WEST and GAFF 1976).

A leaf hung on the detecting microphone in an acoustic chamber and illuminated with an incandescent bulb produces a succession of clicks which develop to a crescendo, then die away as the leaf becomes more wilted in a few hours. Clicks and weight of the leaf are recorded. The leaf is then supplied with water in a humid chamber and restored to maximum water content, and allowed to lose water under the same conditions a second time. At intervals the balancing pressure is measured using a pressure bomb and the dry weight is measured last. Finally, through the intermediary of weight,the tensions at which cavitation occurs can be computed (Fig. 5.2). Up to now only a few species have been studied extensively using this method. Table 5.1 shows striking similarities be-

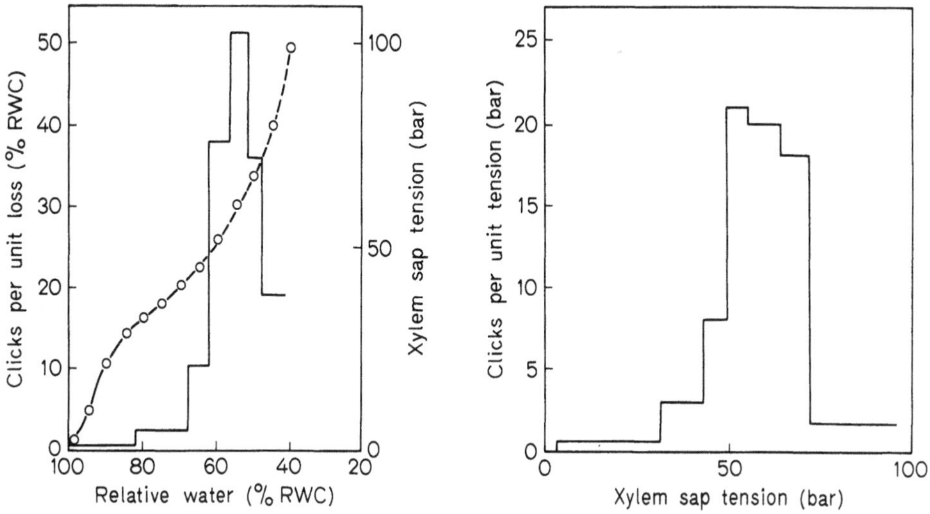

Fig. 5.2. Relationship between xylem sap tensions and cavitation in Manna ash (*Fraxinus ornus*) compound leaf with petiole. *Left Histogram* shows relationship between clicks produced per unit relative water content. *Curve* shows equivalent xylem tension inferred from pressure bomb extraction of water to give different relative water contents. (*Broken line* is extrapolated from the pressure-volume curve) (From MILBURN, TYREE, LO GULLO and SALLEO unpublished results). *Right* Numbers of the xylem conduits disrupted by cavitation (clicks) as xylem sap tensions increase. Most cavitation occurs from 30 to 70 bar tension

Table 5.1. Approximate ranges of xylem sap tensions encountered in a range of plants growing under relatively natural fluctuating water regimes. The ranges of sap tensions at which cavitation occurs in most conduits, as indicated by acoustic detection, are only slightly greater than those encountered naturally in most instances. The correlation between plant stature and cavitation susceptibility seems slight. All tensions were measured with a pressure bomb except two[a] measured by psychrometer. *Fraxinus ornus* grew in Sicily; *F. excelsior* in Scotland. (Unpublished results marked with an asterisk*)

Plant	Xylem Pressure (Approx. range) (bar)	Cavitation Pressure (Approx. range) (bar)	Source
Tomato (*Lycopersicum*)	0 to −5	−6 to −9	NONHEBEL and MILBURN*
Plantain (*Plantago*)	0 to −15	−5 to −15	MILBURN and MCLAUGHLIN (1974)
Castor bean (*Ricinus*)	−2 to −10	−10 to −15	DODOO and MILBURN*
Coconut (*Cocos*)	−2 to −13	−13	MILBURN and DAVIS (1973)
Larch (*Larix*)	−2 to −14	−5 to −15	MILBURN and MCLELLAN*
Apple (*Malus*)[a]	ca. −6	−12 to −25	WEST and GAFF (1976)
Gumtree (*Eucalyptus*)[a]	?	−28 to −40	CROMBIE and DOLEY*
Ash (*Fraxinus excelsior*)	ca. −9	−30 to −70	CROMBIE and MILBURN*
Ash (*Fraxinus ornus*)	−30 to −45	−30 to −70	MILBURN et al.*

tween temperate and tropical herbaceous species. However, trees in arid regions contrast markedly with similar species grown in temperate regions. We do not yet know the extent to which this reflects plasticity of development under the appropriate conditions or genetic differences in the cultivars concerned.

5.3.3 The Question of Reversal of Cavitation Damage

If sufficient root pressure can be maintained over a period of time, gas bubbles entrained in the conduits of small herbs, e.g., grasses, will dissolve. Not only does the positive pressure tend to dissolve the gas bubbles and restore full liquid continuity, but the surface tension of the sap also produces an excess gas pressure. This gas pressure increases in supra atmospheric pressure as the size of the bubble diminishes according to the formula $P=\frac{2T}{r}$ where P is the excess pressure, T is the surface tension of sap and r is the radius of the bubble concerned. The question now arises if this pressure could be sufficient to restore liquid continuity in tall trees because with each 1 m elevation a negative pressure of 0.1 bar must be counteracted by the excess pressure required for dissolution.

There is no way the trivial root pressures in trees could assist in the reversal of cavitation, but if rain water were to gain access at a sufficient rate this would become a strong possibility. Assisting in this process would be hydraulic bottlenecks which have been shown to exist in trees at various points (LARSON and ISEBRANDS 1978; ZIMMERMANN 1978c) which may prevent the water from gravitating downward too rapidly. Despite the fact that cavitation embolisms seem difficult to remove in tall trees, there is a considerable body of evidence suggesting that they may in fact be removed. GIBBS (1958) and others produced extensive evidence showing that not only did the water content of many trees decrease by about 40% in the summer months, but that it also increased once more in the autumn and winter. The extent of this change in water content seems too great to be accounted for by mere changes in cellular water content and implies embolization and subsequent refilling of xylem conduits. Similar results have also been obtained more recently by WARING and FRANKLIN (1979). MILBURN and MCLELLAN (unpublished) showed that the sap pressure of *Larix* twigs decreased markedly (from ca. 15 to ca. 0.3 bar) when they were sprayed in simulated rain water to the point at which emboli would redissolve. This suggests that gas emboli can be redissolved under certain circumstances, but the final resolution of this issue awaits further investigations.

5.3.4 Water Storage in Xylem Fibers

There may be a special mechanism of sap storage peculiar to xylem fibers of certain trees. This seems to be particularly pronounced in the genus *Acer* and is most evident in the leafless state. On warm days from autumn to spring leafless *Acer* trees exude xylem sap when wounded. Even twigs cut from the trees exude when warmed. Exudation appears to be caused by gas pressure in xylem fibers. Conversely, liquid is taken up when the gases are cooled. Gas bubbles can be seen in the xylem fibers but not in the xylem conduits (SAUTER 1974). Pressurization can increase very greatly as a consequence of freezing (O'MALLEY and MILBURN, unpublished). The volume of sap involved is ca. 1–3% of the total water content of the tree.

5.3.5 Water Storage in Bark

Early evidence suggested that whole tree trunks shrink by day and expand by night as the xylem conduits are influenced by changes in xylem sap tension (FRIEDRICH 1897; MACDOUGALL et al. 1929). More recently, however, it has been discovered that it is primarily the cells of the bark and less the wood which are responsible for the dendrographic records obtained (e.g., KLEPPER et al. 1971; HINCKLEY et al. 1974). Apparently bark has a hydraulic capacitance which responds to changes in xylem sap tension with a delay of up to several hours before equilibrium is reached. It seems therefore that dendrographic records register changes in living cell water balance of the same kind observed in succulent or fleshy leaves or fruits. It may be recalled that the role of fruits was shown long ago (see KOZLOWSKI 1972). MAXIMOV (1929) in his book showed that leafy twigs without fruits (lemons) wilted rapidly but remained turgid at the expense of water from the fruits when they were left attached, an observation first reported by BARTHOLOMEW (1926). When dendrographs were placed at different heights on the same tree they showed that the upper parts of the stem shrink before the lower parts when transpiration begins in the morning (HUBER and SCHMIDT 1936; DOBBS and SCOTT 1971). Such results can be plotted as hysteresis curves. Such curves are the result of the respective longitudinal and radial resistances in the stem. They have been obtained in many other ways, e.g., by plotting transpiration vs. absorption (results of KRAMER 1937), transpiration rate vs. water potential, etc.

5.3.6 Water Storage in Primary Tissue

It seems that the most immediate store of water must reside in the cells forming a sheathing layer around the vascular bundles. Studies with linear displacement transducers show that the water relations of cells adjacent to xylem can change in hydraulic capacitance within minutes or even seconds (e.g., CHANEY and KOZLOWSKI 1971; STANSELL et al. 1973). In secondarily thickened stems the radial hydraulic resistance has been reported to increase across the cambium, although this is to some extent offset by conduction through ray cells (MOLZ and KLEPPER 1972).

During adverse conditions, roots have been reported to decrease by more than half their diameter, which tends to separate them deleteriously from the soil (see KOZLOWSKI 1972). The value of this store must be questionable.

5.4 Regulation of Water Flow

While the magnitude of water storage is not very great in most plants, its significance may be greater than immediately apparent. Plants function as hydraulic units in which the different stores are connected via hydraulic resistances which can have the effect of isolating water sources and sinks.

The mechanism of transport of water through living cell catenas, especially in roots and leaves, has been a controversial issue. The hydraulic conductance of cell protoplasts is most unlikely to remain constant because of the changes in solute composition and the well established changes in cytoplasmic viscosity. Cell walls are in some measure compressible, which affects the dimension of the matrix capillaries and hence also their hydraulic capacitance (e.g., KLEMM 1956). Even the xylem, a seemingly simple physical transport tissue, is likely to change rapidly in hydraulic conductance if it becomes embolized through cavitation. Tylose formation introduces yet another factor which may decrease the hydraulic conductance of the tissue, although it normally follows embolism (ZIMMERMANN 1979).

5.5 Conclusions

Xylem anatomy is of fundamental importance in the ascent of sap. Negative pressures can only survive in the xylem of plants for long periods of time because water is confined to millions or even billions of small compartments (tracheids and vessels). Tracheid-to-tracheid and vessel-to-vessel movement of water is facilitated by bordered pits in which large primary-wall areas are exposed so that resistance to flow from one compartment to the next is relatively low. The secondary wall arches over these pit areas in such a way that mechanical strength of the conducting channels is maintained, thus preventing them from collapse. The pores in pit membranes (primary walls) are small enough (probably of the order of 20 nm) that a gas-water interface cannot easily pass from an injured compartment to a functioning one. Water-conducting compartments thus become isolated as soon as they become nonfunctional by entry of air through an injury, or by appearance of water vapor by cavitation, and damage remains confined. The presence of many short vessels in xylem with long vessels (Fig. 5.1), the three-dimensional network of vessels and the varying "spiral grain" of the stem assures distribution of water along the stem. Water stored in plants in different specialized organs fulfils a number of roles. In only a few species does water in living cells serve as a significant reservoir of water during drought. Nevertheless, the smaller stores of readily available water are probably of great significance in cushioning the physiology of the plant from sudden environmental changes. These stores cause diurnal hysteresis in sap fluxes. Significant changes in water storage occur seasonally and also to a small extent diurnally, but their major role lies in maintaining xylem sap transport. Decreasing conductivities in peripheral plant parts, and conductivity constrictions at junctions to branches, twigs, and petioles assures survival of tensile water of the main stem in case of extreme stress.

References

Baas P (1976) Some functional and adaptive aspects of vessel member morphology. Leiden Bot Ser 3:157–181

Bailey IW (1933) The cambium and its derivative tissues. VIII. Structure, distribution and diagnostic significance of vestured pits in dicotyledons. J Arnold Arbor Harv Univ 14:259–283

Barbour MG, Cunningham G, Oechel WC, Bamberg SA (1977) Growth and development, form and function. In: Mabry TJ, Hunziker JH, DiFeo DR (Jr) (eds) Creosote bush, biology and chemistry of *Larrea* in New World deserts. Dowden, Hutchinson and Ross, Stroudsburg, pp 48–91

Bartholomew ET (1926) Internal decline of lemons III. Water deficit in lemon fruits caused by excessive leaf evaporation. Am J Bot 13:102–117

Braun HJ (1959) Die Vernetzung der Gefäße bei *Populus*. Z Bot 45:421–434

Bristow JM (1975) The structure and function of roots in aquatic vascular plants. In: Torrey JG, Clarkson DT (eds) The development and function of roots. Academic Press, London New York, pp 221–236

Carlquist S (1975) Ecological strategies of xylem evolution. Univ Cal Press. Berkeley Los Angeles London

Chaney WR, Kozlowski TT (1971) Water transport in relation to expansion and contraction of leaves and fruits of Calamondin orange. J Hortic Sci 46:71–78

Dixon HH (1914) Transpiration and the ascent of sap in plants. MacMillan, London

Dobbs RC, Scott DRM (1971) Distribution of diurnal fluctuations in stem circumference of Douglas fir. Can J For Res 1:80–83

Friedrich J (1897) Über den Einfluß der Witterung auf den Baumzuwachs. Zentralbl gesamte Forstwes 23:471–495

Gibbs RD (1958) Patterns of the seasonal water content of trees. In: Thimann KV (ed) The physiology of forest trees. Ronald, New York, pp 43–69

Graaf Van der NA, Baas P (1974) Wood anatomical variation in relation to latitude and altitude. Blumea 22:101–121

Hanscom Z, Ting IP (1978) Responses of succulents to plant water stress. Plant Physiol 61:327–330

Hellkvist J, Richards GP, Jarvis PG (1974) Vertical gradients of water potential and tissue water relations in Sitka spruce trees measured with the pressure chamber. J Appl Ecol 11:637–668

Hinckley TM, Chambers JL, Brukerhoff DN, Roberts JE, Turner J (1974) Effect of mid-day shading on net assimilation rate, leaf surface resistance, branch diameter, and xylem potential in a white oak sapling. Can J For Res 4:296–300

Huber B (1928) Weitere quantitative Untersuchungen über das Wasserleitungssystem der Pflanzen. Jahrb Wiss Bot 67:877–959

Huber B, Schmidt E (1936) Weitere thermo-elektrische Untersuchungen über den Transpirationsstrom der Bäume. (Further thermo-electric investigations on the transpiration stream in trees). Tharandter Forstl Jahrb 87:369–412 (1936) (Xerox copies of English translation available from National Translation Center, 35 West 33rd Street, Chicago, Illinois 60616)

Jarvis PG (1975) Water transfer in plants. In: Vries de DA, Alfen van NK (eds) Heat and mass transfer in the environment of vegetation. Scripta, Washington DC, pp 369–394

Jeje A, Zimmermann MH (1979) Resistance to water flow in xylem vessels. J Exp Bot 30:817–827

Klemm G (1956) Untersuchungen über den Transpirationswiderstand der Mesophyllmembranen und seine Bedeutung als Regulator für die stomatäre Transpiration. Planta 47:547–587

Klepper B, Browning VD, Taylor HM (1971) Stem diameter in relation to plant water status. Plant Physiol 48:683–685
Kozlowski TT (1972) Shrinking and swelling of plant tissues. In: Kozlowski TT (ed) Water deficits and plant growth III. Plant responses in control of water balance. Academic Press, London New York, pp 1–64
Kozlowski TT, Winget CH (1963) Patterns of water movement in forest trees. Bot Gaz 124:301–311
Kramer PJ (1937) The relation between the rate of transpiration and the rate of absorption of water in plants. Am J Bot 24:10–15
Kursanov AL (1957) The root system as an organ of metabolism. International conference of radioisotopes in scientific research, UNESCO/NS/RIC/128. Pergamon, New York London
Larson PR, Isebrands JG (1978) Functional significance of the nodal constricted zone in *Populus deltoides* Bartr. Can J Bot 56:801–804
Lutz HJ (1952) Occurrence of clefts in the wood of living white spruce in Alaska. J For 50:99–102
MacDougal DT, Overton JB, Smith GM (1929) The hydrostatic-pneumatic system of certain trees: movements of liquids and gases. Carnegie Inst Washington Publ 397
Maximov NA (1929) The plant in relation to water. English translation. Yapp RH (ed). Allen and Unwin, London
Milburn JA (1973a) Cavitation in *Ricinus* by acoustic detection: Induction in excised leaves by various factors. Planta 110:253–265
Milburn JA (1973b) Cavitation studies on whole *Ricinus* plants by acoustic detection. Planta 112:333–342
Milburn JA (1979) Water flow in plants. Longman, London New York
Milburn JA, Davis TA (1973) Role of pressure in xylem transport of coconut and other palms. Physiol Plant 29:415–420
Milburn JA, Johnson RPC (1966) The conduction of sap. II. Detections of vibrations produced by sap cavitation in *Ricinus* xylem. Planta 69:43–52
Milburn JA, McLaughlin ME (1974) Studies of cavitation in isolated vascular bundles and whole leaves of *Plantago major* L. New Phytol 73:861–871
Milburn JA, McLellan JA (1978) The detection of cavitation in plants at a cytological level. Proc Fed Eur Soc Plant Physiol, Edinburgh (Abstr) 369–370
Molz FJ, Klepper B (1972) Radial propagation of water potential in stems. Agron J 64:469–473
Ohtani J, Ishida S (1976) Study on the pit of wood cells using scanning electron microscopy. 5. Vestured pits in Japanese dicotyledonous woods. Res Bull College Exp For Hokkaido Univ 33:407–436
Passioura JB (1972) The effect of root geometry on the yield of wheat growing on stored water. Aust J Agric Res 23:745–752
Renner O (1915) Theoretisches und experimentelles zur Kohäsionstheorie der Wasserbewegung. Jahrb Wiss Bot 56:617–667
Richter H (1976) The water status in the plant – experimental evidence. In: Lange OL, Kappen L, Schulze E-D (eds) Water and plant life. Ecol Stud. Vol 19. Springer, Berlin Heidelberg New York, pp 42–58
Ritchie GA, Hinckley TM (1975) The pressure chamber as an instrument for ecological research. In: MacFadyen A (ed) Advances in ecological research Vol IX. Academic Press, London New York, pp 165–254
Running SW, Waring RH, Rydell RA (1975) Physiological control of water flux in conifers: a computer simulation model. Oecologia 18:1–18
Sauter JJ (1974) Maple. In: McGraw-Hill yearbook of science and technology
Scholander PF (1968) How mangroves desalinate seawater. Physiol Plant 21:251–261
Scholander PF, Hammel HT, Bradstreeet ED, Hemmingsen EA (1965) Sap pressures in vascular plants. Science 148:339–346
Skene DS, Balodis V (1968) A study of vessel length in *Eucalyptus obliqua* L'Hérit. J Exp Bot 19:825–830

Spomer GG (1968) Sensors monitor tensions in transpiration streams of trees. Science 161:484–485

Stansell JR, Klepper B, Browing VD, Taylor HM (1973) Plant water status in relation to clouds. Agron J 65:677–678

Tyree MT, Caldwell C, Dainty J (1975) The water relations of hemlock (*Tsuga canadensis*). V. The localization of resistances to bulk water flow. Can J Bot 53:1078–1084

Vité JP, Rudinsky JA (1959) The water-conducting system in conifers and their importance to the distribution of trunk-injected chemicals. Contrib Boyce Thompson Inst 20:27–38

Waring RH, Franklin JF (1979) Evergreen coniferous forests of the Pacific northwest. Science 204:1380–1386

Waring RH, Running SW (1976) Water uptake, storage and transpiration by conifers: A physiological model. In: Lange OL, Kappen L, Schulze E-D (eds) Water and plant life – problems and modern approaches. Ecol Stud Vol 19 Springer, Berlin Heidelberg New York, pp 189–202

Waring RH, Running SW (1978) Sapwood water storage: Its contribution to transpiration and effect upon water conductance through the stems of old growth Douglas fir. Plant Cell Environ 1:131–140

West DW, Gaff DF (1976) Xylem cavitation in excised leaves of *Malus sylvestris* Mill. and measurement of leaf water status with the pressure chamber. Planta 129:15–18

Zimmermann MH (1971) Dicotyledonous wood structure made apparent by sequential sections. Film E 1735. Inst Wiss Film, Nonnenstieg 72, 34 Göttingen, Germany. (Film data and summary available as a reprint)

Zimmermann MH (1978a) Vessel ends and the disruption of water flow in plants. Phytopathology 68:253–255

Zimmermann MH (1978b) Structural requirements for optimal water conduction in tree stems. In: Tomlinson PB, Zimmermann MH (eds) Tropical trees as living systems. Univ Press, Cambridge, pp 517–532

Zimmermann MH (1978c) Hydraulic architecture of some diffuse-porous trees. Can J Bot 56:2286–2295

Zimmermann MH (1979) The discovery of tylose formation by a Viennese lady in 1845. Int Assoc Wood Anat Bull 2–3:51–56

Zimmermann MH, Brown CL (1971) Trees: structure and function, 3rd edn 1977. Springer, Berlin Heidelberg New York

Zimmermann MH, Jeje A (1981) Vessel-length distribution in the stem of some American woody species. Can J Bot 59:1882–1892

Zimmermann MH, Tomlinson PB (1966) Analysis of complex vascular systems in plants: optical shuttle method. Science 152:72–73

Zweypfennig RC VJ (1978) A hypothesis on the function of vestured pits. Int Assoc Wood Anat Bull 1:13–15

6 Resistance of Plant Surfaces to Water Loss: Transport Properties of Cutin, Suberin and Associated Lipids

J. SCHÖNHERR

CONTENTS

6.1 Introduction

The driving force of transpiration is the water potential gradient between leaf cells and the surrounding atmosphere. The water potential of the atmosphere decreases rapidly with decreasing humidity and may exceed −2,000 bar. This is a hostile environment for plant cells having a water potential of about −5 to −50 bar, and in adapting to terrestrial growing conditions, aerial parts of plants have evolved water-saving devices in the form of cuticle and periderm. All primary aerial parts of higher plants and ferns are covered by a cuticle, retained in submerged leaves and stems of the higher water plants (ARBER 1920), and there are indications that at least some mosses have a cuticle (PRIESTLY 1943). In secondary organs the cuticle is replaced by a periderm.

BRONGNIART (1830) was probably the first to observe and isolate plant cuticles. Since that time cuticles have been the subject of numerous investigations. The early literature on cuticular transpiration was reviewed by STÅLFELT (1956). An extensive and comprehensive review of chemistry, structure and function of cuticles and periderms is found in MARTIN and JUNIPER (1970). Penetration of cuticles by solutes applied to the foliage has been studied extensively and was reviewed by HULL (1970). There are a number of reviews on chemistry and biochemistry of cutin, suberin, and associated waxes, some of the most recent being by HOLLOWAY et al. (1972) and KOLATTUKUDY (1975, 1980). The present review focuses on the magnitude of permeability coefficients for water transport across cuticles and periderms and on the relationship between permeability of the membranes and their chemistry and structure. For this purpose unambiguous permeability coefficients are needed. These can be obtained only by working with isolated membranes, and this review is therefore restricted to these studies. The cuticular transpiration measured using whole leaves or plants is a complex function of different permeabilities and resistances which are affected by many variables. Some of them cannot be controlled adequately when using whole leaves and uncertainties arise due to the presence of stomata and trichomes.

6.2 Theory and Experimental Methods

6.2.1 Flux Equations

Permeation of liquids or vapors across polymer membranes is usually described and analyzed in terms of the following parameters: The diffusion coefficient (D) which reflects the mobility of the species in the membrane; the solubility or the distribution coefficient (S) which reflects the concentration of the species in the membrane; the permeability coefficient (P) which reflects both mobility and solubility (CRANK and PARK 1968).

The flux (J) of any species, as given by Fick's law, is proportional to D and the driving force dC/dx where C is the concentration and x is the space coordinate in direction of flow. In the steady state

$$J=\frac{-D\,(\bar{C}_i-\bar{C}_o)}{l} \qquad (6.1)$$

where l is the membrane thickness and the $\bar{C}$'s are the concentrations of the species in the membrane just at the outer and inner membrane surfaces.

If the surface concentrations are not known Eq. (6.1) may be rewritten in terms of a permeability coefficient P:

$$J=\frac{P\,(p_i-p_o)}{l} \qquad (6.2)$$

where (p_i-p_o) is the driving force in terms of the partial pressures of the vapor at the opposite sides of the membrane and

$$P=DS, \qquad (6.3)$$

provided the membrane is homogeneous and the diffusion coefficient is independent of the driving force. The permeability coefficient often includes the membrane thickness when it is unknown or difficult to measure (SCHÖNHERR 1976a, b).

The permeability coefficient is easily obtained by measuring the steady-state flux due to a given driving force and S may be obtained from the sorption isotherm. With P and S known, D can be calculated from Eq. (6.3). Alternatively, D may be obtained from the extrapolated time lag t_e, the intersection of the straight line portion of a plot total amount diffused vs. time (CRANK and PARK 1968):

$$t_e=\frac{l^2}{6D}. \qquad (6.4)$$

Water permeation of porous membranes is often analyzed in terms of total pore area A^p or fractional pore area (A^p/A^m), where A^m is the membrane area. Equation (6.3) may than be written

$$P=\frac{D\,A^p/A^m}{l} \qquad (6.5)$$

and water permeability can be analyzed in terms of number of pores (n), average pore radius (r^p), and average path length (l) (SOLOMON 1968; SCHÖNHERR 1976a).

6.2.2 Experimental Methods

6.2.2.1 Membranes

Since the relationship between membrane structure, membrane composition, and membrane permeability is to be studied, enzymatically isolated membranes that lack stomata and trichomes must be used. Isolation with pectinase and cellulase at 35 °C does not change membrane structure (HOCH 1975) or permeability (SCHÖNHERR 1976b; Schönherr and MÉRIDA 1981). High temperatures cause irreversible changes in membrane permeability and structure (SCHÖNHERR et al. 1979) and must be avoided.

Components of cuticles will operationally be classified according to their solubility in lipid solvents such as chloroform. Nonextracted cuticles will be called cuticular membranes (CM), the soluble components will be called soluble lipids or soluble cuticular lipids (SCL) and the insoluble remainder will be called polymer matrix (MX). Layers of periderms obtained by enzymatic isolation (potato tuber periderm) or by stripping (birch trunk) are called periderm membranes (PM).

6.2.2.2 Types of Flux Measurements

Diffusion

Permeability coefficients for diffusion of water (P_d) are obtained using the system water/membrane/water with identical solutions on both sides, such that no gradient of water potential or water activity acts across the membrane. Under these conditions the only possible mechanism of water transfer is diffusion. Since there is no net water transport the diffusional flux is measured using tritiated water (THO). Under steady-state conditions the permeability coefficient is calculated from Eq. (6.6)

$$P_d = \frac{J^*}{\Delta[\mathrm{THO}]} \tag{6.6}$$

where J^* is the tracer flux (in dpm $m^{-2}\,s^{-1}$) and Δ[THO] is the concentration gradient of THO across the membrane (in dpm m^{-3}). Using these dimensions P_d has the dimension $m\,s^{-1}$ and contains implicitly the membrane thickness.

Volume Flow of Water

The volume flow of water (J_v in $m^3\,m^{-2}\,s^{-1}$) caused by a gradient of osmotic pressure (or more generally by a gradient of water potential) can be obtained using the system water/membrane/water. Again, solutions on both sides are identical, except that on one side a solute is added. In the steady state the osmotic permeability coefficient (P_f) is calculated from Eq. (6.7)

$$P_f = \frac{J_v}{\bar{v}\,\Delta C_s} \tag{6.7}$$

where $\bar{v}$ is the partial molar volume of water (in $m^3\,mol^{-1}$) and C_s ($mol\,m^{-3}$) is the concentration gradient of solute across the membrane. Using these dimensions P_f has the dimensions $m\,s^{-1}$. If a water potential gradient acts across the membrane, the mechanism of transport may no longer be diffusion alone. P_f is equal to P_d only if water molecules move individually across the membrane. However, when liquid continuity exists across the membrane (porous membranes) a gradient of osmotic pressure or water potential causes a laminar flow that increases with increasing pore size (SOLOMON 1968; SCHÖNHERR 1976a). Thus, in porous membranes $P_f > P_d$.

Transpiration

Ordinarily, the inner surface of the cuticle is in contact with the water of the epidermal wall and the outer surface is exposed to the atmosphere. This situation is mimicked by measuring water permeability using the set-up water/membrane/water vapor (SCHÖNHERR and SCHMIDT 1979; SCHÖNHERR and LENDZIAN 1981).

Permeability coefficients of transpiration (P_{tr}) have been measured using either THO (SCHÖNHERR and SCHMIDT 1979; SCHÖNHERR and MÉRIDA 1981) or by using a gravimetric method (SCHÖNHERR and LENDZIAN 1981).

Using THO as tracer J* is measured in the steady state and P_{tr} is calculated from Eq. (6.6). It should be noted that the magnitude of P_{tr} usually depends on the relative humidity on the outer surface of the cuticle (SCHÖNHERR and SCHMIDT 1979). The maximum value of P_{tr} is measured when $a_{wv}=1$ (relative humidity 100%). The P_{tr} values usually decrease with decreasing a_{wv}. The cause for this phenomenon will be discussed later.

From P_{tr} the actual transpirational flux (J_{tr}) can be calculated for any gradient of water activity ($a_w - a_{wv}$) from Eq. (6.8)

$$J_{tr} = P_{tr}\,(a_w - a_{wv}) \tag{6.8}$$

provided P_{tr} is independent of a_{wv} or the dependence is known. In Eq. (6.8) a_w is the activity of the liquid phase (usually close to 1) and a_{wv} is the water activity of the vapor phase as defined by Eq. (6.9)

$$a_{wv} = \frac{p_{wv}}{p^0_{wv}} = \frac{RH}{100} \tag{6.9}$$

where p^0_{wv} is the saturation vapor pressure at a given temperature, p_{wv} is the actual vapor pressure and RH is the relative humidity.

The gravimetric method gives J_{tr} (in kg m^{-2} s^{-1}) directly and P_{tr} is calculated from Eq. (6.10)

$$P_{tr} = \frac{J_{tr}}{\rho\,(a_w - a_{wv})} \tag{6.10}$$

where the specific gravity of water (ρ in kg m^{-3}) is needed to obtain P_{tr} in the usual dimension (ms^{-1}).

6.3 Water Permeability of Cuticular Membranes

6.3.1 Diffusion

Permeability coefficients for diffusion of water across CM are around 10^{-9} m s^{-1} (Table 6.1). One of the lowest values was found for CM from onion bulb scales (*Allium cepa*) which is also the thinnest membrane studied.

Table 6.1. Permeability coefficients for transport of water across isolated cuticular membranes from various species

Species	Organ	pH	Cation		Coefficient ($m\ s^{-1}$)	CV	Membrane thickness (m)	Number of membranes
Citrus aurantium[c]	Leaf	6.0	Ca^{2+}	P_d	1.08×10^{-9}	(72)	2.5×10^{-6}	69
Citrus aurantium[a]	Leaf	3.0	Na^+	P_d	1.01×10^{-9}	(36)	2.5×10^{-6}	27
Citrus aurantium[a]	Leaf	9.0	Na^+	P_d	1.75×10^{-9}	(44)	2.5×10^{-6}	27
Pyrus communis[a]	Leaf	5.0	Na^+	P_d	0.86×10^{-9}	(37)	2.0×10^{-6}	3
Pyrus communis[e]	Leaf	6.0	Ca^{2+}	P_{tr}	0.49×10^{-9}	(22)	2.0×10^{-6}	10
Hedera helix[e]	Leaf	6.0	Ca^{2+}	P_{tr}	0.10×10^{-9}	(35)	1.5×10^{-6}	10
Lycopersicon esc.[e]	Fruit	6.0	Ca^{2+}	P_{tr}	1.89×10^{-9}	(40)	4.0×10^{-6}	10
Solanum melongena[b]	Fruit	6.0	Ca^{2+}	P_d	2.30×10^{-9}	(31)	2.0×10^{-6}	6
Clivia miniata[f]	Leaf	6.0	Ca^{2+}	P_{tr}	0.10×10^{-9}	(23)	6.0×10^{-6}	10
Allium cepa[d]	Bulb	6.0	Ca^{2+}	P_d	0.43×10^{-9}	(43)	0.8×10^{-6}	30
Potamogeton lucens[a]	Leaf	7.0	Na^+	P_d	2.55×10^{-6}	(13)	0.2×10^{-6}	3

Membrane thickness represents rough estimates obtained either from the weight per cm^2 of membrane or from electron micrographs. CV is the coefficient of variation. Number of membranes refers to the number of membranes studied individually.
[a] SCHÖNHERR 1976b; [b] SCHÖNHERR and SCHMIDT 1979; [c] HAAS and SCHÖNHERR 1979; [d] SCHÖNHERR and MÉRIDA 1981; [e] SCHÖNHERR and LENDZIAN 1981; [f] SCHÖNHERR unpublished

The large variability (compare standard deviations) of P_d values even if membranes are taken from the same leaf, fruit, or bulb scale is a typical characteristic of CM. It has been shown that this variability is due to the soluble cuticular lipids associated with cuticles (HAAS and SCHÖNHERR 1979). Due to this variability the effects of pH, counterion, or water activity on permeability can only be studied using the same membranes for all treatments. With CM no significant effects of pH and counterions were found (SCHÖNHERR 1976b). There is no obvious correlation between P_d and membrane thickness, as predicted by Fick's law [Eq. (6.1)]. Permeability of onion CM is just as low as that of *Clivia* CM, even though the latter is more than ten times as thick. Interestingly, the CM of *Potamogeton* leaves are orders of magnitude more permeable than the CM of terrestrial organs, as the submerged leaves of *Potamogeton* do not need protection against water loss.

6.3.2 Effect of Vapor Pressure on Permeability of Cuticular Membranes

Permeability of CM depends on the vapor pressure of the atmosphere in contact with the outer surface of the CM (Fig. 6.1). The effect is most pronounced with *Citrus* CM and egg plant fruit CM, while little effect is observed with onion bulb scale CM. In fact, many CM of onion did not respond at all to a_{wv} (SCHÖNHERR and SCHMIDT 1979). In view of the large variability between membranes it must be stressed that the effect of a_{wv} on P_{tr} was investigated

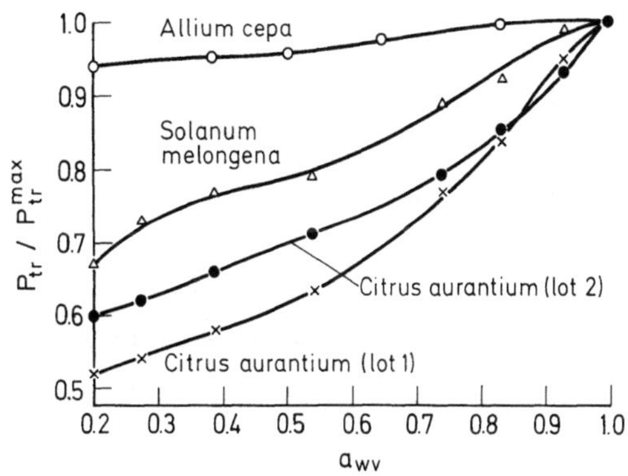

Fig. 6.1. Effect of water activity of the vapor phase (a_{wv}) on permeability coefficients of transpiration P_{tr} of cuticular membranes. P_{tr}^{max} was determined at $a_{wv}=1$. The maximum permeability coefficients were for *Citrus aurantium* 5.17×10^{-10} m s^{-1} (lot 1, Na^+-form pH 3) and 3.33×10^{-9} m s^{-1} (lot 2, Ca^{2+}-form, pH 6); *Allium cepa* 4.30×10^{-10} m s^{-1} (Ca^{2+}-form, pH 6); *Solanum melongena* 2.5×10^{-9} m s^{-1} (Ca^{2+}-form, pH 6). Values are averages of 5 to 30 membranes (Schönherr and Schmidt 1979, and unpublished results)

using the same membranes at all values of a_{wv}. Since experimental error was smaller than 3% the effect of a_{wv} on P_{tr} is highly significant.

6.3.3 Cuticular Resistances

The parameter most frequently used to characterize water permeability of cuticles is the resistance (r). The resistance is defined as driving force per unit flux, thus P^{-1} is a resistance, its dimension is s m^{-1}. Unfortunately, the driving force of transpiration which is used most frequently to calculate r is not the gradient of vapor pressure or the gradient of water activity used to calculate permeability coefficients. Instead, the gradient of absolute humidity (ΔC_{wv}) (kg water vapor per m^3 of air) is employed, and the resistance (r) is calculated from Eq. (6.11):

$$r = \frac{\Delta C_{wv}}{J_{tr}}. \tag{6.11}$$

Again, the dimension of (r) is s m^{-1} which may lead to confusion. If resistances are calculated according to Eq. (6.11) it is assumed that the gradient of water vapor concentration is the driving force of transpiration. It is further assumed that the inner side of the cuticle is in equilibrium with saturated vapor (Slatyer 1967). This implies that the liquid-vapor transition occurs prior to entry of water into the cuticle and that single individual water vapor molecules move across the cuticle.

Cuticular resistances quoted in the literature (Cowan and Milthorpe 1968) range from 2,000 to 5,000 s m^{-1} for mesophytes and 5,000 to 40,000 for xerophytes. Holmgreen et al. (1965) reported cuticular resistances of 8,500 (*Acer platanoides*), 8,300 (*Betula verrucosa*) and 38,000 s m^{-1} (*Quercus robur*).

In order to compare these conventional resistances determined using whole leaves with permeability coefficients as defined by Eq. (6.6), (6.8) and (6.10),

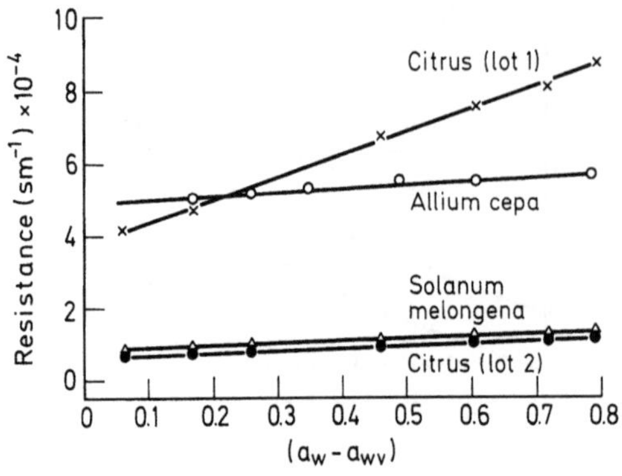

Fig. 6.2. Resistances of cuticular membranes as function of $a_w - a_{wv}$. Data taken from Fig. 6.1. See text for calculation

Table 6.2. Comparison of water permeability of cuticular membranes with permeability of artificial hydrophobic polymer membranes at 25 °C

Type of membrane			Permeability coefficient $m\ s^{-1}\ cm^{-1}$
Artificial membranes	Kel-F	(polychlorotrifluorinated ethylene)	4.12×10^{-14}
	Saran	(polyvinylidenechloride)	6.39×10^{-14}
	Manolene	(polyethylene)	8.33×10^{-14}
	Profax	(polypropylene)	24.20×10^{-14}
Cuticular membranes	*Hedera helix*	(leaf CM)	2.00×10^{-14}
	Allium cepa	(bulb scale CM)	3.44×10^{-14}
	Pyrus communis	(leaf CM)	9.80×10^{-14}
	Citrus aurantium	(leaf CM)	27.00×10^{-14}
	Solanum melongena	(fruit CM)	46.00×10^{-14}
	Lycopersicon esculentum	(fruit CM)	75.60×10^{-14}

Permeability coefficients are expressed per cm thickness of membrane. Data for artificial membranes taken from SIVADJIAN and RIBEIRO (1964), data for cuticular membranes are taken from Table 6.1

the permeability coefficients must be converted to resistances by combinding Eq. (6.8) with Eq. (6.11) which yields:

$$r = \frac{\Delta C_{wv}}{P_{tr}\,(a_w - a_{wv})}. \tag{6.12}$$

Resistances range from approximately $10^4\ s\ m^{-1}$ to $6.8 \times 10^4\ s\ m^{-1}$ (Fig. 6.2) which is in fair agreement. However, all resistances are dependent on humidity. Between 94% and 21% humidity resistances increased by a factor of 1.87 (*Citrus*, Na^+-form), 1.73 (*Citrus* in Ca^{2+}-form), 1.59 (egg plant fruit) and 1.12 (onion). It has been argued that the dependence of r on a_{wv} shows that liquid continuity exists across the cuticles and therefore C_{wv} is not the proper driving force of transpiration (SCHÖNHERR and SCHMIDT 1979).

6.3.4 Water Permeability of Cuticular Membranes as Compared to Artificial Polymer Membranes

Before attempting to analyze the chemical and physical bases of cuticular resistance to water transfer it is interesting to compare them with artificial polymer membranes of low permeability and known composition. For this purpose permeability coefficients are expressed per cm thickness of membrane. This procedure is legitimate for homogeneous polymer membranes. With cuticles, however, the calculated values should be used only for purposes of comparison, as Fick's law is not obeyed, as was pointed out already.

Water permeability of plant cuticles is of the same order of magnitude as permeability of hydrophobic synthetic polymer membranes (Table 6.2). Interestingly, the thinnest CM (*Hedera* and *Allium*) proved to be the least permeable ones. The polymer membranes included in Table 6.2 are the least permeable to water manufactured so far. Given the fact that the building materials available to plants for making a hydrophobic barrier are simply carbon and hydrogen, it appears that evolution selected the optimum solution to the problem.

6.4 Water Permeability of Periderm Membranes

Limited information is available on water permeability of isolated periderm membranes. Permeability of periderm membranes stripped off the trunks of *Betula pendula* has been investigated by SCHÖNHERR and ZIEGLER (1980). Permeability of potato tuber periderm membranes (*Solanum tuberosum*) isolated enzymatically was studied by VOGT in this laboratory (unpublished results).

6.4.1 Water Permeability of *Betula* Periderm

Permeability coefficients are proportional to the reciprocal of membrane thickness or the reciprocal of membrane weight (Fig. 6.3) as predicted by Fick's law [Eq. (6.1)]. Thus the individual periderm layers act as resistances in series and the permeability of the periderm covering the trunk can be calculated from Fig. 6.3. This periderm layer is usually about 1 mm thick, which gives a P_d of 1.6×10^{-10} m s^{-1}. This is a very high level of protection if compared with cuticular membranes (Table 6.1). However, about 3% of the total periderm area is covered by lenticels that have a much higher permeability. The problem of lenticels has been discussed elsewhere (SCHÖNHERR and ZIEGLER 1980).

The permeability coefficient of a hypothetical periderm layer having 1 cm thickness would be 1.60×10^{-11} m s^{-1} which is about two orders of magnitude higher than the permeability of cuticles (cf. Table 6.2). Thus, on a unit thickness basis birch periderm is quite permeable. Permeability coefficients depend on humidity (Fig. 6.4). The effect of a_{wv} on P_{tr} is more pronounced than with cuticular membranes.

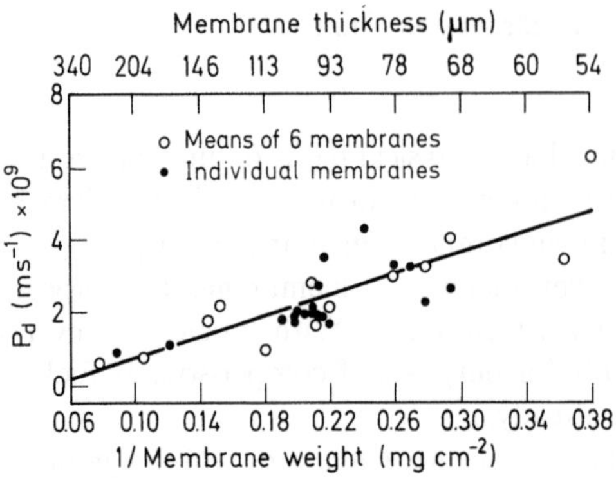

Fig. 6.3. Permeability of *Betula* periderm as affected by membrane thickness (SCHÖNHERR and ZIEGLER 1980)

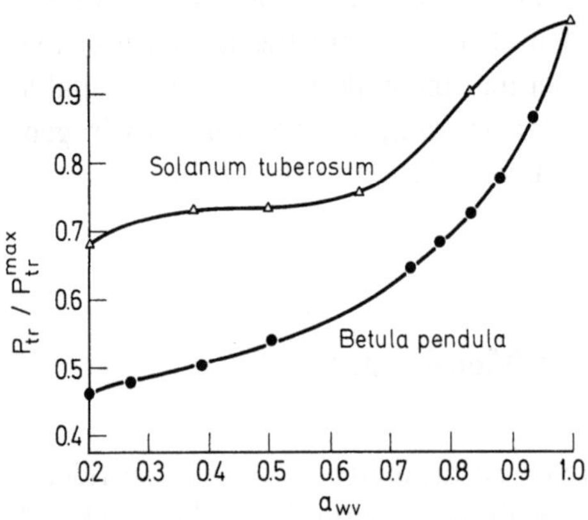

Fig. 6.4. The effect of water activity of the vapor phase (a_{wv}) on permeability coefficients of transpiration of potato tuber and birch trunk periderm. P_{tr}^{max} (determined at $a_{wv}=1$) for potato periderm 2.5×10^{-9} m s^{-1} (SD 56%, means of six membranes) and for *Betula* periderm 9.55×10^{-9} m s^{-1} (SD=39%, means of six membranes) [SCHÖNHERR and ZIEGLER 1980, and unpublished results (potato)]

6.4.2 Permeability of Potato Tuber Periderm

Potato tuber periderm layers are only three to five cell layers thick, which amounts to about 30 to 50 μm. Permeability coefficients of these membranes range from 10^{-9} to 10^{-10} m s^{-1}, providing a level of protection comparable to that of cuticular membranes. There are indications that permeability decreases rapidly after harvesting and storage in dry air. Permeability coefficients decrease with decreasing humidity (Fig. 6.4) (unpublished results).

6.5 Interpretation of Permeability Coefficients

Plant cuticles and periderms have both been shown to have an extremely low water permeability. Permeability coefficients range from 10^{-9} to 10^{-10} m s^{-1} and it appears that this is the level of protection necessary for terrestrial plant growth in temperate zones. The challenge is now to relate permeability coefficients to chemistry and structure of cuticles and periderms.

The permeability to water of a polymer membrane is proportional to its water content and to the mobility of water in the membrane [Eq. (6.3)]. It follows that the permeability coefficient for water should be higher for the more polar polymers than for the nonpolar ones. This prediction is generally observed (BARRIE 1968).

Permeability coefficients are simply proportionality coefficients that relate the fluxes to the driving forces. Information concerning the effects of structure and composition on cuticular permeability can be obtained on the basis of certain transport models. In developing these models I shall rely on studies with artificial polymer membranes, as these membranes have a lower degree of complexity, both in composition and in structure.

6.5.1 Membrane Models Based on Composition of Membranes

Polymers with low permeability to water have a number of properties in common: they consist of carbon chains with no hydrophilic substituents and there is lateral symmetry on each carbon atom in the chain (BARRIE 1968). Regularity of structure which enables close packing and the absence of polar groups are prerequisites for low water permeability. Even though crystallinity reduces water permeability, it is not essential, as amorphous polymers in the glass state can be excellent barriers (BENNET et al. 1962). High permeabilities are usually found for polar polymers and polymers having a high segmental mobility.

Polymers may be classified according to the effect of vapor pressure on permeability. Hydrophobic polymers such as polyethylene, polypropylene, and polyethylenetetraphthalate have permeability coefficients that are independent of water vapor pressure. Fluxes are proportional to the driving force, the vapor pressure gradient. Solubility is proportional to vapor pressure, that is a linear sorption isotherm is observed (YASUDA and STANNETT 1962). In these polymers solubility is extremely low. For example, in polyethylene only 2.8×10^{-5} g of water vapor are sorbed (at 25 °C and 100% humidity) per cm^3 of polymer. At all vapor pressures water permeates as individual molecules and water clusters do not form. Hence, over the entire range of p/p_o there is only water-polymer interaction and no water-water interaction in the polymer. Permeability coefficients are constant over the entire range of p/p_o (Fig. 6.5, polyethylene) and the system vapor/membrane/vapor yields the same values as the system water/membrane/water vapor (YASUDA and STANNETT 1962). These types of membranes will be referred to as solubility membranes (SCHÖNHERR and SCHMIDT 1979). If the membrane is exposed to water on one side all hydrogen bonds must be broken before the molecule can enter the membrane (Fig. 6.6). In this specific case it is permitted to use the gradient of water vapor concentration as driving force [Eq. (6.11)] in calculating permeability coefficients, as the phase transition liquid water–water vapor occurs prior to the entry of water molecules into the membrane.

In a membrane containing polar substituents Henry's law (sorption proportional to vapor pressure) is not obeyed and solubility is not directly proportional to vapor pressure. The sorption isotherm increases with vapor pressure (BARRIE 1968; YASUDA and STANNETT 1962), such that often a BET type isotherm is observed. With increasing vapor pressure the number of water molecules sur-

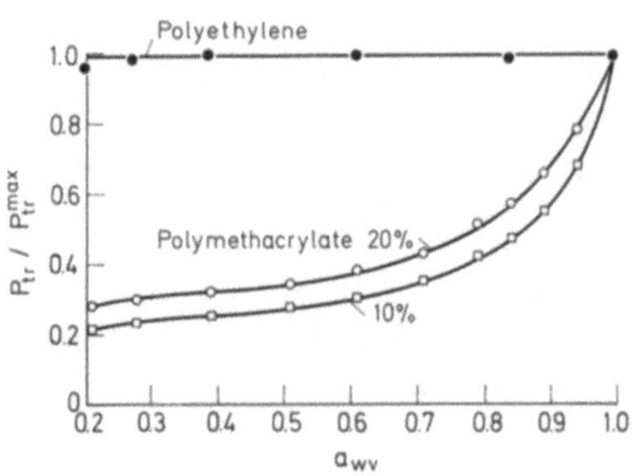

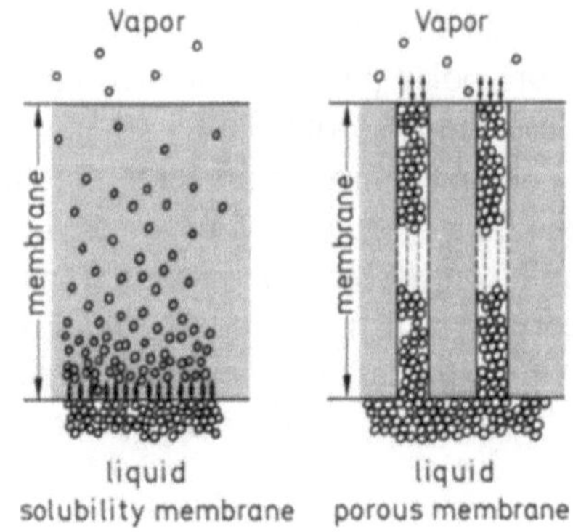

Fig. 6.5. The effect of the activity of water vapor (a_{wv}) on water permeability of polyethylene and polymethacrylate membranes in Ca^{2+}-form at pH 6. Maximum permeabilities (at a_{wv}= 1) were 6.2×10^{-11} m s^{-1} (polyethylene); 4.23×10^{-7} m s^{-1} (polymethacrylate, 20% cross-linking) and 4.30×10^{-7} m s^{-1} (10% cross-linking) (SCHÖNHERR and ZIEGLER 1980)

Fig. 6.6. Transport models for lipid membranes. *Model I* solubility membrane. *Model II* porous membrane. Liquid/vapor interface indicated by *double arrows* (SCHÖNHERR and SCHMIDT 1979)

rounding the polar groups increases and as a consequence both diffusion and permeability coefficients usually increase with increasing water content of the polymer (BARRIE 1968).

In polar polymers water sorption leads to the formation of water clusters or to continuous water-filled channels, provided the polar groups are sufficiently close to each other. Membranes of this type, having permeability coefficients that depend on vapor pressure (such as polymethacrylate in Fig. 6.5) have been termed porous membranes (SCHÖNHERR 1976a; SCHÖNHERR and SCHMIDT 1979). In these membranes water continuity exists across the entire membrane when it is exposed to water on one side and vapor on the other and the liquid/vapor transition takes place at the membrane/vapor interphase. Therefore ΔC_{wv} is not the driving force of water across the membrane (Fig. 6.6).

6.5.2 Membrane Models Based on Structure of Membranes

The membrane matrix may be amorphous or crystalline. Within a given membrane both amorphous and crystalline regions may occur. The presence of crystallites reduces the cross-sectional area for diffusion and increases the effective path length.

Analysis of transport in heterogeneous membranes is difficult (BARRER 1968). In the simplest case water permeability of polymers containing crystalline regions (nylon, polyethylene-tetraphthalate) was analyzed using the equation

$$P = P_a \, v_a \, k \tag{6.13}$$

where P_a and v_a are the permeability coefficient and the volume fraction of the amorphous phase and k is a structure factor which is a function of v_a

(LASOSKI and COBBS 1959; KLUTE 1959a, b). In this case the crystalline phase is practically impermeable to water and is dispersed within a continuous amorphous phase.

A heterogeneous membrane may also consist of two (or more) interpenetrating phases which are continuous, or they may consist of laminates of the type AB, ABA, ABAB etc. when they consist of only two phases. For this latter case explicit relations between diffusion coefficients of the two phases and the hold-up time have been worked out for the special case of D being independent of vapor pressure and a linear sorption isotherm (BARRER 1968). For all other cases theory is incomplete and rather involved (see discussion by BARRER 1968).

In a heterogeneous membrane consisting of alternating layers of A and B normal to the direction of flux (resistances in series) the total permeability of a membrane P of thickness l is related to the permeabilities and thicknesses of the laminates by the equation

$$\frac{l}{P}=\frac{l_A}{P_A}+\frac{l_B}{P_B}+\cdots \tag{6.14}$$

provided D is independent of vapor pressure and the sorption isotherm is linear (BARRER 1968).

6.5.3 Chemistry and Structure of Cuticular Membranes

According to these models analysis of cuticular permeability to water must focus on the following points: (1) chemistry of cuticles and their structure, (2) arrangement of components of cuticles within the membrane, (3) mobility of water within the components and solubility of water in the components. A complete analysis of cuticular permeability cannot be given at the present time because water vapor sorption of cuticles (or periderms) has not been studied so far. Therefore D and S have not been studied independently and only the product of both, namely P, has been measured. The discussion can therefore only be a qualitative one.

6.5.3.1 Chemistry of Cuticles

The polymer matrix makes up the bulk of the CM. Unfortunately, little is known concerning the intact polymer matrix, as practically all investigations focused on the degradation products obtained by hydrolysis or hydrogenolysis (MARTIN and JUNIPER 1970; KOLATTUKUDY 1975, 1980). The insoluble fraction of cuticles is never pure cutin, it also contains various nonlipid components such as cellulose, polyuronic acids, proteins, and phenolic compounds (MEYER 1938; ROELOFSEN 1952; HUELIN 1959; MARTIN and JUNIPER 1970; KOLATTUKUDY 1975, 1980; SCHÖNHERR and HUBER 1977; HUNT and BAKER 1980). In isolated cuticles the cutin ranges from 20 to 1,030 μg cm^{-2} or from 20% to 84% by weight (HOLLOWAY et al. 1972; BAKER and PROCOPIOU 1975). According to HOLLOWAY et al. (1972) plant cutins may be classified depending on the major monomer acid found after hydrolysis. They distinguish the hexadecanoic acid, the dihydroxyhexadecanoic, and the trihydroxyhexadecanoic acid type. Since cutin monomers are cross-linked by ester, peroxide, and ether bridges (CRISP 1965) the presence of high amounts of di- and trihydroxyfatty acid may affect the degree of cross-linking and therefore swelling in water. However, this relationship has not been investigated.

The soluble cuticular lipids (SCL or so-called waxes) are a very complex mixture, comprising from 6 to 380 μg cm^{-2} or from 2% to 30% by weight (NORRIS 1974; BAKER and PROCOPIOU 1975; BAKER et al. 1975; HAAS and SCHÖNHERR 1979). Of the CM used in permeability studies only *Citrus* leaf CM has been quantitatively investigated so far (HAAS and SCHÖNHERR 1979) and this contains fatty acids (C_{16} to C_{26}), fatty alcohols (C_{26} to C_{34}), esters (C_{38} to C_{44}) and paraffins (C_{16} to C_{33}). The total amount of SCL was about 10 μg cm^{-2}, or 3% by weight. With respect to water permeability it must be stressed that all SCL have long hydrophobic chains, the major homologs have no side chains, and some have a polar functional group.

Water permeability of lipids that form monolayers has been investigated (LAMER et al. 1964; GAINES 1968). The resistance to evaporation of a stearic acid or hexadecanol monolayer at high surface pressures is about 20 s m^{-1}. Interestingly the resistance is not a linear function of chain length. Rather, it is the logarithm of the resistance that increases linearly with chain length, such that for every 2 $-CH_2$ groups added resistance doubles (ARCHER and LAMER 1955; LAMER et al. 1964). For example, the resistance of monolayers of fatty alcohols 16, 18, 20, and 22 carbon atoms long is 20, 40, 80, and 160 s m^{-1}, respectively. Thus, a monolayer of a C_{22}-alcohol has the same resistance as eight monolayers of hexadecanol in series. To build the same resistance only 0.22 μg cm^{-2} are needed in the first case and 13 μg cm^{-2} in the second. This fact provides a very interesting hint as to the selective advantage of using long chain lipids in the cuticle to build a transpiration resistance. For the log relationship to be effective it is necessary that the SCL molecules are oriented parallel to the normal of the membrane. This type of arrangement has been proposed for CM from studies using polarized light (MEYER 1938; ROELOFSEN 1952).

6.5.3.2 Structure of Cuticular Membranes

Chemical analysis shows cuticles to be heterogeneous membranes. Studies into the fine structure of CM using polarized light or electron microscopy essentially confirm this. In describing the fine structure of cuticles the terminology proposed by ROELOFSEN (1952), SITTE and RENNIER (1963) and WATTENDORFF and HOLLOWAY (1980) will be used. The outer epidermal wall consists of cuticular membrane and cell wall. In many plant species the CM is composed of a thin, outer cuticle proper (CP) and a thicker cuticular layer (CL). The CM of *Clivia* leaves (Figs. 6.7–6.8) and onion bulb scales (Fig. 6.9) belong to this class. In other species, however, such a distinction is not possible, for example in *Citrus* CM (Fig. 6.10), where a fine electron-dense reticulum or fibrillae extend throughout the entire CM.

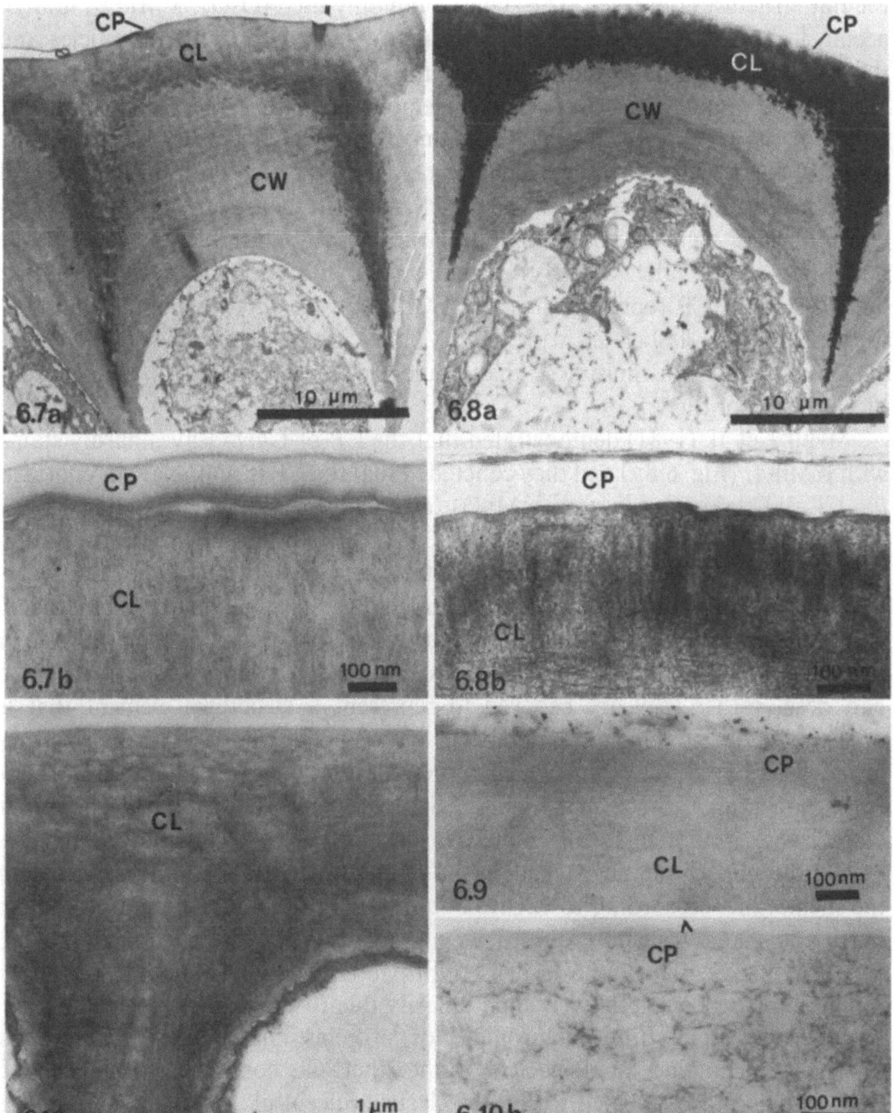

Figs. 6.7–6.10. Electron micrographs of plant cuticles. The upper cuticular membrane of *Clivia miniata* leaves fixed with glutaraldehyde and osmium tetroxide (**6.7**) and with glutaraldehyde and potassium permenganate (**6.8**); **6.7b** and **6.8b** are high magnification images of the cuticle proper. **6.9** Cuticular membrane of the inner bulb scale of *Allium cepa* fixed with glutaraldehyde and osmium tetroxide. **6.10** Cuticular membrane (**6.10a**) and the outer layer of the CM (**6.10b**) of the upper surface of *Citrus aurantium* leaves fixed with glutaraldehyde and osmium tetroxide. *CP* cuticle proper; *CL* cuticular layer; *CW* cell wall

With regard to permeability the key problem is the chemical identity of the fine structures observed. The stains commonly used (OsO_4, $KMnO_4$, uranylacetat, lead citrate) are not specific for any of the components of cuticles and the problem cannot be solved at the present time. Using a histochemical reaction, WATTENDORFF and HOLLOWAY (1980) demonstrated that the electron opaque layers of the CP of *Agave* leaves and most of the CL consist of cutin. The fibrils of the CL gave a polysaccharide reaction but it could not be precluded that other substances present in the cuticle may also contribute to the electron-dense fibrillae seen following osmium fixation and staining with other heavy metals. Indirect evidence was presented to the effect that the electron-lucent lamellae of the CP represent layers of SCL. However, it was pointed out that most of the SCL associated with the CM must be located elsewhere, as *Agave* leaves contain much more SCL than could be accommodated in the electron-lucent layers of the CP.

MÉRIDA et al. (1981) demonstrated that the CP of *Clivia* leaves hardly stains with $KMnO_4$ (Fig. 6.8a) and they concluded that the CP is nonpolar. In contrast, the CL did stain heavily with $KMnO_4$, especially in young, expanding leaves, showing it to be more polar than the CP (Fig. 6.8). Following post-fixation with osmium and staining with heavy metals, the CP showed a lamellar appearance (Fig. 6.7b). More lamellae were observed if the specimen had been extracted with chloroform at room temperature prior to fixation with osmium, and it was suggested that the electron-lucent layers are not simply voids where SCL had been, rather they are believed to represent a structural element that is nonextractable; possibly a nonpolar type of cutin. The thickness of the electron-lucent layers is about 5 nm, which represents the length of chain with about 35 carbon atoms. Since many of the SCL are of this length, this nonpolar cutin layers would be ideally suited to serve as a matrix within which the SCL molecules could be packed in an orientation parallel to the normal of the membrane. According to MÉRIDA et al. (1981) the CP represents the main barrier to water transport due to its nonpolar nature, even though the CL is much thicker. The electron-dense fibrillae in the CL were interpreted to represent channels of relatively high polarity due to the presence of polar functional groups from cutin, polyuronic acids, proteins, and cellulose. These polar channels are not incrusted by SCL, rather the SCL contained within the CL were assumed to be located within the electron-lucent globules of the CL.

This picture of the cuticle of *Clivia* tallies with earlier results of studies using polarized light (MEYER 1938; ROELOFSEN 1952). The CL of *Clivia* leaves is anisotropic and appears negatively birefringent when cross-sections are viewed with polarized light. Birefringence disappeared on heating to 90 °C and reappeared on cooling. Extraction with lipid solvent removed birefringence irreversibly. MEYER (1938) concluded that negative birefringence was due to SCL embedded in the CL. The CL exhibited form double refraction and it was suggested that the SCL are contained in tangentially oriented pores of the cutin, with the long axes of the molecules arranged parallel to the normal of the membrane. MEYER (1938) found the CP of *Clivia* to be isotropic, when cross-sections were viewed. However, using very thin sections and high magnification ROELOFSEN (1952) found the CP to be weakly negatively birefringent. After extraction with chloroform or pyridine the CP was weakly positively birefringent. This positive double refraction could not be eliminated by heating or imbibition with solvents matching the refractive index of the CP. Hence, it cannot be form birefringence and was attributed to a nonextractable structural element of the CP. The CP does not contain cellulose and it is most likely that the positive double refraction is due to cutin. Birefringence due to embedded SCL has been observed with all CM investigated so far (Frey-Wyssling 1959; Roelofsen 1959). Sitte and Rennier (1963) investigated a wide variety of cuticles using polarized light, staining techniques, and electron microscopy. In most species the outermost layer of the CM was negatively birefringent and the innermost layer was positive. In thick CM (*Agave americana, Ficus elastica, Picea excelsa, Prunus laurocerasus*) more than two layers differing in optical properties were observed. The negative birefringence of the outer layer of the CM was attributed to embedded crystalline SCL, as it disappeared on heating and on extraction with hot pyridine. In many instances, a residual birefringence was found even in extracted cuticles imbibed with suitable solvents. This form birefringence was attributed to the presence of voids in the polymer matrix orientated parallel to the surface of the cuticles. In very young leaves of *Ficus elastica* form birefringence was found even in nonextracted cuticles which disappeared during development. Possibly, SCL are deposited in preformed voids in the cutin during leaf evelopment. Occasionally, the authors observed a weak birefringence due to cutin itself.

Lamellation of the CP has been observed in the CP of other species as well (Chafe and Wardrop 1973; Jarvis and Wardrop 1974; Sargent 1976; Heide Jørgensen 1978; Wattendorff and Holloway 1980). However, with many other species lamellae could not be found (see Wattendorff and Holloway 1980 for a discussion). Possibly not all cuticles have a lamellar CP, or standard procedures of sample preparation are not suited to reveal the lamellar character of some types of cuticles. Lamellae could not be detected in the CM of *Citrus aurantium* (Figs. 6.10a and b) by transmission electron microscopy. On the other hand, Eckl and Gruler (1980) found lamellae in *Citrus* leaves by a freeze fracturing technique. Freeze fracturing revealed lamellae in the CP of tobacco leaves, *Tilia* buds and hairs from ovules of cotton (Wilson 1980).

The lamellar structure of *Citrus* CM was confirmed by electron spin resonance (ESR) spectroscopy (Eckl and Gruler 1980). At ambient temperatures the hydrophobic SCL occur in distinct layers orientated parallel to the membrane surface and the long axis of the molecules is also parallel to the surface. The soluble lipids are in the solid state. The hydroxy fatty acid monomers of cutin are not randomly orientated, rather they have a preferred orientation normal to the surface below 16 °C, and between 16° and 38 °C they are slightly tilted.

This brief discussion shows that the fine structure of cuticles is not well understood at the present time. Polarized light microscopy clearly shows that SCL occur in a highly ordered state. This was confirmed by ESR-spectroscopy, which also demonstrates a considerable degree of order within cutin (Eckl and Gruler 1980). However, the exact location and orientation of the SCL within the cuticle is not yet known.

There is also conflicting evidence concerning the fine structure of extracuticular lipids (also called surface waxes) observed on the CM of most plant species (MARTIN and JUNIPER 1970). These extracuticular lipids usually exhibit a characteristic crystalline appearance. Using a model system JEFFREE et al. (1975) demonstrated that the type of crystallites formed depends on the composition of the extracuticular lipids. Recently, RENTSCHLER (1979) argued that the crystallites are preparation artifacts. The role of extracuticular lipids in permeability is unknown as it is not possible to remove them without disturbing intracuticular SCL.

6.5.4 Transport Properties of Polymer Matrix Membranes

6.5.4.1 Ion Exchange

The polymer matrix contains a considerable amount of dissociable groups; that is it contains strong dipoles. The MX is a polyelectrolyte with isoelectric point around 3 (SCHÖNHERR and HUBER 1977). Above pH 3 the MX carries a net negative charge due to dissociable $-COOH$ groups contributed by cutin, polyuronic acids, and proteins (SCHÖNHERR and BUKOVAC 1973). The total exchange capacity of polymer matrix obtained from various leaves and plants ranged from 0.25 to 0.50 Eq kg^{-1}. The MX from tomato fruit contains large amounts of dissociable $-OH$ groups from phenolic compounds covalently bound to the polymer matrix (SCHÖNHERR and BUKOVAC 1973; HUNT and BAKER 1980).

The $-COOH$ groups have a pronounced selectivity for Ca^{2+} over monovalent alkali metal ions. Thus, in vivo the cuticle is probably in the Ca^{2+}-form. The high selectivity for divalent ions indicates that the fixed charges are not randomly distributed. They may form clusters or bands. In fact, the MX usually shows a reticulated structure when viewed with the EM and stained with heavy metals (HOCH 1975; MAIER 1968). However, distribution of charges across the matrix appears to be uniform as asymmetry potentials were small or nonexistent (SCHÖNHERR and HUBER 1977).

In addition to $-COOH$ and $-OH$ groups all cuticles investigated contained $-NH_3^+$ groups contributed by basic amino acids of protein(s) contained in the MX. The fixed positive charges amount to 12 to 25 mEq. kg^{-1}. The identity of the protein is unknown but possibly it is the enzyme(s) implicated in cutin biosynthesis (KOLATTUKUDY 1980).

6.5.4.2 Water Permeability of Polymer Matrix Membranes

The polymer matrix contains polar functional groups, some of which are dissociable. As a consequence permeability of MX-membranes depends on pH and counterions (SCHÖNHERR 1974, 1976a) and on water activity (SCHÖNHERR and SCHMIDT 1979). Permeability coefficients of MX-membranes are around 10^{-7} m s^{-1} (Table 6.3), that is, they are one to three orders of magnitude larger than those of CM.

Table 6.3. Water permeability of polymer matrix membranes and the effect of extraction of soluble cuticular lipids on water permeability

Species		P_{tr} (MX) $m\ s^{-1}$	P_{tr} (MX)/P_{tr} (CM)
Lycopersicon esculentum	fruit[a]	3.76×10^{-8}	20
Clivia miniata	leaf[c]	1.66×10^{-8}	171
Citrus aurantium	leaf[a]	15.60×10^{-8}	188
Pyrus communis	leaf[a]	10.80×10^{-8}	218
Hedera helix	leaf[a]	2.65×10^{-8}	265
Allium cepa	bulb scale[b]	66.30×10^{-8}	1542

[a] SCHÖNHERR and LENDZIAN 1981; [b] SCHÖNHERR and MÉRIDA 1981;
[c] SCHÖNHERR unpublished

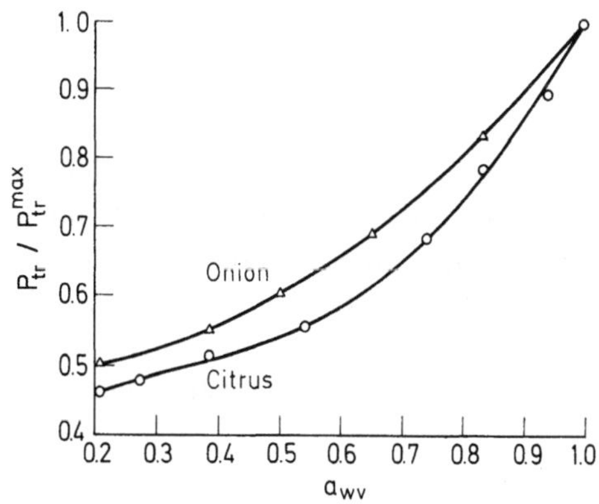

Fig. 6.11. The effect of the water activity of the vapor phase on permeability coefficients of polymer matrix membranes in Ca^{2+}-form at pH 6. Maximum permeability (at $a_{wv}=1$) of *Citrus* MX was $1.55 \times 10^{-7}\ m\ s^{-1}$ and of onion MX was $8.55 \times 10^{-7}\ m\ s^{-1}$. (SCHÖNHERR and SCHMIDT 1979, and SCHÖNHERR and MÉRIDA 1981)

Water permeability of MX-membranes in Na^+-form increases with increasing pH while in the Ca^{2+}-form pH has little effect (SCHÖNHERR 1976b). Apparently, the polarity of the –COOH group is similar to its Ca^{2+}-salt. Permeability also depends on ionic strength: P_d increases with increasing NaCl concentrations up to 0.1 M; higher salt concentrations reduce P_d due to deswelling of the MX (SCHÖNHERR 1974).

Polymer matrix membranes are porous membranes due to the presence of polar functional groups. The dependence of P_{tr} on a_{wv} is pronounced (Fig. 6.11) but it is not as strong as for polymethacrylic acid (Fig. 6.5), which has many more polar groups. Furthermore, increasing the degree of cross-linking decreases P_{tr} and the effect of a_{wv} on P_{tr} (Fig. 6.5). From the ratio P_f/P_d the average pore radius was estimated to be 0.45 nm and 0.41 nm for *Citrus* and onion MX, respectively (SCHÖNHERR 1976a, b). Thus, the pores are quite narrow, only about three water molecules or one glucose molecule can be accommodated in the cross-section.

The effects of humidity, pH, and counterions on permeability of MX-membranes may be due to their effects on water content and water mobility [Eq. (6.3)]. At the present time it is not possible to decide to what degree S and D are affected by the experimental variables, as vapor sorption isotherms have not been determined. A quantitative analysis of water permeability of MX-membranes in terms of S and D is therefore not possible at the moment.

6.5.5 Transport Properties of Soluble Cuticular Lipids and Cuticular Membranes

Transport properties of SCL cannot be determined separately. Instead they must be deduced from comparison of CM with MX-membranes. As shown, good evidence exists that cuticular membranes are laminated. Extraction of SCL from CM increases water permeability by a factor of 20 to 1542, even though SCL make up only a small percentage of the total mass. It is not possible to explain the effect of SCL on permeability by assuming the polymer matrix to be a continuous phase within which domains of SCL exist that are crystalline and practically water-impermeable, as implied by Eq. (6.13). In addition, such a model could not explain the differences found between CM and MX-membranes in their response to a_{wv} (Figs. 6.1 and 6.11). The alternative model – resistances arranged in series – [Eq. (6.14)] agrees with both structural evidence and transport behavior. The permeability of CM is completely determined by the layer(s) of the crystalline, hydrophobic SCL (SCHÖNHERR 1976a; SCHÖNHERR and SCHMIDT 1979). The assumption of a simple bilaminate membrane (SCHÖNHERR 1976a) of the type MX-SCL is probably too simple in view of the EM evidence. However, the conclusions drawn from this model are still valid for a multilaminated membrance since

$$\frac{1}{P(SCL)_{total}}=\frac{l_1}{P(SCL_1)}+\frac{l_2}{P(SCL_2)}+\cdots. \qquad (6.15)$$

However, only $P(SCL)_{total}$ can be estimated because SCL layers cannot be removed successively. At least in some cases the SCL layers seem to develop successively and centripetally (SARGENT 1976) and they may differ in composition and permeability. If large differences in composition exist, the individual SCL layers should have different transition temperatures and the Arrhenius plot ln P vs. 1/T should bear this out (Fig. 6.12). However, only one transition is found in the temperature range of 5° to 65 °C for *Citrus* and onion CM at 44° and 48 °C, respectively. In contrast, no sudden increase in water permeability of MX-membranes is found (Fig. 6.11). In the Ca^{2+}-form ln P of MX-membranes departs from the straight line at higher temperatures as the temperature-dependent increase in permeability slows down. This may be due to a lower water content of the MX at higher temperatures. The fact that only one transition was found in *Citrus* CM has been taken as indication that only one SCL layer exists or that (in case more than one SCL layer should exist) all layers have the same composition and transition temperature (SCHÖNHERR et al. 1979).

More recent work has shown that the situation is much more complicated. On the basis of combined calorimetry, ESR-spectroscopy and by measuring thermal expansion of CM, ECKL and GRULER (1980) found phase transitions

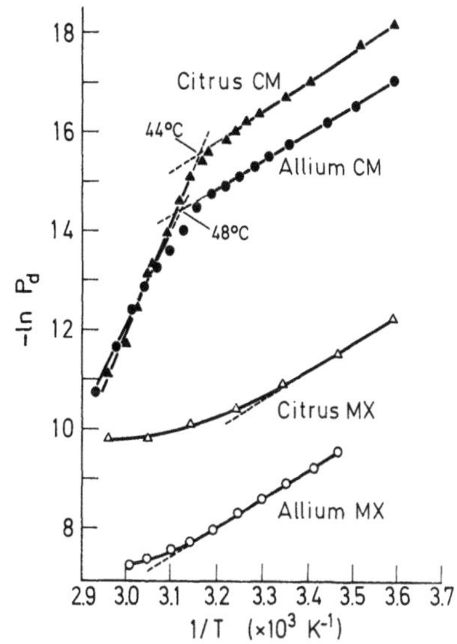

Fig. 6.12. The effect of temperature on water permeability of *Citrus* and onion CM and MX-membranes in Ca^{2+}-form. (SCHÖNHERR et al. 1979 and SCHÖNHERR and MÉRIDA 1981)

at 16 °C (MX), 38 °C (SCL), and 44 °C (MX). The sudden increase in water permeability of *Citrus* CM is probably due to changes in properties of both polymer matrix and SCL. The transition of the SCL at 38 °C is probably of the type solid/liquid crystalline. Between 38 °C and 44 °C the average molecular orientation of the SCL is still perpendicular to the normal of the membrane. Above 44 °C there is a change in molecular orientation of SCL. The main orientation is now parallel to the normal and the sudden increase in water permeability is probably due to the higher mobility of the barrier and possibly to the formation of polar holes in it. The change in orientation caused by temperatures higher than 44 °C is not reversible on cooling, even though the SCL become solid again. This is in good agreement with the observation that water permeability is irreversibly increased by temperatures above 44 °C (SCHÖNHERR et al. 1979; SCHÖNHERR and MÉRIDA 1981). There is no lipid loss from the CM by high temperatures, provided the membranes are in the Ca^{2+}-form. From membranes in the Na^{+}-form fatty acids are lost at high temperatures (HAAS and SCHÖNHERR 1981).

The results obtained with onion CM are quite similar to those with *Citrus*, except that the transition occurred at 48 °C. Unfortunately, additional information needed for interpretation (SCL amounts and composition, calorimetry ESR-spectra etc.) is not available at the present time.

The difference between *Citrus* and onion cuticular membranes must be stressed. In *Citrus* the effect of a_{wv} on P_{tr} is similar for both CM and MX-membranes. It is clear that the barrier formed by *Citrus* SCL has polar pores. In *Citrus* SCL fatty acids and alcohols dominate (HAAS and SCHÖNHERR 1979). In contrast, water permeability of onion CM responded little or not at all to changes in a_{wv}, indicating a much more hydrophobic barrier lacking polar functional groups. In fact, preliminary data indicate that alkanes amount to 70% of the total SCL in onion CM (HAAS personal communication). Interestingly, comparing permeability and composition of *Citrus* CM of different years, HAAS and SCHÖNHERR (1979) found permeability to be highest in years when "polar" SCL (fatty acids and alcohols) were highest and alkanes were lowest. They also pointed out the need to study both permeability and composition on the same membranes, otherwise differences will not be statistically significant.

In spite of these difficulties encountered in interpretation two facts have clearly emerged: (1) soluble cuticular lipids determine water permeability of CM. This can be attributed to their long hydrophobic chains which are unbranched and highly oriented. Future research must show to what extent the type of SCL and their amounts affect permeability. The interaction between cutin and SCL is another problem that must be solved. (2) There is no relationship between thickness of CM and permeability. There are a number of indications that the cuticle proper is the main barrier to water and that the cuticular layers are relatively permeable.

6.5.6 Transport Properties of Periderm Membranes

Periderm membranes are cellular membranes. Secondary walls of the individual cells are free of cellulose and consist of suberin, polyphenols, and soluble lipids (SITTE 1955, 1962, 1975; WATTENDORFF 1974; FALK and EL-HADIDI 1961; HOLLOWAY et al. 1972; KOLATTUKUDY 1975, 1980). The secondary walls are birefringent and show numerous light bands in the EM, believed to represent soluble lipids (SITTE 1975; FALK and EL-HADIDI 1961; SOLIDAY et al. 1979). Thus the secondary walls of phellem cells are the barriers to water transport and the role of the hydrophobic lamellae has been pointed out (SITTE 1975). Phellem cells are connected by their mutual middle lamellae and primary walls.

From the point of view of water permeability the key questions are: What is the composition of middle lamellae and primary walls? Are they polar or apolar?

In *Betula* periderm middle lamellae and primary walls are very polar, they stain readily with toluidine blue and if $AgNO_3$ is added as tracer of water flux, silver precipitates form abundantly in middle lamellae and primary walls but not in secondary walls (SCHÖNHERR and ZIEGLER 1980). It was concluded that in *Betula* periderm water flows across the network formed by middle lamellae and primary walls. Consistent with this view are the following observations: extraction with chloroform causes a weight loss of 50% but has little effect on water permeability; P_{tr} strongly depends on a_{wv} and this dependence is little altered by solvent extraction; no phase transition was found between 5° and 65 °C (SCHÖNHERR and ZIEGLER 1980). In *Betula* periderm tangential secondary walls contain numerous pores in radial direction believed to be the remains of plasmodesmata. They are accessible to water as they always contain silver precipitates (SCHÖNHERR and ZIEGLER 1980).

The lumina of phellem cells are probably filled with water when the membranes are submerged in water.

If subjected to dry air the lumina will drain (SCHÖNHERR and ZIEGLER 1980), the secondary hydrophobic walls will function as excluded volume [Eq. (6.13)] and water will flow through the hydrophilic middle lamellae and primary walls. Since per unit length birch periderm is quite water-permeable, a relatively thick layer (1 mm) is needed to build a barrier that provides sufficient protection.

A similar analysis cannot be given for potato tuber periderm for lack of detailed information. It is clear, though, that potato tuber periderm differs in many respects from birch periderm. The effect of a_{wv} on P_{tr} is smaller and chloroform extraction increases water permeability markedly. This observation agrees with a recent report according to which water permeability of potato wound periderm depends on synthesis of alkanes and alcohols but not on synthesis of suberin (SOLIDAY et al. 1979). Attempts to trace the path of water across potato tuber periderm failed; no silver could be found anywhere, not even after 5 days of exposure at 40 °C. This shows that middle lamellae and primary walls differ markedly from those of *Betula* periderm. The secondary walls also differ in that no plasmodesmata could be found. In flowing across potato tuber periderm water must cross numerous laminated and hydrophobic barriers arranged in series. This agrees with a observed hold-up time of 6 to 12 h. A detailed analysis must await further experimentation.

6.6 Concluding Remarks

6.6.1 The Problem of Adaptation to Water Stress

Most xerophytic plant species have much thicker cuticles than mesophytic, hygrophytic, or even hydrophytic ones. This observation is probably responsible for the common belief that plants respond to water stress with the formation of thick cuticles that reduce passive water loss from the plant. Data do not support this belief. The thin CM of onion bulb scales, pear or ivy leaves are just as efficient water barriers as the thick *Clivia* CM. Water permeability coefficients for CM or periderm membranes all range from 10^{-9} to 10^{-10} m s^{-1}, and it appears that this is the level of protection necessary for terrestrial plant growth. Thus, water permeability coefficients do not differ greatly between species. The real difference lies in the way plants accomplish this goal. Some species build a very thin CM (*Allium, Pyrus, Hedera*), others a rather thick one (*Clivia, Lycopersicon*). By the same token, the thin potato tuber periderm is just as efficient a water barrier as the very heavy *Betula* trunk periderm. Why different plants have adopted such widely different strategies is not known. It is clear that SCL represent the major water barrier. However, the amounts and composition of SCL differ widely among species that differ only little in water permeability such as *Allium, Clivia,* and *Citrus*. Also, environmental conditions are known to affect composition of SCL but at the present time it is not known whether or not water permeability is also affected. Combined studies of lipid composition and water permeability have not been performed. At the present time there is insufficient evidence to show that there is an adaptation to water stress at the level of the cuticle.

6.6.2 Relationship Between Permeability, Composition, and Structure of Cuticles

Water permeability of cuticles of terrestrial plants is very low. As water barriers, cuticles compare favorably with the least permeable artificial polymeric membranes known.

Cuticles are heterogeneous membranes. They consist of a polymer matrix that is insoluble in lipid solvents, and of soluble lipids. Evidence is presented to show that the components are arranged in laminas normal to the direction of water flow.

The polymer matrix represents the bulk of the total mass of the membranes studied. It has a relatively high water permeability due to the presence of strong dipoles which are hydrated in the presence of water. Therefore, water permeability of polymer membranes depends on water activity.

Water permeability of cuticles is determined by the soluble lipids associated with the polymer matrix. Soluble lipids amount only to a small fraction of the total membrane mass, therefore water permeability of cuticles is not proportional to membrane thickness or weight. The low water permeability of the lamina formed by soluble cuticular lipids can be mainly attributed to two facts: (a) they have long hydrophobic chains and (b) they occur as crystallites at ambient temperatures. In some types of cuticles the polar functions of the soluble lipids appear to form continuous polar channels.

It is suggested that the main function of the polymer matrix is to provide a mechanically (and possibly microbiologically) resistant matrix for the soluble lipids to be embedded in. While it is clear that the soluble lipids determine water permeability, two important questions remain unanswered: (a) What is the role of the numerous different kinds of soluble lipids ordinarily found in cuticles? (b) What is the relationship between the amounts of soluble lipids and water permeability?

Only when these questions have been answered can an attempt be made to analyze the role of cuticles in adaptation to water stress.

Water permeability of potato tuber periderm is also quite low. However, the structural and chemical bases of water permeability cannot be analyzed at the present time for lack of detailed information.

Betula trunk periderm has a high water permeability on a unit thickness basis. However, the periderm layer is thick, thus providing a level of protection against water loss comparable to that of cuticles. In *Betula* periderm middle lamellae and primary walls are not incrusted with lipophilic substances and they are the main channels for water flow.

Acknowledgements. An early version of the manuscript was reviewed by Drs. K. Lendzian (Munich), C.B. Osmond (Canberra), P. Sitte (Freiburg) and H. Ziegler (Munich). I greatfully acknowledge their helpful advice. Electron micrographs have been prepared by Drs. T. Mérida (Caracas) and W.H. Schmidt (Munich) to whom I am much indebted.

References

Arber A (1920) Water plants. Univ Press, Cambridge

Archer RJ, LaMer VK (1955) The rate of evaporation of water through fatty acid monolayers. J Phys Chem 59:200–208

Baker EA, Procopiou J (1975) The cuticle of *Citrus* species. II. Composition of intracuticular lipids of leaves and fruits. J Sci Food Chem 26:1347–1352

Baker EA, Procopiou J, Hunt GM (1975) The cuticle of *Citrus* species. Composition of leaf and fruit waxes. J Sci Food Chem 26:1093–1101

Barrer RM (1968) Diffusion and permeation in heterogeneous media. In: Crank J, Park GS (eds) Diffusion in polymers. Academic Press, London New York, pp 165–217

Barrie JA (1968) Water in polymers. In: Crank J, Park GS (eds) Diffusion in polymers. Academic Press, London New York, pp 259–313

Bennett B, Cooper CW, Luttinger M (1962) The preparation and gas permeability of vinylidene chloride copolymer films. J Appl Polym Sci 6:S23–S24

Brongniart A (1830) Ann Sci Nat Bot 1 Ser 21:420– (quoted from Martin and Juniper 1970)

Chafe SC, Wardrop AB (1973) Fine structural observations on the epiderms. II. The cuticle. Planta 109:39–48

Cowan IR, Milthorpe FL (1968) Plant factors influencing the water status of plant tissues. In: Kozlowski TT (ed) Water deficits and plant growth Vol I. Academic Press, London New York, pp 137–193

Crank J, Park GS (1968) Methods of measurement. In: Crank J, Park GS (eds) Diffusion in polymers. Academic Press, London New York, pp 1–39

Crisp CE (1965) The biopolymer cutin. Doct Diss, Univ Calif

Eckl K, Gruler H (1980) Phase transitions in plant cuticles. Planta 150:102–113

Falk VH, El-Hadidi MN (1961) Der Feinbau der Suberinschichten verkorkter Zellwände. Z Naturforsch Teil B 16:134–137

Frey-Wyssling A (1959) Die pflanzliche Zellwand. Springer, Berlin Göttingen Heidelberg

Gaines GL (1968) Insoluble monolayers at liquid-gas interfaces. Interscience, New York London Sydney

Haas K, Schönherr J (1979) Composition of soluble cuticular lipids and water permeability of cuticular membranes from *Citrus* leaves. Planta 146:399–403

Haas K, Schönherr J (1981) Temperature dependent effects of buffer treatments on soluble cuticular lipids of isolated cuticles. Biochem Biophys Pflanz 176:101–104

Heide Jørgensen HS (1978) The xeromorphic leaves of *Hakea suaveolens* R.B.: Structure of epidermal cells, cuticle development and ectodesmata. Bot Tidsskr 72:227–244

Hoch HC (1975) Ultrastructural alterations observed in isolated apple leaf cuticle. Can J Bot 53:2006–2013

Holloway JP, Baker EA, Martin JT (1972) The chemistry of plant cutins and suberins. An Quim 68:905–916

Holmgreen P, Jarvis PG, Jarvis MS (1965) Resistance to carbon dioxide and water vapor transfer in leaves of different plant species. Physiol Plant 18:557–573

Huelin FE (1959) Studies on the natural coatings of apples. IV. The nature of cutin. Aust J Biol Sci 12:175–180

Hull HM (1970) Leaf structure as related to absorption of pesticides and other compounds. Residue Rev 31:1–155

Hunt GM, Baker EA (1980) Phenolic constituents of tomato fruit cuticles. Phytochemistry 19:1415–1419

Jarvis LR, Wardrop AB (1974) The development of the cuticle *Phormium tenax*. Planta 119:101–112

Jeffree CE, Baker EA, Holloway PJ (1975) Ultrastructure and recrystallization of plant epicuticular waxes. New Phytol 75:539–549

Klute CH (1959a) Diffusion of small molecules in semicrystalline polymers: water in polyethylene. J Appl Polym 1:340–346

Klute CH (1959b) Diffusion of small molecules in semicrystalline polymers. II. The permeability of unoriented polymer films. J Polym Sci 41:307–317

Kolattukudy PE (1975) Biochemistry of cutin, suberin and waxes, the lipid barriers on plants. In: Calliard T, Mercer EI (eds) Recent advances in the chemistry and biochemistry of lipids. Academic Press, London New York, pp 203–246
Kolattukudy PE (1980) Biopolyester membranes of plants: cutin and suberin. Science 208:990–1000
LaMer VK, Healy TW, Aylmore LAG (1964) The transport of water through monolayers of long-chain n-paraffinic alcohols. J Colloid Sci 19:673–684
Lasoski Jr SW, Cobbs Jr WH (1959) Moisture permeability of polymers. I. Role of crystallinity and orientation. J Polym Sci 36:21–33
Maier U (1968) Dendritenartige Strukturen in der Cuticularschicht von *Lilium candidum*. Protoplasma 65:243–246
Martin JT, Juniper BE (1970) The cuticle of plants. Arnold, London
Mérida T, Schönherr J, Schmidt HW (1981) Fine structure of plant cuticles in relation to water permeability: The fine structure of the cuticle of *Clivia miniata* Reg leaves. Planta 152:259–267
Meyer M (1938) Die submikroskopische Struktur der kutinisierten Zellmembranen. Protoplasma 29:552–586
Norris RF (1974) Penetration of 2,4-D in relation to cuticle thickness. Am J Bot 61:74–79
Priestly JB (1943) The cuticle of angiosperms. Bot Rev 9:593–616
Rentschler I (1979) Zur Präparation der Wachsschicht von Blättern für die Elektronenmikroskopie. Microsc Acta 82:47–52
Roelofsen PA (1952) The submicroscopic structure of cell walls. Acta Bot Neerl 1:99–114
Roelofsen PA (1959) The plant cell wall. Borntraeger, Berlin
Sargent C (1976) The occurence of a secondary cuticle in *Libertia elegans* (Iridaceae). Ann Bot (London) 40:355–359
Schönherr J (1974) The nature of the pH effect on water permeability of plant cuticles. Ber Dtsch Bot Ges 87:389–402
Schönherr J (1976a) Water permeability of isolated cuticular membranes: The effect of pH and cations on diffusion, hydrodynamic permeability and size of polar pores in the cutin matrix. Planta 128:113–126
Schönherr J (1976b) Water permeability of isolated cuticular membranes: The effect of cuticular waxes on diffusion of water. Planta 131:159–164
Schönherr J, Bukovac MJ (1973) Ion exchange properties of isolated tomato fruit cuticular membrane: Exchange capacity, nature of fixed charges and cation selectivity. Planta 109:73–93
Schönherr J, Huber R (1977) Plant cuticles are polyelectrolytes with isoelectric points around three. Plant Physiol 59:145–150
Schönherr J, Lendzian K (1981) A simple and inexpensive method of measuring water permeability of isolated plant cuticular membranes Z Pflanzenphysiol 102:321–327
Schönherr J, Mérida T (1981) Water permeability of plant cuticular membranes: The effects of humidity and temperature on permeability of non-isolated cuticles of onion bulb scales. Plant Cell Environ 4:349–354
Schönherr J, Schmidt HW (1979) Water permeability of plant cuticles. Dependence of permeability coefficients of cuticular transpiration on vapor pressure saturation deficit. Planta 144:391–400
Schönherr J, Ziegler H (1980) Water permeability of *Betula* periderm. Planta 147:345–354
Schönherr J, Eckl K, Gruler H (1979) Water permeability of plant cuticles: The effect of temperature on diffusion of water. Planta 147:21–26
Sitte P (1955) Der Feinbau verkorkter Zellwände. Mikroskopie 10:178–200
Sitte P (1962) Zum Feinbau der Suberinschichten im Flaschenkork. Protoplasma 54:555–559
Sitte P (1975) Die Bedeutung der molekularen Lamellenbauweise von Korkmembranen. Biochem Physiol Pflanz 168:287–297
Sitte P, Rennier R (1963) Untersuchungen an cuticularen Zellwandschichten. Planta 60:19–40
Sivadjian J, Ribeiro D (1964) Comparative studies on the permeability to water and to water vapor of plastic materials by the hygrophotographic technique. J Appl Polym Sci 8:1403–1413

Slatyer RO (1967) Plant water relationships. Academic Press, London New York
Soliday CL, Kolattukudy PE, Davis RW (1979) Chemical and ultrastructural evidence that waxes associated with the suberin polymer constitute the major diffusion barrier to water vapor in potato tuber (*Solanum tuberosum* L.). Planta 146:607–614
Solomon AK (1968) Characterization of biological membranes by equivalent pores. J Gen Physiol 51:335s–346s
Stålfelt MG (1956) Die kutikuläre Transpiration. In: Ruhland W (ed) Handbuch der Pflanzenphysiologie Vol III. Springer, Berlin Göttingen Heidelberg, pp 342–350
Wattendorff J (1974) Ultrahistochemical reactions of the suberized cell walls in *Acorus, Acacia* and *Larix*. Z Pflanzenphysiol 73:214–225
Wattendorff J, Holloway PJ (1980) Studies on the ultrastructure and histochemistry of plant cuticles: The cuticular membrane of *Agave americana* L. in situ. Ann Bot (London) 46:13–28
Wilson JHM (1980) Freeze fractured cuticles of plants. Plant Sci Lett 18:121–126
Yasuda H, Stannett V (1962) Permeation, solution and diffusion of water in some high polymers. J Polym Sci 57:907–923

7 Stomatal Responses, Water Loss and CO_2 Assimilation Rates of Plants in Contrasting Environments

E.-D. SCHULZE and A.E. HALL

CONTENTS

7.1 Introduction

Stomatal apertures are the major pathway for the movement of CO_2 from the atmosphere into the mesophyll of leaves. The presence of this pathway for the movement of gases also results in water loss from the hydrated surfaces within leaves to the atmosphere. Stomatal aperture appears to be controlled by complex mechanisms which operate to maintain a variable balance between allowing CO_2 uptake to proceed, while restricting the loss of water vapor, and preventing leaf desiccation. Recent reviews have examined the physiological bases of stomatal function (RASCHKE 1979; ALLAWAY and MILTHORPE 1976) and stomatal responses to environment (SHERIFF 1979; BURROWS and MILTHORPE 1976; HALL et al. 1976). Analyses which integrated stomatal effects on CO_2 exchange, transpiration, and energy balance were developed based upon theory (COWAN 1977), which have led to hypotheses concerning optimal stomatal function (see Chap. 17, this Vol.; COWAN and FARQUHAR 1977). However, information concerning the simultaneous responses of stomata, water loss, and CO_2 assimilation rates has not been reviewed for plants in natural environments.

The following chapter attempts a comparative analysis of the variation in the operational range of leaf conductance and CO_2 assimilation rate, and the regulation of these processes within the operational range, including an analysis of the relationships which exist between stomata, plant water loss and plant water status. Finally, consideration is given to the responses of plants in contrasting natural environments, and levels of leaf conductance, water loss, and CO_2 assimilation rate are examined in relation to maximal levels for natural conditions.

For this review, studies were analyzed in which simultaneous measurements of leaf conductance and CO_2 assimilation rate had been determined. Many studies of this type have been conducted in controlled environments but few data are available for naturally varying field conditions. Data obtained by micrometerological methods were not included because they usually lack precision and are difficult to interpret due to the complexity of plant canopies. When evaluating relationships between stomata and plant water status, data obtained with viscous flow porometers were generally not included because they tend to be inaccurate.

Leaf conductance was used rather than leaf resistance because conductance is more linearly related to transpiration, CO_2 assimilation, and stomatal aperture (BURROWS and MILTHORPE 1976; HALL et al. 1976). Leaf conductance values refer to water vapor and were calculated from the transpiration rate and the difference in vapor pressure between leaf and air. Conductances are presented – unless explicitly stated – on a projected leaf area basis and are expressed in the same units as the vapor and CO_2 fluxes using the definition of COWAN (1977) described in Chapter 8, this Volume. Leaf conductances refer to the combined effects of boundary layers, stomata, and cuticles on both sides of the leaf. Intercellular CO_2 concentrations were calculated from independent measurements of CO_2 flux and leaf conductance.

7.2 Operational Range of Leaf Conductance and CO_2 Assimilation Rate

Maximum leaf conductance for water vapor is mainly determined by the density and geometric properties of open stomata, and the boundary layer. Minimum leaf conductance for water vapor is determined by the conductance of the cuticle and the extent of stomatal closure. KÖRNER et al. (1979) examined maximum leaf conductances of 13 groups of morphologically and/or ecologically comparable plant species and related these data to estimates of photosynthetic capacity mainly obtained from other sources. They proposed that, for a wide range of C_3 plants, photosynthetic capacity was linearly correlated with maximum leaf conductance and that C_4 grasses had greater photosynthetic capacity than C_3 grasses, but similar maximum leaf conductances. The way in which CO_2 assimilation (A) and leaf conductance (g) are correlated appears to have great ecological significance. If during variations of environmental or plant internal conditions, A and g change proportionally in a straight line relationship which passes through the origin, the intercellular CO_2 concentration, c_i remains constant. To the extent that leaf temperature and atmospheric vapor pressure do not vary, water use efficiency also remains constant. In this case stomatal movements do not change the conditions of CO_2 fixation in the mesophyll. In contrast, curvilinear correlations between g and A and linear relations which do not pass through the origin indicate that either g or A responds more strongly than the other parameter to changes of environmental or plant internal conditions. In this case c_i and water use efficiency are not maintained constant; c_i decreases whenever g is more affected by changes of environmental or plant internal conditions than A. In the following section we shall examine this correlation between the range of leaf conductance and the photosynthetic capacity for plants subjected to different environments during growth.

In contrast to KÖRNER et al. (1979), in the following study leaf conductances (g) were examined for conditions where CO_2 assimilation rate was maximal, and with natural concentrations of CO_2 and O_2 (A_{max}), as suggested by HALL and SCHULZE (1980a). This upper reference level for the range of leaf conductance (g at A_{max}) is probably lower, in many circumstances, than the maximal leaf conductance which could have been achieved in a leaf by special experimental treatments. But, few data are available where maximum leaf conductances have been deliberately sought, and measurements of maximum leaf conductance may be difficult because stomata may not respond in a steady-state manner under such conditions, e.g. at high temperatures. Instead, leaf conductance at which CO_2 assimilation is maximal appears to have greater ecological significance for understanding plant adaptations to their environment.

Several data sets are available where correlations had been examined between A and g in attempts to explain stomatal effects on photosynthesis. These data are evaluated in an inverse manner to determine the extent to which the operational range of leaf conductance is linearly or curvilinearly correlated with photosynthetic capacity. Long-term effects of plant climate and effects of leaf age, plant conditioning, and genotype are investigated.

7.2.1 Long-Term Effects of the Level of Visible Radiation During Growth

Plants adapted to shaded habitats have a lower capacity for CO_2 assimilation than plants adapted to, and grown in, habitats where they are exposed to high levels of solar radiation. These differences in capacity are associated with many differences between sun and shade leaves in photosynthesis and respiration, in leaf anatomy, and in leaf conductance to water vapor (BJÖRKMAN 1973; see Chap. 3, Vol. 12 A). *Atriplex patula* normally grows in exposed habitats, but it exhibits shade-type responses when grown in the shade (Fig. 7.1). BJÖRKMAN (1973) calculated the dashed lines in Fig. 7.1 from CO_2 response curves

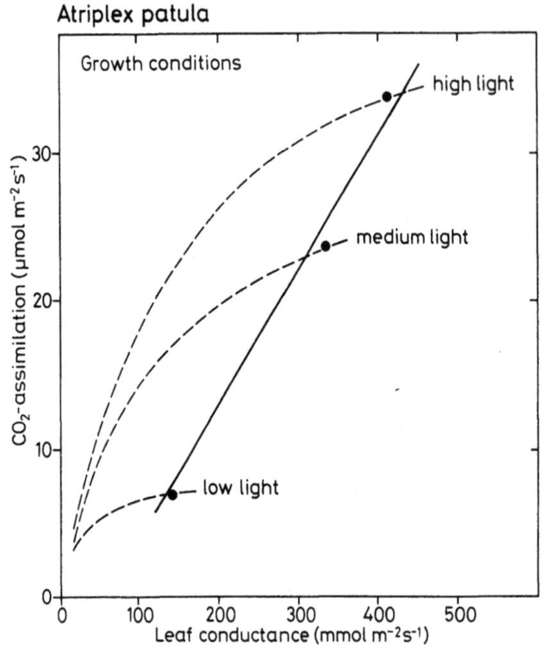

Fig. 7.1. CO_2 assimilation of *Atriplex patula* grown in different light climates as related to leaf conductance. The data points are CO_2 assimilation rates measured under saturating light, optimal temperature, and humidity, and natural concentrations of CO_2 and O_2, and leaf conductance to water vapor measured concurrently. The *dashed lines* are predicted effects of leaf conductance on CO_2 assimilation rate calculated using measured responses of assimilation to CO_2 concentration. (After BJÖRKMAN 1973)

and used them to demonstrate that the lower CO_2 assimilation rates of the shade-grown plants were not solely due to their smaller leaf conductances. It is apparent that g (at A_{max}) was linearly correlated with A_{max}.

Comprehensive investigations of correlations between stomatal conductance and photosynthetic capacity were conducted by WONG (1979) and WONG et al. (1979). Also their data exhibit linear regression lines between g (at A_{max}) and A_{max} for plants of several species, as they varied due to several factors including long-term effects of the radiation environment during plant growth. For maize, the varying characteristics in leaves of both sun- or shade-grown plants are well approximated by one regression line passing through the origin (Fig. 7.2). DOWNES (1971) studied a wide range of C_4 genotypes within the genus *Sorghum;* and their characteristics of g (at A_{max}) and A_{max} yielded also a regression line with a slope similar to that of *Zea mays* (Fig. 7.2). In this and subsequent figures, it will be apparent that substantial differences exist between C_3 and C_4 plants in the slope of the regression line of g (at A_{max}) with A_{max}. This is a corollary of the fact that c_i is lower in C_4 than in C_3 plants. A smaller slope of the regression line indicates that higher CO_2 assimilation rates would be obtained at similar leaf conductances; and, therefore, small slopes are associated with high intrinsic water use efficiency.

Large differences are also apparent in the slopes of the regression lines obtained with different C_3 species. The study of HOLMGREN (1968) with sun and shade ecotypes of *Solidago virgaurea* grown in both sun and shade environments, exhibit a linear regression between g (at A_{max}) and A_{max} which had a much lower slope than the regression line of g (at A_{max}) and A_{max} in *Fagus silvatica.* BJÖRKMAN (1968) demonstrated that substantial differences were present amongst ecotypes of *Solidago virgaurea* in their photosynthetic and respirato-

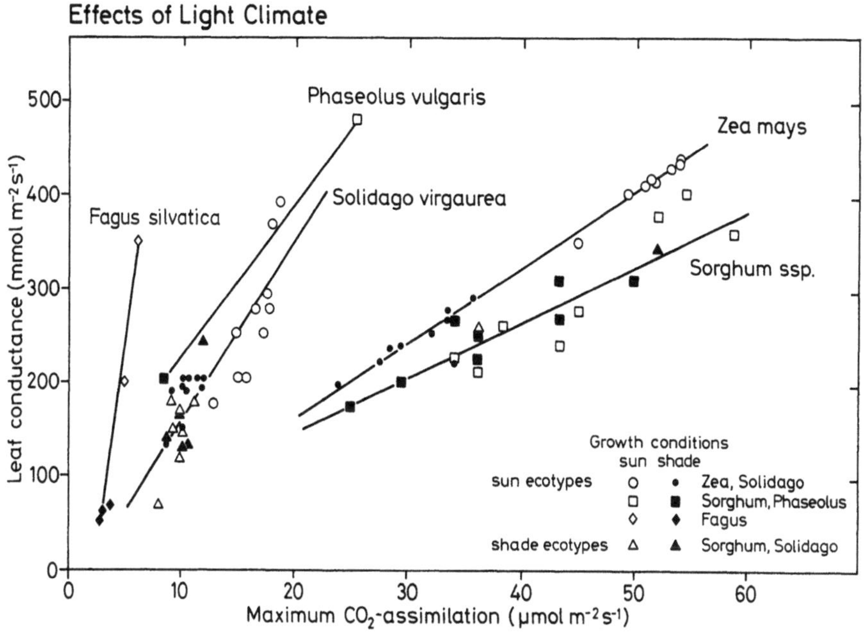

Fig. 7.2. The effects of light climate on the relation between leaf conductance and maximum CO_2 assimilation at optimal temperatures and humidity, and natural concentrations of CO_2 and O_2. The variation resulted from differences among leaves of plants grown under sun and shade conditions. *Fagus silvatica:* SCHULZE (1970), *Phaseolus vulgaris* and *Zea mays:* WONG (1979), *Solidago virgaurea:* HOLMGREN (1968), *Sorghum ssp.:* DOWNES (1971)

ry metabolism. Consequently, the linear regression in Fig. 7.2 is not simply due to stomatal effects on CO_2 uptake rate. The data on *Fagus silvatica* were obtained on the same tree and contain the long-term effects of different levels of radiation in the exposed and more shaded part of the forest canopy.

7.2.2 Influence of Inorganic Nutrition

In many habitats, the supply of plant nutrients, especially nitrogen, strongly limits plant growth and photosynthetic capacity. Linear regressions passing through the origin were observed between g (at A_{max}) and A_{max} by WONG (1979) for C_3 and C_4 plants subjected to variations in the supplies of nitrogen and phosphate (Fig. 7.3). Similar correlations are apparent in the data of GOUDRIAAN and VAN KEULEN (1979) for the effects of nitrogen deficiency on *Zea mays;* however, their data for *Helianthus annuus,* subjected to nitrogen deficiencies, exhibit only poor correlations between g (at A_{max}) and A_{max}. In addition, the data of MEDINA (1972) for *Atriplex patula* exhibit no association between g and A at two levels of nitrogen supply. The lack of correlation between g (at A_{max}) and A_{max} that has been observed on some occasions may indicate that conductance is not influenced by the metabolism of the leaf mesophyll.

It is unlikely that the linear correlations described by WONG (1979) can be solely explained by nutrient deficiencies affecting g, and g affecting A. Defi-

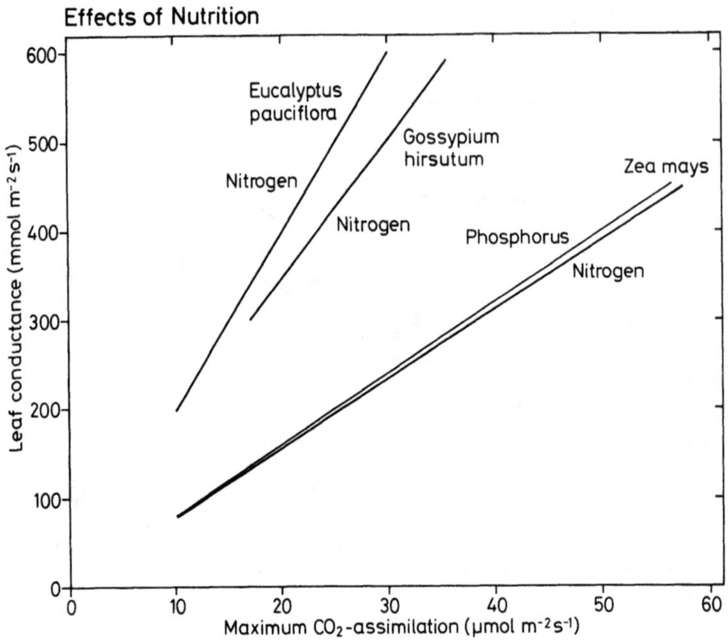

Fig. 7.3. The effects of nutrition on the relation between leaf conductance and maximum CO_2 assimilation at optimal temperatures and humidity, natural concentrations of CO_2 and O_2, and irradiance levels sufficient to achieve near maximal assimilation. The variation resulted from deficiencies of nitrogen and phosphate. (After WONG 1979)

ciencies of nitrogen usually result in lower levels of photosynthetic enzymes per unit area, as occurred in the studies of WONG (1979). For C_3 species, the responses of A to independent variations in g are strongly curvilinear (refer to the dashed line in Fig. 7.1), as may be demonstrated by subjecting plants to short-term variations in humidity (FARQUHAR et al. 1980).

7.2.3 Leaf Age Effects

Variations in photosynthetic capacity and leaf conductance with leaf age reflect developmental changes, and effects due to intraplant competition for nutrients and long-term changes in radiation level. During the short period of leaf expansion, photosynthetic capacity increases, after which A_{max} remains relatively constant and then steadily declines as the leaf senesces (SCHULZE 1970; LUDLOW and WILSON 1971; WOODWARD and RAWSON 1976). Under constant measurement conditions, linear correlations have been observed between concurrent measurements of g (at A_{max}) and A_{max} as they varied with leaf age (Fig. 7.4). The C_4 plants, *Sorghum almum* and *Pennisetum purpureum*, had a smaller regression slope than the C_3 plants, as previously reported for studies of the long-term effects of radiation and plant nutrition. Data on variation in leaf conductance and photosynthetic capacity obtained by LUDLOW and WILSON (1971) for several tropical legumes exhibited poorer correlations than the data for tropical grasses presented in Fig. 7.4.

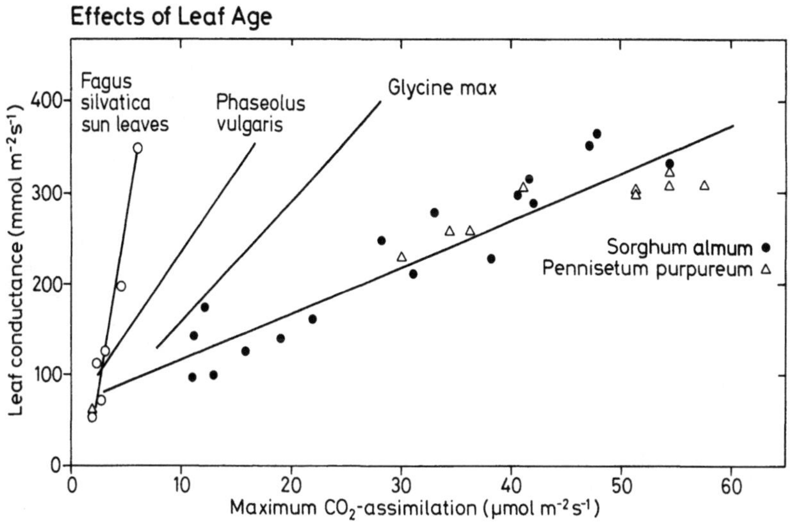

Fig. 7.4. The effects of leaf age on the relation between leaf conductance and maximum CO_2 assimilation made under near optimal conditions for assimilation, with variation due to leaf age observed on either the same day with different leaves or on different days with the same leaves. *Fagus silvatica:* SCHULZE (1970), *Glycine max:* WOODWARD and RAWSON (1976), *Phaseolus vulgaris:* DAVIS and MCCREE (1978), *Sorghum almum* and *Pennisetum purpureum:* LUDLOW and WILSON (1971)

7.2.4 Influences of Plant Conditioning and Genotype

There are several ways, in addition to the radiation environment and nutrient supplies in the soil, in which prior conditions during growth are known to influence photosynthetic capacity. Pretreatment at different temperatures (BAUER 1978) and assimilate source-sink manipulations (HALL and MILTHORPE 1978; RAWSON et al. 1976) have resulted in changes in photosynthetic capacity. They also resulted in concomitant changes in leaf conductance. While it cannot be stated that all factors which have long-term effects on photosynthetic capacity also influence the range of leaf conductance, there are many experiments where associations between these parameters are apparent. Some additional examples are presented which have relevance to genotypic influences on the association between g (at A_{max}) and A_{max}.

Viruses may be considered as having conditioning effects on their host plants which are either physiological or genetical in origin. For healthy and beet yellows virus-infected plants of *Beta vulgaris,* g (at A_{max}) was correlated with A_{max} (Fig. 7.5). Measurements with five provenances of *Picea sitchensis* exhibited linear regressions between g (at A_{max}) and A_{max} (dashed line in Fig. 7.5). Presumably, this variation in g and A was due to genotypic differences; and it is of interest that these data (the dashed line) followed the same regression line as the seasonal variation of g (at A_{max}) and A_{max} obtained with shoots from one provenance at different times of year (data points of LUDLOW and JARVIS 1971 in Fig. 7.5). An association between g (at A_{max}) and A_{max} of genetic origin is also apparent in the data of BENNETT and ROOK (1978) for two contrasting

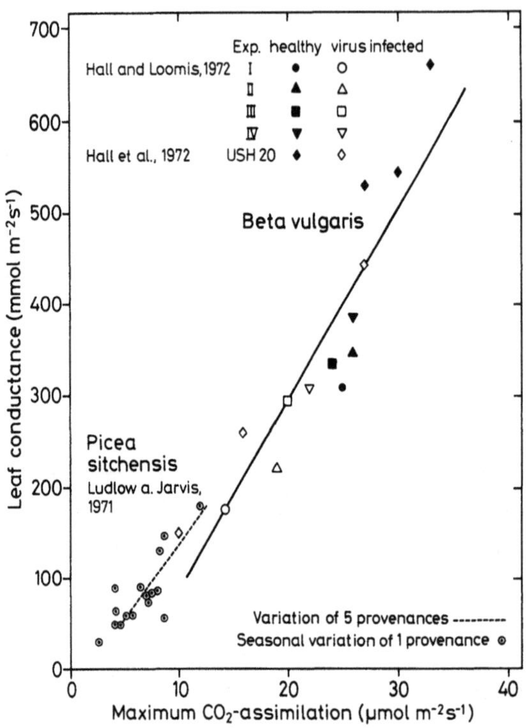

Fig. 7.5. The effects of genotype and virus infection on the relation between leaf conductance and maximum CO_2 assimilation, made under near optimal conditions for assimilation. The *data points* for *Picea sitchensis* indicate seasonal variation, whereas the *regression line* indicates genotypic variation. *Beta vulgaris:* HALL and LOOMIS (1972) and HALL et al. (1972), *Picea sitchensis:* LUDLOW and JARVIS (1971)

clones of *Pinus radiata*. Apparently, genetic variation within a species can result in variation in the range of leaf conductance which exhibits the same relationship with photosynthetic capacity as may occur due to long-term environmental effects on one genotype. The studies with ecotypes of *Solidago virgaurea* (Fig. 7.2) also support this observation.

Interspecific effects on the relationship between g (at A_{max}) and A_{max} are evident in studies with *Citrus* species (KHAIRI and HALL 1976a, b). *Citrus limon,* which is thought to be native to a warm subhumid region, was compared with the hybrid between *Citrus limon* and *Eremocitrus glauca;* the latter species is thought to be native to a hot semi-arid region. The relationship between the curves of g (at A_{max}) versus A_{max} for the species and the hybrid (Fig. 7.6) is similar to that already described for C_3 versus C_4 species. A higher ratio of A_{max} to g, as exhibited by the hybrid, should result in higher water use efficiency and improved adaptation to arid habitats compared with *Citrus limon.* In another study, *Citropsis gabunensis,* which is thought to be native to shade environments in the humid tropics, was compared with *Citrus sinensis,* which is thought to be native to the subhumid subtropics. Compared with *Citrus sinensis,* the data for *Citropsis gabunensis* are displaced toward the origin as might be expected for a shade species, and are rotated away from the A axis, which is consistent with an environment which has lower evaporative demands (Fig. 7.6).

A schematic description of possible relationships between the operational range of leaf conductance and the capacity for photosynthesis which summarizes

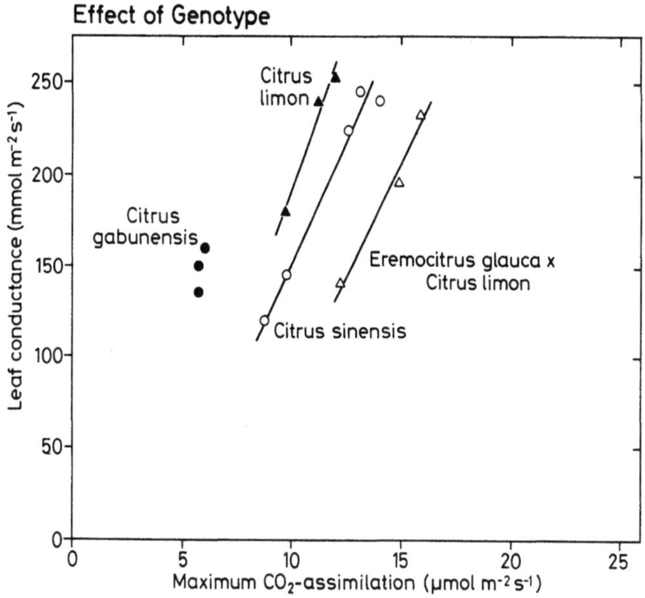

Fig. 7.6. The effects of genotype within the genus *Citrus* and related genera on the relation between leaf conductance and maximum CO_2 assimilation made under near-optimal conditions for assimilation. The variation resulted from day-to-day changes in conductance and capacity for CO_2 assimilation which may have been due to source-sink effects. (After Khairi and Hall 1976a, b)

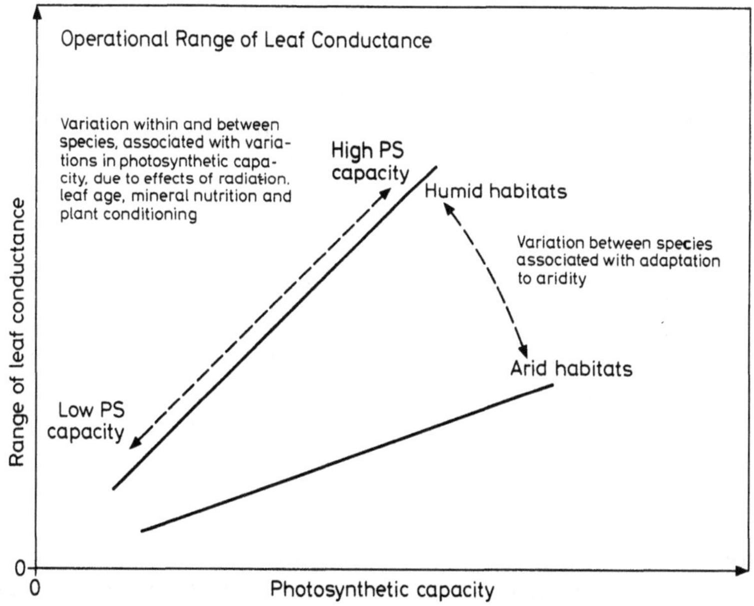

Fig. 7.7. Hypothetical relations between the operational range of leaf conductance and photosynthetic capacity as indicated by maximal CO_2 assimilation rate at near-optimal conditions for assimilation

also the previous figures is presented in Fig. 7.7. It is hypothesized that environmental and plant conditions which result in a lower photosynthetic capacity (and a lower investment) per unit leaf area, may result in a smaller range of leaf conductance. FARQUHAR (1978a) demonstrated that a linear association between g and the capacity for CO_2 assimilation is consistent with the concept of optimal stomatal function as proposed by COWAN and FARQUHAR (1977). The slopes of the regression line of g (at A_{max}) versus A_{max} varied substantially among C_3 species, although C_3 plants may not reach a slope as low as in C_4 plants. In the present study *Fagus silvatica* showed the steepest slope of all C_3 species and *Gossypium hirsutum* the lowest slope. Smaller slopes, as exhibited by some C_3 species and most C_4 species, signify higher intrinsic water use efficiency and lower c_i, and improved adaptation to arid habitats. It is hypothesized that genotypes within the same species will tend to exhibit relations between g (at A_{max}) and A_{max} which have similar regression lines, whereas substantial differences in slope of the regression line will normally require substantial differences in genotype.

7.3 Regulation of Leaf Conductance and CO_2 Assimilation Rate Within the Operational Range

During diurnal courses, stomatal conductance varies in response to short-term changes in the level of visible radiation, humidity, and temperature; and to a variable extent, these changes are influenced by effects on stomata of variations in CO_2 concentration inside the leaf. These stomatal responses are substantially reversible, except where plants are subjected to extreme variations in environment. In the following sections, the short-term reversible responses of stomata to radiation, humidity, and temperature are discussed; and the regulation of leaf conductance is examined in relation to the balance between water loss and CO_2 assimilation.

7.3.1 Short-Term Effects of Visible Radiation

Stomatal responses to light have been examined on many occasions in controlled environments. Leaf conductance usually exhibits a hyperbolic response to photon flux (BURROWS and MILTHORPE 1976), which is similar to the response of CO_2 assimilation rate, resulting on some, but not all, occasions in a linear regression between g and A (WONG 1979; RAMOS 1981). Gain analyses based upon the methods developed by COWAN (1977) and FARQUHAR et al. (1978) have been used to determine the extent to which stomatal responses to light are indirect and mediated by internal CO_2 concentration. For some C_4 species, g response to internal CO_2 concentration may be an important factor in g response to light (RAMOS 1981; RASCHKE et al. 1978). In contrast, the response of stomata in some C_3 species is either a direct response to photon flux or a response to some product of photosynthesis in the mesophyll tissue (WONG et al. 1978, 1979), and stomatal response to internal CO_2 concentration is relatively insignificant.

The photon flux required for 95% of maximal conductance varies substantially (100 to 2,000 μmol photon $m^{-2} s^{-1}$) depending upon the species, leaf age, and preconditioning. On many occasions, greater photon fluxes are required for full stomatal opening in C_4 than in C_3 species. In addition, plants with grass-type stomata from the families Poaceae and Cyperaceae may respond differently to blue light than plants from other families (JOHNSSON et al. 1976). On occasions, plants have exhibited only small responses of leaf conductance to light; e.g., sunflower in the studies of GOUDRIAAN and VAN LAAR (1978), but not in the studies of RAMOS (1981). These variations in stomatal sensitivity to light have not been explained. Responses obtained with *Larix decidua* in natural conditions by BENECKE et al. (1981) are described in Fig. 7.8. Stomata of older leaves were less responsive to reductions in photon flux than younger leaves. These studies also demonstrated that there are cases where the short-term relations between A and g are different from the long-term relations between A_{max} and g (at A_{max}).

7.3.2 Responses to Humidity and Temperature

Stomatal responses to humidity may reflect a mechanism which prevents extreme rates of plant water loss and extreme plant water deficits when plants are subjected to high evaporative demands (HALL et al. 1976). The mechanism for stomatal response to humidity has not been clearly established (SHERIFF 1979). LANGE et al. (1971) proposed that stomatal response to humidity was mediated by cuticular transpiration of the outer epidermis, whereas SHERIFF and MEIDNER (1975) proposed that it may be a consequence of the evaporation from the inner walls of the guard and subsidiary cells. It should be recognized, however, that the latter mechanism could not result in steady-state decreases in transpiration with increases in evaporative demand (COWAN 1977; FARQUHAR 1978; TYREE and YIANOULIS 1980).

Stomatal responses to humidity have been observed with virtually all species that have been examined (some exceptions are discussed by LÖSCH 1980). Stomatal responses to humidity were first comprehensively described for plants in desert habitats (SCHULZE et al. 1972a), but they have also been observed with plants from less arid habitats (CAMACHO-B et al. 1974), arctic zones (JOHNSON and CALDWELL 1976), temperate zones (JARVIS 1980), and even humid tropical zones (KHAIRI and HALL 1976b). In addition, stomatal response to humidity has been observed at night with CAM plants (LANGE and MEDINA 1979; OSMOND et al. 1979).

The sensitivity of stomata to humidity has not been associated with either habitat conditions or leaf type. Generally, stomata of plants with larger maximal conductances are more sensitive to humidity, whether the differences in maximal conductance are genetic in origin (Fig. 7.9) or due to long-term drought (HALL and SCHULZE 1980a; LUDLOW 1980). However, on a relative scale, the responses

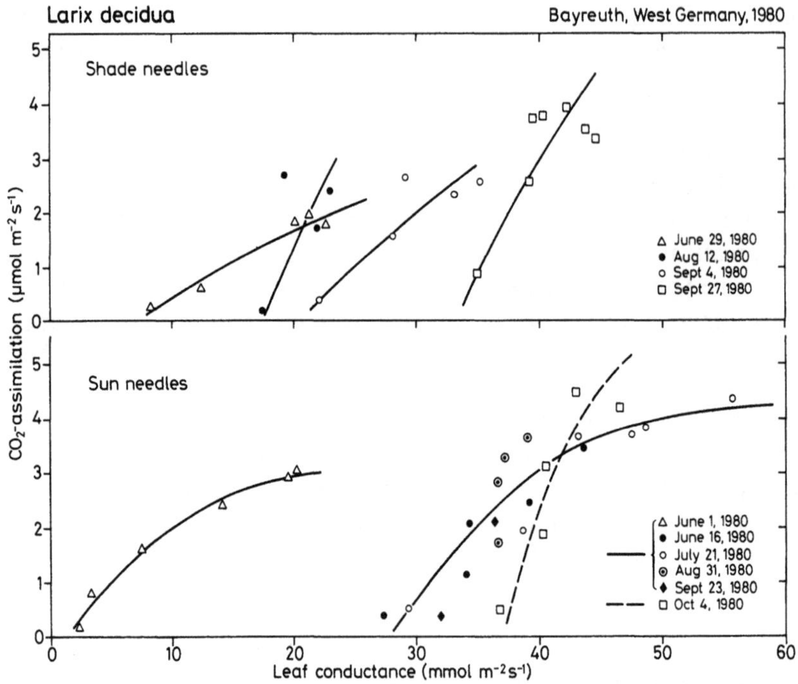

Fig. 7.8. Short-term relation between CO_2 assimilation at optimal temperature and high air humidity and leaf conductance of sun and shade needles of *Larix decidua* in different seasons. The *data points* show steady state conditions at different light intensities. All rates are related to total surface area (ratio of total/projected area is about 2.3). (After BENECKE et al. 1981)

may be similar with an increase in vapor pressure difference from 10 to 20 mbar resulting in a 50% decrease in stomatal conductance. Pretreatment with dry air may inhibit stomatal response to humidity (RAWSON et al. 1977).

Substantial regulation of transpiration by sensitive stomatal responses to humidity has been observed with several conifers (BENNETT and ROOK 1978; JARVIS 1980), and some herbaceous angiosperms (DRAKE et al. 1970; HALL and KAUFMANN 1975; HALL and SCHULZE 1980b; LUDLOW 1980). For *Macroptilium atropurpureum* and *Vigna unguiculata,* sensitive regulation of water loss was associated with extreme drought avoidance. However, studies with several woody angiosperms indicated that stomatal effects on transpiration during responses to humidity were smaller, and substantial increases in transpiration rate occurred with decreases in external humidity (KHAIRI and HALL 1976b; OSONUBI and DAVIES 1980).

In natural environments, the regulation of transpiration is strongly dependent upon stomatal responses to both vapor pressure difference and temperature. When the vapor pressure deficit of the air decreases during the day time, it is usually due to increasing temperature because ambient vapor pressures frequently remain relatively constant during the day. Stomatal responses to temperature per se have often been confounded with responses to vapor pressure difference. When the effects of temperature and humidity are separated, leaf

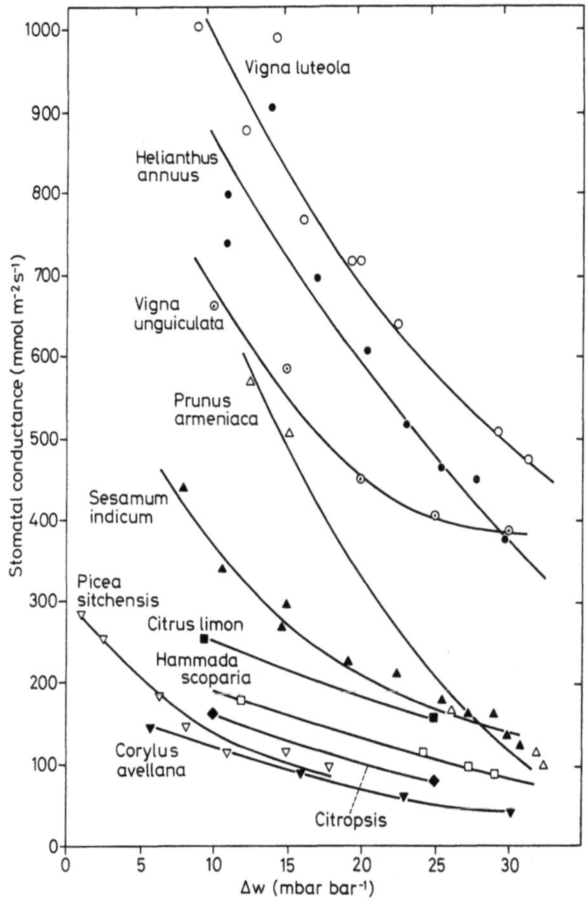

Fig. 7.9. Stomatal conductance as related to the relative leaf/air vapor pressure difference, Δw, with constant leaf temperature and radiation, *Hammada scoparia* (rates related to total surface area of the round twigs), *Helianthus annuus, Prunus armeniaca, Sesamum indicum, Vigna luteola:* after HALL et al. (1976), *Corylus avellana:* SCHULZE and KÜPPERS (1979), *Citrus limon, Citropsis gabunensis:* KHAIRI and HALL (1976b), *Picea sitchensis:* JARVIS (1980), *Vigna unguiculata:* HALL and SCHULZE (1980a)

conductance increases with temperature at temperatures above the optimum for photosynthesis (HALL et al. 1976).

Transpiration influences leaf energy balance; in plants adequately supplied with water, stomata may regulate temperature, either maintaining leaf temperature near the optimum for processes such as photosynthesis, or preventing high temperatures that would result in tissue necrosis.

7.3.3 The Balance Between Water Loss and Assimilation

A comprehensive and elegant hypothesis for the regulation of leaf conductance within the operational range has been developed by COWAN (1977) and COWAN and FARQUHAR (1977). Detailed discussions of this hypothesis have been pre-

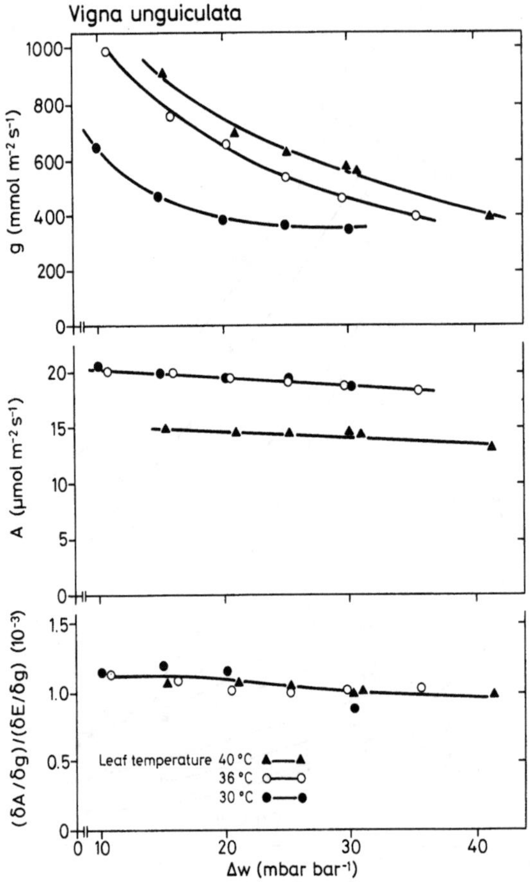

Fig. 7.10. Leaf conductance to water vapor, g, CO_2 assimilation, *A*, and the ratio for the sensitivity of assimilation and transpiration to changes in leaf conductance, $(\partial A/\partial g)/(\partial E/\partial g)$, as related to changes in relative leaf/air vapor pressure difference. The experiment was conducted with step decreases in humidity at three constant leaf temperatures on 3 days with the same attached leaflet and photon flux of 900 µmol m^{-2} s^{-1}, and external CO_2 relative partial pressure of 300 ± 5 µbar bar^{-1}. (After HALL and SCHULZE 1980a)

sented in Chapter 17, this Volume (see also Chap. 1, this Vol.). However, some discussion is warranted in this chapter because this hypothesis integrates stomatal effects on energy balance and the transport of H_2O and CO_2 between leaf and air.

If daily CO_2 assimilation is to be maximized at a given level of daily transpiration, stomatal conductance should vary during the day in a manner which keeps the gain ratio $\frac{\partial A}{\partial g}/\frac{\partial E}{\partial g}$ constant. These partial differentials describe the sensitivity of CO_2 assimilation rate (A) and transpiration rate (E) to changes in leaf conductance (g). Recent studies under steady-state conditions in controlled environments demonstrated that stomatal responses to humidity (FARQUHAR et al. 1980), and to humidity and temperature (HALL and SCHULZE 1980a) resulted in relatively constant gain ratios. The constancy of the gain ratio with different temperatures was quite remarkable (Fig. 7.10, also see Table 1 in HALL and SCHULZE 1980a). It is of interest to determine how well these data fit alternative hypotheses for stomatal function. FARQUHAR et al. (1980) demonstrated that their data for stomatal response to humidity by *Nicotiana glauca* did not exhibit sufficient feedback control of plant water relations to keep transpiration constant nor sufficient feedback control of internal CO_2 concentra-

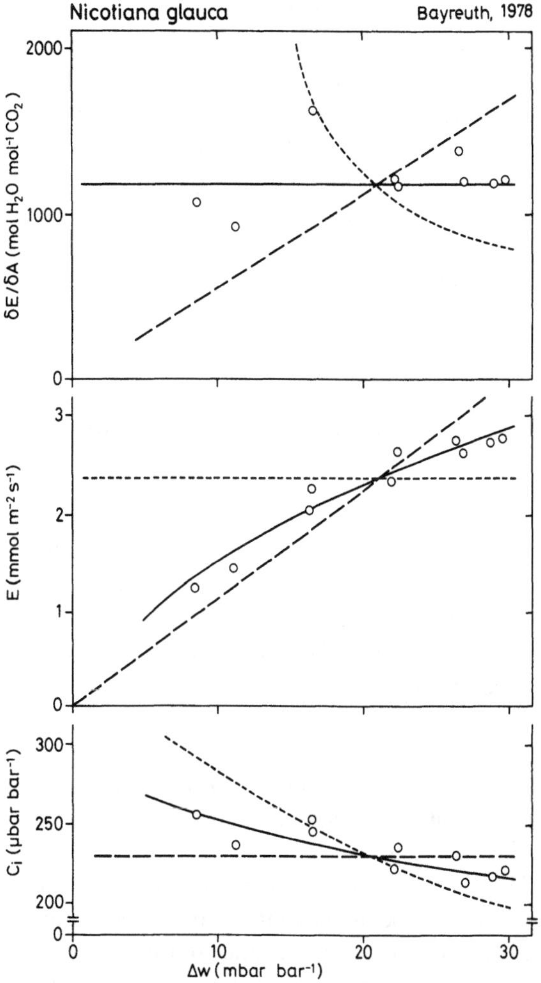

Fig. 7.11. The ratio for the sensitivity of transpiration and assimilation to changes in leaf conductance, $\partial E/\partial A$, the rate of transpiration, E, and the intercellular mole fraction of CO_2, c_i, as related to the relative leaf/air vapor pressure difference, Δw. The *circles* represent the measured data. The relationships represented by the various lines were derived assuming constant $\partial E/\partial A$ (1,200 mol H_2O mol^{-1} CO_2 *drawn line*), constant E (2.36 mmol m^{-2} s^{-1} *dotted line*), and constant c_i (231 µbar bar^{-1} *dashed line*). (After FARQUHAR et al. 1980)

tion to keep CO_2 assimilation rate constant (Fig. 7.11). The data in Fig. 7.10 are consistent with partial regulation of leaf temperature by stomata; but this is, also, an integral part of the hypothesis that $\partial A/\partial E$ is kept constant with diurnal variations in air temperature and humidity. At this time, it is not known how $\partial A/\partial E$ is influenced by short-term variations in photon flux or by the multiple variations in environment that occur during the day in natural conditions. If the daily assimilation of canopies is to be maximized at a given level of daily transpiration, all leaves must exhibit the same constant gain ratio during the day (COWAN and FARQUHAR 1977). The predictions of the model of FARQUHAR (1978a) may be interpreted as indicating that, where g (at A_{max}) and A_{max} vary in a proportional manner, $\partial A/\partial E$ may be relatively constant. This indicates that the long-term effects of radiation and mineral nutrition during leaf growth and of leaf age may result in only small variations in $\partial A/\partial E$ within the canopy (except where short-term variations in photon flux cause g and A to vary in a nonproportional manner). Higher values of $\partial A/\partial E$ are generally to be expected where the slope of g (at A_{max}) versus A_{max} is smaller.

Smaller slopes have been observed for C_4 versus C_3 species (Figs. 7.2, 7.3, and 7.4) and for some C_3 species which have been subjected to drought (HALL and SCHULZE 1980a and Fig. 7.16 in Sect. 7.4). In addition, correlation coefficients closer to one, for g (at A_{max}) versus A_{max}, may indicate more precise regulation of $\partial A/\partial E$. It is apparent that data concerning relationships between the ranges of leaf conductance and photosynthetic capacity may provide an approach toward defining the optimality of stomatal function, which may allow comparisons between species in different environments at a more general level than was previously possible.

7.4 Stomatal Responses and Plant Water Status

Leaf conductance responds to changes in plant water status but it also influences leaf water potential through effects on transpiration. Under natural conditions the effect of leaf conductance on transpiration and leaf water potential can often be detected. However, it is usually difficult to separate the effects on stomata of changes in plant water status from effects due to changing atmospheric conditions.

7.4.1 Leaf Conductance and the Control of Water Loss and Plant Water Status

The flow of water through the soil-plant-atmosphere continuum has been described as being proportional to the water potential gradient, and inversely proportional to the resistance to flow imposed by the soil, roots, vascular system, leaves, and canopy (COWAN and MILTHORPE 1968; COWAN 1965; see Chap. 1, this Vol.). However, it can be demonstrated that this model is only approximately valid and it is not consistent with some experimental observations (Chap. 8, this Vol.).

The dominant resistance to flow is in the vapor phase. Consequently, leaf water potential should be regarded as the dependent variable, responding to changes in the flow rate of water (as determined by atmospheric conditions and leaf and canopy resistance), and upon the resistance to flow of liquid water in the soil plant system (COWAN and MILTHORPE 1968; ELFVING et al. 1972; KAUFMANN 1976). When the flux of water through plants is much greater than the rate of change in plant water content, it is possible to describe diurnal changes in leaf water potential and water flux by a sequence of steady states even under variable natural conditions (COWAN 1965).

A schematic description of possible short-term and long-term relations between leaf water potential, Ψ, and transpiration, E, is presented in Fig. 7.12a. When transpiration rates are low, such as during the night, leaf water potentials approach a maximum value which is related to the soil water potential in the root zone. Leaf water potential usually decreases after sunrise and approaches

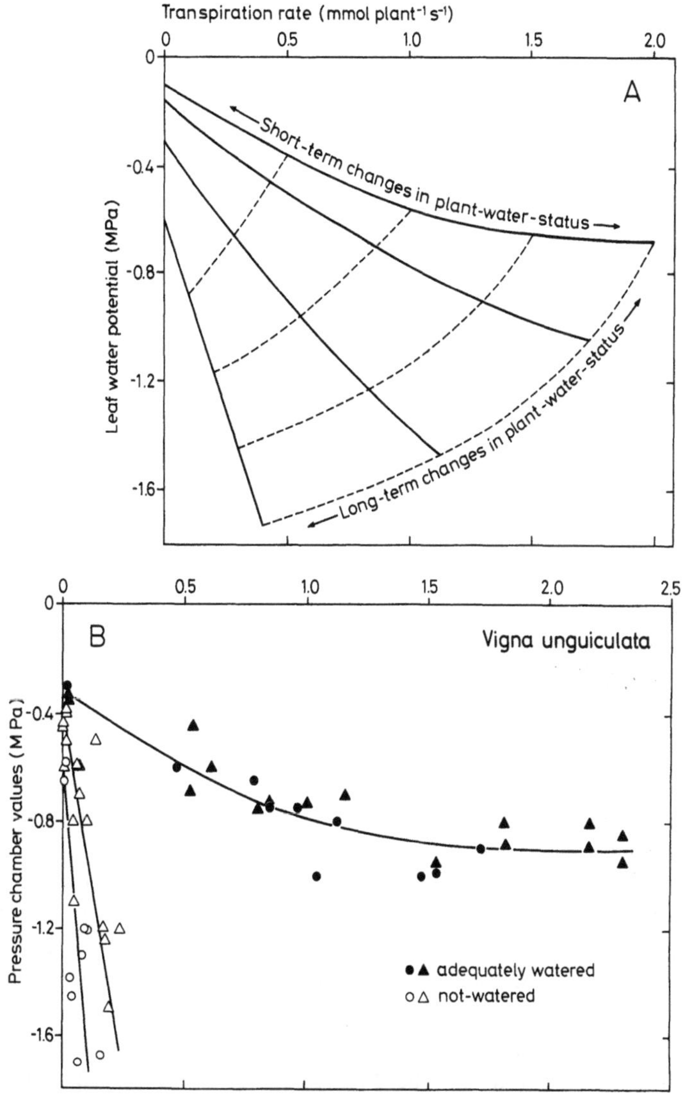

Fig. 7.12. A Schematic representation of short-term and long-term relations between leaf water potential and transpiration under different equal increments of evaporative demand. **B** Xylem pressure potential as a function of transpiration per plant of *Vigna unguiculata* in controlled environments. The plants in the dry treatments had not been supplied with water for 25 days, whereas the other plants had been irrigated on the evening prior to the day on which measurements were made. (After HALL and SCHULZE 1980b)

a minimum value at the time of maximum transpiration. In Fig. 7.12A this change of Ψ by variations in E is referred to as "short-term changes in plant-water-status". The plant may operate at different slopes for these short-term changes in plant-water-status depending on the conductance in the pathway of water from the soil into the leaf. Soil drought results in both a decrease in pre-dawn leaf water potential and increases in resistances to flow in the soil and plant (see Chaps. 1 and 3, this Vol.). In addition, other factors such as dying of root tips, hardwood formation, or cavitation of xylem vessels will decrease the conductance of the pathway of water from the soil into the leaf (see Chap. 5, this Vol.). Only some of these factors, e.g., the flow of water from the surroundings of the root tip may be reversible in the short term. In Fig. 7.12A those factors which change the slope of the Ψ/E relationship are referred to as "long-term changes in plant-water status". The dependence of Ψ on changes in E has provided explanations for diurnal and seasonal variations in leaf water potential (ELFVING et al. 1972; KAUFMANN 1976; KAPPEN et al. 1975), and descriptions of species differences in the efficiency of the water transport system from soil to leaves (CAMACHO-B et al. 1974). This conceptual model demonstrates that the same level of leaf water potential may occur in dry and in well-watered plants but at different levels of transpirational water loss. However, plants subjected to drying soils usually exhibit lower leaf water potentials than well-watered, but genetically similar plants.

Substantial differences have been observed among plants in the extent to which leaf water potential decreases when plants are subjected to short-term changes in atmospheric conditions (e.g., CAMACHO-B et al. 1974; SÁNCHEZ-DÍAZ and MOONEY 1979; LUDLOW 1980). Several woody species have exhibited larger changes in leaf water potential with variations in atmospheric conditions than herbaceous species, possibly due to differences in resistances to flow within the conducting system between soil and leaves.

Seasonal changes in leaf water potential resulting from changes in soil water status have often been greater for xerophytes and woody species than for herbaceous mesophytes. Some herbaceous legumes and succulents have exhibited extremely small changes in leaf water potential under long-term drought. Differences between plants in leaf water potential under long-term drought are mainly due to differences in rooting density and depth and in the regulation of transpiration at the stomata and canopy level. Transpiration rates of the whole plant determine both the rapidity and extent to which soil water is depleted in the root zone, and the extent to which leaf water potential decreases during the day. The strong influence of the regulation of transpiration on the leaf water status of *Vigna unguiculata* is shown in Fig. 7.12B for both adequate soil water and soil drought. In this example, the differences in transpiration rates were due to differences in leaf area, leaf orientation, and average leaf conductance (HALL and SCHULZE 1980b).

In the schematic description of possible relations between leaf water potential and transpiration (Fig. 7.12A), the maximal transpiration rate which can be achieved in an individual plant is described as decreasing with long-term drought, and this is clearly apparent in Fig. 7.12B. Possible mechanisms for the relationship between stomatal conductance and plant water status are examined in the next section.

7.4.2 Effects of Short-Term and Long-Term Changes in Plant Water Status on Leaf Conductance and CO_2 Assimilation

When plants are subjected to constant atmospheric conditions and soil water is not replenished, plant water deficits develop and leaf conductance decreases.

However, the relationship between stomatal conductance and plant water deficits is not simple.

STÅLFELT (1929) described leaves as a hydraulic feedback system in which turgor pressure changes occur in guard cells and subsidiary cells, due to passive and active processes, and cause stomatal responses. This hypothesis is consistent with the transient stomatal opening and subsequent closing when leaves are excised, and the dynamic stomatal responses described for excised leaves by RASCHKE (1970), who proposed for these conditions that "stomatal aperture is controlled by the water potential in the water conducting system". Close coupling of stomatal aperture with leaf water potential is also apparent with intact plants during stomatal oscillations (COWAN 1972). In most studies with intact plants, however, close coupling between stomatal conductance and leaf water potential has not been observed.

In natural environments leaf water potential, temperature, and ambient humidity change in synchrony and it is not known which of these factors is responsible for the changes in stomatal conductance that also occur. SCHULZE and KÜPPERS (1979) separated these factors using a system in which attached leaves were maintained in constant temperature and humidity, while plant water potential was varied by changing the evaporative demand on the remaining part of the shoot. They established for adequately watered plants and for plants subjected to soil drought, that for *Corylus avellana* the changes in leaf water potential which occur in the natural environment (up to 8 bar) would have negligible influence on stomatal conductance of this plant even at −24 bar total water potential.

It has also been proposed that insensitivity to changes in leaf water potential only occurs up to a threshold level, beyond which stomata abruptly close (TURNER 1974; LUDLOW 1980). Threshold responses have been observed when the water supply to the leaves has been drastically changed by excision or by manipulating the root environment in an artificial manner (HALL and HOFFMAN 1976). Evidence for threshold responses under natural conditions is not strong. It is of two types. Firstly, correlations have been made between leaf resistance (mainly measured with transit-time porometers) and leaf water potential using data obtained during diurnal courses. These correlations may also be explained by reduced humidity causing both partial stomatal closure and decreases in leaf water potential (through effects on transpiration, see Fig. 7.12). Furthermore, when data are expressed as resistance rather than conductance a linear relation may appear curvilinear. Secondly, correlations have been made between leaf conductance and leaf water potential using the day-to-day changes in values which occur during a drying cycle. Some examples are presented in Fig. 7.13, and the correlations indicate progressive changes in leaf conductance over a range leaf water potentials. For example, leaf conductance of *Glycine max* decreased rapidly when leaf water potential was between −15 and −16 bar. However, these data were obtained at midday on different days, and diurnal courses for the same days show that the response was not a simple and direct reaction to changes in water potential. The diurnal courses of leaf conductance and leaf water potential of *Glycine max* on the 13th and 33rd days after irrigation are presented in Fig. 7.14. On both days leaf conductance

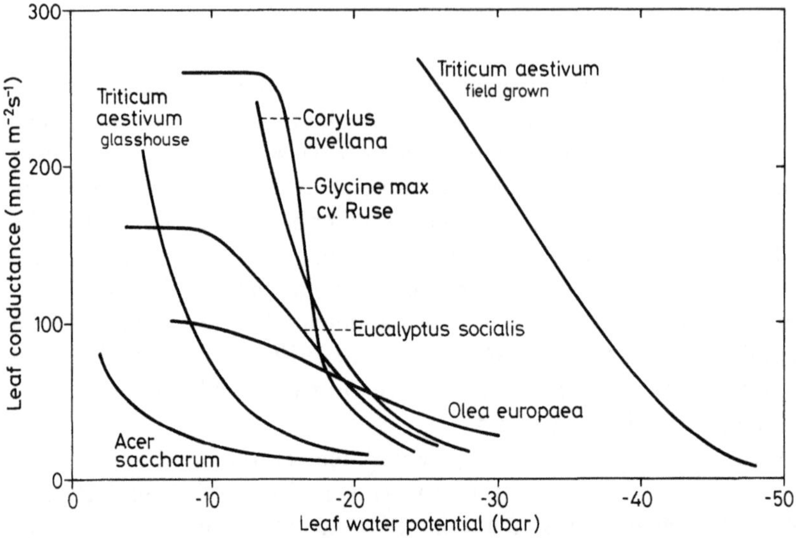

Fig. 7.13. Leaf conductance as related to changes in leaf water potential during progressive drought. *Acer saccharum:* DAVIES and KOZLOWSKI (1977), *Corylus avellana:* SCHULZE and KÜPPERS (1979), *Eucalyptus socialis:* COLLATZ et al. (1976), *Glycine max* cv. *Ruse:* TURNER et al. (1978), *Olea europaea:* LARCHER et al. (1981), *Triticum aestivum* glass house: LAWLOR (1976), *Triticum aestivum* field grown: R.A. FISHER, personal communication

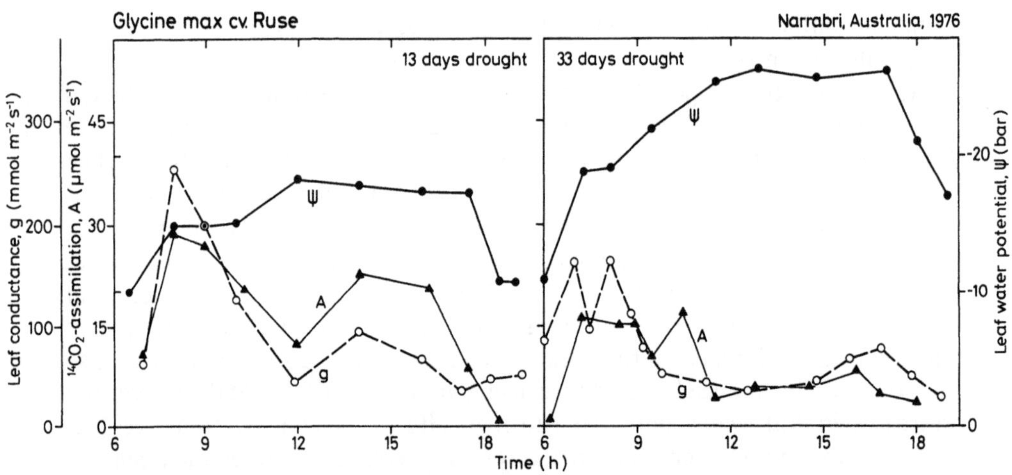

Fig. 7.14. Diurnal course of leaf conductance, g, CO_2 assimilation, A, and leaf water potential, Ψ, after 13 and 33 days of drought. (After TURNER et al. 1978)

increased substantially in the late afternoon even though leaf water potentials were at or far below the threshold. For these data there was no clear relationship between leaf conductance and leaf water potential.

Differences among the absolute range in which leaf conductance is correlated to leaf water potential for the same genotypes grown in laboratory and field conditions (e.g., *Triticum aestivum* in Fig. 7.13), have been interpreted as result-

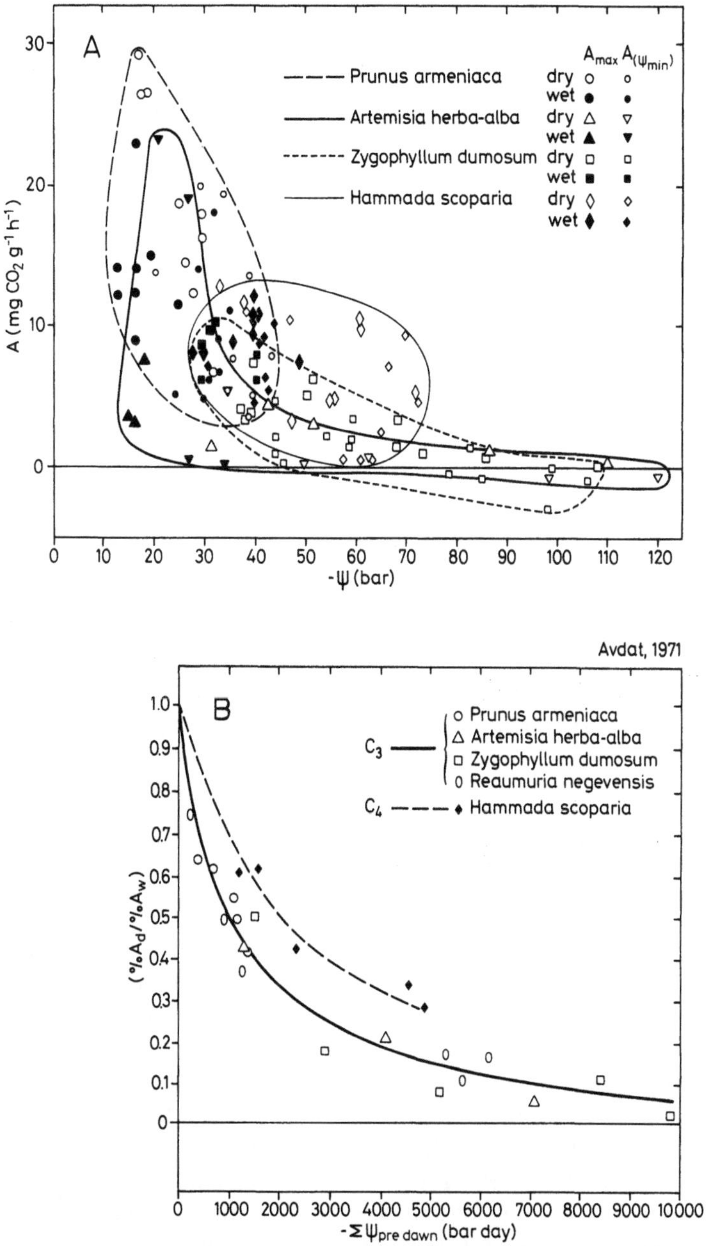

Fig. 7.15. A CO_2 assimilation, A, of nonirrigated and irrigated species as related to simultaneous measurements of water potential, Ψ. The *data points* show maximal rates of CO_2 uptake (A_{max}) during selected diurnal courses of a dry season as well as assimilation rates observed at the time of day when water stress was maximal ($A_{(\Psi_{min})}$) (After SCHULZE et al. 1980a) **B**. Normalized maximal net photosynthesis rates ($\%A_d/\%A_w$) of nonirrigated and irrigated plants during the course of a dry season under desert conditions as related to cumulative predawn water potentials since the time of last rainfall ($-\Sigma\Psi_{pre\text{-}dawn}$). The rates of CO_2 uptake were initialized to be 100% at the time of last rainfall. (After SCHULZE et al. 1980a)

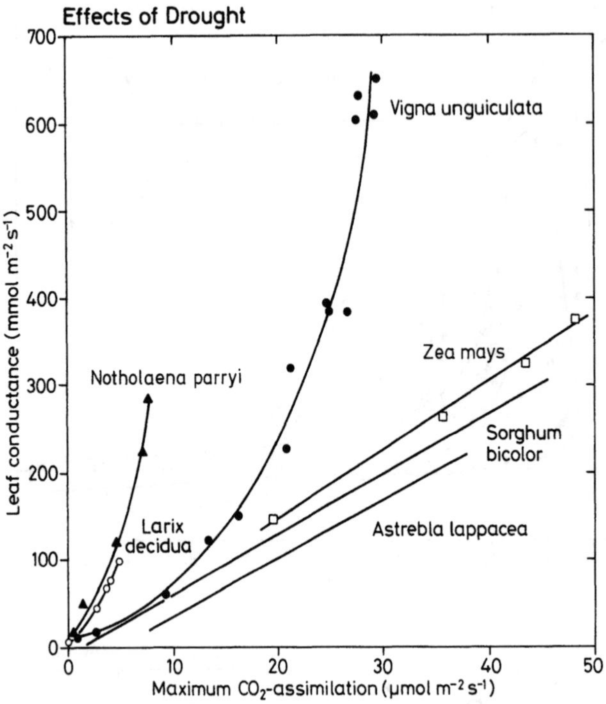

Fig. 7.16. The effects of drought on the relation between leaf conductance and maximum CO_2 assimilation, made under near-optimal conditions for assimilation. C_4 plants: *Astrebla lappacea* (DOLEY and TRIVETT 1974), *Sorghum bicolor* (JONES 1979), *Zea mays* (WONG 1979); C_3 plants: *Larix decidua* with rates related to total needle area (HAVRANEK and BENECKE 1978), *Larrea divaricata* (MOONEY et al. 1976), *Notholaena parryi* (NOBEL 1978), *Vigna unguiculata* (HALL and SCHULZE 1980a)

ing from osmotic adjustment (TURNER 1979). Decrease in osmotic potential results in maintenance of bulk leaf turgor at lower leaf water potentials (JONES and TURNER 1978). However, JONES and RAWSON (1979) observed changes in the slopes of lines correlating leaf conductance with leaf water potential with different rates of drying for *Sorghum bicolor,* and they were not related to differences in bulk leaf osmotic potential. Consequently, it is concluded that the evidence for threshold responses of stomata to changes in bulk leaf water status is weak for intact plants in natural environments. But, even with very slow drying out and repeated drying out cycles, it is a fact that stomata close with drought. It remains open which water relations parameter is responsible for this regulation.

Further progress may be made by testing for the existence of long-term effects of plant water status on maximal stomatal conductance which are mediated by the drought experienced by plants over days and months or through changes in the root (BATES and HALL 1981) rather than the prevailing leaf water status as discussed in Chapter 8, this Volume. The data of NOBEL (1978) indicate that daily maximum leaf conductances of *Notholaena parryi* were correlated with soil water potential during a drying cycle. Daily decreases in soil water potential are usually associated with daily decreases in pre-dawn leaf water potential. RUNNING (1976, 1980) reported that maximum daily leaf conductances of *Pinus contorta* and *Pseudotsuga menziesii* were correlated with pre-dawn pressure chamber values. In contrast, the variation in maximum daily CO_2 assimilation rates, for different species under long-term drought in the Negev desert, was not related to either pre-dawn or prevailing leaf pressure chamber values (Fig. 7.15A). Maximum daily CO_2 assimilation rates of these species were related to their drought experience, expressed as cumulative daily pre-dawn pressure chamber values (Fig. 7.15B). It remains to be determined whether long-term responses of leaf conductance are correlated with some measure of drought experience, soil water potential or pre-dawn leaf water potential.

It has been reported for *Zea mays* (WONG 1979) that variations in leaf conductance with long-term drought exhibit the same linear correlation with maximal CO_2 assimilation rate that was also observed with long-term responses to radiation, inorganic nutrition and leaf age (Fig. 7.16). However, paired measurements of daily maximal CO_2 assimilation rate and leaf conductance for two C_3 species under long-term drought exhibited curvilinear relations (Fig. 7.16). In contrast with Figs. 7.1 to 7.7 drought affected the stomata of the C_3 species to a relatively greater extent than it affected photosynthetic metabolism. This response to drought by the C_3 species may have adaptive significance since intrinsic water use efficiency increased as the soil water supply decreased. It is not known whether the curves in Fig. 7.16 are generally representative at least with respect to shape of the responses to drought of C_3 and C_4 plants.

7.5 Responses of Plants in Natural Environments

Stomatal responses are the principle mechanism in plants for short-term regulation of carbon gain and water loss. This regulation operates within specific ranges and depends on the naturally variable conditions of the habitat. For arid conditions STOCKER (1956) distinguished between various daily patterns of gas exchange, namely one-peaked and two-peaked daily courses, and he associated these types and various intermediate forms with morphological adaptations in different species and with environmental conditions such as light, humidity, and water stress. This analysis is expanded in the following discussion to plant responses in contrasting environments. The variations in CO_2 assimilation rates and leaf conductances in natural habitats are also discussed in relation to A_{max} and g (at A_{max}).

7.5.1 Diurnal and Seasonal Courses of Leaf Conductance and CO_2 Assimilation

7.5.1.1 Dry Environments

Under arid conditions daily courses of gas exchange may vary substantially for different species due to differences in plant structure and metabolism

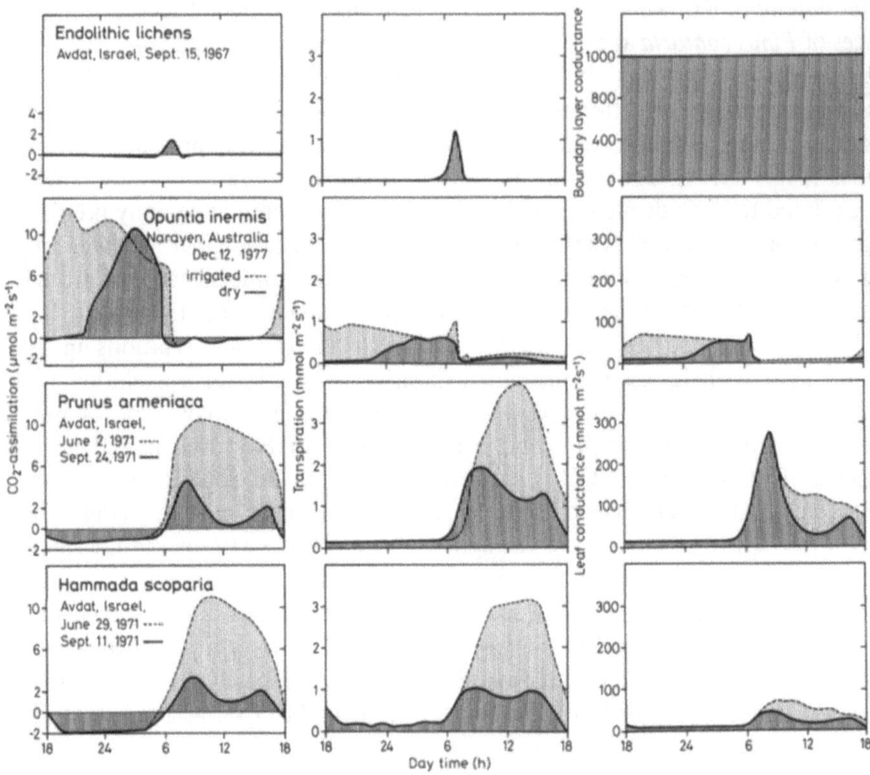

Fig. 7.17. Diurnal courses (dark and light shaded areas) of CO_2 assimilation, transpiration, and leaf or boundary layer conductance in different plant types in a desert environment: Poikilohydric plants (endolithic lichens: LANGE et al. 1970b), CAM plants (*Opuntia inermis:* OSMOND et al. 1979), C_3 plants (*Prunus armeniaca:* SCHULZE et al. 1980a, 1974, 1975), C_4 plants (*Hammada scoparia* with rates related to total surface area of the round assimilating leafless twigs: SCHULZE et al. 1980a; KAPPEN et al. 1976, and original data of the Israel 1971 investigation of O.L. LANGE)

(Fig. 7.17). Many arid regions are characterized by the presence of poikilohydric lichens. Their water relations are completely different from homoiohydric plants since they have no stomata. Without regulation, water status is governed by the humidity of the surrounding air and by the boundary layer. In a dynamic equilibrium with their surrounding, lichens gain water at night with humid air and dew fall (see Chap. 4, this Vol.). This activates respiration at night and allows a short period of CO_2 uptake in the morning before the thallus dries out again due to evaporative water loss (LANGE et al. 1970a, b). Water uptake from the atmosphere is generally restricted to poikilohydric plants, since the cuticle of homoiohydric plants which is effective in restricting water loss is also an effective barrier for water uptake. Hydration of vascular plants by dew may occur to a small extent, and it is still a controversial topic (e.g., SLATYER 1956; KERR and BEARDSELL 1975), as discussed in Chapter 4, this Volume.

Different patterns of CO_2 and water vapor exchange are exhibited by succulents during Crassulacean acid metabolism (Chap. 15, this Vol.). CO_2 uptake occurs at night through open stomata when air humidity is high and the transpiration rate is low. During the day stomata are fully closed and transpiration is very low even though evaporative demands may be high. The result is a high intrinsic water use efficiency. Leaf conductance of CAM plants appears to be partially controlled at night by stomatal response to humidity (OSMOND et al. 1979; LANGE and MEDINA 1979). The rate of CO_2 exchange of CAM plants is highly dependent on night-time temperatures and on the water status of the plant (LANGE et al. 1975; VON WILLERT et al. 1980). CAM plants exhibit changes in the pattern of CO_2 uptake from day-time CO_2 uptake, to night-time CO_2 uptake, and even to internal recycling of CO_2 when stomata are permanently closed (SZAREK and TING 1974).

Figure 7.17 gives examples of one-peaked and two-peaked daily courses of CO_2 and water vapor exchange and leaf conductance for C_3 and C_4 plants as they have been observed depending on environmental conditions and on the long-term effects of drought. It should be made clear, however, that a distinction of various types of daily patterns is convenient, but that all these patterns are part of a continuum, in which gas exchange depends on a considerable number of factors.

One-peaked daily courses of C_3 and C_4 plants with high rates of CO_2 and water vapor exchange are frequently observed with well-watered plants. One-peaked diurnal courses of leaf conductance were found in *Phragmites communis*, growing in a moist habitat but under extremely high evaporative demands, and leaf temperatures were up to 8 °C below ambient air temperatures due to transpirational cooling (PEARCY et al. 1974a). Irrigated *Tidestromia oblongifolia* maintained one-peaked daily courses of CO_2 uptake even at temperatures close to 50 °C (BJÖRKMAN et al. 1972b). However, other species may show irregular one-peaked patterns even when adequately irrigated. *Prunus armeniaca* (Fig. 7.17) showed greater stomatal opening in the morning and a steady decline during the day. The pattern of CO_2 and water vapor exchange varies with increases in atmospheric and soil drought. But it appears that C_4 plants, e.g., *Hammada scoparia*, may maintain one-peaked courses of gas exchange under more extreme conditions of atmospheric and soil drought than C_3 plants growing in the same habitat, e.g., *Zygophyllum dumosum* or *Artemisia herba-alba* (SCHULZE et al. 1980b).

Plants sometimes exhibit one-peaked patterns of gas exchange at very low rates of CO_2 uptake and water vapor loss as a result of long-term drought (STOCKER 1956). *Atriplex hymenelytra* maintained daily leaf conductance and CO_2 uptake at almost constant but low levels in the dry season (PEARCY et al. 1974b). With more extreme conditions of soil drought a short morning peak of CO_2 uptake was observed followed by CO_2 losses in the light (SCHULZE et al. 1980). In this case transpiration rate was low and mainly due to cuticular water loss. Under natural environmental conditions one-peaked patterns of A at low rates of CO_2 uptake may be due to high day-time temperatures, and leaf conductance may be low due to the effects of low air humidity in addition to severe soil water stress. But even under conditions of constant

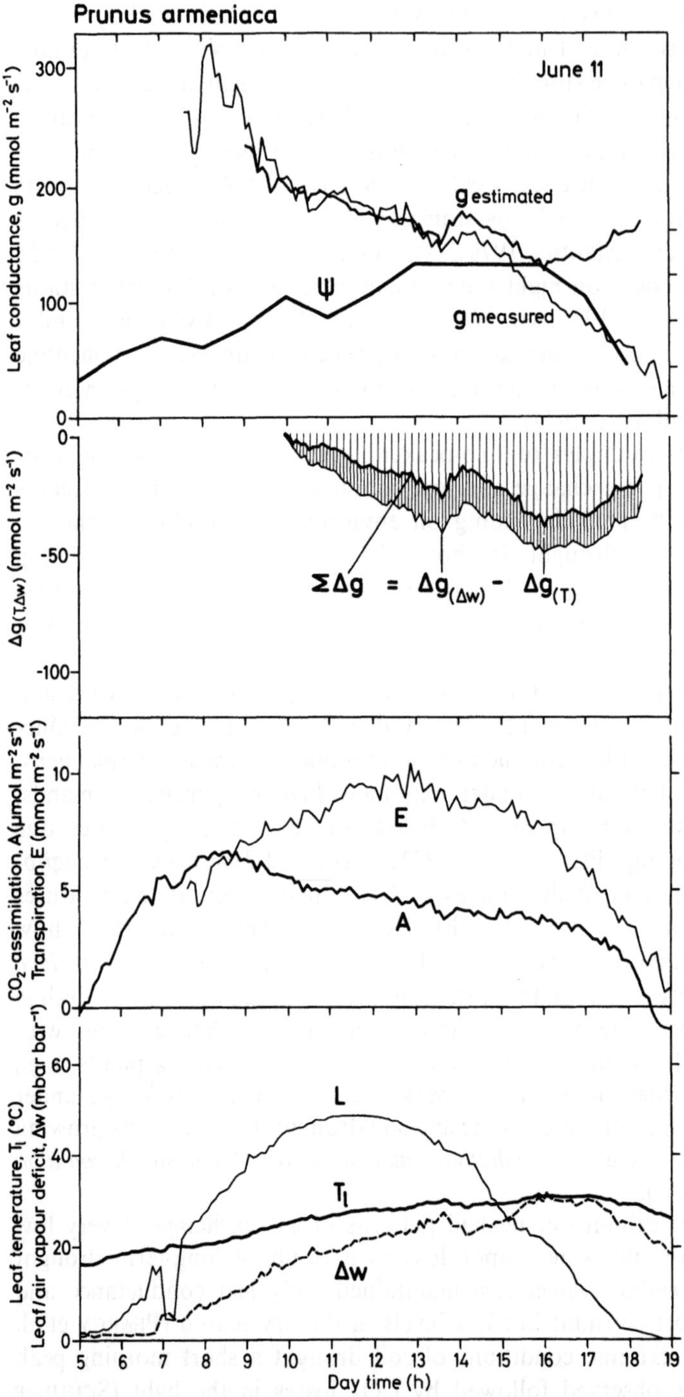

Fig. 7.18. Diurnal courses of leaf water potential (Ψ), measured and predicted leaf conductance for water vapor, g, change in leaf conductance caused by changes in the leaf/air vapor pressure concentration difference ($\Delta g_{(\Delta w)}$), change in leaf conductance caused by changes in leaf temperature ($\Delta g_{(T)}$), total change in leaf conductance caused by changes in humidity and leaf temperature ($\Sigma\Delta g$), CO_2 assimilation, A, transpiration, E, leaf tempera-

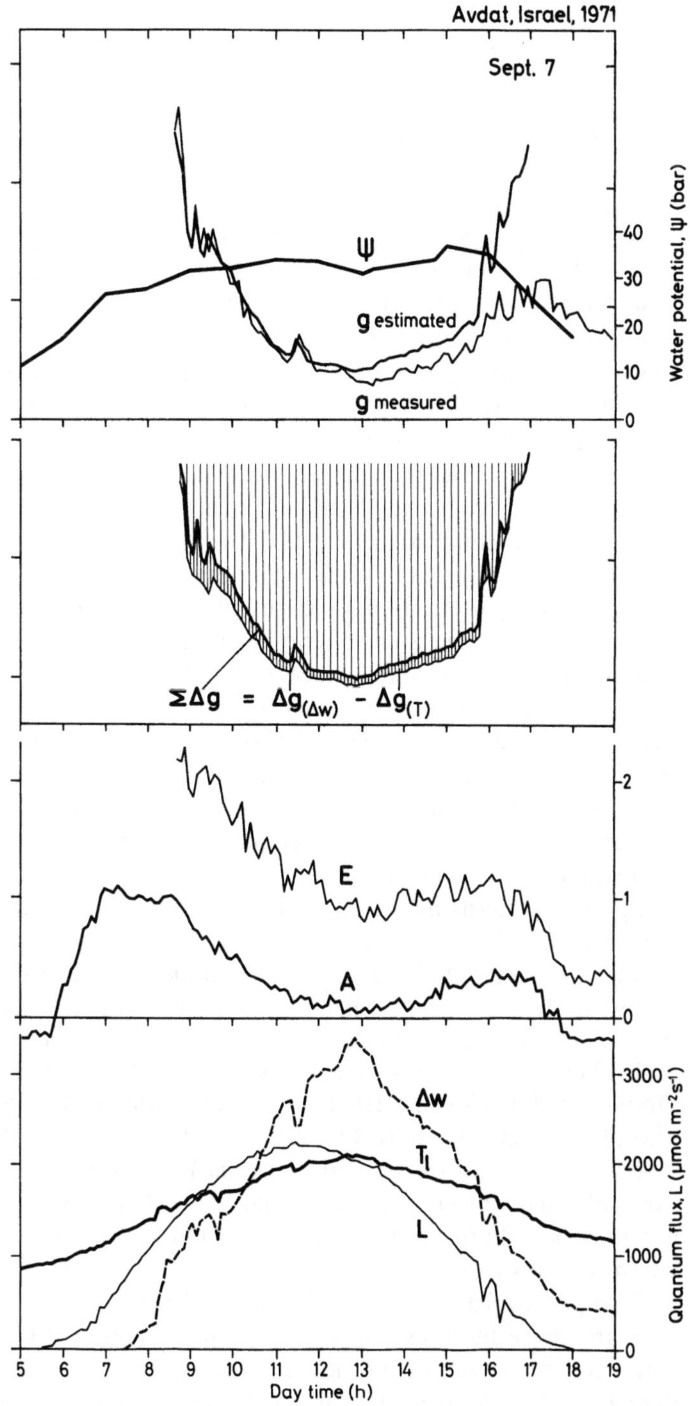

ture, T_l, leaf/air vapor concentration difference, Δw, and quantum flux, L. The changes in leaf conductance with temperature and humidity were calculated from independent steady-state experiments with only one factor variable. The calculations are based on the measured early morning leaf conductance and the measured changes in temperature and humidity from that time on. (After SCHULZE et al. 1972, 1973, 1974, 1975) temperatures in the range of the temperature optimum of CO_2 uptake DE PUIT and CALDWELL (1973) observed a change in daily pattern from a one-peaked pattern with high assimilation rates and leaf conductances at noon to a one-peaked pattern with low assimilation rates and leaf conductances and a morning peak of CO_2 uptake. In this case plant water stress affected both leaf conductance and the capacity of CO_2 assimilation in a correlated manner as discussed in Section 7.4.2.

Two-peaked daily courses of H_2O and CO_2 exchange have been observed in C_3 and in C_4 plants (Fig. 7.17). They occur in arid regions (LANGE et al. 1969; CONNOR et al. 1971; TURNER et al. 1978; PEARCY et al. 1974), in mediterranean climates (MOONEY et al. 1974; TENHUNEN et al. 1981), in temperate forests (BENECKE et al. 1981; SCHULZE 1970) and even in arctic environments (TIESZEN 1978). During a two-peaked diurnal course, CO_2 uptake increases to high rates in the early morning, but stomata close at noon to such a degree that CO_2 uptake decreases. Transpiration may decrease also, but this may not be generally observed since the increased difference in vapor concentration between leaf and air at noon compensates for the effect of stomatal closure. Stomata open again in the late afternoon, resulting in a second peak of CO_2 uptake.

Two-peaked patterns of daily gas exchange by *Prunus armeniaca* were mainly due to changes in the atmospheric environment (SCHULZE et al. 1974, 1975). Daily courses of leaf conductance (Fig. 7.18) were predicted using a model which was based upon independent experiments where only one atmospheric factor was varied (SCHULZE et al. 1974). Effects of variation in air temperature and air humidity accounted for the daily patterns of leaf conductance during the period of light saturation. The cool and humid air conditions early in the wet season resulted in one-peaked patterns of CO_2 and water vapor exchange, whereas the dry and hot conditions in summer resulted in two-peaked patterns. One-peaked patterns have been artificially induced by high air humidity even at the end of a dry season when two-peaked patterns were common in natural weather (SCHULZE et al. 1972) and two-peaked patterns have been observed at low air humidity even when plants were adequately irrigated (TENHUNEN et al. 1980). Similar responses were found in *Zygophyllum dumosum* (SCHULZE et al. 1980b) and in forest trees (BENECKE 1980). For mediterranean climates, MOONEY et al. (1974) correlated two-peaked patterns of *Heteromeles arbutifolia* with plant water potentials. In contrast, *Arbutus unedo* in Portugal, which has a very similar leaf structure to *Heteromeles arbutifolia,* shows a very strong stomatal response to changes in humidity and temperature, which could result in two-peaked diurnal patterns of leaf conductance similar to those of *Prunus armeniaca* (TENHUNEN et al. 1981).

Two-peaked patterns of only CO_2 assimilation may be observed, if temperatures are above the optimum for CO_2 uptake (Fig. 7.19). In *Artemisia herba-alba* a continuous decrease of leaf conductance and transpiration rate occurred during the day. At the same time the daily pattern of CO_2 uptake was two-peaked. CO_2 metabolism was affected by temperatures above the optimum at noon (SCHULZE 1972b). In this case it remains unclear to what extent stomata were directly responding to the environment and to what extent they responded directly or indirectly to plant water status. With long-term drought, *Artemisia*

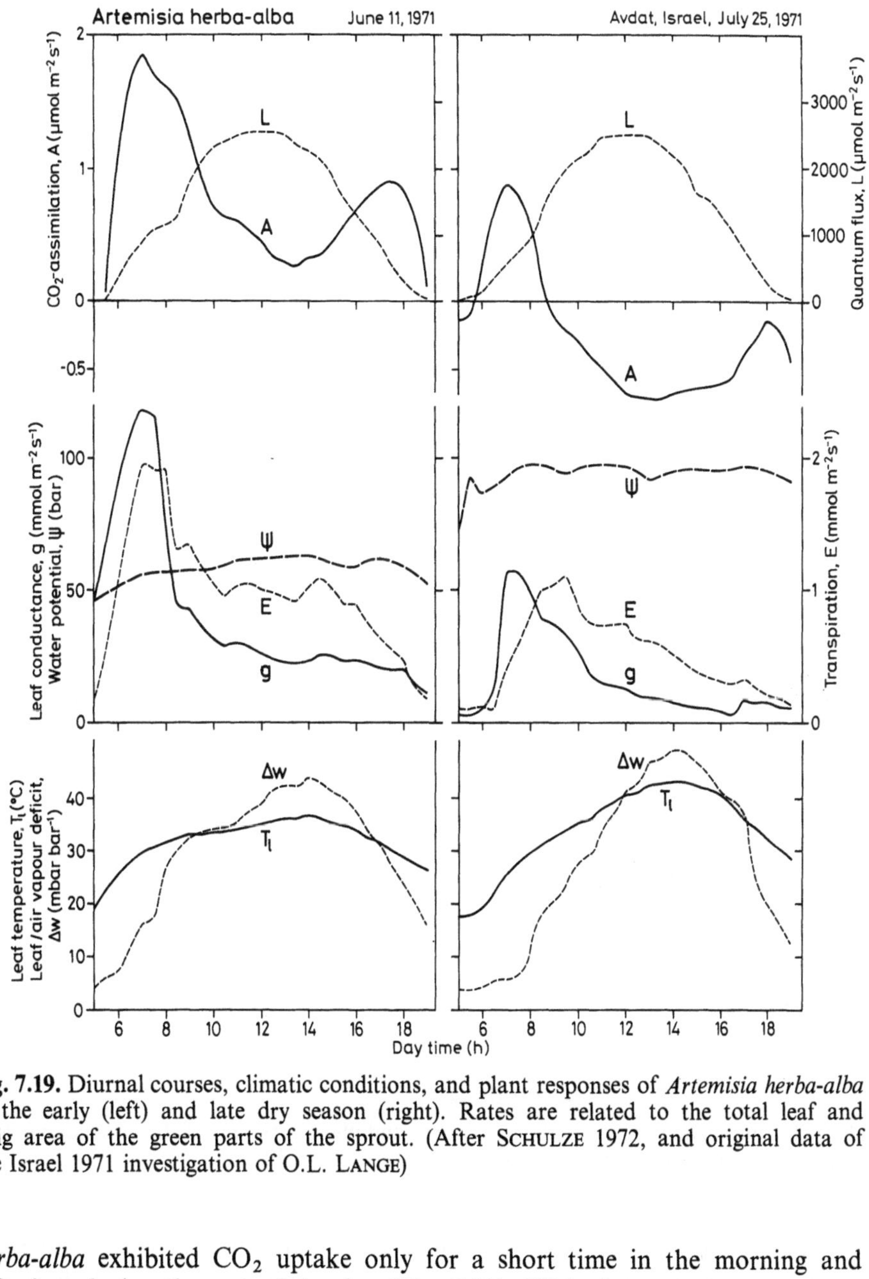

Fig. 7.19. Diurnal courses, climatic conditions, and plant responses of *Artemisia herba-alba* in the early (left) and late dry season (right). Rates are related to the total leaf and twig area of the green parts of the sprout. (After SCHULZE 1972, and original data of the Israel 1971 investigation of O.L. LANGE)

herba-alba exhibited CO_2 uptake only for a short time in the morning and CO_2 loss during the rest of the day (Fig. 7.19). With decreasing temperatures in the afternoon this CO_2 loss decreased until the compensation point was reached where CO_2 exchange was negligible. The result is a two-peaked pattern but in the range of CO_2 loss.

Under arid conditions plants have evolved additional means for controlling leaf temperature and water balance apart from stomatal control, namely leaf pubescence and leaf movement. The presence of leaf pubescence (leaf hairs) in *Encelia farinosa* reduced the absorptance of irradiance by as much as 56% compared with the nonpubescent plant *Encelia californica*. This reduction in

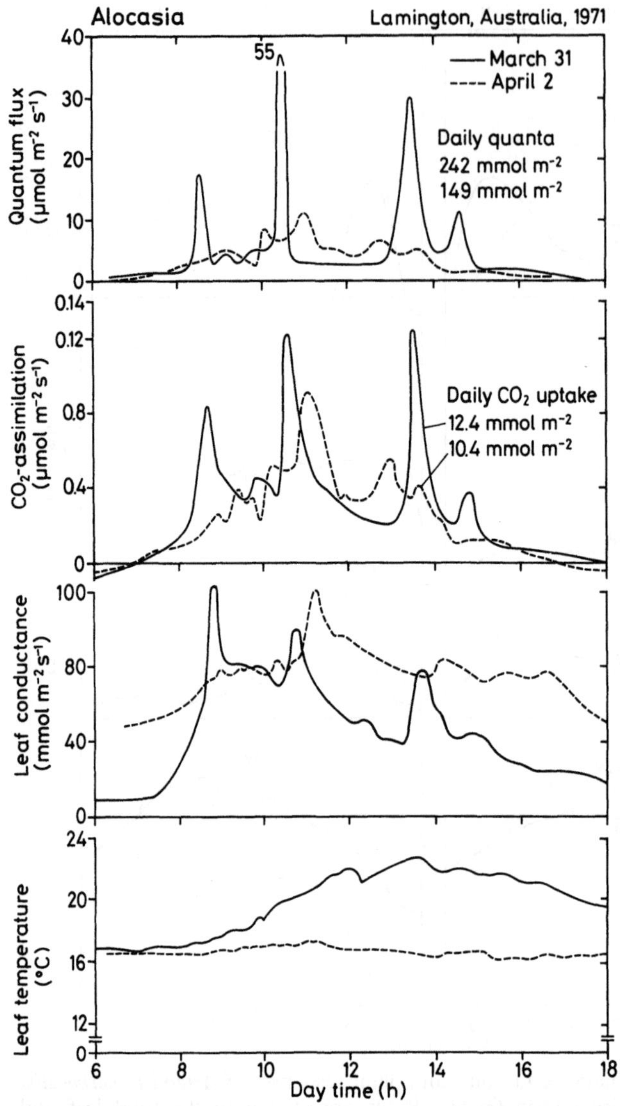

Fig. 7.20. Diurnal courses of weather conditions and plant responses of *Alocasia macrorrhiza* on the forest floor of a rain forest. (After BJÖRKMAN et al. 1972a)

absorptance resulted in lower leaf temperatures and lower transpiration rates (EHLERINGER et al. 1976). In addition, leaf hairs reflect quanta which would tend to reduce the rate of photosynthesis. However, for a high radiation climate with high air temperatures EHLERINGER and MOONEY (1978) predicted that the reductions in leaf temperature, bringing it closer to the optimum, would more than counterbalance the effect on photosynthesis of decreased light. Leaf movement may change its mode on the same plant individual and even at different times of the day. The most contrasting modes of leaf movement are sun-tracking and sun-avoiding. SHACKEL and HALL (1979) provided experimental evidence for the effects on leaf temperature of sun-tracking and sun-avoiding leaf movements and MOONEY and EHLERINGER (1978) discussed the effect on photosynthesis of sun-tracking compared with stationary leaves.

7.5.1.2 Environments with Low and Variable Irradiance

Light limitation of net photosynthesis is known for temperate and humid tropical environments. These habitats are generally characterized by high air humidity and the absence of prolonged dry periods, compared to semi-arid and arid regions. The forest floor of tropical rain forests has extremely low light levels and high air humidity. *Alocasia macrorrhiza* (Fig. 7.20) has an extremely low light saturation point for higher plants of 80–100 µmol photons $m^{-2}\,s^{-1}$. But even this photon flux is rarely reached in its natural habitat. CO_2 assimilation rate closely followed changes in photon flux during the day. The diurnal course of leaf conductance was very similar to that of light and CO_2 uptake. Ecologically, the important result is that in an environment of low evaporative demand and high water supply, plants appear to lose water only at a rate which is correlated with the capacity of the mesophyll for fixing CO_2. COWAN (1977) showed for this experiment that at no time was g limiting A to any appreciable extent.

The effects of low and variable light on leaf conductance and CO_2 assimilation are illustrated by the performances of sun and shade leaves of *Fagus silvatica* (Fig. 7.21). The comparison contains a long-term effect of the level of visible radiation during growth on the maximal rate of CO_2 exchange and on leaf conductance, as well as short-term effects of changing weather conditions during the day on the diurnal course of CO_2 and water vapor exchange. The long-term effect of the level of irradiance during growth can be detected from the highest rates of CO_2 uptake during the day, which were very similar in young sun and shade leaves, but which differed considerably in summer between old sun and shade leaves. In summer, shade leaves achieved half the assimilation rates of sun leaves but with only 1/10 of the irradiance of sun leaves. Leaf conductance appeared to be associated with the level of CO_2 uptake for sun and shade leaves in the long-term manner discussed in Section 7.2.1 for g (at A_{max}) and A_{max}. The short-term performance of CO_2 assimilation and leaf conductance within their operational ranges can be seen from the diurnal courses in response to weather conditions. The rates of CO_2 uptake were strongly dependent on light intensity. *Fagus silvatica* has a broad temperature optimum for photosynthesis between 10° and 24 °C (SCHULZE 1970), and only was limited by temperature on the day in May (Fig. 7.21). The diurnal course of leaf conductance varied in a similar way to CO_2 uptake and light. Additionally, for sun leaves of *Fagus silvatica*, stomata close at midday in response to high vapor pressure deficits (SCHULZE 1970). Similar results were found in other tree species (ROBERTS et al. 1979; ZOBEL and LIU 1980; BENECKE et al. 1981). For apple leaves daily changes of leaf conductance were modeled based upon responses to variations in light and humidity only (THORPE et al. 1980). In addition, the influence of plant internal water relations on stomata must be considered. Field observations with tall trees of *Pinus contorta* indicated correlations between maximal leaf conductance and pre-dawn leaf water potentials but the daily

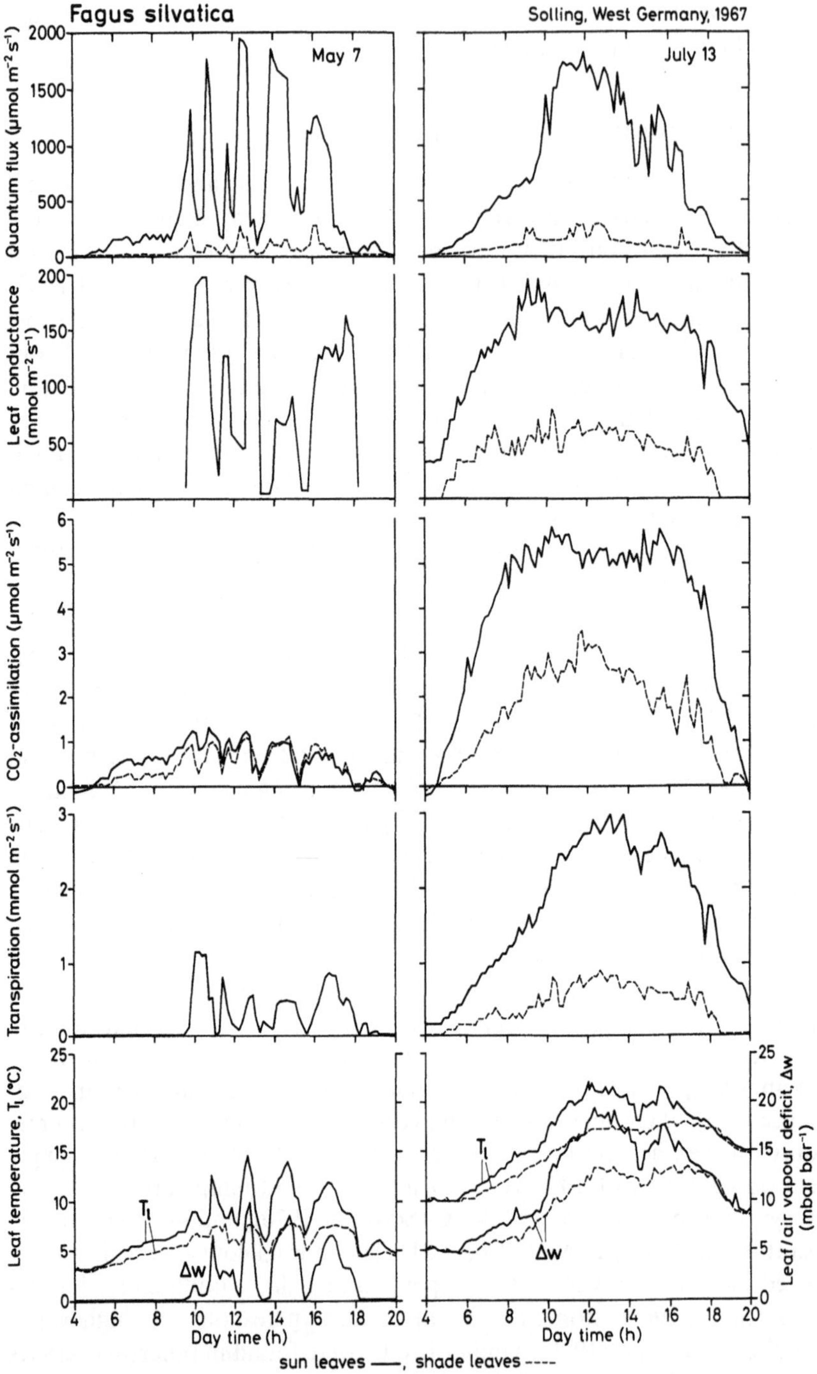

Fig. 7.21. Diurnal courses of weather conditions and plant responses of sun and shade leaves of *Fagus silvatica* shortly after bud break (May 7) and during summer (July 13). (After SCHULZE 1970)

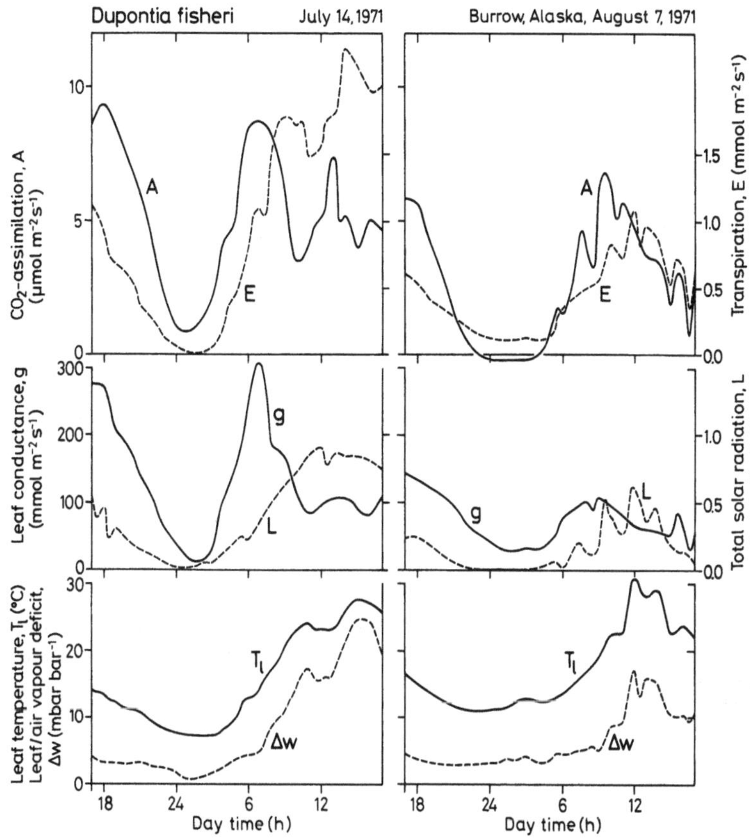

Fig. 7.22. Diurnal courses of weather conditions and plant responses of *Dupontia fisheri* in a tundra habitat. (After TIESZEN 1978)

course of leaf conductance was related to changes in humidity difference between leaf and air (RUNNING 1980).

7.5.1.3 Cold Environments

Arctic and alpine environments are characterized by low temperatures and a short growing season (WALTER 1968). Night-time temperatures may drop below freezing even during summer (TIESZEN 1978; TRANQUILLINI 1957), although daytime temperatures may be quite high, especially in plant forms like cushion plants (LARCHER 1977). In *Dupontia fisheri* (Fig. 7.22) CO_2 uptake closely followed irradiance and continued for long periods due to the midnight sun of the arctic summer (TIESZEN 1978). Leaf conductance closely followed irradiance and CO_2 assimilation at low light intensity, but lower leaf conductances were observed in full sunlight. According to TIESZEN (1978), this midday closure of stomata resulted from water stress under high irradiance and high leaf temperature conditions. However, during the course of the day, there was also a pronounced and concurrent increase of the water vapor pressure difference between leaf and air. Therefore it is not clear whether the stomata responded to water stress, temperature, or humidity, especially since midday temperatures were within the range of optimum CO_2 uptake. Strong light and humidity response of leaf conductance have been observed with *Pinus mugo* at the alpine timberline of New Zealand (BENECKE et al. 1976), and changes in temperature did not substantially affect photosynthesis of *Picea mariana* during the summer months at the subarctic tree line (VOWINKEL et al. 1975). For several arctic species CO_2 assimilation rate was relatively independent of temperature even at 3 °C (TIESZEN 1975). Slight inhibition of photosynthesis was observed at lower temperatures, but CO_2 uptake was positive as low as −5 °C and as high as 35 °C in *Dupontia fisheri* (TIESZEN 1978). Temperatures below freezing of the needle strongly affected the photosynthetic capacity of *Pinus cembra* (TRANQUILLINI 1957), but the effects of such low temperatures on stomatal conductance have not been reported for natural arctic or alpine field environments.

Although tundras are not generally considered as water-limited environments, development of plant water stress and subsequent effects on productivity have been suggested for a number of habitats in both arctic and alpine regions. CALDWELL et al. (1978) proposed that water stress may even play a key role in the distribution of tundra species along microtopographic gradients. EHLERINGER and MILLER (1975) and MILLER et al. (1978) reported minimal seasonal water potentials for some tundra species as low as −30 bar (*Deschampsia*) and −40 bar (*Kobresia*). JOHNSON and CALDWELL (1976) described considerable effects of long-term water stress – as produced in the laboratory by soil drought – on both leaf conductance and CO_2 assimilation rate. But the effects of water stress on plant responses have not been studied in the field in arctic environments.

7.5.2 Variations of CO_2 Assimilation Rate and Leaf Conductance Within Their Operational Ranges Under Natural Field Conditions

The previous sections discussed A_{max} and g (at A_{max}) in relation to habitat and plant internal factors. The extent to which the daily average rates of CO_2 assimilation ($\bar{A}$) and leaf conductance ($\bar{g}$) fall below their maximal levels in natural environments is not known, although this should have significance to the adaptation of plants. In the following, the average performance of A and g will be examined under prolonged periods of natural weather conditions, and the climatic factors which account for differences between $\bar{A}$ and A_{max} will be estimated.

In the following, first, the way in which A changes with g will be established for representative weather conditions in contrasting climates, second, the arithmetic averages of daytime changes in A and g will be depicted, third, $\bar{A}$ will be related to A_{max} using an empirical model (SCHULZE et al. 1976), which predicts the extent to which $\bar{A}$ is reduced below A_{max} due to suboptimal levels of light, temperature, or humidity. This analysis gives information concerning the climatic factors which strongly influence A and g in contrasting environments. The

C_4 plant *Hammada scoparia* of arid regions and the C_3 plants *Fagus silvatica* and *Prunus armeniaca* growing in humid temperate and in arid environments, respectively, are examined. In these experiments care was taken in measuring A_{max}, but it was not always possible to accurately measure g (at A_{max}) and g at low light during humid morning hours or during rain in the field because of water vapor condensation in the chambers.

Paired values of A and g of *Fagus silvatica* increased until a plateau was reached which probably represents the level at which A was saturated with respect to g (Fig. 7.23 solid line). Additional values were observed which had a lower A/g ratio. They occurred in the field following rapid changes in light,, e.g., by clouds, when A dropped instantaneously but stomata remained open. At low rates of CO_2 uptake, shade leaves reached higher levels of A for the same degree of stomatal opening than sun leaves, but the highest values of A and g were lower in shade than in sun leaves, presumably due to long-term effects of light during leaf expansion. The daily average rates of A and g as calculated from 14 continuous daily recordings are very close to the solid line in Fig. 7.23. Thus, the occasions when stomata were widely open at low rates of A were rare with respect to the total time. The data points in Fig. 7.23 do not represent all half-hourly means of the indicated time period (July 12–28), since it was not always possible to accurately measure g during rain and in the early morning hours (Fig. 7.26 contains the total time comparison). $\bar{g}$ was calculated from measured values in similar climatic conditions. If the stomata of *Fagus silvatica* were adjusted to a certain light climate then they appeared to open to such a degree that further increases in g would result in small increases in A compared with the increases in water loss.

Figure 7.24 and Fig. 7.25 shows paired values of half-hourly means of A and g for *Prunus armeniaca* and *Hammada scoparia* during diurnal courses over weekly periods from the end of the wet season to late in the dry season. Despite considerable scatter, there is a general increase of A with increases in g. The rates of A decreased from the wet to the dry season due to long-term effects of drought. In the morning and in the evening at low light and humid air, stomata were frequently widely open at low rates of CO_2 uptake (Fig. 7.24: dashed lines with arrow heads indicate the change of rates with time). The relation between A and g was curvilinear during humidity-induced stomatal closure at optimum temperature (Fig. 7.24: dotted line). For *Prunus armeniaca* water-stressed plants reach higher rates of CO_2 uptake for the same level of leaf conductance than earlier in the dry season when the plants had adequate water. When comparing absolute rates it must be recognized that A was based upon total surface area for the leafless stems of *Hammada scoparia*, and projected leaf area for *Prunus* and *Fagus*.

Of the three species which were examined only *Fagus* achieved rates of CO_2 uptake under natural weather conditions which reached A_{max}. In arid climates A_{max} was reached neither in the late wet season nor in the dry season. Generally in the morning, the photon flux was too low and for the C_4 plant, additionally, it was too cold. During times of light saturation it was too dry for stomatal opening and for the C_3 plant it was too hot for optimal CO_2 uptake. A_{max} was measured only in special experiments where all atmospheric

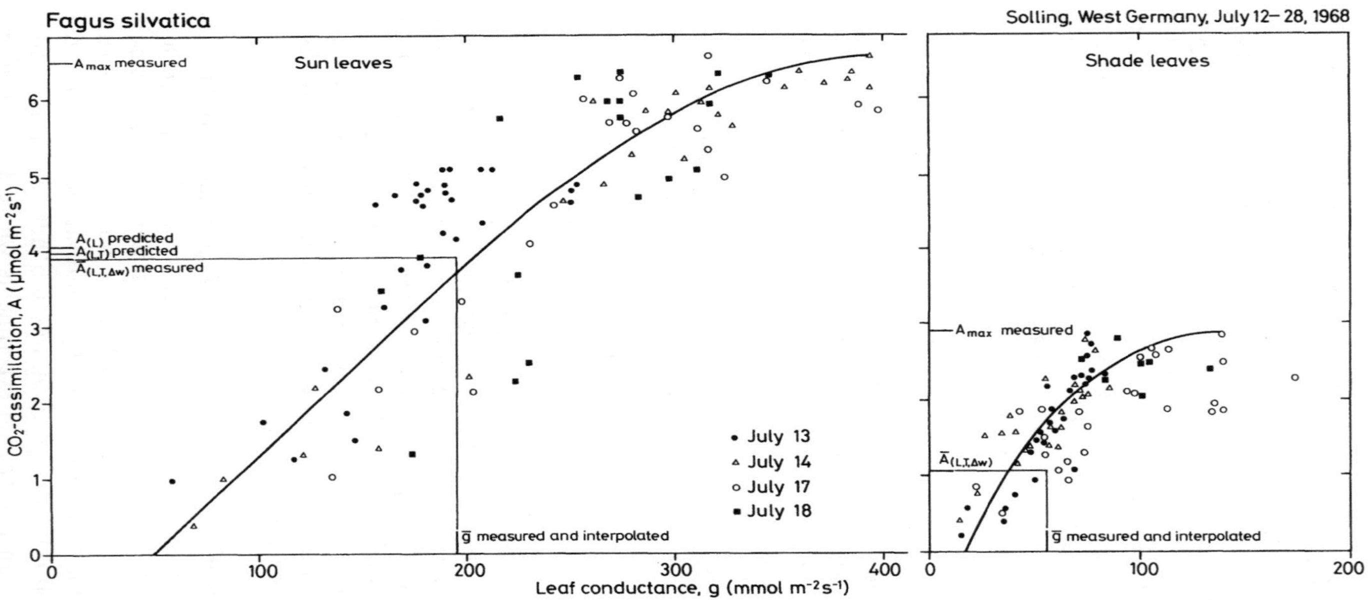

Fig. 7.23. Short-term changes of CO_2 assimilation, *A*, during diurnal courses of a week's period (half-hourly means) as related to concurrently measured leaf conductances, *g*, in sun and shade leaves of *Fagus silvatica*. The *data points* do not represent the full time record. In early morning hours after dew and during rainfall only A was measured but not g. The average rate of CO_2 assimilation, $\bar{A}$, is based on a full record of measurements, whereas the average rate of leaf conductance, $\bar{g}$, contains measured and interpolated values. A_{max} was measured in special steady state experiments with humid air, optimal temperatures and light saturation. $A_{(L)}$ indicates the predicted average assimilation rate if temperature and humidity were optimal during the day from dawn to dusk. $A_{(L,T)}$ indicates the predicted average assimilation rate if only humidity were optimal during the day from dawn to dusk. (Original data of the Solling investigation of SCHULZE 1970)

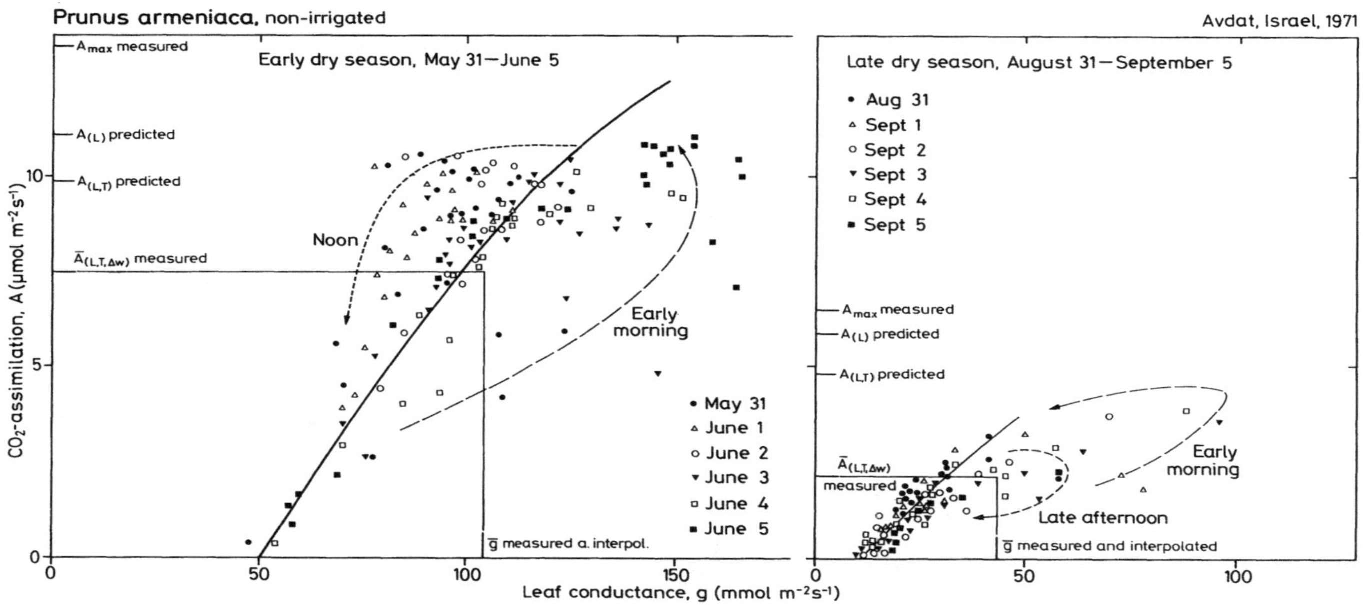

Fig. 7.24. Same as Fig. 7.23 but for *Prunus armeniaca* in the early and late dry season (original data of the Israel 1971 investigation of O.L. LANGE). *Dashed lines* indicate changes of A with g at changing light during dawn and dusk. *Dotted lines* indicate changes of A with g during diurnal changes in air humidity

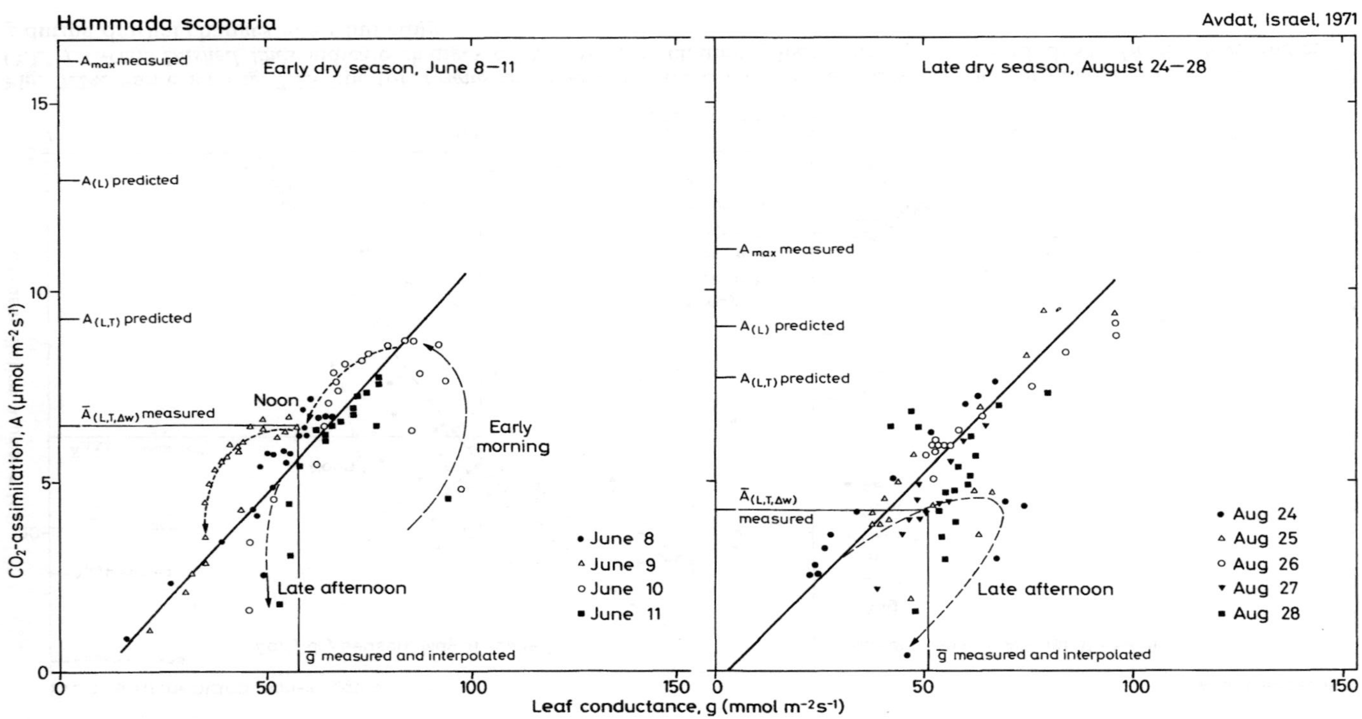

Fig. 7.25. Same as Fig. 7.24 but for *Hammada scoparia* in the early and late dry season. Rates are related to total surface area of the round, assimilating, leafless twigs (original data of the Israel 1971 investigation of O.L. LANGE)

Table 7.1. Climatic conditions and the deviation of average daily CO_2 assimilation from A_{max} under contrasting environmental conditions

	Temperate zone		Arid zone			
	Fagus silvatica		C_3 plant *Prunus armeniaca*		C_4 plant *Hammada scoparia*	
	Sun leaves	Shade leaves	Early dry	Late dry	Early dry	Late dry
Climatic conditions						
Rel. portion of total daytime (%):						
1. below light saturation	67	95	39	31	60	42
2. >10% off temperature optimum	25	29	37	55	51	52
3. leaf/air vapor pressure difference >5 mbar	9	<1	33	79	90	95
Rel. portion of time with light saturation:						
1. >10% off temperature optimum	20	4	5	81	12	32
2. leaf/air vapor pressure difference >5 mbar	20	3	96	96	100	100
CO_2 assimilation						
Daily average A/A_{max} (%)	62	38	68	32	40	42
Depression of daily average A below A_{max} is due to (%):						
1. Nonsaturated light	96	99	42	15	34	29
2. Nonoptimal temperature	4	1	19	26	20	21
3. Low air humidity	<1	<1	39	63	46	50

factors were kept artificially optimal. The highest rates of A under natural weather conditions for the plants in the arid zone were 15%–40% below A_{max}.

In all climatic conditions daily average rates of A were significantly below A_{max} (Table 7.1: 32%–68% below A_{max}). But the reasons for this depression of $\bar{A}$ below A_{max} were different in humid temperate and arid climates. In the temperate zone, light was below saturation during $^2/_3$ of the total daytime in sun leaves (95% in shade leaves) but temperatures and humidities were quite favorable for gas exchange. In contrast, in the arid region the leaf/air vapor pressure difference was greater than 5 mbar during more than 95% of

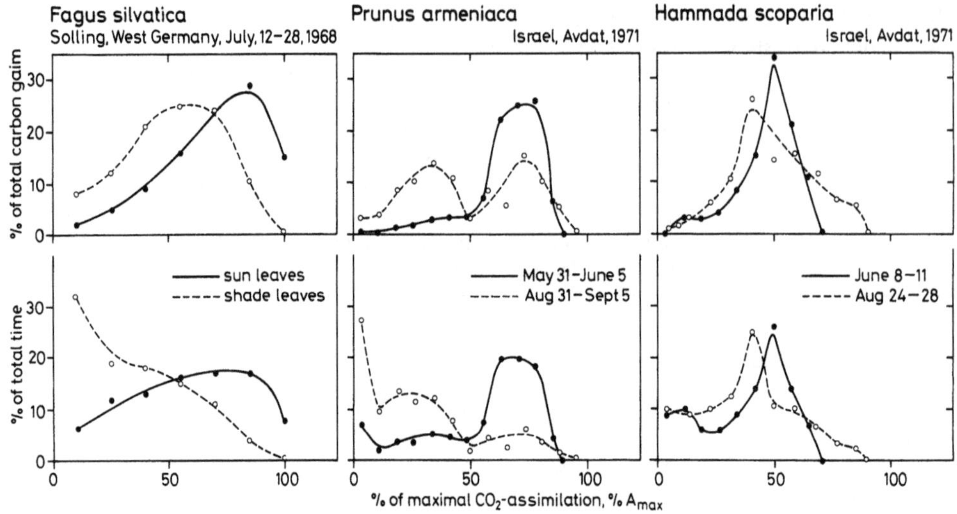

Fig. 7.26. The relative proportions of the total carbon gain which are assimilated at various rates of CO_2 assimilation, and the relative portion of the total daytime during which certain rates of assimilation were observed under natural conditions of *Fagus silvatica* sun and shade leaves, and *Prunus armeniaca* and *Hammada scoparia* leaves early and late in the dry season

the time of light saturation, but light was below saturation only during 30%–40% of the time for C_3 plants and 46%–50% for C_4 plants (Table 7.1).

Using an empirical model of CO_2 assimilation (SCHULZE et al. 1976), calculations were made keeping either light, temperature, or humidity at optimal levels throughout the daylight time while varying all the other factors as they occur in the natural environment. Thus interpolations were made between $\bar{A}$ and A_{max} for conditions which were partially optimal. In all simulations the effects of light, humidity, and temperature explained more than 90% of the measured variation in A. In *Fagus silvatica* light appears to be the single dominant factor reducing the rate of A below A_{max} in the natural environment. The conditions were similar for *Picea abies* (FUCHS et al. 1977) and for herbaceous plants on the forest floor (SCHULZE 1972b). It would be interesting to know the extent to which a high rate of A_{max} is adaptive in conditions of low average irradiance, recognizing the investment cost for maintaining a high level of A_{max}. Figure 7.26 shows that a relatively large proportion of carbon gain was assimilated during the short period of light saturation in sun leaves of *Fagus silvatica*. Therefore CO_2 uptake at A_{max} significantly contributed to the carbon balance. In contrast, however, in shade leaves the contribution of CO_2 assimilation rates at A_{max} to the daily carbon gain was negligible. Experiments with *Larix decidua* and various broad-leaved shrubs, such as *Acer, Crataegus,* and *Rosa* indicate that the results of *Fagus* may be representative for the temperate region.

In arid regions, the effect of nonsaturated light was predicted as being low but the effects of temperature (too cool or too hot for *Prunus* and too cool for *Hammada,* LANGE et al. 1978) and air humidity were substantial (Table 7.1). For *Prunus armeniaca* the effect of air humidity was dominant, demonstrating the limitations to CO_2 assimilation of this species in a desert environment, where it can exist only under conditions of run-off farming (EVENARI et al. 1977). It is obvious from Fig. 7.26 for *Prunus armeniaca* that a relatively large proportion of the total carbon gain was assimilated during the cool and humid morning hours at low light at 75% of A_{max}. A second large proportion of the total carbon gain was assimilated at 25% of A_{max} during the time of stomatal closure at noon. This is in contrast to the desert species *Hammada scoparia* in which the largest portion of total carbon gain was assimilated at 50% of A_{max} and the short periods which allowed higher rates of CO_2 uptake did not contribute significantly to the carbon balance.

7.6 Conclusions

The environmental and plant factors that influence the range of stomatal conductance were examined. Linear correlations were apparent between maximal CO_2 assimilation rates (A_{max} was determined at optimal temperature, light and humidity, and natural concentrations of CO_2 and O_2), and concurrent levels of leaf conductance (g), as they both varied due to long-term effects of light, mineral nutrition, and leaf age. In some cases, however, g (at A_{max}) was not correlated with A_{max}; and it is likely that the linear correlations were not due to a close physiological coupling of these plant parameters. For C_3 species, the correlation is not simply due to environmental influences on g which, in turn, influence A through effects on the diffusion of CO_2 into leaves. Apparently, the range of stomatal conductance varies such that it matches the photosynthetic capacity as it is determined by the long-term effects of several environmental and plant factors.

The linear regression of g (at A_{max}) versus A_{max} was relatively constant for plants from the same species but varied among and between C_3 and C_4 plants from different species. Smaller regression slopes, indicating higher intrinsic water use efficiency, were apparent in C_4 species and some C_3 species that are adapted to more arid habitats.

The regulation of leaf conductance and CO_2 assimilation rate within the operational range is strongly dependent upon the short-term responses of stomata to light, humidity, and temperature. Stomatal responses to light may result from indirect effects of light on CO_2 assimilation rate which operate via stomatal response to internal CO_2 concentration or some product of CO_2 assimilation in the mesophyll tissue, or direct effects of light on guard cell metabolism. Stomatal response to internal CO_2 concentration is substantial in some C_4 species, whereas stomatal response to light by some C_3 species cannot be explained by their responses to internal CO_2 concentration.

Leaf conductance response to humidity is present in many species with contrasting adaptation, including species from humid habitats. Increases in temperature result in increases in leaf conductance over a substantial range of temperatures when the vapor pressure difference between leaf and air is kept constant. Evidence has been obtained which indicates that stomatal responses to humidity and temperature may maximize daily water use efficiency as described by the optimization hypothesis of COWAN and FARQUHAR (1977). This approach for defining and analyzing optimal function has general applicability to many plant processes and should provide a useful conceptual base for future research.

The effect of water stress is complicated by the interaction of effects of transpiration on plant water potential due to stomatal opening and air humidity and effects of plant water status on leaf conductance. It is useful from a conceptual and an experimental standpoint to separate the long-term and short-term effects of plant water status on stomatal function. Evidence for the occurrence of short-term effects under natural conditions is controversial. It is suggested that stomata of intact plants, under natural conditions frequently do not respond to changes in bulk leaf water status in either a coupled or a threshold manner even with extreme soil drought. However, leaf water potential is strongly dependent upon the transpiration rate. The sensitivity with which leaf water potential changes with transpiration is affected by resistances to liquid water flow from the soil to the leaf, and is strongly influenced by water depletion in the root zone. Long-term effects of water stress determine the maximum capacity of the stomata to open and the sensitivity of their responses to atmospheric conditions, and this appears to be associated with long-term effects of drought on the stomata themselves and possibly long-term effects of drought on the capacity of the leaf for fixing CO_2. The relation between g (at A_{max}) and A_{max} during water stress was not linear for some C_3 plants. This indicated that the effect of long-term water stress on stomata was greater than on CO_2 assimilation. Separate studies indicate that maximal leaf conductance was correlated with pre-dawn leaf water potentials, whereas CO_2 assimilation was correlated with the drought experience as expressed by cumulative pre-dawn water potentials. The differential response of g and A causes a rotation of the g/A_{max} regression, indicating greater intrinsic water use efficiency with progressive depletion of soil water.

Although mechanisms of plant responses have been studied successfully in the laboratory, the integrated effect of habitat conditions and the species-specific adaptations of plants to certain environments can only be studied under natural conditions. The diurnal courses of CO_2 assimilation rate and water vapor exchange provide fundamental information on plant responses and adaptations to natural environments. But it is difficult to determine the mechanisms responsible for diurnal courses if several environmental factors change simultaneously. In this case an interpretation appears to be only possible if measurements under naturally variable conditions are supplemented by field experiments where individual environmental factors are changed. The diurnal courses of CO_2 assimilation and leaf conductance describe the short-term effects of mainly atmospheric conditions on A and g within their operational range. The operational ranges reflect the long-term effects of habitat conditions during growth. In humid and arid, as well as in hot and cold climates, daily changes of stomatal response and CO_2 assimilation appear to be closely associated with short-term changes in light, humidity, and temperature. Even on the forest floor of a rain forest, plants appear to adjust stomatal aperture at a level which is related to the rate of CO_2 uptake. But this short-term relation is not unique, e.g., during the morning hours of a desert environment, or during light saturation with humid air in a temperate climate, g might increase more than the short-term association between A and g would suggest. Air humidity appears to be the next most important atmospheric parameter in humid, arid, and arctic environments, whereas some plants appear to be quite well adapted to the temperature regimes of their habitats. At the present time, the effects of plant water status during diurnal courses appear not to be adequately understood. A continual decrease of leaf conductance during the day, which was observed not only in diurnal courses at low rates of CO_2 and water vapor exchange, but also under constant environmental conditions, may be related to long-term effects of drought, which may partially operate through changes in the root which cause concurrent changes in photosynthetic capacity and leaf conductance.

During short-term changes in environment the association between A and g is modified by differential effects of light, humidity, and temperature on CO_2 uptake and leaf conductance. Average daily rates of CO_2 uptake were related to A_{max} for three species from humid temperate and arid regions. Ecologically it seems important that the highest rates of CO_2 assimilation which were measured under natural weather conditions were 15% to 40% below A_{max} in arid regions. The photon flux was too low in the morning and later in the day the air was too dry for maximal CO_2 uptake. The daily average rate of CO_2 assimilation was only 32% to 68% of A_{max} in humid temperate and arid regions. But the reasons for the depression of $\bar{A}$ below A_{max} were different for humid and arid environments. More than 95% of the depression was due to non-saturated light in the humid temperate region, whereas air humidity (39–63% of the depression) followed by light (15–42% of the depression) were the most significant factors in arid climates. Ecologically, it is not clear to what extent a higher rate of A_{max} is adaptive, when $\bar{A}$ is more than 50% below A_{max} recognizing the high investment costs of maintaining a high level of A_{max}.

It is quite obvious that the foregoing discussion on plant responses in contrasting environments is based on a relatively small number of species which have been studied thoroughly. Understanding plant adaptation to contrasting environments will require information on mechanisms obtained in the laboratory and in the field and descriptions of the responses of plants under natural conditions. A more balanced approach appears to be essential in future research. The large number of response curves for single variables, which have been obtained in laboratories, have not been adequately complemented by studies of plant responses under naturally varying conditions in the field, not even for agricultural plants. For understanding stomatal function and regulation it appears necessary to study the concomitant variations of leaf conductance and CO_2 assimilation rate. Investigations on plant responses under natural conditions should be given priority in which measurements are made of diurnal courses of CO_2 and water vapor exchange under naturally varying weather but over prolonged periods of time, as well as of response curves of single variables under experimentally controlled constant conditions in the field. In these latter experiments A_{max} and g (at A_{max}) should be determined. The extent to which the daily average rates of CO_2 assimilation and leaf conductance fall below their maximal levels appear to provide quantitative information on the adaptation of plants in natural habitats, and on the factors which have a major influence on the performance of plants in contrasting environments.

Acknowledgements. The work was supported by the Deutsche Forschungsgemeinschaft SFB 137. We thank Prof. Dr. I.R. COWAN for reviewing the manuscript.

References

Allaway WG, Milthorpe FL (1976) Structure and functioning of stomata. In: Kozlowski TT (ed) Water deficits and plant growth Vol IV. Academic Press, London New York, pp 57–102

Bates LM, Hall AE (1981) Stomatal closure with soilwater depletion not associated with changes in bulk leaf water status. Oecologia 50:62–65

Bauer H (1978) Photosynthesis of ivy leaves (*Hedera helix*) after heat stress. I. CO_2 gas exchange and diffusion resistance. Physiol Plant 44:400–406

Benecke U (1980) Photosynthesis and transpiration of *Pinus radiata* D. Don under natural conditions in a forest stand. Oecologia 44:192–198

Benecke U, Havranek W, McCracken I (1976) Comparative study of water use by tree species in a mountain environment. Proc Soil Plant Water Symp, Palmerston North

Benecke U, Schulze E-D, Matyssek R, Havranek M (1981) Environmental control of CO_2 assimilation and leaf conductance in *Larix decidua* Mill. I. A comparison of contrasting natural environments. Oecologia 50:54–61

Bennett KJ, Rook DA (1978) Stomatal and mesophyll resistances in two clones of *Pinus radiata* D. Don known to differ in transpiration and survival rate. Aust J Plant Physiol 5:231–238

Björkman O (1968) Further studies on differentiation of photosynthetic properties in sun and shade ecotypes of *Solidago virgaurea*. Physiol Plant 21:84–99

Björkman O (1973) Comparative studies in photosynthesis in higher plants. In: Giese A (edd) Current topics in photobiology and photochemistry, Vol VIII. Academic Press, London New York, pp 1–63

Björkman O, Ludlow MM, Morrow PA (1972a) Photosynthetic performance of two rainforest species in their native habitat and analysis of their gas exchange. Carnegie Inst Washington Yearb 71/72:94–102

Björkman O, Pearcy RW, Harrison AT, Mooney HA (1972b) Photosynthetic adaptation to high temperatures: a field study in Death Valley, Calif. Science 175:786–789

Burrows FJ, Milthorpe FL (1976) Stomatal conductance in the control of gas exchange. In: Kozlowski TT (ed) Water deficits and plant growth, vol IV. Academic Press, London New York, pp 103–152

Caldwell MM, Johnson DA, Fareed M (1978) Constraints on tundra productivity: photosynthetic capacity in relation to solar radiation utilization and water stress in arctic and alpine tundras. In: Tieszen LL (ed) Vegetation and production ecology of an Alaskan arctic tundra. Ecol Stud, Vol 29. Springer, Berlin Heidelberg New York, pp 323–342

Camacho-B SE, Hall AE, Kaufmann MR (1974) Efficiency and regulation of water transport in some woody and herbaceous species. Plant Physiol 54:169–172

Collatz J, Ferrar PJ, Slatyer RO (1976) Effects of water stress and differential hardening treatments on photosynthetic characteristics of a xeromorphic shrub, *Eucalyptus socialis* F. Muell. Oecologia 23:95–105

Connor DJ, Tunstall BR, Driessche Van den (1971) An analysis of photosynthetic response in a Brigalow forest. Photosynthetica 5:218–225

Cowan IR (1965) Transport of water in the soil-plant-atmosphere system. J Appl Ecol 2:221–239

Cowan IR (1972) Oscillations in stomatal conductance and plant functioning associated with stomatal conductance: Observations and a model. Planta 106:185–219

Cowan IR (1977) Stomatal behaviour and environment. Adv Bot Res 4:117–228

Cowan IR, Farquhar GD (1977) Stomatal function in relation to leaf metabolism and environment. In: Integration of activity in the higher plant. Soc Exp Biol Symp 31. Univ Press, Cambridge, pp 471–505

Cowan IR, Milthorpe FL (1968) Plant factors influencing the water status of plant tissues. In: Kozlowski TT (ed) Water deficits and plant growth Vol I. Academic Press, London New York, pp 137–193

Davies WJ, Kozlowski TT (1977) Variations among woody plants in stomatal conductance and photosynthesis during and after drought. Plant Soil 46:435–444

Davis SD, McCree KJ (1978) Photosynthetic rate and diffusion conductance as a function of age in leaves of bean plants. Crop Sci 18:280–282

De Puit EJ, Caldwell MM (1973) Seasonal pattern of net photosynthesis of *Artemisia tridentata*. Am J Bot 60:426–435

Doley D, Trivett NBA (1974) Effects of low water potentials on transpiration and photosynthesis in Mitchell Grass (*Astrebla lappacea*) Aust J Plant Physiol 1:539–550

Downes RW (1971) Relationship between evolutionary adaptation and gas exchange characteristics of diverse sorghum taxa. Aust J Biol Sci 24:843–852

Drake BG, Raschke K, Salisbury FB (1970) Temperatures and transpiration resistance of *Xanthium* leaves as affected by air temperature, humidity and wind speed. Plant Physiol 46:324–330

Ehleringer JR, Miller PC (1975) Water relations of selected plant species in the alpine tundra, Colorado. Ecology 56:370–380

Ehleringer JR, Mooney HA (1978) Leaf hairs: effects on physiological activity and adaptive value to a desert shrub. Oecologia 37:183–201

Ehleringer JR, Björkman O, Mooney HA (1976) Leaf pubescence: Effects on absorptance and photosynthesis in a desert shrub. Science 192:376–377

Elfving DC, Kaufmann MR, Hall AE (1972) Interpreting leaf water potential measurements with a model of the soil plant atmosphere continuum. Physiol Plant 27:161–168

Evenari M, Lange OL, Schulze E-D, Kappen L, Buschbom U (1977) Net photosynthesis, dry matter production and phenological development of apricot trees (*Prunus armeniaca* L.) cultivated in the Negev highlands (Israel). Flora 166:383–414

Farquhar GD (1978a) Carbon assimilation in relation to transpiration and fluxes of ammonia. In: Marcelle R (ed) Internal control of photosynthesis. Junk, The Hague, pp 321–328

Farquhar GD (1978b) Feedforward responses of stomata to humidity. Aust J Plant Physiol 5:787–800

Farquhar GD, Dubbe DR, Raschke K (1978) Gain of the feedback loop involving carbon dioxide and stomata. Plant Physiol 62:406–412

Farquhar GD, Schulze E-D, Küppers M (1980) Responses to humidity by stomata of *Nicotiana glauca* L. and *Corylus avellana* L. are consistent with the optimization of carbon dioxide uptake with respect to water loss. Aust J Plant Physiol 7:315–327

Fuchs M, Schulze E-D, Fuchs MI (1977) Spatial distribution of photosynthetic capacity and performance in a montane spruce forest of Northern Germany. II. Climatic control of carbon dioxide uptake. Oecologia 29:329–340

Goudriaan J, Keulen van H (1979) The direct and indirect effects of nitrogen shortage on photosynthesis and transpiration in maize and sunflower. Neth J Agric Sci 27:227–234

Goudriaan J, Laar van HH (1978) Relations between leaf resistance, CO_2 concentration and CO_2 assimilation in maize, beans, Lalang grass and sunflower. Photosynthetica 12:241–249

Hall AE, Hoffman GJ (1976) Leaf conductance response to humidity and water transport. Agron J 68:876–881

Hall AE, Kaufmann MR (1975) Stomatal response to environment with *Sesamum indicum* L. Plant Physiol 55:455–459

Hall AE, Loomis RS (1972) An explanation for the difference in photosynthetic capabilities

of healthy and beet yellows virus infected sugar beets (*Beta vulgaris* L.) Plant Physiol 50:576–580
Hall AE, Schulze E-D (1980a) Stomatal responses to environment and a possible interrelation between stomatal effects on transpiration and CO_2 assimilation. Plant Cell Environ 3:467–474
Hall AE, Schulze E-D (1980b) Drought effects on transpiration and leaf water status of cowpea in controlled environments. Aust J Plant Physiol 7:141–147
Hall AE, Hunt WF, Loomis RS (1972) Variations in leaf resistance, net photosynthesis, and tolerance to the beet yellow virus among varieties of sugar beet (*Beta vulgaris* L.) Crop Sci 12:558–561
Hall AE, Schulze E-D, Lange OL (1976) Current perspectives of steady state stomatal responses to environment. In: Lange OL, Kappen L, Schulze E-D (eds) Water and plant life. Ecol Stud Vol 19. Springer, Berlin Heidelberg New York, pp 169–188
Hall AJ, Milthorpe FL (1978) Assimilate source-sink relationships in *Capsicum annuum* L. III. The effects of fruit excision on photosynthesis and leaf and stem carbohydrates. Aust J Plant Physiol 5:1–13
Havranek WM, Benecke U (1978) The influence of soil moisture on water potential, transpiration and photosynthesis of conifer seedlings. Plant Soil 49:91–103
Holmgren P (1968) Leaf factors affecting light saturated photosynthesis in ecotypes of *Solidago virgaurea* from exposed and shaded habitats. Physiol Plant 21:676–698
Jarvis PG (1980) Stomatal response to water stress in conifers. In: Turner NC, Kramer PJ (eds) Adaptation of plants to water and high temperature stress. Wiley and Sons, New York, pp 105–122
Jones MM (1979) Physiological responses of sorghum and sunflower to leaf water deficits. Ph D Thesis, Austr Natl Univ
Jones MM, Rawson HM (1979) Influence of rate of development of leaf water deficits upon photosynthesis, leaf conductance, water use efficiency, and osmotic potential in *Sorghum*. Physiol Plant 45:103–111
Jones MM, Turner NC (1978) Osmotic adjustment in leaves of *Sorghum* in response to water deficits. Plant Physiol 61:122–126
Johnson DA, Caldwell MM (1976) Water potential components, stomatal function, and liquid phase water transport resistances of four arctic and alpine species in relation to moisture stress. Physiol Plant 36:271–278
Johnsson M, Issaias S, Brogårdh T, Johnsson A (1976) Rapid blue-light-induced transpiration response restricted to plants with grass-like stomata. Physiol Plant 36:229–232
Kappen L, Oertli JJ, Lange OL, Schulze E-D, Evenari M, Buschbom U (1975) Seasonal and diurnal courses of water relations of the arido-active plant *Hammada scoparia* in the Negev desert. Oecologia 21:175–192
Kappen L, Lange OL, Schulze E-D, Evenari M, Buschbom U (1976) Distributional pattern of water relations and net photosynthesis of *Hammada scoparia* (Pomel) Iljin in a desert environment. Oecologia 23:323–334
Kaufmann MR (1976) Water transport through plants. Current perspectives. In: Wardlaw IF, Passioura JB (eds) Transport and transfer processes in plants. Academic Press, London New York, pp 313–327
Kerr JP, Beardsell MF (1975) Effect of dew on leaf water potentials and crop resistances in a paspalum pasture. Agron J 67:596–599
Khairi MMA, Hall AE (1976a) Temperature and humidity effects on net photosynthesis and transpiration of *Citrus*. Physiol Plant 36:29–34
Khairi MMA, Hall AE (1976b) Comparative studies of net photosynthesis and transpiration of some *Citrus* species and relatives. Physiol Plant 36:35–39
Körner CH, Scheel JA, Bauer H (1979) Maximum leaf diffusive conductance in vascular plants. Photosynthetica 13:45–82
Lange OL, Medina E (1979) Stomata of the CAM plant *Tillandsia recurvata* respond directly to humidity. Oecologia 40:357–363
Lange OL, Koch W, Schulze E-D (1969) CO_2-Gaswechsel und Wasserhaushalt von Pflanzen in der Negev-Wüste am Ende der Trockenzeit. Ber Dtsch Bot Ges 82:39–61

Lange OL, Schulze E-D, Koch W (1970a) Experimentell-ökologische Untersuchungen an Flechten der Negev-Wüste. CO_2-Gaswechsel und Wasserhaushalt von *Ramalina maciformis* (Del.) Bory am natürlichen Standort während der sommerlichen Trockenperiode. Flora 159:38–62

Lange OL, Schulze E-D, Koch W (1970b) Ecophysiological investigations on lichens of the Negev Desert. III. CO_2 gas exchange and water relations of crustose and foliose lichens in their natural habitat during the summer dry period. Flora 159:525–528

Lange OL, Lösch R, Schulze E-D, Kappen L (1971) Responses of stomata to changes in humidity. Planta 100:76–86

Lange OL, Schulze E-D, Kappen L, Evenari M, Buschbom U (1975) CO_2 exchange pattern under natural conditions of *Caralluna negevensis,* a CAM plant of the Negev Desert. Photosynthetica 9:318–326

Lange OL, Schulze E-D, Evenari M, Kappen L, Buschbom U (1978) The temperature-related photosynthetic capacity of plants under desert conditions. III. Ecological significance of the seasonal change of the photosynthetic response to temperature. Oecologia 34:89–100

Larcher W (1971) Ergebnisse des IBP-Projektes „Zwergstrauchheide Patscherkofel" Sitzungsberichte Österr Akad Wiss Math Naturwiss K1 Abt I 186:301–371

Larcher W, Moares De JAPV, Bauer H (1981) Adaptive responses of leaf water potential, CO_2 gas exchange and water use efficiency of *Olea europaea* during drying and rewatering. Proc Saloniki Symp Productivity of Medit Ecosystems

Lawlor DW (1976) Water stress induced changes in photosynthesis, photorespiration, respiration and CO_2 compensation concentration of wheat. Photosynthetica 10:378–387

Lösch R (1980) Stomatal responses to changes in air humidity. In: Sen DN, Chawan DD, Bunsal RP (eds) Structure, function and ecology of stomata. Dehra Dun, pp 189–216

Ludlow MM (1980) Adaptive significance of stomatal responses to water stress. In: Turner NC, Kramer PJ (eds) Adaptation of plants to water and high temperature stress. Wiley and Sons, New York, pp 123–138

Ludlow MM, Jarvis PG (1971) Photosynthesis in sitka spruce (*Picea sitchensis* (Bong) Carr) I. General characteristics. J Appl Ecol 8:925–953

Ludlow MM, Wilson GL (1971) Photosynthesis of tropical pasture plants. I. Illuminance, carbon dioxide concentration, leaf temperature and leaf-air vapour pressure difference. Aust J Biol Sci 24:449–470

Medina E (1972) Photosynthetic capacity and carboxydismutase activity in *Atriplex patula* leaves as determined by nitrogen nutrition and light intensity during growth. In: Forti G, Buron M, Metzner H (eds) Proc 2nd int congr photosynthes res. Junk, The Hague, pp 2527–2536

Miller PC, Stoner MA, Ehleringer JR (1978) Some aspects of water relations of arctic and alpine regions. In: Tieszen LL (ed) Vegetation and production ecology of an Alaskan arctic tundra. Ecol Stud Vol 29. Springer, Berlin Heidelberg New Yrok, pp 343–358

Mooney HA, Ehleringer JR (1978) The carbon gain benefits of solar tracking in a desert annual. Plant Cell Environ 1:307–311

Mooney HA, Harrison AT, Morrow PA (1974) Environmental limitations of photosynthesis on a California evergreen shrub. Oecologia 19:293–301

Mooney HA, Björkman O, Collatz GJ (1976) Photosynthetic acclimation to temperature and water stress in the desert shrub *Larrea divaricata.* Carnegie Inst Washington 76:328–335

Nobel PS (1978) Microhabitat, water relations, and photosynthesis of a desert fern, *Notholaena parryi.* Oecologia 31:293–309

Osmond CB, Ludlow MM, Davis R, Cowan IR, Powles SB, Winter K (1979) Stomatal responses to humidity in *Opuntia inermis* in relation to control of CO_2 and H_2O exchange patterns. Oecologia 41:65–76

Osonubi O, Davies WJ (1980) The influence of plant water stress on stomatal control of gas exchange at different levels of atmospheric humidity. Oecologia 46:1–6

Pearcy RW, Berry JA, Bartholomew B (1974a) Field photosynthetic performance and leaf temperature of *Phragmites communis* under summer conditions in Death Valley, California. Photosynthetica 8:104–108

Pearcy RW, Harrison AT, Mooney HA, Björkman O (1974b) Seasonal changes in net photosynthesis of *Atriplex hymenelytra* shrubs growing in Death Valley, Calif. Oecologia 17:111–121

Ramos C (1981) Relationship between leaf conductance, CO_2 partial pressure and CO_2 uptake rate in leaves, and a test of a portable instrument for its measurement in the field. Ph D Thesis, Univ Calif Riverside

Raschke K (1970) Stomatal response to pressure changes and interruptions in the water supply of detached leaves of *Zea mays* L. Plant Physiol 45:415–423

Raschke K (1979) Movements of stomata. In: Haupt W, Feinleib ME (eds) Physiology of movements. Encyclopedia of plant physiology new series Vol VII. Springer, Berlin Heidelberg New York, pp 383–441

Raschke K, Hanebuth WF, Farquhar GD (1978) Relationship between stomatal conductance and light intensity in leaves of *Zea mays* L. derived from experiments using the mesophyll as shade. Planta 139:73–78

Rawson HM, Gifford RM, Bremner PM (1976) Carbon dioxide exchange in relation to sink demand in wheat. Planta 132:19–23

Rawson HM, Beeg JE, Woodward RG (1977) The effects of atmospheric humidity on photosynthesis, transpiration and water use efficiency of leaves of several plant species. Planta 134:5–10

Roberts SW, Knoerr KR, Strain BR (1979) Comparative field water relations of four co-occurring tree species. Can J Bot 57:1876–1882

Running SW (1976) Environmental control of leaf water conductance in conifers. Can J For Res 6:104–112

Running SW (1980) Environmental and physiological control of water flux through *Pinus contorta*. Can J For Res 10:82–91

Sánchez-Díaz ÜF, Mooney HA (1979) Resistance to water transfer in desert shrubs native to Death Valley, Calif. Physiol Plant 46:139–146

Schulze E-D (1970) Der CO_2-Gaswechsel der Buche (*Fagus silvatica* L.) in Abhängigkeit von den Klimafaktoren im Freiland. Flora 159:177–232

Schulze E-D (1972a) A new type of climatized gas exchange chamber for net photosynthesis and transpiration measurements in the field. Oecologia 10:243–251

Schulze E-D (1972b) Die Wirkung von Licht und Temperatur auf den CO_2-Gaswechsel verschiedener Lebensformen aus der Krautschicht eines montanen Buchenwaldes. Oecologia 9:235–258

Schulze E-D, Küppers M (1979) Short-term and long-term effects of plant water deficits on stomatal responses to humidity in *Corylus avellana* L. Planta 146:319–326

Schulze E-D, Lange OL, Buschbom U, Kappen L, Evenari M (1972a) Stomatal responses to changes in humidity in plants growing in the desert. Planta 108:259–270

Schulze E-D, Lange OL, Koch W (1972b) Eco-physiological investigations on wild and cultivated plants in the Negev Desert. III. Daily courses of net photosynthesis and transpiration at the end of the dry period. Oecologia 9:317–340

Schulze E-D, Lange OL, Kappen L, Buschbom U, Evenari M (1973) Stomatal responses to changes in temperature at increasing water stress. Planta 110:20–42

Schulze E-D, Lange OL, Evenari M, Kappen L, Buschbom U (1974) The role of air humidity and leaf temperature in controlling stomatal resistance of *Prunus armeniaca* L. under desert conditions. I. A simulation of the daily course of stomatal resistance. Oecologia 17:159–170

Schulze E-D, Lange OL, Kappen L, Evenari M, Buschbom U (1975a) Physiological basis of primary production of perennial higher plants in the Negev desert. In: Cooper JP (ed) Photosynthesis and productivity in different environments, IBP Vol III. Univ Press, Cambridge, pp 107–119

Schulze E-D, Lange OL, Kappen L, Evenari M, Buschbom U (1975b) The role of air humidity and leaf temperature in controlling stomatal resistance of *Prunus armeniaca*

L. under desert conditions. II. The significance of leaf water status and internal carbon dioxide concentration. Oecologia 18:219–233

Schulze E-D, Lange OL, Evenari M, Kappen L, Buschbom U (1976) An empirical model of net photosynthesis for the desert plant *Hammada scoparia* (Pomel) Iljin. I. Description and test of the model. Oecologia 22:355–372

Schulze E-D, Hall AE, Lange OL, Evenari M, Kappen L, Buschbom U (1980a) Long-term effects of drought on wild and cultivated plants in the Negev desert. I. Maximal rates of net photosynthesis. Oecologia 45:11–18

Schulze E-D, Lange OL, Evenari M, Kappen L, Buschbom U (1980b) Long-term effects of drought on wild and cultivated plants in the Negev desert. II. Diurnal patterns of net photosynthesis and daily carbon gain. Oecologia 45:19–25

Shackel KA, Hall AE (1979) Reversible leaflet movements in relation to drought adaptation of cowpea, *Vigna unguiculata* (L.) Walp. Aust J Plant Physiol 6:265–276

Sheriff DM, Meidner H (1975) Water movement into and through *Tradescantia virginiana* L. leaves. J Exp Bot 26:897–902

Sheriff DW (1979) Stomatal aperture and the sensing of the environment by guard cells. Plant Cell Environ 2:15–22

Slatyer RO (1956) Absorption of water from atmospheres of different humidity and its transport through plants. Aust J Biol Sci 9:552–558

Stålfelt MG (1929) Die Abhängigkeit der Spaltöffnungsreaktionen von der Wasserbilanz. Planta 8:287–340

Stocker O (1956) Die Abhängigkeit der Transpiration von den Umweltfaktoren. In: Ruhland W (ed) Encyclopedia of plant physiology Vol III. Springer, Berlin Göttingen Heidelberg, pp 436–488

Szarek SR, Ting IP (1974) Seasonal patterns of acid metabolism and gas exchange in *Opuntia basilaris*. Plant Physiol 54:76–81

Tenhunen JD, Meyer A, Lange OL, Gates DM (1980) Development of a photosynthesis model with emphasis on ecological applications. V. Test of the applicability of a steady state model to description of net photosynthesis of *Prunus armeniaca* under field conditions. Oecologia 45:147–155

Tenhunen JD, Lange OL, Braun M, Meyer A, Lösch R, Pereira JS (1981) Midday stomatal closure in *Arbutus unedo* leaves in a natural macchia and under simulated habitat conditions in an environmental chamber. Oecologia 47:365–367

Thorpe MR, Warrit B, Landsberg JJ (1980) Responses of apple leaf stomata: A model for single leaves and a whole tree. Plant Cell Environ 3:23–28

Tieszen LL (1975) CO_2 exchange in the Alaskan arctic tundra: Seasonal changes in the rate of photosynthesis of four species. Photosynthetica 9:376–390

Tieszen LL (1978) Photosynthesis in the principal Barrow, Alaska, species: A summary of field and laboratory responses. In: Tieszen LL (ed) Vegetation and production ecology of an Alaskan arctic tundra. Ecol Stud Vol 29. Springer, Berlin Heidelberg New York, pp 241–268

Tranquillini W (1957) Standortklima, Wasserbilanz und CO_2-Gaswechsel junger Zirben (*Pinus cembra* L.) an der alpinen Waldgrenze. Planta 49:612–661

Turner NC (1974) Stomatal response to light and water under field conditions. In: Bieleski RL, Ferguson AR, Cresswell MM (eds) Mechanisms of regulation of plant growth. Roy Soc NZ Bull 12:423–432

Turner NC (1979) Drought resistance and adaptation to water deficits in crop plants. In: Mussell H, Staples R (eds) Stress physiology in crop plants. Wiley and Sons, New York, pp 344–372

Turner NC, Begg JE, Rawson HM, English SD, Hearn AB (1978) Agronomic and physiological responses of soybean and sorghum crops to water deficits. III. Components of leaf water potential, leaf conductance, $^{14}CO_2$ photosynthesis, and adaptation to water deficits. Aust J Plant Physiol 5:179–194

Tyree MT, Yianoulis P (1980) The site of water evaporation from substomatal cavities, liquid path resistance and hydroactive stomatal closure. Ann Bot (London) 46:175–193

Vowinkel T, Oechel WC, Boll WG (1975) The effect of climate on the photosynthesis of *Picea mariana* at the subarctic tree line. 1. Field measurements. Can J Bot 53:604–620

Walter H (1968) Die Vegetation der Erde. II. Die gemäßigten und arktischen Zonen. Fischer, Stuttgart

Willert von DJ, Brinckmann E, Scheitler B, Schulze ED, Thomas DA, Treichel S (1980) Ökophysiologische Untersuchungen an Pflanzen der Namib-Wüste. Naturwissenschaften 67:21–28

Wong SC (1979) Stomatal behaviour in relation to photosynthesis. Ph D Thesis, Aust Natl Univ, Canberra

Wong SC, Cowan IR, Farquhar GD (1978) Leaf conductance in relation to assimilation in *Eucalyptus pauciflora* Sieb. ex Spreng: influence of irradiance and partial pressure of carbon dioxide. Plant Physiol 62:670–764

Wong SC, Cowan IR, Farquhar GD (1979) Stomatal conductance correlates with photosynthetic capacity. Nature (London) 282:424–426

Woodward RG, Rawson HM (1976) Photosynthesis and transpiration in dicotyledonous plants. II. Expanding and senescing leaves of soybean. Aust J Plant Physiol 3:257–267

Zobel DB, Liu VT (1980) Effects of environment, seedling age, and seed source on leaf resistance of three species of *Chamaecyparis* and *Tsuga chinensis*. Oecologia 46:412–419

8 Mathematical Models of Plant Water Loss and Plant Water Relations

A.E. HALL

CONTENTS

8.1 Introduction

Quantitative descriptions of the functioning of individual organisms or plant communities in different environments are a major objective of physiological research. Mathematical models of plant water loss and plant water relations represent attempts to quantify these aspects of plant function. In addition, mathematical models can be objective and efficient systems for integrating and summarizing available information, and they are useful for expressing and testing hypotheses. However, the values and limitations of mathematical models of biological systems have been underestimated. The development and use of mathematical models should follow rational guidelines analogous to the conceptual constraints that are widely accepted as being appropriate in experimental studies. Unfortunately, guidelines for the development, evaluation, and use of mathematical models of ecological systems are only just beginning to emerge. Guidelines will be briefly discussed in the section on objectives and procedures of modeling to illustrate some of the limitations of mathematical models.

In this chapter the physiological and physical concepts on which models of plant water loss may be based will be examined. The transport of water in the soil-plant system is examined mainly in relation to models of water loss. More detailed treatments of water flow in the soil-plant-atmosphere continuum are provided in Chapters 1, 3, and 5, this Volume.

Mathematical models have been described as being mechanistical, phenomenological, or empirical (it should be recognized, however, that a model which is mechanistic at one level of organization, such as the cellular level, is not mechanistic at lower levels of organization, such as the organelle or enzyme levels). The development of phenomenological and empirical models that may be useful for pursuing objectives of physiological ecology will be discussed. Only *deterministic models* are considered. The other type of model, *stochastic*, is used where the objective is to predict the probability of specific outcomes, and where probabilities are assigned to input variables and parameters. The reader is referred to JONES (1980) for a description of a *stochastic model* of water relations, and a discussion of the importance and uses of stochastic modeling. However, few stochastic models of water relations have been developed, and difficult problems arise when attempting to develop stochastic models of biological systems, due to the presence of homeostatic mechanisms (refer to the review by LOOMIS et al. 1979).

This analysis is restricted to homeohydric plants (those having stomata), and does not consider plants having Crassulacean acid metabolism which are discussed in Chapter 15, this Volume. The intent is to provide a conceptual basis for the development of models that reliably predict water loss from plants. Current models are discussed in relation to this objective.

8.1.1 Objectives and Procedures of Modeling

Mathematical models of biological systems are works of art combining both fact and fiction, and they only occasionally and partially simulate reality (PASSIOURA 1973). These models are only gross representations of the elegantly complex systems that are called plants. It would be prudent to assume that we will *never* be able to develop mathematical models that *perfectly* mimic biological systems. *However, it is possible to usefully model, for specific purposes, mental constructs of the subsystems within plants and ecosystems* (LOOMIS et al. 1979, see also the discussion of general questions of modelling in Chap. 4, Vol. 12 D).

Consequently, the mandatory initial step in modeling is to define precisely the objectives of the intended study. As with experimental studies, this frequently involves testing hypotheses. Some examples of objectives of earlier models illustrate the diverse possibilities in modeling plant water relations.

Hypotheses concerning stomatal mechanics have been subjected to considerable analysis using mathematical models (COOKE et al. 1976; DEMICHELE and

SHARPE 1973; SHARPE and WU 1978; SHOEMAKER and SRIVASTAVA 1973) but the complexity of the analyses and paucity of experimental data precluded simple or general interpretation of the results that were obtained. Consequently, including submodels of stomatal mechanics in models of the regulation of water loss may not, at this time, improve predictions or explanations concerning the effects of environment on transpiration by leaves or canopies.

Models have been developed for evaluating stomatal responses to environment by applying differential calculus and classical feedback theory (FARQUHAR et al. 1978). This approach has been used to evaluate the extent to which stomatal responses to internal [CO_2] are responsible for stomatal opening with increases in irradiance (WONG et al. 1978). An extension of this approach has been used to describe the contributions of stomatal water loss (feedback) and cuticular water loss (feedforward) to stomatal response to humidity (FARQUHAR 1978). This approach can be used in a general manner both to determine and quantify the effects of environment on stomata, and provides a fundamental basis for understanding and modeling the regulation of water loss by stomata.

Models have been used to explore relations between the regulation of water loss and plant water status (COWAN and MILTHORPE 1968; COWAN 1972), including the effects of water storage in plants (RUNNING et al. 1975; WARING and RUNNING 1978). These models have the potential capability of relating water transport to plant function as mediated by water stress.

Models of water and solute transport in plants (DALTON et al. 1975; FISCUS and KRAMER 1975) indicate the errors that may occur when water flow in the soil-plant system is treated without considering effects of solute transport. Consequently, when modeling water transport it is necessary to consider whether submodels of ion uptake and translocation should be included.

Relationships between stomatal regulation of transpiration and carbon dioxide assimilation have been modeled (COWAN and FARQUHAR 1977), and are examined in detail in Chapter 17, this Volume. Models of this type illustrate the need for considering both of these processes when relating stomata to plant function.

Another type of objective, that has significance for both agriculture and for understanding natural systems, is the prediction of plant water use (SLABBERS 1980) and seasonal balances of water (FITZPATRICK and NIX 1969; HALL and DANCETTE 1978; ROSENTHAL et al. 1977). For these objectives, models must describe the regulation of water loss at the canopy or community levels. Ultimately, predictions of water and carbon balances should provide information that explains certain aspects of plant adaptation and competition (EHLERINGER and MOONEY 1978; JONES 1980; MOONEY and EHLERINGER 1978).

It is unlikely that any single model could accomplish all of these objectives. However, when models are developed, consideration should be given to the possibility that they may also be useful as submodels of more comprehensive models. For example, LANDSBERG and BUTLER (1980) linked a model of stomatal response to environment, developed by THORPE et al. (1980), to the model of energy balance and aerodynamic transfer described by MONTEITH (1965), and examined the implications to transpiration of stomatal response to humidity. This example illustrates the hierarchic levels within ecosystems and the use of models to predict or explain the function of upper levels (i.e., the canopy) by integrating information obtained at a lower level (i.e., the leaf).

Model development should include testing. A *predictive* model must provide reasonable predictions over the range of environmental and plant conditions that are relevant to the objectives for which the model was developed. *Explanatory* models, which are developed for analyzing hypotheses concerning the internal functioning or regulation of systems, must be tested with respect to both internal consistency and gross prediction. The extent of testing depends upon the quality of submodels. Certain aspects of mass transport and energy relations may be considered as being firmly based upon established laws of physics and require little additional testing. In contrast, submodels based upon biological assumptions will usually require at least some tests of their validity with the appropriate species and environmental conditions as determined by the objectives of the study. The quantitative and qualitative variations in results of experimental studies of soil-plant water relations and stomatal physiology indicate that for these systems specific tests are essential procedures in model development at this time. Plant water relations and stomatal responses are strongly dependent upon previous growth conditions, present environmental conditions, and the methods used for measurement, and in some cases the causal factors for major differences in response have not been adequately identified. Some of the contrasts and similarities that have been observed in stomatal responses to environment are discussed in Chapter 7, this Volume.

Lastly, it should be expected that all models will have deficiencies. Comprehensive analyses of model deficiencies should be conducted and reported, so that the constraints concerning model use and interpretation of model responses are obvious. Models, like interpretations from experimental studies, should be recognized as only occupying some position on a continuum between the states of general ignorance and partial understanding.

8.1.2 The Significance of the Linkage Between Plant Water Loss and Plant Water Relations

The boundaries of models normally should not exclude processes which strongly interact with the processes described by the model. In this context, when modeling plant regulation of water loss, it would appear necessary to consider possible relations with plant water status and liquid water transport in the plant. It may be anticipated that a coherent system would include interactions between the regulation of water loss and tissue water relations which act to prevent extreme dehydration of the protoplasm or to maintain plant turgor (COWAN 1977a; DAVIES et al. 1980). Differences do exist, however, among different plant life forms in the extent to which transpiration is regulated in response to changing environmental conditions (FISCHER and TURNER 1978).

Changes in plant water status influence plant water loss through effects on stomatal function and canopy characteristics. These interactions should be included in comprehensive models and are discussed in Sections 8.2.2.2 and 8.4. Additional linkages between plant water status and water loss, which may be important in specific circumstances, are discussed in the remainder of this section.

Changes in water potential have small effects on the equilibrium vapor pressure. For example, vapor pressure would be decreased only 1%, 2%, and 10% by decreases in water potential from 0 to −14, −28 and −145 bar, respectively (SLATYER 1967). In addition, variation in water potential over this range would have little effect on the latent heat of vaporization (COWAN 1977a). Consequently, if the water potential of the solutions at the evaporating surfaces within leaves is similar to bulk leaf water potential, it may be assumed that decreases in leaf water potential have little *direct* effect on the rate of water loss from leaves (COWAN 1977a; WEATHERLEY 1970).

However, JARVIS and SLATYER (1970) proposed that the water potential at the sites of evaporation within the leaf may be substantially lower than the bulk leaf water potential at high rates of transpiration. The studies of FARQUHAR and RASCHKE (1978) did not confirm this observation and they concluded that "under conditions relevant to studies of stomatal behavior, the water vapor pressure at the sites of evaporation is equal to the saturation vapor pressure". At this time it is not known whether this conclusion is generally valid and further studies should be conducted, especially with xeromorphic plants, and plants subjected to drought.

Additional nonstomatal features which influence water loss include wax tubes in the stomatal antechamber of some conifers (JEFFREE et al. 1971), resinous layers over stomata, such as may be observed on the leaves of *Eriodictyon trichocalyx*, and the various leaf characteristics that influence boundary layer and cuticular resistances. Leaf-rolling (O'TOOLE and CRUZ 1980), leaf pubescence, and changes in leaf size, due to drought, influence boundary-layer resistance. In addition, drought-induced changes in leaf orientation (SHACKEL and HALL 1979) and leaf pubescence (EHLERINGER and MOONEY 1978) influence transpiration through effects on radiation balance and leaf temperature.

The linkages present between the regulation of plant water loss and plant water relations have implications for experimental approaches. It would appear prudent to study stomata as they function as integral parts in whole plant systems. Excised leaves and epidermal strips may have value, as simplified systems, for studying specific mechanisms but in some of these systems stomatal responses may have differed from their responses in intact plants. The following sections of this review mainly consider experimental evidence obtained with intact plants to avoid the possibility of artifacts of this type.

8.1.3 The Significance of the Linkage Between Regulation of Transpiration and Carbon Dioxide Assimilation

There are several reasons why it is useful to include submodels of carbon dioxide assimilation in some models of the regulation of transpiration. Photosynthesis and transpiration are intimately linked through mutual dependencies on gaseous exchanges between leaf and air, and on leaf energy balance. In addition, stomatal apertures of some species are strongly dependent upon the concentration of CO_2 inside the leaf (RASCHKE 1976). Modeling the CO_2 concentration inside leaves requires a submodel of carbon dioxide assimilation. Recently, it has been proposed that stomatal conductance may be influenced by a product from photosynthesis in the mesophyll tissue (WONG et al. 1979) or by the energy status of the mesophyll (WONG 1979; see also Chap. 7, this Vol.). If relations of this type are present between photosynthesis and stomata a more complex submodel of photosynthesis and respiration would be required to describe stomatal responses to the internal and external environment of the leaf. Alternatively, the regulation of water loss may be modeled based upon hypothetical relationships between transpiration and CO_2 assimilation (COWAN and FARQUHAR 1977; HALL and SCHULZE 1980a) as will be discussed in Section 8.2.2.3.

Where the objective of models of water loss is to predict plant performance or adaptation, it may be useful to incorporate a submodel of CO_2 assimilation, although more empirical models relating plant performance to plant water status (TURK and HALL 1980a, b) or relative water use (TURK and HALL 1980c) are more appropriate in some circumstances.

8.2 Models of Leaf Water Loss and Stomatal Response to Environment

Consideration is given to quantitative aspects of water loss at the leaf level of organization (Sect. 8.2.1). Detailed analyses of stomatal responses to environment are presented in Section 8.2.2. A brief analysis of canopy water loss is presented in Section 8.4 to illustrate approaches that may be used to relate leaf function to the functioning of isolated plants or plant communities.

8.2.1 Physical Aspects of the Influence of the Leaf and the Environment on Water Loss from Plants

This discussion is designed as a supplement to the comprehensive treatment of the transfer of H_2O and CO_2 between leaf and air by JARVIS (1971) which describes methods widely used at this time. However, these widely used methods can result in errors of interpretation in some circumstances, and mathematical treatments are available to overcome these problems.

Methods should clearly specify and differentiate between the effects of the leaf epidermis and the environment on the regulation of water loss. To facilitate explanation, but with no loss in rigor, it is assumed, at this stage, that the boundary layer resistance is negligible and the system is isothermal. The widely used approach considers

$$J_i = \Delta C_i / r_i \tag{8.1}$$

where J_i is the flux (mass area^{-1} time^{-1}) of a specific gas (i), ΔC_i is the difference in concentration (mass volume^{-1}) of the gas along the flow path, and r_i (time length^{-1}) is a resistance to diffusion. The characteristics of this resistance may be ascertained by comparing Eq. (8.1) with Fick's law which shows that:

$$r_i = L/D' \tag{8.2}$$

where L is the effective path length, and D′ is an effective diffusion coefficient. D′ may be related to conventional diffusion coefficients (D_i) by the following equation:

$$D' = D_i/K^* \tag{8.3}$$

where K* is a dimensionless coefficient describing the influences of stomatal densities, and aperture dimensions on gaseous diffusion. However, diffusion coefficients are dependent upon temperature and atmospheric pressure as described by the experimentally determined function of FULLER et al. (1966):

$$D_h = D_h^\circ \, (T/T_o)^{1.75} \, (P_o/P) \tag{8.4}$$

where D_h° is the diffusion coefficient for water vapor in air at reference levels of absolute temperature (T_o) and atmospheric pressure (P_o). $D_h^\circ = 0.25\ cm^2\ s^{-1}$ at 293 K and 1.01 bar. D_h is the same diffusion coefficient but at ambient temperature (T) and atmospheric pressure (P). The characteristics of the commonly used resistance to water vapor (r_h) may be specified by combining Eqs. (8.2), (8.3) and (8.4).

$$r_h = L\,K^*/[D_h^\circ \, (T/T_o)^{1.75} \, (P_o/P)] \tag{8.5}$$

Consequently, a leaf with constant epidermal properties, as specified by L and K*, would have an apparent resistance to water vapor (r_h) that is 19% greater at 10 °C than at 40 °C, and 44% greater at sea level compared with r_h at 3,000 m elevation (atmospheric pressures are approximately 1.01 bar at sea level and 0.7 bar at 3,000 m elevation). Influences of pressure on gaseous fluxes that are similar to theoretical expectations have been demonstrated using physical models and leaves by GALE (1973). Fortunately, the errors due to variations in pressure and temperature are not large enough to invalidate the conclusions of most earlier studies of plant regulation of water loss based upon Eq. (8.1). However, the errors are large enough to justify development of an improved system, especially for studies along altitudinal gradients (GALE 1972; SMITH and GELLER 1979) or where a large range of temperature is encountered.

A method has been described by COWAN (1977b) which appears to overcome some of the error due to effects of temperature and pressure on diffusivities.

$$J_h = \frac{\Delta e}{r^t\,P} \tag{8.6}$$

In this method the flux of water vapor (J_h) is related to the difference in vapor pressure (Δe) divided by the atmospheric pressure (P), and the total resistance to diffusion of water vapor (r^t) has the units 1/transpirational flux. The relation between r^t and the terms in the empirical Eq. (8.5) may be ascertained by assuming isothermal conditions and applying the ideal gas equation:

$$r^t = \frac{R\,T_o^{1.75}}{M\,D_h^o\,P_o\,T^{0.75}}\,L\,K^*, \tag{8.7}$$

r^t is a function of the geometric properties of the epidermis (L and K*) and several constants (R, T_o, M, D_h^o and P_o) where R is the universal gas constant and M is the molecular weight of water, but it also has a small dependence upon absolute temperature ($1/T^{.75}$). However, epidermises of leaves may not have this particular temperature dependence. The movement of water vapor through stomata involves additional complexities including incomplete molecular slip of diffusing molecules at surfaces and the effects of temperature gradients (COWAN 1977b). These complexities have not been adequately quantified for leaves but the definition proposed by COWAN (1977b) in Eq. (8.6) has advantages and should be used instead of the commonly used definition of resistance [Eq. (8.1)]. Resistances based upon the old system (r) may be converted to the system of COWAN using the following equation:

$$r^t = r\,V_o\,\frac{P_o\,T}{P\,T_o} \tag{8.8}$$

where V_o is the molar volume of an ideal gas at T_o and P_o ($V_o = 0.0224\ m^3\ mol^{-1}$ at $T_o = 273$ K and $P_o = 1.01$ bar).

Gaseous transport equations are also useful for estimating the carbon dioxide concentration inside the leaf as described by JARVIS (1971). It is appropriate to continue using concentrations that are on a volume basis because at low concentrations, volume fractions are equivalent to relative partial pressures (e.g., ppm is equivalent to μbar partial pressure/bar atmospheric pressure). Resistances to the diffusion of one gas are commonly converted to apply to another gas, by using the ratios of the diffusion coefficients as described by JARVIS (1971). One uncertainty in this approach is that at any time several gases are diffusing between leaf and air, and for a complete analysis all molecular collisions must be considered. JARMAN (1974) provided equations which describe the effects of these collisions, and concluded that estimates of the difference in CO_2 concentration between the external air and intercellular spaces based upon the old system (JARVIS 1971) may be in error by 0% to 80%. However, JARMAN (1974) pointed out that the large relative errors only occur at the light or CO_2 compensation points where the absolute gradients and absolute errors are small. Estimates of internal CO_2 concentration may be improved by assuming that CO_2 transfer in the cuticular pathway is negligible and basing the calculation on flow in the stomatal pathway.

It is useful to separate the influence of the epidermis on gaseous transfer from the effects of the boundary layer which depend upon leaf dimensions, surface roughness, wind speed, and air-flow characteristics. A detailed analysis of the measurement and properties of boundary layer resistances has been presented by JARVIS (1971) and only supplementary information will be presented here. Boundary layer effects on water vapor loss from a single leaf surface may be described as follows:

$$J_h = \frac{\Delta e}{(r^a + r^e)\,P} \tag{8.9}$$

where r^a is the boundary layer resistance, and r^e is the resistance of the leaf epidermis. When leaves are isolateral (equal resistances on both adaxial and abaxial surfaces) Eq. (8.9) is also appropriate for partitioning epidermal and boundary layer effects on both surfaces using a total flux measurement. However, leaves are often anisolateral and errors occur if r^e is estimated from total flux measurements using Eq. (8.9) (HALL, Appendix A 1970). The most straightforward approach is to measure gaseous fluxes separately at each surface of the leaf, but this increases the equipment and effort required for controlled environment studies, and may put serious constraints on cuvette design. An alternative approach was suggested by HALL (Appendix A 1970).

The level of anisolaterality (K) is defined as:

$$K = \frac{r^u}{r^l} \tag{8.10}$$

where r^u and r^l are the resistances of the upper and lower epidermises. The combined resistance of the two surfaces (r^*) may be obtained from

$$\frac{1}{r^*} = \frac{1}{r^u} + \frac{1}{r^l}. \tag{8.11}$$

It may be shown that the estimate of leaf resistance (r^e) obtained from whole leaf fluxes and Eq. (8.9) is related to r^* as determined from separate measurements on both leaf surfaces by

$$r^e = \frac{r^a\, r^*\, (K-1)^2}{r^*\,(K+1)^2 + 4\, r^a\, K} + r^*. \tag{8.12}$$

The error term is a function of r^a, K and r^*. The error approaches zero when r^a is small and does not exceed the magnitude of r^a. Consequently, epidermal and boundary layer effects on gaseous fluxes can be reasonably partitioned using measurements of whole leaf fluxes, providing the boundary layer resistance is maintained at a low level by using high wind speeds and turbulent air flow. In addition, it is easier to resolve changes in epidermal resistance when the boundary layer resistance is small, and the high wind speeds and turbulence result in only small gradients in gaseous concentrations within cuvettes.

When comparing the effects of phenomena that operate in a catenary sequence, such as epidermal and boundary layer effects, it is convenient to use resistances. However, conductance is generally more useful when evaluating plant responses to environment. Transpiration, leaf water potential, and net photosynthesis are more linearly related to leaf conductance than to leaf resistance. Arithmetic means may be taken of multiple measurements of leaf conductance in canopies because the system consists of parallel pathways. In addition, the relationship between stomatal aperture and conductance is more linear than the relationship with resistance (BURROWS and MILTHORPE 1976). This preference for conductance is not trivial in that erroneous interpretations, including some of those supporting "threshold response hypotheses", have arisen because stomatal properties were expressed in terms of resistance.

8.2.2 Steady-State Models of Stomatal Function

A *mechanistic model* of stomatal responses to environment would require adequate descriptions of the following submodels: the mechanics of guard cell wall deformation and physical relations with adjacent cells; the biochemistry and biophysics of the control of osmotic potential within guard cells, and relations with other epidermal and mesophyll cells; and the characteristics of water, solute, and possibly hormonal transport within leaves and throughout plants. I consider that, given the present state of understanding of these systems, *mechanistic models* of stomatal response to environment would be cumbersome and highly unreliable, but ALLAWAY and MILTHORPE (1976) are more optimistic. However, *predictive models* capable of providing useful information could be developed for specific circumstances. Three types of *predictive models* are considered: multiple correlation models, phenomenological models, and models based upon hypothetical systems of stomatal regulation.

Substantial simplification may be achieved if the diurnal course of stomatal conductance may be treated as a sequence of steady states. It has been suggested that this assumption may be valid for the treatment of water flow in some plants (COWAN and MILTHORPE 1968), and steady-state analysis may provide adequate descriptions of a substantial proportion of stomatal responses in natural environments. The hysteresis observed on some occasions for stomatal responses to humidity and radiation may be approximately treated by steady-state models if the direction of change is included as a variable. Steady-state models would not be appropriate for strongly dynamic systems such as where stomatal oscillations occur.

8.2.2.1 Multiple Correlation Models

The simplest model of stomatal response to environment would include the independent variables: radiation, leaf temperature, some measure of plant water status, and possibly some aspect of atmospheric humidity and internal CO_2 concentration. Even with only four or five independent variables, it is likely that multiple correlation models, for which the definition and relations of variables have been determined empirically, will have little generality for treating different environments. HINCKLEY et al. (1975) developed multiple regression equations relating leaf resistance to solar irradiance, air temperature, vapor pressure deficit, and xylem pressure potential, but they concluded that these equations gave poor predictions. An intrinsically linear, logarithmic multiple regression analysis was applied by PALLARDY and KOZLOWSKI (1979). They obtained empirical relationships between leaf resistance and solar radiation, vapor pressure deficit, temperature, and leaf water potential. Surprisingly, the partial correlation coefficient for leaf water potential implied stomatal opening with greater leaf water deficits.

It is apparent that caution should be exercised in using empirical models to estimate the contribution of individual factors to stomatal responses in natural environments. Generally, highly empirical models should only be used for predicting the outcome of complex processes, and not for analyzing the mechanisms involved. For example, multiple correlation models may be used to predict missing values in diurnal and seasonal courses so that improved estimates of daily and annual balances may be obtained. An important principle is that empirical models are more reliable when used to interpolate rather than to extrapolate outside of the range of values used in developing the model. Empirical approaches may be considerably strengthened by developing *phenomenological models* of the type described by JARVIS (1976), with subsequent use of least squares regression analysis to obtain input parameters.

8.2.2.2 Phenomenological Models

The reader is referred to the paper by JARVIS (1976) which provides an excellent description and example of a *phenomenological model* of stomatal responses to environment. In the development of phenomenological models of stomatal responses to environment there are six basic steps. First, the external and internal variables that substantially influence stomatal conductance must be determined. Second, the manner in which individual variables influence stomatal conductance must be established at appropriate levels of the other variables. Third, equations must be developed that describe the effects of individual variables and interactions with the other variables. Fourth, values for the input parameters in these equations must be ascertained. Fifth, the predictions of the model must be tested. Sixth, it is likely that the predictions will not be perfect and it will be necessary to analyze the model to search for the sources of error, and correct them by returning to previous steps and repeating the process.

A detailed analysis of information that is relevant to each of these steps is provided, beginning with an examination of the manner in which stomata responded to light, humidity, temperature, CO_2 concentration and plant water status (see the detailed discussion in Chap. 7, this Vol.).

1. *Light.* Variations in solar radiation or visible light below the levels required for maximal stomatal opening result in changes in stomatal conductance. Three hypotheses for the mechanism of stomatal response to radiation have been proposed: that the response is a direct response of the guard cells to radiation; or that the response is an indirect consequence of radiation effects on mesophyll photosynthesis, either via changes in internal CO_2 concentration or changes in leaf energy status. However, a submodel based only upon stomatal responses to internal CO_2 concentration would not adequately describe the stomatal responses to light of many species (WONG et al. 1978; WONG 1979; RAMOS 1981). Stomatal responses to light may be modeled as depending upon the photon

flux, but the action spectrum should be considered. For some species, blue light appears to have specific effects on stomata, and the action spectrum differs from that of photosynthesis (RASCHKE 1979). In cases where the major response to light of stomata is associated with photosynthesis, the appropriate variable for modeling this effect would be the photon flux between 400 and 700 nm.

The responses of stomatal conductance to photon flux approximate a rectangular hyperbola, although other functions may be more appropriate for some species (BURROWS and MILTHORPE 1976). The photon flux (between 400 to 700 nm) required for 95% of maximal conductance varies from low values to full sunlight (100 to 2,000 μmol photon m^{-2} s^{-1}) depending upon the species, leaf age, and environment (TURNER 1974; BURROWS and MILTHORPE 1976). These reviews demonstrate that the concept of stomata opening fully at a low light threshold applies to only a few species and conditions. The use of resistance, and transit-time porometers, which are imprecise when stomatal conductances are moderate to large were partially responsible for the threshold hypothesis. Stomatal responses to photon flux may vary for opening responses in the morning and closing responses in the evening with some species (HINCKLEY et al. 1975; RAMOS 1981). Larger stomatal conductances were observed in the morning than in the evening at the same intermediate photon fluxes. These data may be explained by either hysteresis in opening and closing responses or the effects of endogenous diurnal rhythms. In addition, hysteresis may be observed in the opposite direction if stomatal responses to photon flux are determined over a short period of time (JARVIS 1980). This indicates that in some circumstances different parameters for stomatal responses to photon flux may be needed with increasing and decreasing fluxes or at different times during the photoperiod. For environments with variable cloud cover or where sunflecking is present, the use of steady-state models may result in some error because the half-time for stomatal response to radiation is several minutes, and stomatal conductances may reflect an integrated, earlier experience and not prevailing light levels (JARVIS 1980).

2. *Humidity*. Variations in atmospheric humidity frequently result in changes in stomatal conductance independently of effects due to changes in bulk leaf water status (HALL et al. 1976; HALL and HOFFMAN 1976; SCHULZE and KÜPPERS 1979; LÖSCH 1979; OSONUBI and DAVIES 1980a). It has been hypothesized that the response is related to changes in the water status of the epidermis through influences of humidity on cuticular water loss (LANGE et al. 1971; LÖSCH 1977); consequently, the vapor pressure difference between leaf and air (or relative vapor pressure difference) appears to be the most appropriate characterization of "humidity".

Mechanisms for stomatal response to "humidity" have been discussed in relation to water transport within leaves by SHERIFF (1979) and LÖSCH (1979). A model of stomatal response to humidity has been developed which assumes that the response is related to water loss from the external surfaces of the stomatal apparatus and from inside the leaf (FARQUHAR 1978). This model provides an excellent basis for further studies, but it has not been rigorously tested due to present inability to measure cuticular water loss from surfaces containing stomata. Empirical, exponential (BURROWS and MILTHORPE 1976) and linear (THORPE et al. 1980) models have been proposed for conductance response to vapor pressure difference, but these models only approximately predict observed responses. Stomatal conductance responses to the combined effects of vapor pressure difference and photon flux have been modeled by THORPE et al. (1980).

Some complexities of stomatal response to humidity warrant discussion. Stomatal opening responses to increasing humidity are slower than closing responses to decreasing humidity, and incomplete opening may occur after stomata are subjected to very low humidities (LANGE et al. 1971; LÖSCH 1977, 1979; SCHULZE and KÜPPERS 1979). In addition, at very high humidities, leaf conductance may decrease when humidity is increased (ASTON 1976). It is possible that part of leaf conductance response to humidity is nonstomatal, but much of the response is stomatal (ASTON 1976) and its significance to adaptation necessitates its inclusion in predictive models.

3. *Temperature*. Stomata respond to temperature, but until recently the characteristics of the responses have been varied and subject to dispute (HALL et al. 1976; LÖSCH 1979). It has been shown that much of the variation may have been due to measuring stomatal responses under unspecified, varying, and different levels of vapor pressure difference between leaf and air (HALL et al. 1976).

Measurements of stomatal conductance response to temperature, in which vapor pressure difference was kept constant, gave either the optimum-type response (JARVIS 1980) typical of many biological responses to temperature (SHARPE and DEMICHELLE 1977) or progressive increases in conductance over a substantial range of temperature (HALL et al. 1976). The shape of the curve relating stomatal conductance to temperature may change with drought (SCHULZE et al. 1973) or may remain the same (OSONUBI and DAVIES 1980b).

Irreversibility in stomatal responses to temperature causes difficulties for both measurement and modeling. Plants that were exposed for less than 1 h to extremely high (BAUER 1978) or extremely low (NEILSON and JARVIS 1975) temperatures had leaf conductances that remained low for several hours or days. SCHULZE et al. (1974) simulated the diurnal course of stomatal resistance of *Prunus armeniaca* with a model that assumes additive effects of leaf temperature and vapor pressure difference. The model reasonably predicted much of the diurnal variation in leaf resistance, but during the afternoons of hot dry days the observed stomatal resistances were larger than predicted values. The authors suggested that this partial stomatal closure in the afternoons may have been due to after-effects of high temperature or hysteresis in stomatal response to humidity.

4. *CO_2 Concentration*. Different regions of the stomatal apparatus are exposed to CO_2 concentrations varying between external ambient levels and the concentration within leaves. It is usually assumed that stomata only respond to the internal CO_2 concentration, which is defined as the average concentration at the evaporating surfaces within leaves, but this has not been rigorously established. Alternatively, stomata may respond to the gradient in CO_2 concentration between the external and internal walls of guard cells. In this case, they would respond differently to variations in internal concentration due to

changes in external concentration, than to those due to changes in internal processes such as photosynthesis. The only studies that I am aware of on this point are the unpublished observations of I.R. COWAN and G.D. FARQUHAR reported by WONG (1979, p 162), which do not support the hypothesis that stomata respond to the gradient in CO_2 concentration.

Models of stomatal responses to CO_2 concentration may contribute to predictions concerning the consequences to plants of the expected increases in global CO_2 concentration (BAES et al. 1977). However, the inclusion of submodels for describing stomatal responses to natural variations in internal CO_2 concentration may not be necessary in natural plant habitats. There are many experiments with intact, attached leaves in which external CO_2 concentrations were varied to simulate natural variations in internal CO_2 concentration, which only had small effects on leaf conductance (BURROWS and MILTHORPE 1976; HALL et al. 1976; JARVIS 1980). However, stomata of some species respond strongly to changes in CO_2 concentration (RASCHKE et al. 1978; WONG 1979; RAMOS 1981). When evaluating and modeling stomatal responses to internal CO_2 concentration it is useful to apply gain analysis methodology as described by WONG et al. (1978) to separate the effects of different factors on stomatal conductance and internal CO_2 concentration.

RASCHKE (1976) has hypothesized that "stress" may sensitize stomata to internal CO_2 concentration. This hypothesis is particularly important as a possible conceptual basis for modeling stress effects on stomata, and for modeling and understanding the relationships between regulation of water loss and CO_2 assimilation. Unfortunately, few rigorous tests of this hypothesis (with intact plants) have been reported, and there are some indications that water stress does not influence stomata of certain conifers (JARVIS 1980) or *Zea mays* in this manner (WONG 1979).

Several models have included assumed relations between stomatal conductance and internal CO_2 concentration (PENNING DE VRIES 1972; VAN BAVEL et al. 1973; WAGGONER 1969) or external CO_2 concentration (JARVIS 1976). Little evidence was provided to justify these submodels, but most authors stated that the effect of CO_2 was either small or unknown. It is apparent that further experimental studies are needed of stomatal responses to CO_2 with intact plants grown in different environments.

5. *Plant Water Status.* It is well known that stomatal conductances tend to decrease when plants are subjected to drought, but general relationships between stomatal conductance and plant water status have not been discovered. Present differences of opinion illustrate the controversy that prevails concerning relations between stomatal conductance and plant water status. TURNER and BEGG (1978) stated "It is now generally recognized that the stomata do not respond to changes in leaf water potential ··· until a critical threshold level ··· is reached, and then stomatal closure occurs over a narrow range of leaf water potential ···. ··· Many of the differences in the leaf water potential for stomatal closure can be accounted for by differences in the osmotic potential or osmotic adjustment of plants". In contrast, JONES and RAWSON (1979) stated "in sorghum, leaf conductance declines continuously over a range of leaf water potential from -15 to -30 bar. The steepness of the initial decline was depen-

dent upon the rate of change of leaf water potential so that although at high rates of stress (7 and 12 bar/day) the response of leaf conductance to water potential resembles the threshold response reported elsewhere for sorghum (TURNER 1974), at lower rate of stress (1.5 bar/day) the response extends over an approximately 15 bar range of leaf water potential. ··· osmotic adjustment in sorghum is not responsible for the higher values of leaf conductance at low leaf water potentials in slowly stressed plants ··· it appears that in sorghum the relationship between leaf conductance and leaf turgor potential is not unique ···". They also suggested that the threshold concept "may be a result of combination of factors such as the use of diffusion porometry, the use of stomatal resistance units, and the imposition of very rapid rate of stress, rather than being a reflection of an inherent mechanism involved in guard cell movement". Evidence which supports both sets of statements is available in the literature.

I will present an analysis of the influence of plant water status on stomata that partially reconciles these apparently conflicting points of view, and provides a conceptual basis for the development of a general model. It should be recognized that a priori we should not expect stomatal conductance to be uniquely related to either bulk leaf water potential or bulk leaf pressure potential. Stomatal apertures are probably related to the turgor pressures of guard cells and adjacent cells, but it cannot be assumed that the osmotic potentials of these cells are either constant or uniquely related to bulk leaf osmotic potential (RASCHKE 1979; BEADLE et al. 1978). It is also useful to distinguish between effects on stomata of short-term (minutes) and long-term (days) changes in plant water status. When the water status of leaves of several species was rapidly disturbed by detaching them or manipulating the osmotic potential of the root zone, leaf conductance exhibited threshold responses to bulk leaf water potential (EHLIG and GARDNER 1964; HALL and HOFFMAN 1976). These responses resemble the threshold model of TURNER and BEGG (1978). In contrast, EHLIG and GARDNER (1964) observed progressive decreases in relative transpiration for whole plants with day-to-day decreases in leaf water potential as the water in the root zone was depleted which do not fit the threshold model. It should be noted that under field conditions, environmental factors and plant water status vary during the day, and leaf water potential depends upon the transpiration rate. Consequently, it is difficult under these conditions to determine separately the effects on stomata of leaf water status and the environmental factors that also influence the evaporative demand. In an ingenious experiment, SCHULZE and KÜPPERS (1979) modified the water status of individual attached leaves of *Corylus avellana* as much as 8 bar, during the day, without changing the environment of these leaves or the root zone environment (they varied the evaporative demand on the remaining part of the shoot). They established that the changes in leaf water potential that could occur during a diurnal course had negligible influence on stomatal conductance. ACKERSON et al. (1977) also observed little change in stomatal resistance of cotton with decreases in leaf water potential to as low as −25 bar under field conditions. Consequently, it is possible that progressive decreases in transpiration with progressive decreases in leaf water potential are not due to a close coupling between leaf conductance and leaf water potential.

It is hypothesized that short-term (minutes) changes in leaf water status have a threshold effect on stomatal conductance, but that in some natural conditions and with some species the threshold leaf water potential for stomatal closure may not be reached. The results of JONES and RAWSON (1979) and others which are not consistent with the threshold hypothesis may be partially explained by a hypothesis for long-term (days) effects of changes in plant water status on stomata. Long-term changes in plant water status influence maximum daily leaf conductance (HALL and SCHULZE 1980a) and may also modulate stomatal response to other environmental variables (HALL et al. 1976). Progressive decreases in leaf conductance were observed with decreases over 2 weeks in soil water potential in the root zone of a desert fern (NOBEL 1978). Pre-dawn plant water potential provides another measure of day-to-day changes in plant water status. Correlations have been reported between pre-dawn pressure chamber values and mean daily leaf conductance (RUNNING et al. 1975) or maximum morning conductance (RUNNING 1980a). It is also possible that plant water potential values integrated over time may provide a better measure of long-term changes in plant water status than pre-dawn water potential. SCHULZE et al. (1980) have shown that drought-induced changes in photosynthesis were related to cumulative daily plant water potential, but this type of analysis has not been applied to the effects of drought on leaf conductance. A model which includes a time-integrated measure of drought experience could also incorporate the effects of rate of drying described by JONES and RAWSON (1979) and possibly after-effects of water stress on stomatal opening potential of the type described by FISCHER et al. (1970).

A comprehensive model would include submodels for both the long-term effects of drought experience, and short-term, possibly threshold, relations with leaf water status. The only interpretation of leaf conductance data that I am aware of, which follows this scheme, is that of RUNNING (1976). He described leaf conductance responses of conifers as including both a relation between early-morning maximum leaf conductance and pre-dawn water potential, and threshold relations with leaf water potential during the rest of the day. It should be recognized, however, that it is difficult to determine from diurnal courses whether leaf conductance is responding to leaf water potential. The decreases in leaf conductance during the day, observed by RUNNING (1976), could have been due to the effects of one or more of several covarying factors, including the humidity difference between leaf and air (RUNNING 1980a).

Methods for quantifying the responses of stomata to environment are considered. Ideally the development of equations for stomatal responses would be based upon leaf conductance responses to individual variables in controlled environment cuvettes in the field (SCHULZE et al. 1974) or in growth chambers (THORPE et al. 1980). Next, the extent of interactions between variables must be determined so that the individual equations may be linked in a coherent general function. Individual functions may be combined by multiplication (JARVIS 1976), by addition (SCHULZE et al. 1974), or by using reciprocals (BURROWS and MILTHORPE 1976). The most effective method for combining functions is not known at this time.

A reference stomatal conductance may be included in models as a parameter which is influenced by the long-term effects of drought, temperature preconditioning, leaf age etc. (HALL and SCHULZE 1980a). The short-term effects of radiation, humidity, and temperature are then considered as having proportionate effects (multiplicative model) or additive effects in relation to the reference conductance. The reference conductance is defined as the stomatal conductance which occurs when the CO_2 assimilation rate is maximized by optimizing temperature and humidity, and using saturating photon flux and normal air (21% $[O_2]$ and 330 ppm $[CO_2]$). An alternative procedure would be to include maximum conductance as a parameter, but its measurement may require use of conditions which result in instability of the stomatal system such as high temperatures (and also moderately high humidity, saturating photon flux, and possibly low CO_2 concentration). The values of input parameters may be determined from studies where variables are individually changed in steps (SCHULZE et al. 1974; THORPE et al. 1980) or from field data, where several variables were changing at the same time, by simultaneous solution of equations using nonlinear least squares regression analysis (JARVIS 1976).

Ideally, model tests should be conducted with a set of field data which includes an adequate range of values for important variables, and which was not used for obtaining input parameters. If the tests indicate that the model is not adequate, investigations of the sources of error could include sensitivity tests using reasonable ranges of parameter values, but excessive juggling of parameter values in attempts to obtain a reasonable fit is not recommended. To improve the model it may be necessary to go back and determine whether the assumptions, environmental variables, and functions are appropriate for the system and objectives of the model.

8.2.2.3 Models Based Upon Hypothetical Systems of Stomatal Regulation

The complexity of stomatal responses to environment described in the previous section indicates that developing phenomenological models of this system will be a difficult task. An alternative approach has evolved from recent research that may contribute to the development of predictive models. In this approach it is assumed that the complex responses of stomata have certain end results, such as maintenance of leaf turgor (DAVIES et al. 1980) or internal CO_2 concentration (WONG 1979) or specific types of water-use optimization.

The elegant theory of COWAN and FARQUHAR (1977) provides a basis for modeling stomatal function in relation to the optimization of water-use (see also Chaps. 1, 7, and 17, this Vol.). They pointed out that, within the constraints set by the maximal and minimal stomatal conductances that are possible for a particular leaf: on any day, a specific diurnal course of stomatal conductance would result in maximum daily CO_2 assimilation for a specific level of daily water use. They hypothesized, based upon the calculus of variations, that this optimal function would be achieved if stomatal apertures varied so that $\partial A/\partial E$ remained constant during the day: where ∂A is the change in CO_2 assimilation rate and ∂E is the change in transpiration rate that would result from an infinitesimal change in stomatal conductance (δg). $\partial A/\partial E$ may be more explicitly described by the ratio of partial derivatives $(\partial A/\partial g)/(\partial E/\partial g)$. For this hypothesis to provide a basis for effective predictive models of stomatal response to environment, two aspects must be evaluated. First, the extent to which $\partial A/\partial E$ is constant during the day must be determined for different species and environmental conditions. Second, the plant and environmental factors which cause short-term and long-term changes in the value of $\partial A/\partial E$ and the response functions for these effects must be determined.

It has been established that $\partial A/\partial E$ remained reasonably constant with variations in atmospheric humidity at constant leaf temperature for *Nicotiana glauca* and *Corylus avellana* (FARQUHAR et al. 1980), and with variations in both humidity and temperature for *Vigna unguiculata* (HALL and SCHULZE 1980a). However, $\partial A/\partial E$ was not constant when stomata did not fully recover with humidification after exposure to low humidities, and in one set of studies with a plant subjected to drought (HALL and SCHULZE 1980a). In addition, provisional studies indicated that $\partial A/\partial E$ was not constant when photon flux was varied, but stomatal conductance was linearly correlated with CO_2 assimilation rate (C. RAMOS personal communication). Linear associations between stomatal conductance and CO_2 assimilation rate, with short-term changes in photon flux, have been reported for several species (GOUDRIAAN and VAN LAAR 1978; LOUWERSE and ZWEERDE 1977; WONG et al. 1978, 1979; WONG 1979).

An approach for modeling diurnal courses of stomatal conductance is proposed based upon the observations described in the previous paragraph. It is assumed that the effects of light, at constant humidity and temperature, may be modeled by assuming a relation between stomatal conductance and CO_2 assimilation rate. Stomatal responses to humidity and temperature would be modeled by assuming that $\partial A/\partial E$ remains constant at a particular photon flux. Operationally, it would be necessary to determine the stomatal conductance that satisfies these constraints. Stomatal conductances that result in constant $\partial A/\partial E$ could be determined by computerized iteration procedures to simultaneously solve the equations in Appendix I of COWAN and FARQUHAR (1977), which were also presented by HALL and SCHULZE (1980a), and equations for modeling photosynthesis and respiration. A model of leaf photosynthesis and respiration was developed by HALL (1979) for applications of this type. Models of photosynthesis are discussed in Chapter 16, this Volume. The relation between stomatal conductance and CO_2 assimilation rate with changes in photon flux could be determined empirically.

For modeling long-term (day-to-day) changes in stomatal response to environment it would be necessary to understand and quantify the day-to-day changes in level of $\partial A/\partial E$ or reference stomatal conductance that may occur (as mentioned earlier the reference stomatal conductance is that which occurs when CO_2 assimilation rate is maximized but at normal $[O_2]$ and $[CO_2]$). The studies of HALL and SCHULZE (1980a) with *Vigna unguiculata* indicated that day-to-day values of $\partial A/\partial E$ increased and reference stomatal conductance decreased with progressive drought. However, for *Corylus avellana* there was a tendency for $\partial A/\partial E$ to decrease with progressive drought (FARQUHAR et al. 1980). Obviously, more studies are needed of the levels of $\partial A/\partial E$ that occur especially in natural environments, but these measurement require sophisticated technology (HALL and SCHULZE 1980a). It is possible that the levels of $\partial A/\partial E$, the reference stomatal conductance, and the capacity for assimilation CO_2 are all interrelated (HALL and SCHULZE 1980a). As discussed in Chapter 7, this Volume, there is abundant evidence that day-to-day decreases in capacity for assimilating CO_2 are accompanied by decreases in the reference level of conductance irrespective of the factor responsible for the decreases in capacity for assimilating CO_2. The extent to which these associations between the reference conductance and capacity for assimilating CO_2 indicate a relatively constant $\partial A/\partial E$, as suggested by FARQUHAR (1979), is not known.

The advantages and disadvantages of modeling stomatal responses to environment based upon assumptions concerning the levels of $\partial A/\partial E$ and relations with reference stomatal conductance and CO_2 assimilation capacity are considered. Requiring an effective model of CO_2 assimilation is not a disadvantage for most applications because this model is also necessary for predicting the consequences of stomatal conductances to photosynthesis and adaptation. A major advantage of the $\partial A/\partial E$ hypothesis is that it integrates stomatal effects on the gaseous transfer of both water vapor and CO_2, the effects of energy balance on transpiration and leaf temperature, and the effects of internal CO_2 concentration and leaf temperature on internal leaf capacity for assimilating CO_2. The model does not include possible relations between leaf conductance and short-term effects of changes in leaf water status. However, for cases where short-term effects of leaf water status do occur, they could be modeled by assuming a threshold response of leaf conductance to bulk leaf water potential or that a system exists for maintaining bulk leaf pressure potential above a particular level. Modeling long-term effects by assuming that there is a relationship between reference stomatal conductance and CO_2 assimilation capacity would be extremely effective if the relationship is generally valid for specific genotypes or species and independent of the factor responsible for the long-term effects. The disadvantage of this approach is that it is constrained so that, if it does not work adequately, it will be difficult to modify.

8.2.3 Dynamic Models of Stomatal Function

Models which describe phenomena that continuously change with time and which contain time as an independent variable are defined as being dynamic. When steady-state models are applied in discrete-time-step simulations there is frequently the appearance but not the substance of dynamic modeling. Discrete-step simulation models which include dynamic submodels describing endogenous rhythms or the long-term effects of drought may be considered hybrids with both steady-state and dynamic characteristics.

Superficial evaluation would indicate that all biological processes are dynamic, and that dynamic models would be more powerful than steady-state or hybrid models. In practice, the type of model that would be more powerful would depend upon the objectives of the exercise, and a trade-off between the increased predictive and analytical capabilities of dynamic models, and their increased complexity which approximates an order of magnitude greater than that of steady-state models of the same system.

A dynamic model has been developed for analyses of oscillations in stomatal conductance (COWAN 1972). At this time, it is not known whether stomatal oscillations occur in natural environments, and what their adaptive significance might be (COWAN 1977b), although the relevance of stomatal oscillations to water-use efficiency and plant growth has been discussed by COWAN and TROUGHTON (1971). A powerful quantitative method has been developed for predicting the environmental conditions that promote the occurrence of stomatal oscillations (FARQUHAR and COWAN 1974). This model predicts that the lack of oscillations in some controlled environment systems may be due to lack of control of external humidity or control of leaf temperature, both of which are artificial compared with natural environments and would tend to dampen oscillations. The study of stomatal oscillation may ultimately lead to insights into the mechanisms and regulation (CRAM 1976) of stomatal response to environment and water transport in plants. But the development of comprehensive, dynamic models of stomatal function requires a better understanding of the system than is available at this time.

8.3 Integrated Models of Plant Water Loss and Water Relations

The models of plant water loss which were discussed in Section 8.2 should be linked with models of liquid water transport in the soil-plant system and of plant water relations as suggested in Section 8.1.2. A major problem with this approach is that even though simple models are available of liquid water transport in the soil-plant system, there are indications that these models and current understanding have substantial flaws (as will be discussed in this section). Also, special problems arise when attempting to link certain threshold models for the response of leaf conductance to leaf water potential with models for liquid water transport in the soil-plant system. A threshold model which has only two states, open versus fully closed, cannot be linked to a steady-state model for liquid water transport if discontinuities are to be excluded. A threshold model of this type could be linked to a dynamic model of liquid water transport but, depending upon the time constants of the system, it would tend to predict oscillations of leaf conductance and these rarely occur in natural conditions. A threshold model which includes the possibility of progressive stomatal closure over a particular range of leaf water potentials would not necessarily exhibit these problems.

8.3.1 Steady-State Models of Plant Water Loss and Water Relations

Submodels describing the short-term (diurnal) changes and long-term (seasonal) changes in leaf water status are both necessary if a comprehensive model of plant water loss and water relations is to be developed. The diurnal course of leaf water potential may be modeled as mainly depending upon transpiration

(ELFVING et al. 1972; JARVIS 1976) and as a sequence of steady states (COWAN and MILTHORPE 1968), providing the rate of change in plant water content is small compared with the flow through the system. This approach may be valid for most herbaceous species and for some woody species (KAUFMANN 1976), but not for some succulents and many conifers where water loss from foliage may be sustained for several days or weeks by water stored in the plants (RUNNING et al. 1975; WARING and RUNNING 1978; RUNNING 1980b). A hydrologic balance model would also be needed to predict day-to-day changes in soil water conditions in the root zone (HALL and DANCETTE 1978).

However, relations between leaf water potential and transpiration rate frequently do not fit simple models. They have varied substantially depending upon the species, conditions of growth and measurement, and laboratory in which the studies were conducted (KAUFMANN 1976; COWAN 1977b). There are several areas where further research may resolve some of the controversy.

First, theoretical analyses indicate that liquid water transport may be intimately linked with solute transport (DALTON et al. 1975; FISCUS and KRAMER 1975). A cycling of solutes within plants, due to active pumping of ions into xylem tissue and active removal from xylem tissue, could provide the energy necessary for water to move against a water potential gradient in the whole-plant system (HALL and HOFFMAN 1976). The observation that roots excised from transpiring plants exuded sap (MEYER and RITCHIE 1980) supports this interpretation. However, the contribution of solute cycling to water transport may be small, especially at high transpiration rates.

Second, substantial interface resistance between roots and soil may occur in specific circumstances (COWAN and MILTHORPE 1968; FAIZ 1973; HERKELRATH 1975). This may resolve some of the controversy concerning whether the major resistance to water flow is in the root or the soil (ROSE et al. 1976). Under conditions where roots shrink away from soil particles, the major resistance would be neither in the soil nor in the root per se but between them.

Third, the liquid water pathway between soil and roots may exhibit irreversibility. Root systems frequently penetrate both dry and moist layers in the soil profile; yet they probably do not normally act as conduits cycling substantial quantities of water from the moist zone to the dry zone. An exceptional case of water cycling within root systems has been hypothesized for trees in an arid region of Chile where deep ground water is present (MOONEY et al. 1980). Adequate explanations for this irreversibility have not been provided, and this phenomenon has not been incorporated into models of the soil-plant continuum of water.

Fourth, the expected equilibrium between plant and soil water potentials, when transpiration is slow, has not been observed in some circumstances. Leaf water potential of some xerophytes and woody species have remained at low values (several bars negative) even after irrigation (SCHULZE et al. 1980; SYVERTSEN et al. 1975). Associations between lack of equilibrium at zero transpiration and rapid leaf enlargement have been described by BOYER (1974). But it is unlikely that the flow of water involved in growth, by itself, could account for the large deviations from equilibrium exhibited by some xerophytes and woody species. Lack of equilibrium also occurs when the liquid water continuum between soil and plant is broken. This has been observed with certain CAM plants which maintained extremely high water potentials even though the soil was very dry. The plant-soil system becomes coupled soon after rain, due to the rapid growth of new roots. The extent to which uncoupling occurs among different species and environments is not known.

It should be apparent that current controversy and lack of understanding of water transport in the soil-plant system seriously constrains model development. Even the basis for evaluating components of water potential has been questioned (SPANNER 1973; PASSIOURA 1980).

8.3.2 Dynamic Models of Plant Water Loss and Water Relations

There are some plants which have low rates of water loss during certain seasons and sufficient water within the plant to sustain this water loss for several days or weeks, even though water uptake from the soil is negligible. Examples of this phenomenon are provided by evergreen conifers during winter, succulent CAM species in dry soils, and bottle trees (e.g., *Brachychiton australis*). By analogy with electrical circuits this phenomenon may be modeled as being dynamic and equivalent to capacitance (COWAN and MILTHORPE 1968). Charge and discharge of water stored in the plant could exhibit diurnal cycles superimposed upon seasonal cycles. The occurrence of substantial capacitance may partially explain hysteresis in relations between leaf water potential and transpiration with water potentials being lower in the afternoon than in the morning at a particular transpiration rate (JARVIS 1976; KAPPEN et al. 1975). A detailed discussion of the quantitative treatment of the capacitance of different plant tissues has been presented by JARVIS (1975). A dynamic model of water transport in plants has been developed by COWAN (1972), which incorporates effects due to capacitance, and which was used for analyzing oscillations in stomatal conductance. The influence of capacitance on the water balance of conifers has been examined using a discrete-step (daily increment) steady-state model (RUNNING et al. 1975).

Water flow in the soil-plant system as a whole may be influenced by the capacitance of specific tissues. The relationship between root water potential and root shrinkage would influence water flow through effects on the root-soil interface resistance. Stomatal apertures may be influenced by water flow and water deficits in the epidermal and mesophyll tissues (SHERIFF 1979; RUNNING 1980b), and consequently, by the capacitance of the epidermis or mesophyll. Limited information concerning values of parameters for these phenomena makes it difficult to model them at this time.

It is speculated that dynamic characteristics of water transport in plants may be due to linkages between oscillations in plant water status and stomatal apertures, and solute pumping across membranes. Evidence has been obtained indicating that ion pumping across membranes may be influenced by turgor pressure (CRAM 1976; COSTER et al. 1977). In an analogous manner increased tension in the xylem vessels of roots may cause increased ion pumping into the xylem. A mechanism of this type, if coupled with regulated removal of solutes from the xylem vessels in the shoot, could keep xylem pressure potential and leaf water potential relatively constant with changes in transpirational flow (as was observed by HALL and SCHULZE 1980b; KAUFMANN 1976; WEATHERLEY 1970) or result in water flow in the direction of higher water potentials (HALL and HOFFMAN 1976; SHARP and DAVIES 1979). The development of realistic dynamic models of plant loss and water relations will require a more complete understanding of the water transport and water relations of the soil-plant system than is currently available.

8.4 Integrated Models of Canopy Water Loss and Water Relations

The water use of the dense populations of uniform plants commonly used in mechanized agriculture may be interpreted on a unit land area basis. In contrast, the water use of sparse populations may be interpreted on a unit plant basis. For dense communities containing different species it may be appropriate to determine the water-use of the community as a whole per unit land area, and attempt to quantify the contribution of each species in the community. A review of integrated models of canopy water loss has been presented by JURY (1979) and additional information has been presented by SLABBERS (1980).

Modeling canopies on a unit leaf area basis must be considered a monumental task because the structural and environmental characteristics of canopies are extremely complex. Canopies have been modeled as discrete strata containing leaves having a specific leaf area index and average angle (DUNCAN et al. 1967), but the substantial input requirements of models of this type preclude many uses, even with uniform populations of plants. Gaseous transfer characteristics within canopies are even more complex and may contain mesoscale mass flow of air (LEMON et al. 1973). Comprehensive models have been developed for simulating the soil-plant-atmosphere continuum of water, and CO_2 exchange of field crops (LEMON et al. 1973; VAN BAVEL 1974), but for input they require a detailed description of the canopy structure, and this is usually not available.

Some guidelines are presented for phenomenological models of canopy water loss and water relations which have modest input requirements. Evaporation from soil surfaces has been successfully modeled as being dependent upon the evaporative demand, until the surface dries, after which it is assumed to be dependent upon the water transfer properties of the soil (RITCHIE 1972). Soil evaporation may be considered as only occurring from the land surface that is not covered by the canopy (HALL and DANCETTE 1978) or as depending upon leaf area index (RITCHIE 1972) or the net radiation at the soil surface (TANNER and JURY 1976). Percent cover is by far the most convenient of these parameters for partitioning evaporation and transpiration. Transpiration per unit land area has been modeled as being proportional to the proportion of land area covered by the canopy (HALL and DANCETTE 1978), but this neglects micro-scale advection that may be important in some cases (HANKS et al. 1971). In addition a submodel for the canopy conductance to water vapor loss may be developed and incorporated into a model of canopy energy balance and gaseous transfer such as the model of MONTEITH (1965). In field studies, canopy conductances have been estimated by a model consisting of summing leaf conductance × leaf area index for different strata in the canopy (BISCOE et al. 1976). They found that this model gave canopy conductances which were correlated with values obtained using micrometeorological data and the equation of MONTEITH (1965).

Linking canopy resistance to soil water deficits is a major problem. First, it should be recognized that the development and maintenance of ground cover (and leaf area index) will be influenced by long-term drought (TURK and HALL 1980a, b). For canopies that completely cover the land surface, drought will mainly influence canopy water loss through effects on leaf conductances. A threshold hypothesis has been proposed for this phenomenon because changes in canopy conductance were not observed until more than 75% of the available water in the root zone was depleted (RITCHIE 1973). This hypothesis has been used in hydrologic balance models where the regulation of transpiration was linked to available soil water (ROSENTHAL et al. 1977). In addition, this threshold hypothesis for the regulation of canopy water loss has been linked to the threshold hypothesis for relations between stomatal conductance and leaf water potential in a model developed by SLABBERS (1980). In contrast, TURK and HALL (1980c) did not observe a threshold response, they observed an approximately linear relation between canopy conductance and the amount of available water in the root zone. By analogy with the hypothesis presented in Section 8.2.2.2 for relations between stomatal conductance and drought, it is proposed that long-term drought may have progressive effects on canopy conductance as observed by TURK and HALL (1980c), whereas more rapidly induced soil water deficits may result in threshold responses of the type described by RITCHIE (1973).

Submodels relating the development of ground cover (or leaf area index), and canopy conductance to soil water status in the root zone have the advantage that soil water status can be approximately modeled by hydrologic budget simulations (HALL and DANCETTE 1978). Linking the development of ground cover (or leaf area index), and canopy conductance to plant water status has the disadvantage that it would be necessary to develop a submodel of canopy water status. Models of canopy water status may have to include the effects of branched pathways (RICHTER 1973) and differential transpirational fluxes within the canopy.

It may be anticipated that developing models of water loss and water relations of canopies formed by communities of different species will be difficult. One approach to this complex system would involve a three-dimensional assemblage of unit cubes describing the root and shoot distribution and characteristics of component species in the community (personal communication by R.S. LOOMIS).

8.5 Conclusions: Contributions of Mathematical Models to Research

Mathematical modeling of biological systems is a relatively new approach to obtaining improved understanding and it has been criticized. Some of the criticism is warranted and may be partially due to the primitive state of ecological modeling and the fact that guidelines concerning procedures to be used in modeling ecological systems are only just evolving. I suggest, however, that much is to be gained from close integration between modeling and experimental studies.

The complexity of ecosystems requires integrated modeling and experimental approaches. Experimental studies can only provide limited quantities of information concerning the hydrologic budget of ecosystems because of the infinite number of soil-plant-atmosphere systems that are of interest to ecologists, agriculturalists, and others. Mathematical models can be used in an interpolative manner to reconstruct certain aspects of the functions of ecosystems over a period of days, months, or years using only isolated sets of experimental data (VAN BAVEL 1974).

Evaluating the effects of individual environmental variables on the soil-plant-atmosphere system represents another contribution of modeling. Individual response curves may be determined in controlled environments. Mathematical models based upon this information may then be used to predict, and are tested against, responses observed in natural environments in which many variables are changing over the same time period. In this way, mathematical models provide an important link between controlled environment studies and field studies (VAN BAVEL 1974). However, the quality of the link depends upon the adequacy of the controlled environment studies. Much of our present knowledge of water transport and plant response to water deficits has been obtained with plants in small pots of sterilized root media or even worse, with hydroponics or excised roots or shoots. It is likely that in many cases the responses may not be relevant to plant responses in most natural habitats.

Mathematical models coupled with controlled environment studies can provide information that cannot be obtained in the field at this time. For example, it is possible that atmospheric CO_2 concentrations may double during the next century (BAES et al. 1977). It would appear important for man to obtain some estimates of the consequences of this drastic disturbance to the ecosphere and to the habitats of man and the other biological organisms. These estimates could be obtained with controlled environment studies, to obtain isolated data sets describing plant responses to elevated CO_2 concentrations, and by using mathematical models to predict plant responses in hypothetical field environments. Another example is illustrated by the work of LANGE et al. (1978). They evaluated the adaptation of different plant species to the thermal regime of a location in the Negev desert by examining deviations between optimal temperatures for CO_2 assimilation and actual diurnal temperatures. Mathematical models are well suited to studies of this type. First, it is necessary to have a model that is able to predict the functions of a particular ecotype in its habit during the appropriate season. The model may then be used to predict ecotype performance in moderately different environments (e.g., warmer or cooler) to obtain information on the adaptation of the ecotype to its present habitat.

Plant breeders frequently attempt to develop genotypes of crop species with improved adaptation to specific habitats. A major problem with respect to water-limiting environments is that a priori it is not known whether it will be necessary to select for increased or decreased rates of particular processes such as root growth (HALL et al. 1979). It is likely that some intermediate level will be optimal and this level will depend upon both the genetic background and the habitat. Mathematical models may be used to predict the consequences to plant performance of genetic manipulation of specific plant parameters. This may help plant breeders to establish priorities among the infinite possibilities open to them for genetic manipulation. Mathematical models of hydrologic balance have been used to predict optimal cycle lengths for annual crops in semiarid environments (DANCETTE and HALL 1979; HALL and DANCETTE 1978) which may be used to guide plant breeding programs.

Possibly the most important present and future applications of mathematical modeling of the systems described in this chapter concern water management including: water sheds, planning and operating water supply systems, and irrigated agriculture. Mathematical models have made major contributions in these areas but in one area it can only be assumed that modern science and human societies have failed. Irrespective of considerable research on soil-plant-atmosphere water relations, the timing of irrigations in much of the world is still based upon a combination of empirical experience and artistry. It is hoped that a combination of experimental studies, mathematical modeling, and development of new technologies will soon provide effective and practical scientific methods for determining when different crops should be irrigated in different environmental conditions.

I will conclude by pointing out a unique contribution that mathematical modeling can provide for ecology. Studies with mathematical models can, in principle, make contributions to understanding that are not possible at this time through experimental studies. The history of the physical sciences has many examples of predictions based upon mathematical models, some of which, with the passage of time and development of technology, have been supported or disproved by experimental studies. Mathematical models will probably play this same role in plant water relations and ecology as a whole.

Acknowledgements. I appreciate the comments provided by Professors C.B. OSMOND, O.L. LANGE and E.-D. SCHULZE, and Dr. J.B. PASSIOURA concerning earlier drafts of this review.

References

Ackerson RC, Krieg DR, Haring CL, Chang N (1977) Effects of plant water status on stomatal activity, photosynthesis, and nitrate reductase activity of field grown cotton. Crop Sci 17:81–84

Allway WG, Milthorpe FL (1976) Structure and functioning of stomata. In: Kozlowski TT (ed) Water deficits and plant growth Vol IV. Academic Press, London New York, pp 57–102

Aston MJ (1976) Variation of stomatal diffusive resistance with ambient humidity in sunflower (*Helianthus annuus*). Aust J Plant Physiol 3:489–502

Baes CF Jr, Goeller HE, Olson JS, Rotty RM (1977) Carbon dioxide and climate: the uncontrolled experiment. Am Sci 65:310–320

Bauer H (1978) Photosynthesis of ivy leaves (*Hedera helix*) after heat stress 1. CO_2-gas exchange and diffusion resistances. Physiol Plant 44:400–406

Bavel Van CHM (1974) Soil water potential and plant behavior: a case modeling study with sunflowers. Oecol Plant 9:89–109
Bavel Van CHM, DeMichele DW, Ahmed J (1973) A model of gas and energy exchange regulation by stomatal action in plant leaves. Tex Agric Exp Stn Publ MP-1078:28
Beadle CL, Turner NC, Jarvis PC (1978) Critical water potential for stomatal closure in sitka spruce. Physiol Plant 43:160–165
Biscoe PV, Cohen Y, Wallace JS (1976) Community water relations. Daily and seasonal changes of water potentials in cereals. Philos Trans R Soc London Ser B 273:565–580
Boyer JS (1974) Water transport in plants: mechanism of apparent changes in resistance during absorption. Planta 117:187–207
Burrows FJ, Milthorpe FL (1976) Stomatal conductance in the control of gas exchange. In: Kozlowski TT (ed) Water deficits and plant growth Vol IV. Academic Press, London New York, pp 103–152
Cooke JR, Debaerdemaeker JG, Rand RH, Mang HA (1976) A finite element shell analysis of guard cell deformation. Trans ASAE 19:1107–1121
Coster HGL, Steudle E, Zimmermann U (1977) Turgor pressure sensing in plant cell membranes. Plant Physiol 58:636–643
Cowan IR (1972) Oscillations in stomatal conductance and plant functioning associated with stomatal conductance: observations and a model. Planta 106:185–219
Cowan IR (1977a) Water use in high plants. In: McIntyre AK (ed) Water: planets, plants and people. Aust Acad Sci, Canberra, pp 71–107
Cowan IR (1977b) Stomatal behaviour and environment. Adv Bot Res 4:117–228
Cowan IR, Farquhar GD (1977) Stomatal function in relation to leaf metabolism and environment. In: Integration of activity in the higher plant. Soc Exp Biol Symp XXXI. Univ Press, Cambridge, pp 471–505
Cowan IR, Milthorpe FL (1968) Plant factors influencing the water status of plant tissues. In: Kozlowski TT (ed) Water deficits and plant growth Vol I. Academic Press, London New York, pp 137–193
Cowan IR. Troughton JH (1971) The relative role of stomata in transpiration and assimilation. Planta 97:325–336
Cram WJ (1976) Negative feedback regulation of transport in cells. The maintenance of turgor, volume and nutrient supply. In: Lüttge U, Pitman MG (eds) Transport in plants II. Part A. Cells. Encyclopedia of plant physiology new series Vol II. Springer, Berlin Heidelberg New York, pp 284–316
Dalton FN, Raats PAC, Gardner WR (1975) Simultaneous uptake of water and solutes by plant roots. Agron J 67:334–339
Dancette C, Hall AE (1979) Agroclimatology applied to water management in the Sudanian and Sahelian zones of Africa. In: Hall AE, Cannell GH, Lawton HW (eds) Agriculture in semi-arid environments. Ecol Stud Vol 34. Springer, Berlin Heidelberg New York, pp 97–118
Davies WJ, Wilson JA, Sharp RE, Osonubi O (1980) Control of stomatal behaviour in water-stressed plants. In: Jarvis PG, Mansfield TA (eds) Stomatal physiology. Univ Press, Cambridge
DeMichele DW, Sharpe PJH (1973) An analysis of the mechanics of guard cell motion. J Theor Biol 41:77–96
Duncan WG, Loomis RS, Williams WA, Hanau R (1967) A model for simulating photosynthesis in plant communities. Hilgardia 38:181–205
Ehleringer JR, Mooney HA (1978) Leaf hairs: effects on physiological activity and adaptive value to a desert shrub. Oecologia 37:183–200
Ehlig CF, Gardner WR (1964) Relationship between transpiration and the internal water relations of plants. Agron J 56:127–130
Elfving DC, Kaufmann MR, Hall AE (1972) Interpreting leaf water potential measurements with a model of the soil-plant-atmosphere continuum. Physiol Plant 27:161–168
Faiz SMA (1973) Soil-root water relations. Ph D Thesis, Univ Aberdeen
Farquhar GD (1978) Feed forward responses of stomata to humidity. Aust J Plant Physiol 5:787–800
Farquhar GD (1979) Carbon assimilation in relation to transpiration and fluxes of ammonia.

In: Marcelle R, Clijsters H, Poucke van M (eds) Photosynthesis and plant development. Junk, The Hague, pp 321–328
Farquhar GD, Cowan IR (1974) Oscillations in stomatal conductance the influence of environmental gain. Plant Physiol 54:769–772
Farquhar GD, Raschke K (1978) On the resistance to transpiration of the sites of evaporation within the leaf. Plant Physiol 61:1000–1005
Farquhar GD, Dubbe DR, Raschke K (1978) Gain of the feedback loop involving carbon dioxide and stomata. Plant Physiol 62:406–412
Farquhar GD, Schulze E-D, Küppers M (1980) Responses to humidity by stomata of *Nicotiana glauca* L. and *Corylus avellana* L. are consistent with the optimization of carbon dioxide uptake with respect to water loss. Aust J Plant Physiol 7:315–327
Fischer RA, Turner NC (1978) Plant productivity in the arid and semiarid zones. Annu Rev Plant Physiol 29:277–317
Fischer RA, Hsiao TC, Hagan RM (1970) After-effects of water stress on stomatal opening potential. I. Techniques and magnitudes. J Exp Bot 21:371–385
Fiscus EL, Kramer PJ (1975) General model for osmotic and pressure-induced flow in plant roots. Proc Natl Acad Sci USA 72:3114–3118
Fitzpatrick EA, Nix HA (1969) A model for simulating soil water regime in alternating fallow-crop systems. Agric Meteorol 6:303–319
Fuller EN, Schettler PD, Giddings JC (1966) A new method for prediction of binary gas-phase diffusion coefficients. Ind Eng Chem 58(5):19–27
Gale J (1972) Elevation and transpiration: some theoretical considerations with special reference to mediterranean-type climate. J Appl Ecol 9:691–702
Gale J (1973) Experiment evidence for the effect of barometric pressure on photosynthesis and transpiration. In: Plant response to climate factors. Proc Uppsala Symp, 1970 UNESCO, pp 289–294
Goudriaan J, Laar Van HH (1978) Relations between leaf resistance, CO_2-concentration and CO_2-assimilation in maize, beans, Lalang grass and sunflower. Photosynthetica 12:241–249
Hall AE (1970) Photosynthetic capabilities of healthy and beet yellows virus-infected sugar beets (*Beta vulgaris* L.). Ph D Thesis, Univ California, Davis
Hall AE (1979) A model of leaf photosynthesis and respiration for predicting carbon dioxide assimilation in different environments. Oecologia 143:299–316
Hall AE, Dancette C (1978) Analysis of fallow-farming systems in semi-arid Africa using a model to simulate the hydrologic budget. Agron J 70:816–823
Hall AE, Hoffman GJ (1976) Leaf conductance response to humidity and water transport in plants. Agron J 68:876–881
Hall AE, Schulze E-D (1980a) Stomatal responses to environment and a possible interrelation between stomatal effects on transpiration and CO_2 assimilation. Plant Cell Environ 3:467–474
Hall AE, Schulze E-D (1980b) Drought effects on transpiration and leaf water status of cowpea in controlled environments. Aust J Plant Physiol 7:141–147
Hall AE, Schulze E-D, Lange OL (1976) Current perspectives of steadystate stomatal responses to environment. In: Lange OL, Kappen L, Schulze E-D (eds) Water and plant life. Ecol Stud Vol 19. Springer, Berlin Heidelberg New York, pp 169–188
Hall AE, Foster KW, Waines JG (1979) Crop adaptation to semi-arid environments. In: Hall AE, Cannell GH, Lawton HW (eds) Agriculture in semi-arid environments. Ecol Stud Vol 34. Springer, Berlin Heidelberg New York, pp 148–179
Hanks RJ, Allen LH, Gardner H (1971) Advection and evapotranspiration of wide-row sorghum in the Central Great Plains. Agron J 62:520–527
Herkelrath WN (1975) Water uptake by plant roots. Ph D Thesis, Univ Wisconsin, Madison
Hinckley TM, Schroeder MO, Roberts JE, Bruckerhoff DN (1975) Effects of several environmental variables and xylem pressure potential on leaf surface resistance in white oak. For Sci 22:201–211
Hsiao TC, Allaway WG (1973) Action spectra for guard cell Rb^+ uptake and stomatal opening in *Vicia faba*. Plant Physiol 51:82–88
Jarman PD (1974) The diffusion of carbon dioxide and water vapor through stomata. J Exp Bot 25:927–936

Jarvis PG (1971) The estimation of resistances to carbon dioxide transfer. In: Sestak Z, Catsky J, Jarvis PG (eds) Plant photosynthetic production manual of methods. Junk, The Hague, pp 566–631

Jarvis PG (1975) Water in plants. In: Vries de DA, Afgan NH (eds) Heat and mass transfer in the biosphere I. Transfer processes in the plant environment. Halsted, Washington, pp 369–394

Jarvis PG (1976) The interpretation of the variations in leaf water potential and stomatal conductance found in canopies in the field. Philos Trans R Soc London Ser B 273:593–610

Jarvis PG (1980) Stomatal response to water stress in conifers. In: Turner NC, Kramer PJ (eds) Adaptation of plants to water and high temperature stress. Wiley-Interscience, New York, pp 105–122

Jarvis PG, Slatyer RO (1970) The role of the mesophyll cell wall in leaf transpiration. Planta 90:303–322

Jeffree CE, Johnson RPC, Jarvis PG (1971) Epicuticular wax in the stomatal antechamber of Sitka spruce and its effects on the diffusion of water vapor and carbon dioxide. Planta 98:1–10

Johnsson M, Issaias S, Brogardh T, Johnsson A (1976) Rapid blue-light-induced transpiration response restricted to plants with grass-like stomata. Physiol Plant 36:229–232

Jones HG (1980) Interaction and integration of adaptive responses to water stress: the implications of an unpredictable environment. In: Turner NC, Kramer PJ (eds) Adaptation of plants to water and high temperature stress. Wiley-Interscience, New York, pp 353–365

Jones MM, Rawson HM (1979) Influence of rate of development of leaf water deficits upon photosynthesis, leaf conductance, water use efficiency, and osmotic potential in sorghum. Physiol Plant 45:103–111

Jury WA (1979) Water transport through soil, plant, and atmosphere. In: Hall AE, Cannell GH, Lawton HW (eds) Agriculture in semi-arid environments. Ecol Stud Vol 34. Springer, Berlin Heidelberg New York, pp 180–199

Kappen L, Oertli JJ, Lange OL, Schulze E-D, Evenari M, Buschbom U (1975) Seasonal and diurnal courses of water relations of the aridoactive plant *Hammada scoparia* in the Negev desert. Oecologia 21:175–192

Kaufmann MR (1976) Water transport through plants: current perspectives. In: Wardlaw IF, Passioura JB (eds) Transport and transfer processes in plants. Academic Press, London New York, pp 313–327

Landsberg JJ, Butler DR (1980) Stomatal responses to humidity: implications for transpiration. Plant Cell Environ 3:29–33

Lange OL, Lösch R, Schulze E-D, Kappen L (1971) Responses of stomata to changes in humidity. Planta 100:76–86

Lange OL, Schulze E-D, Evenari M, Kappen L, Buschbom U (1978) The temperature-related photosynthetic capacity of plants under desert conditions. III. Ecological significance of the seasonal changes of the photosynthetic response to temperature. Oecologia 34:89–100

Lemon ER, Stewart DW, Shawcroft RW, Jensen SE (1973) Experiments in predicting evapotranspiration by simulation with a soil-plant-atmosphere model (SPAM). In: Bruce RR et al. (eds) Field soil water regime. Soil Sci Soc, Am, Inc SSSA Spec Publ No 5, Madison, pp 57–76

Lösch R (1977) Responses of stomata to environmental factors-experiments with isolated epidermal strips of *Polypodium vulgare* 1. Temperature and humidity. Oecologia 29:85–97

Lösch R (1979) Stomatal responses to changes in air humidity. In: Sen DN, Chawan DD, Bansal RP (eds) Structure, function and ecology of stomata. Dehra Dun, pp 189–216

Loomis RS, Rabbinge R, Ng E (1979) Explanatory models in crop physiology. Annu Rev Plant Physiol 30:339–367

Louwerse W, Zweerder WVD (1977) Photosynthesis, transpiration and leaf morphology of *Phaseolus vulgaris* and *Zea mays* grown at different irradiances in artificial and sunlight. Photosynthetica 11:11–21

Meyer WS, Ritchie JT (1980) Resistance to water flow in the sorghum plant. Plant Physiol 65:33–39
Monteith JL (1965) Evaporation and environment. In State and movement of water in living organisms. Soc Exp Biol Symp XIX, Univ Press, Cambridge, pp 205–234
Mooney HA, Ehleringer JR (1978) The carbon gain benefits of solar tracking in a desert annual. Plant Cell Environ 1:307–311
Mooney HA, Gulmon SL, Rundel PW, Ehleringer J (1980) Further observations on the water relations of *Prosopis tamarugo* of the northern Atacama desert. Oecologia 44:177–180
Neilson RE, Jarvis PG (1975) Photosynthesis in sitka spruce (*Picea sitchensis* (Bong.) Carr.) VI. Response of stomata to temperature. J Appl Ecol 12:879–891
Nobel PS (1978) Microhabitat, water relations, and photosynthesis of a desert fern *Notholaena parryi*. Oecologia 31:293–309
Osonubi O, Davies WJ (1980a) The influence of plant water stress on stomatal control of gas exchange at different levels of atmospheric humidity. Oecologia 46:1–6
Osonubi O, Davies WJ (1980b) The influence of water stress on the photosynthetic performance and stomatal behaviour of tree seedlings subjected to variation in temperature and irradiance. Oecologia 45:3–10
O'Toole JC, Cruz RT (1980) Response of leaf water potential, stomatal resistance, and leaf rolling to water stress. Plant Physiol 65:428–432
Pallardy SG, Kozlowski TT (1979) Relationship of leaf diffusion resistance of *Populus* clones to leaf water potential and environment. Oecologia 40:371–380
Passioura JB (1973) Sense and nonsense in crop simulation. J Aust Inst Agric Sci 39:181–183
Passioura JB (1980) The meaning of matric potential. J Exp Bot 31:1161–1169
Penning de Vries FWT (1972) A model for simulating transpiration of leaves with special attention to stomatal functioning. J Appl Ecol 9:57–77
Ramos C (1981) Relationship between leaf conductance, CO_2 partial pressure and CO_2 uptake rate in leaves, and a test of a portable instrument for its measurement in the field. Ph D Thesis, Univ California, Riverside
Raschke K (1976) How stomata resolve the dilemma of opposing priorities. Philos Trans R Soc London Ser B 273:551–560
Raschke K (1979) Movements of stomata. In: Haupt W, Feinleib ME (eds) Physiology of movements. Encyclopedia of plant physiology new series Vol VII. Springer, Berlin Heidelberg New York, pp 383–441
Raschke K, Hanebuth WF, Farquhar GD (1978) Relationship between stomatal conductance and light intensity in leaves of *Zea mays* L., derived from experiments using the mesophyll as shade. Planta 139:73–78
Richter H (1973) Frictional potential losses and total water potential in plants: a reevaluation. J Exp Bot 24:983–994
Ritchie JT (1972) Model for predicting evaporation from a row crop with incomplete cover. Water Resour Res 8:1204–1213
Ritchie JT (1973) Influence of soil water status and meteorological conditions on evaporation from a corn canopy. Agron J 65:893–897
Rose CW, Byrne GF, Hansen GK (1976) Water transport from soil through plant to atmosphere: a lumped parameter model. Agric Meteorol 16:171–184
Rosenthal WD, Kanemasu E, Raney RJ, Stone LR (1977) Evaluation of an evapotranspiration model for corn. Agron J 69:461–464
Running SW (1976) Environmental control of leaf water conductance in conifers. Can J For Res 6:104–112
Running SW (1980a) Environmental and physiological control of water flux through *Pinus contorta*. Can J For Res 10:82–91
Running SW (1980b) Relating plant capacitance to the water relations of *Pinus contorta*. For Ecol Manage 2:237–252
Running SW, Waring RH, Rydell RA (1975) Physiological control of water flux in conifers. Oecologia 18:1–16
Schulze E-D, Küppers M (1979) Short-term effects of plant water deficits on stomatal response to humidity in *Corylus avellana* L. Planta 146:319–326

Schulze E-D, Lange OL, Kappen L, Buschbom U, Evenari M (1973) Stomatal responses to changes in temperature at increasing water stress. Planta 110:29–42

Schulze E-D, Lange OL, Evenari M, Kappen L, Buschbom U (1974) The role of air humidity and leaf temperature in controlling stomatal resistance of *Prunus armeniaca* L. under desert conditions. I. A simulation of the daily course of stomatal resistance. Oecologia 17:159–170

Schulze E-D, Hall AE, Lange OL, Evenari M, Kappen L, Buschbom U (1980) Long-term effects of drought on wild and cultivated plants in the Negev Desert. Oecologia 45:11–18

Shackel KA, Hall AE (1979) Reversible leaflet movements in relation to drought adaptation of cowpeas. *Vigna unguiculata* (L.) Walp. Aust J Plant Physiol 6:265–276

Sharp RE, Davies WJ (1979) Solute regulation and growth by roots and shoots of water-stressed maize plants. Planta 147:43–49

Sharpe PJH, De Michelle DW (1977) Reaction kinetics of poikilotherm development. J Theor Biol 64:649–670

Sharpe PJH, Wu H (1978) Stomatal mechanics: volume changes during opening. Plant Cell Environ 1:259–268

Sheriff DW (1979) Stomatal aperture and the sensing of the environment by guard cells. Plant Cell Environ 2:15–22

Shoemaker EM, Srivastava LM (1973) The mechanics of stomatal opening in corn (*Zea mays* L.) leaves. J Theor Biol 42:219–225

Slabbers PJ (1980) Practical prediction of actual evapotranspiration. Irrig Sci 1:185–196

Slatyer RO (1967) Plant-water relationships. Academic Press, London New York

Smith WK, Geller GN (1979) Plant transpiration at high elevations: theory, field measurements and comparisons with desert plants. Oecologia 41:109–122

Spanner DC (1973) The components of the water potential in plants and soils. J Exp Bot 24:816–819

Syvertsen JP, Cunningham GL, Feather TV (1975) Anomalous diurnal patterns of stem xylem water potentials in *Larrea tridentata*. Ecology 56:1423–1428

Tanner CB, Jury WA (1976) Estimating evaporation and transpiration from a row crop during incomplete cover. Agron J 68:239–243

Thorpe MR, Warrit B, Landsberg JJ (1980) Responses of apple leaf stomata: a model for single leaves and a whole tree. Plant Cell Environ 3:23–27

Turk KJ, Hall AE (1980a) Drought adaptation of cowpea. II. Influence of drought on plant water status and relations with seed yield. Agron J 72:421–427

Turk KJ, Hall AE (1980b) Drought adaptation of cowpea. III. Influence of drought on plant growth and relations with seed yield. Agron J 72:428–433

Turk KJ, Hall AE (1980c) Drought adaptation of cowpea. IV. Influence of drought on water use, and relations with growth and seed yield. Agron J 72:434–439

Turner NC (1974) Stomatal response to light and water under field conditions. In: Bieleski RL, Ferguson AR, Cresswell MM (eds) Mechanisms of regulation of plant growth. Roy Soc NZ Bull 12:423–432

Turner NC, Begg JE (1978) Responses of pasture plants to water deficits. In: Wilson JR (ed) Plant relations in pastures. CSIRO, Melbourne, pp 50–60

Waggoner PE (1969) Predicting the effect upon net photosynthesis of changes in leaf metabolism and physics. Crop Sci 9:315–321

Waring RH, Running SW (1978) Sapwood water storage: its contribution to transpiration and effect upon water conductance through the stems of old-growth Douglas fir. Plant Cell Environ 1:131–140

Weatherley PE (1970) Some aspects of water relations. Adv Bot Res 3:171–206

Wong SC (1979) Stomatal behavior in relation to photosynthesis. PhD, Thesis, Aust Natl Univ

Wong SC, Cowan IR, Farquhar GD (1978) Leaf conductance in relation to assimilation in *Eucalyptus pauciflora* Sieb. ex Spreng. Influence of irradiance and partial pressure of carbon dioxide. Plant Physiol 62:670–674

Wong SC, Cowan IR, Farquhar GD (1979) Stomatal conductance correlates with photosynthetic capacity. Nature (London) 282:424–426

9 Physiological Responses to Moderate Water Stress

K.J. Bradford and T.C. Hsiao

CONTENTS

9.1 Introduction

On a global basis, water is a paramount factor in determining the distribution of species, and the responses and adaptations of a species to water stress are critical for its success in any environmental niche. Numerous studies have reported a myriad of changes elicited by water stress. The changes observed are dependent on the species under study and on the severity, duration, and time course of the stress. Before reviewing the changes in detail, we will first present an overview of stress and responses using the stress–strain concept of physics. Next, we will discuss specific water-related parameters for quantifying plant water status and briefly consider how changes in the parameters may affect plant functions. This is followed by the main body which first reviews and analyzes selected responses to water stress and then examines the integrated adaptive behavior of whole plants.

The intent of this chapter is to discuss selected topics from the personal viewpoint of the authors. For more comprehensive coverage, other recent reviews should be consulted (PALEG and ASPINALL 1981; TURNER and KRAMER 1980).

9.2 Stress and Responses: an Overview

The numerous physiological responses of plants to water deficits generally vary with the severity as well as the duration of the stress. Only the most sensitive processes are altered by a very mild stress. As the stress increases, these changes intensify, and additional processes become affected in accordance to their relative sensitivities to the stress. If the stress is prolonged, there is more time for the initial effects to lead to secondary and tertiary responses. These changes with time, taken as a whole, often represent modulation of the system to meet the demand of the altered environment. An example is leaf senescence during drought, discussed in Section 9.8.3. Very few responses to water stress represent true lesions, especially when the stress is not extremely severe. One exception is the cavitation of water in the xylem under high tension (MILBURN 1979; see Chap. 5, this Vol.), which constitutes outright damage to the water transport system. In many other cases, the changes effected by water stress which appear to be damaging, probably represent operation of control mechanisms to adjust the performance of the organism in accordance with the environmental constraint. Selected examples of this behavior are discussed in the following sections.

A general conceptual framework of stress physiology may help to formalize some of the aforementioned points and aid in the detailed examinations of plant responses to water stress to follow. The framework is the physical model of stress–strain relationships (Fig. 9.1). It should be pointed out at the onset, however, that this simple analogy is useful only to a point. The term "stress" is in common use in the physiological literature but its meaning has been controversial at times (KRAMER 1980; LEVITT 1980). In physics and mechanics, stress refers to the force applied to a body and strain is used to describe

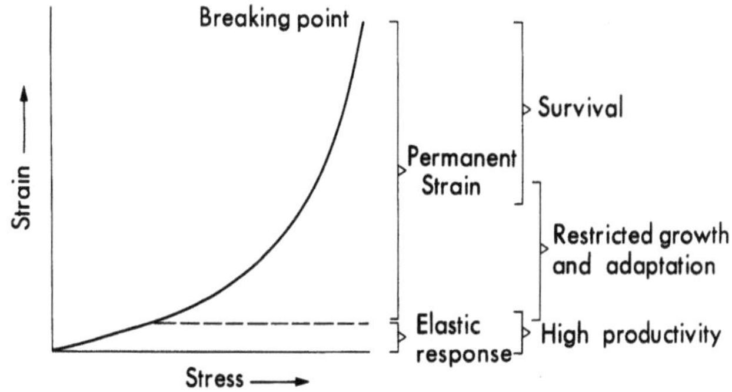

Fig. 9.1. Generalized stress–strain diagram of physical material, as a model for plant responses to water stress. Overall plant responses in terms of growth and productivity, corresponding to the regions of physical strain, are given on the *right margin* of the diagram

the deformation of the body resulting from the stress. In accordance with that usage, plant water deficiency would be the stress, and responses to the deficiency, the strain. In physics, stress and strain are usually introduced first under the topic of elasticity. In engineering, they are treated in depth under the topic of strength of materials. The names of these topics suggest strength and ability to resist on the one hand, and yielding and resilience on the other, behavior characteristics very pertinent to consideration of plant responses and adaptation to stresses.

The analogy can be extended to include magnitudes of stress and strain. Strain increases with each increment of stress, as seen in stress–strain diagrams (Fig. 9.1). A small amount of stress, when applied to a body, causes limited and reversible deformation. The original form is fully recovered when the stress is removed. As shown in the figure, this perfectly elastic response to stress holds only within the elastic limit. When deformation or strain is pushed above this limit by additional stress, the body is permanently deformed more and more as stress continues to rise. The final point on the diagram is reached when the body can strain no further and breaks under yet another increment of stress. In terms of environmental stresses and plant responses, the region of elastic strain represents stress levels which have little or no enduring effect on plant performance and productivity. An example could be the mild water stress encountered daily at midday by plants growing on well-watered soil in sunny weather. Diminution of the stress in the late afternoon and recovery overnight may be such that the daily overall growth is not inhibited. As stress becomes more than just very mild, however, the effects on plants are at least partly permanent. This is represented by the region of permanent strain above the elastic limit.

In the lower portion of the permanent strain region, growth and productivity are somewhat restricted, and adaptive responses and adjustment to the stress set in. In the upper portion of the permanent strain region, growth is virtually stopped, extensive damage due to the stress is obvious, and the plant is struggling to survive. The breaking point of strain is reached when stress is sufficiently severe to kill the plant. In our view, the interval of elastic strain on the ordinate is very narrow for plants subjected to water stress. Most of the strain under water stress would fall in the permanent strain region. Permanent strain should reflect largely the physiological or developmental plasticity a plant exhibits under water stress. The greater the interval of permanent strain on the ordinate a plant can undergo before the breaking point, the wider is the range of stress to which the plant is adaptable. Adaptation to more severe stress entails severe reductions in growth and productivity, but ensures survival.

The behavior of the physical model is time-dependent, at least in certain cases. When stressed by a force just below the elasticity limit for a short period, materials deform and regain their original shape when the stress is removed. On the other hand, if the same force is applied for a very long time, some materials (e.g., wood and rubber) begin to be permanently deformed or strained. This parallels the dependence of plant responses to stress on the duration of the stress, except that the time dependence is much more marked in the case of the plant.

The analogy between physics and physiology should not be carried much further. Plants are infinitely more complicated, with feedback and feedforward loops linking different parts of the plant and diverse processes within cells. The operation of homeostatic mechanisms in plants under stress results in a modified organism after sufficient time lapse; that organism exhibits a different set of response curves to stress than the original. On the other hand, it is precisely the homeostatic behavior that implements the adjustment and adaptation to stress and provides the means to cope with environmental adversity. Another disparity between the model and the actual plant is the shape of the stress–strain curves. Figure 9.1 is common for physical material, but the shape for plant responses to environmental stress can be markedly different, depending on the kind of stress and the parameter chosen to indicate stress intensity. In certain cases, the curve is likely to be convex upward instead of concave.

The physical concept of stress–strain relationships has long been used by LEVITT (1980) to describe environmental stresses and plant responses. His treatment differs substantially from ours and involves more detailed and often different analogies betwen the physical model and the plant system. We believe some of his analogies are not warranted and may lead to confusion. Nonetheless, we acknowledge his leading role in generalizing concepts in stress physiology.

Regardless of the limitations in drawing parallels between the physical and biological systems, stress–strain functions, schematized in Fig. 9.1, are central to stress physiology. Most stress studies implicitly or explicitly seek to define such functions, with the ordinate specified and often restricted to a single process or parameter (for example, Fig. 9.2 A). These parameters can be for any level of organization, ranging from molecules to plant communities. Examples are the rates of respiration for the cellular level and biomass accumulation over time for the whole-plant level. Of course, the behavior of any organism is the integrated sum of the function for each underlying process. As for the dependence of responses on the duration and history of stress, one may visualize the expansion of any given function into a time series. The shape of the curve changes with time, as dictated by the stress level during the time interval, as well as by the stress-response functions preceding the interval. Thus, the stress–strain relationship is represented by three-dimensional surfaces, with stress severity and time on two of the three axes as independent variables. All in all, the value of knowing the time course of stress severity and of responses cannot be overstated in stress studies.

9.3 Levels of Water Stress and Its Indicators

With stress severity being one of the critical factors in determining plant responses, what parameter should be used to measure water stress in studies of plant responses to water deficits? Tissue water potential Ψ has been used widely as a measure of water stress in the last decade. No doubt the concept of water potential is well based on thermodynamics, and Ψ gradient is all important for water transport (see Chaps. 1, 2, and 5, this Vol.). It is not certain, however, that Ψ, per se, is critical for physiological functions. Other parameters closely reflecting plant water status which may also serve as indicators of water stress are: chemical activity of the tissue water, the abolute or relative water content (RWC) of the tissue, turgor pressure in cells (Ψ_p), and the total concentration of all solutes in cells (as indicated by Ψ_π). The nature of these parameters and their significance in plant functions are now briefly discussed.

Chemical activity of water (a_w) is related fundamentally to Ψ and can be used interchangeably with Ψ[1]. The reason for not being able to clearly relate Ψ directly to physiological functions is that because of their logarithmic relationship, a_w is reduced only very slightly when Ψ is lowered 10 or 20 bar. For example, a reduction in Ψ of 20 bar corresponds to a reduction in a_w of only 1.5%. Such small changes are unlikely to cause substantial perturbations in reactions involving water as reactants or products. The lack of changes in certain metabolic events when Ψ was raised by increasing hydrostatic pressure (FLOWERS and HANSON 1969; DHINDSA and CLELAND 1975b) supports this conclusion.

Absolute water content is perhaps the most direct measure of tissue water status. As an indicator of water stress, however, it is confounded by tissue size and dry matter content. The use of relative water content (RWC) avoids this difficulty but requires the accurate determination of water content of the tissue at saturation or full turgor ($\Psi=0$). This, however, involves serious uncertainties because the saturation point is not clear-cut (CATSKY 1974). With plants of high water status, marked changes in physiological processes can be caused by a loss of 1% or 2% of the tissue water. Because of the uncertainties involved in its determination, no significant changes in RWC could be measured in such studies, in spite of obvious effects of the stress (HSIAO 1970; HUFFAKER et al. 1970). Experimentally, water content is related to Ψ in the classical Höfler

[1] $\Psi=(RT/\bar{v}_w)\ln a_w$, where R is the universal gas constant, T is absolute temperature, and $\bar{v}_w$ is the partial molal volume of water

diagram (see Chap. 2, this Vol.). Ψ is dependent on water content because both of its key components, Ψ_π and Ψ_p, are theoretically functions of water volume in the protoplasm. Water content can be related conceptually to plant responses to water stress. For example, sufficient reduction in water content causes herbaceous plants to lose their normal shape and form, the maintenance of which is dependent on cell inflation by water. With sufficient reduction in cellular water volume, metabolites become significantly more concentrated and spatial relationships among organelles and membranes are substantially altered.

Turgor pressure is another parameter closely tied to tissue water status, but only in the water status region above the zero turgor point (Chap. 2, this Vol.). When nearly saturated tissue (RWC approaches 1.0) loses small amounts of water, Ψ_p drops steeply, as is the case for Ψ. Turgor pressure is extremely difficult to measure directly. The common procedure of estimating Ψ_p by $\Psi - \Psi_\pi$ is not without pitfalls, especially where Ψ and Ψ_π are measured on separate subsamples and using different techniques, each with its own inherent errors. Turgor pressure is known to be critical for expansive growth (see below). Although by definition Ψ_p is a unique function of the pressure in the water with atmospheric pressure as datum, from the viewpoint of physiological functions the key is not the absolute pressure Ψ_p, but the pressure differential ($\Delta\Psi_p$) across the plasmalemma-cell wall boundary. This point was made clear earlier in a detailed analysis (HSIAO 1973a). Consideration of Ψ_p instead of $\Delta\Psi_p$ is adequate in most natural situations, however, because the pressure outside the cell wall and in the intercellular air space is usually atmospheric.

Osmotic potential (Ψ_π) is yet another possible indicator of water stress. Its relationship to water volume is described by Morse's equation, making it a rather insensitive indicator because changes in water volume having functional significance are usually only a small fraction of the total. More importantly, Ψ_π obviously is also a function of the amount of solutes present. The latter quantity can be markedly affected by osmotic adjustment to water stress or salinity, or by variations in other environmental parameters such as light (ACEVEDO et al. 1979).

The foregoing discussion makes it clear that no single parameter is totally sufficient as the indicator of water stress, except for the most restricted situations. In the vast majority of the cases, a combination of Ψ and RWC or Ψ and Ψ_p provides much more insight than the use of any one of the three parameters. A major part of the current literature, however, still reports only Ψ.

Regardless of the indicator, it is helpful to identify a general reference point of water stress. The most frequently mentioned reference point is the level of water stress that just induces obvious wilting in leaves low in sclerenchyma tissue, such as those of annual crops. This level also corresponds approximately to the point of nearly zero turgor in cells (TURNER 1974) and the point when stomata close pronouncedly in response to the water stress in the bulk leaf tissue. Needless to say, CO_2 assimilation is also substantially restricted at that point. For many field-grown herbaceous crop species without a significant history of water stress, the wilting point falls in a Ψ range of -13 to -16 bar. When subjected to the preconditioning effect of water stress, however, herbaceous crop species capable of osmotic adjustment would show wilting only at substantially lower Ψ values, in some cases perhaps as low as -28 to -30 bar. In many tree and wild species, wilting as a visual symptom of water stress may not be observed due to the mechanical strength of the leaves. The Ψ value representing the point of near-zero turgor and threshold for stomatal closure may also be much lower than -16 bar, reflecting more solutes in the tissue (FERERES et al. 1978). When RWC is used as an indicator, however, there appears to be more unity among species and in the same species with varying growth history – the near-zero turgor point for the vast majority of leaves appears to be in the RWC range of 0.75 to 0.85.

As will be amplified below, many physiological processes are altered by water stress levels milder than those causing wilting and stomatal closure. The level of water stress that triggers perturbation in plant functions depends on the processes in question, as well as on the duration of the stress and preconditioning. In many cases, the thresholds correspond to high values of Ψ and remarkably small reductions in tissue water status may cause distinct deficiency responses. Indeed, for some processes such as expansive growth the threshold may not exist.

9.4 Effects of Stress on Growth and Underlying Metabolism

The ultimate goal of plant ecophysiology is to understand how environment and physiological processes interact to determine the growth and reproduction of plants. Just as the availability of water is a dominant factor of the physical environment, expansive growth (irreversible cell enlargement) is an overriding physiological event. On the short term, cellular expansion and utilization of substrates are intimately linked through feedback channels to all aspects of metabolism. On the long term, the expansion of leaves determines the active photosynthetic surface area and sets limits on the future growth potential of the plant. Thus, expansive growth can be viewed as an integrator of the environmental and metabolic events which influence overall plant productivity. For this reason, the various short-term physiological responses to water stress will be considered within the context of expansive growth. This will involve an outline of the physics of growth, followed by current ideas on the metabolic mechanisms underlying cellular expansion.

9.4.1 A Physical View of Expansive Growth

Expansive growth has long been known to be highly sensitive to variations in Ψ (see HSIAO 1973a). Various lines of evidence suggest that the initial effect of reduced Ψ on expansion is physical and is mediated by changes in Ψ_p. The best of such evidence is that obtained with giant internodal cells of *Nitella* whose Ψ_p could be monitored directly (GREEN 1968; GREEN et al. 1971). Sudden stepwise increases or decreases in Ψ_p caused corresponding increases or decreases in growth, at least for a short duration. The evidence for higher plants is less direct. Rewatering of maize plants under mild water stress permitted the resumption of growth within seconds (ACEVEDO et al. 1971), suggesting that the increase in Ψ immediately promotes cell expansion with little or no intermediate metabolic reactions. Stepwise increases in Ψ (by increasing hydrostatic pressure) around the roots of rice resulted in proportional acceleration of the leaf elongation rate (CUTLER et al. 1980b). Applications of external pressure of around 2 bar to sunflower leaves, though elevating their Ψ, actually stopped their expansion (BOYER 1968). When maize leaf elongation was markedly slowed for about 1 h by mild stress, release of the stress effected a transitory phase of growth more rapid than the unstressed steady-state rate so there was no net reduction in elongation for the total period (ACEVEDO et al. 1971). These data appear to fit best the view that the initial effect of water stress on growth is physical and mediated by changes in Ψ_p. Implicit in this statement, however, is the view that when stress continues for some time, metabolic factors come into play, and Ψ_p then is only one of the parameters determining growth rates.

From a physical point of view, expansive growth can be analyzed in terms of cell volume, water uptake and Ψ gradients, turgor pressure, Ψ_π, and cell extensibility. The ways in which these factors interact to determine the rate of growth are discussed, along with the effects of water stress. This discussion is limited, as a similar treatment is presented in Chapter 2, this Volume, though with somewhat different emphasis and viewpoints.

One view of instantaneous growth is to consider it as the irreversible stretching of the cell wall induced by pressure. Described as a simple equation (LOCKHART 1965; GREEN et al. 1971),

$$dV/Vdt = E_g\,(\Psi_p - \Psi_{p,th}) \qquad (9.1)$$

where V is the cell volume and dV/Vdt is the rate of volume increase (cell expansion) relative to the total volume. E_g is a coefficient of gross extensibility and is the slope of the curve relating growth rate to Ψ_p, and refers only to the "plastic" or irreversible extensibility of the cell. $\Psi_{p,th}$ indicates that a minimum Ψ_p must be exceeded before growth occurs and is the intercept on the Ψ_p axis of the curve. It should be pointed out at the onset that though parts of a physical description, E_g and $\Psi_{p,th}$ should be highly dependent on metabolic events especially in the cell wall, as will be discussed below.

The existence of a threshold turgor below which no extension occurs has been observed in a variety of plant tissues (BOYER 1968; GREEN et al. 1971, 1977; HSIAO et al. 1976a; MEYER and BOYER 1972), with values for $\Psi_{p,th}$ ranging from 1 to 6.5 bar. The instantaneous growth rate is therefore sensitive only to that increment of Ψ_p in excess of $\Psi_{p,th}$, which may be only a very few bars if $\Psi_{p,th}$ is high. The relationship between cell volume and Ψ_p dictates that when Ψ of quite turgid cells is reduced, Ψ_p falls much more rapidly than Ψ_π. Thus, a small reduction in Ψ can prevent growth (BOYER 1968; ACEVEDO et al. 1971), presumably because it erases the difference between Ψ_p and $\Psi_{p,th}$.

For sustained growth, there must be continuous water uptake and volume enlargement of the cell. To maintain a Ψ gradient for the uptake as cellular volume increases, more solutes must be accumulated or generated within the cell. Otherwise, the Ψ gradient will be dissipated as the internal solutes are diluted. So steady-state growth implies steady-state maintenance of osmolarity (GRENETZ and LIST 1973). The role of water uptake and volume increments can be accounted for by combining Eq. (9.1) with $\Psi = \Psi_p + \Psi_\pi$ and the fact that water uptake rate is jointly determined by the Ψ gradient and the conductance of the tissue to water. The resulting equations are (HSIAO et al. 1976a; LOCKHART 1965):

$$\frac{dV}{Vdt} = \frac{E_g C}{E_g + C} (\Delta \Psi + \Psi_p - \Psi_{p,th}) \tag{9.2a}$$

$$= \frac{E_g C}{E_g + C} (\Psi^{out} - \Psi_\pi - \Psi_{p,th}) \tag{9.2b}$$

where C is hydraulic conductance which incorporates pathway geometry (usually unknown) of the tissue, $\Delta \Psi$ is the Ψ gradient from the water source to the inside of the cell, and Ψ_π is the osmotic potential in the growing cell. The special interaction between C and E_g in determining growth rate and the need for Ψ gradients in the growing zone, shown explicitly by the equations, are further discussed in Chapter 2, this Volume. The equations are useful for examining the effect of Ψ of the medium on steady-state or near steady-state expansive growth. Inhibition of leaf elongation in rice due to osmotic stress in the root medium can be rapidly reversed by applying hydrostatic pressure to the solution to raise its Ψ (CUTLER et al. 1980b). The resumption of growth is apparently a direct response to an increase in leaf Ψ, as NULSEN et al. (1975) have shown that altering Ψ in the rooting medium via changes in pressure rapidly leads to variations in leaf Ψ of similar magnitude. Thus, when certain conditions are met (e.g., E_g, C, and $\Psi_{p,th}$ remain constant), effects on growth of changes in Ψ gradient for water uptake may be reasonably predicted by Eq. (9.2).

Equations (9.1) and (9.2), because of their apparent simplicity, are useful for conceptual analysis of growth responses to water stress. Actually, their application is simple only when it is implicitly assumed that E_g, C, $\Psi_{p,th}$, and Ψ_π are constants. While those assumptions appear valid momentarily after tissue water status is suddenly changed, there are ample data indicating that these parameters may shift with time thereafter, making the use of the equations less than straightforward. On the other hand, it is precisely these shifts which provide the means for the plant to adjust and adapt to changing water status. Either a reduction in $\Psi_{p,th}$ or an increase in E_g would allow continuation of expansive growth at lower values of Ψ_p. There is evidence that both of these adjustments can occur in response to water stress (GREEN et al. 1977; HSIAO et al. 1976a; CUTLER et al. 1980a). GREEN and coworkers (1977) view the cell wall as a self-stabilizing system which tends to restore the steady-state growth rate following shifts in Ψ and Ψ_p. The gradual resumption of leaf elongation after the sudden imposition of water stress on the roots of maize and the deceleration of growth rate from the hyper-normal to normal following sudden stress release (ACEVEDO et al. 1971), are in accordance with this view. Also in apparent agreement is the finding of BUNCE (1977) which showed that the average growth rates of soybean leaves were similar in the growth chamber, greenhouse, and field despite the different values of Ψ and Ψ_p under these conditions. Nonetheless, the validity of that view is probably limited to certain situations. When water stress lasts more than a few hours and is the only treatment differential, there is evidence that E_g and $\Psi_{p,th}$ may be changed in such a way to restrict growth at any given Ψ_p. That is particularly true when the growing tissue undergoes substantial osmotic adjustment in response to the water stress (CUTLER et al. 1980a; MEYER and BOYER 1972; MICHELENA and BOYER 1980; SHARP et al. 1979). Systematic studies of effects of the duration of stress on expansive growth as a function of Ψ_p are badly needed. The apparently contradicting patterns of changes in E_g and $\Psi_{p,th}$ to either maintain or restrict growth under water stress are further examined in the section on long-term and integrated behavior (Sect. 9.8).

The maintenance of Ψ_π in the face of dilution by cell enlargement is a critical factor in sustaining growth, as is explicit in Eq. (9.2b). When Ψ^{out} is declining under conditions of water stress, even more solutes must be accumulated in the cells to maintain $\Delta\Psi$. Osmotic maintenance of the growing zone and some aspects of osmotic adjustment in response to water stress are discussed in the following section on the mechanisms underlying expansive growth. More comprehensive treatments of osmotic adjustment and regulation are found elsewhere in this volume (Chapters 2 and 3, Volume 12 C).

9.4.2 Mechanism of Growth

9.4.2.1 Wall Extensibility

As defined here [Eq. (9.1)], cell wall extensibility (E_g) represents the slope of a plot of dV/Vdt versus Ψ_p. It is therefore determined by measuring growth rate and Ψ_p of a tissue over time as Ψ is varied. E_g is a steady-state parameter which incorporates the various aspects of physiology and metabolism which regulate the rate of growth for a given Ψ_p. Plastic, or irreversible, wall extension can also be measured in killed tissues following treatments which influence the growth rate (CLELAND 1967b). Extensibility determined in this way represents a static measure of the ability of the wall to extend plastically at the time the tissue was killed. Thus, the two measures of extensibility are not directly comparable, although they both are indications of the ability of the wall to extend when Ψ_p exceeds $\Psi_{p,th}$ (CLELAND 1967a). The term E_g will be reserved for cases where growth rate and Ψ_p have been measured. When plastic extensibility has been estimated from growth rates only or from in vitro techniques, the term "extensibility" will be used.

Agents which can accelerate the rate of growth, such as indoleacetic acid (IAA) or gibberellins (GA), also increase the cell wall extensibility (ADAMS et al. 1975; CLELAND 1967a). The "acid-growth hypothesis" has been proposed to account for hormonal effects on wall extensibility (for review, see CLELAND and RAYLE 1978). In essence, this hypothesis postulates that the growth-initiating event is the stimulation of H^+ extrusion from the plasma membrane, causing acidification of the cell wall, which in turn activates the enzymes reponsible for wall loosening. There is considerable experimental evidence for this hypothesis, primarily in excised *Avena* coleoptiles and *Pisum* stem segments, but also in intact *Phaseolus* leaves (VAN VOLKENBURGH and CLELAND 1980). Support for the acid-growth hypothesis is far from unanimous, however (see review by PENNY and PENNY 1978). For example, STUART and JONES (1978b) found no evidence for the involvement of proton secretion or acidification in the promotion of elongation of lettuce hypocotyls by GA. In the water plant *Ranunculus sceleratus*, in which stem elongation is promoted by both IAA and ethylene, IAA stimulated proton extrusion while ethylene inhibited it (CRAKER et al. 1978). Both growth and proton extrusion are inhibited by ABA (RAYLE 1973), but the effect on growth persists even in low pH buffers (REHM and CLINE 1973), suggesting additional modes of action for the hormone. In view of these multiple mechanisms for growth regulation, hope of identifying a single key event which is sensitive to water deficit may be unrealistic.

However, in oat coleoptiles and pea stems, where the acid growth hypothesis does seem to apply, a 0.2 M osmotic stress prevented elongation but increased the rate of proton extrusion (CLELAND 1975; MARRÉ et al. 1973). A reduction in Ψ of only about -1.5 bar was sufficient to promote proton extrusion from wheat coleoptiles (GÖRING et al. 1978). Since the stress in these experiments was imposed osmotically, it is important to note that the effects on acidification occurred without plasmolysis, which is known to inhibit proton transport (MASUDA et al. 1978; RUBINSTEIN 1977). Cell wall acidification in response to decreased Ψ would be expected to increase extensibility, and therefore act to restore the growth rate. On the short term, such compensation can occur (ACEVEDO et al. 1971), suggesting that proton extrusion may represent one biochemical component of the self-stabilizing system governing wall extensibility envisioned by GREEN and coworkers (1977).

On a time scale of days or weeks, E_g may show variable responses to water stress, as was mentioned above. A problem with most estimates of E_g is that Ψ_p and elongation rate are usually not measured on the same leaf. It has recently become evident that Ψ_p may be quite different in mature versus expanding leaves during water stress (MICHELENA and BOYER 1980; MUNNS et al. 1979). The elongation zone may not be lacking in turgor pressure, but growth is still restricted, indicating an increase in $\Psi_{p,th}$ or a decrease in E_g with prolonged stress. The recently developed pressure probe technique for direct measurement of Ψ_p in higher plant cells may be of use in further investigations of this phenomenon (HÜSKEN et al. 1978), although there are still many problems associated with its application in expanding cells.

Cell wall relaxation is associated with metabolism of hemicellulose constituents which serve to cross-link cellulose microfibrils. This metabolism is thought to reflect the breakage and reformation of bonds which allow the microfibrils to reorient during expansion (LABAVITCH 1981). In pea stems, a xyloglucan is released from cell walls coincident with the initiation of elongation following IAA application (LABAVITCH and RAY 1974). Solubilization of the xyloglucan still occurred when elongation was prevented by hypertonic solutions. IAA-induced release of hemicellulose components from cell walls of *Avena* coleoptiles changes little or may even increase as Ψ_p falls until $\Psi_p \leqq \Psi_{p,th}$, then it rapidly ceases (LOESCHER and NEVINS 1973; SAKURAI and MASUDA 1977). A change in wall metabolism apparently takes place when Ψ_p becomes less than $\Psi_{p,th}$ which results in a decrease in extensibility (MASUDA 1978). It is interesting to speculate that for small turgor reductions ($\Psi_p \geq \Psi_{p,th}$), promotion of proton extrusion and wall metabolism may increase extensibility and maintain the growth rate. After more severe stress, wall metabolism may cease and extensibility decrease, thus accounting for the reduction in E_g observed in plants subjected to repeated or severe drought stress. It would be useful to determine the relationship of wall metabolism to Ψ_p and growth in tissues which have experienced varying degrees of water stress.

9.4.2.2 Osmotic Maintenance

When the cell wall relaxes, an immediate consequence is that Ψ_p falls slightly, and Ψ^{in} becomes less than Ψ^{out}. Water then enters the cell down the Ψ gradient. So long as the wall continues to relax and Ψ_π in the cell is maintained [see Eq. (9.2b)], water enters the cell and growth continues. The maintenance of Ψ_π as the cell volume increases requires a net accumulation of solutes. It appears that malate and its counterions are involved, as they are in the processes of pH regulation, stomatal movement, and ion uptake (HSIAO 1976; OSMOND 1976; SMITH and RAVEN 1979). Dissociation of malic acid provides H^+ for exchange with extra-cellular K^+, and the K^+-malate is sequestered in the vacuole. DHINDSA and coworkers (1975) found that K^+-malate in growing cotton fibers represented over 50% of the total solutes accumulated. Furthermore, the fluctuation in estimated Ψ_p could be almost completely accounted for by increases and decreases in the K^+-malate content. When growth of *Avena* coleoptiles was stimulated by IAA, CO_2 fixation into malate and K^+ uptake were also enhanced, though the total osmotic contribution of the accumulated K^+-malate was not estimated (HASCHKE and LÜTTGE 1975a, b, 1977; JOHNSON and RAYLE 1976; STOUT et al. 1978). It is not known whether malate synthesis is triggered by the dilution of cellular solutes, or by ion transport and charge balance requirements resulting from the stimulation of H^+ extrusion by IAA.

It is well established that giant algal cells can sense decreases in turgor pressure and respond by actively accumulating ions or synthesizing organic solutes (BISSON and GUTKNECHT 1980; KAUSS 1977; ZIMMERMANN 1978). In higher plant cells, however, solute uptake seems to be only indirectly related to Ψ_p. During GA-induced growth of lettuce hypocotyls, Ψ_π (and therefore Ψ_p) declined roughly in proportion to the increase in tissue volume if additional salts were not supplied (STUART and JONES 1978a). The provision of KCl in the bathing medium counteracted this decline in Ψ_π and also promoted elongation. However, GA did not stimulate uptake of KCl in the absence of elongation, and after elongation, Ψ_π of GA-treated tissues was consistently less negative than that of the respective controls. A contrasting case was reported for cucumber hypocotyls, where GA caused a reduction in Ψ_π regardless of whether elongation was stimulated, the effect being greater when exogenous substrates were supplied (KATSUMI and KAZAMA 1978). This seems to be related to the specific mechanism of GA action in this tissue, rather than to turgor regulation per se. In *Avena* coleoptiles, neither turgor reduction nor IAA promoted solute uptake in the absence of growth (STEVENSON and CLELAND 1981). REED and BONNER (1974) reported that although ABA would inhibit K^+ uptake into *Avena* coleoptiles, there was no simple quantitative relationship between this effect and the degree of inhibition of growth, either by ABA or by osmotic stress. The conclusion from these studies is that solute accumulation in growing tissues is more closely related to the concentration of solutes available for uptake than it is to Ψ_p. In the absence of expansion, solutes will accumulate to some equilibrium level inside the cell. When dilution occurs due to growth, more solutes will be taken up, but this seems to depend more on the transport properties of the membrane (possibly hormone-modulated) than it does on Ψ_p. If solutes are not available or cannot be generated, growth will continue only until Ψ_p has been reduced below $\Psi_{p,th}$.

If this is indeed the case, then the crucial factor in osmotic maintenance during growth is the rate of supply of solutes to the elongating region. MCNEIL (1976) showed that the rate of accumulation of solutes, and particularly K^+, in expanding sunflower hypocotyls was greatest in the region where growth was most rapid. During water stress, import of solutes to the expansion zone can continue and turgor be maintained, even though growth is inhibited (BARLOW et al. 1980; MICHELENA and BOYER 1980). The relative insensitivity of phloem transport to low Ψ, and the resulting accumulation of solutes, may be important for the survival of the meristematic tissues during water stress.

There is some evidence that in nongrowing tissues, ion accumulation is sensitive to Ψ_p. MCNEIL (1976) reported a doubling of the rate of ^{86}Rb uptake due to osmotic stress (130 mM mannitol) in nongrowing sunflower hypocotyls, but the total change in Ψ_π was not measured. Carrot root slices will accumulate K^+ and Cl^- independently of turgor changes, while in beet root slices, Cl^- uptake is stimulated by turgor reduction and may be involved in turgor regulation. CRAM (1980) suggests that this represents a fundamental difference between glycophilic (carrot) and halophilic (beet) tissues, the latter accumulating ions to maintain constant turgor, the former to maintain constant internal ion concentration. In both tissues, however, ion accumulation is sensitive to other feedback mechanisms, such as the intracellular Cl^- concentrations, and is more likely related to nutrient uptake than to a turgor maintenance mechanism.

9.4.2.3 Hydraulic Conductance

The hydraulic conductance, C, indicates the ability of the tissue to supply water to the growing cells. The permeability of cell membranes to water has generally been assumed to be quite high, so that for single cells or slowly growing tissues, water transport would not limit growth (RAY et al. 1972). Preliminary results with the pressure-probe technique indicate that this assumption is probably valid (STEUDLE and COSGROVE 1981). In rapidly growing shoot tissues, however, where some cells are distant from the water source (i.e., the xylem vessels), considerable Ψ gradients may occur even when transpiration is negligible (BOYER 1968). Recent analyses estimate that although individual cells in a growing tissue will be close to Ψ equilibrium with their surroundings, overall Ψ gradients within the tissue may range from -1 to -2 bar (MOLZ and BOYER 1978; SILK and WAGNER 1980). The major uncertainty in these estimates is the magnitude of C, which may change during growth. For auxin-induced growth of soybean hypocotyls, BOYER and WU (1978) calculated that increases in E_g and C contributed about equally to the stimulation of growth rate. The change in C presumably resulted from an increase in membrane permeability to water, which is known to be sensitive to a variety of chemical and environmental stimuli (KUIPER 1972; LEE-STADELMANN and STADELMANN 1976).

The importance of C in regulating growth rates during water stress is a complex question (see Chap. 2, this Vol.). The magnitude of C determines the Ψ gradient within the tissue for a given rate of water flow, or growth. The importance of the auxin-induced increase in C may be to offset the larger $\Delta\Psi$ that would be required to sustain a higher flow rate. A similar increase in C during water stress would allow maintenance of the growth rate with a smaller $\Delta\Psi$ between the xylem and the expanding cells. Alternatively, if prolonged water deficit limits growth through effects on $\Psi_{p,th}$ or E_g, the magnitude of C will be less important as flow rates will be slow and $\Delta\Psi$ will be small. Much more experimental information is required before we can realistically evaluate the importance of C in determining the overall growth rate during water stress.

9.4.2.4 Cell Wall Synthesis

In addition to the factors discussed above, synthesis of new cell wall material is an essential component of steady-state growth. There is generally a close parallel between the incorporation of glucose into cell wall material and the rate of elongation (BAKER and RAY 1965; RAY 1962). In certain cases, however, wall synthesis can be divorced from its usual relation to growth. For example, elongation can occur in excised tissues without corresponding wall synthesis if additional substrates are not supplied. This is accompanied by an apparent thinning of the cell walls (RAY 1962). Certain inhibitors of elongation, such as 20 mM Ca^{2+}, do not prevent wall synthesis, leading to an increase in wall material and presumably thicker walls (BAKER and RAY 1965). It is possible that when elongation is prevented by water stress, wall synthesis might continue, producing small cells with thicker walls, as was reported for water-stressed cotton leaves (CUTLER et al. 1977). However, experiments using osmotic stress to reduce Ψ have found that wall synthesis is inhibited in parallel with growth, and wall thickness remains constant (BAYLEY and SETTERFIELD 1957; BAKER and RAY 1965), suggesting the existence of tight feedback links between wall expansion and wall synthesis.

9.4.2.5 Protein Synthesis

As with cell wall synthesis, protein synthesis is an obvious requirement for sustained cell enlargement. The myriad of functions of proteins in cellular physiology as enzymes, structural and membrane components, etc., makes it unlikely that expansive growth could continue for very long without simultaneous protein synthesis. The rapidity with which inhibitors of protein synthesis will arrest growth has led to the concept of growth-limiting proteins (CLELAND 1971; PENNY 1971). These are postulated to be rapidly turned over so that when new synthesis is blocked, these essential proteins become depleted and growth ceases. More recent results have tended to confirm this hypothesis (BATES and CLELAND 1979; LADO et al. 1977).

On the other hand, mild water stress is known to inhibit protein synthesis as well as growth (reviewed by HSIAO 1973a). In rapidly expanding *Zea mays* leaves, a brief desiccation treatment that reduced Ψ only 2 or 3 bar caused a dramatic shift of polyribosomes to monoribosomes, which was interpreted to reflect a decrease in overall protein synthesis (HSIAO 1970). Similar results have been reported for several other species (MORILLA et al. 1973; BARLOW et al. 1977). DHINDSA and CLELAND (1975b) found that the application of pressure to osmotically stressed *Avena* coleoptiles did not reverse the inhibitory effect on protein synthesis, indicating that the response was not due to a reduction in tissue Ψ.

HSIAO (1970) originally suggested, on the basis of the sensitivity of growth and polyribosome changes to reductions in Ψ, that inhibition of protein synthesis may represent a feedback effect of slowed growth rates. Recently, SCOTT and coworkers (1979) compared polyribosome levels of mature and elongating wheat leaves as Ψ varied from -8 to -30 bar. Water stress caused a decline in the polyribosome percentage in the elongating leaves but had no significant effect in mature leaves. In the young tissue, the shift from polyribosomes to monoribosomes occurred concomitantly with the cessation of elongation. These data, confirmed by unpublished work of HSIAO on maize, indicate that there is a close relationship between expansion and protein synthesis in growing cells. These regulatory links must be very tight, as the reciprocal effects of water stress on growth and protein synthesis, and of protein synthesis inhibitors on growth, can both be expressed within a matter of minutes. How slowed growth rates affect the activity of ribosomes remains an intriguing question.

The inhibition of protein synthesis during water stress is not indiscriminate. Synthesis of some proteins is promoted by water stress while that of others is reduced (DHINDSA and CLELAND 1975a). BRADY and coworkers (1974) were able to show by a dual-labeling technique that water stress (91% RWC for 2 h) reduced the rate of synthesis of Fraction I protein in young wheat leaves. This included effects on both subunits of ribulose bisphosphate carboxylase, and there was a preferential loss of chloroplast polyribosomes. Interestingly, a similar decrease in the synthesis of Fraction I protein relative to other soluble proteins was induced by ABA (10^{-6} M). In contrast, DHINDSA and CLELAND (1975b) reported that ABA indiscriminately inhibited protein synthesis in *Avena* coleoptiles, which presumably synthesize very little RuBPCase. These provocative results have yet to be followed up in detail using modern molecular biology techniques.

As would be expected from the effects of water stress on protein synthesis, the activities of a large number of enzymes have been found to vary in response to stress (see TODD 1972; HSIAO 1973a). These are too numerous to catalog here, but in general, enzyme activities are lowered by water stress except those involved in hydrolysis or catabolism, which often increase. Two well-studied enzymes exhibiting this pattern are nitrate reductase (NR) and ribonuclease (RNase), the former decreasing rapidly during water stress, the latter increasing in activity. These enzymes will be used as examples to illustrate the control mechanisms operating to regulate enzyme levels during water stress.

Slight reductions in water content (HUFFAKER et al. 1970) or Ψ (MORILLA et al. 1973) resulted in a rapid decline in NR activity in barley and maize seedlings. MORILLA and coworkers (1973) concluded that inhibition of enzyme synthesis, coupled with rapid turnover of the enzyme, was responsible for the loss of activity. Correlations between polyribosome levels and NR activity during stress and recovery, the prevention of restoration of NR activity by cycloheximide after rewatering, as well as the insensitivity of the rate of decline in NR to water stress, support this hypothesis. Others (PLAUT 1974; HEUER et al. 1979) have suggested that at least part of the inhibition of NR activity by water stress results from inactivation of existing enzyme, but more specific information on the rates of turnover is needed to establish this point.

Since NR is inducible by nitrate (HEWITT 1975), it is possible that water stress effects are an indirect result of reduced nitrate uptake. SHANER and BOYER (1976) showed that the level of NR in the tissue was related to the flux of nitrate from the roots, rather than to the tissue nitrate content or to Ψ. Reduced nitrate uptake by stressed roots in conjunction with stomatal closure and lessened transpiration resulted in a severe restriction of nitrate flux to the shoot. Apparently, nitrate in the leaf is compartmentalized so that only that in the transport path is effective in inducing NR. This example illustrates the integrative nature of plant response to water stress, with root ion uptake, stomatal control of transpiration, and leaf protein synthesis interacting to alter the activity of a key assimilative enzyme.

In contrast to NR, RNase activity increases as a result of water stress, but only when stress is quite severe (MARANVILLE and PAULSEN 1972). In isolated tobacco protoplasts, synthesis of the enzyme is induced by increases in the osmotic strength of the medium (PREMECZ et al. 1977), indicating that changes in protoplast volume or concentration of internal solutes is the triggering mechanism. RNase is also sensitive to modification by plant growth regulators, with ABA increasing and kinetin decreasing the activity in turgid plants (ARAD et al. 1973; SODEK and WRIGHT 1969). During water stress, however, these hormonal relationships can be reversed due to contrasting effects on leaf water balance. When barley leaves were pretreated with either ABA or kinetin and then stressed by desiccation or salinity, RNase activity was determined by the water status, rather than the hormonal treatment (ARAD et al. 1973; ARAD and RICHMOND 1976). ABA pretreatment tended to reduce the severity of the fall in Ψ, and also resulted in lowered RNase activity. Kinetin, on the other hand, enhanced the loss of tissue water and stimulated RNase activity. The hormonal effects were therefore reversed from those found in the absence of water stress. It appears that the specific hormonal controls operating in turgid tissue can be overridden when water deficit becomes excessive.

9.5 Synthesis and Transport of Growth Regulators in Response to Stress

As virtually all aspects of plant growth and development are influenced by both water stress and growth regulators, it is not surprising that the major plant hormones have been investigated with respect to water stress. The literature is too voluminous to be covered comprehensively here. Emphasis will be placed on the changes in endogenous levels of growth regulators in response to water stress and the possible consequences of these changes with respect to plant water balance, growth, and development. Other viewpoints can be found in previous reviews (ASPINALL 1980; DAVIES et al. 1980; ITAI and BENZIONI 1976; MANSFIELD et al. 1978; VAADIA 1976; WALTON 1980; WRIGHT 1978).

9.5.1 Indoleacetic Acid

Information on the effect of water stress on endogenous indoleacetic acid (IAA) levels is scarce and often conflicting. Diffusable auxin content of both *Helianthus annuus* and *Anastatica hierochuntica* decreased with decreasing soil water content (HARTUNG and WITT 1968). IAA in *Vicia faba* leaves, on the other hand, increased considerably during a drying cycle, but the response was relatively slow compared to the changes in ethylene and ABA levels (HALL et al. 1977). These studies suffer from poor characterization of Ψ and its components, and the degree and duration of stress varied widely. DARBYSHIRE (1971) measured a linear increase in IAA oxidase activity with decreasing Ψ between -2 and -15 bar in tomato, and suggested that leaf IAA content would be reduced as a result. This expectation was not supported in work with a wilty mutant of tomato, *flacca*, which contains higher than normal amounts of IAA-like activity, apparently due to a stimulation of IAA synthesis rather than to reduced degradation (TAL et al. 1979). However, it is not clear in this mutant which effects are caused by water stress and which result from altered hormonal relationships, particularly with respect to ABA. At present, no conclusions can be drawn concerning a specific relationship between Ψ and endogenous IAA levels.

There is evidence that basipetal auxin transport is inhibited by water stress ($\Psi < -10$ bar) (KALDEWEY et al. 1974; DAVENPORT et al. 1977, 1980). Ethylene synthesis may be involved here, as the gas is known to reduce auxin transport (BEYER and MORGAN 1971). The combination of ethylene production and reduced auxin transport is thought to be involved in the promotion of leaf abscission by water stress (see Sect. 9.8.2).

9.5.2 Gibberellins

Little is known of the effect of water stress on metabolism of gibberellins (GA). Endogenous GA activity declined rapidly in stressed lettuce leaves, and rose again during a subsequent recovery period (AHARONI et al. 1977). The fall in GA activity preceded the rise in ABA content of the same tissue. These results are complicated by the use of detached leaves, which are simultaneously senescencing. Nonetheless, water stress (10% reduction in RWC) accelerated the rate of decline in GA (Aharoni and Richmond 1978). Gibberellins are known to be synthesized in roots as well as leaves (Torrey 1976), but effects of water stress on the sites of synthesis and transport of GA have not been studied. The paucity of information on endogenous GA during water stress prevents speculation as to their importance in physiological responses.

9.5.3 Ethylene

Increased ethylene production is a general response of plant tissues to environmental stresses, including water stress (Yang and Pratt 1978). Since ethylene is a gas, it has been suggested that stomatal closure during water stress might limit diffusion from the plant, leading to increased internal concentrations (El-Beltagy and Hall 1974). Further work has shown, however, that ethylene readily diffuses from leaves and stems, and that similar results are obtained whether ethylene emanation or internal accumulation are measured (Wright 1980; Zeroni et al. 1977). This implies that water stress increases the rate of ethylene synthesis. The biosynthetic pathway of ethylene has recently been deduced, including the identification of a new intermediate, aminocyclopropanecarboxylate (ACC) (Adams and Yang 1979; Lürssen et al. 1979). ACC is derived from S-adenosylmethionine (SAM), and is in turn converted into ethylene. The rate-limiting step in the sequence is the production of ACC, and most factors which accelerate ethylene synthesis act to stimulate the conversion of SAM to ACC (see Yang 1980 for review). As expected, when wheat leaves were allowed to lose 10% fresh weight, synthesis of both ACC and ethylene increased markedly (Apelbaum and Yang 1981). The mechanism by which water stress stimulates the activity of the ACC-forming enzyme remains unknown.

Ethylene has generally been considered to act near its site of synthesis, since long-distance transport of the gas is unlikely. In response to waterlogging or root anaerobiosis, however, ACC synthesis in the roots of tomato plants is accelerated, but conversion of ACC to ethylene is blocked (Bradford and Yang 1980, 1981). ACC accumulates in the roots and is transported in the xylem to the shoot, where in the presence of oxygen it is rapidly converted to ethylene. Thus, "ethylene" may be mobile in the plant in the form of its precursor, ACC. Waterlogging and water deficit are quite different stresses (Bradford and Yang 1981), and it is not known whether ACC transport is involved in the stimulation of ethylene production by reduced Ψ. It should, however, be considered as a possible mechanism for hormonal communication of stress responses within the plant.

9.5.4 Cytokinins

Information on how endogenous cytokinins (CK) are affected by water stress is quite limited, despite considerable speculation as to their role in plant responses to drought (ITAI and BENZIONI 1976; VAADIA 1976). Cytokinins are thought to be synthesized primarily in the roots, although they can be metabolized throughout the plant (VAN STADEN and DAVEY 1979). Xylem sap from droughted (ITAI and VAADIA 1965) or osmotically stressed (ITAI et al. 1968) roots contained considerably less CK activity per unit volume than that from control roots. When the stress consisted of an enhanced evaporative demand which caused only transient wilting of the leaves, CK in both root exudate and leaf extracts were reduced by 50% (ITAI and VAADIA 1971). ITAI et al. (1968) proposed that a reduction in CK export from the roots could be responsible for leaf senescence and protein synthesis inhibition in water-stressed plants. Unfortunately, subsequent work in this area has been virtually nil. Key data needed to establish the validity of the hypothesis, such as characterization of the degree and duration of stress required, the kinetics of CK transport during stress, the levels of CK in the leaf, and the specific CK involved, are not available.

The notion that a drop in CK export from the roots acts as a "signal" to the shoot of root distress remains provocative. ASPINALL (1980) has noted that "such a 'signal' would seem to be redundant during water stress, since shoot water potential falls more rapidly than root potential during stress". This depends to a certain degree, however, on how stomata act to regulate leaf Ψ during soil water depletion (BATES and HALL 1981). Stomata can also respond to other root stresses while leaf Ψ remains unaffected (BRADFORD and YANG 1981; SOJKA and STOLZY 1980). Whether CK are involved in these stomatal responses is unknown, but the concept of root-to-shoot communication of root status is still useful (JONES 1980). The physiology of CK in water-stressed plants deserves further experimental consideration.

9.5.5 Abscisic Acid

In mesophytic plants, the endogenous levels of abscisic acid (ABA) rise markedly in response to moderate water stress, apparently due to de novo synthesis (MILBORROW 1974; WALTON 1980). The increase is most dramatic in the leaves, but has also been observed in roots, stems, and xylem and phloem sap (HOAD 1975, 1978; WALTON et al. 1976; ZEEVAART 1977). Following the discovery that stomatal closure can be induced by ABA, considerable effort has been directed toward elucidating the relationship between the increase in ABA levels and stomatal behavior during water stress, as will be discussed below (Sect. 9.6.1). The present section emphasizes the regulation of ABA levels in the plant and possible nonstomatal effects of ABA.

In some cases, as when leaves are sampled at various times in a drying cycle, ABA content changes little until a "threshold" Ψ is reached, then it accumulates rapidly with a further fall in Ψ (ZABADAL 1974). If Ψ of detached leaves is reduced to various levels and then held constant to allow time for ABA accumulation, a more gradual rise in ABA content may be observed (WRIGHT 1977). At least part of the reason for the apparent threshold Ψ for ABA synthesis may be due to increasing stress duration at the same time that

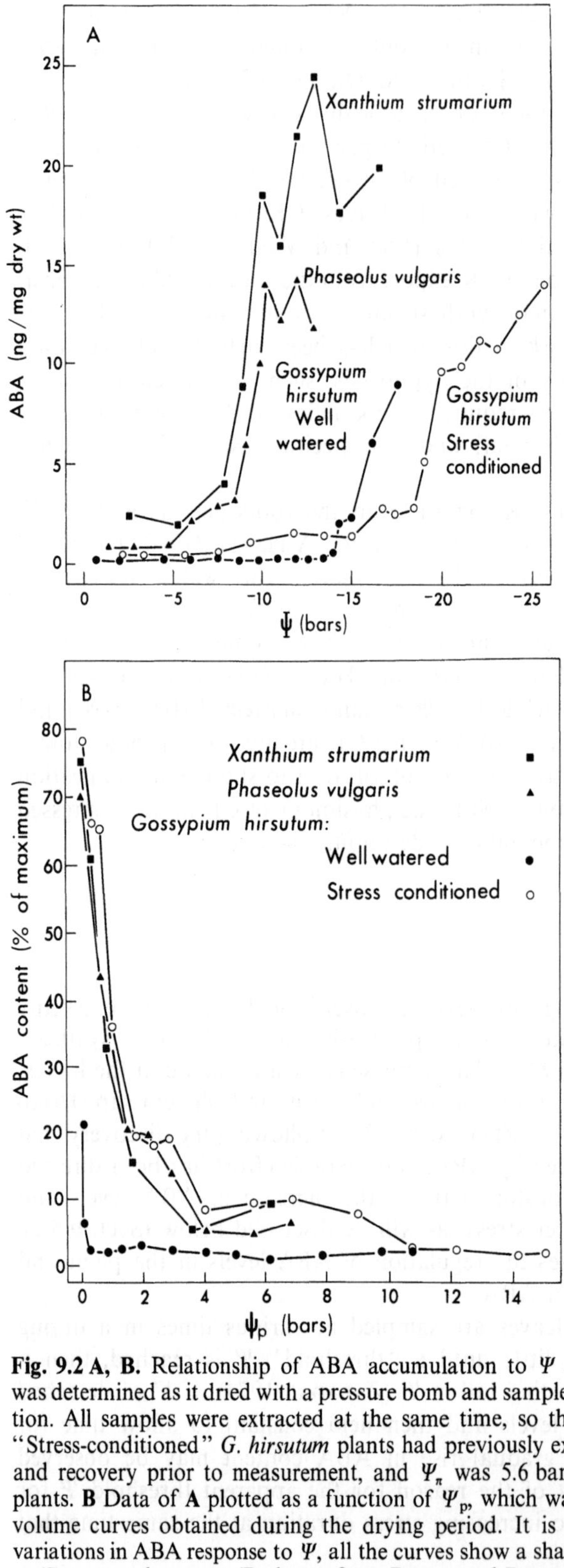

Fig. 9.2 A, B. Relationship of ABA accumulation to Ψ and Ψ_p. **A** Ψ of an excised leaf was determined as it dried with a pressure bomb and samples were taken for ABA determination. All samples were extracted at the same time, so that stress duration was identical. "Stress-conditioned" *G. hirsutum* plants had previously experienced eight cycles of wilting and recovery prior to measurement, and Ψ_π was 5.6 bar lower than in the well-watered plants. **B** Data of **A** plotted as a function of Ψ_p, which was determined from the pressure-volume curves obtained during the drying period. It is apparent that, despite the large variations in ABA response to Ψ, all the curves show a sharp increase in ABA accumulation as Ψ_p approaches zero. (Redrawn from PIERCE and RASCHKE 1980)

Ψ is falling in rapid desiccation experiments. However, even when precautions are taken to exclude this complication, a plot of ABA content versus leaf Ψ still shows a marked change in slope as Ψ_p approaches zero (Fig. 9.2). If the Ψ corresponding to zero turgor is lowered by osmotic adjustment, the point of rapid ABA accumulation is also shifted (Fig. 9.2 A). It is likely, therefore, that the very large (up to 40-fold) increases in ABA content during water stress are triggered by the loss of leaf turgor.

The gradual rise in ABA content when turgor is low but above zero (Fig. 9.2 B) may be due to heterogeneity in Ψ_π or volumetric elastic modulus among leaf cells, broadening the range of Ψ over which individual cells would reach zero turgor, or to inherent differences in the turgor sensitivity of different cells. It is these smaller changes in ABA content which can occur before Ψ_p reaches zero which should perhaps attract our attention, as the combination of osmotic adjustment and stomatal closure often prevents the complete loss of turgor in field-grown plants (TURNER and JONES 1980). DAVIES and LAKSO (1978) found that the ABA content of apple leaves increased linearly with a decline in Ψ_p during a slowly developing water stress, even though Ψ_p did not fall below 5 bar. ACKERSON (1980) reported that repeated stress and recovery cycles resulted in an increase in the ABA content of turgid leaves, but had little effect on the amount of ABA accumulated upon wilting. It is interesting in this regard that well-watered cotton plants showed virtually no increase in ABA until Ψ_p was less than 1 bar, while plants which had experienced several stress cycles began to accumulate ABA when Ψ_p was still at least 9 bar (Fig. 9.2 B). Consequently, modification of the turgor sensitivity of ABA synthesis (or degradation) may be more closely related to adaptation to water stress than is the rate of ABA production or degree of accumulation when Ψ_p reaches zero.

The notion that small changes in endogenous ABA concentration might be physiologically significant has been invoked to explain kinetic discrepancies between increases in leaf ABA content and stomatal closure (see Sect. 9.6.1). These hypotheses suggest that the initial response to water deficit may be release of ABA from a "bound" or "compartmentalized" form as an initial response to stress, followed by stimulation of synthesis as a subsequent event (RASCHKE 1975). It is unlikely that free ABA is released by hydrolysis of ABA conjugates (probably ABA-glucosyl ester), as the conjugated forms represent only a small fraction of the total ABA, and they increase rather than decrease during wilting (MILBORROW 1978; ZEEVAART 1980).

Compartmentation remains a distinct possibility, both among cell types and within a given cell, and there is evidence that ABA is not uniformly distributed within cells. It has been thought that this might result from ABA synthesis occurring primarily in the chloroplasts (LOVEYS 1977; MILBORROW 1974), but this is now controversial (HARTUNG et al. 1981; WALTON 1980). Regardless of the site(s) of synthesis, the distribution of ABA among the cellular compartments seems to be determined primarily by pH gradients. HEILMANN and coworkers (1980) found that up to 80% of the total ABA content of illuminated spinach leaves can be contained in the chloroplasts, which they explained on the basis of the equilibration of a weak acid (pK_a of ABA=4.8) between two compartments of differing pH, assuming that the chloroplast membrane is permeable only to the undissociated species. Further results indicate that this assumption is probably valid, both for the chloroplast envelope and the plasma membrane, and that the transport of ABA does not seem to be carrier mediated (GIMMLER et al. 1981; KAISER and HARTUNG 1981). A difference in uptake of ABA into chloroplasts in light or darkness could be accounted for by the pH gradient which exists across the outer membrane under the two conditions (HEILMANN et al. 1980). The hypothesis also predicts that ABA concentration will be very low in the vacuole, since it is slightly acidic, and MILBORROW (1979) found that the "vacuolar" sap of citrus fruit vesicles did contain considerably less ABA than the remaining "cytoplasmic" fraction. Intracellular ABA concentration should therefore be greatest in the chloroplasts, intermediate in the cytoplasm, and least in the vacuole, the differences being accentuated in the light (COWAN et al. 1982).

If water stress causes ABA redistribution, then it should affect either the pH gradients or the membrane permeability to anionic ABA. HARTUNG and coworkers (1981) reported that osmotic stress caused release of ABA from isolated chloroplasts, which they attributed to a decrease in stromal pH due to water stress. In intact cells, however, decreases in Ψ largely affect Ψ_p, not cell volume or Ψ_π. It remains to be shown that changes in hydrostatic pressure will alter chloroplast pH or ABA compartmentation. Others have proposed that the initial effect of water stress is to increase the permeability of chloroplast membranes to ABA (MANSFIELD et al. 1978; MILBORROW 1979). In light of the pH-dependent distribution model, this would require an increase in permeability to anionic ABA, as the membrane is already highly permeable to the protonated species. Such measurements have not been made on chloroplasts obtained from water-stressed plants. While there is now a testable hypothesis for the intracellular distribution of ABA, we still do not know how a fall in Ψ_p acts to alter ABA compartmentation or synthesis.

ABA is metabolized to phaseic acid (PA), dihydrophaseic acid (DPA), and ABA-glucosyl conjugates (see WALTON 1980, for review). The ABA concentration in the leaf therefore reflects the relative rates of ABA synthesis and metabolism. HARRISON and WALTON (1975) concluded that both synthesis and metabolism were accelerated during water deficit, leading to a new, higher steady-state level of ABA in bean leaves. ZEEVAART (1980) also found that metabolites continued to accumulate even when ABA levels remained constant in stressed *Xanthium* leaves. When the stress was relieved, ABA levels fell after a 1-h lag, and the PA accumulation rate increased. The rate of ABA catabolism increases when even slight turgor is restored following stress (PIERCE and RASCHKE 1981). The synthesis and degradation of ABA seem to be finely balanced around the point of zero turgor to regulate the cellular ABA content.

The high level of ABA present in water-stressed leaves raises the question of whether transport to other parts of the plant occurs. ZEEVAART (1977) presented evidence for the export of ABA and its metabolites in phloem exudate from stressed *Ricinus* leaves. Tracer studies in turgid plants also support the notion that ABA is mobile in the phloem (BELLANDI and DÖRFFLING 1974). Very high concentrations of ABA (up to 0.1 mM) were reported in phloem sap exuding from cut pods of water stressed *Lupinus albus* (HOAD 1978). Steam girdling of petioles prevented export of ^{14}C-ABA applied to the leaves (HARTUNG 1976) and led to an increase in endogenous ABA levels (SETTER et al. 1980). In addition to compartmentation and rates of synthesis and degradation, translocation must also be considered as a mechanism for internal regulation of ABA levels.

Although HOAD (1975) concluded that ABA appearing in the xylem sap of stressed sunflower plants had originated in the leaves, roots of several species have been shown to synthesize ABA following water loss or osmotic stress (BARR 1973; MILBORROW and ROBINSON 1973; WALTON et al. 1976). Thus, it is likely that ABA levels increase in roots during water stress, but perhaps less markedly than in leaves. A major resistance to water transport in the plant occurs in the roots. Various authors have reported that treatment with ABA increases the hydraulic conductance (C) of the root system (GLINKA 1980; TAL and IMBER 1971), which would reduce the Ψ gradient required for water transport and maintain a higher leaf Ψ. Others, however, found either no effect (CRAM and PITMAN 1972) or a decrease (MARKHART et al. 1979) in C following ABA application. Even in cases where hydrostatic pressure (MARKHART et al. 1979) or suction (GLINKA 1977) were used to minimize the contribution of ion transport to the estimation of C, contrasting effects were observed. As the results of such experiments have been shown to be highly dependent upon growth conditions, methods of measurement, duration of treatment, species, and other factors (PITMAN et al. 1974a), such differences are perhaps not surprising.

A major source of variation, however, may reside in the widely differing concentrations of ABA employed. For example, KARMOKER and VAN STEVENINCK (1978) found maximum stimulation of volume flow through bean root systems at 0.5 μM ABA, with decreasing effects as the ABA concentration was increased to 10 μM. PITMAN and WELLFARE (1978) reported an initial stimulation of volume flow in barley roots followed by inhibition, but the ABA concentration used was 50 μM, which is at least 100-fold greater than the reported maximum ABA content of roots even following an episode of water stress (WALTON et al. 1976). Although the data for endogenous ABA levels in roots are limited, it would seem that studies with applied ABA should be restricted to concentrations in the range of 1 μM or less if the results are to have physiological significance. There is a clear need for more information on the endogenous ABA levels (and compartmentation?) in roots and how these change in response to water stress.

Estimations of C involve measurements of ion uptake and transport to the xylem, and these processes are also sensitive to ABA and water stress. Both water stress (from 98% to 88% RWC) and ABA (1 μM) had little effect on ion uptake into barley roots, but transport from roots to shoots was reduced, resulting in an accumulation of ions in the root (PITMAN et al. 1974b). But again, there are conflicting reports of stimulation of ion uptake and transport by ABA (COLLINS and KERRIGAN 1974; GLINKA 1980). Regardless of the fine details, all reports agree that ABA can markedly influence the ionic relations of roots. ABA could therefore play a role in the accumulation of solutes which occurs in roots subjected to water stress (GREACEN and OH 1972; SHARP and DAVIES 1979).

The restriction of shoot growth and continuation of root growth are both important adaptations to water stress (see Sect. 9.3). Since ABA is a potent inhibitor of shoot growth, accumulation of ABA during water stress may participate in this developmental adaptation. QUARRIE and JONES (1977) compared the effects of water stress or ABA applications on the growth and development of wheat. To a large degree, artificially elevating ABA levels in well-watered plants mimicked such effects of water stress as reduced leaf growth, small epidermal cell size, fewer stomata per leaf, increased differentiation of trichomes, and fewer spikelets reduce per ear. DAVIES and coworkers (1980) cited unpublished observations that ABA applications to well-watered plants resulted in a stimulation of root growth relative to shoot growth.

These studies implicate ABA in controlling certain developmental changes during stress, but the relationships are not always clear cut. For example, ABA levels can be higher in young, growing leaves than in mature leaves (RASCHKE and ZEEVAART 1976), and developing soybean seeds contain very high concentrations of ABA at a time when the growth rate is greatest (QUEBEDEAUX et al. 1976). Presumably the ABA is compartmentalized or does not have its usual inhibitory effect in these tissues. On the other hand, the inhibition of leaf growth during water stress, even though osmotic adjustment has restored turgor in the elongating region itself (see Sect. 9.2), indicates that regulatory factors other than Ψ_p act to limit growth.

A wide variety of environmental conditions in addition to water stress will cause an increase in leaf ABA levels, including nutrient deficiency or toxicity, salinity, chilling, and waterlogging (RIKIN et al. 1976; HIRON and WRIGHT 1973; BOUSSIBA and RICHMOND 1976; BOUSSIBA et al. 1975; MIZRAHI et al. 1972). Some of these, such as nutrient deficiency and waterlogging, need not involve reductions in Ψ (BOUSSIBA and RICHMOND 1976; BRADFORD and YANG 1981). Furthermore, applications of ABA can increase the tolerance of plants to a subsequent stress (RIKIN and RICHMOND 1976; MIZRAHI and RICHMOND 1972). In some cases, such as wilting induced by salination or cessation of aeration of the root system, the increased tolerance can be attributed to stomatal closure. However, the ability of ABA to delay or prevent chilling injury implicates membrane changes related to ion permeability or hydraulic conductance (RIKIN and RICHMOND 1976; RIKIN et al. 1976). Determining how ABA affects cellular membranes, whether in stomata or in root cells, is certainly a central topic for future research.

9.6 Stomatal Behavior and Photosynthesis

Photosynthesis has long been known to be partially or completely suppressed by water stress if it is sufficiently severe. Research on this topic early in the century has been reviewed by BOYER (1976). Generally, in short-term experiments, assimilation is not sensitive to reductions in leaf Ψ until a threshold

Ψ value is reached. Further reductions in Ψ below that value, however, cause first gentle then very steep drops in CO_2 assimilation, to the extent of reaching a negative net photosynthesis rate (due to respiration). The threshold Ψ varies widely (e.g., -5 to -25 bar) for different species and experimental conditions; data bearing on this point have been tabulated (BOYER 1976). The influence of growth history and osmotic adjustment due to prior water stress on the shape and position of the photosynthetic response curves to leaf Ψ is discussed in Chapters 7 and 18, this Volume.

Almost always, transpiration is suppressed by water stress concurrently with the suppression of photosynthesis. Assimilation and transpiration have much in common. In its influx into leaves to the sites of carboxylation, CO_2 travels essentially the entire pathway for transpiration, though in a direction opposite to that of the movement of water vapor (see Chaps. 7 and 17, this Vol.). In addition, the necessary energy for transpiration (heat of vaporization) usually also derives from radiation, as does the energy for photosynthesis. Effects of water stress on size and geometry of the foliage, therefore, would impinge similarly on both processes (HSIAO and BRADFORD 1982). Thus, it is not surprising that the means evolved by plants to control transpiration and avoid excessive dehydration when water-stressed also result in reductions in CO_2 assimilation, though perhaps in different proportions. Three varieties of transpirational control have evolved which can be categorized as reduced effective leaf surface area, reduced radiation absorption per unit effective surface area, and increased resistances to water vapor in the transport pathway. The first category is elaborated in Section 9.8.3, and the remaining two have been discussed to various extents in Chapters 1 and 2, Volume 12 A and in Chapters 7 and 8, this Volume. The considerations here are therefore limited mainly to the underlying mechanisms, especially those for stomatal responses.

Also discussed are effects other than those arising out of transpirational control, as CO_2 assimilation may be inhibited independently of diffusional limitations. These responses do not appear to have a clear adaptive role and may be viewed as "lesions" or damage. In terms of the resistance network analysis for gas exchange, these effects are reflected mostly in increases in intracellular resistance, as distinguished from changes in epidermal and boundary layer resistances, the latter being parts of the pathway shared by CO_2 and water vapor.

9.6.1 Stomatal Responses to Water Stress and Humidity

Generally, the aforementioned decline in CO_2 assimilation associated with reductions in leaf Ψ has been attributed primarily to stomatal closure and the consequent increase in leaf epidermal resistance (r_e). As discussed elsewhere in this Volume (Chap. 17), stomata may have evolved as regulatory valves to minimize water loss for a given amount of carbon gained. A different view, probably more applicable to crop species, is that stomata operate to meet two often conflicting demands: maximizing assimilation while preventing the reduction of plant water status to levels substantially damaging to the tissue. When water status in a leaf falls below a threshold value, stomata respond by closing, with the consequent reduction in transpiration as well as assimilation. The threshold behavior is most common in herbaceous crop species (BEADLE et al. 1973; DAVIES et al. 1981; TURNER 1974; HSIAO 1973a; LUDLOW 1980), but is also often observed in woody species (JARVIS 1980). It is consistent with the concept of stomata serving as an overriding protective mechanism against excessive tissue water loss. Midday stomatal closure in many crops in the field when soil water depletes sufficiently is an example of such a response. For many herbs growing in the field under high irradiance and favorable environment, the threshold for stomatal closure in recently matured leaves in light lies in the range of -12 to -16 bar, provided that the plants have had no recent history of substantial water stress. With a history of stress and osmotic adjustment, the threshold can be much lower (BEGG and TURNER 1976; LUDLOW 1980). The threshold may also differ for leaves of different ages and prior exposure to radiation (JORDAN et al. 1975).

Stomata also close partially in many cases when air humidity is low and the evaporative demand is high, although the water status of the bulk leaf tissue is above the closure threshold. This response has been covered in detail in Chapter 7 this Volume. Since this behavior results in the immediate reduction in CO_2 assimilation and the conservation of water in spite of the absence of tissue water stress, it may be advantageous only in locations where the rainfall probability diminishes as the season progresses and competing species which may use up the water conserved in the soil are absent. Behavior of this kind may be called "pessimistic" (JONES 1980; Y. VAADIA personal communication), as it appears to anticipate the worst water supply situation in the future. Not surprisingly then, the literature leaves the impression that crop species selected for high productivity tend to show minimal or no stomatal response to humidity while trees and wild herbs often have marked responses. The interplay between stomatal responses to bulk leaf water status and to humidity have been analyzed by LUDLOW (1980).

Stomatal closure due either to low leaf water status or high evaporative demand is based on changes in turgor balance between guard cells and surrounding epidermal cells. Because the mechanism is turgor-mediated, the notion is still held by some that excessive loss of leaf water, or differential loss of water between guard cells and their surroundings, can lead directly to changes in turgor balance and stomatal closure. Actually, the mechanism is much more complex, and is, in the words of STÅLFELT, not just "hydropassive" but "hydroactive". The former term refers to stomatal movements arising as the result of the gain or loss of only water by cells, whereas the latter refers to stomatal movements resulting from gain or loss of solutes by cells followed by changes in water volumes. Many years ago, STÅLFELT (1955) had repeatedly observed that as water stress developed in leaves, solute loss from guard cells, reduction in guard cell volume, and stomatal closure occurred simultaneously. The necessity of solute gain and loss from guard cells for more than just transient stomatal movement is demonstrated in the elegant analysis by RASCHKE (1979) of the mechanics of stomata. Here we will use a simpler approach similar to that of BOYER (1976) to make the same point. For stomata to open substantially, turgor in guard cells (Ψ_p^{gc}) must be much higher than turgor in subsidiary cells (Ψ_p^{sc}) (MEIDNER and EDWARDS 1975; RASCHKE 1979). On the other hand, because of their adjacent positions and the lack of an impervious barrier to water transport, Ψ in guard cells and subsidiary cells must be in near equilibrium, in the absence of sudden stepwise changes in water status. Since $\Psi = \Psi_p + \Psi_\pi$,

$$\Psi_p^{gc} + \Psi_\pi^{gc} = \Psi_p^{sc} + \Psi_\pi^{sc}.$$

It follows that for $\Psi_p^{gc} \gg \Psi_p^{sc}$, $\Psi_\pi^{gc} \ll \Psi_\pi^{sc}$. That is, for stomata to open substantially, osmotic potential must be much more negative and solutes much more concentrated in guard cells than in subsidiary cells. Conversely, for stomata to close as a consequence of water stress or high humidity gradient, the turgor difference and hence solute difference between guard cells and subsidiary cells must be markedly reduced. Merely a redistribution of water between guard cells and their neighbors without an underlying movement of solutes would be insufficient to cause the changes in turgor needed to effect closure. Generally, the changes in Ψ_π and hence Ψ_p in guard cells associated with full opening and closing are in the range of 10 to 40 bar (HSIAO 1976; RASCHKE 1979).

STÅLFELT's early work did not identify the solute species involved in the hydroactive closing of stomata under water stress. Since K^+ and associated anions account for virtually all of the solutes being transported in and out of guard cells in stomatal movements elicited by changes in CO_2 concentration and light (reviewed by HSIAO 1976), it is no surprise that K^+ was also found to be the cationic solute underlying stomatal movement in response to changes in bulk leaf water status (HSIAO 1973b; EHRET and BOYER 1979). As the leaf undergoes water stress, stomata close concurrently with losses in stainable K^+ from guard cells. Changes in stomatal aperture and guard cell K^+ are well correlated, either as stomata close during water stress or as they reopen when stress is released (HSIAO 1973b). When water status of the leaf is either raised or lowered rapidly, stomatal movement lags behind the change in tissue water content but appears to be synchronous with variations in stainable K^+ in guard cells, as shown in Fig. 9.3. As the leaf tissue rapidly lost water during a 10-min drying period, stomatal aperture and stainable K^+ in guard cells declined only slightly. Subsequent to the drying, the tissue was kept in an air stream of 100% relative humidity for 30 min. During that time there was no further drying and the tissue maintained its water content, although stomata continued to close in parallel with K^+ loss from the guard cells. In fact, most of the decline in K^+ and aperture took place during the period when there was no further drop in tissue water content. These data suggest that the hydropassive component is insignificant in stomatal movement induced by variations in leaf water status. Stress apparently triggers the efflux of K^+ and anions from guard cells with the consequent diminution of turgor differences between guard cells and their neighbors, leading to stomatal closure.

Having verified the hydroactive nature of the stomatal closing elicited by water stress and demonstrated the key role for K^+, there remains still the question of how a reduction in leaf water status leads to K^+ efflux. A popular hypothesis, that water stress causes abscisic acid (ABA) accumulation which in turn leads to the efflux of K^+ from guard cells, was recently reviewed by

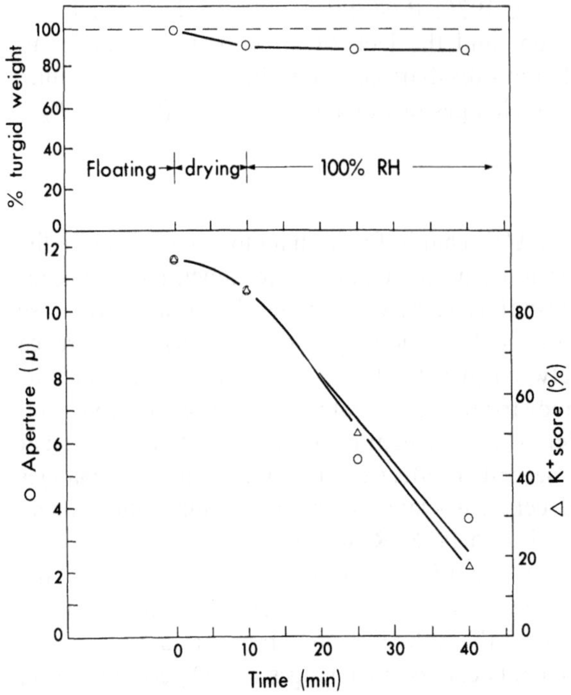

Fig. 9.3. Changes in stomatal aperture and guard cell K^+ in leaf discs of *Vicia faba* in response to water stress. *Above* Fresh weight of leaf discs as percent of turgid weight when floating on water, after drying in moving air, and while being kept in saturated moving air. *Below* Stomatal aperture and stainable K^+ score in guard cells of the leaf discs. Replicates of discs were taken from recently mature leaves and floated on water under moving air of controlled humidity and temperature. At time zero, the discs were removed from water and allowed to dry in the same moving air for the indicated duration, then transferred to moving saturated air to prevent further desiccation. Discs were removed at various times, two epidermal strips were taken from each disc, and the apertures in one strip under oil were measured microscopically. The other strip was stained for K^+ with cobaltinitrite, darkened by sulfide, and scored microscopically as percent of guard cell area occupied by the stain. FISCHER (1971) has shown a linear relationship between K^+ content and percent of guard cell area stained. Data are means of 4 discs, with 20 stomata measured for aperture and 4 microscope fields scored for K^+ per disc. (Unpublished data of T.C. HSIAO)

RASCHKE (1979). While most of the evidence indeed indicates a mediating role for ABA in stomatal behavior in general and under conditions of gradually intensifying water stress in particular, some data are inconsistent with the hypothesis. These inconsistencies were pointed out in an earlier review (HSIAO 1973a) and are basically discrepancies between the time course of variations in ABA levels and stomatal behavior. When suddenly stressed, the accumulation of ABA in leaves can lag behind stomatal closure (BEARDSELL and COHEN 1975; HENSON 1981; WALTON et al. 1977). Generally ABA buildup in leaves becomes measurable only after 20 min to 1 h of stress (e.g., WRIGHT 1969; PIERCE and RASCHKE 1980a) but stomatal closure begins sooner (e.g., Fig. 9.3), often within 5 min from the start of the reduction in tissue water status (GRAZ-

IANI and LIVNE 1971; HENSON 1981; HSIAO unpublished). Upon rewatering of stressed plants, leaf ABA declines ahead of recovery in stomatal opening (BEARDSELL and COHEN 1975; LOVEYS and KRIEDEMANN 1973; LUDLOW et al. 1980). These discrepancies commonly are attributed to preferential localization of ABA in or near guard cells. Since guard cells constitute only about 1% of leaf volume (RASCHKE 1979), ABA content of bulk leaf tissue is a poor indicator of its concentration at the stomatal site. Recent results with a highly sensitive radioimmunoassay for ABA indicate that guard cell protoplasts are capable of synthesizing ABA in response to an osmotic stress (WEILER et al. 1982). On the other hand, there is some indication that ABA elicited by water stress is produced largely in the mesophyll and may be transported from there to the guard cells (LOVEYS 1977). Changes in pH within the mesophyll cells due to stress could result in an increase in ABA in the apoplast (COWAN et al. 1982), but experimental data on whether this occurs in vivo are lacking. Also, gaps exist as to how ABA acts to change the rate and direction of ion transport between guard cells and their neighbors. Further discussion of these points is found elsewhere (DAVIES et al. 1981; HSIAO 1976; RASCHKE 1979; WALTON et al. 1977).

Even less is known about the mechanism underlying stomatal response to humidity at relatively high leaf water status. LÖSCH and SCHENK (1978) found some correlation between stomatal aperture of, and stainable K^+ in, guard cells of *Valerianella locusta* as air humidity was varied. Upon a reduction or elevation in humidity, aperture decreased or increased respectively, but always ahead of decreases or increases in guard cell K^+. This is in contrast to the close correlation found between aperture and K^+ in the same plant when illumination was varied. The authors were unable to conclude whether the initial aperture change was hydropassive followed by stabilization and adjustment via K^+ transport, or whether the initial response was underlaid by changes in solutes other than K^+. It is not known if ABA plays a role in the humidity response.

9.6.2 Nonstomatal Effects of Water Stress on Photosynthesis

In the resistance network analysis of CO_2 assimilation, the intracellular resistance (r_i) represents all the "resistances" encountered by CO_2 in moving from the intercellular air spaces to the carboxylation sites in the chloroplasts. Since this includes dissolution of CO_2 in the cell wall water, movement across the wall, the membranes, and the cytoplasm to the sites of the carboxylase, as well as the biochemical reaction of carboxylation, r_i should be clearly distinguished from the epidermal resistance discussed above which involves only gaseous diffusion. Examination of r_i should provide a gross view of water stress effects on the photosynthetic processes downstream from stomata. It is unquestionable that under very severe stress, when tissue RWC is only a fraction of that under well-watered conditions, the photosynthetic apparatus will be damaged and the processes other than gaseous CO_2 transport inhibited. Therefore, the important question is not whether r_i can be increased by stress, but rather what stress severity will cause r_i to rise, what mechanisms underlie changes in r_i, and what the importance of these changes is to plant adaptation to water stress conditions.

Early work indicated that the response of r_i to water stress can be quite variable, sometimes increasing at about the same water status as the threshold for stomatal closure, and other times remaining unchanged until very severe stress had developed (reviewed by HSIAO 1973a). More recent studies, while using improved gas exchange techniques and broadening the range of species tested, have still not uncovered the causes for the different behaviors of r_i under water stress (PEARCY 1982). In addition to possible inherent differences among species, certainly part of the variability must lie in the rates and degrees of stress development. Studies demonstrating little or no effect of stress on r_i down to the range of 60%–70% RWC often involved rapid imposition of stress, in a matter of less than one hour to a few hours (MEDERSKI et al. 1975; SLATYER 1973; TROUGHTON and SLATYER 1969). On the other hand, stress usually developed over a period of several days to weeks in studies showing marked effects on r_i (JONES 1973b; COLLATZ 1977; COLLATZ et al. 1976). There are contrasting data, however, particularly when osmotic adjustment is involved. For example, r_i of the perennial desert shrub *Larrea divaricata* increased about proportionally with the increase in stomatal resistance as stress developed rapidly in potted plants (MOONEY et al. 1977). In the natural environment, however, r_i stayed nearly constant down to a leaf Ψ of -49 bar and was relatively independent of stomatal behavior. JONES and RAWSON (1979) found a similar effect of the rate of stress development on r_i of *Sorghum*. In both cases, osmotic adjustment occurred during the slowly developing stress to maintain some turgor and presumably prevent alteration of the photosynthetic machinery. But even osmotic adjustment does not always insure an unaltered r_i. Cotton which had osmotically adjusted to drought assimilated at higher rates at low leaf Ψ, but at lower rates at high leaf Ψ, compared to unadjusted plants (ACKERSON and HEBERT 1981; OLIVEIRA and HSIAO 1980). Since stomatal conductance was higher in the adjusted plants than in the controls at any given leaf Ψ (ACKERSON and HEBERT 1981), r_i must have increased in the adjusted leaves in spite of the better maintenance of leaf turgor.

This latter example illustrates a conceptual problem which may underlie much of the controversy over water stress effects on r_i. If r_i is taken to represent all factors influencing the rate of assimilation of CO_2 from the intercellular spaces, then the conclusion that r_i increased in the stressed leaves is justified. However, for the resistance analog model to be valid, the photosynthetic rate must be proportional to the intercellular CO_2 concentration (c_i), which generally means making measurements at CO_2 levels well below ambient (JONES 1973a). Intercellular resistance is then estimated from the inverse of the assimilation versus c_i curve. At ambient CO_2 concentrations, other factors, such as electron transport rate or ability to regenerate RuP_2, may co-limit the assimilation rate, which will no longer be a linear function of c_i (Chap. 16, this Vol.). Thus, photosynthesis can be limited by nonstomatal factors even though the calculated r_i (determined at low c_i) remains unchanged.

If, in the example cited, water stress affected primarily the CO_2-saturated rate of photosynthesis, then the measured r_i (at low CO_2) could have remained unaltered although the apparent r_i (at ambient CO_2) had increased. Assumption of the linear model for r_i leads to the conclusion that most of the reduction in photosynthesis due to water stress results from stomatal closure (SLATYER 1973). However, there are data suggesting that the degree to which stomata limit photosynthesis remains constant or even decreases during water stress due to associated alterations in the photosynthetic machinery (FARQUHAR and SHARKEY 1982; JONES 1973b; VON CAEMMERER 1981). This may even result in an increase in c_i in stressed plants, which has been reported to be the case for cotton (RADIN and ACKERSON 1981). The relative importance of stomatal versus nonstomatal limitations to photosynthesis may depend upon the rate of imposition, degree, and duration of water deficit. Systematic comparisons of the responses of stomata and r_i of selected species at various rates of stress development, and at both low and high c_i, should clear up much of the present confusion.

The biochemistry of photosynthesis is often divided into the light reactions and the dark reactions. Studies of the effects of water stress on the dark reactions have focused on RuP_2 carboxylase/oxygenase. The activity of this enzyme extracted from several C_3 crop species was found to decrease as stress developed over a period of several days or longer (JONES 1973b; JOHNSON et al. 1974; O'TOOLE et al. 1976). Although CO_2 assimilation followed the same general trend, r_i was only loosely related to RuP_2 carboxylase activity (JONES 1973b; O'TOOLE et al. 1976). With Sitka spruce (*Picea sitchensis*), the assayed activity of the enzyme remained about constant while r_i increased during the drying cycle, down to a needle Ψ of -30 bar (BEADLE and JARVIS 1977). The other enzymes reduced by slowly developing stress are carbonic anhydrase in cotton (JONES 1973b), and ribulose-5-phosphate kinase and PEP carboxylase in barley (HUFFAKER et al. 1970). Other photosynthetic enzymes are yet to be examined.

Water stress readily inhibits light-induced chlorophyll formation in etiolated leaves but has little or no effect on chlorophyll contents in green leaves of crop species (BOYER 1976; JONES 1973a; reviewed by HSIAO 1973a). Chlorophyll in Sitka spruce did not change significantly until needle Ψ reached -21 bar after slow drying (BEADLE and JARVIS 1977).

Early studies, conducted mostly in the 1960's on crop species, indicated that reactions in the chloroplasts, especially the light reactions, were inhibited by very severe stress (e.g., $<50\%$ RWC) but probably little affected by mild to moderate stress (reviewed by HSIAO 1973a). BOYER and BOWEN (1970), however, reported convincing data indicating that even moderate stress on sunflower leaves inhibited the activity of Photosystem II in chloroplast fragments isolated from these leaves. More severe stress, on the other hand, was necessary to affect chloroplast fragments from pea leaves. The inhibition of Photosystem II in sunflower was well correlated with the inhibition of CO_2 assimilation by intact leaves for the range of leaf Ψ from the beginning of to complete stomatal closure. FRY (1970) first reported an inhibition of electron transport, measured as Hill activity, for cotton chloroplast fragments isolated from severely stressed leaves. Later, FRY (1972) demonstrated that the inhibition was also present in the Ψ range expected for the stomatal threshold for cotton. BEADLE and JARVIS (1977) found, however, that electron transport in fragmented chloroplasts, measured as uncoupled Photosystems I and II activities, remained about constant as Sitka spruce underwent slow stress from a needle Ψ of -7 bar to about -28 bar, when assimilation became nil and the needles neared the point of shedding due to desiccation. Unfortunately, the data were quite variable from date to date and hence would not have shown gentle trends or small changes. MOHANTY and BOYER (1976) found that the quantum yields of photosynthesis in intact leaves as well as of Photosystem II activity in isolated chloroplast fragments were reduced by stressing the plant. Quantum yield of leaf photosynthesis was also observed to be reduced in stressed *Larrea* by MOONEY et al. (1977), but the reduction appeared to have accounted for only a part of the measured increase in r_i. These data indicate that basic aspects of light-harvesting and energy conversion were altered by stress. Interestingly, quantum yield of *Larrea* adapted to the Death Valley environment was not affected by stress down to leaf Ψ of -49 bar (MOONEY et al. 1977).

Most exciting are the recent results of GOVINDJEE et al. (1981) on chlorophyll *a* fluorescence transients, showing reduced ratio of maximum to minimum fluorescence (P/O) in intact leaves of *Nerium oleander, Atriplex triangularis,* and *Tolmiea menziesii* under water stress. They interpreted the data to be indicative of a blockage of electron flow on the water side of Photosystem II since Q, the electron acceptor of Photosystem II, must remain in a largely oxidized state for chlorophyll *a* fluorescence to be low under excitation. The sensitivity of the response was such that the authors actually proposed the use of the P/O ratio as an indicator of leaf Ψ, at least in the range of -9 to -40 bar. The changes were evident whether the stress developed in 1 h or over a period of days and were readily reversible upon rewatering. Unfortunately, no accompanying data on assimilation or other aspects of photosynthesis were presented.

In sunflowers, various features of electron transport and cyclic and noncyclic phosphorylation of isolated chloroplasts were shown to be inhibited by rapid stress in the range affecting stomata and CO_2 assimilation (KECK and BOYER 1974). Both phosphorylation systems were reduced to zero at -16 to -18 bar of leaf Ψ, about 5 bar below the zero turgor point and a stress level which still permitted some net photosynthesis in those leaves (BOYER 1976). Inhibition of Photosystems I and II activities was also only partial at these stress levels, indicating an uncoupling of phosphorylation from electron flow (KECK and BOYER 1974). In a later study conducted by Boyer and associates (YOUNIS et al. 1979) on spinach, however, photophosphorylation by chloroplasts was reduced only somewhat at -15 bar and was still about 40% of the control rate at -25 bar. At -25 bar, the spinach leaves contained only about one third of the water compared to the control (YOUNIS et al. 1979), and probably would have lost viability if kept at that water content for a number of hours.

By varying the water content of isolated chloroplasts drastically, SANTARIUS and ERNST (1967) concluded that marked dehydration leads to reversible uncoupling of photophosphorylation from electron transport. YOUNIS et al. (1979) succeeded in isolating coupling factor CF_1 from chloroplasts of rapidly stressed and control spinach leaves. CF_1 from stressed leaves exhibited less ATPase activity and much less tendency to bind the fluorescent analogs of ADP and ATP than CF_1 from control leaves. Circular dichroism spectra indicated that stress altered the conformation of CF_1 protein, probably in the random coil region. Results with reconstitution experiments suggested that both CF_1 and the chloroplast membrane were altered by stress. The effects of stress were obvious at leaf Ψ of -16 to -18 bar, more so at still lower Ψ values, and were reversible in vivo but not in vitro. To date, these findings represent the only definitive evidence of a macromolecular structural alteration of functional significance induced by stress.

Reversibility represents a puzzling aspect of stress effects on chloroplast functions. Isolation and assay of chloroplast activities were usually carried out in identical media of relatively high Ψ whether the chloroplasts originated from stressed or control leaves. Thus, the effects of stress are residual and cannot be reversed in vitro, yet rewatering of the stressed plant effected quick recovery (BOYER 1976; YOUNIS et al. 1979). When chloroplasts from well-watered plants were stressed in osmotic solutions, similar effects on electron transport and photophosphorylation were observed but at reductions of chloroplastic volume much larger than those required if stressed in vivo (BOYER 1976; HSIAO 1973a; SANTARIUS and ERNST 1967). Further, the effects were reversed completely when the chloroplasts were placed back in the standard assay medium. In the case of the coupling factor, the CF_1 protein studied by YOUNIS et al. (1979) was stored in 2 M $(NH_4)_2SO_4$, Ψ of which should be less than -100 bar. Yet, CF_1 from the control leaves remained highly active, in contrast to the low activity of CF_1 from the stressed. All these observations point to some unique feature of the cytoplasmic environment playing a key role in mediating the effects of stress and of rewatering. What is more, the mediating role appeared to be all-encompassing, perhaps impinging upon the whole complement of light reactions.

A discussion of causal mechanisms of water stress effects on photosynthesis would not be complete without mentioning the recent proposals of OSMOND et al. (1980). As these authors note, environmental conditions conducive to water stress are virtually always associated with high light intensity. If leaves are exposed to very low CO_2 levels in the light, "photoinhibition" may occur due to damage to the photosynthetic apparatus. It is possible that when stomata close due to rapid water stress in the light, the reduction in c_i, coupled perhaps with some unknown mechanism induced by water stress, brings about photoinhibition, acounting for the reduction in quantum yield and increase in r_i. When stress develops gradually, plants may adjust both stomatal and nonstomatal components of photosynthesis in a coordinated manner, in accordance with the hypothesis of COWAN and FARQUHAR (see Chaps. 7 and 17, this Vol.). Some of the biochemical aspects of photoinhibition have been discussed recently by LORIMER and ANDREWS (1980) in a lucid review, along with alternative points of view. Experimental data are as yet too sparse for a proper evaluation of the importance of photoinhibition during water stress, but it remains a provocative possibility. The notion of coordinated adjustments under slow stress is well worth considering. Some early data already hinted at a concerted response of several components of the CO_2-assimilation complex (HSIAO 1973a, p. 531). It is likely that slow stress results in modified leaves, clearly distinct from control leaves in their photosynthetic characteristics, but with their functional units still tightly integrated. Further discussion of coordinated modulation of plant responses under stress is found in Section 9.8.

9.7 Other Processes Affected by Water Stress

Although photosynthesis and growth have been emphasized, various other physiological processes are also affected by water stress. Of these, respiration, accumulation and transport of specific metabolites, and nutrient uptake will be briefly discussed due to their relevance to our overall theme. More comprehensive coverage can be found in other reviews (HSIAO 1973a) and volumes (MUSSELL and STAPLES 1979; TURNER and KRAMER 1980).

9.7.1 Respiration

In general, dark respiration declines moderately as Ψ falls when stress is not too severe (HSIAO 1973a). Apparent respiration in the light also showed a similar trend (BOYER 1971). GREENWAY and LEAHY (1970) suggested from experiments with osmotic stresses that a fall in Ψ_p might be responsible for inhibition of respiration, but osmotic stress resulting in plasmolysis may not be identical with desiccation. In cases where the stress was induced by air drying, Ψ_p was not determined (KOEPPE et al. 1973). Recent evidence obtained with water-stressed wheat apices suggests that the relationship between Ψ_p and respiration is not quantitative (PHELOUNG and BARLOW 1981). Turgor pressure in the apex fell initially when water was withheld, then remained constant at 1 bar due to osmotic adjustment as Ψ decreased to -51 bar. Respiration, however, continued to decline almost linearly with Ψ. Substrates for respiration were not limiting, as import continued and carbohydrate content actually increased during stress. Osmotic stress produced very similar results, but only when slowly penetrating osmotica were used and cell volume decreased. Respiration of mitochondria isolated from water-stressed leaves was also inhibited, although respiratory control was not lost (BELL et al. 1971). This seems not to be due to the lower activity of water, but could be an effect of increasing concentration of specific solutes (CAMPBELL et al. 1976). The decline in respiration during water stress is therefore only qualitatively related to Ψ_p. The initial drop in respiration rate may be due to reduced demand as elongation is inhibited, but the continued reduction seems to be more closely related to the decrease in cell volume.

9.7.2 Proline and Betaine Accumulation and Transport

Hydrolysis of proteins and accumulation of amino acids is a well-known response to water stress, although early studies were often inconclusive due to the severe and poorly characterized stresses employed (BARNETT and NAYLOR 1966; PETRIE and WOOD 1938). One amino acid, proline, is synthesized at an accelerated rate during water stress while its oxidation is inhibited, resulting in large accumulations of proline in water-stressed tissues (BOGGESS et al. 1976; STEWART et al. 1977). It has been advocated that this is advantageous to the plant in coping with drought and that proline accumulation be used as an indicator in selecting for drought resistance in crop breeding (SINGH et al. 1973). Recent work, however, suggests that the opposite may be true – proline accumulation is indicative of the stress damage (HANSON et al. 1977, 1979; reviewed by STEWART and HANSON 1980). Proline accumulation generally begins only after water deficit has become severe enough to prevent growth and cause stomatal closure (e.g., MCMICHAEL and ELMORE 1977). Except for certain marine organisms and halophytes (FLOWERS et al. 1977), proline accumulation is more closely associated with survival in the face of a severe environmental stress than it is with adaptation for continued growth.

Betaine is another compound which accumulates in plants in response to salinity or drought (STOREY et al. 1977). Unlike proline, which is rapidly metabolized upon rewatering (e.g., ASPINALL 1980), betaine appears to be a metabolically inert end product (HANSON and NELSEN 1978). It may be useful as an indicator of previous stress history, since the plant content reflects the cumulative effect of prior stress (LADYMAN et al. 1980). Its physiological role is uncertain, although it may be a compatible cytoplasmic solute for osmotic adjustment (see Chap. 2, Vol. 12 C).

Regardless of the metabolic functions of the compounds, studies of proline and betaine transport during water stress have provided useful information. Export of both compounds from mature barley leaves to younger leaves continued at moderate levels of water stress (LADYMAN et al. 1980; TULLY et al. 1979; reviewed by HANSON and HITZ 1982). Only very severe stress or steam girdling of the leaf prevented export and promoted accumulation within the leaf. This lends further support to the notion that phloem transport to the expanding and meristematic tissues may be relatively insensitive to water stress.

9.7.3 Nutrient Uptake

It has been known for many years that water stress reduces plant absorption of inorganic nutrients. A slowing of the transpiration stream alone often reduces uptake by roots as well as movement to the shoot. In addition, water stress apparently inhibits the mechanisms of uptake and of nutrient unloading into the xylem more directly. The early data bearing on these points have been summarized (HSIAO 1973a). The special case of inhibition of symbiotic nitrogen fixation by water stress has been reviewed by SPRENT (1981) and will not be considered here.

A reduction in transpiration rate may not result in a proportional decrease in the movement of nutrients from the roots to the shoot. Loading of nutrients into the xylem tends to remain steady, and xylem sap becomes more concentrated when transpiration is low. In one study (SHANER and BOYER 1976), nitrate content in the xylem stream increased sufficiently to compensate completely for a 50% reduction in transpiration (effected by increasing air humidity). Water stress, however, inhibited both nitrate flow to the shoot and transpiration concurrently, suggesting that ion transport processes may have been affected as well. In the absence of transpiration flow, the uptake of Cl^- into excised roots was inhibited by prior stresses of various severities (COLE and ALSTON 1974). The effect became detectable when root Ψ was reduced (to -3 or -4 bar) by drying in an air stream for 30 min. Recovery of uptake after the stress required at least several hours. Similar effects were reported for Rb^+ uptake, but only for stress levels of −9 to −11 bar, achieved by immersing the roots in mannitol or polyethylene glycol 6000 solutions (PARRONDO et al. 1975). Low Ψ may directly affect uptake, or an accumulation of nutrients in the xylem when transpiration is low could impede ion pumping upstream at the epidermal and cortical cells. Transport of nutrients to the shoot may also be affected by stress independently of uptake. When roots which had been pulse-labeled with radioactive phosphate were subsequently stressed in mannitol solution of −6 bar, movement of the prior-accumulated phosphate to the shoot was markedly inhibited (GREENWAY and KLEPPER 1968), suggesting an effect of stress on xylem loading.

The mechanisms underlying stress effects on ion transport in higher plants are obscure. Recently much interest has been aroused by work on turgor sensing and regulation in algal cells. Ion fluxes are closely linked to $\Delta\Psi_p$ across the plasmalemma-cell wall complex, with very small reductions in $\Delta\Psi_p$ and turgor leading to net ion influxes, possibly sensed via changes in membrane conformation (BISSON and GUTKNECHT 1980; ZIMMERMANN 1978). With higher plants, ion transport appears to be insensitive to changes in turgor of at least several bars (see also Sect. 9.4.2.2). More importantly, reductions in turgor led to inhibition of transport, not stimulation. It appears doubtful that the conceptual framework developed for algal systems can be applied to higher plants without drastic modification.

For plants growing in the soil, additional suppressive effects on ion uptake due to factors associated with soil drying can be dominant. These factors have been analyzed quantitatively by NYE and TINKER (1977). It suffices to say that often the transport of the nutrients from the bulk soil to the root surface plays a major role in limiting uptake.

In spite of the effects discussed above, it is probable that nutritional deficiencies may not be aggravated or induced by water stress in many cases. Suppressed growth under water stress may reduce the demand for nutrients to a level commensurate with the reduced accumulation. In a long-term experiment with maize grown in soil in pots with and without water stress, VERASAN and PHILLIPS (1978) found the cumulative uptake of N, K, P, Ca, and Mg, as well as biomass production, to be linearly related to cumulative evapotranspiration. The relative slopes of the regression lines for the nutrients, however, were lower than that for the biomass. When we calculated the content of nutrients per unit biomass from their regression lines, we found the percent of each nutrient to be higher in the stressed than in the unstressed plants. In fact, the percentage contents of some nutrients were more than doubled by stress. In a study on *Phaseolus vulgaris* using ^{15}N, uptake of ammonium or nitrate ions into roots and shoots over periods of hours was reduced markedly by placing the roots in −4 bar polyethylene glycol 1540 (FROTA and TUCKER 1978a). In spite of the lower uptake, more ^{15}N accumulated in the stressed tissue as inorganic ions and as soluble nitrogenous compounds, particularly as α-amino nitrogen. On the other hand, much less ^{15}N was incorporated into proteins in the stressed tissue (FROTA and TUCKER 1978b). These results indicated that stress reduced the demand for nitrogen (for protein synthesis) below the level of supply so that the soluble nitrogen pool was expanded.

The aforementioned results notwithstanding, it is also probable that in certain cases stress does induce nutritional deficiencies, as indicated by lower percentage composition of the nutrients in the tissue. The problem is most likely to occur in the field when stress is developing slowly and nutrients such as nitrogen are confined to the upper soil layer which dries out during a period of drought (Bataglia 1980).

9.8 Integrated and Long-Term Behavior

In an early review (Hsiao 1973a), it was pointed out that regarding metabolic changes, the sequence of events upon the onset of water stress is important in determining which, among the many perturbations, are primary events and which are secondary or tertiary. It follows that attention should be focused on the changes at the onset of stress and shortly thereafter and that stress application should be sudden. These comments are still valid, as far as research on mechanisms underlying responses to stress is concerned. What is clear, however, is that an accurate understanding of the integrated behavior of the whole organism cannot be had by only probing into its parts. The difference between organisms and chemical reactions is the existence of the higher orders of organization that coordinate and orchestrate events drastically different in nature and separated spatially and often temporally as well. The effects of prolonged water stress on a plant may be pronounced, yet changes in any single process at a given time during the stress may only be very minor and difficult to measure. This is probably due to amplification with time as well as coordinated modulations of a large number of processes, as will be further discussed below.

In nature, water stress in higher plants usually does not develop and become severe in a matter of minutes or hours. Rather, it is a gradual process taking days to weeks, durations which provide sufficient time for coordinated responses. Many of these appear to result in adaption and enhance survival under stress.

9.8.1 Cause Versus Effect – Blurred Distinction

Reduced to the biochemical level, various parts of metabolism are controlled and coordinated via feedback loops. The continuous and cyclic nature of such a loop makes it difficult to distinguish whether changes in any segment are the cause or the effect, once the perturbation starts. This characteristic is well known in metabolic studies and is equally applicable to the higher levels of organization – to behavior of cells, organs, and the whole plant. A good example is the behavior of stomata, whose various control loops have been discussed by Raschke (1979). Here we attempt to recast the parameters underlying expansive growth, detailed in Section 9.4, as a flow chart with several loops to illustrate the interlocking nature of the various components of the process and

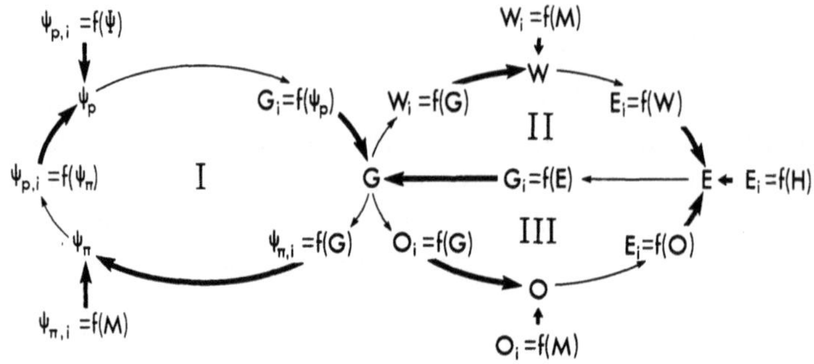

Fig. 9.4. Interdependence of expansive growth and underlying parameters. *G* expansive growth; *M* general metabolism (substrates, energy); *E* gross extensibility; *W* cell wall synthesis and metabolism; *O* synthesis of macromolecules other than cell wall; *H* hormonal effects. Each parameter is taken as a function of a number of other parameters. For example, E=f(W, O, H). The component functions E_i=f(W), E_i=f(O), and E_i=f(H) together determine the total E. *Heavy arrows* (→) indicate the "summing" of the components or partials to make the whole. *Light arrows* (→) indicate that the whole parameter in turn serves as the variable on which the partial of another function depends. Completion of loops is indicative of the interlocking nature of the various processes. Loops are labeled *I, II,* and *III* for easy reference. See text for further explanations

how it may be futile to attempt to find a "limiting" step once changes set in and are propagated with time. Some speculations and simplifying assumptions are necessary in constructing this chart, but the overview should be reasonably valid and certainly illustrates the integrated nature of the system.

In the chart (Fig. 9.4), the starting point, as before, is that the instantaneous growth rate (G) is a function of turgor pressure and gross extensibility of the wall (E). Following the mathematical concept of total derivative being the sum of the partials, we diagramatically show that G results from or is a sum of its partials or components (G_i, where i=1, 2), one being the rate as a function of Ψ_p, and another, the rate as a function of E. Implicit in the equations for components [e.g., $G_i=f(\Psi_p)$] is the fact that all variables other than those explicitly shown are held constant. The figure shows that the independent variable Ψ_p in the equation for G_i, in turn, is dependent on tissue Ψ_π and Ψ. Further on, Ψ_π is in turn dependent on metabolism (M) and on G, thus completing a loop involving G, Ψ_p, and Ψ_π. The dependence of Ψ_π on metabolism is obvious, either for solutes generated internally or for those transported into the growing tissue. The dependence of Ψ_π on G is due to dilution by growth-associated increase in volume, as discussed earlier.

The other two loops in Figure 9.4 involve cell building blocks: cell wall and other macromolecules. In Section 9.4, it was pointed out that E is a function of hormonal activity (H) and cell wall synthesis and metabolism (W). W, in turn, is dependent on G, as well as on general metabolism (M). This completes loop II in the diagram. The third loop, connecting G, E, and synthesis of macromolecules other than cell wall (O), involves the most uncertainty. The most serious is that it is not known whether synthesis and metabolism of proteins and lipids directly affect extensibility of cells or indirectly influence it by affecting W. The loop is drawn arbitrarily through G, O, and E; it could well be through

G, O, W, and E. The evidence suggesting that protein synthesis (a part of O) under water stress is partly a function of G has been discussed (Sect. 9.4.2.4).

In Fig. 9.4, Loop I shows that a decrease in leaf Ψ can lead to reduction in Ψ_p and then in G. The reduction in G, in turn, leads to less volume dilution and hence a partial reestablishment of Ψ_p, thus allowing for resumption of growth, as observed with *Nitella* (GREEN 1968). Loop II shows that simultaneously with the adjustments in Loop I arising out of the initial reduction in Ψ, changes in wall metabolism and hence extensibility may take place. The reduced G may make the wall more extensible, as is found with *Nitella,* coleoptile systems (Sect. 9.4.2.1), and maize leaves (ACEVEDO et al. 1971) in cases of very short stress intervals. On the other hand, longer stress may lead to the opposite effect of apparent reduction in E (MEYER and BOYER 1972; MICHELENA and BOYER 1980; SHARP et al. 1979; HSIAO unpublished), giving rise to the phenomena of slower growth in spite of higher Ψ_p. Thus, with prolonged stress, the function $W_i(G)$ or $E_i(W)$ has apparently undergone a sign change. It is also conceivable that the apparent reduction in E may have arisen from modification of nonwall components in Loop III. There is virtually no information bearing on these points. The significance of "wall stiffening" with longer stress durations in adaptation is also not clear. We may speculate that once stress sets in and persists, it would be important for the plant to restrict expansive growth of leaves while it undergoes osmotic adjustment to maintain turgor for other purposes. Continued fast leaf growth would lead to the undesirable result of a larger surface for transpiration and greater water expenditure. One test of this hypothesis would be to compare effects of stress on extensibility of leaves with that of roots, as the argument of restricting expansion as an adaptive advantage would not hold for roots.

The above discussion, though considering only a small segment (expansive growth) of the whole, makes it clear that once stress persists beyond the initial period of minutes or a few hours, the apparent tight coordination among the various processes in the plant makes questions on cause–effect relations moot for many functions operating near a state of dynamic equilibrium. Only when the various control loops have just been perturbed by a nearly stepwise change in one of the parameters is the causal question pertinent.

9.8.2 Hormonal Interactions in Developmental Responses to Water Stress

The preceding discussion is particularly pertinent to hormonal interactions during water stress. Contrasting and counteracting effects of different growth regulators on a given developmental process are certainly the rule rather than the exception. The analysis or application of a single growth regulator often yields results which can only be interpreted in the light of possible alterations in synthesis or activity of other hormones. In addition, environmental stress can alter these functional relationships and their developmental consequences. This makes generalization difficult, particularly since regulatory mechanisms vary widely among species. Our approach here will be to describe some cases which have been examined in detail as examples of the types of interactions that

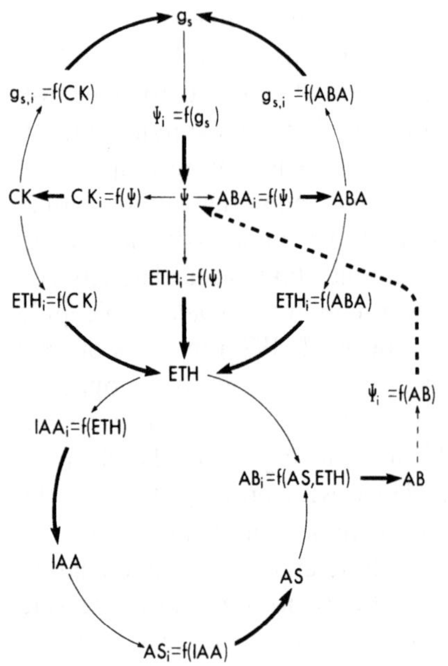

Fig. 9.5. Hormonal interactions during water stress. *ETH* ethylene synthesis rate; *ABA* abscisic acid level; *CK* cytokinin level; g_s stomatal conductance; *IAA* IAA transport rate to abscission zone; *AS* abscission zone sensitivity to ethylene; *AB* extent of abscission. The notation is consistent with that in Fig. 9.4. The *upper part* of the figure illustrates interactions between Ψ, ABA, and CK which influence the rate of ethylene synthesis, and the possible feedback loop via stomatal conductance. The *lower part* of the graph indicates the role of ethylene in regulation of leaf abscission during water stress. The *dotted lines* symbolize the more indirect nature of the feedback loop linking ethylene, abscission, and Ψ. See text for further explanation

can be expected. Integrating these into an overall scheme of plant adaptation requires some imagination, speculation, and extrapolation from the existing data, especially with reference to wild plants for which detailed information is often lacking. The schemes discussed below should be considered as working hypotheses which will undoubtedly be amended in the face of new evidence.

A tacit assumption underlying our concepts of growth regulation is that the response evoked is proportional to the endogenous hormonal concentration at a site of action. So our starting point should be how water stress affects the synthesis, compartmentation, translocation, and metabolism of the various growth regulators, which was discussed in Section 9.5. However, one hormone may dramatically influence the rate of synthesis of another, leading to complex interactions. Such a case has been described for water stress-induced ethylene synthesis in wheat leaves (WRIGHT 1977, 1979, 1980). After a lag of about 2 h, the ethylene synthesis rate increases roughly in proportion to the decrease in Ψ between -8 and -11 bar. Unfortunately, values for Ψ_p have not been reported, but the sharp increase in ABA synthesis and the early wilting symptoms which appear at about -9 bar suggest that Ψ_p is probably zero when $\Psi = -9$ bar. These relationships have been diagrammed using the notation described above (Fig. 9.5). The endogenous levels of ABA and ethylene are both functions of Ψ, rising as Ψ falls. Cytokinin levels in response to stress seem not to have been measured in this tissue, but in analogy with other reports (AHARONI et al. 1977; Sect. 9.5.4), they are also given as a function of Ψ, presumably decreasing at lower Ψ. Applied ABA inhibited ethylene synthesis moderately at physiological concentrations, and almost completely at very high concentrations. Thus, the ethylene level is represented as a function of ABA, with ethylene decreasing as ABA accumulates. Exogenous CK greatly stimulated ethylene production during water stress, and counteracted the inhibitory effect of ABA.

The increase in ABA and decline in CK would both tend to offset the stimulation of ethylene synthesis by low Ψ. The ethylene synthesis rate at any time is therefore influenced by Ψ, CK, and ABA as these three parameters fluctuate. This could not be properly called a feedback loop, as no modulation of the other inputs by ethylene is shown. In lettuce leaves, the presence or absence of ethylene did not influence ABA accumulation during water stress (AHARONI 1978), and data are not available for ethylene effects on CK levels. A possible indirect feedback effect of ethylene will be discussed below.

The control loop can be closed in another way, however, by postulating hormonal influences on stomatal conductance (g_s). While there are still many questions here (Sect. 9.6), the assumption is useful for illustrative purposes. ABA would be expected to cause stomatal closure, while CK have been found to promote opening in various graminaceous species (e.g., BENGTSON et al. 1979; JEWER and INCOLL 1980). As ABA rises and CK fall during water stress, stomatal closure would occur and g_s would be reduced. The lower g_s would limit transpiration, allowing Ψ to rise, and thereby completing the feedback loop involving ABA and CK (Fig. 9.5). Ethylene has no direct effect on stomatal behavior in species which have been tested (BRADFORD and HSIAO unpublished; PALLAGHY and RASCHKE 1972). Even this simple model demonstrates the complex interactions which can underlie an individual parameter which we have chosen to measure, such as ethylene synthesis rate. It also serves to highlight the extent of our ignorance, as this represents one of the most well-characterized cases of hormonal interactions during water stress, yet much of the scheme is still speculative, particularly regarding CK involvement.

What are the possible long-term consequences of these hormonal fluctuations with respect to adaptation to drought? Stomatal closure to reduce transpiration has been mentioned above. However, this may not be sufficient for plant survival if transpirational losses continue to occur through the cuticle, depleting soil water with little gain to assimilation. An alternative is to abscise leaves and reduce the transpirational area. The success of drought-deciduous plants in arid areas attests to the effectiveness of this strategy. The hormonal mechanisms regulating leaf abscission in cotton have been well studied, and will be used to illustrate hormonal control of a developmental response to water stress.

It is well known that leaves and young bolls of cotton often abscise following an episode of drought. Although abscission may not occur until after rewatering, the final extent of abscission was found to be linearly related to the predawn Ψ between -8 and -25 bar (MCMICHAEL et al. 1973). Ethylene production by intact cotton petioles was stimulated by the same range of Ψ (MCMICHAEL et al. 1972). The ability of the petioles to transport IAA also declines as Ψ falls below -8 bar (DAVENPORT et al. 1977, 1980), possibly due to ethylene inhibition of the transport process (BEYER and MORGAN 1971). In Fig. 9.5, this is represented as the IAA transport rate (or IAA level in the abscission zone) being a function of the ethylene synthesis rate. In cotton, ethylene treatment alone causes relatively little abscission in well-watered plants, but abscission following water stress is potentiated by the gas (JORDAN et al. 1972). The increased sensitivity of the cells of the abscission zone to ethylene is a consequence of a reduced supply of IAA from the leaf blade (JACKSON et al. 1973), which may represent an additional direct effect of water stress on IAA synthesis or transport not shown in Fig. 9.5. Abscission zone sensitivity to ethylene (AS) is therefore a function of IAA. Once the cells of the abscission zone are responsive to ethylene, the final extent of abscission (AB) will be dependent upon the ethylene synthesis rate at that time. The correspondence between minimum Ψ and extent of abscission following rewatering may reflect the speed with which the interactions governing ethylene synthesis act to reduce production of the hormone when Ψ increases. After abscission has occurred, the transpirational demand on the plant is reduced, allowing Ψ of the remaining (usually youngest) leaves to increase, representing an indirect feedback on Ψ by ethylene (Fig. 9.5.). Supporting evidence for this overall scheme comes from the work of AHARONI (1978), who found that water stress elicited quite distinct patterns of ethylene production in leaves that do, or do not abscise. A broader survey of stress-induced ethylene production in drought-deciduous or evergreen species from arid regions is needed to establish the generality of this physiological mechanism. An important aspect of developmental regulation, that the sensitivity of the target tissue to a hormone may be as important as the hormonal concentration, is well illustrated by this example. While the complete scheme presented in Fig. 9.5 is constructed from data on a monocot (wheat) and a dicot (cotton), and may not be strictly applicable to either, it does indicate the type of detailed information that is needed to understand the mechanisms of regulation of responses to water stress.

Tissue sensitivity and hormone transport may also interact to result in specific responses to water stress, such as interference in reproductive physiology. KING and EVANS (1977) found that a brief (8 h, −37 bar) water stress during the first inductive long day considerably inhibited flowering in *Lolium temulentum*. ABA increased rapidly in the leaves during water stress, then declined soon after rewatering. ABA also accumulated in the apex, but only after a lag of 9 to 22 h. As applied ABA will inhibit flower induction in this species (EVANS 1966), it is likely that ABA transported from the leaves to the apex was responsible for the stress effect. HOAD (1973) had earlier reported that in *Ricinus*, stem apices accumulated ABA during water stress only if they were attached to the plant, and not if excised prior to dehydration, and had concluded that phloem transport from mature leaves was the probable source of ABA in the stressed apices. A similar study by MORGAN (1980) implicated ABA in stress-induced male sterility of wheat. Spikelets from plants with wilted leaves ($\Psi_p=0$) contained three times the ABA found in spikelets from turgid plants, even though the spikelets themselves osmotically adjusted during stress to maintain constant turgor ($\Psi_p=6$ bar). Seed set was reduced from 55% to 4% due to the stress, and application of ABA to the flag leaf of turgid plants reproduced the effects of water stress on pollen morphology and viability. Tracer studies have confirmed that ABA can be transported from the leaves to the grain in wheat (GOLDBACH and GOLDBACH 1975; DEWDNEY and MCWHA 1978). It should be mentioned that during normal development, wheat grains accumulate high levels of ABA which is probably synthesized in situ (KING 1976; DEWDNEY and MCWHA 1978), emphasizing the importance of the sensitivity of the target organ in determining the ultimate effect of an alteration in growth regulator levels. Excessive ABA at the time of microsporogenesis or photoperiodic induction can reduce fertility, but a high level of ABA may be a normal part of seed development.

These examples have highlighted the aspects of growth regulation which need further clarification with respect to water stress: the rates of production and accumulation of growth substances, their transport, the physiological consequences, and the interactions among all these processes. The most useful approach to achieve this is to document the effects of water stress on as many of the growth regulators and physiological processes as possible for a given response. Only by viewing the overall pattern can we begin to penetrate the maze of interlocking processes which ultimately result in the characteristic developmental or adaptive adjustments to water deficit.

9.8.3 Time Course of Gross Responses and Adaptation

Having focused mainly on specific processes in considering the integrated responses, we now discuss the overall changes a plant undergoes when subjected to gradually intensifying stresses. Species differ in their responses, the extent of their adaptation to drought, and the detailed mechanisms involved. Nonetheless, general patterns may be described which are applicable to a large number of species, particularly of crop plants. Most plants have evolved with water stress as one of the major environmental constraints. The overall responses to water stress almost always lead to adjustments in water use and supply such as to aid in the completion of the life cycle and ensure the propagation of the species. The various gross changes elicited as water stress develops and intensifies are viewed below in the light of this overriding consideration.

Luxuriant vegetative growth is characteristic of plants well supplied with water, when other factors are not limiting. The first sign of water shortage in the field usually is a restriction in foliage growth (FISCHER and HAGAN 1965; JORDAN 1981), attributable to the high sensitivity of expansive growth to water stress discussed earlier. When foliar coverage of the ground is incomplete, a slowing down in leaf expansive growth results in less radiation interception by the plant and less transpiration compared to plants receiving ample water (HSIAO et al. 1976b). In the absence of neighboring strong competitors for water, the water saved in the soil would be stretched out to meet the demands of a longer ontogenetic span. The trade-off is the "permanent strain" of reduced plant size and productivity.

Recently, more quantitative data were obtained on this response and showed that a very mild water stress is sufficient to have substantial effects on canopy size if the stress lasts for some time. The phenomenon may be considered an example of positive feedback behavior or amplification with time. In these cases, a slight alteration in a process leads in time to differences much larger than just proportional to the alteration. Classically, the time course of plant growth or biomass production exhibits an exponential phase in the early stage (WILLIAMS 1975). In other words, biomass production follows first-order kinetics during that period and accumulates at a rate proportional to the existing biomass

(BLACKMAN 1919). A plausible basis for this behavior is that light absorption by the foliage is approximately proportional to the size of the canopy, and if canopy size is proportional to the plant biomass and CO_2 assimilation is proportional to the amount of light absorbed, then the rate of biomass production would be proportional to the existing biomass. Also, the rate of increase in light interception by the canopy would be proportional to the light interception by the canopy. Of course, this consideration only holds for early stages of development when most of the biomass is in the foliage and not in the nonphotosynthetic organs, and when there is little mutual shading among plants so that light interception and absorption is proportional to canopy size. The integrated equation for first-order kinetics of biomass accumulation is

$$m = m_o e^{kt}$$

where m is the biomass per plant at time t, m_o is m at time zero, e is the base of natural logarithm, and k is a proportionality constant representing biomass produced per unit of time per unit of existing biomass. An analogous equation can also be written for light interception by the canopy, with m changed to the amount of light intercepted per plant per unit of time and k changed correspondingly (see also Chap. 1, Vol. 12 A). Although this approach involves a number of simplifications, it appears to be applicable to experimental data. As shown in Fig. 9.6, the interception of photosynthetically active radiation (PAR) by bean plants in rows at midday during the early growth stage follows first-order kinetics and progresses exponentially with time. PAR interception is given on the basis of per m^2 land area, but can easily be converted to the per plant basis. Data for the three treatments all fit exponential curves with high regression coefficients. The irrigated high-density planting (HI) and conventional density planting (CI) differed in the amount of light intercepted but showed similar values (0.068 vs 0.063) for the k of Eq. (9.3), suggesting that their difference lies primarily in m_o or its equivalent for light interception. In contrast, the unirrigated high-density planting (HD) growing on only water stored in the soil exhibited a significantly lower k value (0.053). Although the lack of irrigation reduced k only about 22%, the reduction in light intercepted on day 40 is considerably larger (36%) because of the compounding effect. There was a corresponding difference in biomass of the HI and HD treatment (ARRUDA and HSIAO unpublished).

The difference in leaf Ψ at midday between HI and HD was very small and varied between 1 to 2 bar for the period shown in Fig. 9.6. Since expansive growth is so sensitive to small reductions in Ψ, this difference in Ψ probably accounted for the difference in the k's of the two treatments. There was no detectable difference in epidermal resistance of the leaves. Because of the exponential dependence of m or its equivalent on k and t, a small difference in k is amplified with time. In this case, the compounding was over a period of only about 20 days. Even then the amplification effect on the difference between HI and HD at the end of the period is quite evident. Equation (9.3) shows that for longer time intervals, the disparity would be even more marked. For example, a time lapse of 40 days would result in a 45% reduction in

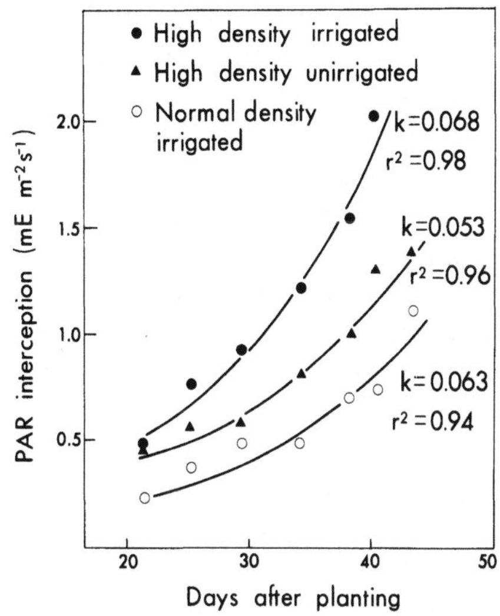

Fig. 9.6. Interception of photosynthetically active radiation (PAR) by a bean canopy in the field at noon as influenced by planting density and irrigation. k is a proportionality constant similar to k in Eq. (9.3) but for PAR interception, and r is the regression coefficient for exponential functions of PAR interception versus time after planting. High density planting was at 38 plants m^{-2}, and normal density at 19 plants m^{-2}. Irrigated treatments received water about weekly by sprinkler to replenish the evapo-transpiration loss. Unirrigated treatment grew solely on the water stored in the soil, wetted through the profile at planting. Leaf Ψ at midday were −7.7, −7.8, and −8.8 bar, respectively, for the irrigated normal density, irrigated high density, and unirrigated normal density treatments on day 27; −5.9, −5.9, and −8.4 bar, respectively, on day 31; and −6.7, −7.2, and −7.4 bar, respectively, on day 43. Leaf diffusion resistances ranged from 1.6 to 2.0 s cm^{-1} during this period and were not different between treatments. (Unpublished data of F. ARRUDA and T.C. HSIAO)

light interception for the 22% difference in k. We suggest that the often-observed pronounced difference in plant size between stressed and control plants when there is little or no shading by neighbors is due to this phenomenon. Such observations had often been puzzling in cases when measured Ψ differences were small. The large difference in plant sizes had seemed to be out of proportion with the small difference in Ψ. Further, we also suggest that the concept, though based on data on crops, is applicable to plants in their native habitats as long as the community population is sparse and light interception by foliage is partial.

While the slowed canopy development restricts the transpiration surface area and conserves water in the soil, other responses in the plant may be initiated which tend to increase the water supply to the plant. Increases in roots relative to shoots have often been observed when water is limiting (EL NADI et al. 1969; PEARSON 1966). While the change may be attributed mostly to reduction in shoot growth, there are instances when water stress effected an increase in the absolute root biomass (HSIAO and ACEVEDO 1974; SHARP and DAVIES 1975). More significant in terms of adaptation is the fact that root growth under stress is confined to regions of relatively moist soil, such as the deeper layers (HSIAO and ACEVEDO 1974; KLEPPER et al. 1973; ELLIS et al. 1977). Thus, the beneficial effect on water supply is more than just proportional to the change in the total root system.

There is uncertainty regarding the stress level necessary to cause the shift in root–shoot balance. In a short-term pot study of maize, SHARP and DAVIES (1975) found the change in root–shoot ratio to start in the same two-day period as when root and leaf Ψ began to decline and leaf elongation became notably suppressed by water stress. It is possible that root growth is favored over shoot growth mostly by stress levels not completely inhibitory to photosynthesis. Such stresses, by restricting leaf expansion, may leave more assimilates available for root growth. When stress is severe enough to suppress photosynthesis substantially, root growth would also be reduced.

The mechanism accounting for the favored root growth under water stress is not known. The situation is made complex by the fact that the mechanical strength of most soils increases dramatically as the soil dries, offering greater and greater impedance to root elongation. To overcome this impedance as well as the lower water potential, the roots osmotically adjust to continue growth in the drying soil (GREACEN and OH 1972). The adjustment is more pronounced in the growing region of the root tip than in the mature basal region of the root (SHARP and DAVIES 1979). This parallels the ability of the elongating regions of leaves to osmotically adjust more effectively than the mature leaves (BARLOW et al. 1980; MICHELENA and BOYER 1980), but why root growth continues while leaf growth is inhibited remains puzzling. The hypothesized role of ABA in modifying root–shoot balance under water stress has been discussed in Section 9.5.5. As was pointed out, some data are inconsistent with the hypothesis and more studies are needed.

Osmotic adjustment throughout the plant in response to water stress constitutes another adaptive strategy to enhance available water in many soils. By building up osmotica in the cells, turgor and physiological functions can be maintained at low tissue Ψ. The ability to withstand low tissue Ψ permits the establishment of sufficient Ψ gradient within the plant to continue water absorption from soils of low Ψ values. How much additional water can be extracted from the soil is dependent on the degree of osmotic adjustment as well as on the moisture release characteristics of the soil. If the soil is coarse-textured and releases little water in the Ψ range between the Ψ values of the unadjusted and osmotically adjusted roots, then little additional water can be gained by the plant through the adjustment. If the soil is fine-textured and releases significant amounts of water in that Ψ range, the plant would increase its water supply considerably by osmotic adjustment. The extent of osmotic adjustment is highly variable among species, and is also affected by the speed of development of water stress (see Chap. 2, Vol. 12 C).

If the saving in water achieved by restricting canopy development and the maintenance or even enhancement of water supply through root development and osmotic adjustment are insufficient to prevent leaf Ψ from dropping substantially with time, sooner or later leaf Ψ will reach the threshold for stomatal closure. Initially, stomatal closure will be confined to the midday periods of peak irradiance and evaporative demand. As stress becomes more severe, however, stomata may open only briefly in the morning and close for much of the remaining part of the day (MACDOWALL 1963; SAUGIER 1976). In as much as they close and open in a matter of minutes, stomata constitute a dynamic means of controlling water loss and CO_2 assimilation. The direct response of stomata to humidity and postulated optimization of carbon assimilation and water use are discussed in Chapters 7, 17, and 18 of this Volume.

Loss of turgor and wilting often coincide with the point when leaf Ψ reaches the threshold value for stomatal closure. Wilting reflects the loss of mechanical strength of the tissue. When wilted, the lamina of many dicots droops and its petiole bends downward, resulting in a smaller projected area on the horizontal plane for the leaf (TURNER 1974). If the leaf area index (LAI) is not very high, a droopy canopy would intercept less radiation than a turgid one when the sun is overhead, leading to a smaller radiation energy load and less evaporative demand on the crop at midday (LUXMOORE et al. 1971; RAWSON 1979). Thus, transpiration is reduced beyond that achieved solely by stomatal closure (RAWSON 1979).

Stress-induced changes in leaf shape may involve specialized anatomical features. Leaf rolling in some grass species under water stress is thought to be due to volume changes in the bulliform or motor cells distributed longitudinally in rows along the adaxial side of the leaf. When severely stressed, the leaves can be tightly rolled into a needle shape and held nearly vertical. Rolling appears to start at about the same leaf Ψ as stomatal closure (BEGG 1980; JORDAN 1981; O'TOOLE and CRUZ 1980) and contributes to reducing transpiration independently of stomatal closure (O'TOOLE et al. 1979). Radiation load on the crop decreases due to the more erect leaves and smaller light interception surface (JORDAN 1982), and the transpiring surface area is reduced due to the overlapping lamina in the leaf cylinder.

If stress sufficiently severe to cause wilting and stomatal closure persists or intensifies, senescence of older leaves usually becomes evident in a matter of days (BOYER and MCPHERSON 1975; FISCHER 1973). Senescence generally starts with the oldest and the most shaded leaves and advances sequentially to the youngest as stress progresses. In most dicots, senescence culminates in abscission of the leaf. Shedding of leaves is well known as a means of adaptation to drought in native communities (KOZLOWSKI 1976). Nutrients such as nitrogen, phosphorus, and potassium in the senescing leaves are mobilized for use in the remaining parts of the plant (THOMAS and STODDART 1980). Senescing leaves have lower stomatal conductance (FRIEDRICH and HUFFAKER 1980; WITTENBACH et al. 1980) and their stomata are more sensitive to water stress (JORDAN et al. 1975). The more important effect in the long run, however, is that the green leaf area is reduced, accounting for much of the diminution in transpiration (FISCHER and KOHN 1966).

The stress level which triggers leaf senescence is probably dependent on nitrogen nutrition and the stage of development, though definitive data are lacking. It is well known that nitrogen deficiency alone induces early senescence of older leaves, and water stress appeared to accelerate that further in one field study (BATAGLIA 1980). Senescence may also be more readily triggered if water stress occurs in the late stage of plant development, when the demands for nutrients by reproductive organs are strong (Jordan 1982) and may be met by remobilization from the older leaves (Sinclair and de Wit 1975).

When leaves are wilted and their Ψ already very low, a sudden increase in evaporative demand due to weather changes in some cases can cause death of leaves by dehydration. Leaves exposed to the sun and wind are most susceptible, regardless of age. In contrast to the well-programmed senescence process, there is no recycling of the leaf constituents in such death. Nonetheless, the transpiring surface area would be reduced and plant water balance is improved.

The success of a species in a stressful habitat is determined by its reproduction and propagation. The reproductive processes of plants are highly responsive to water stress throughout the sequence of stress development. The responses, however, appear to be mostly indirect and mediated by fluctuations in the supply of assimilates (Fischer 1980; but see Sect. 9.8.2). Fruit or grain yields under water stress are highly correlated with the size of the assimilate-manufacturing plant (Fischer 1979; Jordan 1982). At floral initiation, the number of florets to form is partly determined by assimilate source size. If stressed during initiation, growth of florets may be slowed by a lack of turgor (Barlow et al. 1980) and some florets may not develop fully, probably because of a lack of assimilates (Fischer 1973; Chinnici and Peterson 1979). Severe stress at or near anthesis can prevent fertilization (Jordan 1982; Hsiao unpublished), but this is not a common problem in the field. After anthesis, the number of seeds can be reduced by water stress-induced abortion, which is probably a consequence of insufficient assimilate supply (Hsiao unpublished). Reproductive sink size is constantly adjusted during stress to result in a reasonable balance between plant size and fruit load and ensure that at least some seeds will mature, though their number may be drastically reduced.

All the responses described above are costly to the plant in terms of its growth and productivity and can be viewed as permanent strains. Some are more costly than others. For example, leaf senescence represents a direct loss of carbon, some mineral nutrients, and photosynthetic capacity, while stomatal closure and wilting represent only a temporary slowing down in photosynthesis. On the other hand, senescence is less costly than leaf death by desiccation in that some of the nutrients are saved by translocation during senescence. The shift in the partition of carbon in favor of the roots over shoots results in a direct cost as well as a more subtle compounding depressive effective on shoot growth if the canopy is incomplete. Quantification of these costs as well as the adaptive benefits has only begun and remains as a major task ahead in stress physiology.

Some of the responses are quickly reversible and many are essentially irreversible. The most straightforward assumption would be that the readily reversible responses constitute the first "line of defense" when stresses are mild, with the irreversible responses coming into play only when the situation becomes desperate. It is therefore surprising that a nearly irreversible restriction of canopy size is one of the most sensitive responses to water stress for many species. This behavior must reflect the dominant combinations of rainfall distribution and soil water storage characteristics during evolution. Of course, species differ in their adaptive strategies to drought. The individual responses discussed above

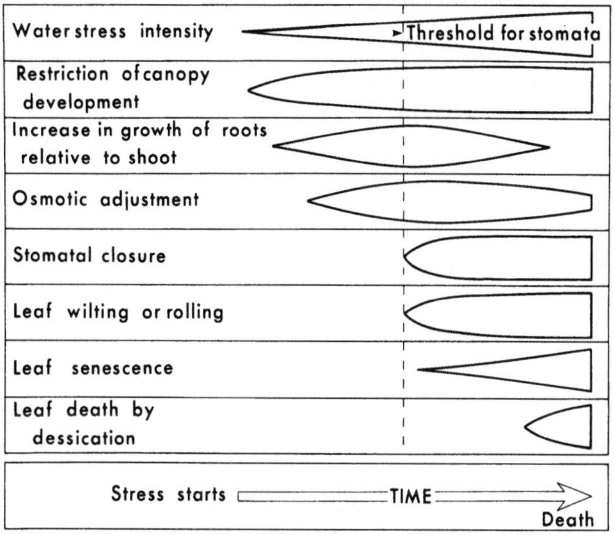

Fig. 9.7. Generalized time course of gross and adaptive changes in crop plants in response to the gradual development of water stress in the field. Width of the band at a given time represents the relative magnitude of the given response. The starting position of each band on the time scale is also indicative of the threshold water stress level for eliciting the response. Band shape reflects variations of the response with increasing stress intensity and duration. For example, preferential growth of roots relative to shoot is depicted to be nearly maximum near the threshold for stomatal closure, but declines as the soil dries out further and soil mechanical impedance rises

may be used to varying extent or not at all by a given species. For brevity, several well-known adaptive responses, particularly the photonastic movement and changes in pubescence of leaves, and the enhanced expression of CAM, have been left out. More complete treatments of divergent adaptive strategies are found in Chapters 10, 11, 15, and 18 of this Volume.

9.9 Conclusion

In conclusion, the time course of adaptive changes as water stress develops and intensifies is summarized as a diagram (Fig. 9.7). It is obvious that the generalization is based on incomplete information and often only qualitative observations. The details in the diagram will improve as more information develops. The diagram makes clear the sequence of events and emphasizes the fact that several important adaptive changes occur well ahead of stomatal closure. Stomatal closure triggered by reductions in bulk leaf water status is one of the last "lines of defense". In the last score of years, stomatal control of transpiration and associated changes in photosynthesis have been heavily emphasized when considering productivity and adaptation to drought. The other equally important responses, especially those occurring ahead of stomatal closure, have received general recognition in the physiological literature only in the last few years (e.g., Begg 1980; Turner 1979), although some authors have called attention to them much earlier (e.g., Fischer and Hagan 1965; Hsiao and Acevedo 1974; Hsiao et al. 1976b). A more balanced approach in water stress physiology in the years to come should lead to a better understanding of the system as a whole.

Acknowledgments. The authors express their appreciation for the assistance of Ms. Wendy Hall and Ms. Val Rawlings in preparing the manuscript. Partial funding was provided by National Science Foundation Grant No. PCM 78-2371 to Wendy K. Silk and T.C. Hsiao.

References

Acevedo E, Hsiao TC, Henderson DW (1971) Immediate and subsequent growth responses of maize leaves to changes in water status. Plant Physiol 48:631–636

Acevedo E, Fereres E, Hsiao TC, Henderson DW (1979) Diurnal growth trends, water potential, and osmotic adjustment of maize and sorghum leaves in the field. Plant Physiol 64:476–480

Ackerson RC (1980) Stomatal response of cotton to water stress and abscisic acid as affected by water stress history. Plant Physiol 65:455–459

Ackerson RC, Hebert RR (1981) Osmoregulation in cotton in response to water stress. I. Alterations in photosynthesis, leaf conductance, translocation, and ultrastructure. Plant Physiol 67:484–488

Adams DO, Yang SF (1979) Ethylene biosynthesis: Identification of l-aminocyclopropane-l-carboxylic acid as an intermediate in the conversion of methionine to ethylene. Proc Natl Acad Sci USA 76:170–174

Adams PA, Montague MJ, Tepfer M, Rayle DL, Kuma HI, Kaufman PB (1975) Effect of gibberellic acid on the plasticity and elasticity of *Avena* stem segments. Plant Physiol 56:757–760

Aharoni N (1978) Relationship between leaf water status and endogenous ethylene in detached leaves. Plant Physiol 61:658–662

Aharoni N, Richmond AE (1978) Endogenous gibberellin and abscisic acid content as related to senescence of detached lettuce leaves. Plant Physiol 62:224–228

Aharoni N, Blumenfeld A, Richmond AE (1977) Hormonal activity in detached lettuce leaves as affected by leaf water content. Plant Physiol 59:1169–1173

Apelbaum A, Yang SF (1981) Biosynthesis of stress ethylene induced by water deficit. Plant Physiol 68:594–596

Arad S, Richmond AE (1976) Leaf cell water and enzyme activity. Plant Physiol 57:656–658

Arad S, Mizrahi, Richmond AE (1973) Leaf water content and hormone effects on ribonuclease activity. Plant Physiol 52:510–512

Aspinall D (1980) Role of abscisic acid and other hormones in adaptation to water stress. In: Turner NC, Kramer PJ (eds) Adaptation of plants to water and high temperature stress. Wiley and Sons, New York, pp 155–172

Baker DB, Ray PM (1965) Relation between effects of auxin on cell wall synthesis and cell elongation. Plant Physiol 40:360–368

Barlow EWR, Munns R, Scott NS, Reisner AH (1977) Water potential, growth and polyribosome content of the stressed wheat apex. J Exp Bot 28:909–916

Barlow EWR, Munns RE, Brady CJ (1980) Drought responses of apical meristems. In: Turner NC, Kramer PJ (eds) Adaptation of plants to water and high temperature stress. Wiley and Sons, New York, pp 191–205

Barnett NM, Naylor AW (1966) Amino acid and protein metabolism in Bermuda grass during water stress. Plant Physiol 41:1222–1230

Barr ML (1973) Biosynthesis of abscisic acid: Incorporation of radioactivity from ^{14}C mevalonic acid by intact roots. Plant Physiol Suppl 51:47

Bates GW, Cleland RE (1979) Protein synthesis and auxin-induced growth: Inhibitor studies. Planta 145:437–442

Bates LM, Hall AE (1981) Stomatal closure with soil water depletion not associated with changes in bulk leaf water status. Oecologia 50:62–65

Bayley ST, Setterfield G (1957) The influence of mannitol and auxin on growth of cell walls in *Avena* coleoptiles. Ann Bot 21:633–641

Bataglia OC (1980) Effects of nitrogen-water-relations on maize productivity. Ph D Dissertation, Univ California, Davis

Beadle CL, Jarvis PG (1977) Effects of shoot water status on some photosynthetic partial processes in Sitka spruce. Physiol Plant 41:7–13

Beadle CL, Stevenson KR, Neumann HH, Thurtell GW, King KM (1973) Diffusive resistance, transpiration, and photosynthesis in single leaves of corn and sorghum in relation to leaf water potential. Can J Plant Sci 53:537–544

Beardsell MF, Cohen D (1975) Relationships between leaf water status, abscisic acid levels, and stomatal resistance in maize and sorghum. Plant Physiol 56:207–212

Begg JE (1980) Morphological adaptations of leaves to water stress. In: Turner NC, Kramer PJ (eds) Adaptation of plants to water and high temperature stress. Wiley and Sons, New York, pp 33–42

Begg JE, Turner NC (1976) Crop water deficits. Adv Agron 28:161–217

Bell DT, Koeppe DE, Miller RJ (1971) The effects of drought stress on respiration of isolated corn mitochondria. Plant Physiol 48:413–415

Bellandi PM, Dörffling K (1974) Transport of abscisic acid-2-^{14}C in intact pea seedlings. Physiol Plant 32:365–368

Bengston C, Falk SD, Larsson S (1979) Effects of kinetin on transpiration rate and abscisic acid content of water stressed young wheat plants. Physiol Plant 45:183–188

Beyer EM, Morgan PW (1971) Abscission: The role of ethylene modification of auxin transport. Plant Physiol 48:208–212

Bisson MA, Gutknecht J (1980) Osmotic regulation in algae. In: Spanswick RM, Lucas WJ, Dainty J (eds) Plant membrane transport: Current conceptual issues. Elsevier/North Holland, Amsterdam, pp 131–146

Blackman VH (1919) The compound interest law and plant growth. Ann Bot 33:353–360

Boggess SF, Stewart CR, Aspinall D, Paleg LG (1976) Effect of water stress on proline synthesis from radioactive precursors. Plant Physiol 58:398–401

Boussiba S, Richmond AE (1976) Abscisic acid and the aftereffect of stress in tobacco plants. Planta 129:217–219

Boussiba S, Rikin A, Richmond AE (1975) The role of abscisic acid in cross-adaptation of tobacco plants. Plant Physiol 56:337–339

Boyer JS (1968) Relationship of water potential to growth of leaves. Plant Physiol 43:1056–1062

Boyer JS (1971) Nonstomatal inhibition of photosynthesis in sunflower at low leaf water potentials and high light intensities. Plant Physiol 48:532–536

Boyer JS (1976) Water deficits and photosynthesis. In: Kozlowski TT (ed) Water deficits and plant growth Vol IV. Soil water measurement, plant responses, and breeding for drought resistance. Academic Press, London New York, pp 153–190

Boyer JS, Bowen BL (1970) Inhibition of oxygen evolution in chloroplasts isolated from leaves with low water potentials. Plant Physiol 45:612–615

Boyer JS, McPherson HG (1975) Physiology of water deficits in cereal crops. Adv Agron 27:1–23

Boyer JS, Wu G (1978) Auxin increases the hydraulic conductivity of auxin-sensitive hypocotyl tissue. Planta 139:227–237

Bradford KJ, Yang SF (1980) Xylem transport of l-aminocyclopropane-l-carboxylic acid, an ethylene precursor, in waterlogged tomato plants. Plant Physiol 65:322–326

Bradford KJ, Yang SF (1981) Physiological responses of plants to waterlogging. Hort Science 16:25–30

Brady CJ, Scott NS, Munns R (1974) The interaction of water stress with the senescence

pattern of leaves. In: Bieleski RL, Ferguson AR, Creswell MM (eds) Roy Soc N Z, Bull, 12:403–409
Bunce JA (1977) Leaf elongation in relation to leaf water potential in soybean. J Exp Bot 28:156–161
Caemmerer von S (1981) On the relationship between chloroplast biochemistry and gas exchange of leaves. Ph D Dissertation, Aust Natl Univ Canberra
Campbell LC, Raison JK, Brady CJ (1976) Response of plant mitochondria to media of high solute content. J Bioenerg 8:121–129
Catsky J (1974) Water content. In: Slavík B (ed) Methods of studying plant water relations. Ecol Stud Vol IX. Springer, Berlin Heidelberg New York, pp 121–156
Chinnici MF, Peterson DM (1979) Temperature and drought effects on blast and other characteristics in developing oats. Crop Sci 19:893–897
Cleland RE (1967a) A dual role of turgor pressure in auxin-induced cell elongation in *Avena* coleoptiles. Planta 77:182–191
Cleland RE (1967b) Extensibility of isolated cell walls: Measurement and changes during cell elongation. Planta 74:197–209
Cleland RE (1971) Instability of growth-limiting proteins of the *Avena* coleoptile and their pool size in relation to auxin. Planta 99:1–11
Cleland RE (1975) Auxin-induced hydrogen ion excretion: correlation with growth, and control by external pH and water stress. Planta 127:233–242
Cleland RE, Rayle DL (1978) Auxin, H^+-excretion and cell elongation. Bot Mag Tokyo Spec Issue 1:125–139
Cole PJ, Alston AM (1974) Effect of transient dehydration on absorption of chloride by wheat roots. Plant Soil 40:243–247
Collatz GJ (1977) Influence of certain environmental factors on photosynthesis and photorespiration in *Simmondsia chinensis*. Planta 134:127–132
Collatz J, Ferrar PJ, Slatyer RO (1976) Effects of water stress and differential hardening treatments on photosynthetic characteristics of a xeromorphic shrub, *Eucalyptus socialis* F. Muell. Oecologia 23:95–105
Collins JC, Kerrigan AP (1974) The effect of kinetin and abscisic acid on water and ion transport in isolated maize roots. New Phytol 73:309–314
Cowan IR, Raven JA, Hartung W, Farquhar GD (1982) A possible role for abscisic acid in coupling stomatal conductance and photosynthetic carbon metabolism in leaves. Aust J Plant Physiol in press
Craker LE, Cookson C, Osborne DJ (1978) Control of proton extrusion and cell elongation by ethylene and auxin in the water-plant *Ranunculus sceleratus*. Plant Sci Lett 12:379–385
Cram WJ (1980) Chloride accumulation as a homeostatic system: Negative feedback signals for concentration and turgor maintenance differ in a glycophyte and a halophyte. Aust J Plant Physiol 7:237–249
Cram WJ, Pitman MA (1972) The action of abscisic acid on ion uptake and water flow in plant roots. Aust J Biol Sci 25:1125–1132
Cutler JM, Rains DW, Loomis RS (1977) The importance of cell size in the water relations of plants. Physiol Plant 40:255–260
Cutler JM, Shahan KW, Steponkus PL (1980a) Influence of water deficits and osmotic adjustment on leaf elongation in rice. Crop Sci 20:314–318
Cutler JM, Steponkus PL, Wach MJ, Shahan KW (1980b) Dynamic aspects and enhancement of leaf elongation in rice. Plant Physiol 66:147–152
Darbyshire B (1971) Changes in indoleacetic acid oxidase activity associated with plant water potential. Physiol Plant 25:80–84
Davenport TL, Morgan PW, Jordan WR (1977) Auxin transport as related to leaf abscission during water stress in cotton. Plant Physiol 59:554–557
Davenport TL, Morgan PW, Jordan WR (1980) Reduction of auxin transport capacity with age and internal water deficits in cotton petioles. Plant Physiol 65:1023–1025
Davies FS, Lakso AN (1978) Water relations in apple seedlings: changes in water potential components, abscisic acid levels and stomatal conductances under irrigated and non-irrigated conditions. J Am Soc Hortic Sci 103:310–313

Davies WJ, Mansfield TA, Wellburn RR (1980) A role for abscisic acid in drought endurance and drought avoidance. In: Skoog F (ed) Plant growth substances 1979. Springer, Berlin Heidelberg New York, pp 242–253
Davies WJ, Wilson JA, Sharp RE, Osonubi O (1981) Control of stomatal behaviour in water-stressed plants. In: Jarvis PG, Mansfield TA (eds) Stomatal physiology. Univ Press, Cambridge, pp 163–185
Dewdney SJ, McWha JA (1978) The metabolism and transport of abscisic acid during grain fill in wheat. J Exp Bot 29:1299–1308
Dhindsa RS, Cleland RE (1975a) Water stress and protein synthesis. I. Differential inhibition of protein synthesis. Plant Physiol 55:778–781
Dhindsa RS, Cleland RE (1975b) Water stress and protein synthesis. II. Interaction between water stress, hydrostatic pressure, and abscisic acid on the pattern of protein synthesis in *Avena* coleoptiles. Plant Physiol 55:782–785
Dhindsa RS, Baesley CA, Ting IP (1975) Osmoregulation in cotton fiber: Accumulation of potassium and malate during growth. Plant Physiol 56:394–398
Ehret DL, Boyer JS (1979) Potassium loss from stomatal guard cells at low water potential. J Exp Bot 30:225–234
El-Beltagy AS, Hall MA (1974) Effect of water stress upon endogenous ethylene levels in *Vicia faba*. New Phytol 73:47–60
Ellis FB, Elliot JG, Barnes BT, Howse KR (1977) Comparison of direct drilling, reduced cultivation and ploughing on the growth of cereals. II. Spring barley on a sandy loam soil: soil physical conditions and root growth. J Agric Sci 89:631–642
El Nadi AH, Brouwer R, Locher J Th (1969) Some responses of the root and the shoot of *Vicia faba* plants to water stress. Neth J Agric Sci 17:133–142
Evans LT (1966) Abscisin II: Inhibitory effect on flower induction in a long-day plant. Science 151:107–108
Farquhar GD, Sharkey TD (1982) Stomatal conductance and photosynthesis. Annu Rev Plant Physiol 33:317–345
Fereres E, Acevedo E, Henderson DW, Hsiao TC (1978) Seasonal changes in water potential and turgor maintenance in sorghum and maize under water stress. Physiol Plant 44:261–267
Fischer RA (1971) Role of potassium in stomatal opening in the leaf of *Vicia faba*. Plant Physiol 47:555–558
Fischer RA (1973) The effect of water stress at various stages of development on yield processes in wheat. In: Slatyer RO (ed) Plant response to climatic factors. Proc Uppsala Symp UNESCO, Paris
Fischer RA (1979) Growth and water limitation to dryland wheat yield in Australia: a physiological framework. J Aust Inst Agric Sci 45:83–93
Fischer RA (1980) Influence of water stress on crop yield in semiarid regions. In: Turner NC, Kramer PJ (eds) Adaptation of plants to water and high temperature stress. Wiley and Sons, New York, pp 323–340
Fischer RA, Hagan RM (1965) Plant water relations, irrigation management and crop yield. Exp Agric 1:161–177
Fischer RA, Kohn GD (1966) The relationship between evapotranspiration and growth in the wheat crop. Aust J Agric Res 17:255–267
Flowers TJ, Hanson JB (1969) The effect of reduced water potential on soybean mitochondria. Plant Physiol 44:939–945
Flowers TJ, Troke PF, Yeo AR (1977) The mechanism of salt tolerance in halophytes. Annu Rev Plant Physiol 28:89–121
Friedrich JW, Huffaker RC (1980) Photosynthesis, leaf resistance and ribulose-1,5-bisphosphate carboxylase degradation in senescing barley leaves. Plant Physiol 65:1103–1107
Frota JNE, Tucker TC (1978a) Absorption rates of ammonium and nitrate by red kidney beans under salt and water stress. Soil Sci Soc Am J 42:753–756
Frota JNE, Tucker TC (1978b) Salt and water stress influences nitrogen metabolism in red kidney beans. Soil Sci Soc Am J 42:743–746
Fry KE (1970) Some factors affecting the Hill reaction activity in cotton chloroplasts. Plant Physiol 45:465–469

Fry KE (1972) Inhibition of ferricyanide reduction in chloroplasts prepared from water-stressed cotton leaves. Crop Sci 12:698–701
Gimmler H, Heilmann B, Demmig B, Hartung W (1981) The permeability of the plasmalemma and the chloroplast envelope of spinach mesophyll cells for phytohormones. Z Naturforsch 36C:672–678
Glinka Z (1977) Effects of abscisic acid and of hydrostatic pressure gradient on water movement through excised sunflower roots. Plant Physiol 59:933–935
Glinka Z (1980) Abscisic acid promotes both volume flow and ion release to the xylem in sunflower roots. Plant Physiol 65:537–540
Goldbach H, Goldbach E (1975) Abscisic acid translocation and influence of water stress on grain abscisic acid content. J Exp Bot 28:1342–1350
Göring H, Bleiss W, Kretschmer H (1978) Stimulated elongation growth of coleoptile segments as a consequence of activated H^+ secretion after temporary turgor reduction. Biochem Physiol Pflanz 173:373–376
Govindjee, Downton WJS, Fork DC, Armond PA (1981) Chlorophyll *a* fluorescence transient as an indicator of water potential of leaves. Plant Sci Lett 20:191–194
Graziani Y, Livne A (1971) Dehydration, water fluxes, and permeability of tobacco leaf tissue. Plant Physiol 48:575–579
Greacen EL, Oh JS (1972) Physics of root growth. Nature (London) New Biol 235:24–25
Green PB (1968) Growth physics in *Nitella*: a method for continuous in vivo analysis of extensibility based on a micro-manometer technique for turgor pressure. Plant Physiol 43:1169–1184
Green PB, Erickson RO, Buggy J (1971) Metabolic and physical control of cell elongation rate – in vivo studies in *Nitella*. Plant Physiol 47:423–430
Green PB, Bauer K, Cummins WR (1977) Biophysical model for plant cell growth: auxin effects. In: Jungreis AM, Hodges TK, Kleinzeller A, Schultz SG (eds) Water relations in membrane transport in plants and animals. Academic Press, London New York, pp 30–45
Greenway H, Klepper B (1968) Phosphorus transport to the xylem and its regulation by water flow. Planta 83:119–136
Greenway H, Leahy M (1970) Effects of rapidly and slowly permeating osmotica on metabolism. Plant Physiol 46:259–262
Grenetz PS, List A JR (1973) A model for predicting growth response of plants to changes in external water potential: *Zea mays* primary root. J Theor Biol 39:29–45
Hall MA, Kapuya JA, Sivakumaran S, John A (1977) The role of ethylene in the response of plants to stress. Pestic Sci 8:217–223
Hanson AD, Hitz WD (1982) Water deficit and the nitrogen economy. In: Taylor H, Jordan W, Sinclair T (eds) Efficient water use in crop production. Am Soc Agron, Madison, Wisconsin, in press
Hanson AD, Nelsen CE (1978) Betaine accumulation and [^{14}C] formate metabolism in water-stressed barley leaves. Plant Physiol 62:305–312
Hanson AD, Nelsen CE, Everson EH (1977) Evaluation of free proline accumulation as an index of drought resistance using two contrasting barley cultivars. Crop Sci 17:720–726
Hanson AD, Nelsen CE, Pedersen AR, Everson EH (1979) Capacity for proline accumulation during water stress in barley and its implications for breeding for drought resistance. Crop Sci 19:489–493
Harrison MA, Walton DC (1975) Abscisic acid metabolism in water stressed bean leaves. Plant Physiol 56:250–256
Hartung W (1976) Effect of water stress on transport of [2-^{14}C] abscisic acid in intact plants of *Phaseolus coccineus* L. Oecologia 26:177–183
Hartung W, Witt J (1968) Über den Einfluß der Bodenfeuchtigkeit auf den Wuchsstoffgehalt von *Anastatica hierochuntica* und *Helianthus annuus*. Flora (Jena) Abt B 157:603–614
Hartung W, Heilmann B, Gimmler H (1981) Do chloroplasts play a role in abscisic acid synthesis? Plant Sci Lett 22:235–242
Haschke H-P, Lüttge U (1975a) Interaction between IAA, potassium and malate accumulation, and growth in *Avena* coleoptile segments. Z Pflanzenphysiol 76:450–455

Haschke H-P, Lüttge U (1975b) Stoichiometric correlation of malate accumulation with auxin-dependent K^+-H^+ exchange and growth in *Avena* coleoptile segments. Plant Physiol 56:696–698

Haschke H-P, Lüttge U (1977) Action of auxin on CO_2 dark fixation in *Avena* coleoptile segments as related to elongation growth. Plant Sci Lett 8:53–58

Heilmann B, Hartung W, Gimmler H (1980) The distribution of abscisic acid between chloroplasts and cytoplasm of leaf cells and the permeability of the chloroplast envelope for abscisic acid. Z Pflanzenphysiol 97:67–78

Henson IE (1981) Changes in abscisic acid content during stomatal closure in pearl millet (*Pennisetum americanum* (L.) Leeke). Plant Sci Lett 21:121–127

Heuer B, Plaut Z, Federman E (1979) Nitrate and nitrite reduction in wheat leaves as affected by different types of water stress. Physiol Plant 46:318–323

Hewitt EB (1975) Assimilatory nitrate – nitrite reduction. Annu Rev Plant Physiol 26:73–100

Hiron RW, Wright STC (1973) The role of endogenous abscisic acid in the response of plants to stress. J Exp Bot 24:769–781

Hoad GV (1973) Effect of moisture stress on abscisic acid levels in *Ricinus communis* L. with particular reference to phloem exudate. Planta 113:367–372

Hoad GV (1975) Effect of osmotic stress on abscisic acid levels in xylem sap of sunflower. Planta 124:25–29

Hoad GV (1978) Effect of water stress on abscisic acid levels in white lupin (*Lupinus albus* L.) fruit, leaves and phloem exudate. Planta 142:287–290

Hsiao TC (1970) Rapid changes in levels of polyribosomes in *Zea mays* in response to water stress. Plant Physiol 46:281–285

Hsiao TC (1973a) Plant responses to water stress. Annu Rev Plant Physiol 24:519–570

Hsiao TC (1973b) Effects of water deficit on guard cell potassium and stomatal movement. Plant Physiol Suppl 51:9

Hsiao TC (1976) Stomatal ion transport. In: Lüttge U, Pitman MG (eds) Encyclopedia of plant physiology new ser vol II. Transport in plants II, Part B, tissues and organs. Springer, Berlin Heidelberg New York, pp 193–221

Hsiao TC, Acevedo E (1974) Plant responses to water deficits, water use efficiency and drought resistance. Agric Meteorol 14:59–84

Hsiao TC, Acevedo E, Fereres E, Henderson DW (1976a) Stress metabolism: water stress, growth, and osmotic adjustment. Philos Trans R Soc London Ser B 273:479–500

Hsiao TC, Fereres E, Acevedo E, Henderson DW (1976b) Water stress and dynamics of growth and yield of crop plants. In: Lange, OL, Kappen L, Schulze E-D (eds) Water and plant life: problems and modern approaches. Ecol Stud Vol 19. Springer, Berlin Heidelberg New York, pp 281–305

Hsiao TC, Bradford KJ (1982) Physiological consequences of cellular water deficits. In: Taylor HM, Jordan WR, Sinclair TR (eds) Efficient water use in crop production. Am Soc Agron, Madison, Wisconsin, in press

Huang C-Y, Boyer JS, Vanderhoef LN (1975) Limitation of acetylene reduction (nitrogen fixation) by photosynthesis in soybean having low water potentials. Plant Physiol 56:228–232

Hüsken D, Steudle E, Zimmermann U (1978) Pressure probe technique for measuring water relations of cells in higher plants. Plant Physiol 61:158–163

Huffaker RC, Radin T, Kleinkopf GE, Cox EL (1970) Effects of mild water stress on enzymes of nitrate assimilation and of the carboxylative phase of photosynthesis in barley. Crop Sci 10:471–474

Itai C, Benzioni A (1976) Water stress and hormonal response. In: Lange OL, Kappen L, Schulze E-D (eds) Water and plant life: problems and modern approaches. Ecol Stud Vol 19. Springer, Berlin Heidelberg New York, pp 176–188

Itai C, Vaadia Y (1965) Kinetin-like activity in root exudate of water-stressed sunflower plants. Physiol Plant 18:941–944

Itai C, Vaadia Y (1971) Cytokinin activity in water-stressed shoots. Plant Physiol 47:87–90

Itai C, Richmond A, Vaadia Y (1968) The role of root cytokinins during water and salinity stress. Isr J Bot 17:187–195

Jackson MB, Hartley CB, Osborne DJ (1973) Timing abscission in *Phaseolus vulgaris*

L. by controlling ethylene production and sensitivity to ethylene. New Phytol 72:1251–1260
Jarvis PG (1980) Stomatal response to water stress in conifers. In: Turner NC, Kramer PJ (eds) Adaptation of plants to water and high temperature stress. Wiley and Sons, New York, pp 105–122
Jewer PC, Incoll LD (1980) Promotion of stomatal opening in the grass *Anthephora pubescens* Nees by a range of natural and synthetic cytokinins. Planta 150:218–221
Johnson KD, Rayle DL (1976) Enhancement of CO_2 uptake in *Avena* coleoptiles by fusicoccin. Plant Physiol 57:806–811
Johnson RR, Frey NM, Moss DN (1974) Effect of water stress on photosynthesis and transpiration of flag leaves and spikes of barley and wheat. Crop Sci 14:728–731
Jones HG (1973a) Limiting factors in photosynthesis. New Phytol 72:1089–1094
Jones HG (1973b) Moderate-term water stresses and associated changes in some photosynthetic parameters in cotton. New Phytol 72:1095–1105
Jones HG (1980) Interaction and integration of adaptive responses to water stress: The implications of an unpredictable environment. In: Turner NC, Kramer PJ (eds) Adaptation of plants to water and high temperature stress. Wiley and Sons, New York, pp 353–365
Jones MM, Rawson HM (1979) Influence of rate of development of leaf water deficits upon photosynthesis, leaf conductance, water use efficiency, and osmotic potential in sorghum. Physiol Plant 45:103–111
Jordan WR (1982) Whole plant response to water deficits: An overview. In: Taylor H, Jordan W, Sinclair T (eds) Limitations to efficient water use in crop production. Am Soc Agron, Madison, Wisconsin, in press
Jordan WR, Miller FR (1980) Genetic variability in sorghum root systems: implications for drought tolerance. In: Turner NC, Kramer PJ (eds) Adaptation of plants to water and high temperature stress. Wiley and Sons, New York, pp 383–400
Jordan WR, Morgan PW, Davenport TL (1972) Water stress enhances ethylene-mediated leaf abscission in cotton. Plant Physiol 50:756–758
Jordan WR, Brown KW, Thomas JC (1975) Leaf age as a determinant in stomatal control of water loss from cotton during water stress. Plant Physiol 56:595–599
Kaiser WM, Hartung W (1981) Uptake and release of abscisic acid by isolated photoautotrophic mesophyll cells, depending on pH gradients. Plant Physiol 68:202–206
Kaldewey H, Ginkel U, Wawczyniak G (1974) Auxin transport and water stress in pea (*Pisum sativum* L.). Ber Dtsch Bot Ges 87:563–576
Karmoker JL, Steveninck van RFM (1978) Stimulation of volume flow and ion flux by abscisic acid in excised root systems of *Phaseolus vulgaris* cv. Redland Pioneer. Planta 141:37–43
Katsumi M, Kazama H (1978) Gibberellin control of cell elongation in cucumber hypocotyl sections. Bot Mag Tokyo Spec Issue 1:141–158
Kauss H (1977) Biochemistry of osmotic regulation. In: Northcote DH (ed) Int Rev Biochem, Plant Biochem II, vol XIII. Univ Park Press, Baltimore, pp 119–140
Keck RW, Boyer JS (1974) Chloroplast response to low leaf water potentials. III. Differing inhibition of electron transport and photophosphorylation. Plant Physiol 53:474–479
King RW (1976) Abscisic acid in developing wheat grains and its relationship to grain growth and maturation. Planta 132:43–51
King RW, Evans LT (1977) Inhibition of flowering in *Lolium temulentum* L. by water stress: a role for abscisic acid. Aust J Plant Physiol 4:225–233
Klepper B, Taylor HM, Huck MG, Fiscus EL (1973) Water relations and growth of cotton in drying soil. Agron J 65:307–310
Koeppe DE, Miller RJ, Bell DT (1973) Drought-affected mitochondrial processes as related to tissue and whole plant response. Agron J 65:566–569
Kozlowski TT (1976) Water supply and leaf shedding. In: Kozlowski TT (ed) Water deficits and plant growth, vol IV. Academic Press, London New York, pp 191–231
Kramer PJ (1980) Drought, stress, and the origin of adaptations. In: Turner NC, Kramer PJ (eds) Adaptation of plants to water and high temperature stress. Wiley and Sons, New York, pp 7–20

Kuiper PJC (1972) Water transport across membranes. Annu Rev Plant Physiol 23:157–172
Labavitch JM (1981) Cell wall turnover in plant development. Annu Rev Plant Physiol 32:385–406
Labavitch JM, Ray PM (1974) Relationship between promotion of xyloglucan metabolism and induction of elongation by indoleacetic acid. Plant Physiol 54:499–502
Lado P, Rasi-Caldogno F, Colombo R (1977) Effect of cycloheximide on IAA- or FC-induced cell enlargement in pea internode segments. Plant Sci Lett 9:93–101
Ladyman JAR, Hitz WD, Hanson AD (1980) Translocation and metabolism of glycine betaine by barley plants in relation to water stress. Planta 150:191–196
Lahiri AN (1980) Interaction of water stress and mineral nutrition on growth and yield. In: Turner NC, Kramer PJ (eds) Adaptation of plants to water and high temperature stress. Wiley and Sons, New York, pp 341–352
Lee-Stadelmann OY, Stadelmann EJ (1976) Cell permeability and water stress. In: Lange OL, Kappen L, Schulze E-D (eds) Water and plant life: Problems and modern approaches. Ecol Stud Vol 19. Springer, Berlin Heidelberg New York, pp 268–280
Levitt J (1980) Stress and strain terminology. In: Levitt J (ed) Responses of plants to environmental stresses, Vol I, 2nd edn. Academic Press, London New York, pp 3–19
Lockhart JA (1965) An analysis of irreversible plant cell elongation. J Theor Biol 8:264–276
Loescher WA, Nevins DJ (1973) Turgor-dependent changes in *Avena* coleoptile cell wall composition. Plant Physiol 52:248–251
Lorimer GH, Andrews TJ (1980) The C_2 photo- and chemorespiratory carbon oxidation cycle. In: Hatch MD, Boardman NK (eds) The biochemistry of plants, vol VIII. Academic Press, London New York, pp 329–374
Lösch R, Schenk B (1978) Humidity responses of stomata and the potassium content of guard cells. J Exp Bot 29:781–787
Loveys BR (1977) The intracellular location of abscisic acid in stressed and non-stressed leaf tissue. Physiol Plant 40:6–10
Loveys BR, Kriedemann PE (1973) Rapid changes in abscisic acid-like inhibitors following alterations in vine leaf water potential. Physiol Plant 28:476–479
Ludlow MM (1980) Adaptive significance of stomatal responses to water stress. In: Turner NC, Kramer PJ (eds) Adaptation of plants to water and high temperature stress. Wiley and Sons, New York, pp 123–138
Ludlow MM, Ng TT, Ford CW (1980) Recovery after water stress of leaf gas exchange in *Panicum maximum* var. *trichoglume*. Aust J Plant Physiol 7:299–313
Lürssen K, Naumann K, Schröder R (1979) l-Aminocyclopropane-l-carboxylic acid – a new intermediate of ethylene biosynthesis. Naturwissenschaften 66:264–265
Luxmoore RJ, Millington RJ, Marcellos H (1971) Soybean canopy structure and some radiant energy relations. Agron J 63:111–114
MacDowall FDH (1963) Midday closure of stomata in aging tobacco leaves. Can J Bot 41:1289–1300
Mansfield TA, Wellburn AR, Moreira TJS (1978) The role of abscisic acid and farnesol in the alleviation of water stress. Philos Trans R Soc London Ser B 284:471–482
Maranville JW, Paulsen GM (1972) Alteration of protein composition of corn (*Zea mays* L.) seedlings during moisture stress. Crop Sci 12:660–663
Markhart AH III, Fiscus EL, Naylor AW, Kramer PJ (1979) Effect of abscisic acid on root hydraulic conductivity. Plant Physiol 64:611–614
Marré E, Lado P, Rasi Caldogno F, Colombo R (1973) Correlation between cell enlargement in pea internode segments and decrease in the pH of the medium of incubation. I. Effects of fusicoccin, natural and synthetic auxins and mannitol. Plant Sci Lett 1:179–184
Masuda Y (1978) Auxin-induced cell wall loosening. Bot Mag Tokyo Spec Issue 1:103–123
Masuda Y, Sakurai N, Tazawa M, Shimmen T (1978) Effects of osmotic shock on auxin-induced cell extension, cell wall changes and acidification in *Avena* coleoptile segments. Plant Cell Physiol 19:857–867
McMichael BL, Elmore CD (1977) Proline accumulation in water-stressed cotton leaves. Crop Sci 17:905–908
McMichael BL, Jordan WR, Powell RD (1972) An effect of water stress on ethylene production by intact cotton petioles. Plant Physiol 49:658–660

McMichael BL, Jordan WR, Powell RD (1973) Abscission processes in cotton: induction by plant water deficit. Agron J 65:202–204
McNeil DL (1976) The basis of osmotic pressure maintenance during expansion growth in *Helianthus annuus* hypocotyls. Aust J Plant Physiol 3:311–324
Mederski HJ, Chen LH, Curry RB (1975) Effect of leaf water deficit on stomatal and nonstomatal regulation of net carbon dioxide assimilation. Plant Physiol 55:589–593
Meidner H, Edwards M (1975) Direct measurement of turgor pressure potentials of guard cells. J Exp Bot 26:319–320
Meyer RF, Boyer JS (1972) Sensitivity of cell division and cell elongation to low water potentials in soybean hypocotyls. Planta 108:77–87
Michelena VA, Boyer JS (1980) Growth and osmotic adjustment in maize leaves having low water potentials. Plant Physiol Suppl 65:8
Milborrow BV (1974) The chemistry and physiology of abscisic acid. Annu Rev Plant Physiol 25:259–307
Milborrow BV (1978) The stability of conjugated abscisic acid during wilting. J Exp Bot 29:1059–1066
Milborrow BV (1979) Antitranspirants and the regulation of abscisic acid content. Aust J Plant Physiol 6:249–254
Milborrow BV, Robinson DR (1973) Factors affecting the biosynthesis of abscisic acid. J Exp Bot 24:537–548
Milburn JA (1979) Water flow in plants. Longman, London New York
Mizrahi Y, Richmond AE (1972) Hormonal modification of plant response to water stress. Aust J Biol Sci 25:437–442
Mizrahi Y, Blumenfeld A, Richmond AE (1972) The role of abscisic acid and salination in the adaptive response of plants to reduced root aeration. Plant Cell Physiol 13:15–21
Mohanty P, Boyer JS (1976) Chloroplast response to low leaf water potentials. IV. Quantum yield is reduced. Plant Physiol 57:704–709
Molz FJ, Boyer JS (1978) Growth-induced water potentials in plant cells and tissues. Plant Physiol 62:423–429
Mooney HA, Björkman O, Collatz GJ (1977) Photosynthetic acclimation to temperature and water stress in the desert shrub *Larrea divaricata*. Carnegie Inst Yearb 76:328–335
Morgan JM (1980) Possible role of abscisic acid in reducing seed set in water-stressed wheat plants. Nature (London) 285:655–657
Morilla CA, Boyer JS, Hageman RH (1973) Nitrate reductase activity and polyribosomal content of corn (*Zea mays* L.) having low leaf water potentials. Plant Physiol 51:817–824
Munns R, Brady CJ, Barlow EWR (1979) Solute accumulation in the apex and leaves of wheat during water stress. Aust J Plant Physiol 6:379–389
Mussell H, Staples RC (1979) Stress physiology in crop plants. Wiley-Interscience, New York, 510 p
Nulsen RA, Turtell GW, Stevenson FR (1975) Response of leaf water potential to pressure changes at the root surface of corn plants. Agron J 69:951–954
Nye PH, Tinker PB (1977) Solute movement in the soil-root system. Stud Ecol, vol IV Univ California Press, Berkeley, 342 p
Oliveira EC, Hsiao TC (1980) Osmotic adjustment of cotton to water stress: time course and associated changes in growth and assimilation. Plant Physiol Suppl 65:6
Osmond CB (1976) Ion absorption and carbon metabolism in cells of higher plants. In: Lüttge U, Pitman MG (eds) Transport in plants II. Encyclopedia of plant physiology new ser vol II, part A, Springer, Berlin Heidelberg New York, pp 347–372
Osmond CB, Winter K, Powles SB (1980) Adaptive significance of carbon dioxide cycling during photosynthesis in water-stressed plants. In: Turner NC, Kramer PJ (eds) Adaptation of plants to water and high temperature stress. Wiley and Sons, New York, pp 139–154
O'Toole JC, Cruz RT (1980) Response of leaf water potential, stomatal resistance and leaf rolling to water stress. Plant Physiol 31:44–53
O'Toole JC, Crookston RK, Treharne KJ, Ozbun JL (1976) Mesophyll resistance and carboxylase activity. A comparison under water stress conditions. Plant Physiol 57:465–468

O'Toole JC, Cruz RT, Singh TN (1979) Leaf rolling and transpiration. Plant Sci Lett 16:111–114

Paleg LG, Aspinall D (eds) (1981) Physiology and biochemistry of drought resistance in plants. Academic Press, Sydney, 492 p

Pallaghy CK, Raschke K (1972) No stomatal response to ethylene. Plant Physiol 49:275–276

Parrondo RT, Smith RC, Lazurick K (1975) Rubidium absorption by corn root tissue after a brief period of water stress and during recovery. Physiol Plant 35:34–38

Pate JS (1980) Transport and partitioning of nitrogenous solutes. Annu Rev Plant Physiol 31:313–340

Pearcy RW (1982) Non-stomatal inhibition of photosynthesis by water stress. In: Taylor HM, Jordan WR, Sinclair TR (eds) Efficient water use in crop production. Am Soc Agron, Madison, Wisconsin, in press

Pearson RW (1966) Soil environment and root development. In: Pierre WH, Kirkham D, Pesek J, Shaw R (eds) Plant environment and efficient water use. Am Soc Agron, Soil Sci Soc Am, Madison, Wisconsin, pp 95–126

Penny P (1971) Growth-limiting proteins in relation to auxin-induced cell elongation in lupin hypocotyls. Plant Physiol 48:720–723

Penny P, Penny D (1978) Rapid responses to phytohormones In: Letham DS, Goodwin PB, Higgins TJV (eds) Phytohormones and related compounds – a comprehensive treatise, vol II. Elsevier, Amsterdam, pp 537–597

Petrie AHK, Wood JG (1938) Studies on the nitrogen metabolism of plants. I. The relation between the content of proteins, amino acids, and water in the leaves. Ann Bot 2:33–60

Pheloung P, Barlow EWR (1981) Respiration and carbohydrate accumulation in the water-stressed wheat apex. J Exp Bot 32:921–931

Pierce M, Raschke K (1980) Correlation between loss of turgor and accumulation of abscisic acid in detached leaves. Planta 148:174–182

Pierce M, Raschke K (1981) Synthesis and metabolism of abscisic acid in detached leaves of *Phaseolus vulgaris* L. after loss and recovery of turgor. Planta 153:156–165

Pitman MG, Wellfare O (1978) Inhibition of ion transport in excised barley roots by abscisic acid; relation to water permeability of the roots. J Exp Bot 29:1125–1138

Pitman MG, Lüttge U, Läuchli A, Ball E (1974a) Action of abscisic acid on ion transport as affected by root temperature and nutrient status. J Exp Bot 25:147–155

Pitman MG, Lüttge U, Läuchli A, Ball E (1974b) Effect of previous water stress on ion uptake and transport in barley seedlings. Aust J Plant Physiol 1:377–385

Plaut Z (1974) Nitrate reductase activity of wheat seedlings during exposure to and recovery from water stress and salinity. Physiol Plant 30:212–217

Premecz G, Olah T, Gulyas A, Nyitrai A, Palfi G, Farkas GL (1977) Is the increase in ribonuclease level in isolated tobacco protoplasts due to osmotic stress? Plant Sci Lett 9:195–200

Quarrie SA, Jones HG (1977) Effects of abscisic acid and water stress on development and morphology of wheat. J Exp Bot 28:192–203

Quebedeaux B, Sweetser PB, Rowell JC (1976) Abscisic acid levels in soybean reproductive structures during development. Plant Physiol 58:363–366

Radin JW, Ackerson RC (1981) Water relations of cotton plants under nitrogen deficiency III. Stomatal conductance, photosynthesis, and abscisic acid accumulation during drought. Plant Physiol 67:115–119

Raschke K (1975) Stomatal action. Annu Rev Plant Physiol 26:309–340

Raschke K (1979) Movements of stomata. In: Haupt W, Feinleib ME (eds) Physiology of movements. Encyclopedia of plant physiology new ser vol VII. Springer, Berlin Heidelberg New York, pp 383–441

Raschke K, Zeevaart JAD (1976) Abscisic acid content, transpiration, and stomatal conductance as related to leaf age in plants of *Xanthium strumarium* L. Plant Physiol 58:169–174

Rawson HM (1979) Vertical wilting and photosynthesis, transpiration, and water use efficiency of sunflower leaves. Aust J Plant Physiol 6:109–120

Ray PM (1962) Cell wall synthesis and cell elongation in oat coleoptile tissue. Am J Bot 49:928–939

Ray PM, Green PB, Cleland R (1972) Role of turgor in plant cell growth. Nature (London) 239:163–164
Rayle DL (1973) Auxin-induced hydrogen ion secretion in *Avena* coleoptiles and its implication. Planta 114:63–73
Reed NR, Bonner BA (1974) The effect of abscisic acid on the uptake of potassium and chloride into *Avena* coleoptile sections. Planta 116:173–185
Rehm MM, Cline MG (1973) Inhibition of low pH-induced elongation in *Avena* coleoptiles by abscisic acid. Plant Physiol 51:946–948
Rikin A, Richmond AE (1976) Amelioration of chilling injuries in cucumber seedlings by abscisic acid. Physiol Plant 38:95–97
Rikin A, Blumenfeld A, Richmond AE (1976) Chilling resistance as affected by stressing environment and ABA. Bot Gaz 137:307–312
Rubinstein B (1977) Osmotic shock inhibits auxin-stimulated acidification and growth. Plant Physiol 59:369–371
Sakurai W, Masuda Y (1977) Effect of indole-3-acetic acid on cell wall loosening: changes in mechanical properties and noncellulosic glucose content of *Avena* coleoptile cell wall. Plant Cell Physiol 18:587–594
Santarius KA, Ernst R (1967) Hill reaction and photophosphorylation of isolated chloroplasts in relation to water content. I. Removal of water by means of concentrated solutions. Planta 73:91–108
Saugier B (1976) Sunflower. In: Monteith JL (ed) Vegetation and the atmosphere, vol II. Case studies. Academic Press, London New York, pp 87–119
Scott NS, Munns R, Barlow EWR (1979) Polyribosome content in young and aged wheat leaves subjected to drought. J Exp Bot 30:905–911
Setter TL, Brun WA, Brenner ML (1980) Effect of obstructed translocation on leaf abscisic acid, and associated stomatal closure and photosynthesis decline. Plant Physiol 65:1111–1115
Shaner DL, Boyer JS (1976) Nitrate reductase activity in maize (*Zea mays* L.) leaves. II. Regulation by nitrate flux at low leaf water potential. Plant Physiol 58:505–509
Sharp RE, Davies WJ (1975) Solute regulation and growth by roots and shoots of water-stressed maize plants. Planta 147:43–49
Sharp RE, Osonubi O, Wood WA, Davies WJ (1979) A simple instrument for measuring leaf extension in grasses, and its application in the study of the effects of water stress on maize and sorghum. Ann Bot 44:35–45
Silk WK, Wagner KK (1980) Growth-sustaining water potential distributions in the primary corn root: A noncompartmental continuum model. Plant Physiol 66:859–863
Sinclair TR, Wit de CT (1975) Photosynthate and nitrogen requirements for seed production by various crops. Science 189:565–567
Singh TN, Paleg LG, Aspinall D (1973) Stress metabolism. III. Variations in response to water deficit in the barley plant. Aust J Biol Sci 26:65–76
Slatyer RO (1973) Effects of short periods of water stress on leaf photosynthesis. In: Slatyer RO (ed) Plant response to climatic factors. Proc Uppsala Symp 1970. UNESCO, Paris, pp 271–276
Smith FA, Raven JA (1979) Intracellular pH and its regulation. Annu Rev Plant Physiol 30:289–311
Sodek L, Wright STC (1969) The effect of kinetin on ribonuclease, acid phosphatase, lipase and esterase levels in detached wheat leaves. Phytochemistry 8:1629–1640
Sprent JI (1981) Nitrogen fixation. In: Paleg LG, Aspinall D (eds) Physiology and biochemistry of drought resistance in plants. Academic Press, Sydney, pp 131–143
Sojka RE, Stolzy LH (1980) Soil-oxygen effects on stomatal response. Soil Sci 130:350–358
Stålfelt MG (1955) The stomata as a hydrophotic regulator of the water deficit of the plant. Physiol Plant 8:572–593
Staden van J, Davey JE (1979) The synthesis, transport and metabolism of endogenous cytokinins. Plant Cell Environ 2:93–106
Steudle E, Cosgrove D (1981) Water relations of growing pea epicotyl. Abstr XIII Int Bot Congr, Sydney, August 21–28, p 247

Stevenson TT, Cleland RE (1981) Osmoregulation in the *Avena* coleoptile in relation to auxin and growth. Plant Physiol 67:749–753
Stewart CR, Hanson AD (1980) Proline accumulation as a metabolic response to water stress. In: Turner NC, Kramer PJ (eds) Adaptation of plants to water and high temperature stress. Wiley and Sons, New York, pp 173–189
Stewart CR, Boggess SF, Aspinall D, Paleg LG (1977) Inhibition of proline oxidation by water stress. Plant Physiol 59:930–932
Storey R, Ahmad N, Wyn Jones RG (1977) Taxonomic and ecological aspects of the distribution of glycine betaine and related compounds in plants. Oecologia 27:319–332
Stout RG, Johnson KD, Rayle DC (1978) Rapid auxin and fusicoccin-enhanced Rb^+ uptake and malate synthesis in *Avena* coleoptile sections. Planta 139:35–41
Stuart DA, Jones RL (1978a) Role of cation and anion uptake in salt-stimulated elongation of lettuce hypocotyl sections. Plant Physiol 61:180–183
Stuart DA, Jones RL (1978b) The role of acidification in gibberellic acid- and fusicoccin-induced elongation growth of lettuce hypocotyl sections. Planta 142:135–145
Tal M, Imber D (1971) Abnormal stomatal behavior and hormonal imbalance in *flacca*, a wilty mutant of tomato III. Hormonal effects on the water status in the plant. Plant Physiol 47:849–850
Tal M, Imber D, Erez A, Epstein E (1979) Abnormal stomatal behavior and hormonal imbalance in *flacca*, a wilty mutant of tomato. V. Effect of abscisic acid on indoleacetic acid metabolism and ethylene evolution. Plant Physiol 63:1044–1048
Thomas H, Stoddart JL (1980) Leaf senescence. Annu Rev Plant Physiol 31:83–111
Todd GW (1972) Water deficits and enzymatic activity. In: Kozlowski TT (ed) Water deficits and plant growth. Vol III. Academic Press, London New York, pp 177–216
Torrey JA (1976) Root hormones and plant growth. Annu Rev Plant Physiol 27:435–484
Troughton JH, Slatyer RO (1969) Plant water status, leaf temperature, and the calculated mesophyll resistance to carbon dioxide of cotton leaves. Aust J Biol Sci 22:815–827
Tully RE, Hanson AD, Nelsen CE (1979) Proline accumulation in water-stressed barley leaves in relation to translocation and the nitrogen budget. Plant Physiol 63:518–523
Turner NC (1974) Stomatal behavior and water status of maize, sorghum and tobacco under field conditions. Plant Physiol 53:360–365
Turner NC (1979) Drought resistance and adaptation to water deficits in crop plants. In: Mussel H, Staples RC (eds) Stress physiology in crop plants. Wiley-Interscience, New York, pp 343–372
Turner NC, Jones MM (1980) Turgor maintenance by osmotic adjustment: A review and evaluation. In: Turner NC, Kramer PJ (eds) Adaptation of plants to water and high temperature stress. Wiley and Sons, New York, pp 87–103
Turner NC, Kramer PJ (eds) (1980) Adaptation of plants to water and high temperature stress. Wiley and Sons, New York, 382 p
Vaadia Y (1976) Plant hormones and water stress. Philos Trans R Soc London Ser B 273:513–522
Verasan V, Phillips RE (1978) Effects of soil water stress on growth and nutrient accumulation in corn. Agron J 70:613–618
Volkenburgh van E, Cleland RE (1980) Proton excretion and cell expansion in bean leaves. Planta 148:273–278
Walton DC (1980) Biochemistry and physiology of abscisic acid. Annu Rev Plant Physiol 31:453–489
Walton DC, Harrison MA, Cote P (1976) The effects of water stress on abscisic acid levels and metabolism in roots of *Phaseolus vulgaris* L. and other plants. Planta 131:141–144
Walton DC, Galson E, Harrison MA (1977) The relationship between stomatal resistance and abscisic acid levels in leaves of water-stressed bean plants. Planta 133:145–148
Weiler EW, Schnabl H, Hornberg C (1982) Stress-related levels of abscisic acid in guard cell protoplasts of *Vicia fabe* L. Planta 154:24–28
Williams RF (1975) The shoot apex and leaf growth. Univ Press, Cambridge, 256 p

Wittenbach VA, Ackerson RC, Giaquinta RT, Hebert RR (1980) Changes in photosynthesis, ribulose bisphosphate carboxylase, proteolytic activity and ultrastructure of soybean leaves during senescence. Crop Sci 20:225–231
Wright STC (1969) An increase in the "inhibitor-β" content of detached wheat leaves following a period of wilting. Planta 86:10–20
Wright STC (1977) The relationship between leaf water potential (Ψ leaf) and the levels of abscisic acid and ethylene in excised wheat leaves. Planta 134:183–189
Wright STC (1978) Phytohormones and stress phenomena. In: Letham DS, Goodwin PB, Higgins TJV (eds) Phytohormones and related compounds: A comprehensive treatise, vol II. Phytohormones and the development of higher plants. Elsevier, Amsterdam, pp 497–536
Wright STC (1979) The effect of 6-benzyladenine and leaf ageing treatment on the levels of stress-induced ethylene emanating from wilted wheat leaves. Planta 144:179–188
Wright STC (1980) The effect of plant growth regulator treatments on the levels of ethylene emanating from excised turgid and wilted wheat leaves. Planta 148:381–388
Yang SF (1980) Regulation of ethylene biosynthesis. HortScience 15:238–243
Yang SF, Pratt HK (1978) The physiology of ethylene in wounded plant tissues. In: Kahl G (ed) Biochemistry of wounded plant tissues. De Gruyter, Berlin, pp 595–622
Younis HM, Boyer JS, Govindjee (1979) Conformation and activity of chloroplast coupling factor exposed to low chemical potential of water in cells. Biochim Biophys Acta 548:328–340
Zabadal TJ (1974) A water potential threshold for the increase of abscisic acid in leaves. Plant Physiol 53:125–127
Zeevaart JAD (1977) Sites of abscisic acid synthesis and metabolism in *Ricinus communis* L. Plant Physiol 59:788–791
Zeevaart JAD (1980) Changes in levels of abscisic acid and its metabolites in excised leaf blades of *Xanthium strumarium* during and after water stress. Plant Physiol 66:672–678
Zeroni M, Jerie PH, Hall MA (1977) Studies on the movement and distribution of ethylene in *Vicia faba* L. Planta 134:119–125
Zimmermann U (1978) Physics of turgor and osmoregulation. Annu Rev Plant Physiol 29:121–148

10 Desiccation-Tolerance

J.D. Bewley and J.E. Krochko

CONTENTS

10.1 Introduction

Productivity and geographical distribution of plant species are limited as much, if not more, by water stress than by any other environmental factor. Fully one-third of the exposed land mass is classified as arid or semi-arid while the remainder is subjected to seasonal or local variations in water supply. It is hardly surprising that a strong evolutionary impetus toward improved drought-resistance has resulted in a considerable array of adaptations at all levels of organization (ecological, morphological, physiological, and biochemical). Within this variety three basic strategies are recognizable: drought-evasion, drought-avoidance, and drought-tolerance (Cloudsley-Thompson and Chadwick 1964). Drought evasion (or escape) is seen most easily in the ephemeral annuals. Due to an abbreviated life-cycle these plants complete their vegetative growth and reproductive cycle during periods of moisture availability, surviving the intervening dry periods as desiccation-tolerant seeds. Drought-avoidance is achieved primarily through adaptations that retard water loss and/or increase water absorption. Plants in which these features are well-developed effectively forestall internal water deficits when exogenous supplies are scarce. Drought-tolerance is a feature of plants capable of experiencing protoplasmic dehydration without permanent injury. Except for the seed stage, this latter mechanism is poorly developed in the majority of higher plants. Losses greater than 40%–90% of normal water content (LEVITT 1972), or equilibration with relative humidities of 91% to 98% (LEVITT et al. 1960) are lethal for most. By contrast, many lower plants (and some angiosperms) are remarkable in the ability of their vegetative tissues to withstand extreme drought, i.e., desiccation. Known also as poikilohydrous plants (WALTER 1955), poikiloxerophytes (HENCKEL and PRONINA 1968), or resurrection plants, these desiccation-tolerant plants are loosely defined by their unique ability to revive from an air-dry condition (the air being of low relative humidity). Plants that succumb to desiccation are called desiccation-sensitive, desiccation-intolerant or homoiohydrous (WALTER 1955) plants. Plants that can survive water deficits only by avoiding or resisting desiccation are desiccation-resistant plants, and these fall into the desiccation-intolerant category.

By far the majority of research on water stress in plants has involved desiccation-intolerant species. This is quite understandable, considering that poikilohydrous plants comprise a very limited proportion of native vegetation (even in the regions where they are found) and that they make an imperceptible contribution to agriculture. Generally, desiccation-tolerant plants have sacrificed productivity for survival in the regions to which they are adapted.

Much information is now available regarding the physiological and biochemical changes occurring in desiccation-intolerant species of plants during and after exposure to sub-lethal levels of water stress. Yet very little is understood of the factors responsible for a plant's susceptibility to severe water deficits, or of those that confer greater tolerance. Equally little is known about the basis for desiccation-tolerance per se. Nevertheless, some interesting information on the metabolism of desiccation-tolerant plants under stress has been gained over the past few years, which has necessitated a re-evaluation of established concepts of desiccation-tolerance and the proposal of new ones. These will be elaborated upon in subsequent sections. Some comparisons are made with desiccation-intolerant plants throughout this chapter. However, the reader should refer to others in this volume for a more detailed discussion of the effects of drought on homoiohydrous species.

Desiccation-tolerance in plants has been reviewed elsewhere recently (BEWLEY 1979; GAFF 1980), and the reader is directed to these publications for aspects not covered here.

10.2 Taxonomic Origins of Desiccation-Tolerance in Higher Plants

In addition to numerous examples of fungi, algae (DAVIS 1972), bryophytes and lichens the list of known desiccation-tolerant plants now includes approximately 60–70 species of ferns and fern-allies, and at least 60 species of angiosperms (Table 10.1). GAFF (1980), citing unpublished data, reported that 83 species of tolerant seed plants have been identified (see also Chaps. 2 and 4, this Vol.).

During the evolution of higher plants, the differentiation of vascular tissue and concomitant increases in plant size and complexity have led to the development of predominantly drought-avoidance mechanisms in species adapted to xeric regions. These kinds of adaptation have the advantage of enabling plants to continue growing during times of restricted water supply. As is now obvious, higher plants have not been restricted to this course despite the intricacies of their structure and physiology. Most are genetically capable of producing desiccation-tolerant reproductive propagules (seeds). The absence of gymnosperms from the survey (Table 10.1) (VIEWEG and ZIEGLER 1969; GAFF 1972) suggests that limitations on size for the successful re-establishment of water continuity in the xylem after drying may be important. The angiosperm examples are all herbs or small shrubs, rarely exceeding 1 m in height. The tallest reported are specimens of *Xerophyta eglandulosa* at 1.7 m (GAFF 1977). Drought-tolerance limits determined in the laboratory reveal that, as a group, the reviviscent angiosperms are more hardy than the ferns, although somewhat less so than either lichens or bryophytes (GAFF 1971, 1977). The disorganization of subcellular constituents that occurs in desiccation-tolerant seed plants during drying frequently is more severe than that observed in rehydrated mosses and ferns, or homoiohydrous plants subjected to sub-lethal levels of water stress (FELLOWS and BOYER 1978; GAFF and HALLAM 1974; GAFF et al. 1976; TUCKER et al. 1975). At least one-third of the desiccation-tolerant monocot species are poikilochlorophyllous (Table 10.1); they suffer the loss of the majority of their chlorophyll while retaining viability during drying.

Examples of desiccation-tolerant pteridophytes comprise greater than 0.65% of the recognized species in this taxon, and are found in a high proportion of families [10 out of 15 – after WILLIS (1973)]. Their distribution is pancontinental (Table 10.1). At present, desiccation-tolerance has been described for approximately 0.1% of monocot species and 0.01% of dicots, but it is very probable that more cases will be documented. Most of the desiccation-tolerant angiosperms are native to southern Africa (39 of the species included in Table 10.1), with fewer from other regions, and none from North America. The most drought-tolerant seed plant from North America is probably the desert shrub, *Larrea divaricata.* Its buds and immature leaves (but not mature leaves) survive considerable dehydration (RUNYON 1934) although the tolerance limitations of these organs have not been fully established. GAFF and LATZ (1978) have suggested that the relative abundance of tolerant species from southern Africa possibly reflects a longer duration of continuous aridity there; according to KOCH (1964) the Namib desert (which is drier than any area in Australia) has existed since the Cretaceous period.

Examples of desiccation-tolerant angiosperms are found in eight families belonging to five orders in this classification (after CRONQUIST, in RADFORD et al. 1974) (Table 10.1). The lack of close relationships among many of these suggests that the ability to withstand air-drying has evolved independently on

Table 10.1. Desiccation-tolerant vascular plants

				Genus [c]	Location [d]	References
Pteridophyta [a]	Lycopsida	Selaginellales	Selaginellaceae	Selaginella (10)	N. Am., S. Am., Afr.	1, 2, 3, 6, 8, 12, 20, 24, 28
	Filicopsida	Pteridales	Sinopteridaceae	Cheilanthes (21)	N. Am., Afr., Aust.	3, 7, 8, 13, 20, 28
				Notholaena (6)	N. Am., S. Afr.	17, 19, 20, 26, 28
				Pellaea (7)	N. Am., S. Afr., Aust.	3, 6, 7, 8, 28
				Doryopteris (2)	S. Afr., Aust.	2, 3, 7
			Actiniopteridaceae	Actiniopteris (2)	S. Afr.	3
			Hemionitidaceae	Gymnopteris	N. Am.	28
				Paraceterach	Aust.	7
			Adiantaceae	Adiantum	S. Afr.	3
		Aspidiales	Aspleniaceae	Asplenium (4)	Afr.	3, 20
				Ceterach (2)	Eur., Med., S. Afr.	3, 20
				Pleurosorus	Aust.	7
		Schizaeales	Schizaeaceae	Mohria	S. Afr.	2, 3
		Davalliales	Oleandraceae	Arthropteris	S. Afr.	3
		Polypodiales	Grammitidaceae	Ctenopteris	Aust.	7
			Polypodiaceae	Polypodium (2)	N. Am., S. Afr.	3, 8, 23, 26
				Platycerium	W. Afr.	12, 13
Spermatophyta Angiospermae [b]	Dicotyledonae	Hamamelidales	Myrothamnaceae	Myrothamus (2)	S. Afr.	2, 6, 8, 14, 16, 24, 25, 26, 28, 29
		Scrophulariales	Scrophulariaceae	Craterostigma (4)	S. Afr.	2, 3, 6, 8, 24
				Chamaegigas [e]	S. Afr.	2, 6
				Illysanthes (2)	S. Afr.	3, 6
			Gesneriaceae	Boea	Aust.	6, 7, 8
				Haberlea [f]	Med.	6, 27
				Ramonda [f] (3)	Eur., Med.	6, 8, 20, 24, 27
		Lamiales	Labiatae	Satureja	S. Am.	30
	Monocotyledonae	Cyperales	Cyperaceae	*Coleochloa (2)	S. Afr.	2, 3, 5, 6, 8, 24
				*Afrotrilepis	W. Afr.	6, 12, 13, 22
				*Carex [g]	C. Asia	6, 12

		*Fimbristylis[h]	Aust.	6, 7
		*Cyperus[h]	S. Afr.	3, 5, 6
		*Mariscus[h]	S. Afr.	5, 6
		*Kyllinga[h]	S. Afr.	3, 5, 6
	Gramineae	Oropetium	S. Afr.	2, 5, 6, 8
		Microchloa (2)	S. Afr.	2, 5, 6
		*Eragrostis[h] (3)	S. Afr.	5, 6, 24
		*Sporobolus[h] (4)	Aust., S. Afr.	2, 5, 6, 7, 8, 24
		Brachyachne	S. Afr.	5, 6
		Micraira (6)	Aust.	6, 7
		Tripogon (2)	S. Afr., Aust.	5, 6, 7, 8
		*Eragrostiella	Aust.	7
		Poa	Aust., Med.	6, 7, 27
Liliales	Liliaceae	*Borya (3)	Aust.	4, 6, 7, 8, 9, 24
	Velloziaceae	Talbotia	S. Afr.	6, 8, 11, 24
		*Xerophyta (13)	Afr., S. Am.	2, 3, 6, 8, 10, 18, 21, 24

[a] Classification according to WILLIS (1973)

[b] Orders and Families according to CRONQUIST (in RADFORD et al. 1974)

[c] For genera containing more than one desiccation-tolerant species, the number is indicated in brackets. Genera denoted by an asterisk (*) contain at least one poikilochlorophyllous example

[d] Geographical distribution or collection site of desiccation-tolerant species: N. Am. (North America), S. Am. (South America), Aust. (Australia, New Zealand), C. Asia (central Asia), Eur. (Europe), Med. (southern Europe and/or Mediterranean), Afr. (Africa) or more specifically S. Afr. (southern Africa) and W. Afr. (West Africa)

[e] Only submerged leaves and immature floating leaves are desiccation-tolerant

[f] Desiccation-tolerance disputed by HENCKEL and PRONINA (1973)

[g] Desiccation-tolerance considered ambiguous by WALTER and STADELMANN (1968)

[h] Includes species in which desiccation-tolerance is confined to basal portion of leaf [and also bulbils (*Poa bulbosa*) or immature leaves (*Sporobolus lampranthus*)]

References: 1. EICKMEIER (1979); 2. GAFF (1971); 3. GAFF (1977); 4. GAFF and CHURCHILL (1976); 5. GAFF and ELLIS (1974); 6. GAFF and HALLAM (1974); 7. GAFF and LATZ (1978); 8. GAFF and MCGREGOR (1979); 9. GAFF et al. (1976); 10. HALLAM and GAFF (1978a); 11. HALLAM and GAFF (1978b); 12. HAMBLER (1961); 13. HAMBLER (1964); 14. HENCKEL and PRONINA (1969); 15. HENCKEL and PRONINA (1973); 16. HOFFMANN (1968); 17. ILJIN (1957); 18. MEGURO et al. (1977); 19. NOBEL (1978); 20. OPPENHEIMER and HALEVY (1962); 21. OWOSEYE and SANFORD (1972); 22. RICHARDS (1957); 23. STUART (1968); 24. TYMMS and GAFF (1979); 25. VIEWEG and ZIEGLER (1969); 26. WALTER (1955); 27. WALTER and STADELMANN (1968); 28. WALTER and STADELMANN (1974); 29. ZIEGLER AND VIEWEG (1970); 30. MONTENEGRO et al. (1979)

a number of occasions. Desiccation-tolerance has been described for most (or all) species in some genera [e.g., *Myrothamnus* (2/2), *Chamaegigas* (1/1), *Haberlea* (1/1), *Ramonda* (3/4), *Borya* (3/3)], a high proportion in others [e.g., *Craterostigma* (4/20), *Coleochloa* (2/7), *Afrotrilepis* (1/2), *Microchloa* (2/4), *Xerophyta* (13/55)], and for very few (or only one) in others [e.g., *Boea* (1/25), *Carex* (1/1500), *Fimbristylis* (1/300), *Cyperus* (1/550), *Kyllinga* (1/60), *Oropetium* (1/5), *Eragrostis* (3/300), *Brachyachne* (1/10), *Tripogon* (2/20), *Eragrostiella* (1/7)] (Table 10.1; from WILLIS 1973). The majority of desiccation-tolerant grasses belong to three tribes (Eragrosteae, Chlorideae, Sporoboleae) that HUTCHINSON (1973) considers are closely related. Yet, in general, this trait occurs sporadically and infrequently in genera of the Gramineae and Cyperaceae (see above; GAFF and ELLIS 1974; GAFF and LATZ 1978). For many of the monocot examples listed in Table 10.1 only the basal portion of the leaf, or immature leaves, survives desiccation. It frequently has been reported (for desiccation-intolerant species) that young, meristematic tissues are more drought-tolerant than mature, differentiated tissues on the same plant (reviewed in LEVITT 1972). On the basis of these observations, GAFF and ELLIS (1974) have suggested that the basal-meristematic pattern of growth in the grass and sedge leaves might allow for a gradual extension of extreme drought-tolerance developed in the basal meristem to progressively older areas of the leaf. Species in which desiccation-tolerance is confined to the lower portions of the leaf (e.g., *Kyllinga alba, Eragrostis paradoxa*) may be at an earlier stage of evolutionary adaptation than species in which the whole leaf is tolerant (e.g., *Sporobolus stapfianus, Coleochloa setifera*) (GAFF and ELLIS 1974).

Because of the diversity of morphological form and taxonomic origins of desiccation-tolerant plants it is difficult to assign common characteristics to them. Nevertheless, there are certain features that appear to be of general applicability to a variety of tolerant species, and these are outlined in Table 10.2. Further details on some of these features are to be found in subsequent sections.

10.3 Desiccation-Tolerance: a Constitutional or Conditional Attribute? Some Ecological Considerations

One of the most intriguing questions concerning desiccation-tolerant plants is whether their tolerance is an inherited, constant feature of the protoplasm, or whether it develops in response to imposed water deficits, as environmental acclimation (hardening). The answer is by no means clear-cut, for some tolerant plants appear to be able to survive desiccation conditions more severe than those imposed by their environment, some only when subjected to the conditions of their particular habitat in which they are found, and others show obvious changes in tolerance with changes in ambient conditions (i.e., acclimation).

For bryophytes (e.g., BROWN and BUCK 1979; CLAUSEN 1952, 1964; DILKS and PROCTOR 1974, 1976a; LEE and STEWART 1971; NÖRR 1974a; OCHI 1952)

Table 10.2. Some general characteristics of desiccation-tolerant terrestrial plants

I. Morphology	
1. Life forms	Lichens, bryophytes, ferns, herbs and shrubs
2. Morphology	Extremely wide range of growth forms. Maximum height 1.7 m
3. Leaf form	Often small, or leathery or needle like. Usually rolls or curls tightly during water loss
II. Life History	
4. Longevity of vegetative phase	Long to very long
5. Phenology of reproductive structure production	No general relationship with season
6. Frequency of reproductive structure production	Intermittent over a long life history
7. Proportion of annual production devoted to reproduction	Small
8. Growth period	Not seasonal. Opportunistic, determined by water availability
III. Physiology	
9. Relative growth rate	Slow and spasmodic
10. Response to stress	Morphogenetic responses slow and small in magnitude
11. Metabolic integrity	Metabolism suspended when dry. Strong conservation of metabolic systems which are reactivated when wet. Cellular damage under stress limited to extensive, but repair or restitution effected relatively quickly when wet
12. Photosynthesis and productivity	Opportunistic. Photosynthesis not obviously linked to vegetative growth
IV. Miscellaneous	
13. Litter	Sparse, sometimes persistent
14. Palatability to herbivores	Low

and lichens (e.g., LECHOWICZ and ADAMS 1974; BUCK and BROWN 1979; KERSHAW 1972; SMITH 1962; LANGE 1953) there are good correlations between drought-tolerance and the water status of the habitats where particular species (or races within species) are found. Similarly, in intertidal algae, the vertical disposition of species relative to the shoreline reflects their abilities to recover from dehydration (BIEBL 1962; BÉRARD-THERRIAULT and CARDINAL 1973; QUADIR et al. 1979; ZANEFELD 1969; DROMGOOLE 1980; SCHONBECK and NORTON 1978). These groups contain species that show seasonal variations in drought-tolerance which correspond in a general way to the patterns of desiccation stress encountered in the field (CLAUSEN 1952; DILKS and PROCTOR 1976a; HOSOKAWA and KUBOTA 1957; LECHOWICZ and ADAMS 1974; NÖRR 1974a; SCHONBECK and NORTON 1979). CLAUSEN (1952) found for hepatics that it was

Table 10.3. Desiccation-tolerance in *Fucus spiralis* collected from the upper and lower parts of its zone, measured as growth rate after 48 h exposure at 15–22 °C. (After Schonbeck and Norton 1979)

	Linear growth (mm per 15 days)	Percent weight gain after 15 days
Plants from upper part of zone		
Desiccated	0.8 ± 0.6[a]	6.5 ± 3.6
Control	5.6 ± 0.4	32.3 ± 3.4
Plants from lower part of zone		
Desiccated	-0.3 ± 0.4[b]	-3.8 ± 2.1[b]
Control	6.3 ± 0.5	42.7 ± 3.2

[a] Mean $\pm$ S.E., n=5 for each treatment
[b] Plants killed by desiccation

predominantly the species from intermediate habitats that exhibited these season-related changes; those from very moist or very dry sites possessed relatively immutable tolerance limits. Similar trends occur in intertidal algae. In a comparison of two species occupying adjacent positions on the high-shore, Schonbeck and Norton (1979) found that the lower one (*Fucus spiralis*) hardened and dehardened with the changing seasons to a much greater extent than the more tolerant species (*Pelvetia canaliculata*) growing above it. Grime (1977, 1979) has postulated that greater physiological flexibility is a characteristic of species occupying moderately or mildly stressful habitats; under these conditions interspecific competition phenomena are more important than the physiological influences of the environment for determining success (cf. Chapman 1973). If these observations are an accurate indication of the trends expected, it would seem that species from intermediate habitats are more likely to show forms that differ in drought-tolerance than species adapted to very extreme situations. There is some substantiating evidence for this postulation. Wet and dry races have been identified in species of bryophytes that grow primarily under moderate stress conditions (Clausen 1952; Lee and Stewart 1971). For the alga, *Fucus spiralis*, samples collected from the lower portion of its range were less able to withstand desiccation than plants established higher on the shore (Table 10.3). Comparable differences were not found in collections of a more tolerant species, *P. canaliculata*. Also this latter species showed only insignificant changes in desiccation-tolerance when grown under laboratory conditions designed to induce these. The probable basis for these intraspecific variations in desiccation-tolerance has received little attention. Clausen (1952) observed a reduction in desiccation-tolerance in a dry race of *Lophocolea heterophylla* following brief cultivation on moist peat, and Dilks and Proctor (1976b) have provided evidence for an increased tolerance to desiccation in *Rhytidiadelphus loreus* following a series of alternating wet and dry periods. In *Fucus spiralis*, exposure to subcritical levels of desiccation for 2 to 4 h a day, over 5 days, is sufficient to cause substantial drought-hardening (Schonbeck and Norton 1979). Considering the relatively short duration of these experiments, the alterations are almost certainly the result of acclimation rather than structural changes at the level of the genome. This should not be construed as eliminating the possibility of genetically different races developing in situ over time.

The development of desiccation-tolerance in plants is sometimes caused by environmental factors other than moisture availability. In a green alga, *Spongiochloris typica,* desiccation-tolerance is an ephemeral quality, developing to varying degrees and at different times depending on the nutrient supply of the culture medium (McLEAN 1967). In a liverwort, *Lunularia cruciata,* it is associated with long days (or continuous light) (SCHWABE and NACHMONY-BASCOMB 1963). In the poikilohydrous angiosperm, *Chamaegigas intrepidus,* the ability to survive air-drying is diminished as leaves mature (GAFF 1971).

For the majority of the vascular plants considered to be desiccation-tolerant there are reported no discernable season-related alterations in this ability. During the dry season they will grow vigorously if supplied with water and likewise become quiescent regardless of the month when water is withheld (GAFF et al. 1976; OPPENHEIMER and HALEVY 1962; OWOSEYE and SANFORD 1972). A possible exception is *Borya nitida,* the leaves of which seem to survive desiccation better in the summer months than in winter (GAFF 1980).

For some of the poikilohydrous angiosperms it appears that cytoplasmic adjustments must be made to enable the tissue to survive extreme changes in water content. For example, in *Borya nitida,* which is both poikilohydrous and poikilochlorophyllous (a chlorophyll-loser), desiccation is lethal if drying occurs too rapidly (GAFF and ELLIS 1974; GAFF and HALLAM 1974; GAFF et al. 1976). Survival in this species depends upon the completion of a time-dependent (24 h or longer) tolerance-endowing process at water potentials around 96% R.H. [98% to 90% R.H.; 79% to 34% R.W.C. (relative water content) (GAFF and ELLIS 1974)]. The persistent leaf bases surrounding the stem in this plant probably function under natural conditions to retard drying sufficiently to permit the necessary metabolic events associated with the development of desiccation-tolerance to occur. The cellular processes responsible for the acquirement of desiccation-tolerance in this species have not been extensively studied. It is known, however, that tolerance is associated with extreme losses of chlorophyll, starch and nonsoluble nitrogen and extensive disruption of the internal structure of the cell (GAFF et al. 1976); see also Section 10.5.3.

Observations on the poikilohydrous, poikilochlorophyllous angiosperms *Xerophyta schnitzleinia* (OWOSEYE and SANFORD 1972) and *Afrotrilepis pilosa* (HAMBLER 1961) provide additional evidence that slow drying may be necessary to prevent loss of viability during desiccation. Under natural conditions *X. schnitzleinia* requires at least 5 days to dry while in *A. pilosa* desiccation-tolerance is found only with intact plants or portions of plants with long stem sections (both of these dry more slowly than isolated pieces of this plant). On the other hand, some higher plants can dry out very quickly under field conditions and survive, e.g., *Chamaegigas intrepidus* may desiccate within 1 h (GAFF 1977).

In liverworts (HÖFLER in LEVITT (1956)), mosses (IRMSCHER in LEVITT (1956)) and algae (EVANS 1959) it has been shown that the ability to survive desiccation increases if drying occurs slowly rather than quickly (cf. also ILJIN 1953, 1957). However, in species of Oscillatoriaceae the rate of drying has little effect on subsequent survival [HESS in FOGG (1969)]. For the desiccation-tolerant moss, *Tortula ruralis,* the rate of water loss is irrelevant to the mosses' eventual recovery, although it may alter the physiological condition of these plants (BEWLEY and GWÓŹDŹ 1975; KROCHKO et al. 1979; GWÓŹDŹ and BEWLEY 1975). CLAUSEN (1952) investigated the rate of water loss from isolated pieces of a variety of liverwort species to determine whether a reduced drying rate was associated with structures presumed to accomplish this and with a plants' habitat stress. Those characters which appeared to be important in the laboratory had little relevance in situ. There was no correlation between the rate at which single isolated shoots lost water, the rate of desiccation in the field and a species' desiccation-tolerance.

Some structural features (scales, low surface/volume ratio, mucilaginous sheaths, thick walls, complex internal morphology, pubescence) of poikilohydrous plants have been shown to impede (but not prevent) the rate of water loss during desiccation (LARSON 1979; LARSON and KERSHAW 1976; FOGG 1969; OPPENHEIMER and HALEVY 1962; PARKER 1968). But the importance of these features for survival has not always been demonstrated clearly. Many species of higher and lower plants appear to avoid rapid desiccation by growing in protected or sheltered habitats (see later) or as dense mats which retard water loss, e.g., some bryophytes. The relative importance of structural features in relation to habitat specificity requires more attention: desiccation-tolerance under field conditions is a feature more commonly noted than studied.

There is evidence that poikilohydry has developed to deal not only with water stress, but also with the temperature extremes characteristic of deserts. Metabolic activity in the desert lichens *Chondropsis semiviridis* (ROGERS 1971) and *Ramalina maciformis* (LANGE et al. 1975) is habitually limited to the early morning hours following dew-moistening and/or water vapor uptake. As the morning advances, the elevated temperatures resulting from the extreme levels of solar radiation that impinge upon their exposed habitats cause a rapid desiccation of the plants. Laboratory investigations have shown that when dry the lichens are unharmed by temperatures as high as 65 °C (recovery was determined by the ability to resume photosynthesis), but when fully hydrated they are adversely affected by temperatures greater than 30 °C (*C. semiviridis*) and 38 °C (*R. maciformis*). Since these latter temperatures, and higher, are encountered frequently by the lichens, it would appear that if it were not for their poikilohydrous habit they would require further adaptations to survive. But more recent studies (MACFARLANE and KERSHAW 1978, 1980) have shown that some air-dry thalli of varieties of *Peltigera canina,* and of other lichen species, can be very sensitive to even moderate levels of heat stress, whereas others exhibit a considerable degree of heat tolerance. Thus thermal sensitivity might be an important parameter in the ecology of some poikilohydrous lichens. However, increased resistance to high and low temperatures in the dry state does appear to be a characteristic of some other poikilohydrous tissues (BEWLEY 1973c; FOGG 1969; MALEK and BEWLEY 1978a, b; HAMBLER 1961; LANGE et al. 1975; NÖRR 1974b).

Partial desiccation at certain (usually higher) temperatures may aggravate metabolic imbalances due to water loss, perhaps by accelerating some biochemi-

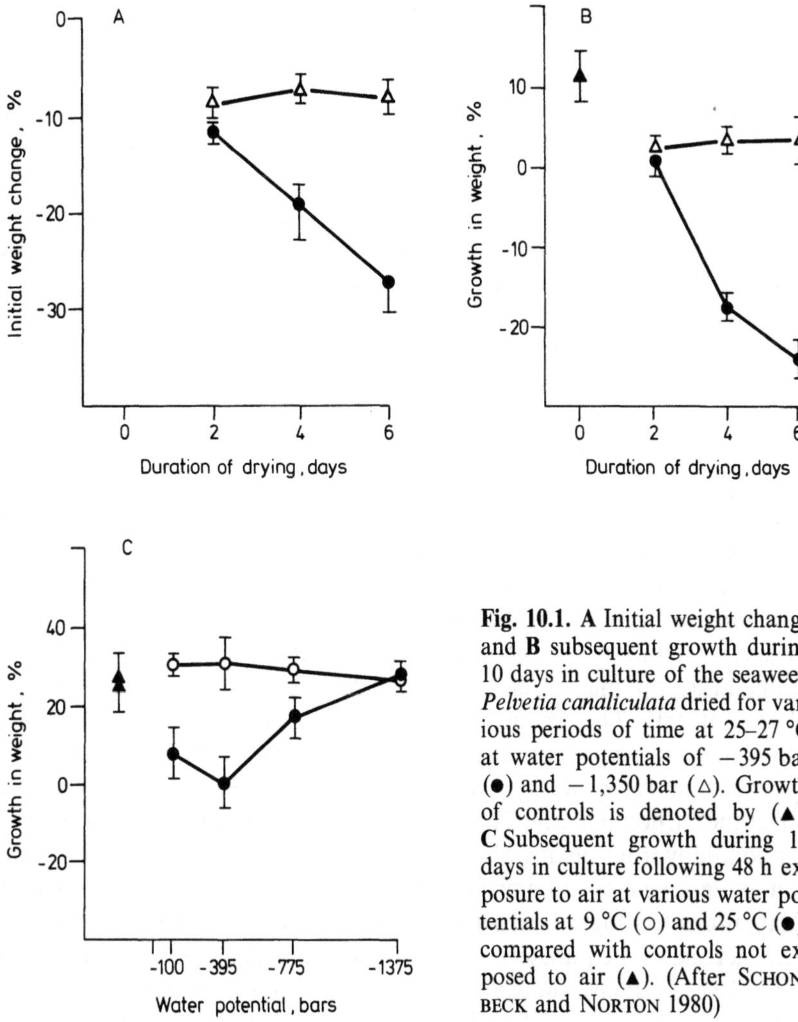

Fig. 10.1. A Initial weight change and **B** subsequent growth during 10 days in culture of the seaweed *Pelvetia canaliculata* dried for various periods of time at 25–27 °C at water potentials of −395 bar (●) and −1,350 bar (△). Growth of controls is denoted by (▲). **C** Subsequent growth during 10 days in culture following 48 h exposure to air at various water potentials at 9 °C (○) and 25 °C (●), compared with controls not exposed to air (▲). (After SCHONBECK and NORTON 1980)

cal processes and inhibiting others. Thus some desiccation-tolerant plants may not only have to survive severe water losses, but also must endure drought events that differ in rates of dehydration and rehydration, and in duration, temperature, and water potential of the dry period. Data on the individual effects of each of these parameters and their interactions may provide considerable insight into plant adaptations to various dry environments. Using *Pelvetia canaliculata* as an example once more, survival of desiccation by this seaweed depends strongly on temperature, duration and water potential of the desiccation event. In laboratory experiments, plants deteriorated gradually and died after 6 days at 25 °C and −395 bar (75% relative humidity), but suffered only slight and nonprogressive injury at −1,350 bar (37% relative humidity) (Fig. 10.1 A and 10.1 B). Damage increased with water potential at 25 °C, but not at 9 °C (Fig. 10.1 C), which suggests that *P. canaliculata* may suffer metabolic imbalances when partially dehydrated at high temperatures. Field observations of this alga have revealed that it occupies a habitat near to its limits of desiccation-tolerance with regard to duration and temperature (SCHONBECK and NORTON 1978), and that the plant exhibits visible injury only after neap tides during warm dry weather result in prolonged exposure. In contrast to this observation, subjecting the terrestrial desiccation-tolerant moss *Tortula ruralis* to 4–15 days partial dehydration at −207 bar (86% relative humidity) at 20°–25 °C is not more damaging than at lower water potentials (SCHONBECK and BEWLEY, unpublished data). This might indicate that severe metabolic imbalances do not develop in partially desiccated *T. ruralis* as they did in *P. canaliculata* under similar conditions. Wide tolerance of temperature in relation to severity of desiccation may permit *T. ruralis* to survive in habitats where drought may last for many weeks and where it may be accompanied by marked fluctuations in temperature and relative humidity.

Elevated temperatures may also have significant consequences for tolerant higher plants during desiccation. A study of water relations in the poikilohydrous fern *Notholaena parryi* have revealed a dramatic decrease in the temperature optimum for net CO_2 exchange in partially dehydrated plants (NOBEL et al. 1978). This has been attributed to (1) the greater fractional stomatal closure at high temperatures, (2) the greater reduction in gross photosynthesis relative to dark respiration plus photorespiration with water stress, and (3) the higher temperature optima for dark respiration plus photorespiration than for gross photosynthesis. The frequent mention in the literature of these contributing physiological components suggests that prolonged metabolic activity in plants experiencing water stress, particularly in a desert environment, may adversely affect the carbon balance. This would have particularly disastrous consequences for poikilohydrous plants, where productivity is already threatened. These effects can be lessened by minimizing the time required for desiccation or through adaptations (ecological, structural, physiological) to prevent excessive temperature increases. The pubescent covering of *Ramondia* (OPPENHEIMER and HALEVY 1962; PARKER 1968), the scales of *Ceterach officinarum* (OPPENHEIMER and HALEVY 1962), the oppressed position of the leaves in *Polytrichum commune* during drying (PARKER 1968), and the armor of dead leaf bases on *Afrotrilepis pilosa* (HAMBLER 1961) may function to prevent excessive insolation in these poikilohydrous plants rather than to reduce the rate of water loss as has commonly been proposed.

From the few ecological observations that have been made on poikilohydrous plants it is evident that many can avoid dramatic changes in temperature and water availability by growing in sheltered or shaded habitats. Many desiccation-tolerant mosses, for example, grow on the north side of trees in the Northern Hemisphere and hence avoid direct exposure to sunlight. In fact, the north side of patches of scrub, hedges, trees, etc., which is in shadow most of the day, provides a microclimate where temperatures are lower than the ambient air, the radiation level is low in red and blue light and liquid water is more readily available than in surrounding exposed areas (STOUTJESDIJK 1974). Desiccation-tolerant plants (such as mosses and liverworts) living in this "open shade" microclimate can avoid frequent desiccation and exposure to elevated temperatures in the partially dried or desiccated state.

In the Chihuahuan, Sonoran, and Mojave deserts, lichens and mosses are scarce and are strongly restricted to north slopes (NASH et al. 1977), or to crevices, gullies, and other shaded situations (ALPERT 1979; NOBEL 1978). In their natural habitat some desert lichens are rarely rewetted by rainfall, but resume their metabolism in atmospheres of high relative humidity and during early morning dew formation (LANGE 1969a). As outlined in Section 10.5.6.2, lichens in the Negev desert moistened in this way resume photosynthesis as dawn breaks and the light intensity increases, and net assimilation is achieved before water is lost to the increasingly drier and warmer air. The longer plants are sheltered from direct sunlight, the greater is their photosynthetic productivity – hence the advantage of the north-facing aspect of their habitat. In the above-mentioned North American deserts, surface temperatures on south-facing slopes may increase by more than 20 °C within the first 30 min after dawn breaks (NASH et al. 1977), resulting in rapid desiccation of any exposed plants, and negligible productivity. Thus the scarcity of mosses and lichens in the desert may be due to the severe restriction on potential net carbon assimilation imposed by rapid desiccation.

Poikilohydrous ferns in the deserts of Arizona, Nigeria, and South Africa are commonly confined to sites exposed to the north and to sites protected by aborescent vegetation (WALTER in OPPENHEIMER and HALEVY 1962; GAFF 1977; HAMBLER 1964). *Pellaea viridis,* from South Africa, has a shade form and a smaller dryland form which grows in exposed regions. *Pellaea calomelanos* frequently grows in full sunlight in one region, but in another region, where aridity and insolation are greater, those portions projecting from the shade of rocks die back in the dry season (GAFF 1977). *Afrotrilepis pilosa* and *Xerophyta schnitzleinia* are both poikilohydric plants found in West Africa as major mat formers, but both have distinctly different habitat tolerances (OWOSEYE and SANFORD 1972). *A. pilosa* develops only as the abundant plant in areas of relatively high humidity which have a short dry season, whereas *X. schnitzleinia* is found only in less humid and more intensely dry areas with a longer-lasting dry season.

It is evident from the examples cited in this section that poikilohydrous plants should not be regarded simply as plants that will grow in any harsh and uncompromising habitat. They survive only where their habitat allows for a net gain in their carbon budget, and in this they do not differ from other plants. As succinctly stated by EICKMEIER (1979) "the ecophysiological strategy of poikilohydric plants represents a complex evolutionary compromise between adaptive costs and benefits related to plant carbon balance in arid environments".

10.4 Theories of Desiccation-Tolerance

The diverse origins and apparent varied preparedness (see Sect. 10.3) of poikilohydrous plants to withstand drying argue against assuming a priori that the mode of survival of desiccation is identical or even similar in all members of this obviously artificial "class" of plants. Several theories have been proposed to explain the extreme drought-tolerance of poikilohydrous tissues. ILJIN has suggested that mechanical injury to the protoplast membranes during drying and remoistening is the prime cause of desiccation sensitivity and that the protoplasm of all plants theoretically is capable of surviving complete dehydration if mechanical stresses are avoided, e.g., by slow drying and gradual rehydration (ILJIN 1953, 1957, reviewed by LEVITT 1956). He concluded that the lethal injuries in plants sensitive to drought are the consequence of tensions that develop as the protoplasm is pulled inward by a shrinking vacuole and outward by a resisting cell wall during drying. In tolerant plants these stresses are minimized through various anatomical and morphological adaptations (reduced cell size, increased cell surface/volume ratio, small or absent vacuoles, lack of plasmodesmata, reduced osmotic pressure, presence of insoluble constituents, easily deformed cell wall). In addition, he considered that gradual drying and remoistening via controlled plasmolysis and deplasmolysis are important for preventing injury in plants not ordinarily tolerant of dehydration.

Other workers also have observed a correlation between desiccation-tolerance in lower plants and reduced cell size (OPPENHEIMER 1960; STADELMANN 1971; WALTER and STADELMANN 1968), small or absent vacuoles (HENCKEL and PRONINA 1968, 1969, 1973; PARKER 1968; WALTER and STADELMANN 1968), solidifying vacuoles (HENCKEL and PRONINA 1968, 1973; OPPENHEIMER 1960; OPPENHEIMER and HALEVY 1962; PARKER 1972; STADELMANN 1971) and the absence of plasmodesmata (HENCKEL and PRONINA 1968, 1969). While it may be that mechanical stresses to the cell are implicated in determining tolerance limits, Iljin's theory is both inadequate and sometimes inaccurate in its description of the features associated with drought-tolerance in many plants. In an examination of drought-tolerant and sensitive mosses, MAHMOUD (1965) found little agreement between desiccation-tolerance and the cellular characteristics described by ILJIN. Additionally, plasmodesmata have been identified in tolerant plants (BEWLEY and PACEY 1978; GAFF et al. 1976; MAHMOUD 1965; TUCKER et al. 1975). It even has been suggested that small cell size simply may represent the result of phylo- and onto-genetically effective scarcity of water during development, rather than being important for desiccation-tolerance (OPPENHEIMER 1960). Iljin's conclusions also have been criticized on the grounds that he determined survival solely on the basis of deplasmolysis and dye retention following remoistening; these phenomena are related to the intactness of the tonoplast rather than to the plasmalemma, and therefore do not distinguish between dead and incompletely killed cells (LEVITT 1956; OPPENHEIMER 1960; OPPENHEIMER and JACOBY 1961). The most common criticism of ILJIN'S theory is that it is too simplistic to adequately explain desiccation-tolerance; this ability is not present (potentially or otherwise) in all plants (LEVITT 1972; OPPENHEIMER 1960; OPPENHEIMER and JACOBY 1961; STADELMANN 1971).

According to STOCKER (1960; reviewed by LEVITT 1972; MAHMOUD 1965; PARKER 1968) the tensions occasioned by desiccation act not on the surface layer of the protoplasm but rather within the internal protoplasm. He believed that the protoplasm consisted of a latticework of threadlike proteins interspersed with water, salts, sugars, and other kinds of proteins. The initial cellular responses to water stress, which he termed the reaction phase, result from an unequal hydration capacity of different parts of this protein framework. This leads to a disorganization and loosening of the protein latticework as water is withdrawn, resulting in a reduction in the viscosity of the protoplasm and an increase in permeability. If drying is gradual enough the cell subsequently enters a restitution or restoration phase during which bonds are strengthened, a reorganization of the protein latticework occurs, and the protoplasmic viscosity increases. These changes result in an overcompensation of the drought reaction and the plant enters a hardened condition. During rapid drying the plant is prevented from completing the restitution phase, ergo an increase in drought-associated injury. In plants able to recover from air-drying STOCKER postulated a reactivation phase following remoistening to repair the damage caused by desiccation.

HENCKEL and PRONINA (1968, 1969, 1973) in a more recent rationalization of drought-tolerance, have attributed to the poikilohydrous plants a specialized metabolism and structure which is neither susceptible to, nor altered by, extreme water stress. Their theory of poikilohydry is based upon the premise "that for this group of plants the dehydration process is not pathological, but a normal reaction" (1968). They suggest that since poikilohydrous plants do not resist dehydration but rather lose water quickly there is no need for cytoplasmic preparations, such as those described by STOCKER (1960). Instead the plant is continually prepared to pass over into a state of anabiosis (latent life) (HENCKEL and PRONINA 1968, 1969). The mechanical stresses engendered by drying are avoided through an absence of plasmodesmata and large vacuoles or a solidification of the vacuolar contents when this structure is present. Metabolic activity of tolerant plants is both persistent and coordinated during desiccation.

STOCKER'S view that the secret of drought-tolerance would be found in the molecular organization of the protoplasm was a significant departure from the theory proposed by ILJIN, in which cell morphology was considered to be of primary importance. One aspect both of these researchers agreed upon was the necessity of accounting for the increased susceptibility of plants to rapid desiccation as compared to slow desiccation. The interaction between drying rate and drought-associated injury was not considered in either the experiments or theory of desiccation-tolerance published by HENCKEL and PRONINA.

In summary, these theories of drought-tolerance lack any type of generalized applicability to such a diverse group as the poikilohydrous plants. Whereas STUART (1968) in his studies of the desiccation-tolerant fern, *Polypodium polypodioides*, found considerable agreement with ILJIN, OCHI (1952) and MAHMOUD (1965) found that the theory of mechanical injury had little applicability to mosses. LEE and STEWART (1971) stated that for bryophytes, desiccation-tolerance is largely physiological and that differences in morphology are of little importance. HÖFLER (cited by LEVITT 1956; OPPENHEIMER 1960) concluded from his studies on liverworts that differences in protoplasmic drought-hardiness are paramount although physical stresses as evidenced by the effects of speed of drying on survival also are involved.

Both ILJIN'S theory of mechanical injury and that proposed by HENCKEL and PRONINA (1968, 1969, 1973) suggest that poikilohydrous plants survive desiccation through the ability of the tissue to avoid damage during drying and rehydration. There are, however, a substantial number of examples in the literature of poikilohydrous plants that do not conform metabolically or structurally as predicted by these theories; their responses to droughting are often very similar to those of homoiohydrous plants subjected to water deficits (although the level of stress required to elicit these changes may differ). While the ability to avoid excessive damage might be important in some desiccation-tolerant plants, it is evident that others do not undergo severe water deficits without exhibiting cellular changes, some of which might be regarded as quite extensive and involve perturbations of organelle structure (Sect. 10.5.3). According to BEWLEY (1979), what may be the critical features of desiccation-tolerance are the abilities (1) to limit damage during desiccation to a reparable level; (2) to maintain physiological integrity in the dry state so that metabolism can be reactivated quickly upon rehydration; and (3) to put a repair mechanism into effect upon rehydration, in particular to retain or regain integrity of membranes and membrane-bound organelles. These features will be examined in the following section.

10.5 Cellular Responses of Tolerant and Intolerant Plants to Desiccation and Rehydration

10.5.1 Plant Water Status

A meaningful assessment of the metabolism of plants of varying sensitivity to drought demands attention to the levels of water loss sustained. Accordingly, the designations defined by HSIAO (1973) will be used for clarification whenever possible. These are (1) mild stress: reduction in Ψ (water potential) by several bar or of RWC by 8–10%, (2) moderate stress: decrease in Ψ by less than 12–15 bar, or lowering of RWC by between 10–20%, and (3) severe stress: Ψ lowered more than 15 bar or RWC more than 20% (see Chap. 9, this Vol.).

Water deficits can affect cellular processes in one or more of the following ways: (1) through a decrease in turgor pressure; (2) by a reduction of the chemical potential or activity of water in cells; (3) by changes in spatial relations in membranes and organelles resulting from a decrease in volume of the protoplasm; (4) by effects of increases in solute concentrations caused by losses of water; (5) by alteration of structure of macromolecules resulting from dehydration (HSIAO 1973). It is considered that (1) [but not (2)] may influence plant metabolism at mild and moderate levels of stress, whereas (3) and (4) would have a greater influence at severe water deficits than at moderate stress levels, and (5) would be important at desiccating water contents. These generalities do not exclude the possibility that there may be cellular reactions or structures particularly sensitive (or resistant) to change via these mechanisms at suboptimal water contents (for detailed discussion of water status in tissues and cells see Chap. 2, this Vol.).

10.5.2 Osmotic Effects

Drought-resistant plants frequently are reported to have lower osmotic potentials than less hardy plants (HOSOKAWA and KUBOTA 1957; ILJIN 1957; LEVITT 1956; PARKER 1968; PATTERSON 1946). The relevance of this correlation to drought-tolerance mechanisms is ambiguous at best; it may simply reflect lowered tissue water content due to restricted water supply. Alternatively, reduced solute potentials may represent a drought-avoidance adaptation for increasing the water-retention capabilities of cells relative to their external environment (KLUGE 1976; LEVITT 1956; OPPENHEIMER 1960). Some plants are capable of actively increasing their stores of solute in response to water deficits (osmoregulation) (HSIAO et al. 1976; JONES and TURNER 1978; KLUGE 1976; MORGAN 1977). In these, avoidance adaptations are also involved if these changes effectively delay the loss of turgor (or restore turgor) to lower water contents than would have occurred otherwise (cf. HSIAO and ACEVEDO 1974).

Increases in free sugars and loss of starch commonly occur in plants [including a poikilohydrous angiosperm, *Xerophyta schnitzleinia* (OWOSEYE and SANFORD 1972)] during water stress (BASKIN and BASKIN 1974; KLUGE 1976; LEVITT 1972), although desiccation and rehydration of the poikilohydrous moss *Tortula ruralis* does not result in any substantial changes in starch or free sugar (sucrose) levels (BEWLEY and THORPE 1974; BEWLEY et al. 1978). Free proline accumulation also occurs in drought-stressed plants (e.g., BARNETT and NAYLOR 1966; BASKIN and BASKIN 1974; PÁLFI et al. 1973; SINGH et al. 1972; STEWART 1972), commencing at moderate to severe levels of water stress. The amounts of this amino acid accumulated have been positively correlated with drought-resistance in varieties and species of desiccation-intolerant plants (SINGH et al. 1972; WALDREN and TEARE 1974), but the validity of such correlations has been questioned (HANSON et al. 1977). Proline levels have been found to increase to different extents in a number of desiccation-tolerant higher plants (POURRAT and HUBAC 1974; TYMMS and GAFF 1979). But no correlations can be drawn between the extreme desiccation-tolerance of resurrection plants and their accumulated proline levels. Moreover, improved survival of desiccation cannot be invoked by exogenous application of proline (TYMMS and GAFF 1979). In the highly tolerant moss, *Tortula ruralis*, proline levels remain very low and unchanged during desiccation and rehydration (BEWLEY unpublished data).

When enzymes, structural proteins, nucleic acids, macromolecular complexes, etc. are desiccated in their native state the integrity of the molecules can be retained if some water remains associated with them to prevent the formation of unfavorable conformations (see reviews by HSIAO 1973; LEVITT 1972; PARKER 1972; TODD 1972), or fragmentation (DARBYSHIRE and STEER 1973). Hence the production or availability of substances (e.g., sugar, polyols, anions, amino acids – including proline) to maintain bound water content could be important features of desiccation-tolerance. Little work has been done to support or reject the possibility. Increased production of polyols by lichens (FARRAR and SMITH 1976) might have some importance in this respect, although some desiccation-tolerant plants (e.g., *T. ruralis*) do not appear to exhibit changes in the pool sizes of low molecular weight compounds such as sugars and amino acids

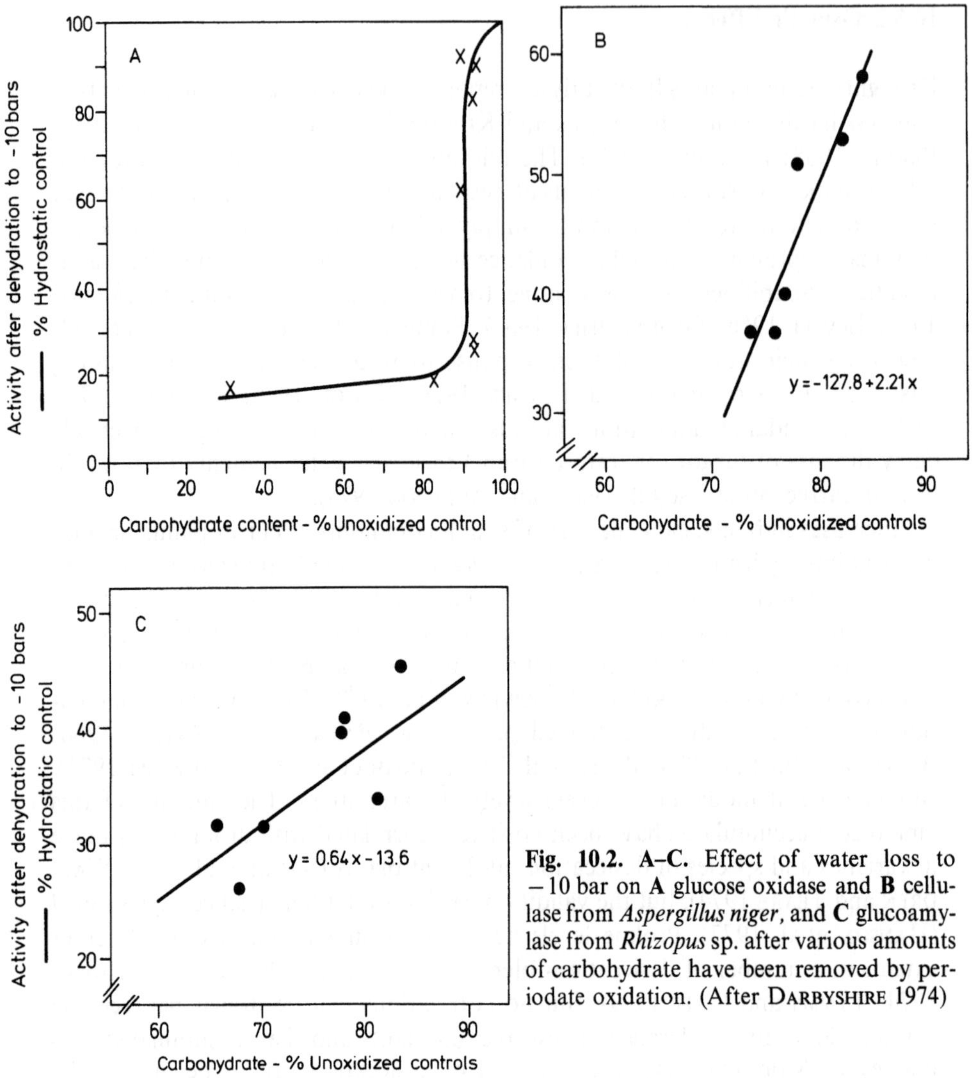

Fig. 10.2. A–C. Effect of water loss to −10 bar on **A** glucose oxidase and **B** cellulase from *Aspergillus niger*, and **C** glucoamylase from *Rhizopus* sp. after various amounts of carbohydrate have been removed by periodate oxidation. (After DARBYSHIRE 1974)

(BEWLEY et al. 1978, BEWLEY unpublished data). Protection of some fungal enzymes against water stress may depend upon their conjugation with carbohydrate as glycosylated proteins (DARBYSHIRE 1974). Removal of 10% or more of the carbohydrate associated with commercially available glucose oxidase, cellulase, and glucoamylase leads to a marked reduction in enzyme activity (Fig. 10.2). The universality of this feature in vivo and in desiccation-tolerant plants remains to be established. Protection against desiccation of chloroplast membrane structures by sugars has been proposed, their purported action being to stabilize proteins (SANTARIUS 1969, 1973). Desiccation leads to the concentration of ions within cells, and the possibility that this results in some damage cannot be ignored. Tolerant plants must be able to neutralize these potentially damaging ion effects, or else accommodate and limit them.

10.5.3 Ultrastructural Changes

There appear to be few remarkable differences in the ultrastructural changes caused by air-drying in desiccation-tolerant plants and the changes observed in desiccation-intolerant plants following treatments at sublethal water deficits. However, there is a complicating factor in the interpretation of many of these studies. When water-stressed or desiccated plant tissues are prepared for microscopic examination, the fixation and embedding procedures frequently result in a partial rehydration of the material. Consequently, even though the water deficit present in the tissue at the time of fixation is known, the published results may represent the condition of cellular structures following the imbibition of additional water. Such an occurrence does not necessarily invalidate the observations, but it is probable that in some cases they represent dehydration–rehydration phenomena.

In desiccation-tolerant mosses after air-drying and a short period of rehydration the chloroplasts are swollen, with distended outer membranes, and the thylakoids are dilated (BEWLEY and PACEY 1978; NOAILLES 1974, 1977, 1978; TUCKER et al. 1975). The pattern of response to desiccation in poikilohydrous angiosperms exhibits one striking variation. Whereas tolerant dicots retain chlorophyll throughout a desiccation/rehydration cycle, in many of the tolerant monocots chlorophyll is degraded during drying. In viable air-dry samples of these "poikilochlorophyllous" plants the thylakoids are indistinct or reduced in number and the grana also are fewer or absent (ZIEGLER and VIEWEG 1970; GAFF and HALLAM 1974; GAFF et al. 1976). In some chlorophyll-retaining species (e.g., *Pellaea calomelanos* and *Talbotia elegans*) chloroplast integrity is maintained during desiccation and remoistening, while in other species (e.g., *Xerophyta villosa*) there is some loss of grana and thylakoids (GAFF and HALLAM 1974; HALLAM and GAFF 1978a, b). Vesicularization within the internal membranes of chloroplasts is common among desiccation-tolerant angiosperms (GAFF and HALLAM 1974; GAFF et al. 1976). However, some of the extreme disruptions observed in micrographs of *B. nitida* are probably artifacts of inadequate fixation (GAFF 1980).

Characteristically mitochondria become swollen and the cristae are indistinct in desiccation-tolerant plants after air-drying and rehydration and in desiccation-intolerant plants exposed to injurious levels of water stress (BEWLEY and PACEY 1978; FREEMAN and DUYSEN 1975; GAFF and HALLAM 1974; GAFF et al. 1976; HALLAM and GAFF 1978a, b; MARINOS and FIFE 1972; NIR et al. 1969; NOAILLES 1974, 1977, 1978; TUCKER et al. 1975; VIEIRA DA SILVA et al. 1974). Also numerous lipid droplets are observed in cells of both groups of plants after these stress treatments (BEWLEY and PACEY 1978; GAFF et al. 1976; NIR et al. 1969; NOAILLES 1977; SWANSON et al. 1976). Chromatin condensation may occur (MARINOS and FIFE 1972; NIR et al. 1969), although the nucleolus and nuclear membrane remain well defined in cells capable of recovery (GAFF et al. 1976; MARINOS and FIFE 1972; NIR et al. 1969; NOAILLES 1977; TUCKER et al. 1975).

The response of the intolerant semi-aquatic moss *Cratoneuron filicinum* to desiccation varies with the speed at which it loses water (BEWLEY and PACEY 1978; KROCHKO et al. 1978). Within the first hour of rehydration of this moss after rapid desiccation the cytoplasm of all cells of the leaves (phyllidia) is disorganized, mitochondria are swollen and have an ill-defined internal structure, the outer chloroplast membrane is lost, and lamellae are dispersed. At the same time of rehydration after very slow desiccation only about 20% of the cells are similarly disrupted – the rest are fairly normal in appearance. After 24 h rehydration all cells of the rapidly desiccated moss are extensively disrupted, but about 50% of the cells of the very slowly dried moss retain their integrity.

The advantages of slow drying in nature have been alluded to in Section 10.3. Perhaps gradual water loss allows for controlled changes in the configuration of macromolecules which favor their stability or which facilitate limited (bio)chemical reactions and hence increases resistance to disruption by dehydration. Presumably even in very slowly dried *C. filicinum* some cells dry out more slowly than others, and these are the ones that survive. As will be evident in succeeding sections, rapid desiccation of *T. ruralis* causes considerably more metabolic disruption than slow desiccation, even though there is complete recovery after both speeds of desiccation. Leaves of many of the higher resurrection plants do not display the extraordinary desiccation-tolerance when detached (which allows them to dry quickly) that they do when dried (more slowly) on intact plants (GAFF 1980). Detached leaves of *Borya nitida* survive 0% relative humidity better if they are initially in the range of ~ −27 to −150 bar for 2 days as they dry.

10.5.4 Leakage and Membrane Integrity, and the "Repair" Concept

Little is known about the changes which the plasmalemma and tonoplast undergo during desiccation and rehydration, although leakage studies indicate that some changes do occur. Many of these studies have been reviewed by SIMON (1974, 1978), who summarizes the evidence that dry viable seeds, lichens, yeasts, pollen, fungi, and their spores all leak solutes when placed in liquid medium. Leaked substances include amino acids, mono-, di- and tri-saccharides, sugar alcohols, organic acids, hormones, phenolics, phosphates, and various fluorescent materials and electrolytes. A common feature is that the major leakage of solutes is a transient phenomenon, often lasting only minutes after the addition of water. Prolonged loss of solutes is characteristic of irreversibly damaged tissues.

Desiccation-tolerant and -intolerant bryophytes exhibit leakage of electrolytes upon rehydration following desiccation (e.g., GUPTA 1976, 1977c, 1979). An index of desiccation-tolerance for bryophytes has been devised based upon potassium retention within cells after storage at high and low relative humidities (BROWN and BUCK 1979). There is a good relationship between this index, tolerance of the plants to stress, and the availability of water in each bryophyte's habitat. An index of recovery has been devised also, which is a useful indication of a bryophyte's ability to survive brief and substantial desiccation stress, but it is less well-correlated to habitat water availability. Desiccation-tolerance and recovery indices have been proposed for lichens also (BUCK and BROWN 1979).

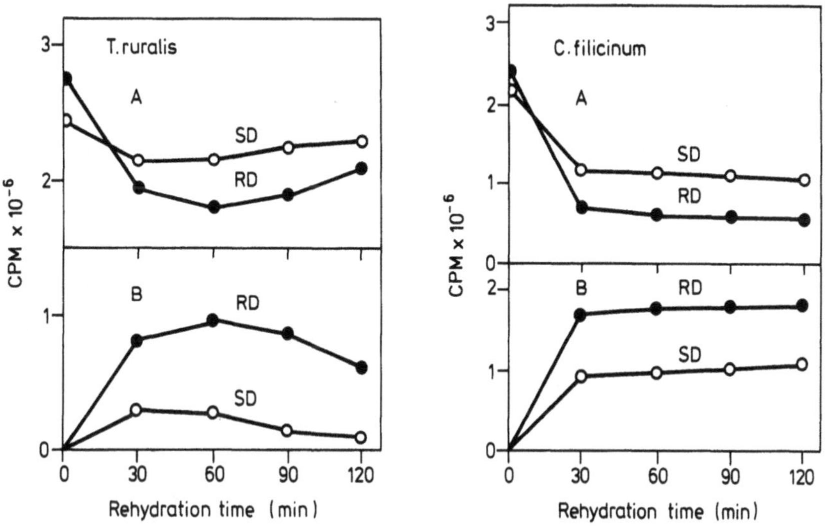

Fig. 10.3 A, B. Leakage of ^{3}H-leucine from the mosses *Tortula ruralis* and *Cratoneuron filicinum* on recovery from rapid desiccation (*RD*) and slow desiccation (*SD*). Changes in total radioactivity in the moss (**A**) and in radioactivity leaked into the bathing solution (**B**) during rehydration are shown. (After DHINDSA and BEWLEY 1977)

The extent of leakage from the one cell-thick phyllidia of the desiccation-tolerant moss *Tortula ruralis* on rehydration is determined by its prior speed of desiccation (Fig. 10.3). After slow drying, leakage is less than half that from rapidly desiccated moss and is similar to that from undried controls. After 1 h the leaked material in the medium is taken up by the moss again. Loss of material from both slowly and rapidly desiccated intolerant *Cratoneuron filicinum* is more extensive than from *T. ruralis* and is not reversed with time. Even so, it is less after slow than after rapid desiccation. The implication from these studies is that membranes of *T. ruralis* undergo certain reversible conformational changes during desiccation to retain their integrity. During slow desiccation these can be accomplished fully, but after rapid desiccation some repair mechanism must be put into effect quickly on rehydration. On the other hand, damage to *C. filicinum* membranes during desiccation and rehydration is apparently too extensive for repair, or the repair mechanisms themselves are damaged, or both.

Leakage from tissues has been explained in terms of changes which take place in the structure of membranes during water loss (SIMON 1974, 1978). Membranes are composed principally of proteins and phospholipids arranged in a fluid bilayer. Individual phospholipid molecules are lined up side-by-side with their polar groups facing the aqueous phase on either side of the membrane and with their hydrocarbon chains forming the hydrophobic central region (Fig. 10.4 A). This molecular organization is stabilized by the relationship between membrane components and water. Low angle X-ray diffraction patterns of isolated membranes have shown that 20%–30% hydration (above fully dried weight) is essential for maintenance of the lipoprotein association in membranes (FINEAN 1969; LUZZATI and HUSSON 1962). At lower water contents the mem-

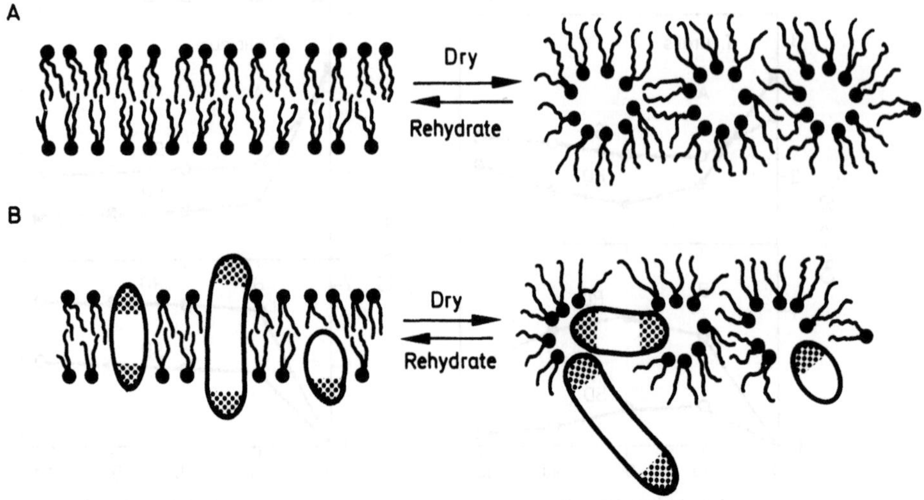

Fig. 10.4. A Diagrammatic representation of the orientation of membrane phospholipids in relation to degree of hydration. *Left* the lamellar configuration with the polar head-groups toward the aqueous layer on either side of the bilayer. *Right* the hexagonal configuration adopted at water levels below 20%. The polar head groups now line narrow water-filled channels. (Based on LUZZATI and HUSSON 1962). **B** Proposed orientation of lipid and protein components in membranes in relation to their degree of hydration. *Left* the membrane under aqueous conditions showing three integral proteins, their polar and ionic moieties *shaded*. *Right* the configuration adopted with less than 20% water, resulting in the displacement of the protein moieties. Upon rehydration these may remain displaced. (Based on SIMON 1978)

brane bilayer becomes rearranged into a hexagonal phase which is largely hydrophobic but is pierced by long water-filled channels lined by the polar heads of the phospholipids (Fig. 10.4 A). The drier the preparations, the narrower the channels become. The structure of membrane proteins appears to be unaffected by desiccation (SCHNEIDER and SCHNEIDER 1972), but they may be displaced from, or within, the membrane as phospholipid components become rearranged (Fig. 10.4 B). Rapid imbibition of dry tissues results in the phospholipid and protein components spontaneously reconstituting the bilayer configuration once a critical degree of hydration is reached. Membrane components may become displaced (or further displaced if already so) by the rapid influx of water. In cells of desiccation-intolerant plants this displacement may be irreversible or too slow to prevent the loss of critical amounts of solute. Displacement also may be greater in some cells of desiccation-tolerant plants (e.g., those rapidly dried, or very dry) than in others (e.g., after slow drying, or with a membrane water content close to the critical 20%–30% minimum). Differences in the degree of disruption of dry membranes could be a consequence of their inherent properties and/or be due to changes occurring during desiccation itself. Even in desiccation-tolerant cells there may be a delay before all displacements are reversed, allowing for solute leakage.

While the above explanation for membrane changes associated with desiccation is an attractive one, it still has to be tested critically. MCKERSIE and

STINSON (1980) have questioned the applicability of the LUZZATI and HUSSON model to plant membranes. They point out that the model was derived from experiments using an ether extract of human brain. The phospholipid and fatty acid composition of this extract might be markedly different from that of membranes found in plants and hence the two phospholipid bilayers might be expected to differ in their hydration requirements. Furthermore, the hexagonal phases as detailed in the LUZZATI and HUSSON model have only been observed in bulk phospholipid extracts and have not been reported in phospholipid bilayers as thin as biological membranes. In the bulk extracts from human brain all of the water was contained within the hexagonal array of channels which on a molecular scale were of great length. Formation of a thin membrane with some water on each side would seem entropically unlikely since a thin slice of the hexagonal model would indicate that both hydrophilic head groups and hydrophobic tails would be exposed to the surrounding water layer. Low and wide angle diffraction studies of bulk phospholipid extracts from *Lotus corniculatus* seeds indicate that the hexagonal phase does not form as a consequence of desiccation, and that the phospholipids are exclusively in the lamellar phase even at 5% water (MCKERSIE and STINSON 1980). Consequently, the observed membrane damage and leakage of solutes from these seeds cannot easily be explained by the formation of a hexagonal phase by membrane phospholipids. Changes in the protein component of the membrane may have to be invoked instead.

Even when its components are reassembled, and leakage is halted, the membrane still may not have regained its functionality. Additional "repair mechanisms" may be necessary to add back essential lipid or protein components. For example, during desiccation mitochondrial membranes may have displaced from or within them certain enzymes, enzyme complexes, or key structural components, which subsequently are replaced over several hours of rehydration. Until such replacement is complete (and the harsher the drying regime the longer this will take) it might be expected that the organelle will show symptoms of its incomplete nature. This indeed appears to be so, as will be detailed in Section 10.5.6.

Only minor changes occur to the fatty acid components of the phospholipid fraction (presumed to be largely membrane) from *T. ruralis* during rapid and slow desiccation (Table 10.4). Upon rehydration after both fast and slow desiccation there is a decline in the higher molecular weight unsaturated fatty acids (18:3, 20:3, 20:4), this decline being appreciably greater upon rehydration after slow desiccation than after rapid desiccation. Nevertheless, in both cases, within 105 min of rehydration the fatty acid composition of the phospholipid fraction returns to that from the undesiccated control moss. This is indicative of a rapid resynthesis of fatty acids and, indeed, there is incorporation of glycerol and acetate into the phospholipid fraction within a short time upon rehydration (Fig. 10.5). Moreover, the synthesis by moss recovering from slow desiccation is greater than that by moss recovering from rapid desiccation, showing a possible correlation between fatty acid levels and rates of their synthesis. There is, however, no obvious correlation between the changing fatty acid composition of the phospholipid fraction during rehydration and the extent

Table 10.4. The relative amounts of the fatty acids in a phospholipid fraction of *Tortula ruralis* upon rehydration after fast and slow drying. (R.R.C. STEWART and BEWLEY unpublished data)

	Palmitic 16:0	Palmitoleic 16:1	Oleic 18:1	Linoleic 18:2	Linolenic 18:3	Eicosatrienoic 20:3	Eicosatetraenoic 20:4
H	100	T	37	35	112	39	27
RD	100	T	38	36	115	36	26
R 15	100	T	36	37	98	29	19
R 45	100	T	31	35	99	27	18
R 75	100	T	34	37	114	39	26
R 105	100	T	37	36	113	34	23
SD	100	T	34	38	118	38	22
S 15	100	T	33	37	99	29	20
S 45	100	T	24	24	71	25	14
S 75	100	T	36	32	84	24	14
S 105	100	T	38	37	116	39	24

H: undesiccated moss; RD: rapidly dried moss within 1 h over silica gel; R 15–105: rapidly dried moss rehydrated from 15–105 min; SD: slowly dried moss over 3 h in an atmosphere of 65% R.H.; S 15–105: slowly dried moss rehydrated from 15–105 min. T: trace amounts

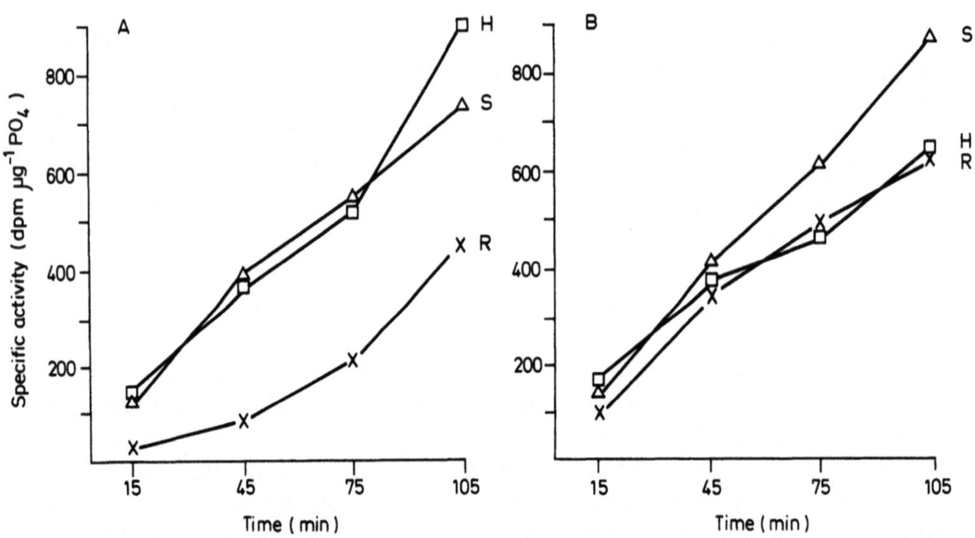

Fig. 10.5 A, B. The time course of incorporation of **A** acetate and **B** glycerol into fatty acids of the phospholipid fraction of *Tortula ruralis* in *H* control, undesiccated moss, *R* moss rehydrated after rapid desiccation and *S* moss rehydrated after slow desiccation. (By R.R.C. STEWART and BEWLEY unpublished data)

of membrane leakage. Rapidly desiccated moss leaks more upon rehydration (Fig. 10.3), but slowly desiccated moss undergoes the greater phospholipid changes (Table 10.4). Probably, then, the phospholipid component of the membrane plays an insignificant role in the control of leakage, and once again we must invoke a role for the unstudied component – the protein. In desiccation-sensitive *C. filicinum* the proportions of 18:3, 20:3, and 20:4 fatty acids in the phospholipid fractions are considerably less than in *T. ruralis*. As in *T. ruralis*, however, there are no changes in fatty acid composition during either rapid or slow desiccation. Upon subsequent rehydration of rapidly dried moss the levels of the above three unsaturated fatty acids decline over the last 2 h, but there are no changes after slow drying (STEWART and BEWLEY, unpublished data). But following both speeds of desiccation there is considerable leakage upon rehydration (Fig. 10.3) – again suggesting a lack of correlation between the status of phospholipids and the extent of leakage upon rehydration.

Alterations to membrane properties by lipid peroxidation could account for some of the observed cellular changes associated with desiccation and rehydration (SIMON 1974). A number of biological oxidations, both enzymatic and spontaneous, generate the free superoxide radical ($O_2^{\cdot -}$) which is cytotoxic, and in turn can react with H_2O_2 to produce singlet oxygen and the hydroxyl radical ($OH\cdot$) – highly potent oxidants (FRIDOVICH 1976). These can induce considerable destruction, particularly to large polymers like nucleic acids, protein and polysaccharides, and to membrane lipids (MEAD 1976). In hydrated tissues free radical production is normally controlled by free radical absorbents or scavenging reactions. One such scavenger is the enzyme superoxide dismutase (SOD) (FRIDOVICH 1975, 1976) which converts $O_2^{\cdot -}$ to H_2O_2, and this in turn can be removed by catalase. It is possible that water loss from cells of desiccation-intolerant plants can upset the balance between free radical producing and scavenging reactions in favor of the former, whereas in desiccation-tolerant plants the balance is maintained.

In desiccation-tolerant *T. ruralis* SOD and catalase activities increase during slow desiccation and lipid peroxidation (as shown by the production of malondialdehyde (MDA), a breakdown product of tri-unsaturated fatty acids with double bonds three carbons apart, e.g., 18:3, 20:3) decreases (Fig. 10.6 A–C). On rehydration lipid peroxidation rises again, but only to control levels, and enzyme levels decline. During rapid desiccation there are no changes, but on rehydration there is a sharp rise in lipid peroxidation, but SOD and catalase levels rise also, sufficiently, presumably, to limit peroxidation damage. Unfortunately, however, it is difficult to draw correlations between peroxidation damage (and the system for scavenging free radicals) and the status of fatty acids within the phospholipid fraction. It is surprising that the levels of MDA rise so quickly and to such an extent upon rehydration of rapidly desiccated moss (Fig. 10.6 C) when the changes in fatty acids (particularly 18:3 and 20:3) are small (Table 10.4). Also, there is little comparison between the small changes in peroxidation products on rehydration after slow desiccation, and the more substantial changes in phospholipid fatty acids. Perhaps, then, the substrates for peroxidation damage and for free radical damage are not to be found within the phospholipid (membrane) fraction, but in some other lipid fraction within the cell. As suggested earlier (Sect. 10.5.3), there is an accumulation of lipid droplets within the cells of some mosses during desiccation (including *T. ruralis*: BEWLEY and PACEY 1978) and it is these which might be the target for peroxidation reactions and protection. On theoretical grounds, the importance of SOD in limiting

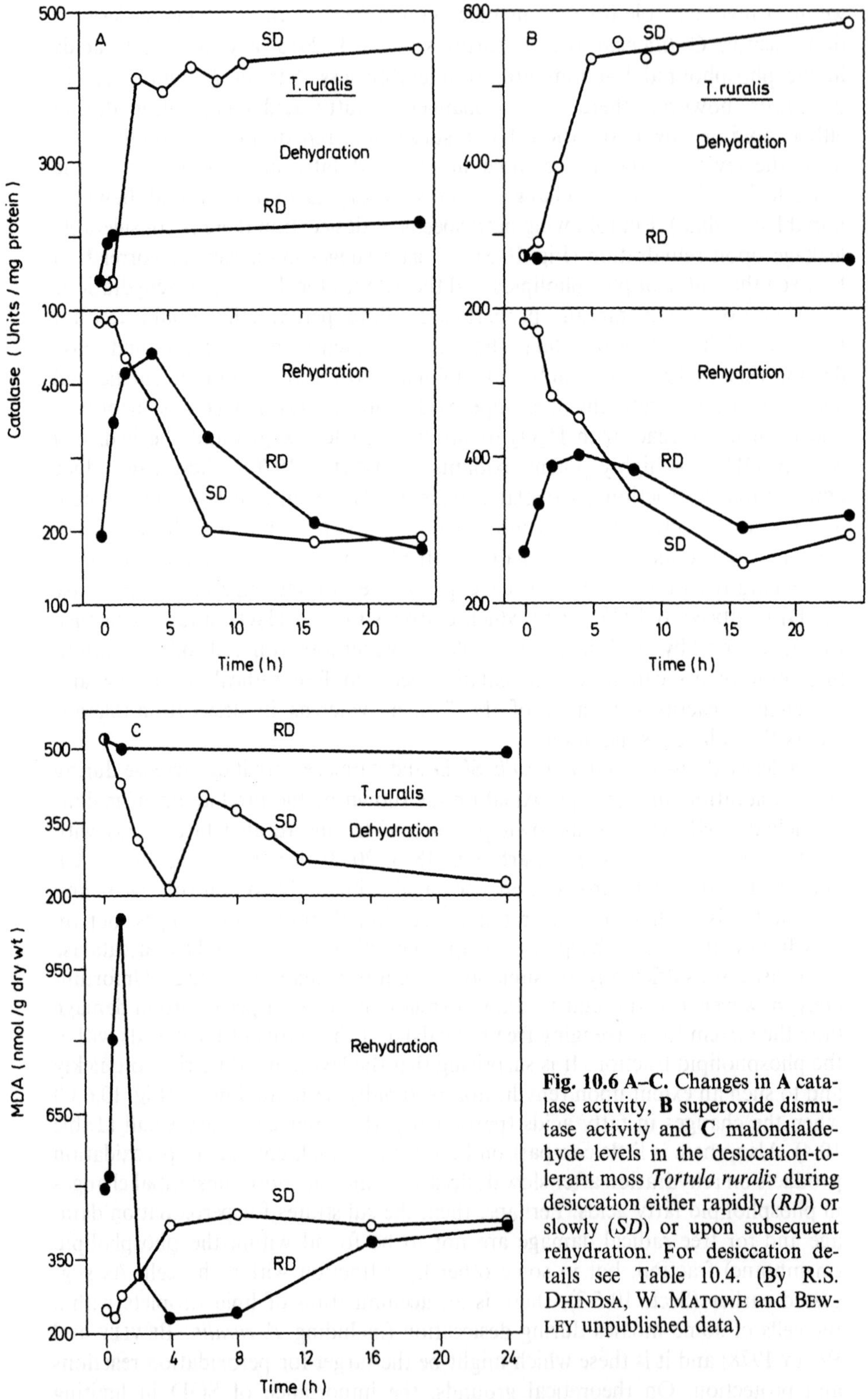

Fig. 10.6 A–C. Changes in **A** catalase activity, **B** superoxide dismutase activity and **C** malondialdehyde levels in the desiccation-tolerant moss *Tortula ruralis* during desiccation either rapidly (*RD*) or slowly (*SD*) or upon subsequent rehydration. For desiccation details see Table 10.4. (By R.S. DHINDSA, W. MATOWE and BEWLEY unpublished data)

membrane peroxidation damage is debatable anyway. SOD is a soluble enzyme (FRIDOVICH 1975) and it may be unable to exert much influence over events taking place in the hydrophobic core of the lipid bilayer.

In desiccation-intolerant *C. filicinum* catalase and SOD decline during drying at either speed, and their activities decline precipitously on rehydration (DHINDSA, MATOWE and BEWLEY unpublished data). Lipid peroxidation, as shown by malondialdehyde production, is greatest on rehydration after slow desiccation, and is 2.5 times greater after 2 h than at the start of rehydration. Also it is 2 times greater than in rehydrated rapidly desiccated moss over the same time period. Yet, the decline in phospholipid fatty acids is considerably greater after rapid than after slow desiccation.

It is evident that desiccation causes perturbations to membranes of both tolerant and intolerant plants, resulting in leakage upon subsequent rehydration. The extent of these perturbations is affected by the speed at which water is lost – rapid desiccation being more deleterious than slow desiccation. Restitution of the membranes occurs upon rehydration of desiccation-tolerant plants, even though there is temporary leakage of low molecular weight solutes until this occurs. This restitution may involve de novo synthesis of membrane components involving the so-called "repair" mechanisms. Limited evidence to date suggests that changes to the phospholipid component due to desiccation and rehydration cannot easily be correlated with the increased leakiness of membranes.

10.5.5 RNA and Protein Synthesis

Very mild to moderate water stress reduces the level of protein synthesis in drought-sensitive vegetative tissues and protein synthesis does not recover in cells subjected to severe water loss (see references in HSIAO 1973; BEWLEY 1981; BEWLEY and LARSEN 1980).

Our understanding of the effects of desiccation on protein and associated RNA synthesis in desiccation-tolerant plants has come largely from studies on the gametophyte of the moss *Tortula ruralis*. There is a recent report that protein synthesis recovers in *Xerophyta villosa* and *Myrothamnus flabellifolia* after extended periods of hydration (GAFF and MCGREGOR 1979 and cited in GAFF 1980), but little work has been done on the higher resurrection plants.

The effects of desiccation of *T. ruralis* on polysomes varies with the speed at which water loss takes place (GWÓŹDŹ et al. 1974). Rapidly desiccated moss contains about half the polysomes of the undesiccated control, but following slow desiccation polysomes are absent (Fig. 10.7). These losses of polysomes could be due to: (1) stress-induced production or activation of ribonucleases which degrade polysomes or, (2) ribosome run-off from mRNA, coupled with failure to reform an initiation complex under stress conditions. That ribosomes simply become detached from mRNA during water loss is an untenable alternative because in rapidly desiccated moss some polysomes are retained (Fig. 10.7).

Ribonuclease activity increases during both rapid and slow desiccation (Fig. 10.7), but it does so only after polysome levels have declined to their

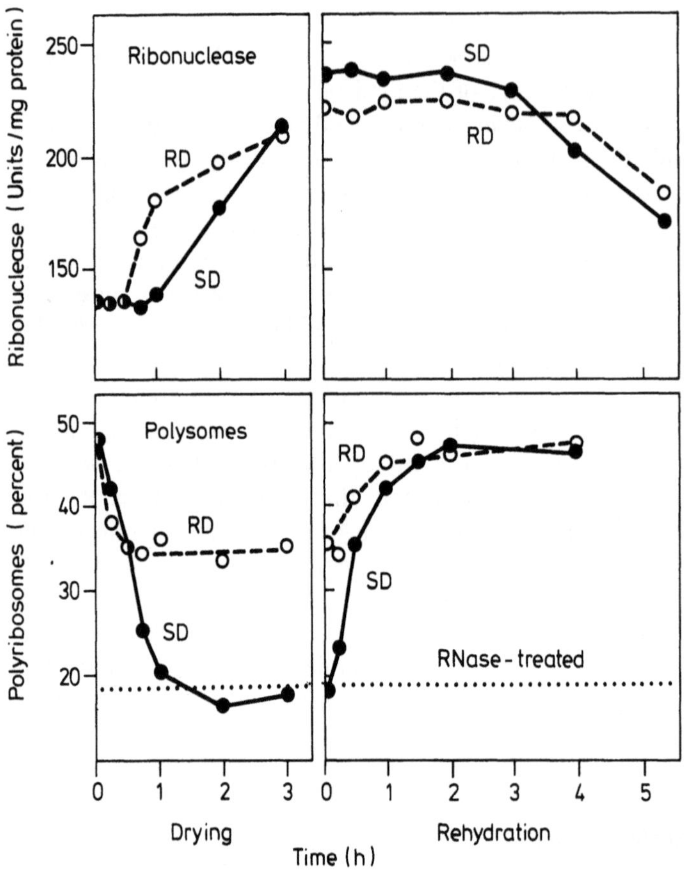

Fig. 10.7. Changes in ribonuclease (*top*) and polyribosomes (*bottom*) during rapid desiccation (*RD*) and slow desiccation (*SD*) of *Tortula ruralis*, and upon subsequent rehydration. For desiccation details see Table 10.4. (After DHINDSA and BEWLEY 1976a)

minimum (which suggests that the enzyme does not arise by de novo synthesis). Moreover, while the final ribonuclease level is the same for mosses dried at both speeds, the final polysome level is much higher after rapid desiccation. Thus the lack of temporal coincidence and quantitative correlation between the increase in enzyme activity and the polysome decrease suggests that ribonuclease is not the causative agent. Polysomes return to control levels within 2 h of rehydration after both speeds of desiccation, and yet ribonuclease levels do not decline until considerably later (Fig. 10.7). This, again, is indicative that ribonuclease is not involved in the regulation of polysome activity. Others (BLEKHMAN 1979) strongly invoke a role for ribonuclease in restricting protein synthesis during water loss from tolerant and intolerant plants, but no temporal correlations between polysome loss and enzyme levels have been made. An increase in ribonuclease as a result of water loss is clearly inadequate evidence for its participation in the cessation of protein synthesis.

During desiccation at both speeds the ribosomal pellet (containing both polysomes and single ribosomes) decreases in its capacity to effect the formation

Table 10.5. Translation in vitro of poly(A)-containing messenger RNA from undesiccated and dry gametophytes of *Tortula ruralis*. (Based on DHINDSA and BEWLEY 1978)

RNA source	^{14}C-Leucine incorporation (cpm)
Fresh moss	9,175
Rapidly desiccated	7,976
Slowly desiccated	7,185
Deletions:	
No mRNA	693
No S-23	85

of peptide bonds, as shown by its increasing inability to utilize puromycin to form peptidyl-puromycin in an in vitro assay system (DHINDSA and BEWLEY 1976a). This is evidence that as polysomes decline the released ribosomes are not complexed with fragments of messenger RNA, whereas they would be if polysomes were degraded by ribonuclease. Similar results were obtained using moss maintained under steady-state water stress conditions (−10 to −60 bar) by polyethylene glycol (DHINDSA and BEWLEY 1976b). Hence ribosome run-off from messenger RNA, coupled with failure to reform an initiation complex, appears to be the primary cause of polysome loss during water loss. Presumably during desiccation at a rapid rate critical water loss occurs before run-off can be completed.

Since loss of polysomes during desiccation of *T. ruralis* is not due to their degradation by ribonuclease it is reasonable to expect that ribosomes exist in slowly desiccated moss, separated from messenger RNA. This is indeed the case (GWÓŹDŹ and BEWLEY 1975). Moreover, cytoplasmic ribosomal RNA (24 S and 17 S) and low molecular weight RNA (4–5 S) are not degraded during desiccation or on subsequent rehydration (TUCKER and BEWLEY 1976).

That messenger RNA also is conserved in the dry moss and is utilized on subsequent rehydration is inferred from the observation that protein synthesis can resume on rehydration of slowly desiccated moss even when most RNA synthesis is inhibited by cordycepin and actinomycin D (DHINDSA and BEWLEY 1978). Other studies on the qualitative aspects of protein synthesis have shown that similar proteins are synthesized before and after desiccation (either rapid or slow) even when RNA synthesis is inhibited on rehydration. This, again, argues for messenger RNA conservation during desiccation (DHINDSA and BEWLEY 1978). Direct and unequivocal evidence for the conservation of messenger RNA in dry *Tortula ruralis* has come from studies in which poly(A)-containing RNA has been extracted from rapidly and slowly desiccated moss, and then made to catalyze the synthesis of polypeptides in vitro (Table 10.5). Extractable poly(A)-messenger RNA activity in slowly desiccated moss is about 80% of that of fully hydrated controls: rapidly desiccated moss contains some 90% messenger RNA activity, although some presumably is conserved on polysomes. Poly(A)-containing RNA also has been extracted from the hydrated and dry gametophyte of *Polytrichum commune* (SEIBERT et al. 1976). Air-drying (claimed

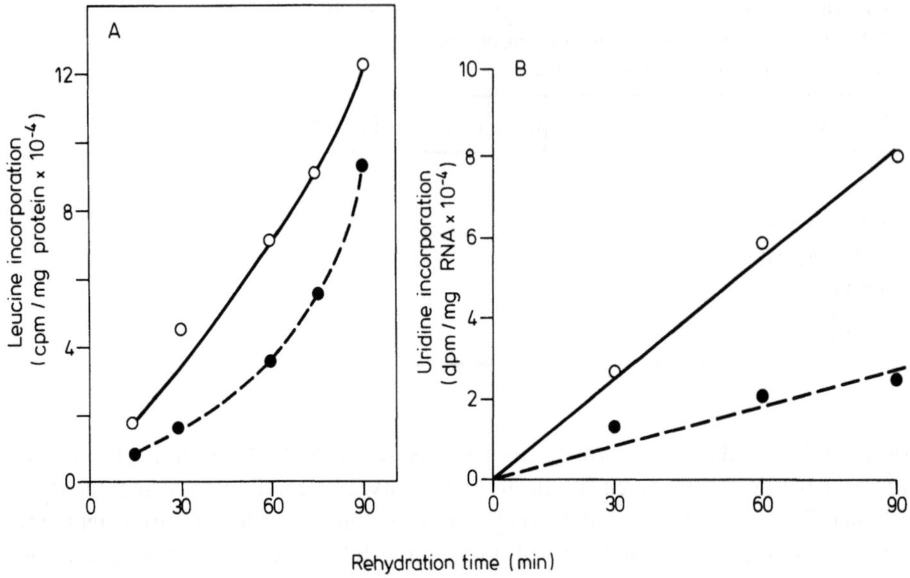

Fig. 10.8 A, B. Protein **A** and RNA **B** synthesis by *Tortula ruralis* on rehydration following rapid (--●--) and slow (-○-) desiccation. For desiccation details see Table 10.4. (**A** after Gwóźdź et al. 1974, **B** by MJ Oliver and Bewley unpublished data)

to be slow drying) of *P. commune* does not result in any degradation of total extractable RNA, but whether or not polysomes are present in the dry moss is not indicated. Hence it is not known if the poly(A)-containing RNA is conserved in the free state, or if it is associated with ribosomes in polysomes. Nor, yet, is it known whether it has messenger RNA-like properties in vitro.

Following desiccation, *T. ruralis* (Bewley 1973a; Gwóźdź et al. 1974), *P. commune* (Seibert et al. 1976) and *Neckera crispa* (Henckel et al. 1977) recommence protein synthesis within a few minutes. In *T. ruralis* polysomes are restored to control levels within 2 h (Fig. 10.7), even after the moss has been stored in the dry state for 10 months (Bewley 1973b). Protein synthesis resumes at a faster rate after slow desiccation than after rapid desiccation (Fig. 10.8 A). The reason for this is not readily apparent, though, because polysomes are present already in rapidly desiccated moss, and resume activity on rehydration (Bewley 1973a), whereas for slowly desiccated moss to restart protein synthesis the recombination of separated messenger RNA and ribosomes must occur.

RNA synthesis in *T. ruralis* also recommences rapidly on rehydration after both speeds of desiccation (Gwóźdź et al. 1974; Dhindsa and Bewley 1978), and more quickly after slow desiccation than after rapid desiccation (Fig. 10.8 B). But the synthesis of neither messenger RNA nor ribosomal RNA appears to be essential for the re-establishment of protein synthesis (Tucker and Bewley 1976; Dhindsa and Bewley 1978). It is claimed that on rehydration of *P. commune* RNA synthesis is delayed by some 30–60 min (Seibert et al. 1976), although insensitivity of the experimental technique used could contribute to this apparent delay.

For comparative purposes, similar studies have been carried out using mosses which are not normally subjected to desiccation in their usual aquatic or semiaquatic habitat. After rapid desiccation *Bryum pseudotriquetrum* does not retain any polysomes, and not only does it not reform any on rehydration but within 24 h even its ribosomes are degraded (BEWLEY et al. 1974). Subsequent studies on *Cratoneuron filicinum* (mis-identified in some publications as *Hygrohypnum luridum*) have shown that this moss retains a reduced number of polysomes in the dry state after rapid desiccation, compared to undesiccated controls (GWÓŹDŹ and BEWLEY 1975), and none following slow desiccation (MALEK and BEWLEY unpublished). Ribosomes increase as polysomes decline in the dried moss, and they retain their activity in vitro. On rehydration, however, activity of both polysomes and ribosomes decreases precipitously (GWÓŹDŹ and BEWLEY 1975), suggesting that it is rehydration, rather than desiccation, which is lethal. Predictably, after both slow and rapid desiccation protein synthesis fails to resume (KROCHKO et al. 1978). Even *C. filicinum,* however, is capable of losing a considerable amount of water before its capacity to resume protein synthesis is impaired. Loss of water to about 45% of original fresh weight is possible without a significant deleterious effect on protein synthesis on rehydration. Moreover protein synthesis can occur at −40 bar and can resume following a 6-h treatment with −60 bar of water potential imposed by polyethylene glycol (DHINDSA and BEWLEY 1977). Although this moss is intolerant of desiccation, and grows almost entirely immersed in water, it has a tolerance of water-stress which is greater than that of many terrestrial higher plants.

What is evident from the studies on *T. ruralis,* a desiccation-tolerant terrestrial moss, is that protein synthesis ceases as water is lost, and resumes again rapidly on rehydration. Protein synthesis itself is quite sensitive to drought stress, being reduced by about 50% at −20 bar water potential although a low amount occurs even at −60 bar (DHINDSA and BEWLEY 1977). Hence protein synthesis may not have a high resistance to drought stress (i.e., it may not continue under severe stress conditions) but it is tolerant of drought, in that the process can resume even after desiccation. Rapid resumption may be aided by the conservation of the components of the protein synthesizing complex in the dry state. This may be particularly important to those plants receiving only temporary hydration (e.g., by morning dew) and which have but a few hours to continue metabolism before becoming desiccated again. The effects of desiccation stress on the protein-synthesizing complex are unknown, although failure of re-initiation under stress is probably an important factor in the cessation of its activity. It may be that the components of the synthetic complex itself have an inherent tolerance of desiccation, for even extracted polysomes can be revived from the air-dry state (GWÓŹDŹ and BEWLEY 1975). It appears, however, that polysomes and ribosomes from desiccation-intolerant *C. filicinum* also retain their activity in the dry moss, as shown by in vitro assay, although they lose activity rapidly on rehydration. Hence it is likely that the possession of a stable protein-synthesizing complex which can withstand severe drought conditions is an important feature of desiccation-tolerant plants, but that other factors are more important in determining the absolute level of tolerance of the cells themselves.

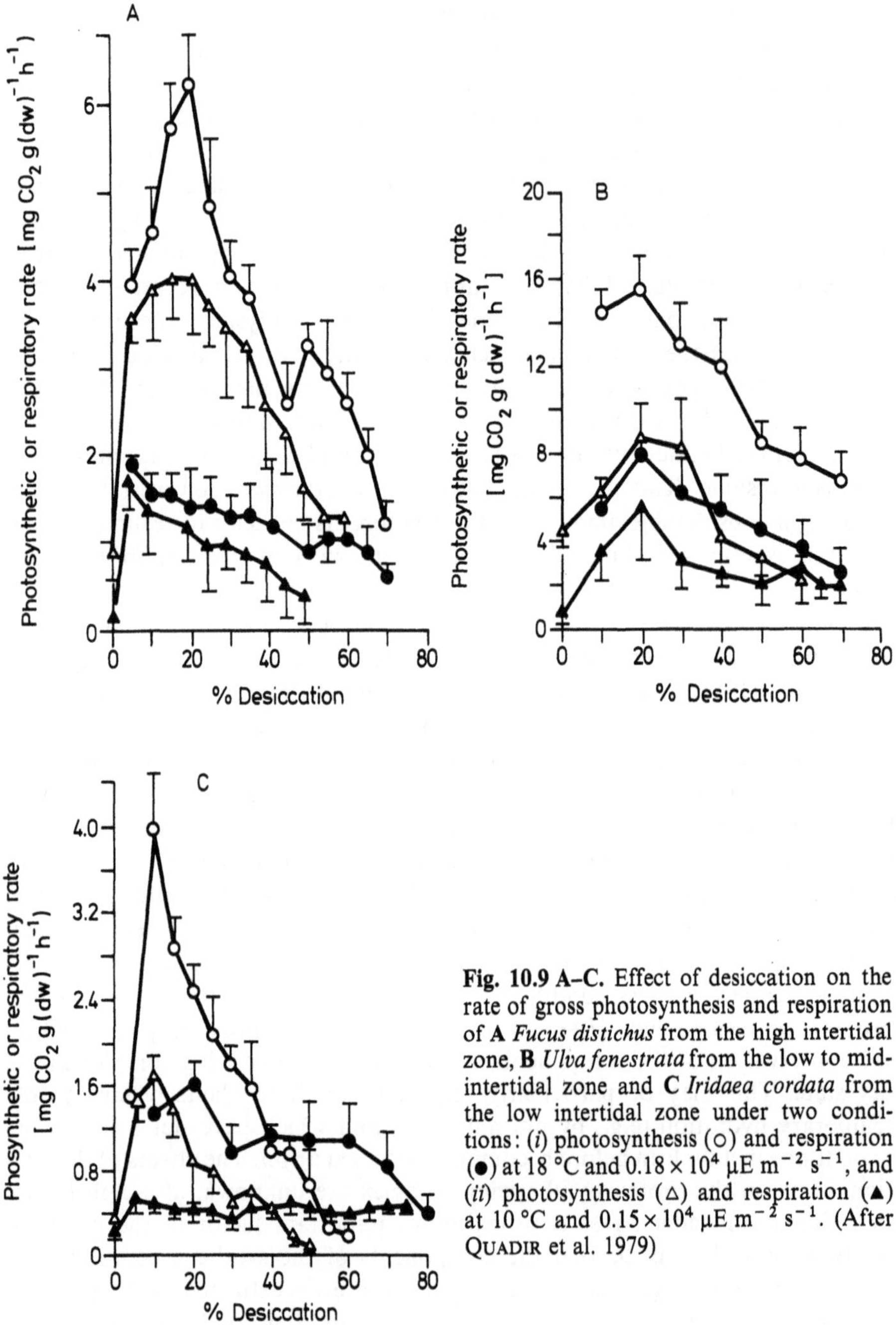

Fig. 10.9 A–C. Effect of desiccation on the rate of gross photosynthesis and respiration of **A** *Fucus distichus* from the high intertidal zone, **B** *Ulva fenestrata* from the low to mid-intertidal zone and **C** *Iridaea cordata* from the low intertidal zone under two conditions: (*i*) photosynthesis (○) and respiration (●) at 18 °C and 0.18×10^4 μE m^{-2} s^{-1}, and (*ii*) photosynthesis (△) and respiration (▲) at 10 °C and 0.15×10^4 μE m^{-2} s^{-1}. (After QUADIR et al. 1979)

10.5.6 Photosynthesis and Respiration

These are by far the most-studied metabolic processes in desiccation-tolerant plants so, for convenience, our discussion is divided into sub-sections, each one dealing with a particular group of plants.

Table 10.6. The survival, photosynthesis (rate of evolution of oxygen in μmol $O_2 \cdot g$ wet $wt^{-1} h^{-1}$), and growth (mean ± S.E.) of three species of fucoid algae after drying for 5 h at 26 °C and 48%–52% R.H. (Based on SCHONBECK and NORTON 1978)

Position in eulittoral zone	Species	Rate of O_2 evolution			Growth	
		Before drying	After 5 h drying and 18 h recovery in sea water	% Original rate	% Gain in wt after 10 days in culture	Condition of thalli after 10 days in culture
Upper fringe	*Pelvetia canaliculata*	23±1	21±1	90	32	Healthy
Upper middle	*Fucus spiralis*	29±3	13±2	45[a]	3	Damaged near tips
Lower	*Fucus serratus*	25±1	0	0	0	Dead

[a] This species will recover fully after longer periods of hydration

10.5.6.1 Algae

Several studies have shown that during drying of eulittoral algae photosynthetic capacity is gradually reduced, although most work has been done with nontolerant species which do not recover after severe water stress (CHAPMAN 1966). Photosynthesis in a variety of seaweeds in the middle and upper zones is maximal after some water loss (Fig. 10.9) [and about 25% loss from *Fucus vesiculosus* and *Ascophyllum nodosum* (BRINKHUIS et al. 1976; JOHNSON et al. 1974)], but then declines with declining fresh weight (Fig. 10.9). It is evident from Fig. 10.9 that the lower zone alga is considerably more sensitive to desiccation than those from the higher zones. Initially water loss from the surface of the fronds may enhance the uptake of CO_2, and hence an increase in photosynthesis, but as water loss progresses the photosynthetic mechanism is presumably disturbed. The kinetics of recovery of carbon fixation after drying of desiccation-tolerant marine algae do not appear to have been studied. Nevertheless, it is evident that the ability to tolerate desiccation and to resume photosynthesis and growth when resubmerged is greatest in species found highest on the shore, and is progressively less in species inhabiting successively lower levels (Table 10.6).

In the red alga *Porphyra* the primary photochemical mechanism of photosynthesis is stable in the dry state and on rehydration is reactivated instantaneously (FORK and HIYAMA 1973). During extensive water loss from desiccation-tolerant marine algae there is complete inactivation of photosynthetic partial processes, the first site that is sensitive to desiccation being the electron transport between photosystems II and I (WILTENS et al. 1978). In contrast, desiccation-intolerant algae do not recover following even limited water loss, suggesting irreversible chloroplast damage.

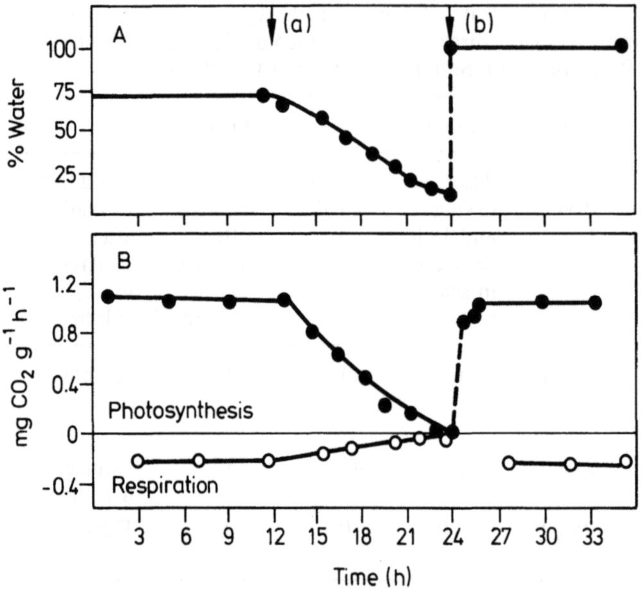

Fig. 10.10. A Water content and **B** carbon dioxide exchange during wetting and drying of the lichen *Ramalina maciformis*. *Arrow* (*a*) indicates the start of exposure to dry air and *arrow* (*b*) the time of rewetting by means of spraying. (After LANGE 1969a)

Respiration in *A. nodosum, Chondrus crispus, Ulva lactuca, F. vesiculosus* and *Gigartina stellata* declines during drying (following a brief rise after some water loss in the last two species (BRINKHUIS et al. 1976; KANWISHER 1957; MATHIESON and BURNS 1971) – see also Fig. 10.9 – and recovery on rehydration can occur after at least 80% water loss (KANWISHER 1957) (with the exception of *A. nodosum*). Similar observations have been made on other marine and freshwater species (OGATA 1968). Predictably, desiccation-tolerant terrestrial algae lose their capacity to respire and photosynthesize during drying, and recover these faculties on subsequent rehydration (BERTSCH 1966; BROCK 1975a; FRAYMOUTH 1928).

10.5.6.2 Lichens

The responses to drying of the whole lichen thallus are a combination of those of the algal (phycobiont) and of the fungal (mycobiont) component. Desiccation-induced changes in photosynthesis, for example, reflect the metabolic state of the algal component which, although it is only 3%–10% of the thallus by weight, is the fixer of carbon, subsequently transported as carbohydrate to the more massive fungal component. This latter is the major site of respiration. Several reviews have been written on the effects of environmental stresses on lichen survival and carbon assimilation (FARRAR 1973, 1976; HARRIS 1976; KAPPEN 1973; LANGE et al. 1975; SMITH 1975) and the reader should consult these for details that are not presented here (see also Chap. 10, Vol. 12 C).

Little is known about the metabolic changes that lichens undergo during desiccation, although LANGE (1969a) has demonstrated that water loss from

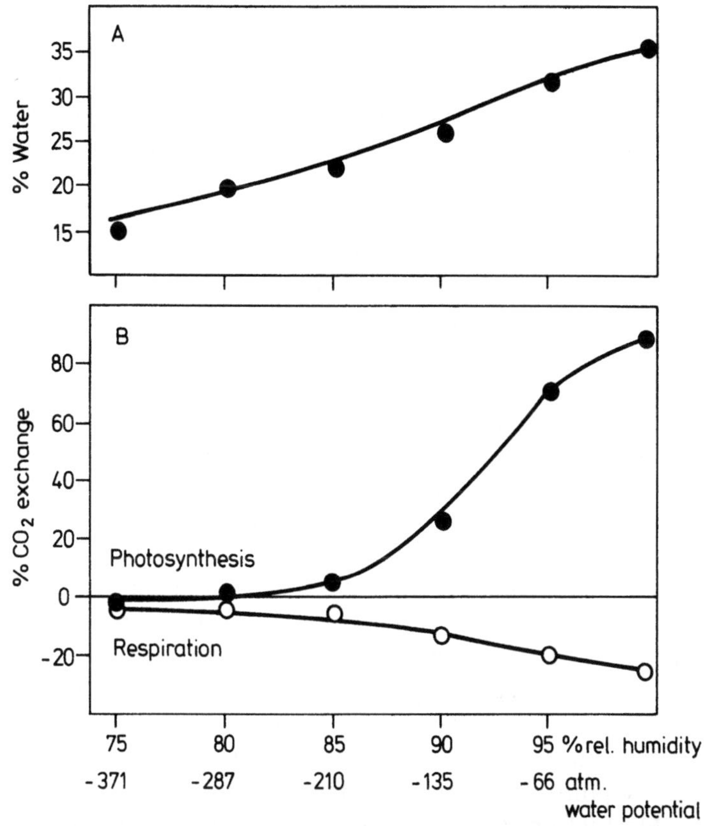

Fig. 10.11. **A** Water content and **B** carbon dioxide exchange of the lichen *Ramalina maciformis* in equilibrium with the water vapor pressure of air having different relative humidities and water potentials. CO_2 exchange at 10,000 lux is expressed as the percentage of that following spraying with water. (After LANGE 1969a)

Ramalina maciformis is paralleled by a continuous decline in CO_2 assimilation and dark respiration (Fig. 10.10). Recovery from desiccation has been studied under both field and laboratory conditions. In their natural habitat some desert lichens are rarely rewetted by rainfall, but rather resume their metabolism in atmospheres of high relative humidity and during early morning dew formation. Dry *Evernia* and *Ramalina* species can absorb sufficient water from moist air to resume CO_2 exchange; in fact, at relative humidities above 80% respiration and photosynthesis increase nonlinearly with increasing moisture content of the ambient air (BERTSCH 1966; LANGE 1969a). These lichens can achieve at least 90% of their maximum activities in water-saturated air, and yet under these conditions (where care is taken to prevent water condensing into droplets) the water content of *R. maciformis*, for example, reaches only 35% (Fig. 10.11). The compensation point of this species is reached at 80% relative humidity (10,000 lux, 10 °C), corresponding to a water content of about 20%, and a water potential close to −300 bar. Fixation of radioactive CO_2 by photosynthesis in *Lepraria membranaceae* has been observed at water potentials as extreme as −450 bar (BROCK 1975b).

When *R. maciformis* is allowed to absorb water from moist air (to a water content of 37%), on subsequent wetting of the thallus with liquid water (to 60% water content) there is an increase in CO_2 assimilation by only 20%. On the other hand, dark respiration more than doubles (LANGE 1969a). The reasons for this are not known, but these observations are of relevance to the environmental conditions to which this lichen is subjected in its natural habitat – the central Negev desert. Here the intervals between winter rainfalls can exceed 12 months, but this and other lichens are often moistened at night by humid air and by early morning dew. Before dawn the water content of the *R. maciformis* thallus may reach 31%, but the night temperatures are too low to allow for much respiration. As dawn breaks and the light intensity increases net assimilation of CO_2 occurs, but photosynthesis then declines within 2–3 h as the thallus loses water to the now drier, and warmer, air. Respiration also resumes, although less rapidly, but by the time optimal temperature conditions are reached the thallus has dried out again (LANGE et al. 1970). It has been calculated that, on a 24 h basis, the mg CO_2 yield of dry weight^{-1} by assimilation is 1.32 and loss by respiration is 0.78, to give a net yield of 0.54 mg g dry weight^{-1}, or an equivalent of 146 µg of organic carbon fixed (summarized by KAPPEN et al. 1975; LANGE et al. 1975). Further calculations, based on the fact that *R. maciformis* will be subjected normally to 198 dew events per annum in the central Negev, suggest a yearly increment of thallus growth of 8.4%. Not all desert-inhabiting lichens resume photosynthesis in response to increasing air humidities (GANNUTZ 1969): *Chondropsis semiviridis*, a native of Australian deserts, probably requires liquid water (dew or rain) (ROGERS 1971).

Most information on the metabolic responses of lichens to rehydration following desiccation has come from laboratory studies in which dried thalli have been rewetted with, and maintained on (or in) liquid water – conditions that might not be always ideal. Immediately on introduction to water there is an intense, but brief (1–2 min) nonmetabolic release of previously adsorbed gases (FARRAR and SMITH 1976; SMITH and MOLESWORTH 1973). This "wetting burst" is a purely physical process also exhibited by nonliving tissues and inert substances, and is of no significance to desiccation-tolerance. It is followed by a period of increased oxygen consumption and CO_2 output over and above basal levels of the undesiccated thallus, and this is known as "resaturation respiration", a phenomenon which may last for from one to many hours, or even days (FARRAR and SMITH 1976; RIED 1953, 1960; SMITH and MOLESWORTH 1973). Even when lichens are not fully dried they exhibit resaturation respiration on rehydration, e.g., *Peltigera polydactyla* after drying to only 40% of full saturation (SMITH and MOLESWORTH 1973). The length and magnitude of this resaturation respiration varies with species, and probably also with conditions of drying and period of desiccation. *C. semiviridis* dried in atmospheres of high relative humidity (hence probably slowly dried) showed very little resaturation respiration on rehydration, basal respiration rates being achieved within an hour (ROGERS 1971). Lichens hydrated by absorbing moisture from the air before being saturated with liquid water, show no resaturation respiration either (BERTSCH 1966; LANGE 1969a). The nature of the resaturation respiration is unknown, but it is claimed that in *Hypogymnia physodes* and *P. polydactyla*, unlike basal respiration, it is cyanide-sensitive (FARRAR and SMITH 1976; SMITH and MOLESWORTH 1973). Whether this implicates mitochondria as the site of resaturation respiration has not been resolved. Resaturation respiration is also azide-sensitive in *P. polydactyla* and DNP-sensitive in *H. physodes*. But these inhibitors have different effects upon metabolism at different concentrations, and so it is unfortunate that the effect of only one concentration was reported. Information on the relationship between oxygen consumption, ATP production and ATP requirements of rehydrated thalli is not available.

The major substrates for respiration (including resaturation respiration) are polyols (e.g., mannitol and arabitol in the fungus, and ribitol in the alga), plus the nonpolyol sucrose (FARRAR and SMITH 1976; SMITH and MOLESWORTH 1973). The highest concentration of polyols is in the fungus, they having been modified from carbon skeletons provided by photosynthetic activity of the alga (SMITH 1975). The net loss of polyols from the thallus of *H. physodes* on rehydration exceeds that which can be accounted for by resaturation respiration alone. The balance is lost due to extensive leakage from alga and fungus into the surrounding medium during the first few minutes of rehydration (FARRAR and SMITH 1976).

Many mesic and xeric lichens cannot survive in a continuously wet habitat, and thrive only under conditions of alternating wet and dry periods. When lichens are rewetted by water vapor there is no leakage of polyols nor any resaturation respiration, and hence (although it has not been established) no depletion of the pools of respirable substrate. But in lichens from temperate climates, where wetting by rain is a common occurrence, there is a potential for depletion of respirable substrates. FARRAR (1976) has hypothesized that the establishment of a sizeable and readily accessible pool of respirable substrate is an important "physiological buffer", so that in times of metabolic stress imposed by alternate wetting and drying cycles substrate may be drawn from the pool, and hence degradation of insoluble cell components (e.g., carbohydrate and protein polymers) does not occur. In time, the pool is replenished by photosynthesis when favorable conditions prevail. If the polyol pool is depleted below a critical level then recovery from desiccation is not possible. Although this concept of physiological buffering is an interesting one, its application to lichens in their natural habitat needs to be established.

Very few studies have been conducted on the metabolism of plants at very low water contents. Using lichens placed in an atmosphere of tritiated water (THO) vapor, COWAN et al. (1979) have shown that different metabolic systems operate at different degrees of desiccation. These are summarized in Fig. 10.12 A and B for *Ramalina celastri* and *Peltigera polydactyla*. It is apparent that events associated with the tricarboxylic acid cycle (b) and related transaminations (a), and with sugar alcohol metabolism (f) operate to produce metabolites at lower water contents than do events associated with carbon fixation, i.e., sugar production (c) and sugar phosphate production (d), and with lipid synthesis (e). Macromolecule synthesis (g) proceeds only at high water contents. The α-oxo acids of the tricarboxylic acid cycle are unstable during desiccation, and it is suggested (COWAN et al. 1979) that amino acid synthesis at very low water contents results in the accumulation of an available store of stable intermediates

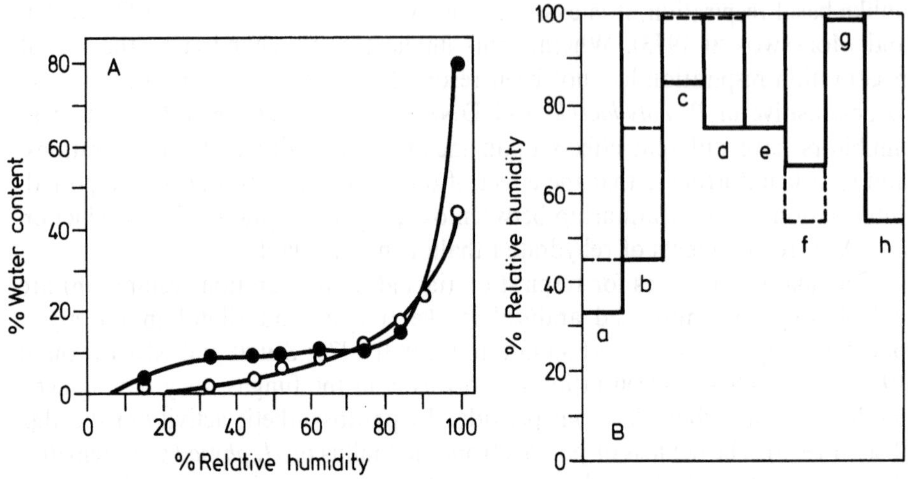

Fig. 10.12. A The relationship between percent relative humidity and percent thallus water content for the lichens *Ramalina celastri* (○) and *Peltigera polydactyla* (●). **B** The relationship between tritium labeling of the major classes of metabolites and the experimental incubation humidity for *R. celastri* (—) and *P. polydactyla* (---). *a* amino acids; *b* organic acids; *c* sugars; *d* sugar phosphates; *e* total lipids; *f* sugar alcohols; *g* soluble macromolecules; *h* insoluble cell residues. (After COWAN et al. 1979)

which feed quickly into the tricarboxylic acid cycle by transamination and deamination on rehydration of the lichen. This allows for rapid resumption of metabolism. It is also suggested that sugar alcohol metabolism assists in the maintenance of a high level of sugar alcohols that play a role as physiological buffers.

10.5.6.3 Bryophytes

Stomata are absent from the gametophytes of bryophytes and gaseous exchange occurs freely at the hydrated cell surfaces. It is evident from a number of studies (BEWLEY et al. 1978; BUSBY and WHITFIELD 1978; DILKS and PROCTOR 1974, 1976a, b, 1979; GUPTA 1977a, b, c; HENCKEL et al. 1970; HENCKEL and PRONINA 1968; HINSHIRI and PROCTOR 1971; KROCHKO et al. 1979; LEE and STEWART 1971; MCKAY 1935; SLAVIK 1965; WILLIS 1964) that decline in water content is accompanied by a decline in ability to photosynthesize and respire, that gaseous exchange does not occur in the desiccated plant, and that in desiccation-tolerant species this resumes rapidly on reintroduction of water. Desiccation-intolerant species, on the other hand, do not recover on subsequent rehydration. Different workers have used different conditions of desiccation and rehydration, and different techniques for measuring gaseous exchange and so, not surprisingly, there are discrepancies in the results obtained. For example, speed of desiccation has a profound effect on the rate of gaseous exchange upon rehydration, as shown in studies of desiccation-tolerant *Tortula ruralis*, and intolerant *Cratoneuron filicinum* (see later). During very slow drying of both species oxygen consumption increases as water is lost, with a sharp increase

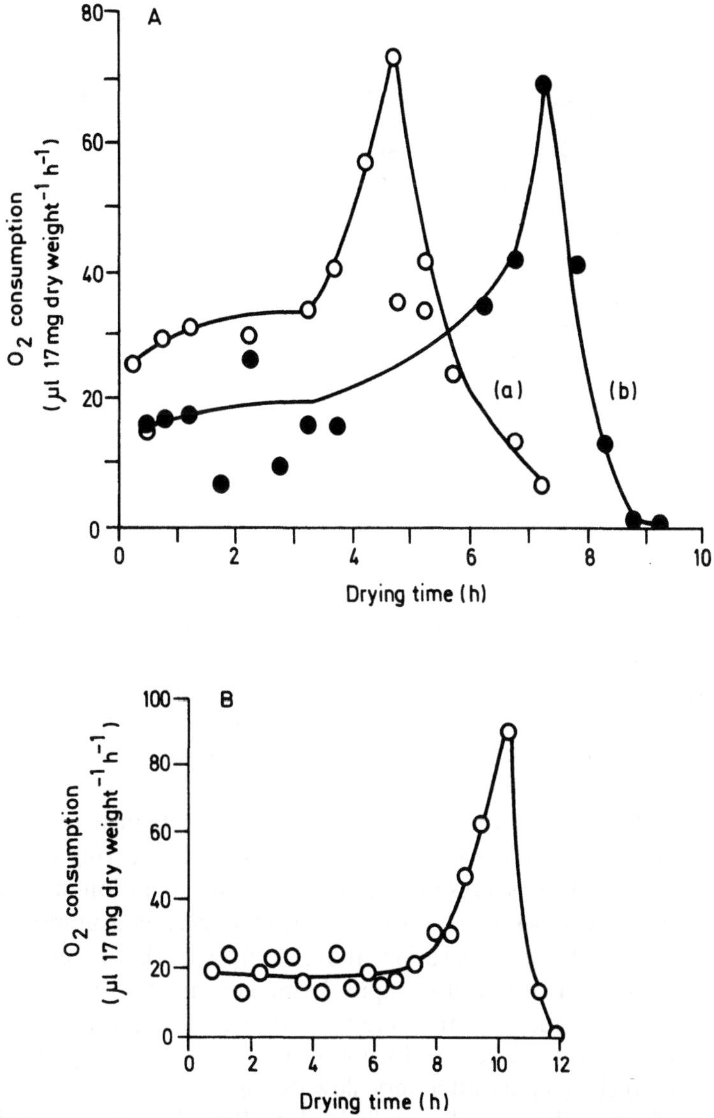

Fig. 10.13 A, B. Oxygen consumption during desiccation of **A** slowly (*a*) and very slowly (*b*) dried *Tortula ruralis* and **B** very slowly dried *Cratoneuron filicinum*. (After KROCHKO et al. 1979)

just prior to dryness (Fig. 10.13). This may not be a consequence only of respiratory activity but of other cellular oxidations, for the pattern of CO_2 evolution in *T. ruralis* is dissimilar from that of oxygen consumption during drying: it declines steadily with water loss (KROCHKO et al. 1979). Oxygen evolution (indicative of photosynthesis) also declines steadily during drying of *T. ruralis* and *C. filicinum* (BEWLEY et al. 1978). In some species there may be an initial elevation of photosynthesis as water is lost (GUPTA 1977b; HENCKEL et al. 1970) which may result from evaporation of excess surface moisture, thus obviating CO_2 diffusion. It has been suggested by HENCKEL and PRONINA (1968)

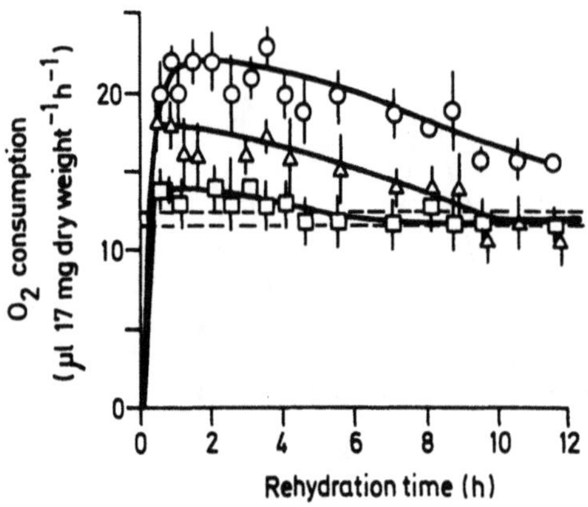

Fig. 10.14. Oxygen consumption in darkness on rehydration of *Tortula ruralis* following rapid desiccation (*RD*) (○), slow desiccation (*SD*) (△) and very slow desiccation (*VSD*) (□). Oxygen consumption by hydrated control moss is shown by the *dashed lines*. (After KROCHKO et al. 1979)

that desiccation-tolerant species can be distinguished from intolerant because in the former respiration (oxygen uptake) falls continually with water loss, whereas in the latter it rises during the period when most water is lost, and then declines sharply. This hypothesis does not have universal applicability, however, because *T. ruralis* and *C. filicinum* exhibit similar patterns of oxygen exchange during desiccation (Fig. 10.13).

On rehydration of *T. ruralis* after rapid drying there is a doubling of oxygen consumption over control levels (cf. "resaturation respiration" in lichens), which are not regained until after about 24 h (Fig. 10.14). Evolution of CO_2 in darkness on rehydration is not elevated by more than 10%, however (KROCHKO et al. 1979). On rehydration after slow drying a moderate increase in oxygen consumption occurs, for some 10–12 h, and after very slow drying the elevation is small and brief. Like basal respiration, resaturation respiration in *T. ruralis* is cyanide-sensitive (KROCHKO 1979). In *C. filicinum* oxygen consumption shows a minor, and relatively brief elevation above control levels on rehydration after slow speeds of drying, but after rapid drying consumption is small (KROCHKO et al. 1979). Ultrastructural studies (KROCHKO et al. 1978) have shown that after rapid drying there is extensive disruption of cellular integrity, but that some cells retain their integrity after very slow drying.

Photosynthesis in *T. ruralis* is also affected by prior speed of desiccation, and by duration of the moss in the dry state. As shown in Fig. 10.15, very rapid drying to −2,080 bar, which reduced tissue water potential from 4 to 0.016 g g dry wt^{-1} in less than an hour, caused a severe but temporary depression of net O_2 evolution upon rehydration. This was more severe after 7 days than after 2 days in the dry state. Slow drying to this, and less extreme (−207 bar: 0.6 g g dry wt^{-1}) stress, or rapid drying to −207 bar resulted in less perturbation and faster recovery (Fig. 10.15). Thus the pattern of gaseous exchange may be affected by the duration that mosses spend in the dry state. In their natural environments most desiccation-tolerant bryophytes may remain dry for several weeks or months between rainfalls, although not all species need to take up liquid water to resume metabolism – some mosses recommence photosynthesis after absorbing water vapor from a moist atmosphere (LANGE 1969b). A dry period of several days to several weeks probably has little effect on the

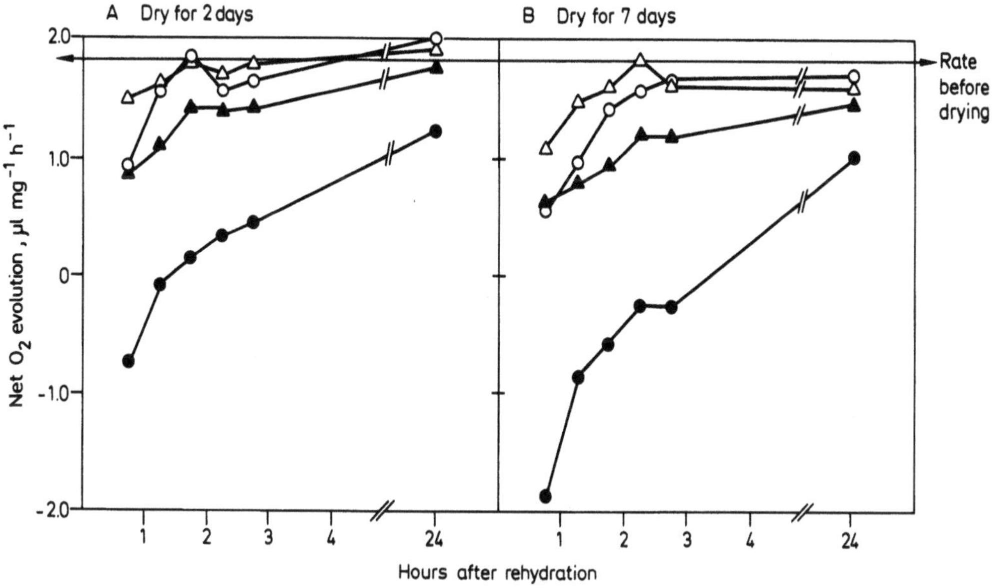

Fig. 10.15 A, B. Net photosynthetic rates of *Tortula ruralis* upon rehydration following rapid (● ▲) or slow (○ △) desiccation to −207 bar (▲ △) or −2,080 bar (● ○) and maintenance in the dry state for **A** 2 days or **B** 7 days. (By MW SCHONBECK and BEWLEY unpublished data)

metabolic recovery of tolerant mosses (Fig. 10.15 and DILKS and PROCTOR 1976a; HINSHIRI and PROCTOR 1971). For example, in *Anomodon viticulosus* and *Porella platyphylla* recovery of net assimilation is complete within 3–4 h of rewetting after maintenance in the dry state for 22 and 60 days, respectively. But after longer time periods net assimilation is initially negative and may never recover. Other factors to be considered are possible seasonal variations in desiccation-tolerance (DILKS and PROCTOR 1976b; LEE and STEWART 1971) and variations between wet and dry races of the same species (LEE and STEWART 1971).

Chlorophyll content of desiccation-tolerant species is largely unaffected by drying and rehydration (BEWLEY et al. 1974; GUPTA 1978), although considerable chlorophyll loss occurs from intolerant species (GUPTA 1978). This presumably is a reflection of general cellular degradation. In *T. ruralis* RuBP carboxylase activity declines by about 30% during desiccation, and recovers on rehydration (SEN GUPTA 1977), but this is unlikely to be the cause of the observed changes in photosynthesis during water loss. Desiccation of the moss *Acrocladium cuspidatum* has little effect on the in vitro activities of nine photosynthetic enzymes (including RuBP carboxylase) even though photosynthesis is completely (and to a considerable extent irreversibly) supressed (LEE and STEWART 1971; STEWART and LEE 1972). One enzyme, though, $NADP^+$-dependent glyceraldehyde-3-phosphate dehydrogenase shows reduced in vitro activity when extracted from dried moss, and this can be restored by the addition of GSH. Perhaps, then, desiccation brings about the oxidation of sulfhydryl residues of the $NADP^+$-G3PDH enzyme, thus accounting, at least in part, for loss of photosynthetic

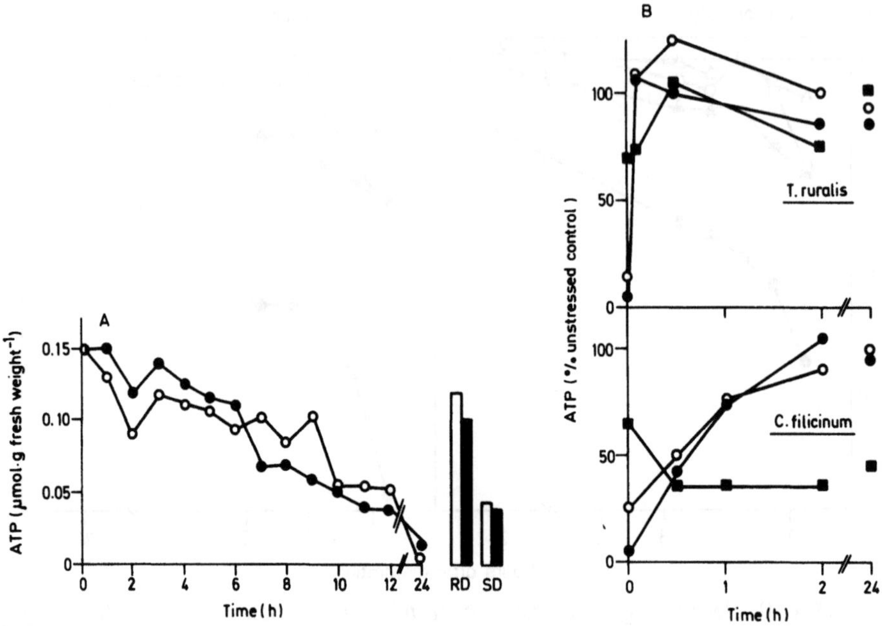

Fig. 10.16. A ATP content during very slow desiccation of *Tortula ruralis* (●) and *Cratoneuron filicinum* (○). Histograms show ATP levels after 24 h in rapidly dried (*RD*) and slowly dried (*SD*) *T. ruralis* (*black*) and *C. filicinum* (*white*), **B** ATP content during rehydration of *T. ruralis* (*top*) and *C. filicinum* (*bottom*) following rapid desiccation (■), slow desiccation (○) and very slow desiccation (●). (After KROCHKO et al. 1979)

capacity. In species where photosynthesis does not resume fully on rehydration this oxidation presumably cannot be reversed in vivo: in tolerant species the enzyme might be protected from oxidation (perhaps the rise in oxygen consumption prior to the completion of desiccation (Fig. 10.13) is related to oxidation of cellular components), or oxidations may be reversible. Loss of enzyme activity from intolerant species is unlikely to be the sole cause of irreversible changes occurring during desiccation (and/or rehydration) for the resultant disorganization of chloroplast integrity (Sect. 10.5.3) could be equally or more important.

HENCKEL and PRONINA (1968) have postulated that the maintenance of control levels of "acid-soluble organic phosphorus with energy rich bonds" (presumably ATP) during drying is an important feature of desiccation-tolerance, to allow their "energy" to be transferred to organelle/membrane structures to facilitate retention of their structural integrity. An apparent decline in ATP levels in desiccation-intolerant *Atrichum undulatum* has been correlated with a rise in ATPase activity, while in the tolerant moss *Neckera crispa* ATP levels remain high and ATPase levels are unaffected by drying (HENCKEL and PRONINA 1968; PRONINA 1972). This hypothesis is questionable, however, and more recent studies on *T. ruralis* and *C. filicinum* clearly do not support it. Changes in ATP levels during desiccation of these two mosses are very similar (Fig. 10.16 A). During very slow drying there is a gradual loss of ATP until in the dried moss negligible amounts remain. After slow drying there is a reduction by

Table 10.7. Percentage of total carbon fixed by nonautotrophic dark CO_2 fixation during slow desiccation of *Tortula ruralis*. (Based on data by SEN GUPTA 1977)

Drying time (h)	Percent Fresh weight	Percent of total C fixed by dark fixation
Undesiccated control	100	4.6
0.25	92	3.0
0.5	87	4.4
1	75	4.0
2	62	4.6
3	54	6.7
4	43	8.4
5	30	64.9
6	23	100
7	17 (dry)	No carbon fixed

about 66% below undesiccated control levels, but after rapid drying ATP levels are undiminished. Rehydration of *T. ruralis* after all speeds of drying results in a rapid return to control levels, usually within 5 min of rewetting (Fig. 10.16 B) (BEWLEY and GWÓŹDŹ 1975). ATP levels in *C. filicinum* increase on rewetting following slow and very slow drying, but not after rapid drying (Fig. 10.16 B). These observations are consistent with those on recovery of respiration. As far as this link between oxygen consumption and ATP synthesis is concerned, there appears to be little correlation between the two during drying of either *T. ruralis* or *C. filicinum*. Nor is there any apparent link during rehydration of *T. ruralis*. Here the amounts of oxygen consumed after rapid and very slow drying are very different (Fig. 10.14), and for many hours, and yet ATP levels are similar (Fig. 10.16 B). It could be that after rapid desiccation ATP production is largely uncoupled from oxygen consumption, but still sufficiently coupled to allow for the maintenance of a constant pool. Certainly mitochondria are considerably swollen after rapid desiccation (TUCKER et al. 1975; BEWLEY and PACEY 1978) which is indicative that their function might be impaired. Alternatively, but seemingly less likely, ATP production and utilization could both be higher, and to an equal extent.

A third aspect of CO_2 exchange, that of dark- or nonautotrophic-CO_2 fixation has been studied in *T. ruralis* and *C. filicinum* (SEN GUPTA 1977). During drying of *T. ruralis* dark fixation of CO_2 declines at a slower speed than does photosynthesis. Whereas it accounts for only 5% of the total CO_2 fixed in hydrated control moss, in nearly-dry moss it accounts for 60%–100% (Table 10.7). Desiccated moss does not fix CO_2, but on rehydration this capacity is regained fully within 5 min, irrespective of prior desiccation speed (Fig. 10.17). Hence, initially upon rehydration dark fixation may account for a considerable percentage of total CO_2 assimilation, and may be important in an anaplerotic sense, enabling TCA cycle intermediates to be directed for synthetic purposes. In *C. filicinum* (the intolerant species) dark fixation of CO_2 is very sensitive to water loss, even more so than is photosynthesis. After rapid desiccation the moss is incapable of light or dark fixation of CO_2, and does so only

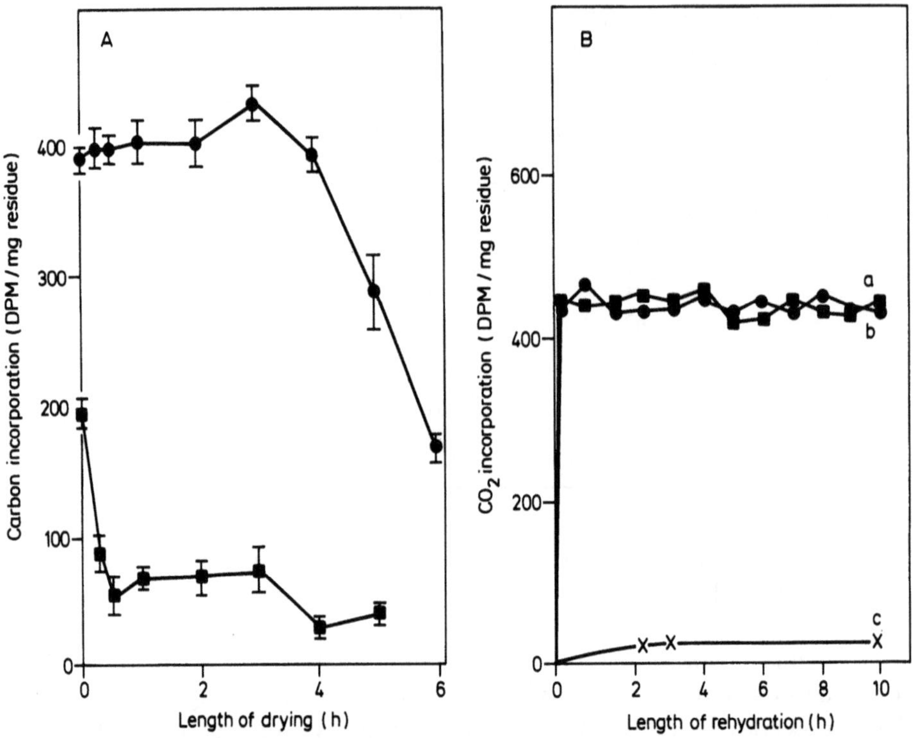

Fig. 10.17. A The effect of slow desiccation on nonautotrophic dark CO_2-fixation in *Tortula ruralis* (●) and *Cratoneuron filicinum* (■). **B** The recovery of dark fixation in *a* slow-dried and **b** fast-dried *T. ruralis* upon rehydration, and *c* in slow-dried *C. filicinum* upon rehydration. (Based on Sen Gupta 1977)

poorly after slower speeds of desiccation (Fig. 10.17). In vitro activities of PEP carboxylase, GOT and MDH do not decline during drying and subsequent rehydration of either *T. ruralis* or *C. filicinum* and so loss of dark-fixation ability is not the result of enzyme destruction within the latter. An alternate, but entirely speculative, possibility is that increasing water deficits cause a decline in acceptor molecules (e.g., PEP) or cofactors.

Dark CO_2 fixation has also recently been reported in the hydrated thalli of two lichen species at rates between 10% and 25% of photosynthesis (Kershaw et al. 1979). Dark fixation is depressed at low thallus water contents, in a similar manner to net photosynthesis.

10.5.6.4 Vegetative Tissues of Vascular Plants

The effects of mild to moderate water stress on CO_2 exchange by desiccation-intolerant higher plants has received considerable attention and there are several extensive reviews (in particular on the effects of drought on photosynthesis) (e.g., Boyer 1976; Cooper 1975; Hsiao 1973; see Chap. 9, this Vol.). Desiccation of intolerant species results in irreversible disruption of cellular organization and metabolism, although some photosynthetic reactions can occur at severe water deficits (Santarius 1967). In the desiccation-tolerant fern *Polypodium polypodioides* both respiration and photosynthesis are reduced as water content declines, and below 35%–40% water content they cease (STUART 1968). Whether reduced photosynthesis due to water loss is mediated through effects on stomata or by changes in the chloroplasts, or both, has not been investigated, although the photosynthetic pigment content of *P. polypodioides* leaves is unaffected by desiccation. On rehydration of this fern, and of *Ceterach officinarum* (OPPENHEIMER and HALEVY 1962) respiration resumes at a higher level than in undesiccated controls. In *C. officinarum* it takes several days for respiration to return to control levels: data are not available for *P. polypodioides*. Photosynthesis resumes soon after rehydration in this latter species (STUART 1968). Desiccation of the poikilohydrous spikemoss *Selaginella lepidophylla* results in a decline in RuBP carboxylase activity by about 40%, but recovery of activity occurs within the first 24 h of rehydration – possibly by de novo synthesis of the enzyme (EICKMEIER 1979). Thus recovery of photosynthesis in this plant after desiccation may be facilitated by the high level of conserved RuBP carboxylase and its ability to quickly restore enzyme levels to normal.

Myrothamnus flabellifolia is a desiccation-tolerant angiosperm which can resume photosynthesis and respiration after long periods in the dry state (HENCKEL and PRONINA 1969; HOFFMANN 1968; VIEWEG and ZIEGLER 1969; ZIEGLER and VIEWEG 1970). A number of important soluble enzymes involved in these processes remain potentially active in the desiccated tissue (GÜNDEL, quoted by KLUGE 1976), and the chlorophyll and carotenoid content of the plant is unchanged during desiccation and rehydration (HOFFMANN 1968). It has been implied (KLUGE 1976) that desiccation-induced changes in membrane components of mitochondria and chloroplasts are responsible for the decline in gaseous exchange, and that their reconstitution on rehydration is essential for resumption of photosynthesis and respiration. Some suggestive evidence has been provided in electron microscope studies (HOFFMANN 1968; ZIEGLER and VIEWEG 1970; WELLBURN and WELLBURN 1976), but more extensive studies are required.

Not all desiccation-tolerant plants retain their photosynthetic pigments within intact chloroplasts during desiccation. The poikilochlorophyllous angiosperms lose both, and reconstitute or resynthesize them on rehydration (GAFF and HALLAM 1974; GAFF et al. 1976; MEGURO et al. 1977). Changes in gaseous exchange during drying and rehydration unfortunately have not been followed in these plants, although electron microscopy studies on organelle integrity have been carried out (see earlier). It is interesting to note that *Borya nitida* only survives desiccation if it is dried slowly, during which chlorophyll is lost, so that the dried plant is yellow. Rapid drying, which does not allow for destruction of chlorophyll, is fatal (GAFF and CHURCHILL 1976).

In concluding this section, it is apparent that in many desiccation-tolerant plants there is a considerable degree of conservation of the components of the respiratory and photosynthetic mechanisms. Perturbations may occur to the structure and contents of both mitochondria and chloroplasts early upon rehydration, but events are soon put in motion to reverse or repair these perturbations. Elucidation of the nature of these repair mechanisms is a fertile field for research.

10.6 Conclusions

While desiccation-tolerance is relatively prevalent in lower plants, this characteristic has been reacquired by a number of higher, vascular plants in unrelated orders and families. Some poikilohydrous plants are able to survive in exposed, uncompromising habitats, that less tolerant plants cannot survive, but others are to be found only in areas where they are protected from the harshest elements of their environment (particularly direct sunlight). Such protection may allow plants to retain water longer, and hence increase their productivity, or to lose water more slowly and thus make the appropriate cytoplasmic adjustments that will allow them to tolerate severe water loss.

There do not appear to be any common morphological features that distinguish poikilohydrous plants from intolerant species – tolerance of desiccation can be attributed largely to properties of the protoplasm. Tolerant plants do not undergo severe water deficits without perturbations to their metabolic and cellular integrity, but critical features of desiccation-tolerance may be the abilities to limit damage due to desiccation to a reparable level and to repair damage quickly upon subsequent rehydration. The integrity of the majority of metabolic systems (or at least of their components) is probably retained in the dry state, and these systems are quickly reactivated upon rehydration. This is probably also an important feature, for hydration of many poikilohydrous plants is a transient phenomenon. Obviously, any plant in any environment can only survive if it is able to achieve a net gain in its carbon status, and the adaptations of poikilohydry can be viewed in relation to this fact (EICKMEIER 1979). Poikilohydric plants can occupy habitats that are unavailable to intolerant plants because they can benefit, relative to potential carbon gain from (1) extreme desiccation (and sometimes temperature) tolerance, (2) utilization of dew or water vapor for hydration (some species only), (3) rapid emergence from the quiescent state and metabolic, including photosynthetic, recovery with the availability of water. On the other hand, these benefits must be considered against the potential carbon costs associated with (1) stored energy and/or time costs for physiological reactivation upon rehydration, (2) physiological deterioration with increasing time in the dry state, (3) low photosynthetic rates, and (4) decline in membrane integrity permitting carbohydrate and nutrient losses during rehydration with liquid water. Different species have developed different ecological, morphological, and (presumably) physiological adaptations to accomplish, ultimately, a net gain in their photosynthetic productivity.

References

Alpert P (1979) Desiccation of desert mosses following a summer rainstorm. Bryologist 82:65–71

Barnett NM, Naylor AW (1966) Amino acid and protein metabolism in Bermudagrass during water stress. Plant Physiol 41:1222–1230

Baskin CC, Baskin JM (1974) Responses of *Astragalus tennesseensis* to drought. Changes in free amino acids and amides during water stress and possible ecological significance. Oecologia 17:11–16

Bérard-Therriault L, Cardinal A (1973) Importance de certains facteurs ecologiques sur la resistance a la desiccation des Fucales (Phaeophyceae). Phycologia 12:41–52

Bertsch A (1966) CO_2-Gaswechsel und Wasserhaushalt der aerophilen Grünalge *Apatococcus lobatus*. Planta 70:46–72

Bewley JD (1973a) Polyribosomes conserved during desiccation of the moss *Tortula ruralis* are active. Plant Physiol 51:285–288

Bewley JD (1973b) Desiccation and protein synthesis in the moss *Tortula ruralis*. Can J Bot 51:203–206

Bewley JD (1973c) The effects of liquid nitrogen temperatures in protein and RNA synthesis in the moss *Tortula ruralis*. Plant Sci Lett 1:303–308

Bewley JD (1979) Physiological aspects of desiccation tolerance. Annu Rev Plant Physiol 30:195–238

Bewley JD (1981) Protein synthesis. In: Paleg LG, Aspinall D (eds) Physiology and biochemistry of drought resistance in plants. Academic Press, London New York, pp 261–282

Bewley JD, Gwóźdź EA (1975) Plant desiccation and protein synthesis. II. On the relationship between endogenous ATP levels and protein synthesizing capacity. Plant Physiol 55:1110–1114

Bewley JD, Larsen KM (1980) Protein synthesis ceases in water-stressed pea roots and maize mesocotyls without loss of polyribosomes. Effects of lethal and non-lethal water stress. J Exp Bot 31:1245–1256

Bewley JD, Pacey J (1978) Desiccation-induced ultrastructural changes in drought-sensitive and drought-tolerant plants. In: Crowe JH, Clegg JS (eds) Dry biological systems. Academic Press, London New York, pp 53–73

Bewley JD, Thorpe TA (1974) On the metabolism of *Tortula ruralis* following desiccation and freezing: Respiration and carbohydrate oxidation. Physiol Plant 32:147–153

Bewley JD, Tucker EB, Gwóźdź EA (1974) The effects of stress on the metabolism of *Tortula ruralis*. In: Bieleski RL, Ferguson AR, Cresswell MM (eds) Mechanisms of regulation of plant growth. Roy Soc NZ Bull 12:395–402

Bewley JD, Halmer P, Krochko J, Winner WE (1978) Metabolism of a drought-tolerant and a drought-sensitive moss. Respiration, ATP synthesis and carbohydrate status. In: Crowe JH, Clegg JS (eds) Dried biological systems. Academic Press, London New York, pp 185–203

Biebl R (1962) Seaweeds. In: Lewin RA (ed) Physiology and biochemistry of algae. Academic Press, London New York, pp 799–815

Blekhman GI (1979) Factors behind changes of ribonuclease activity and characteristics of manifestations of this activity during dehydration of plants. Sov Plant Physiol 26:754–762

Boyer JS (1976) Water deficits and photosynthesis. In: Kozlowski TT (ed) Water deficits and plant growth vol IV. Academic Press, London New York, pp 153–190

Brinkhuis BH, Tempel NR, Jones RF (1976) Photosynthesis and respiration of exposed salt-marsh fucoids. Mar Biol 34:349–360

Brock TD (1975a) Effect of water potential on a *Microcoleus* (Cyanophyceae) from a desert crust. J Phycol 11:316–320

Brock TD (1975b) The effect of water potential on photosynthesis in whole lichens and in their liberated algal components. Planta 124:13–23

Brown DH, Buck GW (1979) Desiccation effects and cation distribution in bryophytes. New Phytol 82:115–125

Buck GW, Brown DH (1979) The effect of desiccation on cation location in lichens. Ann Bot (London) 44:265–277

Busby JR, Whitfield DWA (1978) Water potential, water content, and net assimilation of some boreal forest mosses. Can J Bot 56:1551–1558

Chapman ARO (1973) A critique of prevailing attitudes towards the control of seaweed zonation on the sea shore. Bot Mar 16:80–82

Chapman VJ (1966) The physiological ecology of some New Zealand seaweeds. In: Young EG, McLachlan JL (eds) Proc 5th Int Seaweed Symp Pergamon, Oxford, pp 29–54

Clausen E (1952) Hepatics and humidity, a study on the occurrence of hepatics in a

Danish tract and the influence of relative humidity on their distribution. Dan Bot Ark 15:5–80
Clausen E (1964) The tolerance of hepatics to desiccation and temperature. Bryologist 67:411–417
Cloudsley-Thompson JL, Chadwick MJ (1964) Life in deserts. Defour Editions, Philadelphia
Cooper JP (ed) (1975) Photosynthesis and productivity in different environments. Univ Press, Cambridge
Cowan DA, Green TGA, Wilson AT (1979) Lichen metabolism I. The use of tritium labelled water in studies of anhydrobiotic metabolism in *Ramalina celastri* and *Peltigera polydactyla*. New Phytol 82:489–503
Darbyshire B (1974) The function of the carbohydrate units of three fungal enzymes in their resistance to dehydration. Plant Physiol 54:717–721
Darbyshire B, Steer BT (1973) Dehydration of macromolecules. I. Effect of dehydration-rehydration on indoleacetic acid oxidase, ribonuclease, ribulosediphosphate carboxylase, and ketose-1-phosphate aldolase. Aust J Biol Sci 26:591–604
Davis JS (1972) Survival records in the algae, and the survival role of certain algal pigments, fat and mucilaginous substances. Biologist 54:52–93
Dhindsa RS, Bewley JD (1976a) Plant desiccation: polyribosome loss not due to ribonuclease. Science 191:181–182
Dhindsa RS, Bewley JD (1976b) Water stress and protein synthesis. IV. Responses of a drought-tolerant plant. J Exp Bot 27:513–523
Dhindsa RS, Bewley JD (1977) Water stress and protein synthesis. V. Protein synthesis, protein stability and membrane permeability in a drought-sensitive and a drought-tolerant moss. Plant Physiol 59:295–300
Dhindsa RS, Bewley JD (1978) Messenger RNA is conserved during drying of drought-tolerant *Tortula ruralis*. Proc Natl Acad Sci USA 75:842–846
Dilks TJK, Proctor MCF (1974) The pattern of recovery of bryophytes after desiccation. J Bryol 8:97–115
Dilks TJK, Proctor MCF (1976a) Effects of intermittent desiccation on bryophytes. J Bryol 9:249–264
Dilks TJK, Proctor MCF (1976b) Seasonal variation in desiccation tolerance in some British bryophytes. J Bryol 9:239–247
Dilks TJK, Proctor MCF (1979) Photosynthesis, respiration and water content in bryophytes. New Phytol 82:87–114
Dromgoole FI (1980) Desiccation resistance of intertidal and subtidal algae. Bot Mar 23:149–159
Eickmeier WG (1979) Photosynthetic recovery in the resurrection plant *Selaginella lepidophylla* after wetting. Oecologia 39:93–106
Evans JH (1959) The survival of fresh-water algae during dry periods II. Drying environments. J Ecol 47:55–81
Farrar JF (1973) Lichen physiology: progress and pitfalls. In: Ferry DW, Baddeley MS, Hawksworth DL (eds) Air pollution and lichens. Athlone, London, pp 238–282
Farrar JF (1976) The lichen as an ecosystem: observation and experiment. In: Brown DH, Hawksworth DL, Bailey RH (eds) Lichenology: progress and problems. Academic Press, London New York, pp 385–406
Farrar JF, Smith DC (1976) Ecological physiology of the lichen *Hypogymnia physodes*. III. The importance of the rewetting phase. New Phytol 77:115–125
Fellows RJ, Boyer JS (1978) Altered ultrastructure of cells of sunflower leaves having low water potentials. Protoplasma 93:381–395
Finean JB (1969) Biophysical contributions to membrane structure. Q Rev Biophys 2:1–23
Fogg GE (1969) Survival of algae under adverse conditions. In: Woolhouse HW (ed) Dormancy and survival. Symp Soc Exp Biol vol 23. Univ Press, Cambridge, pp 123–142
Fork DC, Hiyama T (1973) The photochemical reactions of photosynthesis in an alga exposed to extreme conditions. Carnegie Inst Washington Yearb 72:384–388
Fraymouth J (1928) The moisture relations of terrestrial algae. III. The respiration of certain lower plants, including terrestrial algae, with special reference to the influence of drought. Ann Bot (London) 42:75–100

Freeman TP, Duysen ME (1975) The effect of imposed water stress on the development and ultrastructure of wheat chloroplasts. Protoplasma 83:131–145
Fridovich I (1975) Superoxide dismutases. Annu Rev Biochem 44:147–159
Fridovich I (1976) Oxygen radicals, hydrogen peroxide, and oxygen toxicity. In: Pryor WA (ed) Free radicals in biology vol I. Academic Press, London New York, pp 239–277
Gaff DF (1971) Desiccation-tolerant flowering plants in Southern Africa. Science 174:1033–1034
Gaff DF (1972) Drought resistance in *Welwitschia mirabilis* Hook. fil. Dinteria 7:3–7
Gaff DF (1977) Desiccation tolerant vascular plants of Southern Africa. Oecologia 31:95–109
Gaff DF (1980) Protoplasmic tolerance of extreme water stress. In: Turner NC, Kramer PJ (eds) Adaptation of plants to water and high temperature stress. Wiley-Interscience, New York, pp 207–229
Gaff DF, Churchill DM (1976) *Borya nitida* Labill. – an Australian species in the Liliaceae with desiccation-tolerant leaves. Aust J Bot 24:209–224
Gaff DF, Ellis RP (1974) Southern African grasses with foliage that revives after dehydration. Bothalia 11:305–308
Gaff DF, Hallam ND (1974) Resurrecting desiccated plants. In: Bieleski RL, Ferguson AR, Cresswell MM (eds) Mechanisms of regulation of plant growth. Roy Soc NZ Bull 12:389–393
Gaff DF, Latz P (1978) The occurrence of resurrection plants in the Australian flora. Aust J Bot 26:485–492
Gaff DF, McGregor GR (1979) The effect of dehydration and rehydration on the nitrogen content of various fractions from resurrection plants. Biol Plant 21:92–99
Gaff DF, Zee S-Y, O'Brien TP (1976) The fine structure of dehydrated and reviving leaves of *Borya nitida* Labill. – a desiccation-tolerant plant. Aust J Bot 24:225–236
Gannutz TP (1969) Effects of environmental extremes on lichens. Bull Soc Bot Fr Mem 1968. Colloq Lichens 1967, pp 169–179
Grime JP (1977) Evidence for the existence of three primary strategies in plants and its relevance to ecological and evolutionary theory. Am Nat 111:1164–1194
Grime JP (1979) Primary strategies in plants. Trans Bot Soc Edinburgh 43:151–160
Gupta RK (1976) The physiology of the desiccation resistance in bryophytes: nature of organic compounds leaked from desiccated liverwort, *Plagiochila asplenioides*. Biochem Physiol Pflanz 170:389–395
Gupta RK (1977a) An artefact in studies of the responses of respiration of bryophytes to desiccation. Can J Bot 55:1195–1200
Gupta RK (1977b) A note on photosynthesis in relation to water content in liverworts: *Porella platyphylla* and *Scapania undulata*. Aust J Bot 25:363–365
Gupta RK (1977c) A study of photosynthesis and leakage of solutes in relation to the desiccation effects in bryophytes. Can J Bot 55:1186–1194
Gupta RK (1978) The physiology of the desiccation resistance in bryophytes: effect of desiccation on water status and chlorophyll a and b in bryophytes. Indian J Exp Biol 16:354–356
Gupta RK (1979) Leakage of photosynthates from water-stressed liverwort *Scapania undulata* (L.) Dum. Indian J Exp Biol 17:164–166
Gwóźdź EA, Bewley JD (1975) Plant desiccation and protein synthesis. An in vitro system from dry and hydrated mosses using endogenous and synthetic messenger RNA. Plant Physiol 55:340–345
Gwóźdź EA, Bewley JD, Tucker EB (1974) Studies on protein synthesis in *Tortula ruralis*: polyribosome reformation following desiccation. J Exp Bot 25:599–608
Hallam ND, Gaff DF (1978a) Reorganization of fine structure during rehydration of desiccated leaves of *Xerophyta villosa*. New Phytol 81:349–355
Hallam ND, Gaff DF (1978b) Regeneration of chloroplast structure in *Talbotia elegans*: a desiccation-tolerant plant. New Phytol 81:657–662
Hambler DJ (1961) A poikilohydrous poikilochlorophyllous angiosperm from Africa. Nature (London) 191:1415–1416

Hambler DJ (1964) The vegetation of granite outcrops in Western Nigeria. J Ecol 52:573–594
Hanson AD, Nelsen CE, Everson EH (1977) Evaluation of free proline accumulation as an index of drought resistance using two contrasting barley cultivars. Crop Sci 17:720–726
Harris GB (1976) Water content and productivity of lichens. In: Lange OL, Kappen L, Schulze E-D (eds) Water and plant life. Problems and modern approaches. Ecol Stud vol 19. Springer, Berlin Heidelberg New York, pp 452–468
Henckel PA, Pronina ND (1968) Factors underlying dehydration resistance in poikiloxerophytes. Sov Plant Physiol 15:68–74
Henckel PA, Pronina ND (1969) Anabiosis with desiccation of the poikiloxerophytic flowering plant *Myrothamnus flabellifolia*. Sov Plant Physiol 16:745–749
Henckel PA, Pronina ND (1973) The euxerophytic affiliation of *Haberlea rhodopensis* (Family Gesneriaceae). Sov Plant Physiol 20:690–692
Henckel PA, Kurkova EB, Pronina ND (1970) Effect of dehydration on the course of photosynthesis in homeohydrous and poikilohydrous plants. Sov Plant Physiol 17:952–957
Henckel PA, Satarova NA, Shaposhnikova SV (1977) Protein synthesis in poikiloxerophytes and wheat embryos during the initial period of swelling. Sov Plant Physiol 24:737–741
Hinshiri HM, Proctor MCF (1971) The effect of desiccation on subsequent assimilation and respiration of the bryophytes *Anomodon viticulosus* and *Porella platyphylla*. New Phytol 70:527–538
Hoffmann P (1968) Pigmentgehalt und Gaswechsel von *Myrothamnus*-Blättern nach Austrocknung und Wiederaufsättigung. Photosynthetica 2:245–252
Hosokawa T, Kubota H (1957) On the osmotic pressure and resistance to desiccation of epiphytic mosses from a beech forest, southwest Japan. J Ecol 45:579–591
Hsiao TC (1973) Plant responses to water stress. Annu Rev Plant Physiol 24:519–570
Hsiao TC, Acevedo E (1974) Plant responses to water deficits, water-use efficiency, and drought resistance. Agric Meterol 14:59–84
Hsiao TC, Acevedo E, Fereres E, Henderson DW (1976) Stress metabolism. Water stress, growth, and osmotic adjustment. Philos Trans R Soc London Ser B273:479–500
Hutchinson J (1973) The families of flowering plants, 3rd edn. Oxford Univ Press, Oxon
Iljin WS (1953) Causes of death of plants as a consequence of loss of water: conservation of life in desiccated tissues. Bull Torrey Bot Club 80:166–177
Iljin WS (1957) Drought resistance in plants and physiological processes. Annu Rev Plant Physiol 8:257–274
Johnson WS, Gigon A, Gulmon SL, Mooney HA (1974) Comparative photosynthetic capacities of intertidal algae under exposed and submerged conditions. Ecology 55:450–453
Jones MM, Turner HC (1978) Osmotic adjustment in leaves of *Sorghum* in response to water deficits. Plant Physiol 61:122–126
Kanwisher J (1957) Freezing and drying in intertidal algae. Biol Bull 113:275–285
Kappen L (1973) Response to extreme environments. In: Ahmadjian V, Hale ME (eds) The lichens. Academic Press, London New York, pp 311–380
Kappen L, Lange OL, Schulze E-D, Evenari M, Buschbom U (1975) Primary production of lower plants (lichens) in the desert and its physiological basis. In: Cooper JP (ed) Photosynthesis and productivity in different environments. Univ Press, Cambridge, pp 133–143
Kershaw KA (1972) The relationship between moisture content and net assimilation rate of lichen thalli and its ecological significance. Can J Bot 50:543–555
Kershaw KA, Morris T, Tysiaczny MJ, MacFarlane JD (1979) Physiological-environmental interactions in lichens. VIII. The environmental control of dark CO_2 fixation in *Parmelia caperata* (L.) Ach and *Peltigera canina* var. *praetextata* Hue. New Phytol 83:433–444
Kluge M (1976) Carbon and nitrogen metabolism under water stress. In: Lange OL, Kappen L, Schulze E-D (eds) Water and plant life: problems and modern approaches. Ecol Stud vol 19. Springer Berlin Heidelberg New York, pp 243–252

Koch C (1962) The tenebrionidae of southern Africa. XXXI. Comprehensive notes on the tenebrionid fauna of the Namib desert. Ann Transvaal Mus 24:1–98

Krochko JE (1979) Metabolism of a desiccation-tolerant and desiccation-sensitive moss during drying and after rehydration: Respiration, photosynthesis, ATP levels and protein synthesis. MSc Thesis, Univ Calgary

Krochko JE, Bewley JD, Pacey J (1978) The effects of rapid and very slow speeds of drying on the ultrastructure and metabolism of the desiccation-sensitive moss *Cratoneuron filicinum*. J Exp Bot 29:905–917

Krochko JE, Winner WE, Bewley JD (1979) Respiration in relation to adenosine triphosphate content during desiccation and rehydration of a desiccation-tolerant and a desiccation-intolerant moss. Plant Physiol 64:13–17

Lange OL (1953) Hitze- und Trockenresistenz der Flechten in Beziehung zu ihrer Verbreitung. Flora 140:39–97

Lange OL (1969a) Ecophysiological investigations in lichens of the Negev desert. I. CO_2 gas exchange of *Ramalina maciformis* (Del.) Bory under controlled conditions in the laboratory. Flora 158:324–359

Lange OL (1969b) CO_2-gas exchange of mosses following water vapour uptake. Planta 89:90–94

Lange OL, Schulze E-D, Koch W (1970) Experimentell-ökologische Untersuchungen an Flechten der Negev-Wüste. II. CO_2-Gaswechsel und Wasserhaushalt von *Ramalina maciformis* (Del.) Bory am natürlichen Standort während der sommerlichen Trockenperiode. Flora 159:38–62

Lange OL, Schulze E-D, Kappen L, Buschbom U, Evenari M (1975) Adaptations of desert lichens to drought and extrme temperatures. In: Hadley NF (ed) Environmental physiology of desert organisms. Dowden, Hutchinson and Ross, Stroudsburg, pp 20–37

Larson DW (1979) Lichen water relations under drying conditions. New Phytol 82:713–731

Larson DW, Kershaw KA (1976) Studies on lichen-dominated systems. XVIII. Morphological control of evaporation in lichens. Can J Bot 54:2061–2073

Lechowicz MJ, Adams MS (1974) Ecology of *Cladonia* lichens. II. Comparative physiological ecology of *C. mitis, C. rangiferina* and *C. uncialis*. Can J Bot 52:411–422

Lee JA, Stewart GR (1971) Desiccation injury in mosses. I. Intra-specific difference in the effect of moisture stress on photosynthesis. New Phytol 70:1061–1068

Levitt J (1956) The hardiness of plants. Academic Press, London New York

Levitt J (1972) Responses of plants to environmental stresses. Academic Press, London New York

Levitt J, Sullivan CY, Krull E (1960) Some problems in drought resistance. Bull Res Counc Isr Sect D:8 173–179

Luzzati V, Husson F (1962) The structure of the liquid-crystalline phases of lipid-water systems. J Cell Biol 12:207–219

MacFarlane JD, Kershaw KA (1978) Thermal sensitivity in lichens. Science 201:739–741

MacFarlane JD, Kershaw KA (1980) Physiological-environmental interactions in lichens. IX. Thermal stress and lichen ecology. New Phytol 84:669–685

Mahmoud MI (1965) Protoplasmics and drought resistance in mosses. PhD Thesis, Univ California, Davis

Malek L, Bewley JD (1978a) Effects of various rates of freezing on the metabolism of a drought-tolerant plant, the moss *Tortula ruralis*. Plant Physiol 61:334–338

Malek L, Bewley JD (1978b) Protein synthesis related to cold temperatures in the desiccation-tolerant moss *Tortula ruralis*. Physiol Plant 43:313–319

Marinos NG, Fife DN (1972) Ultrastructural changes in wheat embryos during a "presowing drought hardening" treatment. Protoplasma 74:381–396

Mathieson AE, Burns RL (1971) Ecological studies on economic red algae. I. Photosynthesis and respiration of *Chondrus crispus* Stackhouse and *Gigartina stellata* (Stackhouse) Batters. J Exp Mar Biol 7:197–206

McKay E (1935) Photosynthesis in *Grimmia montana*. Plant Physiol 10:803–809

McKersie BD, Stinson RH (1980) Effect of dehydration on leakage and membrane structure in *Lotus corniculatus* L. seeds. Plant Physiol 66:316–320

McLean RJ (1967) Desiccation and heat resistance of the green alga *Spongiochloris typica*. Can J Bot 45:1933–1938

Mead JF (1976) Free radical mechanisms of lipid damage and consequences for cellular membranes. In: Pryor WA (ed) Free radicals in biology vol I. Academic Press, London New York, pp 51–68

Meguro M, Joly CA, Bittencourt MM (1977) Water deficit stress and some aspects of physiological behavior in *Xerophyta plicata* Spring. Velloziaceae. Bol Bot Univ S Paulo 5:27–42

Montenegro G, Hoffmann AJ, Aljaro ME, Hoffmann AE (1979) *Satureja gilliesii*, a poikilohydric shrub from the Chilean mediterranean vegetation. Can J Bot 57:1206–1213

Morgan JM (1977) Differences in osmoregulation between wheat genotypes. Nature (London) 270:234–235

Nash TH III, White SL, Marsh JE (1977) Lichen and moss distribution and biomass in hot desert ecosystems. Bryologist 80:470–479

Nir I, Klein S, Poljakoff-Mayber A (1969) Effect of moisture stress on submicroscopic structure of maize roots. Aust J Biol Sci 22:17–33

Noailles M-C (1974) Comparison de l'ultrastructure du parenchyme des tiges et feuilles d'une mousse normalement hydratée et en cours de desiccation [*Pleurozium schreberi* (Willd.) Mitt.]. CR Acad Sci Ser B 278:2759–2762

Noailles M-C (1977) Quelques aspects cytologiques et physiologiques de la reviviscence chez les Bryophytes. Congres International de Bryologie, Bordeaux. Bryophyt Bibl 13:45–94

Noailles M-C (1978) Etude ultrastructurale de la récupération hydrique aprés une période de secheresse chez une Hypnobryale: *Pleurozium schreberi* (Willd.) Mitt. Ann Sci Nat Bot 19:249–265

Nobel PS (1978) Microhabitat, water relations, and photosynthesis of a desert fern, *Notholaena parryi*. Oecologia 31:293–309

Nobel PS, Longstreth DJ, Hartsock TL (1978) Effect of water stress on the temperature optima of net CO_2 exchange for two desert species. Physiol Plant 44:97–101

Nörr M (1974a) Drought resistance of mosses. Flora 163:371–387

Nörr M (1974b) Heat resistance of mosses. Flora 163:388–397

Ochi H (1952) On the relationship between so-called 'xeromorphic' characters found in mosses and their drought resistance. Bull Soc Plant Ecol (Sendai) 1:183–187

Ogata E (1968) Respiration of some marine plants as affected by dehydration and rehydration. J Shimonoseki Univ Fish 16:89–102

Oppenheimer HR (1960) Adaptation to drought: xerophytism. In: Plant-water relationships in arid and semi-arid conditions. UNESCO, Paris, pp 105–138

Oppenheimer HR, Halevy AH (1962) Anabiosis of *Ceterach officinarum* Lam. et DC. Bull Res Counc Isr D3 11:127–147

Oppenheimer HR, Jacoby B (1961) Does plasmolysis increase the drought tolerance of plant cells? Protoplasma 57:619–627

Owoseye JA, Sanford WW (1972) An ecological study of *Vellozia schnitzleinia*, a drought-enduring plant of northern Nigeria. J Ecol 60:807–817

Pálfi G, Bitó M, Pálfi Z (1973) Free proline and water deficits in plant tissues. Sov Plant Physiol 20:189–193

Parker J (1968) Drought-resistance mechanisms. In: Kozlowski TT (ed) Water deficits and plant growth Vol I. Academic Press, London New York, pp 195–234

Parker J (1972) Protoplasmic resistance to water deficits. In: Kozlowski TT (ed) Water deficits and plant growth Vol III. Academic Press, London New York, pp 125–176

Patterson PM (1946) Osmotic values of bryophytes and problems presented by refractory types. Am J Bot 33:604–611

Pourrat Y, Hubac C (1974) Comparison des mécanismes de la résistance à la sécheresse chez deux plantes desertiques: *Artemisia herba alba* Asso et *Carex pachystylis* (J. Gay) Asch. et Graebn. Physiol Veg 12:135–147

Pronina ND (1972) Effect of dehydration on ATPase activity in poikilohydrous and homeohydrous plants. Sov Plant Physiol 19:731–732

Quadir A, Harrison PJ, Dewreede RE (1979) The effects of emergence and submergence on the photosynthesis and respiration of marine macrophytes. Phycologia 18:83–88
Radford AE, Dickison WC, Massey JR, Bell CR (1974) Vascular plant systematics. Harper and Row, New York
Richards PW (1957) Ecological notes on West African vegetation. I. The plant communities of the Idanre Hills, Nigeria. J Ecol 45:563–577
Ried A (1953) Photosynthese and Atmung bei xerostabilen und xerolabilen Krustenflechten in der Nachwirkung vorausgegangener Entquellungen. Planta 41:436–438
Ried A (1960) Thallusbau und Assimilationshaushalt von Laub- und Krustenflechten. Biol Zentralbl 79:129–151
Rogers RW (1971) Distribution of the lichen *Chondropsis semiviridis* in relation to its heat and drought resistance. New Phytol 70:1069–1077
Runyon EH (1934) The organization of the creosote bush with respect to drought. Ecology 15:128–138
Santarius KA (1969) Der Einfluß von Elektrolyten auf Chloroplasten beim Gefrieren und Trocknen. Planta 89:23–46
Santarius KA (1973) The protective effect of sugars on chloroplast membranes during temperature and water stress and its relationship to frost, desiccation and heat resistance. Planta 113:105–114
Schneider MJT, Schneider AS (1972) Water in biological membranes: Adsorption isotherms and circular dichroism as a function of hydration. J Membr Biol 9:127–140
Schonbeck MW, Norton TA (1978) Factors controlling the upper limits of fucoid algae on the shore. J Exp Mar Biol Ecol 31:303–313
Schonbeck MW, Norton TA (1979) Drought-hardening in the upper-shore seaweeds *Fucus spiralis* and *Pelvetia canaliculata*. J Ecol 67:687–696
Schonbeck MW, Norton TA (1980) The effects on intertidal fucoid algae of exposure to air under various conditions. Bot Mar 23:141–147
Schwabe WW, Nachmony-Bascomb S (1963) Growth and dormancy in *Lunularia cruciata* (L.) Dum. II. The response to daylength and temperature. J Exp Bot 14:353–378
Seibert G, Loris K, Zollner J, Frenzel B, Zahn RK (1976) The conservation of poly-A-containing RNA during the dormant state of the moss *Polytrichum commune*. Nucleic Acids Res 3:1997–2003
Sen Gupta A (1977) Non-autotrophic CO_2 fixation by mosses. MSc Thesis, Univ Calgary
Simon EW (1974) Phospholipids and plant membrane permeability. New Phytol 73:377–420
Simon EW (1978) Membranes in dry and imbibing seeds. In: Crowe JH, Clegg JS (eds) Dry biological systems. Academic Press, London New York, pp 205–224
Singh TN, Aspinall D, Paleg LG (1972) Proline accumulation and varietal adaptability to drought in barley: A potential measure of drought resistance. Nature (London) New Biol 236:188–190
Slavik B (1965) The influence of decreasing hydration on photosynthetic rate in the thalli of the hepatic *Conocephalum conicum*. In: Slavik B (ed) Water stress in plants. Proc Symp Prague 1963, Czech Acad Sci, Prague, p 195
Smith DC (1962) The biology of lichen thalli. Biol Rev 37:537–570
Smith DC (1975) Symbiosis and the biology of lichenised fungi. In: Symbiosis. Symp Soc Exp Biol 29:373–405
Smith DC, Molesworth S (1973) Lichen physiology. XIII. Effect of rewetting dry lichens. New Phytol 72:525–533
Stadelmann CJ (1971) A protoplasmic basis for drought-resistance. A quantitative approach for measuring protoplasmic properties. In: McGinnies WG, Goldman BJ, Paylore P (eds) Food, fibre and arid lands. Univ Ariz Press, Tucson, pp 337–352
Stewart CR (1972) Proline content and metabolism during rehydration of wilted excised leaves in the dark. Plant Physiol 50:679–681
Stewart GR, Lee JA (1972) Desiccation injury in mosses. II. The effect of moisture stress on enzyme levels. New Phytol 71:461–466
Stocker O (1960) Physiological and morphological changes in plants due to water deficiency.

In: Plant-Water Relationships in Arid and Semi-Arid Conditions. UNESCO, Paris, pp 63–94
Stoutjesdijk P (1974) The open shade, an interesting microclimate. Acta Bot Neerl 23:125–130
Stuart TS (1968) Revival of respiration and photosynthesis in dried leaves of *Polypodium polypodioides*. Planta 83:185–206
Swanson ES, Anderson WH, Gellerman JL, Schlenk H (1976) Ultrastructure and lipid composition of mosses. Bryologist 79:339–349
Todd GW (1972) Water deficits and enzymatic activity. In: Kozlowski TT (ed) Water deficits and plant growth Vol III. Academic Press, London New York, pp 177–216
Tucker EB, Bewley JD (1976) Plant desiccation and protein synthesis. III. Stability of cytoplasmic RNA during dehydration, and its synthesis on rehydration of the moss *Tortula ruralis*. Plant Physiol 57:564–567
Tucker EB, Costerton JW, Bewley JD (1975) The ultrastructure of the moss *Tortula ruralis* on recovery from desiccation. Can J Bot 53:94–101
Tymms MJ, Gaff DF (1979) Proline accumulation during water stress in resurrection plants. J Exp Bot 30:165–168
Vieira Da Silva J, Naylor AW, Kramer PJ (1974) Some ultrastructural and enzymatic effects of water stress in cotton (*Gossypium hirsutum* L.) leaves. Proc Natl Acad Sci USA 71:3243–3247
Vieweg GH, Ziegler H (1969) Zur Physiologie von *Myrothamnus flabellifolia*. Ber Dtsch Bot Ges 82:29–36
Waldren RP, Teare ID (1974) Free proline accumulation in drought-stressed plants under laboratory conditions. Plant Soil 40:689–692
Walter H (1955) The water economy and the hydration of plants. Annu Rev Plant Physiol 6:239–252
Walter H, Stadelmann EJ (1968) The physiological prerequisites for the transition of autotrophic plants from water to terrestrial life. BioScience 18:694–701
Walter H, Stadelmann E (1974) A new approach to the water relations of desert plants. In: Brown Jr GW (ed) Desert biology vol II. Academic Press, London New York, pp 213–310
Wellburn, FAM, Wellburn AR (1976) Novel chloroplasts and unusual cellular ultrastructure in the 'resurrection' plant *Myrothamnus flabellifolia* (Myrothamnaceae). Bot J Linn Soc 72:51–54
Willis AJ (1964) Investigations on the physiological ecology of *Tortula ruraliformis*. Trans Br Bryol Soc 4:668–683
Willis JC (1973) A dictionary of flowering plants and ferns, 8th edn. Revised by KH Airy Shaw. Univ Press, Cambridge
Wiltens J, Schrieber U, Vidaver W (1978) Chlorophyll fluorescence induction: an indicator of photosynthetic activity in marine algae undergoing desiccation. Can J Bot 56:2787–2794
Zanefeld JS (1969) Factors controlling the delimitation of littoral benthic marine algal zonation. Am Zool 9:367–391
Ziegler H, Vieweg GH (1970) Poikilohydre Pteridophyta (Farngewächse). Poikilohydre Spermatophyta (Samenpflanzen). In: Walter H, Kreeb K (eds) Die Hydratation und Hydratur des Protoplasmas der Pflanzen und ihre ökophysiologische Bedeutung. Protoplasmatologia vol II, C/6. Springer Wien New York, pp 88–108

11 Frost-Drought and Its Ecological Significance

W. TRANQUILLINI

CONTENTS

11.1 Introduction

The terms frost-drought or frost desiccation (frost-drought has been referred to as "Frosttrocknis" in the German literature since 1915, when the term was introduced by NEGER) are used to indicate the gradual desiccation of above-ground plant organs which occurs during the winter. As a result of low or freezing temperatures, water transport is either severely reduced or completely interrupted in such a manner that water lost via transpiration is inadequately replaced. With continued desiccation, a plant organ reaches a point at which it has lost all the water it can afford to lose without incurring injury. It has then reached its critical water content, or point of minimum osmotic potential. Further desiccation will result in the irreversible damage which is termed frost desiccation injury.

Since a disturbance of plant water economy may occur in the winter even if the ground is not frozen (EBERMAYER 1901), it is preferable to speak of winter desiccation and winter desiccation injury (cf. BANNISTER 1964). Then a clear conceptual distinction is made between this type of injury and actual frost injury. Other terms commonly encountered in the literature are winter browning, winter burn, top dying, red belts, and parch blight.

An extreme water stress experience, characteristic of plants projecting through the snow layer, is a phenomenon occurring in all regions having severe winters. Low temperatures or ice crystal formation in the soil or conducting vessels limit or completely interrupt water transport. Within the temperate zone, this phenomenon occurs increasingly with proximity to the poles and also becomes more pronounced with increasing altitude in mountain areas. The amount of desiccation which trees, shrubs, and other perennial plants experience within their natural distribution areas, however, does not usually tax or exceed the reserve of water which can be lost without incurring injury. Only when plants are cultivated outside of their natural distribution area, or when they occur on the edge of their natural range (e.g., boreal and alpine forest limit) is it likely that the upper limit of their tolerance will be reached, or actually exceeded, with the result that winter desiccation injury occurs.

EBERMAYER (1873, 1901) was the first to attribute extensive winter damage in young pine plantations, which led to a massive shedding of brown needles, to winter desiccation, thereby establishing the frost desiccation theory. This was later challenged by NEGER (1915), NEGER and FUCHS (1915) and MÜNCH (1928a, b, 1933). They maintained that injuries which appeared in forest trees in low-lying areas following a harsh winter were purely freezing injuries. The rapid advance of experimental ecology from about 1920 on made it possible to investigate such phenomena more exactly. As a result, WALTER (1929a, b) and THREN (1934) found large decreases in the osmotic potential of several different plants during the winter in the vicinity of Heidelberg. This indicated that many plants suffer a considerable water deficiency if winters are severe, and that the previously described injuries were due to desiccation. MICHAEL (1963, 1966, 1967a, b) in Germany and SAKAI (1968, 1970) in Japan, very thoroughly tested the validity of the frost desiccation theory by studying trees in low-lying areas. Nevertheless, in more recent, predominantly forestry-oriented literature, injuries are frequently designated as frost desiccation injuries or attributed to the direct influence of frost without experimental verification. Thus, the controversial question first raised around the turn of the century, "frozen or desiccated", is still an important issue today.

At the alpine limit of tree and shrub growth the situation is clearer. Since the work of GOLDSMITH and SMITH (1926), numerous further investigations have shown that the plants for the most part dry out in the winter and that the severe damage seen along the upper limit of existence is due to winter desiccation injury (WARDLE 1971; LARCHER 1977; TRANQUILLINI 1979). Similar conditions are found at the boreal forest limit (HOLTMEIER 1974).

With respect to frost desiccation and its importance for plant distribution, only a few comprehensive descriptions exist (WALTER 1951, 1968; WALTER and KREEB 1970; LARCHER 1957, 1963, 1972; PARKER 1969; TRANQUILLINI 1963, 1976, 1979, 1980; HOLTMEIER 1974; PEARS 1977). Accordingly, this paper will attempt to describe the more recent experimental work on winter desiccation, and to identify those trends which are of particular relevance in clarifying the ecological significance of frost desiccation.

11.2 Water Uptake and Water Conductance

11.2.1 Soil Frost

The soil water easily available to plants, soil pore diameter 0.2–8 μm, freezes, according to LARCHER (1957) and SAKAI (1968), between 0° and −1 °C. At still lower temperatures plant uptake of soil water becomes negligible. LEIK-

OLA and PAAVILAINEN (1972) reported that an upward movement of water took place in tree trunks on warm days in late winter, even though the ground was still frozen. However, this could be explained as resulting from an internal vertical displacement of water following thawing of the conducting tissue (cf. SWANSON 1967).

How long plants remain cut off from the soil water supply depends on the length of time and the depth to which the soil freezes. This in turn depends on the prevailing air temperatures of the habitat, which can be very different from year to year, and on the soil properties, as well as on the depth of the snow cover and the length of time it remains, since this effectively isolates the soil from the influence of the air.

During severe winters in Tharandt, GDR (330 m), the ground froze to a depth of 1 m in areas devoid of snow. It remained frozen to a depth of 30 cm, a depth to which the roots of younger trees penetrate, for 3 months (December to March) (Fig. 11.1). If one considers only the −1 °C isotherm in Fig. 11.1, then the length of time during which plant water uptake could not have occurred is somewhat more than 2 months. Under natural snow conditions during a severe winter, the ground froze to a depth of only about 60 cm, but it likewise remained frozen to a depth of 30 cm for 3 months, and temperatures continued to fall periodically below −1 °C. In other years the soil either did not freeze at all or else froze only for short periods of time (MICHAEL 1967a).

A comparable situation exists along the alpine forest limit where the depth to which the ground freezes depends decidedly on the depth of the snow cover. The latter can be extremely varied within a particular area due to local differences in wind and radiation. In a climatically normal year, the frost penetrated to a depth of only 20 cm in a heavily snow-covered location along the forest limit at Obergurgl (Tyrol) at an altitude of 2,000 m. In no instance did the temperature drop to −1 °C at this depth. At 30 cm depth the temperature remained constantly above 0 °C (Fig. 11.1). In contrast, at a nearby, only lightly snow-covered location, the frost reached a depth of 1 m and remained to a depth of 30 cm from December through April, or 5 months. For 4 months the soil temperature actually remained below −1 °C (AULITZKY 1961). The roots of trees within the forest transition zone penetrate to this depth (cf. WARDLE 1968).

TURNER et al. (1975) and PLATTER (1976) also found conditions which fell within this range of extremes at altitudes of about 2,000 m. They reported that in habitats having little snow cover or in years of low snowfall the soil freezes to a depth of 30 cm, and they estimated that water uptake remains interrupted for a period of several months continuing into April. Interruption of water uptake in any case is always more prolonged at this than at lower altitudes (Fig. 11.1).

Within the boreal spruce forest zone of the Alaskan interior, soil temperature is extremely dependent on both the type of soil and on the vegetation cover. Under a stand of *Picea glauca* the soil temperature at a depth of 30 cm remained below 0 °C for 4 months, in one year from December to the end of March, and in the following year from January until the end of April (VIERECK 1970). Thus conditions here resemble those found along the alpine forest limit (Fig. 11.1).

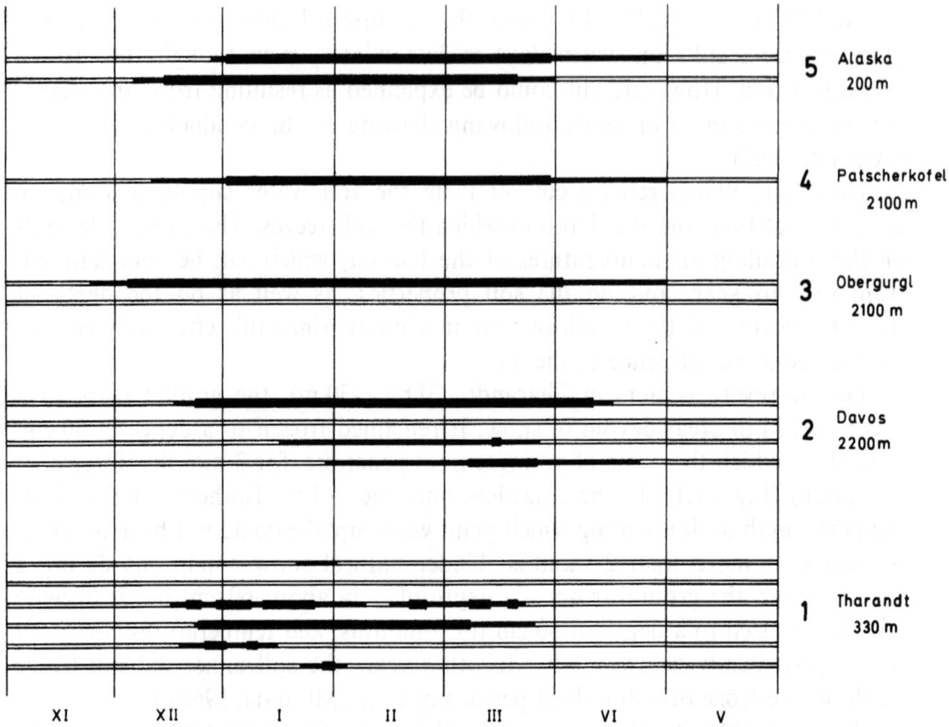

Fig. 11.1. Duration of the soil frost at 30 cm depth in different habitats. *Open double lines* temperature >0 °C; *blackened lines* temperature <0 °C; *thickened lines* temperature <−1 °C. The water available to plants freezes in the region between 0° and −1 °C. *1* Tharandt (GDR), 330 m, for different years (1960–1966). The *three lower pairs of double lines* show years in which the ground was kept artificially snow free; the *upper three pairs* show the soil temperatures occurring under natural snow cover (MICHAEL 1967a). *2* Davos (Switzerland) in winter 1970/71 in different habitats at 2,100–2,200 m. *From bottom to top* heavily snow covered north slope, lightly snow covered east slope, heavily snow covered gully, lightly snow covered ridge (TURNER et al. 1975). *3* Obergurgl (Austria) the average of the winters 1957 and 1958 at 2,070 m. *Lower double lines* a heavily snow-covered horizontal area; *upper double lines* a lightly snow-covered north slope (AULITZKY 1961). *4* Patscherkofel above Innsbruck in winter 1975/1976. A lightly snow-covered west slope in the timberline ecotone at 2,100 m (PLATTER 1976). *5* Alaska (50 km north of Fairbanks), 200 m, under a *Picea glauca* stand in winter 1964/1965 (*lower double lines*) and 1965/1966 (*upper double lines*) (VIERECK 1970)

11.2.2 Ice Crystal Formation in the Water-Conducting System

The impact of ice crystal formation in the water-conducting system on the winter water balance of plants will depend first on the temperature at which ice crystals form within the conducting vessels, and then on whether the supply of water to the peripheral plant parts becomes interrupted, and finally, on whether air bubbles impede water conduction following thawing (LYBECK 1959). Various experiments have shown that the easily displaced water freezes in the conducting vessels of different plant species between 0° and −2 °C (MICHAELIS 1934a; JOHNSTON 1959; ZIMMERMANN 1964; SAKAI 1966; 1968; HAVIS 1971).

According to JOHNSTON (1959) transpiration in *Pinus radiata* rapidly decreased to zero if shoot temperature fell below −2 °C. Several species of broad-leaved trees wilted when their trunks were cooled to between −1° and −2 °C (ZIMMERMANN 1964). In small stems of *Cryptomeria japonica* water transport was already completely interrupted at temperatures below −0.6 °C, and leaf desiccation occurred (SAKAI 1968). According to HAVIS (1971), water first freezes in the conducting vessels of *Kalmia latifolia* and *Ilex opaca* between 0° and −2 °C, but this does not result in full interruption of water transport. This first occurs between −1 °C (Ilex) and −4 °C (Kalmia) when the water in the vessel wall micropores freezes. Possibly a very small amount of water transport may still occur at lower temperatures. HYGEN (1965), for example, found that in completely frozen stems of *Picea abies* dye solutions were still transported at a velocity of 1.5 cm h^{-1}, which is 3% of the normal sap-flow velocity. This slow water movement is thought to occur either within the late wood tracheids or through cell wall cavities.

In contrast to theoretical expectations, repeated freezing does not result in impairment of water movement upon thawing in gymnosperms (HYGEN 1965; HAMMEL 1967). The bordered pits act to isolate the frozen sap water within each tracheid, so that through the expansion of water during freezing, the existing suction forces are not only offset, but rather a positive pressure is generated. As a result, air bubbles are redissolved into solution upon melting (HAMMEL 1967).

Whether or not plant water balance is disturbed during the winter as a result of ice formation in the water-conducting system of branches and stems depends on the time of its occurrence, which varies with habitat. The temperature of thin branches rapidly approaches that of the surrounding air, with the result that such branches freeze and thaw in response to corresponding fluctuations in air temperature. The magnitude of plant temperature fluctuations will be greater still through the added effect of the plant radiation balance. When the needles of coniferous plants are warmed above freezing and the rate of transpiration thereby increases, the portion of the shoot which supplies them usually thaws as well. The situation is different for the thicker branches and trunks of trees which have a greater lag time in response to temperature changes. These remain frozen for a longer time during temporary warm periods in the winter (MICHAELIS 1934a; SAKAI 1966). This is true especially for the base of the tree, at least as long as it is covered with snow (WARDLE 1968). In this case, through the insulating effect of the snow cover, the base can long remain frozen, fully blocking the resupply of water to already rapidly transpiring leaves, even after the ground has thawed (SAKAI 1968).

11.2.3 The Temperature Dependency of Water Uptake and Conduction

In 1901 EBERMAYER had already established that frost is not requisite for the occurrence of winter desiccation, because already at temperatures slightly above the freezing point the water supply becomes insufficient. DÖRING (1935) found, for example, that young *Picea abies* trees absorbed about 35% less water at

Table 11.1. Transpiration rate at 0.5 °C soil temperature in percent of the transpiration rate at 25 °C (KRAMER 1942)

Pinus taeda	13.7	Southern distribution
Pinus caribaea	13.9	
Pinus strobus	37.7	Northern distribution
Pinus resinosa	25.0	

0 °C than at 20 °C, due to increase in root cell resistance to water transport. This leads to a reduction in stomatal pore size and likewise a decrease in transpiration rate. According to KRAMER (1942), the rate of transpiration in different pine species already decreases when soil temperature drops below 25 °C. At temperatures under 15 °C, species having a more southerly distribution area show a greater decrease in the rate of transpiration than do species having a more northerly range of distribution (Table 11.1). KOZLOWSKI (1943) also found that at a soil temperature of 1 °C, the amount of transpiration in *Pinus taeda* and *Pinus strobus* was only 20%–30% of that occurring at 30 °C. According to HAVRANEK (1972), stomatal closure begins at a soil temperature of 5 °C in *Larix decidua* and *Picea abies* and, at 2 °C, transpiration has been reduced by about 50%, compared with the rate at 15 °C. In *Pinus radiata* seedlings, transpiration also decreases as soil temperature drops, due to a reduction in water uptake. The latter occurs in response to a lowering of root membrane permeability which begins just under 15 °C, the temperature at which these membranes are maximally permeable (BABALOLA et al. 1968). According to ANDERSON and MCNAUGHTON (1973) the water content of herbaceous plants from different altitudes was significantly lower at a soil temperature of 3 °C than at 20 °C. For plants from increasingly higher altitudes this difference became decreasingly significant. Thus, mountain plant species appear to be adapted to absorbing water at lower temperatures. This is also supported by the investigations of KAUFMANN (1975, 1977) which indicated that root resistance to water uptake increases with decreasing temperature, due to the greater viscosity of the water, and, at the same time, to changes in root membrane permeability. The latter begin to have a rapid effect under 14 °C in *Pinus radiata,* whereas in tree species of colder regions, such as *Picea engelmannii,* they are first effective under 7 °C.

Root conductivity is not only influenced directly by soil temperature, but also decreases as a result of the hardening processes which take place during the transition to winter dormancy. This transition is initiated by decreasing temperatures and daylength (PARSONS 1978).

11.3 Water Loss

During the frost period in which water transport is diminished or fully interrupted, plants must guard against excessive water loss in order to avoid a rapid increase in water deficit. In evergreen plants this is effected through stomatal closure, whereby water loss occurs only via cuticular transpiration. In deciduous species winter water loss is restricted to transpiring stem axes. With progressive desiccation of plant organs, cuticular and peridermal transpiration can be further reduced through a dehydration of the cutinized layers at the surface which results in an increased diffusion resistance (PISEK and BERGER 1938; see also Chap. 6, this Vol.).

The effect of water loss on the winter water balance of plants, then, will depend on whether or not the stomata are closed in winter, and on the magnitude of cuticular transpiration resistance, and thus ultimately, on the amount of unavoidable water loss occurring under natural conditions, despite these resistances.

11.3.1 Stomatal Movements

One of the causes of stomatal closure in evergreen leaves at the beginning of winter is a rapid increase in the water saturation deficit which occurs when the ground frost penetrates to the root horizon and water ceases to be available (TRANQUILLINI 1957). As shown in Fig. 11.1, this usually occurs in December, at low as well as at high altitudes. The stomata of some plants, however, may already begin to narrow and close during the transition to winter dormancy (LARCHER 1972). According to PARSONS (1978), the stomata of *Cornus stolonifera* close, in the process of frost hardening, when freezing tolerance reaches −12° to −14 °C. This could be brought about through an increase in the concentration of abscisic acid leading to stomatal closure (LITTLE and EIDT 1968; MITTELHEUSER and VAN STEVENINCK 1969). Moreover, in winter the crypts of sunken stomata become sealed with wax, whereby the effectiveness of stomatal closure is increased (JEFFREE et al. 1971; FAHEY 1979).

In the height of winter, during a lasting frost, stomata as a rule remain closed. This is true of plants in lower areas as well as of trees and dwarf shrubs of the subalpine zone (MICHAELIS 1934b; CARTELLIERI 1935; SCHMIDT 1936; TRANQUILLINI 1957; MICHAEL 1967a; TRANQUILLINI and MACHL-EBNER 1971; KÖRNER 1976; LARCHER 1977).

An important question is, when in the spring will an increase in water loss occur due to the reopening of stomata. According to TRANQUILLINI and MACHL-EBNER (1971) stomata of *Pinus cembra* needles open already in winter if the temperature remains at 10 °C for 24 h. Only in February and March will they remain closed under these conditions. In all months, however, a slight stomatal opening will occur if 20 °C temperatures continue for more than 24 h. *Rhododendron* leaves react considerably more rapidly to warm temperatures. CHRISTERSSON (1972) reported that after 4 days at 20 °C, the behavior of *Picea abies* and *Pinus sylvestris* approximated that of nonhardened plants. That is, an increase in transpiration and a normal stomatal reaction to light and dark occurred.

Clearly, in the winter, periods with high temperatures do not last this long. Thus, KÖRNER (1976) found that under actual habitat conditions, the stomata of *Loiseleuria procumbens* and *Calluna vulgaris* at 2,170 m remain continuously closed until the end of March, although on sunny days the leaves are warmed above 20 °C for short periods. In the Rocky Mountains of Wyoming, at 2,900 m, the stomata of *Pinus contorta* first open at the end of April, after 3 warm days and no night frost, if melt water has penetrated into the soil (FAHEY 1979). At this time of the year night frost still induced renewed stomatal closure. *Picea abies* along the forest limit at the northern border of the Alps in Allgäu (FRG) first open their stomata in late winter, if the temperature rises above 8 °C. *Pinus mugo* reacts still more slowly; *Rhododendron,* on the other hand, considerably more rapidly (MICHAELIS 1934b).

11.3.2 Cuticular and Peridermal Transpiration Resistance

Depending on species, the rate of cuticular transpiration occurring under controlled conditions is very different. Rates range relatively from 1 (coarse-needled evergreen conifers) to 80 (thin-leaved herbaceous shade plants) (PISEK and BERGER 1938). The transpiration rate is more important, in terms of surviving frost conditions, than either differences in water reserves or desiccation resistance (PISEK 1956). How effectively many plants can cut back on water loss by curbing stomatal transpiration is shown through comparisons of summer and winter transpiration: the winter water loss of North American conifers is only 1/50 to 1/250 of that occurring in autumn (WEAVER and MOGENSON 1919). In the vicinity of Leningrad (USSR), the winter transpiration of *Pinus sylvestris* was 300 to 400 times less than summer transpiration (IWANOFF 1924). The summer transpiration rate of *Picea abies* and *Pseudotsuga menziesii* located in Tharandt was 65 times the average occurring in the winter when the soil was frozen (MICHAEL 1967a). According to LARCHER (1977) the water use of *Loiseleuria procumbens* at the forest limit in Tyrol was 4.6 g dm^{-2} d^{-1} on good weather days in the summer and, with closed stomata in the winter only 0.05 g, or about 1/100 of the summer rate. The leaf diffusion resistance increased from 3.1 s cm^{-1}, with stomata open, to greater than 100 s cm^{-1} after stomatal closure in winter. The cuticular transpiration resistance in *Picea abies* reached enormously high values after longer periods of desiccation. In isolated annual shoots (water content 120% dry wt.), this resistance increased from around 100 s cm^{-1}, after 4 h of desiccation, to 3,600 s cm^{-1} after 124 h (water content 90% dry wt.) (PLATTER 1976). The leaf conductance (cuticular conductance) of some Alaskan conifers, following stomatal closure in the winter, amounted to 0.013–0.088 cm s^{-1} (*Picea glauca*), <0.001–0.033 (*Pinus contorta*) and <0.001–0.074 (*Pinus banksiana*) (COWLING and KEDROWSKI 1980).

The leafless shoots of deciduous plants also lose water in the winter (KOZLOWSKI 1943). On a surface area basis, these shoots actually transpire considerably more than do shoots with needles. Water reserves are thus more quickly depleted, especially if the leaf excision scars have not yet completely healed, as a consequence of unfavorable climatic conditions in late fall (GORDIAGIN 1930). According to PLATTER (1976), under the same outside conditions, peridermal transpiration from *Larix decidua* axes is approximately equal to that of

Table 11.2. Average cuticular transpiration rate of detached 1-year-old shoots after several days of desiccation at 15 °C, 40% R.H., 10 klx and 3.5 m s^{-1} wind velocity in mg H_2O (g dry wt h)$^{-1}$ (PLATTER 1976)

		October	December
Picea abies	Valley floor	2.30	2.24
	Forest limit	2.56	3.12
	Tree limit	5.90	6.40
Pinus cembra	Forest limit	1.54	1.28
	Tree limit	2.71	3.66
Larix decidua	Valley floor	2.88	2.10
	Forest limit	4.88	2.71
	Tree limit	5.40	4.87

Picea abies shoots with needles, calculated on a dry weight basis. In *Pinus cembra* considerably less transpiration occurs (Table 11.2).

The cuticular transpiration resistance also depends, however, on the outside conditions under which the leaves and shoots developed. In order for complete development of the surface tissues of newly formed shoots to occur, a certain input of warmth and thus a certain amount of time is required. LANGE and SCHULZE (1966) established that, at low altitudes, cutinized layers of *Picea abies* needles require three months from the time of shoot elongation to reach their final thickness. At the forest limit *Picea abies* trees resume growth about 2 months later than in the valley (TRANQUILLINI et al. 1980). If, moreover, the summer is cool so that developmental processes proceed more slowly (HOLZER et al. 1979), then needles and shoots may not be able to mature fully before the onset of winter. BAIG and TRANQUILLINI (1976) as well as PLATTER (1976) were able to demonstrate that a progressive decrease in the thickness of the cuticular layers of all woody species occurs between the valley floor and the "krummholz" limit. Accompanying the reduction in the thickness of the cuticular layers, which may also involve compositional changes, is an exponentially increasing water permeability (cf. ROSSA and LARSEN 1980). The cuticular transpiration rate thus increases with altitude (Table 11.2). After 24 h in an environmental chamber, excised *Picea abies* shoots exhibited only cuticular transpiration. Their water loss was greater, the higher the altitude from which they originated. On the average, water loss in branches from the tree limit (2,100 m) was 2.26 times greater, and in those from the forest limit (1,900 m) 1.70 times greater than that of branches from the valley floor (1,000 m) (Fig. 11.2). That the degree of cuticular maturity decisively affects the magnitude of cuticular transpiration is also shown by the high cuticular transpiration values which are obtained following cool summers and following experimental shortening of the vegetation period (BAIG et al. 1974; TRANQUILLINI 1974; PLATTER 1976).

The cuticular transpiration resistance also varies depending on the origin of the plants. Seedlings of *Pseudotsuga menziesii,* grown from seeds of different altitudinal origin under the same conditions, became desiccated more slowly

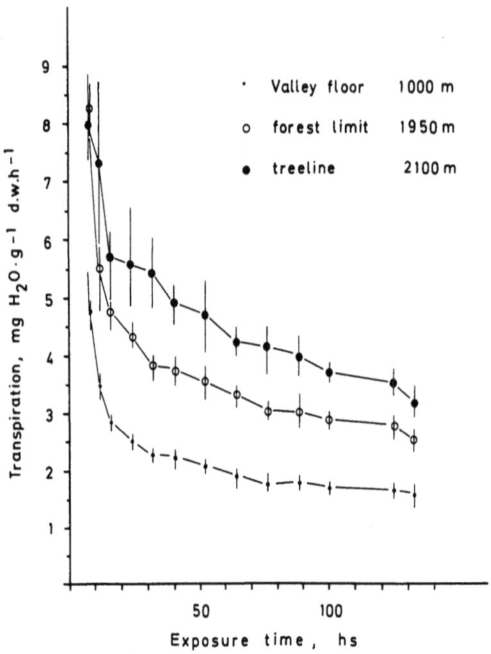

Fig. 11.2. Transpiration in detached 1-year-old shoots of *Picea abies* collected in March from the valley floor (1,000 m), the forest limit (1,950 m) and the tree limit (2,100 m) at Patscherkofel above Innsbruck, in the course of desiccation in a climate-controlled wind tunnel at 10 klx, 15 °C, 40% R.H. and 3.5 m s^{-1} wind velocity. The *steep drop of the curves* at the beginning of the experiment is due to hydroactive stomatal closure. The *smoother portion of the curves* from 20 to 30 h corresponds to the cuticular water loss after stomatal closure. The *vertical lines* indicate the range of variation of the individual measurements (PLATTER 1976)

in an environmental chamber if they came from populations growing at higher altitudes (LARSEN 1978). Trees of higher altitudes are thus better adapted to withstanding desiccation stress than trees of lower altitudes. This adaptation counteracts to some extent the increased cuticular transpiration which is found in high mountain habitats as a result of an incomplete development of the cuticular layers. (For detailed discussion of cuticular resistances see Chap. 6, this Vol.).

11.3.3 Winter Transpiration in Natural Habitats

The amount of cuticular water loss occurring in a particular habitat will depend on the water vapor pressure gradient between the leaf surface and the ambient air (GÄUMANN and JAAG 1936; SLATYER and BIERHUIZEN 1964). This will become steeper when the wind decreases the boundary layer resistance and when the plant organs are heated above air temperature by the sun. In order to evaluate the effects of these two factors on the cuticular transpiration of conifers, the cuticular transpiration rate of *Picea abies* and *Pinus cembra* was measured at different wind velocities and air temperatures during a desiccation experiment. When wind velocity was increased from 4 to 10 m s^{-1}, the water loss in *Picea abies* increased at most only 10%, and in *Pinus cembra* around 21%. With a further increase in wind strength (15 m s^{-1}) the water loss did not increase significantly. If, on the other hand, with a constant dew point of 3 °C, air temperature was increased from 15 °C [9.5 mb vapor pressure deficit (v.p.d.)] to 20° or 25 °C, the transpiration rate at 20 °C (15.8 mb v.p.d.) increased by 30% (*Picea abies*) or almost 60% (*Pinus cembra*). The greatest increase occurred

Table 11.3. The influence of wind and temperature on the cuticular transpiration rate of 1-year-old and 2-year-old branches from *Picea abies* and *Pinus cembra* trees. Change in transpiration rate, in percent, with reference values obtained at 4 m s^{-1} and 15 °C air temperature. The average from trees of different altitudes (BAIG and TRANQUILLINI 1980)

	Picea abies		*Pinus cembra*	
	1-year-old	2-year-old	1-year-old	2-year-old
10 m s^{-1}	2.0	9.7	10.9	21.4
15 m s^{-1}	−1.0	6.6	23.2	25.5
20 °C	25.4	29.8	58.7	52.9
25 °C	129.4	117.1	182.5	173.4

at 25 °C (24.1 mb v.p.d.): 130% (*Picea abies*), 180% (*Pinus cembra*) (Table 11.3). Calculations of the cuticular transpiration resistance indicate that the increase which occurs at higher temperatures results essentially from an increase in the water vapor pressure gradient, and is not the result of stomatal opening.

From this one can see that a strong absorption of radiant energy leading to a heating of plant organs beyond air temperature will have a much greater impact on the rate of cuticular transpiration than will strong wind (BAIG and TRANQUILLINI 1980). However, it is primarily the wind factor which leads to an increase in winter transpiration in regions where strong, dry winds blow throughout the winter, as for example, in the Scottish Highlands (PEARS 1977). This will likewise be the case in the far north where radiation is very weak during the winter. In the subalpine zone, on the other hand, it is the radiative warming of plants that is decisive, above all in late winter. A great deal of impressive evidence demonstrating the occurrence of such radiative warming is available to us (MICHAELIS 1934a; PISEK and SCHIESSL 1946; TRANQUILLINI and TURNER 1961; SAKAI 1966, 1970; PLATTER 1976; LARCHER 1977). According to TRANQUILLINI and TURNER (1961), in April needles of *Pinus cembra* near the surface of the snow may reach a temperature maximum of almost 30 °C. In the extreme case, this is about 21.5 °C higher than the air temperature measured at the same time in a nearby Stevenson screen at 2 m height. Based on the air temperature in the direct vicinity of the branches, sun-exposed needles in the absence of wind had a temperature above air temperature of 17 °C.

Following determination of the rate of winter transpiration at the habitat site, an estimate can be made of the length of time until water reserves are depleted and injury may occur, if no water uptake is possible. In detached annual shoots of *Picea abies* in Tharandt, this time varied, depending on month, between 69 and 21 days (1-year-old shoots) or 47 and 21 days (2-year-old shoots). Water reserves were depleted more rapidly in *Pseudotsuga menziesii* branches, and likewise in branches of *Buxus sempervirens* (Table 11.4).

If one compares these time lengths with the time period during which the soil is continually frozen (Fig. 11.1) in a severe winter (78 days), it is readily apparent that these branches in intact plants must draw upon the water reserves of the adjoining branches and main stem or trunk, if they are to avoid a

Table 11.4. Number of days of desiccation before appearance of injury (MICHAEL 1966)

		January	February	March
Picea abies	1-year-old	69	32	21
	2-year-old	47	24	21
Pseudotsuga menziesii	1-year-old	42	17	20
	2-year-old	25	16	13
Buxus sempervirens	1-year-old	44	20	21

dangerous water deficit and desiccation injury. Young plants are thus very vulnerable to desiccation injury in severe winters and require the protection of the snow layer during the most critical period.

LARCHER (1957) arrived at similar estimates of transpiration for the alpine forest limit based on his measurements of transpiration carried out in the valley floor near Innsbruck. Branches of $^1/_2$ m length, cut in February and March from different plants growing under natural conditions at the forest limit, exhausted their water reserves after 12.5 (*Rhododendron ferrugineum*) to 29 days (*Pinus cembra*). If one considers the water reservoir in the thicker branches and trunks, then, according to Larcher, these life spans in attached branches could be estimated to be about twice as long in the smaller *Pinus cembra* trees, and in the larger trees more than three times as long. Even the *Pinus cembra* trees, which are especially resistant in this respect, do not live longer than 3 months under the conditions at the forest limit, if water uptake remains completely interrupted. Thus, in lightly snow-covered habitats in the region of the forest limit, where the soil remains continuously frozen until April (Fig. 11.1), frost desiccation damage is to be expected. This is especially true for individual trees in the upper transition zone of the forest (timberline ecotone), which, due to a very low cuticular transpiration resistance, use up their water reserves much more rapidly than fully matured trees of lower altitudes. Thus, under environmental chamber conditions, detached *Picea abies* branches from the tree limit (2,140 m) already reach their critical water content, at which desiccation damage first occurs, after 70 h, whereas branches from the forest limit first reach this point after 125 h, and those from the valley floor only after 250 h (Fig. 11.3).

11.4 Water Balance

If, and to what extent winter desiccation occurs in plants in their natural habitats, can be established through series determinations of either water content, osmotic potential or total water potential. If, in addition, one knows at what percent of dehydration desiccation injury begins, in other words, the desiccation resistance (PISEK and LARCHER 1954), then it is possible to determine the desiccation stress, that is, the existing water saturation deficit in percent of the critical water saturation deficit (HÖFLER et al. 1941).

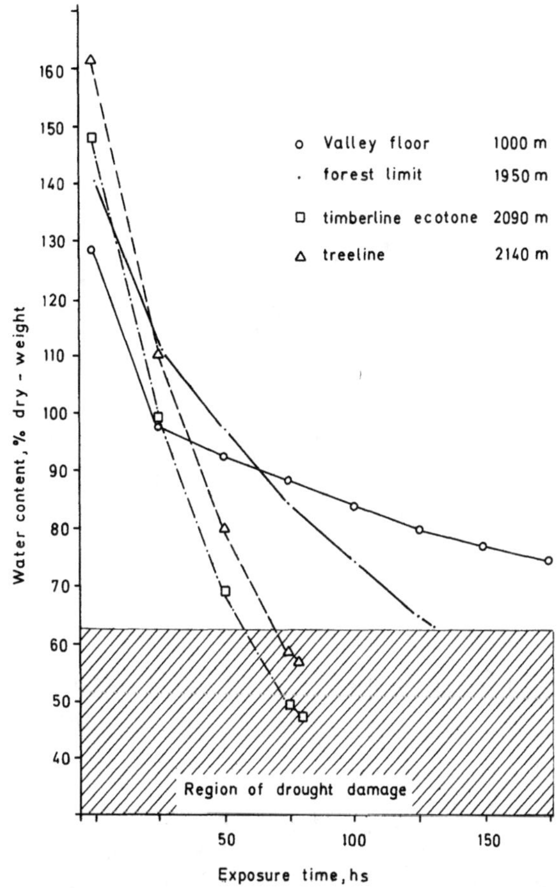

Fig. 11.3. Decrease in water content of detached 1-year-old *Picea abies* shoots collected in December, during desiccation in a climate-controlled wind tunnel at 10 klx, 15 °C, 40% R.H. and 4 m s^{-1} wind velocity. The branches stood in water before the experiment, thus the initial water content corresponds closely to saturation. The shoots were from trees at different altitudes along the Patscherkofel above Innsbruck: valley floor (1,000 m), forest limit (1,950 m), forest transition zone (2,090 m), tree limit (2,140 m). The region in which water contents are first low enough to cause desiccation injury is indicated by *hatched lines*. As can be clearly seen, the higher the elevation from which they originate, the more quickly shoots reach these injurious levels of desiccation, under the same desiccation conditions. This is due to the sharp increase in cuticular conductance which occurs with increasing altitude (cf. Fig. 11.2) (BAIG et al. 1974)

MICHAEL (1967a) very carefully investigated plant water contents in Tharandt. She clearly established that dehydration in older *Picea abies* trees, which reached its maximum in April, never led to injuriously low water contents (Fig. 11.4 A). After longer periods of soil freezing, the water content fell maximally to 75% of the pre-winter value, indicating that only 50% of the safely usable water reserves had been tapped. Similar results were reported by WALTER (1929a, b), THREN (1934), GIBBS (1958), PARKER (1963), ZEIDLER (1964), PLATTER (1976).

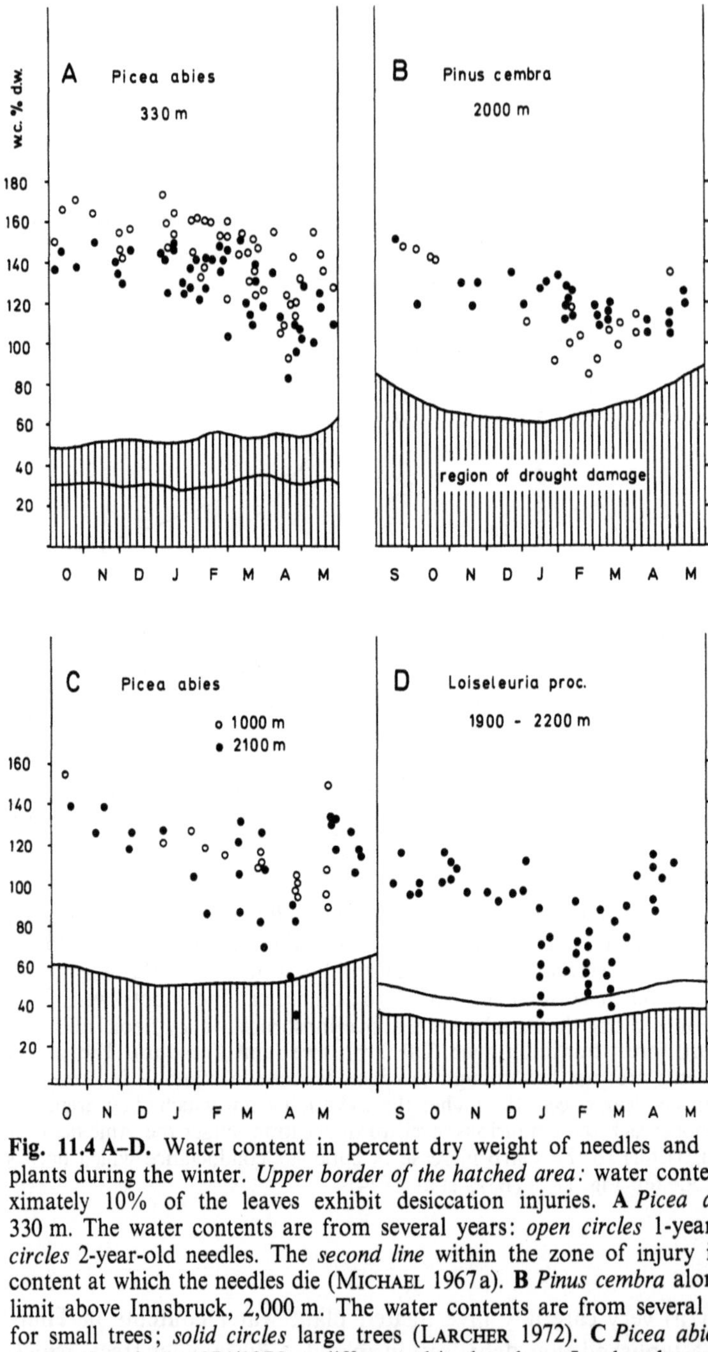

Fig. 11.4 A–D. Water content in percent dry weight of needles and leaves of different plants during the winter. *Upper border of the hatched area:* water content at which approximately 10% of the leaves exhibit desiccation injuries. **A** *Picea abies* in Tharandt, 330 m. The water contents are from several years: *open circles* 1-year-old needles; *solid circles* 2-year-old needles. The *second line* within the zone of injury indicates the water content at which the needles die (MICHAEL 1967a). **B** *Pinus cembra* along the alpine forest limit above Innsbruck, 2,000 m. The water contents are from several years: *open circles* for small trees; *solid circles* large trees (LARCHER 1972). **C** *Picea abies:* Water contents during the winter 1974/1975 at different altitudes above Innsbruck: *open circles* from the valley floor (Igls 1,000 m); *solid circles* from Patscherkofel at 2,100 m (PLATTER 1976). **D** *Loiseleuria procumbens* in the dwarf shrub zone on Patscherkofel. Water content from different years and habitats, 1,900–2,200 m. The *line* above the zone of injury indicates the water content at which the first injuries appear = critical water content (LARCHER 1977)

By way of comparison, the water contents in trees and dwarf shrubs at the alpine forest limit already approach critical levels (Fig. 11.4 B). Here, also, the maximum dehydration occurs in late winter. Smaller *Pinus cembra* trees, which have a smaller amount of water available in their axes, will become more dehydrated than large, adult trees.

Investigations of examples from the timberline ecotone or from the upper dwarf shrub zone, reveal that at these elevations the critical water content is finally exceeded (Fig. 11.4 C and D). Here lethal desiccation damage is repeatedly found (MICHAELIS 1934c; PISEK et al. 1935; STEINER 1935; SCHMIDT 1936; WARDLE 1968; BAIG et al. 1974; KÖRNER 1976; PLATTER 1976). Furthermore, determinations of the osmotic potentials or total water potentials of trees growing at different elevations show that the water stress is greatest in the timberline ecotone (GOLDSMITH and SMITH 1926; MÜLLER-STOLL 1954; TRANQUILLINI 1957; LINDSAY 1971; HANSEN and KLIKOFF 1972; BAIG and TRANQUILLINI 1980).

The distinct and widespread desiccation of trees at the forest limit in the continental part of the Alps is not common to all forest limits. In mountains of regions having milder winters and a stronger oceanic influence no significant decrease in water contents or water potentials is found in trees during the winter (SLATYER 1976; MARCHAND and CHABOT 1978; KINCAID and LYONS 1981).

In contrast, according to BANNISTER (1964), the water contents of *Erica tetralix* and *Calluna vulgaris* in the maritime climate of Scotland sink to their lowest yearly levels in the winter, although the soil is moist and seldom frozen. Here, it could be that the desiccation occurs when water absorption is impeded as a result of lower temperatures to which the plants are not adapted. In other regions of Europe and North America many dwarf shrubs also suffer water deficits in the winter if they are not protected by snow (THREN 1934; HAVIS 1965; HAVAS 1971).

In different regions of Japan (Hokkaido, Middle and East Japan at altitudes above 400–600 m) the water contents of the autochthonous plant, *Cryptomeria japonica* and the introduced plant *Picea glehnii* fall below critical levels during the winter, especially in wind-exposed habitats, and desiccation injury occurs (SAKAI 1968). At higher elevations, *Cryptomeria japonica* reaches its limit of distribution on wind-exposed north slopes due to winter desiccation injury.

Picea abies trees growing in forest plantations in England, which is far outside their natural distribution area, dry out severely in late winter and suffer injury in the form of top dying (DIAMANDIS 1978, 1979).

In Alaska the water content and xylem water potential of autochthonous *Picea glauca* trees dropped sharply in late winter but no injury resulted. A better water balance is maintained in cultivated pine species (*Pinus contorta* and *Pinus banksiana*) as a result of their lower cuticular conductance (see Sect. 11.3.2) (COWLING and KEDROWSKI 1980).

In extremely cold winters with longer periods of soil frost many evergreen species suffer a considerable water loss and a large decrease in osmotic potential due to an insufficient restriction of transpiration and also to an inability to tolerate severe dehydration. To this group belong, above all, plants which have established themselves on the edge or outside of their natural distribution area (WALTER 1929a, b; THREN 1934; ROUSCHAL 1939).

11.5 Winter Injuries

11.5.1 Evidence of the Causes: Desiccation or Freezing Injuries

Climate-induced injuries are found to occur during the winter in different plants in different habitats, sometimes with varying frequency. Ever since EBERMAYER attributed such winter injury in young *Pinus sylvestris* to desiccation, and NEGER and MÜNCH argued in favor of the freezing hypothesis, this controversial question has remained unresolved. Experimental evidence which would substantiate the correctness of one view or the other is difficult to obtain, since low water contents measured at the first appearance of visible damage could have arisen secondarily as a result of injury. Thus, MICHAEL (1966) found that frost-killed branches became desiccated more quickly than living branches, under the same conditions (cf. ZEIDLER 1964). On the other hand, the presence of high water contents at the first appearance of injury does not rule out the possibility that the plants may have been desiccated at an earlier point in time and that the dead organs then became resaturated through a passive movement of water.

For this reason, it has been necessary to rely on indirect evidence as well, in interpreting the causes of winter injury. In many cases a striking parallelism exists between the amount of winter transpiration and the extent of injury. This has been shown to be true for different aged needles in different years, for plants in different habitats, and for different plant species (BAIG et al. 1974). For example, in a study of different conifer species in Hamburg, MEYER (1965) found a close relationship between the degree of winter injury after a cold winter, and the rate of cuticular transpiration determined through laboratory desiccation experiments with isolated needles and shoots. An increasingly greater release of water from these needles and shoots paralleled an increasingly greater severity of injury suffered during the winter by the corresponding plants in the field. According to IWANOFF (1924), winter transpiration is closely correlated with species distribution. Northern species transpire at lower rates than southern species. The high rate of winter transpiration in the latter is thus a primary factor determining the northern limit of their distribution.

Lending further support to the frost desiccation theory is the experimental work from HAVIS (1965). He reported that when the soil is artificially heated to prevent the formation of soil frost, either winter injuries do not occur, or else they are less strongly manifested.

It has been repeatedly claimed, after comparing minimum habitat temperatures with laboratory-determined frost resistances, that in certain habitats the observed winter injury could only have resulted from desiccation stress because habitat temperatures were not low enough for frost injury to occur. This conclusion is also contestable since frost resistance is usually determined on the basis of standardized cooling and thawing velocities which can, under certain conditions, be greatly exceeded in the habitat (TRANQUILLINI and HOLZER 1958). Injuries may then be evoked by only slightly freezing temperatures (WHITE and WEISER 1964; WENTZEL 1965; CHRISTERSSON and SANDSTEDT 1978). Repeated freezing and thawing can also lead to an intensification of injuries (VENN 1979).

Moreover, frost injuries can also arise in late winter–early spring when plant organs become rapidly de-hardened and highly susceptible to freezing damage in a subsequent cold period.

11.5.2 Conclusions: Implications for the Distribution of Plants

At low altitudes all perennial plants within their natural distribution areas are adapted to winter conditions. Some of course require snow protection or else suffer the loss of their semi-evergreen leaves (WALTER 1929a; THREN 1934; ROUSCHAL 1939; HAVIS 1965; HAVAS 1971).

Winter desiccation injuries may occur in forest plantations following severe winters. Young plants are especially endangered, and the question of provenance plays an important role as well, above all in introduced species such as *Pseudotsuga menziesii* in Europe (EBERMAYER 1901; MÜNCH 1928b; KOZLOWSKI 1958; OHNESORGE 1963; ZEIDLER 1964; SCHÖNHAR 1965; WENTZEL 1965; SAKAI 1968; OESCHGER 1973; LARSEN 1978; VENN 1979). That frequent and severe injuries or death occur in cultivations of exotic species following severe winters is understandable, and in many cases results from too little frost resistance (JAHNEL and WATZLAWIK 1956/57).

Various plant species growing near the limits of their natural distribution areas are likewise apt to suffer lethal disturbances in their winter water economies. This is the case for species with an atlantic distribution along their eastern border (WALTER 1929a, b), and for *Quercus robur* along its continental eastern border (GORDIAGIN 1930), as well as for many southern species along their northern border as a result of overly high transpiration rates, e.g., *Salix alba, Ulmus campestris, Tilia platyphyllos, Carpinus betulus* and *Evonymus europaea* (IWANOFF 1924; WALTER 1951).

Winter desiccation is also the primary cause of injury along the boreal forest limit, mainly because new shoots are unable to reach full maturity under the prevailing climatic conditions, and therefore transpire too strongly during the winter (HOLTMEIER 1971, 1974). This applies above all to afforestations composed of species of southern origin, and to the naturally occurring young plants.

Especially distinct, however, are winter desiccation injuries which occur at the upper limits of distribution of trees and dwarf shrubs in all mountains outside of the tropics, but especially in mountains of continental regions. Here the classic frost desiccation conditions prevail – long lasting soil freezing and intense radiative warming of plants. Up to the forest limit and into the lower portion of the dwarf shrub belt, frost desiccation injuries are rare. However, a sudden and dramatic increase occurs in the timberline ecotone and in the upper part of the dwarf shrub belt due, on the one hand, to more extreme climatic conditions, and on the other hand to the shorter time available for development of sufficient protection against cuticular evaporation. Species which require a lower input of warmth to complete this development and which can better restrict their transpiration, are thus in a position to spread to still higher elevations and there to settle in sparsely snow-covered habitats, as well (TRANQUILLINI 1967, 1976, 1979, 1980; WARDLE 1971).

Acknowledgement. I should like to thank Mrs Laura Tenhunen for her clear and expert translation of the text from German to English.

References

Anderson JE, McNaughton SJ (1973) Effects of low soil temperature on transpiration, photosynthesis, leaf relative water content and growth among elevationally diverse plant populations. Ecology 54:1220–1233

Aulitzky H (1961) Die Bodentemperaturverhältnisse an einer zentralalpinen Hanglage beiderseits der Waldgrenze. Arch Meteorol Geophys Bioklimatol Ser B 10:445–532

Babalola O, Boersma L, Youngberg CT (1968) Photosynthesis and transpiration of monterey pine seedlings as a function of soil water suction and soil temperature. Plant Physiol 43:515–521

Baig MN, Tranquillini W (1976) Studies on upper timberline: morphology and anatomy of Norway spruce (*Picea abies*) and stone pine (*Pinus cembra*) needles from various habitat conditions. Can J Bot 54:1622–1632

Baig MN, Tranquillini W (1980) The effect of wind and temperature on cuticular transpiration of *Picea abies* and *Pinus cembra* and their significance in desiccation damage at the alpine tree line. Oecologia 47:252–256

Baig MN, Tranquillini W, Havranek WM (1974) Cuticuläre Transpiration von *Picea abies*- und *Pinus cembra*-Zweigen aus verschiedener Seehöhe und ihre Bedeutung für die winterliche Austrocknung der Bäume an der alpinen Waldgrenze. Centralbl Gesamte Forstwes 91:195–211

Bannister P (1964) The water relations of certain heath plants with reference to their ecological amplitude II. Field studies. J Ecol 52:481–497

Cartellieri E (1935) Jahresgang von osmotischem Wert, Transpiration und Assimilation einiger Ericaceen der alpinen Zwergstrauchheide und von *Pinus cembra*. Jahrb Wiss Bot 82:460–506

Christersson L (1972) The transpiration rate of unhardened, hardened and dehardened seedlings of spruce and pine. Physiol Plant 26:258–263

Christersson L, Sandstedt R (1978) Short-term temperature variation in needles of *Pinus sylvestris* L. Can J For Res 8:480–482

Cowling JE, Kedrowski RA (1980) Winter water relations of native and introduced evergreens in interior Alaska. Can J Bot 58:94–99

Diamandis S (1978) "Top-dying" of Norway spruce *Picea abies* (L.) Karst. with special reference to *Rhizosphaera kalkhoffii* Bubak. III. Moisture content of current and second year needles of Norway spruce. Eur J Pathol 8:357–361

Diamandis S (1979) "Top-dying" of Norway spruce *Picea abies* (L.) Karst. with special reference to *Rhizosphaera kalkhoffii* Bubak. IV. Evidence related to the primary cause of "top-dying". Eur J For Pathol 9:183–191

Döring B (1935) Die Temperaturabhängigkeit der Wasseraufnahme und ihre ökologische Bedeutung. Z Bot 28:305–383

Ebermayer E (1873) Die physikalischen Einwirkungen des Waldes auf Luft und Boden vol I. Krebs, Aschaffenburg

Ebermayer E (1901) Zur Schüttekrankheit der Kiefer. Allg Forst Jagdztg 77 Jhg:309–314

Fahey TJ (1979) The effect of night frost on the transpiration of *Pinus contorta* ssp. *latifolia*. Oecol Plant 14:483–490

Gäumann E, Jaag O (1936) Untersuchungen über die pflanzliche Transpiration. Ber Schweiz Bot Ges 45:411–518

Gibbs RD (1958) Patterns in the seasonal water content of trees. In: Thimann KV (ed) The physiology of forest trees. Ronald, New York, pp 43–69

Goldsmith GW, Smith JH (1926) Some physico-chemical properties of spruce sap. Colo College Publ Sci Ser 13:13
Gordiagin A (1930) Über die winterliche Transpiration einiger Holzgewächse Ostrußlands. Beih Bot Zentralbl 46:93–118
Hammel HT (1967) Freezing of xylem sap without cavitation. Plant Physiol 42:55–66
Hansen DH, Klikoff LG (1972) Water stress in krummholz Wasatch mountains, Utah. Bot Gaz 133:392–394
Havas P (1971) The water economy of the bilberry (*Vaccinium myrtillus*) under winter conditions. Rep Kevo Subarct Res Stn 8:41–52
Havis JR (1965) Desiccation as a factor in winter injury of *Rhododendron*. Proc Am Soc Hortic Sci 86:764–769
Havis JR (1971) Water movement in woody stems during freezing. Cryobiology 8:581–585
Havranek WM (1972) Über die Bedeutung der Bodentemperatur für die Photosynthese und Transpiration junger Forstpflanzen und für die Stoffproduktion an der Waldgrenze. Angew Bot 46:101–116
Höfler K, Migsch H, Rottenburg W (1941) Über die Austrocknungsresistenz landwirtschaftlicher Kulturpflanzen. Forschungsdienst 12:50–61
Holtmeier FK (1971) Waldgrenzenstudien im nördlichen Finnisch-Lappland und angrenzenden Nordnorwegen. Rep Kevo Subarct Res Stn 8:53–62
Holtmeier FK (1974) Geoökologische Beobachtungen und Studien an der subarktischen und alpinen Waldgrenze in vergleichender Sicht. In: Troll C (ed) Erdwiss Forschung Vol 8, Steiner, Wiesbaden
Holzer K, Oberarzbacher P, Unterholzner L, Tranquillini W (1979) Über das Wachstum von Klonen der Fichte in Klimakammern [*Picea abies* (L.) Karsten]. Centralbl Gesamte Forstwes 96:74–86
Hygen G (1965) Water stress in conifers during winter. In: Slavik B (ed) Water stress in plants. Czech Acad Sci, Prague, pp 89–95
Iwanoff L (1924) Über die Transpiration der Holzgewächse im Winter. I, II. Ber Dtsch Bot Ges 42:44–49, 210–218
Jahnel H, Watzlawik G (1956/57) Beobachtungen über die Frosthärte einiger Gehölze im Winter 1955/56 im Forstbotanischen Garten zu Tharandt. Wiss Z Techn Hochsch Dresden 6:543–548
Jeffree CE, Johnson RPC, Jarvis PG (1971) Epicuticular wax in the stomatal antechamber of Sitka spruce and its effects on the diffusion of water vapour and carbon dioxide. Planta 98:1–10
Johnston RD (1959) Control of water movement by stem chilling. Aust J Bot 7:97–108
Kaufmann MR (1975) Leaf water stress in Engelmann spruce. Plant Physiol 56:841–844
Kaufmann MR (1977) Soil temperature and drying cycle effects on water relations of *Pinus radiata*. Can J Bot 55:2413–2418
Kincaid DT, Lyons EE (1981) Winter water relations of red spruce on Mount Monadnock, New Hampshire. Ecology 62:1155–1161
Körner C (1976) Wasserhaushalt und Spaltenverhalten alpiner Zwergsträucher. In: Müller P (ed) Verhandlungen der Gesellschaft für Ökologie Wien 1975. Junk, The Hague, pp 23–30
Kozlowski TT (1943) Transpiration rates of some forest tree species during the dormant season. Plant Physiol 48:252–260
Kozlowski TT (1958) Water relations and growth of trees. J For 56:498–502
Kramer PJ (1942) Species differences with respect to water absorption at low soil temperatures. Am J Bot 29:828–838
Lange OL, Schulze E-D (1966) Untersuchungen über die Dickenentwicklung der kutikularen Zellwandschichten bei der Fichtennadel. Forstwiss Zentralbl 85:27–38
Larcher W (1957) Frosttrocknis an der Waldgrenze und in der alpinen Zwergstrauchheide. Veroeff Mus Ferdinandeum Innsbruck 37:49–81
Larcher W (1963) Zur spätwinterlichen Erschwerung der Wasserbilanz von Holzpflanzen an der Waldgrenze. Der Naturwiss Med Ver Innsbruck 53:125–137
Larcher W (1972) Der Wasserhaushalt immergrüner Pflanzen im Winter. Ber Dtsch Bot Ges 85:315–327

Larcher W (1977) Ergebnisse des IBP-Projekts „Zwergstrauchheide Patscherkofel". Sitzungsber Oesterr Akad Wiss Math Naturwiss Kl Abt I 186:301–371
Larsen JB (1978) Untersuchungen über die winterliche Trockenresistenz von 10 Herkünften der Douglasie (*Pseudotsuga menziesii*). Forstwiss Zentralbl 97:32–40
Leikola M, Paavilainen E (1972) Water uptake of pines from frozen soil. In: Isotopes and radiation in soil-plant relationships including forestry. Proc Sympos Vienna 1971, Int A E A, pp 413–418
Lindsay JH (1971) Annual cycle of leaf water potential in *Picea engelmannii* and *Abies lasiocarpa* at timberline in Wyoming. Arct Alp Res 3:131–138
Little CHA, Eidt DC (1968) Effect of abscisic acid on budbreak and transpiration in woody species. Nature (London) 220:498–499
Lybeck BR (1959) Winter freezing in relation to the rise of sap in tall trees. Plant Physiol 34:482–486
Marchand PJ, Chabot BF (1978) Winter water relations of tree-line plant species on Mt. Washington, N H. Arct Alp Res 10:105–116
Meyer FH (1965) Frosttrocknis bei Koniferen. Ber Dtsch Bot Ges 78:193–202
Michael G (1963) Ein Beitrag zum Frosttrocknisproblem. Naturwissenschaften 50:382
Michael G (1966) Untersuchungen über die winterliche Dürreresistenz einiger immergrüner Gehölze im Hinblick auf eine Frosttrocknisgefahr. Flora (Jena) Abt B156:350–372
Michael G (1967a) Über die Beanspruchung des Wasserhaushaltes einiger immergrüner Gehölze im Mittelgebirge im Zusammenhang mit dem Frosttrocknisproblem. Arch Forstwes 16:1015–1032
Michael G (1967b) Einige Untersuchungsergebnisse zur Frage der Frosttrocknis. Wiss Z Techn Univ Dresden 16:574–575
Michaelis P (1934a) Ökologische Studien an der alpinen Baumgrenze. III. Über die winterlichen Temperaturen der pflanzlichen Organe, insbesondere der Fichte. Beih Bot Zentralbl 52B:333–377
Michaelis P (1934b) Ökologische Studien an der alpinen Baumgrenze. IV. Zur Kenntnis des winterlichen Wasserhaushaltes. Jahrb Wiss Bot 80:169–298
Michaelis P (1934c) Ökologische Studien an der alpinen Baumgrenze. V. Osmotischer Wert und Wassergehalt während des Winters in den verschiedenen Höhenlagen. Jahrb Wiss Bot 80:337–362
Mittelheuser CJ, Steveninck Van RFM (1969) Stomatal closure and inhibition of transpiration induced by RS-abscisic acid. Nature (London) 221:281–282
Müller-Stoll WR (1954) Beiträge zur Ökologie der Waldgrenze am Feldberg im Schwarzwald. In: Janchen E (ed) Festschrift Erwin Aichinger Band II. Angew Pflanzensoziol. Springer, Wien, pp 824–847
Münch E (1928a) Winterschäden an Fichten und anderen Gehölzen. Tharandter Forstl Jahrb 79:276–284
Münch E (1928b) Frostgefährdung wintergrüner Gehölze. Mitt Dtsch Dendrol Ges 40:175–184
Münch E (1933) Frostempfindlichkeit wintergrüner Gehölze. Ber Dtsch Bot Ges 51:(20)–(22)
Neger FW (1915) Rauchwirkung, Spätfrost und Frosttrocknis und ihre Diagnostik. Tharandter Forstl Jahrb 66:195–212
Neger FW, Fuchs J (1915) Untersuchungen über den Nadelfall der Koniferen. Jahrb Wiss Bot 55:608–660
Oeschger HJ (1973) Gefährdung der Douglasie durch Frosttrocknis? Allg Forstz 28:190–191
Ohnesorge B (1963) Frosttrocknis an Koniferen. Allg Forstz 18:471–472
Parker J (1963) Causes of the winter decline in transpiration and photosynthesis in some evergreens. For Sci 9:158–166
Parker J (1969) Further studies of drought resistance in woody plants. Bot Rev 35:317–371
Parsons LR (1978) Water relations, stomatal behavior and root conductivity of red osier dogwood during acclimation to freezing temperatures. Plant Physiol 62:64–70
Pears N (1977) Basic biogeography. Longman, London New York
Pisek A (1956) Der Wasserhaushalt der Meso- und Hygrophyten. In: Ruhland W (ed) Handbuch der Pflanzenphysiologie Vol 3. Pflanze und Wasser. Springer, Berlin Göttingen Heidelberg, pp 825–853

Pisek A, Berger E (1938) Kutikuläre Transpiration und Trockenresistenz isolierter Blätter und Sprosse. Planta 28:124–155
Pisek A, Larcher W (1954) Zusammenhang zwischen Austrocknungsresistenz und Frosthärte bei Immergrünen. Protoplasma 44:30–46
Pisek A, Schiessl R (1946) Die Temperaturbeeinflußbarkeit der Frosthärte von Nadelhölzern und Zwergsträuchern an der alpinen Waldgrenze. Ber Naturwiss Med Ver Innsbruck 47:33–52
Pisek A, Sohm H, Cartellieri E (1935) Untersuchungen über osmotischen Wert und Wassergehalt von Pflanzen und Pflanzengesellschaften der alpinen Stufe. Beih Bot Zentralbl 52B:634–675
Platter W (1976) Wasserhaushalt, cuticuläres Transpirationsvermögen und Dicke der Cutinschichten einiger Nadelholzarten in verschiedenen Höhenlagen und nach experimenteller Verkürzung der Vegetationsperiode. Diss, Univ Innsbruck
Rossa ML, Larsen JB (1980) Die winterlichen Austrocknungsraten verschiedener Herkünfte der Douglasie (*Pseudotsuga menziesii*) und deren Abhängigkeit von der Ausbildung der Cuticula und der Spaltöffnungstiefe. Allg Forst Jagdztg 151:137–147
Rouschal E (1939) Beiträge zum winterlichen Wasserhaushalt von *Cheiranthus cheiri* und anderen wintergrünen Gartenpflanzen. Oesterr Bot Z 88:148–154
Sakai A (1966) Temperature fluctuation in wintering trees. Physiol Plant 19:105–114
Sakai A (1968) Mechanism of desiccation damage of forest trees in winter. Contrib Inst Low Temp Sci Ser B 15:15–35
Sakai A (1970) Mechanism of desiccation damage of conifers wintering in soil-frozen areas. Ecology 51:657–664
Schmidt E (1936) Baumgrenzenstudien am Feldberg im Schwarzwald. Tharandter Forstl Jb 87:1–43
Schönhar S (1965) Frosttrocknis bei Douglasie. Allg Forstz 20:44–45
Slatyer RO (1976) Water deficits in timberline trees in the Snowy Mountains of south-eastern Australia. Oecologia 24:357–366
Slatyer RO, Bierhuizen JF (1964) Transpiration from cotton leaves under a range of environmental conditions in relation to internal and external diffusive resistances. Aust J Biol Sci 17:115–130
Steiner M (1935) Winterliches Bioklima und Wasserhaushalt der Pflanzen an der alpinen Baumgrenze. Bioklim Beibl 2:57–65
Swanson RH (1967) Seasonal course of transpiration of lodgepole pine and Engelmann spruce. In: Sopper WE, Lull HW (eds) Forest hydrology. Pergamon, New York London, pp 419–434
Thren R (1934) Jahreszeitliche Schwankungen des osmotischen Wertes verschiedener ökologischer Typen in der Umgebung von Heidelberg. Z Bot 26:449–526
Tranquillini W (1957) Standortsklima, Wasserbilanz und CO_2-Gaswechsel junger Zirben (*Pinus cembra* L.) an der alpinen Waldgrenze. Planta 49:612–661
Tranquillini W (1963) Climate and water relations of plants in the subalpine region. In: Rutter AJ, Whitehead FH (eds) The water relations of plants. Blackwell, Oxford, pp 153–167
Tranquillini W (1967) Über die physiologischen Ursachen der Wald- und Baumgrenze. Mitt Forstl Bundesversuchsanst Wien 75:457–487
Tranquillini W (1974) Der Einfluß von Seehöhe und Länge der Vegetationszeit auf das cuticuläre Transpirationsvermögen von Fichtensämlingen im Winter. Ber Dtsch Bot Ges 87:175–184
Tranquillini W (1976) Water relations and alpine timberline. In: Lange OL, Kappen L, Schulze E-D (eds) Water and plant life. Ecol Stud Vol 19. Springer, Berlin Heidelberg New York, pp 473–491
Tranquillini W (1979) Physiological ecology of the alpine timberline. Ecol Stud vol 31. Springer, Berlin Heidelberg New York
Tranquillini W (1980) Winter desiccation as the cause for alpine timberline. In: Benecke U, Davis MR (eds) Mountain environments and subalpine tree growth. Proc IUFRO Workshop 1979 Christchurch. N Z For Serv Tech Pap 70, pp 263–267
Tranquillini W, Holzer K (1958) Über das Gefrieren und Auftauen von Coniferennadeln. Ber Dtsch Bot Ges 71:143–156

Tranquillini W, Machl-Ebner I (1971) Über den Einfluß von Wärme auf das Photosynthesevermögen der Zirbe (*Pinus cembra* L.) und der Alpenrose (*Rhododendron ferrugineum* L.) im Winter. Rep Kevo Subarct Res Stn 8:158–166

Tranquillini W, Turner H (1961) Untersuchungen über die Pflanzentemperaturen in der subalpinen Stufe mit besonderer Berücksichtigung der Nadeltemperaturen der Zirbe. Mitt Forstl Bundesversuchsanst Mariabrunn 59:127–151

Tranquillini W, Lechner F, Oberarzbacher P, Unterholzner L, Holzer K (1980) Über das Wachstum von Fichtenklonen in verschiedenen Seehöhen. Mitt Forstl Bundesversuchsanst Wien 129:7–25

Turner H, Rochat P, Streule A (1975) Thermische Charakteristik von Hauptstandorttypen im Bereich der oberen Waldgrenze (Stillberg, Dischmatal bei Davos). Mitt Eidg Anst Forstl Versuchswes 51:95–119

Venn K (1979) Winter vigour in *Picea abies* (L.) Karst. VI. Frost injury to needles of Norway spruce seedlings exposed to experimental freezing during midwinter. Medd Nor Inst Skogforsk 35: –

Viereck LA (1970) Forest succession and soil development adjacent to the Chena river in interior Alaska. Arct Alp Res 2:1–26

Walter H (1929a) Die Winterschäden an unseren immergrünen Pflanzen während der Kälteperiode Januar – März 1929 und ihre Ursachen. Naturwissenschaften 17:854–859

Walter H (1929b) Die osmotischen Werte und die Kälteschäden unserer wintergrünen Pflanzen während der Winterperiode 1929. Ber Dtsch Bot Ges 47:338–348

Walter H (1951) Einführung in die Phytologie Vol III/1. Standortslehre. Ulmer, Stuttgart

Walter H (1968) Die Vegetation der Erde in ökophysiologischer Betrachtung Vol II. Die gemäßigten und arktischen Zonen. Fischer, Jena

Walter H, Kreeb K (1970) Die Hydratation und Hydratur des Protoplasmas der Pflanzen und ihre ökophysiologische Bedeutung, Protoplasmatologia, Vol II C 6, Springer, New York

Wardle P (1968) Engelmann spruce (*Picea engelmannii* Engel.) at its upper limits on the Front Range, Colorado. Ecology 49:483–495

Wardle P (1971) An explanation for alpine timberline. N Z J Bot 9:371–402

Weaver JE, Mogenson A (1919) Relative transpiration of coniferous and broad-leaved trees in autumn and winter. Bot Gaz 68:393–424

Wentzel KF (1965) Die Winterfrostschäden 1962/63 in Koniferenkulturen des Ruhrgebietes und ihre vermutlichen Ursachen. Forstarchiv 36:49–59

White WC, Weiser CJ (1964) The relation of tissue desiccation, extreme cold and rapid temperature fluctuations to winter injury of American arborvitae. Proc Am Soc Hortic Sci 85:554–563

Zeidler G (1964) Die forstlichen Winterschäden 1962/63 in Westfalen-Lippe, ihre Ursachen und Folgerungen. Forstwirtsch Holzwirtsch 19:207–213

Zimmermann MH (1964) Effect of low temperature on ascent of sap in trees. Plant Physiol 39:568–572

12 Water Relations in the Germination of Seeds

D. KOLLER and A. HADAS

CONTENTS

12.1 Introduction: The Germination Process in Seeds

In seeds, the term germination includes those processes that lead to the initiation of growth in the quiescent embryonic sporophyte which they contain. Growth is a measurable and irreversible increase in size that results from cell expansion, whether or not accompanied by cell division. Most seeds contain within their embryos, or in the adjacent endosperm, or perisperm, the substrates that are necessary to sustain the initial growth of the embryo, by which germination is expressed, as well as for the subsequent growth and establishment of their seedlings in their natural habitat. These substrates (polysaccharides, proteins, triglycerides, salts of phytic acid) must first be hydrolyzed from the condensed and insoluble forms in which they are stored. In the dry seed, the metabolic machinery for the hydrolyses, for the subsequent syntheses and for the production of energy is almost entirely absent, inactive, or disassembled. In order for germination to take place, various components of this metabolic machinery must be activated, assembled and synthesized de novo. These processes require energy, but respiratory activity of the dry seed is very low, frequently undetectable. The first requirement for germination is water for hydrolysis, as a medium for translocation by diffusion, for hydration of enzymes, cell membranes and organelles to their operational conformation, and finally to provide the driving force for cell expansion that is initiated by germination.

Absence, or inactivity of any of the essential components of this machinery is sufficient to maintain the embryo in a quiescent state. However, in order for these processes to result in growth, they must be properly integrated, probably by endogenous growth regulators and by the photomorphogenic pigment phytochrome. Research into these aspects has been intensive and was recently reviewed in detail (KHAN 1975; TAYLORSON and HENDRICKS 1977).

A large number of specific metabolic events are known to take place during and after germination. The timing of these events has been and still is of considerable interest. Details of this timetable may differ not only in seeds of different species, but also in different organs of the same embryo and in the adjacent endosperm. These aspects are dealt with in detail in a number of recent reviews and books (CHING 1972; THOMAS 1972; MAYER and SHAIN 1974; MAYER and POLJAKOFF-MAYBER 1975; KHAN 1977; BEWLEY and BLACK 1978).

HABER and LUIPPOLD (1960a, b) showed that initiation of cell division and of cell elongation during germination are controlled separately. The latter is both necessary and sufficient, while the former is neither necessary, nor sufficient for radicle protrusion. It is now generally accepted that radicle protrusion in germination involves primarily cell expansion (HEGARTY 1978), which is not necessarily linked to synthesis of DNA, RNA, or protein (GALLI et al. 1981).

It has also been suggested that in order to germinate seeds must first respire primarily by the pentose phosphate pathway, relative to the Embden-Mayerhof-Parnas pathway of normal respiration, which takes over after germination (ROBERTS and SMITH 1977).

12.1.1 Environmental Requirements for Germination

Because of their stored food, seeds are much more self-contained than the plants which result from their germination. Consequently, the environmental requirements for germination are fewer and simpler, and this relative independence from the environment is carried forward for a considerable period of seedling development. As the embryonic plant does not photosynthesize (although it may contain chlorophyll), its germination requires neither PAR, nor a supply of CO_2. Some seeds do require light for germination, but only in a regulatory capacity, never photosynthetic. The same is true for the early stages of seedling development, which may last several days.

Water Requirements. As long as the seed is not exposed on the soil surface, and as long as the seedling has not emerged above ground, their evaporative losses are negligible, because the evaporative demand of the soil environment is very low. Consequently, the amount of water required for germination itself is that which suffices to bring the relevant tissues of the dry seed to the adequate level of hydration. These are minute amounts of water. The water potential, Ψ, of the dry seed is so low that it can usually take up very rapidly liquid water with which it comes into contact. However, whether or not the final level of seed hydration will suffice for germination may depend on Ψ_o of its immediate environment, and mainly on its partitioning between Ψ_s and Ψ_m (see Sect. 12.4.1).

Water is also required for the terminal stages of germination, to provide the driving force for the extension growth by which germination is expressed (Sect. 12.3.2). All the additional fresh weight of the embryo that results from preautotrophic growth is due to osmotic water uptake in the growing tissues. The amounts of water involved are considerably greater than those required for the initial rehydration, but they are still modest, due to the negligible evaporative losses. During this phase, the gradient in Ψ between the seedling and soil moisture is very small, but whereas during the initial stages of germination all the water must flow toward the soil-seed interface, at this stage the growth of the radicle and of its root hairs into the interstices of the soil matrix reduces the problems of supply. Species and cultivars may differ markedly in water requirements for germination. Differences between cultivars may correspond to the different moisture regimes to which they are adapted (STILES 1948). This is further elaborated in Section 12.4.1.

Oxygen Requirements. Oxygen is required in germination as a terminal acceptor for electrons in respiration, as well as in other oxidative processes of a regulatory nature (ROBERTS and SMITH 1977). Reduction in the availability of oxygen inhibits seed germination in many, but not in all plants (MORINAGA 1926). In oxygen-requiring seeds, such as lettuce, a requirement for increased levels of ATP is established rapidly during the initial stages of germination (imbibition) and this is met by a parallel increase in oxidative phosphorylation and in oxygen consumption (PRADET et al. 1968). This very early supply of metabolic energy is apparently required for the synthesis and assembly of the metabolic machinery that is a prerequisite for germination. There is no reliable information on the ways by which seeds that do not require oxygen obtain their initial supply of metabolic energy in absence of oxidative phosphorylation.

The simultaneous requirements for a supply of water and of oxygen very early in germination conflict with each other, because of the low solubility of oxygen in water. The diffusion of oxygen from the atmosphere to the embryo is therefore very strongly affected by the thickness of the unstirred layer of water around the seed and within it. Furthermore, the geometry of the diffusion pathways through the coats of the dispersal unit changes as they become progressively rehydrated (CÔME and TISSAOUI 1973). This is particularly evident in seeds whose coats become mucilaginous upon wetting (HEYDECKER and ORPHANOS 1968; WITZTUM et al. 1969). If the coats expand more rapidly than the embryo, water may even be sucked into the seed by the resulting pressure differential (ORPHANOS and HEYDECKER 1968).

Temperature Requirements. In order for germination to occur, the temperature of the seed environment must fall within a specific range. The prevailing isothermic environment determines the total germinability, as well as the rate of germination. The temperature range within which germinability is maximal is considerably wider than that which is conducive to maximal rates. The optimal temperature is the range within which both these parameters are maximal. Above this optimum, the progressive decline in germination rate, and eventually also in germinability, probably involve a time-dependent thermal denaturation of proteins (KOLLER 1972) and a phase change of the cell membranes (HENDRICKS and TAYLORSON 1976). These and other thermodynamic aspects of the effects of temperature on germination are discussed by LABOURIAU (1978). The effects of temperature on the kinetics of germination under conditions of increasing soil moisture stress, such as occur in nature, are discussed by KOLLER (1969).

A number of species exhibit a specific requirement for diurnal thermoperiodicity for germination. In general, the high-temperature part of the cycle must exceed a critical level in order for the alternation to be effective and the response is inductive (as to a brief illumination), so that germination subsequently takes place at a (low) constant temperature. Various hypotheses regarding the nature of diurnal thermoperiodicity are reviewed and discussed in the publications listed above.

Mineral Requirements. An exogenous supply of minerals (i.e., from the soil solution) is not required for germination, because all these needs are met from storage in the seed, mainly in the form of salts of phytic acid. A possible exception is the nitrate anion, whose presence enhances seed germination in a large number of species (review by ROBERTS 1972).

12.1.2 Regulatory Mechanisms and Their Role

Dormancy. Seeds of many species exhibit germination that is greatly delayed, sporadic, proceeds at low and erratic rates and reaches low final levels, even when the environmental requirements to sustain subsequent seedling growth are readily available. Such seeds are neither poor in quality, nor low in viability. They merely exhibit the phenomenon of dormancy, which is of prime importance to their survival in the soil (ROBERTS 1972).

After-Ripening. In some dormant seeds this condition gradually passes with age (in dry storage), as they undergo a process which has been termed after-ripening, although the embryos are fully formed and undergo none of the changes associated with ripening. On the functional level, seeds that exhibit after-ripening would be protected against precocious germination at the wrong end of the growing season.

Environmental Regulation. In other dormant seeds, a more or less specific environmental condition, combination, or sequence, induces a dramatic acceleration, uniformity, and high level of germination. In such seeds a key reaction is evidently blocked and this block is removed, or inactivated in response to specific environmental signal(s). Most such seeds that have been investigated respond to environmental signals that occur in the natural habitat of the species only at certain times of the year, in certain localities, or in certain micro-environments underneath the soil surface or on top of it. As such seeds age, their specific requirements for triggering germination generally become less restrictive. This may be a manifestation of after-ripening.

It has been argued that the germination-regulating mechanisms with which these seeds are equipped are used as strategies for survival. In such strategies a number of specific environmental variables, and their combinations, are utilized as signals by which germination is triggered, or synchronized, so as to take advantage of ecological situations that maximize the probability for successful establishment. In the absence of such signals most of the seed population remains to be committed to the irreversible step of germination only with economy, i.e., in small portions and over an extended period (KOLLER 1955, 1969, 1972; MATTHEWS 1976). For instance, germination of a large number of plants is induced by brief exposure(s) of the imbibed seed to white light. This has been interpreted as contributing to the persistence of weeds in cultivated soils (WESSON and WAREING 1969). In other plants, germination is inhibited by exposure of the imbibed seed to the far-red region of the visible spectrum. Green leaves absorb mainly in the blue and the red spectral regions. Light that they transmit in nature is thus relatively enriched in the far-red and highly inhibitory for germination. This response has been interpreted as a means to avoid seedling competition for PAR (SMITH 1973; KING 1975; GORSKI 1975). Germination-regulating mechanisms that involve water relations of the seed and its environment are discussed in Section 12.6.

12.2 Seed Water Relations as Part of Development and Maturation

The developing seed is supplied with water through the conducting system by which it is attached to its mother plant, and loses water by evaporation.

Studies with a number of species (HYDE et al. 1959; BAIN and MERCER 1966; MATTHEWS 1973; BEDFORD and MATTHEWS 1976; ROGERSON and MATTHEWS 1977; VAN ONCKELEN et al. 1980; BARLOW et al. 1980) show that after anthesis seed moisture content changes in three distinct phases. Initiation of each phase appears to be associated with some fundamental change(s) in the seed.

The first phase is initiated by onset of mitosis in the mature ovule, usually following fertilization. Dry matter accumulates at a constant rate, mostly as soluble sugar. Increase in moisture content is rapid, predominantly by osmotic uptake into vacuoles. This maintains a steep $\Delta\Psi$ between the developing seed, which has a low apparent turgor potential, and the rest of the plant (BARLOW et al. 1980). In most seeds the unicellular zygote completes its growth and development into the multi-cellular, fully differentiated embryo during this phase. Despite this intense mitotic and metabolic activity, the seed is incapable of germinating.

Onset of the second phase is characterized by an abrupt and sharp decline in the rate at which the seed accumulates water. This is not caused by inactivation of the conducting system, since dry matter continues to accumulate at the same rate, but by elimination of the $\Delta\Psi$ between the seed (wheat grain) and the rest of the plant (BARLOW et al. 1980). It appears that this phase is initiated when plastids become active sinks for sugars, which they polymerize into osmotically inactive starch. Osmotically active sugars move out of the vacuoles, which thereupon lose water. A sharp decline in respiratory activity takes place. This could be reversed (in peas) by supplying sugars, but not by maintaining high seed moisture content, and is therefore apparently a direct result of the fall in availability of reducing sugars and not at all related to the decrease in moisture content (ROGERSON and MATTHEWS 1977). Concurrently with these changes, the seeds become progressively more germinable, as well as more tolerant of desiccation. Both phenomena have been ascribed to the concurrent increase in integrity of the cell membranes, which is expressed by reduction in leakiness to electrolytes. In beans, the amount of ABA per seed (as well as per unit fresh weight) increases sharply during this phase (VAN ONCKELEN et al. 1980). This may be a response to the progressive desiccation on the one hand, and on the other it may be the cause of the progressive decline, and eventual cessation of growth. Full seedling vigor is reached only toward the end of this phase. In the developing wheat grain Ψ is remarkably unaffected by drastic changes in Ψ^{plant} throughout these two phases. This has been ascribed to high resistance in the water transport pathway into the grain (BARLOW et al. 1980).

The third phase is characterized by onset of a rapid fall in moisture content, which coincides with abrupt cessation in dry matter accumulation. These simultaneous changes indicate that this stage is initiated when the conducting system to the seed is no longer functional. In *Pennisetum americanum* translocation of assimilates into the grain (and consequently also accumulation of dry matter) stops when growth of the embryo into the basal endosperm crushes the transfer cells at the chalazal pad (FUSSELL and DWARTE 1980). In other cases the conducting system may become nonfunctional as a result of the change in the balance of growth regulators in the seed, related to its diminishing growth activity.

Does desiccation have a role in seed development, other than as an aid to dispersal? In developing barley grains germinability increased more rapidly when desiccation was accelerated, by dry weather, or by early harvest, and this has been ascribed to increase in conductance to water by the coats (WELLINGTON and DURHAM 1961; WELLINGTON 1964). An additional, or alternative mechanism is suggested by the finding that exogenous GA fails to induce hydrolytic activity in aleurone of developing grains, unless they were first desiccated, which also made them germinable (EVANS et al. 1975; NICHOLLS 1979). Gibberellin is present in the developing grain and this response to desiccation therefore plays a role in allowing the storage of food during development, preventing its utilization until dispersal. There are indications that desiccation acts in this way by inducing catabolism of ABA (VAN ONCKELEN et al. 1980).

Vital physiological processes take place in the dry seed and their rate depends, among other things, on its equilibrium moisture content. This affects the rate of metabolic processes involved in loss of primary dormancy, as well as the rate at which the coats become modified during after-ripening. Similarly, the rates at which vigor of dry seeds is reduced and their viability is lost as they age in storage are correlated with their moisture content (reviewed by MAGUIRE 1977). Phototransformation of the inactive (red-absorbing) form of phytochrome, P_r, to the active (far-red-absorbing) form, P_{fr}, and vice versa, can take place in dry seeds, but the former requires a higher moisture content. Thus, seed ripening in sunlight would tend to have a low P_{fr}/P_r ratio and require light for germination (MCARTHUR 1978).

12.3 Seed Water Relations During Germination

The kinetics of water uptake by the seed integrate the uptake by its individual parts. The different organs of the embryo, as well as the different tissues of the seed differ in chemical composition, physical organization, physiological activities, and water-retention characteristics (STILES 1948). They may therefore differ in swelling capacity, as well as in the direction of anisotropic swelling. Some of these differences play a role in water uptake and germination.

Uptake of water by the dry seed is characterized by an initial phase of rapid uptake with saturation kinetics, followed by a transition phase with very gradual to negligible uptake, and this is in turn followed by a third phase of rapid and exponentially increasing uptake.

12.3.1 Initial Phase

The kinetics of uptake during this phase are qualitatively the same in live and killed seeds. It has therefore been assumed that this is nonbiological adsorption and capillary uptake. However, this uptake becomes progressively temperature-dependent during the course of this phase (ALLERUP 1958) and this could not be accounted for by effects of temperature on the energy status of water, or on its viscosity. Studies with embryos of Lima bean (Woodstock and Pollock 1965; Pollock and Toole 1966; Klein and Pollock 1968) and soybean (Webster and Leopold 1977; Bramlage et al. 1978) showed that proliferation and reorganization of the membrane system (plasmalemma, ER, mitochondria) starts as early as 20 min after onset of rehydration. Chilling injury results from brief exposures to low (5 °C) temperature during this early part of imbibition, apparently by interfering with membrane reorganization. The rate of this process is probably very temperature-dependent, even in chilling-insensitive seeds, such as *Amaranthus retroflexus* (Taylorson and Hendricks 1972). As the membrane system reorganizes, the relative contribution of osmosis to the overall uptake of water also increases. This enhances the effects of temperature on the later stages of uptake in this phase. Respiratory rate and oxidative phosphorylation also increase rapidly early in the initial phase (Pradet et al. 1968), and this can be ascribed to the early reorganization of the inner mitochondrial membrane system (Webster and Leopold 1977).

Responsiveness of light-sensitive seeds to promotive irradiation increases during the course of incubation. In lettuce, responsiveness increased progressively with initial hydration (Berrie et al. 1974), and in *Chenopodium album* rate of increase in responsiveness correlated well with rate of this rehydration (Karssen 1970). These authors have therefore attributed the increase in responsiveness to the progressive hydration of phytochrome. However, in embryos of *Pinus palustris*, spectrophotometrically detectable levels of phytochrome increased markedly within minutes of contact with water, but physiological changes associated with the phytochrome system did not become evident until several hours later (Tobin and Briggs 1969). Furthermore, in lettuce, phytochrome becomes fully hydrated at a considerably lower water content than that required for full activity of its hypothetical reactant (Hsiao and Vidaver 1971; Loercher 1974; Duke 1978). Results with *Artemisia monosperma* (Koller et al. 1964), *Portulaca oleracea* and *Rumex crispus* (Duke et al. 1977) indicated that increase in responsiveness could not be attributed to increase in level of phytochrome, or its hydration, but to increase in availability of its hypothetical reactant. In *Artemisia* and *Portulaca* responsiveness continued to increase long after the seeds had become fully imbibed.

12.3.2 Transition Phase

Moisture content, respiratory rate, and morphology of the seed remain virtually unchanged throughout this phase, but a variety of metabolic events have been reported to occur, such as changes in levels, or activities of organelles (for instance, polysomes). Stability of respiratory rate during this time indicates that energy is primarily supplied for maintenance. The rate is sufficiently low as not to deplete stored food. This is of obvious advantage, since it is commonplace for seeds to imbibe under conditions that are unfavorable for their germination.

The processes that take place during the transition phase may differ in their operational levels of hydration (Spyropoulos and Lambiris 1980). Such differential requirements for hydration are probably the basis for the technique of "priming" seeds of economically important species for more rapid, or uniform germination, by allowing them to equilibrate with and stay for specific periods and at specific temperatures at controlled water potentials, usually by employing solutions of innocuous, high molecular weight polymers such as polyethylene glycol (HEGARTY 1978; KHAN et al. 1980/81). Priming apparently occurs in nature as well, for instance in some succulent halophytes whose seeds can withstand prolonged immersion in highly saline water and germinate fully upon transfer to pure water (UNGAR 1962; BOORMAN 1968). Priming by different means also occurs in nature. Seeds of many species remain in the transition phase until released by a specific environmental stimulus that either eliminates or inactivates an inhibitory factor, or causes the synthesis or activation of an essential factor. The significance of such responses is discussed in Section 12.1.2. In many cases the seeds can be dried after exposure to the promotive stimulus, yet retain its effects until they are rehydrated. In general, once responsiveness of the seeds to the promotive stimulus is achieved, the seeds remain in the transition phase, "poised" for immediate response (for instance, light-sensitive weed seeds in cultivated soil). In some species, exposure of the seeds to one or more cycles of alternate hydration and dehydration accelerates, or otherwise improves subsequent germination (reviewed by KOLLER 1955). In wheat (HANSON 1973), such treatment was particularly effective for germination under water-stress, or at sub-optimal temperature. Effectiveness depended on duration and temperature of the hydration phase, on conditions of dehydration and on the number of cycles. Treated grains exhibited accelerated activation of α-amylase and higher rates of respiration and of protein synthesis in the aleurone, but not in the embryo.

Among the variety of metabolic and structural changes that take place during the transition phase there may be some that tend to inactivate germination. Physical properties of the coats may become unfavorably modified (NEGBI et al. 1966), essential reactants may become depleted (DUKE et al. 1977), or inhibitory products may accumulate. Under any set of conditions germination would depend on the relative rates of the promotive and inhibitory processes (NEGBI et al. 1966). This may be one cause for the "secondary" dormancy which seeds of certain species enter under conditions that prolong their transition phase (KARSSEN 1980/81). Some of these inhibitory processes appear to require high levels of hydration and are blocked when the transition phase takes place under water stress (NEGBI et al. 1966; KOLLER 1970) and may be reversed by drying (VINCENT and CAVERS 1978). This is probably the underlying mechanism for the promotive effects and after-effects of exposure to water stress in seeds of *Taraxacum kok-saghyz* (LEVITT and HAMM 1943), to high levels of salinity in seeds of certain halophytes, such as *Suaeda depressa* (UNGAR 1962), *Plantago maritima* (BINET 1964) and *Limonium* spp. (BOORMAN 1968).

12.3.3 Growth Phase

The third phase of water uptake occurs only in viable, nondormant seed. Its onset coincides approximately with radicle protrusion and is therefore associated with establishment of cellular extension growth. This accounts for the marked increase in respiratory activity and in oxygen requirements. Growth is initiated prior to radicle protrusion. In the wheat grain the embryo starts to grow during the transition phase. This involves metabolic and energy-requiring processes, but does not depend on supplies from the endosperm, as it takes place equally in excised embryos (on water) and (in intact grains) is not accompanied by increase in embryo dry weight. The increase in embryo fresh weight has been ascribed to water uptake by vacuolation in existing cells, starting with the coleorhiza and epiblast. In certain varieties this growth generates sufficient thrust to overcome the mechanical resistance of the coats. In others, germination is delayed because additional thrust must be generated by mobilizing stored food and this takes place after the scutellum had been rehydrated (WELLINGTON and DURHAM 1961) and had presumably released gibberellin.

12.3.4 Water as a Driving Force

It therefore appears that the growth phase is associated with a fundamental change in the mechanisms controlling seed water relations. Initiation of this phase appears to be the process that is most sensitive to water stress. A lower Ψ_0 is required to suppress radicle elongation after germination than to suppress germination itself (HEGARTY and ROSS 1980/81). This difference may be ascribed to the mechanical resistance of the coats prior to radicle protrusion, which counteracts the thrust generated by the radicle. However, the fact that Ψ of seedlings becomes progressively lower (MCDONOUGH 1975) suggests that this process may indeed have started before radicle protrusion. This mobilization of water provides the driving force for early growth of the embryo, by means of which it overcomes the mechanical resistance of the coats. Several studies, in which the thrust generated by the embryo was compared to the mechanical resistance of the coats, indicate that dormancy may act by blocking the osmotic mobilization of water. In *Juglans* the imbibed embryo achieves the capacity to generate sufficient pressure to fracture the endocarp only during cold treatment (CROCKER et al. 1946). In *Xanthium* the imbibed embryo of the upper (dormant) seed is incapable of generating sufficient thrust to exceed the resistance of its own coat, while that of the lower (nondormant) seed is (ESASHI and LEOPOLD 1968). In light-sensitive lettuce seed, osmotic inhibition of germination could be overcome by promotive irradiation with red light (KAHN 1960). Lower Ψ_0 was required to produce the same inhibition in light than in darkness (REYNOLDS 1975). This could be explained on the basis of results by SCHEIBE and LANG (1965), who showed that the promotive action of light was expressed by increasing the growth capacity of the embryo itself. Furthermore, the P_{fr}-induced increase in the growth rate of the radicle by red light was limited to the initial stages of its growth. The subsequent growth rate was independent of the state of phytochrome (SCHEIBE and LANG 1967). Treatment of the lettuce embryo with red light caused a reduction in its water potential (NABORS and LANG 1971). In subsequent studies, CARPITA et al. (1979a, b) showed that Ψ

in axes of red-irradiated "seeds" was lower by 0.34 MPa than in far-red-treated "seeds" and this difference remained constant during subsequent growth. However, the initial difference in Ψ_S (0.1–0.2 MPa) became progressively less as Ψ_S increased in red-irradiated "seeds", indicating a decrease in Ψ_p. They suggest that the P_{fr}-mediated increase in the driving force for growth of the embryonic axis is an integrated function of cell-wall loosening and increase in osmotic constituents, as a result of activation of a proton pump.

12.4 Statics and Dynamics of Seed-Water Relations

Water uptake by seeds from its surrounding medium (e.g. soil) is controlled by the water energy status within and transportability of water between the various components of the seed-soil system (e.g. seed storage materials, embryonic tissues and coat, soil-seed interface zone and the soil bulk around or in contact with the seed). Initially, water potential in air-dried seeds is extremely low: -50 MPa to -100 MPa (Black 1968; Hegarty 1978). When brought in contact with a moist environment, such as moist soil, humid air, or free water, the seed starts to take up water and the potential of the water in its tissues increases. The relationship between the seed water potential, Ψ^{seed}, and its water content yield characteristic sorption isotherms for different seed species, that depend on the seeds storage materials, as well as on the ambient temperature and solutes in the surrounding medium (Mayer and Poljakoff-Mayber 1975). Similar relations between soil water content and water potential of the soil solution, Ψ^{soil}, depends on texture and structure of the soil and on solute species and concentrations in the soil solution (Childs 1969; Hillel 1971; Baver et al. 1972). Soil texture and structure govern the matric water potential component, Ψ_m, and the solute content and concentrations determine the osmotic water potential component, Ψ_S. Unless the soil is air-dry, water potentials in the soil are generally higher than in the dry seed. Water will therefore move along the potential gradient from the soil to the seed. However, water potential considerations alone can neither explain nor describe the kinetic aspects of water uptake during seed germination (Koller 1972). The kinetic aspects involve the rate of water movement, or water flux along the soil bulk-seed pathway. The water flux toward the seed depends on the water transmission properties of the soil bulk and at the seed-soil contact zone, e.g., soil water diffusivity, or soil capillary conductivity, and seed-soil contact impedance to water flow (Collis-George and Hector 1966; Phillips 1968; Hadas 1969, 1970; Shaykewich and Williams 1971; Williams and Shaykewich 1971; Hadas 1974; Hadas and Russo 1974a, b; Hadas and Stibbe 1973).

These parameters of water potential and water transmission may change with time and soil water content, either by direct relation (e.g., seed metabolism, seed swelling, and seed-water contact) or independently of the process of water uptake (e.g., soil water content, potential and conductance). Effects of these various parameters on water uptake by the seed are discussed in detail below.

12.4.1 Soil Water Potential

It has long been established that the rate and final level of germination decrease with decreasing Ψ^{soil} (DONEEN and MCGILLIVRAY 1943; HUNTER and ERICKSON 1952; COLLIS GEORGE and SANDS 1959). Seeds that exhibit water sensitivity are exceptions to this.

Critical Ψ^{seed}. The ability to germinate and the rate of germination under low Ψ^{soil} depend on the seed species. A seed must absorb a minimal amount of water in order to germinate. This "critical hydration level" is species-specific and corresponds to the "critical Ψ^{seed}", below which germination cannot take place (HUNTER and ERICKSON 1952; MCGINNIES 1960; KAUFMANN 1969; HADAS and STIBBE 1973). HADAS and STIBBE (1973) have suggested that for each seed different critical hydration levels, or critical Ψ^{seed}, exist for various activities or growth processes. Different critical hydration levels probably control cell swelling, imbibition, cell expansion, and cell division (DASBERG 1971; HEGARTY 1978). Attainment of the "critical Ψ^{seed}" depends on the external Ψ^{soil} and seed soil contact. Therefore, it is important to study its dependence on the various soil water potential components contributing to the Ψ^{soil}.

Differential Effects of Ψ_S and Ψ_m. The total Ψ^{soil} is considered to consist of the sum of several components, as follows: gravitational potential, Ψ_g, which depends on the vertical position of the considered point in reference to a given datum; osmotic potential, Ψ_S, derived from the composition and concentrations of the dissolved solute species at the considered point in reference to pure water; pressure potential, Ψ_p, in water-saturated soil, or matric potential, Ψ_m, in unsaturated soil, derived from adsorptive and interfacial (capillary) attraction between soil water and soil matrix and dependent upon soil moisture content.

The two components directly involved in water transport to germinating seeds are Ψ_m and Ψ_S. Thus changes in soil water content and solute composition affect seed germination through changes in both these components of soil water potential. AYERS (1952) and RICHARDS and WADLEIGH (1952) have stated that seeds respond equally to equal changes of Ψ_m and Ψ_S. However, this statement holds as long as the seed coats are selectively permeable to solutes in the osmotic solutions and the seed has perfect contact with the surrounding medium (HADAS and RUSSO 1974a; MANOHAR and HEYDECKER 1974).

HEGARTY (1978) has stated that seeds immersed in water may lose various species of solute to the external solution due to leaky membranes during the imbibition. COLLIS-GEORGE and SANDS (1962) have concluded, from germination behavior, that although Ψ_m and Ψ_S may be of equal nominal free energy value, they do not have the same biological consequences. Their results support the possibility that solutes or ions of inorganic electrolytes move from the osmotic solution into the seed cells. Both reports indicate leaky or selectively permeable membranes in imbibing seeds. Osmotic solutions made of high molecular weight polyelectrolytes, as used by KAUFMANN (1969) and HADAS and RUSSO (1974a, b), show only similarity of equal consequences to germination by both components, namely Ψ_S and Ψ_m. These facts suggest that the contribution of salts commonly found in the soil solution to the total Ψ^{soil} is small or ineffective as long as the seed coat or embryonic cell membranes are leaky or selectively permeable. It is probable that most salt effects on germination are related to ion toxicity. Biological systems differ in their tissue permeabilities to salts and in their susceptibility to stimulatory or toxic effects by different salts and ions found in osmotic solutions or in the soil. Sodium chloride (NaCl) and gypsum ($CaSO_4$) were reported to have toxic effects (UHVITS 1946; COLLIS-GEORGE and SANDS 1959; WIGGANS and GARDNER 1962), while nitrate salts have stimulatory effects on germination (Sect. 12.1.1).

A slight decrease in soil water matric potential, Ψ_m, has a greater effect on water uptake and germination retardation than a corresponding or greater decrease in the soil water osmotic potential, Ψ_S (UHVITS 1946; AYERS and HAYWARD 1948; WIGGANS and GARDNER 1959; COLLIS-GEORGE and SANDS 1959, 1962; MCGINNIES 1960; COLLIS-GEORGE and HECTOR 1966; WILLIAMS and SHAYKEWICH 1971; HADAS and STIBBE 1973; HADAS and RUSSO 1974a, b). A change of 0.01 MPa in Ψ_m was found to cause the same consequences as a 1.0 MPa change in Ψ_S (COLLIS-GEORGE and SANDS 1962). The differences in effectiveness of these soil water potential components are due to the fact that slight changes in the matric water potential involve the recession of water from larger soil pores to smaller ones and are thus accompanied by simultaneous changes in the soil water conductivity and the seed-soil water contact area (SEDGLEY 1963; COLLIS-GEORGE and HECTOR 1966; HADAS 1970; HADAS and RUSSO 1974a, b). Consequently, the "critical Ψ^{seed}" cannot be directly related to the sum of the matric and osmotic components of Ψ^{soil}. The dependence of germination on Ψ^{soil} must be studied separately with respect to changes in the matric component and using high molecular weight substances for the osmotic component. HUNTER and ERICKSON (1952) determined the "critical Ψ_m values" of -1.25, -0.79, -0.66 and -0.35 MPa for seeds of corn, rice, soybeans and sugar beets, respectively. Values of -1.52 to -2.0 MPa, -0.7, -1.2, -0.6, and -0.35 MPa, were reported for sorghum, cotton, chickpea, pea, and clover seeds, respectively, by HADAS (1970), HADAS and STIBBE (1973), and HADAS and RUSSO (1974a, b). Ψ^{soil} values are of the soil bulk under static equilibrium conditions. However, the real values are even lower and should be calculated for the soil-seed interface by using dynamic water flow models (HADAS 1970; HADAS and STIBBE 1973; HADAS and RUSSO 1974b). Such computations yield, for a case of perfect seed soil-water contact, critical Ψ_m values of -1.4, -2.0, -0.45, -1.10, and -1.5 MPa for corn, sorghum, clover, cotton, and chickpea seeds.

Under natural conditions, changes in soil water contents due to precipitation, drainage, evaporation, and uptake by seeds are simultaneously accompanied by changes in Ψ^{soil}. During precipitation, the soil is wetted and its water potential increases, imposing no problems for water uptake by seeds. However, as the soil dries out during imbibition its water potential decreases and this may affect seed germination. Changes in Ψ^{seed}, as inferred from the change with time in water uptake rates by the seeds, were observed by HADAS (1976). Furthermore, germination rate and final level of germination decreased with a continuous decrease in the Ψ_S of osmotic solutions during imbibition (HADAS 1977a, b). A greater total drop, or a greater rate of drop in Ψ_o caused larger decreases in germination rates and in final germination. These effects, which were observed for seeds in perfect contact with the osmotic solution, i.e., without limiting water conductivity, should be even more pronounced for similar changes in total Ψ^{soil}, where conductivity at the seed-soil interface changes with moisture content. HADAS (1977b) found good agreement between projected estimates of seed germination in osmotic solutions in the laboratory and final germination determined under field conditions.

Mechanical Effective Stress. Recently, COLLIS-GEORGE and HECTOR (1966), COLLIS-GEORGE and WILLIAMS (1968) and HADAS (1970, 1977a) examined the possibility that soil Ψ_m affects seed germination by its direct contribution to the mechanical effective stress within the soil matrix, which may physically inhibit the seed from swelling, thus reducing its capacity for imbibition, or for its embryo growth. However, data of HADAS (1970, 1977a) and SHAYKEWICH (1973) have shown that under natural conditions the range of normal stresses of mechanical nature found in the soil cannot impair seed germination.

12.4.2 Water Conductivity and Diffusivity of Soils and Seeds

Water uptake by seeds is a dynamic process and cannot be described, or its limiting factors defined, by an energetical approach alone, but rather by studying the kinetic aspects of the water transport process of the system, namely, the conductance to water across the bulk soil-seed interface and within the seed itself. Obviously, for a given water potential gradient a greater water flux will be attained at a higher soil water conductivity. Reviews by HADAS and RUSSO (1974a, b) show that in most of the reported experimental data relating germination to water uptake rates, the experimental procedures precluded differentiation between the specific effect of changes in Ψ_m, water conductivity, and/or in seed-soil contact area.

Water Conductivities in Soil-Seed System. Water movement in biological systems, across membranes, is best described by diffusion equations, rather than by equations of mass flow systems (PHILLIPS 1968; HADAS 1970; HADAS and RUSSO 1974a, b; see Chap. 17, Vol. 12A). Published values of apparent water conductivity of seeds are lower than those found in moist and wet soils (SHAYKEWICH and WILLIAMS 1971; WILLIAMS and SHAYKEWICH 1971; WARD and SHAYKEWICH 1972). The range of water conductivities of the soils reported was 1.3×10^{-1} m day^{-1} at matric potentials of -0.06 MPa to 3.2×10^{-7} m day^{-1} at -1.50 MPa, whereas the water conductivity of rape and wheat seeds was around 1.1×10^{-7} and 3.2×10^{-7} m day^{-1}, respectively. These values suggest that the soil and the seed conductivities to water are nearly equal only at very low water potentials, indicating that the water conductivity in the bulk soil cannot limit germination. The same conclusion is reached by comparing values of diffusivity to water for seeds and soils.

Diffusivity to Water in Soil-Seed System. Values of soil water diffusivities range from 5×10^7 to $4–5\times10^4$ m^2 day^{-1} for saturated soils to air-dried soils respectively (BRUCE and KLUTE 1956; RIJTEMA 1959; KUNZE and KIRKHAM 1962; DOERING 1965 – for sieved soils, and AMEMIYA 1965; HADAS 1974 – for aggregate soils). Values of water diffusivities for various seeds range from 1.7×10^2 to 1.5×10^{-6} m^2 day^{-1} for rape seeds (SHAYKEWICH and WILLIAMS 1971), 4.2×10^2 to 3×10^{-6} m^2 day^{-1} for wheat (WARD and SHAYKEWICH 1972), 8.9×10^0 to 5.9×10^1 and 2.0×10^1 to 8.4×10^2 m^2 day^{-1} for clover and chickpea seeds (HADAS 1970), 2.6×10^1 to 2.5×10^2 and 1.9×10^1 to 1.6×10^2 m^2 day^{-1} for soybean and corn seeds respectively (PHILLIPS 1968). The average seed water diffusivity increases with both the mean seed water content and soil water content and the range is largest in pure water (PHILLIPS 1968; HADAS 1970; SHAYKEWICH and WILLIAMS 1971; WARD and SHAYKEWICH 1972). Moreover, diffusivities to water of seeds immersed in osmotic solutions of different Ψ_S increase as the Ψ_S increases (HADAS and RUSSO 1974b; HADAS 1976). The data suggest that water distribution within the imbibing seed, as manifested by its relatively low average diffusivity to water, may be a limiting factor controlling imbibition, its rate, and consequently germination.

Soil-Seed Interface. Another rate-limiting process is water flow into the seed across the seed coat-soil interface and through the seed. This deduction is inferred from changes in the average water diffusivities of seeds immersed in osmotic solutions, moist soils and soil aggregates (HADAS 1970, 1976, 1977a).

Seed coat permeability to water is discussed in Sections 12.5.2 and 12.5.3. MORRIS et al. (1968) reported values of seed coat permeability to water for 11 varieties of snap beans. HADAS (1976) has reported that water diffusivity for seed coats of chickpea, pea, and vetch seed ranged from 3×10^0 to 3×10^2, 2.5×10^{-1} to 6×10^0 and 9×10^{-2} to 1.5×10^0 m^2 day^{-1} respectively. These values increased with increasing Ψ_S of the solutions.

Even though reported values of seed coat diffusivity to water are lower than those reported for whole seeds, their influence on water uptake by the seeds was hardly significant (HADAS 1976), but less permeable seed coats may severely impair water uptake. However, as pointed out, the fact that diffusivity to water of the seed coat is lower than that of the soil cannot explain the dependence of the imbibition rate on soil water conductivity, at least for chickpea, pea, and vetch, as reported by HADAS and RUSSO (1974a). In view of their data, as well as of those of others (WILLIAMS and SHAYKEWICH 1971; WARD and SHAYKEWICH 1972), this dependence is caused by change of hydraulic properties at the seed-soil water contact interface. Changes in water flow properties of the contact zone are a consequence of diminishing soil water content at the seed surface and in the soil adjacent to it (HADAS 1970; DASBERG and MENDEL 1971). These effects are more pronounced in a coarse-textured soil such as sand, or in coarsely structured soil (e.g., with large aggregates) in which the changes in the unsaturated soil water conductivity to water are greatest for a small change in the soil water content (AMEMIYA 1965; HADAS 1968).

12.4.3 Impedance to Water Flow at the Seed-Soil Interface

Under natural conditions rate of water uptake by imbibing seeds depends on changes in soil water content and soil hydraulic properties adjacent to the seed surface in contact with the soil particles and soil water (HADAS 1970; HADAS and RUSSO 1974b). These changes and the nonuniformity in water flow at the seed-soil contact interface can be lumped together into a flow parameter which is an index of resistance to water flow to the seed imposed by the seed-soil interface conditions (HADAS 1974, 1976). In most reports of the effects that Ψ^{soil} or hydraulic conductivity have on seed germination, measurements were carried out by using soil plugs, sintered glass, or other porous substrates. Water flow to the seeds was either restricted to a small portion of the seed surface or opposed by pressure of the densely packed material around the seed (COLLIS-GEORGE and SANDS 1959, 1962; COLLIS-GEORGE and HECTOR 1966; DASBERG 1971; SHAYKEWICH and WILLIAMS 1971; WILLIAMS and SHAYKEWICH 1971).

For these reasons the interpretation of the data was criticized by SEDGLEY (1963), HADAS (1970), HADAS and RUSSO (1974a), because the experimental procedures combined the effects of the seed-soil contact impedance to flow with the effects of soil water stress and conductance and normal mechanical stresses on germination. SEDGLEY (1963) has shown that improvement of seed-water contact increased imbibition and hastened germination. COLLIS-GEORGE and HECTOR (1966) calculated the seed coat area in contact with the wet soil and correlated it to the germination rate. They concluded that effects of Ψ_m per se are more important than the indirect control it exerts over the wetted seed-coat area, except at very high matric water potentials. Their analysis was incomplete since in both cases the dynamic aspects of water flow characterization, as given by the hydraulic conductivity of the contact zone, were not included (HADAS 1970; HADAS and RUSSO 1974b).

Contact Impedance. Since the hydraulic conductivity of the seed-soil interface cannot be determined, a procedure has been developed and tested (HADAS and RUSSO 1974a, b) that makes it possible to analyze separately the unique effects that the matric and osmotic components of water potential in the soil bulk, water conductivity, and the seed soil water contact impedance to water flow have on seed imbibition and germination. Their analysis for fully buried, or covered seeds combines the effects that the geometrical configuration, the hydraulic conductivity and Ψ_m at the interface have on water flow to the seed. The impedance factor found shows a good correlation with the wetted area of the seed and the soil conductivity to water. The analyses by HADAS and RUSSO (1974b), and HADAS (1974, 1976) furthermore show that for a given Ψ^{soil} and assuming a perfect seed-soil contact, an increase in the impedance to flow causes a smaller decrease than anticipated in the soil Ψ_m at the seed surface. The contact impedance increases as the seed wetted area, or the soil water conductivity, or both, decrease. Contact impedance for a given Ψ_m increases as the coarseness of the soil texture or structure increase. Thus, for a given seed size and soil Ψ_m the wetted seed coat area will decrease and its contact impedance to flow will increase as the particle coarseness of their immediate surrounding increase. As a consequence, the water uptake rate, as well as the rate and final percentage of germination, will diminish as the proportion of large-sized soil crumbs increases (CURRIE 1973; HADAS 1974, 1977a).

These phenomena may become aggravated for seeds that are partially buried, or lying on the soil surface, since only a small part of the seed coat is in direct contact with the water films.

Contact impedances to water flow at the seed-soil interface of pea and chickpea were neither significantly high to delay water uptake or germination, nor did they increase further during the progress of imbibition (HADAS 1977a). Since these and many other seeds swell during imbibition, the increased volume of the seed compacts the soil adjacent to it and increases the total area in contact with it (HADAS 1970, 1977a). At the same time decrease in water content and diminishing pore sizes at the compacted soil-seed interface cause a marked decrease in water film contact points and hence in water conductivity at the interface. HADAS (1977a) has suggested that the reduction in film contact points and in conductance to water at the seed-soil interface overrides the increase in the total contact area. Swelling seeds on top of a moist soil cannot compact the soil, and thus one may expect a more pronounced increase in the contact impedance, lower rate of water uptake, and delayed germination.

12.5 Pathways for Entry of Water into the Seed

The exterior of the seed (dispersal unit) is never uniform, but exhibits an intricate surface sculpture, as well as specialized regions, such as hilum, micropyle, and chalaza (strophiole in the Leguminosae). The architecture and structure of the seed (dispersal unit) and water-repellant properties of its coats may have a profound influence on the geometry of the pathway for movement of water from the soil into the seed. Using artificial "soil" surfaces, HARPER and BENTON (1966) showed that germination of mucilaginous seeds was least sensitive to Ψ_m, while spiny and reticulate seeds were particularly sensitive. Sensitivity of smooth-coated seeds increased with their size. Dependence of germination on Ψ_m was greatly reduced when evaporative losses were cut. Germination at constant Ψ_m was affected by soil moisture content. However, MOTT (1974) found that features that improve soil-seed contact failed to improve germination on the soil surface. MCWILLIAM and PHILLIPS (1971) have shown that whereas Ψ_S and Ψ_m have equivalent effects on germination of *Lolium perenne* and dehulled *Phalaris tuberosa*, the equivalence no longer holds in nondehulled *Phalaris* because the air gap between the hulls and the pericarp remains vapor-filled at low Ψ_m and causes an increase in resistance to absorption of soil water.

Anatomical studies indicate a structural complexity of the tissues, as well as a specialization of the cell walls from which these coats are made up. Internal to the seed coat are additional envelopes, such as the perisperm and endosperm, with their own physical and chemical peculiarities. Many of these features play a role in the biological functions of the seed, namely in its dispersal and germination. Some of them play a role in water uptake (a) by acting as selectively permeable barriers, and/or (b) by restricting the entry of water through the coats (CHRISTIANSEN and MOORE 1959; MANOHAR and HEYDECKER 1964). Discussion of the anatomical and structural features of the seed coat is outside the scope of the present chapter.

12.5.1 Myxospermy

In species of Plantaginaceae, Linaceae, Brassicaceae, Euphorbiaceae, Onagraceae, Acanthaceae etc., the seed coats become mucilaginous when wet. The mucilage is rapidly formed by uptake of a relatively large volume of water that is then held in readily available form (high Ψ) in the immediate vicinity of the seed. Also, the contact area between the seed and soil water is greatly increased. These features may be of advantage in improving water uptake by the seed, for instance in coarse-textured soils (DEXTER and MIYAMOTO 1959; YOUNG and EVANS 1973), but are of no advantage in the severe water relations at the soil surface (MOTT 1974). On the other hand, myxospermy aggravates oxygen supply. In *Spinacia oleracea* (CAVAZZA 1953) and in *Hirschfeldia incana* (NEGBI et al. 1966) this creates a positive response to reduced Ψ_m, while in *Blepharis persica* germination on the soil surface is inhibited by surface water and promoted by divalent ions, such as Ca^{2+} (WITZTUM et al. 1969). The ecological role of myxospermy is still problematic (GRUBERT 1974).

12.5.2 Selective Permeability

In seeds that exhibit selective permeability, water uptake is controlled by the osmotic, as well as by the matric components of the external water potential (MANOHAR and HEYDECKER 1964), and by the reflection coefficients for the molecular species of the solutes. These considerations must be kept in mind in studies involving osmotic control of seed water relations (NABORS and LANG 1971). Interpretation is complicated by the fact that osmotically active compartments, with different permeability properties, may exist within the seed. SIMON (1974) and SIMON and WIEBE (1975) have shown that selective permeability due to biological membranes might be influenced by environmental factors, either directly, for instance by effects of temperature on fluidity of the lipid phase (see also LABOURIAU 1978), or indirectly, for instance by effects of aeration on availability of energy by oxidative phosphorylation. Selective permeability changes during the course of rehydration, as membrane integrity improves. Selectively permeable coats control leakage of enzymes (DUKE and KAKEFUDA 1981), metabolites, and growth regulators from rehydrating seeds before membrane integrity is established in the embryo and endosperm. Such leakage may not only deplete seed reserves, but also enhance microbial activity in the vicinity of the seed (POLLOCK and TOOLE 1966).

12.5.3 Impermeability to Water

In some seeds the coats are entirely impermeable to water, and only become permeable under the influence of specific environmental complexes. Some of these complexes are clearly related to the ecological niches that the species occupy. This is, for instance, the case with species such as *Rhus ovata* that occupy frequently burnt sites, whose impermeable inner seed coat cracks open above the micropyle and becomes permeable after brief exposures to temperatures that occur at the soil surface during brush fires (STONE and JUHREN 1951). In other cases, the environmental complexes that cause loss of impermeability regulate seed germination in ways that contribute to the successful survival and establishment of the seedlings (KOLLER 1972).

In other seeds, coat impermeability is discontinuous, thus restricting the entry of water to specific pathways, which are either few in number and localized, or more numerous and scattered more uniformly over the seed coat. These features must be taken into account in studies of the effect of the contact area on water uptake by the seed from the soil.

12.5.4 Identification of Pathways

Specific sites for entry of water through otherwise impermeable, or selectively permeable seed coats have been identified by blocking specific regions with water-impermeable material prior to soaking. Other techniques have employed water-soluble stains for imbibition. Sectioning the seeds at intervals during imbibition allows the progress of the wetting front in the seed to be followed. BERGGREN (1963), using such techniques, concluded that the pathway is related to the type of ovule from which the seed had developed: the hilum in seeds from campylotropous ovules, the chalaza – from anatropous, and the micropyle in those from orthotropous ovules. SPURNY (1973) has employed a cinematographic analysis to identify the pathway for entry and to follow the time-course of rehydration of individual regions of the seed coat and organs of the embryo during initial swelling and subsequent germination of pea seeds. Volatile stains (iodine) have been used to identify pathways through which water may enter only in vapor form (HYDE 1954). Seeds with less localized pathways for water entry do not lend themselves to similar procedures for identification of the pathways. Their existence can only be inferred indirectly by the fact that they can open and close, and that when open, the seed can take up water equally well when different regions of its coat are blocked artificially.

12.5.5 Hard-Seededness

Phenomenology. The most extreme, and therefore the most evident manifestation of seed coat impermeability to water is found in "hard" seeds, which take up water very slowly, or not at all when soaked. Such seeds are prevalent in the Papilionaceae, Mimosaceae, Caesalpiniaceae, Malvaceae and Convolvulaceae and exhibit extreme longevity in artificial storage, as well as in nature. Hard-seededness is therefore a major contribution to long-term availability of viable seeds in the soil, and thus to persistence of the species. Characteristically, when such seeds are soaked in water, individual seeds may imbibe and swell more or less abruptly and sporadically over long periods. The time-course of these events ("rate of imbibition") is temperature-dependent (FAYEMI 1957). Water apparently enters the seed very gradually. As the colloidal contents of the seed (up to and including inner layers of the coat) become rehydrated they change their spatial organization. This opens up new pathways for water entry into the seed and/or enlarges existing ones, thus changing a trickle into a flood. This explains the abrupt swelling after long delays.

Softening. Hard seeds can be artificially "softened" by a variety of means that attack the integrity of the coats, without damaging the interior, such as "scarification" by mechanical abrasion, or acid corrosion (PORTER 1949; KOLLER and COHEN 1959), "percussion" (="impaction") treatments (HAMLY 1932; HAGON and BALLARD 1970), or brief immersion in hot (80 °C) water (CHRISTIANSEN and MOORE 1959). Such treatments may attack the coat indiscriminately, but specific regions of the coat may be more susceptible. In nature, permeability is achieved by opening up specific pathways through the otherwise impermeable coats (see Sect. 12.6.4).

Establishment of Seed Coat Impermeability. Seed coat impermeability has been mostly studied in the Papilionaceae (QUINLIVAN 1971), because of their economic importance, but the findings can probably be extended to other families. According to BALLARD (1973) and WERKER et al. (1979), the impermeable layer is that of the palisade (or Malpighian) cells, which contain quinones and have external caps that are very hard, pectinaceous, and embedded in a suberin matrix.

HYDE et al. (1959) have shown that impermeability is absent during early seed development and increases rapidly soon after the rapid phase of moisture uptake by the seed comes to an end. In a more definitive study HYDE (1954) showed that as the seed ripens it loses most of its water rapidly by evaporation over its entire surface. When its water content had fallen (from 150%–200%) to 25%, the coats start to become impermeable. However, water loss continues, although at a much slower rate, by diffusion through the hilum, which has cracked open along the median groove by the tensions created in the drying and shrinking counter-palisade layer. This gradual water loss continues until an equilibrium is reached between Ψ inside the seed and in the surrounding atmosphere. As seed moisture content falls (to 14%) the palisade cells shrink laterally and the coats become impermeable. Thus, as the seeds became dehydrated during ripening their capacity to take up liquid water was equally high as long as their moisture content did not fall below 20% and then decreased markedly and exponentially as moisture content continued to fall to 14%. When ripening took place in atmospheres of controlled humidity the duration of impermeability was negatively correlated with this relative humidity. Thus, when the hilar fissure opens, it allows outward diffusion of water vapor, but prevents entry of liquid water. It was further found that in dry air the hilar fissure opened rapidly and allowed loss of water vapor, while in wet air it closed rapidly and prevented the sorption of water vapor, thus acting as a one-way valve. As a result, seeds exposed to a fluctuating atmospheric humidity could only become progressively more dehydrated, equilibrating with the lowest humidity in the sequence.

Prematuration Conditioning. Germination responses may be quantitatively affected by the parental environment during seed ontogeny, as well as by the position of the seed in the fruit, or inflorescence (review by KOLLER 1972). These effects are also expressed in seed water relations.

Exposure of barley plants to a short heat stress 3 weeks after awn emergence resulted in grains with a thinner coat, that took up water and germinated more readily (KHAN and LAUDE 1969). In hard-seeded Papilionaceae, seed coat impermeability, as well as its susceptibility to summer conditions, differs according to the position of the seed in the pod (KOLLER 1969; HALLORAN and COLLINS 1974). The photoperiodic environment at a late stage in pod development (GUTTERMANN and EVENARI 1972) and length of the growing season of the parent plant apparently affect the ripening process of the seed coat and thus determine the degree of hard-seededness (QUINLIVAN 1965; EVENARI et al. 1966).

12.6 Optimization of Water Supply by Regulation of Germination

The general aspects of germination-regulating mechanisms as strategies for survival are discussed in Section 12.1.2. Water is one of the basic requirements for and plays a central role in the germination process. A considerably larger and more sustained supply of water is required for subsequent growth and survival of the seedling. Yet seeds are small and are therefore able to sample the availability of water only within a small volume of the surrounding soil. Clearly, such a small and localized sampling cannot provide a reliable indication as to whether or not the store of locally available soil moisture is sufficiently large. It thus appears that the probability of survival is actually enhanced by controlling the tendency of quiescent tissue to initiate growth upon rehydration. In this section we shall examine several strategies where the mechanism that regulates germination can be visualized as contributing to ensure the subsequent availability of soil moisture. We shall deal also with mechanisms that regulate the dispersal of the seeds (or more generally, the dispersal unit) in a way that also contributes to ensure the availability of soil moisture.

12.6.1 Optimal Soil Depth for Germination and Emergence

Seeds of most plants land on the soil surface after dissemination. This is where the radiant energy, convection, turbulent transfer, and gravitational flow combine to create rapidly increasing moisture stress, telescoping to a minimum the period during which precipitated water remains available (MCWILLIAM and DOWLING 1970). This is a precarious environment in which to germinate, yet germination in the more hospitable environment of deeper soil encounters difficulties in emergence resulting from the limitations of stored food and from the mechanical impedance by the drying surface layers of the soil (ARNDT 1965a, b). The relative effect of these two conflicting constraints will depend on soil physical conditions and moisture regime on the one hand, and on the nature of the particular seed on the other.

12.6.2 Physiological Depth-Gages

In most natural habitats the soil is undisturbed and its surface is compact. Consequently, the majority of the seed population inhabits a shallow layer at the surface, except where cracks develop. Germination that is triggered, accelerated, and synchronized by environmental conditions which are typical to the soil surface, such as light and/or extreme diurnal temperature fluctuations will help to maximize the utilization of the ephemeral supply of soil moisture at the surface. The energy and daily duration of light (KOLLER et al. 1964; WOOLEY and STOLLER 1978), as well as the amplitude of the diurnal temperature fluctuation, become strongly attenuated with soil depth. These regulatory mechanisms may therefore limit the depth from which the seedlings would have to emerge and the mechanical impedance that they would have to overcome.

A different situation exists in coarse-grained desert soils, where the balance between moisture relations and mechanical impedance at the surface makes the deeper soil layers more hospitable for germination. Inhibition by prolonged irradiation would act to restrict germination to a minimal soil depth in such habitats. This is the case with *Citrullus colocynthis* (KOLLER et al. 1963), whose large seeds are rich in food reserves and whose seedlings exhibit a high growth rate.

12.6.3 Regulation of Depth

In cultivated soil, the optimal depth of planting represents the compromise between the improving water relations and the deteriorating conditions for emergence with depth. The same strategy occurs naturally, unaided by man.

Geocarpy. Some species exhibit geocarpy and actively bury their seeds at a predetermined depth in the soil in the vicinity of the parent plant, while they are still developing. This is achieved in *Faktorovskya aschersoniana* (ZOHARY 1937) and in *Arachis hypogaea* by intercalary growth and positive geotropism of the basal part of the fertilized ovary – the gynophore (SMITH 1956); in *Trifolium subterraneum* by elongation and positive geotropism of the peduncle, followed by burrowing action of successive whorls of sterile calyces (KATZNELSON and MORLEY 1965); in the amphicarpic *Gymnarrhena micrantha* by contraction of the main root prior to transformation of the retracting shoot apex to the reproductive state (KOLLER and ROTH 1964; UKO, unpublished). In addition, geocarpy may provide the rather general advantage of germinating where the parent had successfully completed its own life cycle. Seed burial may also be advantageous by providing some protection against predation.

Hygroscopic Mechanisms. Other species equip their dispersal units with means by which they bury themselves in the soil after dissemination. In *Erodium* spp., the mericarps come equipped with a long, hygroscopic style whose proximal part coils into a tight helix when dry and uncoils when wet (YOUNG et al. 1975). In *Avena sterilis* the same responses occur in the two awns of the dispersal unit (which consists of two fertile florets). In both species the base of the dispersal unit tapers to a sharp tip that carries barbs facing the rear. Diurnal fluctuations in atmospheric humidity result in cycles of coiling and uncoiling that screw the harpoon-like tip of the dispersal unit into the soil. The seed comes to rest at a depth determined on the one hand by the increasing attenuation of the humidity fluctuations and by the increasing mechanical resistance of the soil on the other. In all cases, the rather precise depth at which the seeds come to rest is evidently nonlimiting for emergence.

12.6.4 Seasonal Forecasting

Widespread regions of the earth are characterized by an alternation between a "wet" and a "dry" season, that differ in the abundance, predictability, and pattern of rainfall. With relatively few exceptions (e.g., "summer annuals" that occur where the dry season coincides with summer), the probability for survival and establishment is higher in the wet season, provided over-abundant germination does not increase competition. It is therefore of obvious value for plants to avoid germination in the unreliable soil moisture provided in the dry season and delay it till the onset of the wet season, when supply of moisture is more reliable. Most plants produce their seeds toward the end of the wet season. A requirement for "after-ripening" in such seeds may suffice to reduce the probability of precocious germination, but under field conditions this requirement is usually satisfied fairly early in the dry season. A widespread strategy for further delaying germination until the onset of the following wet season is exhibited by hard-seeded legumes, in which the hilar fissure acts as a one-way valve that under diurnal fluctuations in atmospheric humidity allows only the outward diffusion of water vapor from the seed (Sect. 12.5.5). However, when the atmospheric humidity rises very gradually, the seed becomes progressively rehydrated. Presumably, under such conditions the two superimposed layers that operate the hilar valve have time to become equally hydrated and expand simultaneously, leaving the fissure open (HYDE 1954). Alternatively, colloids inside the seed become hydrated and swell, stretching the coat, and prevent the valve from closing. In the natural environment, seed impermeability is thus progressively lost as the general level of atmospheric humidity gradually rises, with the approach of the wet season.

A different strategy, of high survival value in a mediterranean-type climate (WILLIAMS and ELLIOTT 1960), was observed by QUINLIVAN (1961) in hard seeds of *Trifolium subterraneum, Medicago tribuloides, Lupinus digitatus* and *L. luteus,* whose permeability increased after exposure (in the dry state) to a regime of diurnal temperature fluctuations similar to those experienced on the exposed soil surface during the summer months. Thus in nature impermeability is lost as the wet season approaches. Further studies, with *Lupinus varius* and *Trifolium subterraneum,* showed that the effectiveness of this treatment was determined by the maximal temperature of the cycle, provided the minimal temperature was at least 15 °C lower (QUINLIVAN 1966). In the field, presence of dry litter reduced the maximal temperature at the soil surface and a greater proportion of the seeds remained impermeable (QUINLIVAN 1965). Dry seeds of *Lupinus varius* responded to a regime of diurnal temperature fluctuations only when their moisture content had fallen below a critical level, at which the coats are entirely impermeable and the hilar valve is operative. Seeds with a moisture content above the critical level do not respond at all to temperature fluctuations, but are capable of taking up atmospheric humidity through pathways that are not associated with the hilum or strophiole (QUINLIVAN 1968, 1971). Permeability in response to such temperature alternations is achieved by longitudinal splitting of the Malpighian layer underneath the strophiole ("strophiolar cleft"), which also occurs as a result of percussion. This opening can apparently be resealed by exposure to low atmospheric humidity (HAGON and BALLARD 1970). In *Erodium botrys* permeability similarly increases during exposure to the summer environment. Here, a hilar cap is apparently dislodged by contraction and expansion of the enclosing pericarp in response to the diurnal fluctuations in temperature and humidity (YOUNG et al. 1975).

12.6.5 Integrating Meters of Soil Moisture

As aridity increases, rainfall, even in the rainy season, becomes more erratic in timing, intensity, duration, and localization. Furthermore, after it had reached the surface, its availability in the soil is determined to a large extent by soil surface permeability and by surface topography. In a number of desert plants the dispersal units contain water-soluble inhibitors of germination. The role of such inhibitors may be inferred from results of SORIANO (1953), who showed that samples of desert topsoils, when bottom-irrigated, produced more seedlings if they were previously exposed to (simulated) rainfall. The magnitude of the response was affected by the total amount of precipitation, but more important, also by its intensity. When the dispersal unit is wet, the inhibitors dissolve and diffuse into the surrounding soil water. The process is accelerated when the water flows past the seed, and this occurs in nature by gravitational flow following rainfall. The rate of leaching will be influenced by the flow rate. Consequently, the time required to reduce the inhibitor level sufficiently to allow germination will determine the availability of soil moisture below the seed for subsequent seedling establishment. Water-soluble germination inhibitors thus function as integrating meters of available soil moisture.

12.6.6 Rainfall Sensors for Dispersal

In a number of desert plants the dispersal units are disseminated by mechanisms that are operated by specific environmental water relations. Thus, in the amphicarpic *Gymnarrhena micrantha* the aerial achenes remain enclosed in the dead involucral bracts of the aerial inflorescences until they are released by hygroscopic movements caused by a number of cycles of wetting by dew. In the natural habitat, such cycles are more common toward the end of the dry season, so that germination is delayed until availability of soil moisture is more reliable (KOLLER and ROTH 1964). In *Asteriscus pygmaeus* the achenes are similarly enclosed. The dead involucral bracts open hygroscopically to expose the achenes only when wet by dew, or rain. However, the peripheral achenes on the exposed capitulum are dislodged only under the impact of raindrops (KOLLER and ROTH, unpublished). In *Blepharis persica* the seeds remain locked in the capsule (for years) until they are ejected explosively when raindrops moisten a specialized trigger in the apex of the capsule. They germinate within an hour of landing on the wet soil surface, to which they anchor by specialized mucilaginous hairs that uncoil and poise the emerging radicle at an angle favoring penetration into the soil (GUTTERMAN et al. 1967). Such mechanisms are not only operated by environmental water relations, but also aid in taking advantage of favorable water relations for germination.

References

Allerup S (1958) Effect of temperature on uptake of water in seeds. Physiol Plant 11:99–105

Amemiya M (1965) The influence of aggregate size on moisture content – capillary conductivity relations. Soil Sci Am Proc 29:741–748

Arndt W (1965a) The nature of the mechanical impedance to seedlings by soil surface seals. Aust J Soil Res 3:45–54

Arndt W (1965b) The impedance of soil seals and the forces of emerging seedlings. Aust J Soil Res 3:55–68

Ayers AD (1952) Seed germination as affected by soil moisture and salinity. Agron J 44:82–84

Ayers AD, Hayward HE (1948) A method for measuring the effects of soil salinity on seed germination with observations on several crop plants. Soil Sci Soc Am Proc 13:224–226

Bain JM, Mercer FV (1966) Subcellular organization of the developing cotyledons of *Pisum sativum* L. Aust J Biol Sci 19:49–67

Ballard LAT (1973) Physical barriers to germination. Seed Sci Technol 1:285–303

Barlow EWR, Lee JW, Munns R, Smart MG (1980) Water relations of the developing wheat grain. Aust J Plant Physiol 7:519–525

Baver DA, Gardner WH, Gardner WR (1972) Soil physics. 4th edn. Wiley and Sons, New York

Bedford LV, Matthews S (1976) The effect of seed age at harvest on the germinability and quality of heat-dried pea seeds. Seed Sci Technol 4:275–286

Berggren G (1963) Is the ovule type of importance for the water absorption of the ripe seed? Sven Bot Tidskr 57:377–395

Berrie AMM, Paterson J, West HP (1974) Water content and responsivity of lettuce seed to light. Physiol Plant 31:90–96

Bewley JD, Black M (1978) Physiology and biochemistry of seeds. Vol 1. Springer, Berlin Heidelberg New York

Binet P (1964) Action de la température et de la salinité sur la germination des graines de *Plantago maritima* L. Bull Soc Bot Fr 111:407–411

Black CA (1968) Soil plant relationships. 2nd edn. Wiley and Sons, New York

Boorman LA (1968) Some aspects of the reproductive biology of *Limonium vulgare* Mill. and *Limonium humile* Mill. Ann Bot (London) 32:803–824

Bramlage WJ, Leopold AC, Parrish DJ (1978) Chilling stress to soybeans during imbibition. Plant Physiol 61:525–529

Bruce RR, Klute A (1956) The measurement of soil moisture diffusivity. Soil Sci Soc Am Proc 20:458–462

Carpita NC, Nabors MW, Ross CW, Petretic NL (1979a) The growth physics and water relations of red-light-induced germination in lettuce seeds. III. Changes in the osmotic and pressure potential in the embryonic axes of red- and far-red-treated seeds. Planta 144:217–224

Carpita NC, Nabors MW, Ross CW, Petretic NL (1979b) The growth physics and water relations of red-light-induced germination in lettuce seeds. IV. Biochemical changes in the embryonic axes of red- and far-red-treated seeds. Planta 144:225–233

Cavazza L (1953) L'influenza di bassi tensioni dell'acqua sulla germinazione di alcuni semi. Nuov G Bot Ital 60:759–762

Childs EC (1969) An introduction to the physical basis of soil water phenomena. Wiley-Interscience, New York

Ching TM (1972) Metabolism of germinating seeds. In: Kozlowski TT (ed) Seed biology Vol II. Academic Press, London New York, pp 103–218

Christiansen MJ, Moore PP (1959) Seed coat structural differences that influence water uptake and seed quality in hard seed cotton. Agron J 51:582–584

Collis George N, Hector JB (1966) Germination of seeds as influenced by matric potential and by area of contact between seed and soil water. Aust J Soil Res 4:145–164

Collis George N, Sands JE (1959) The control of seed germination by moisture as a soil physical property. Aust J Agric Res 10:628–637

Collis George N, Sands JE (1962) Comparison of the effects of physical and chemical components of soil water energy on seed germination. Aust J Agric Res 13:575–585

Collis George N, Williams J (1968) Comparison of the effects of soil matric potential and isotropic effective stress on the germination of *Lactuca sativa*. Aust J Soil Res 6:179–192

Côme D, Tissaoui T (1973) Interrelated effects of imbibition, temperature and oxygen on seed germination. In: Heydecker W (ed) Seed ecology. Butterworth, London, pp 157–168

Crocker W, Thornton NC, Schroeder EM (1946) Internal pressure necessary to break shells of nuts and role of shells in delayed germination. Contrib Boyce Thompson Inst 14:173–201

Currie JA (1973) The seed – soil system. In: Heydecker W (ed) Seed ecology. Butterworth, London, pp 463–479

Dasberg S (1971) Soil water movement to germinating seeds. J Exp Bot 22:999–1008

Dasberg S, Mendel K (1971) The effect of soil water and aeration on seed germination. J Exp Bot 22:992–998

Dexter ST, Miyamoto T (1959) Acceleration of water uptake and germination of sugarbeet seed balls by surface coating of hydrophilic colloids. Agron J 51:388–389

Doering EJ (1965) Soil water diffusivity by the one step method. Soil Sci 99:322–326

Doneen LD, McGillivray JH (1943) Germination and emergence of vegetable seeds as affected by different soil moisture conditions. Plant Physiol 18:524–529

Duke SH, Kakefuda G (1981) Role of the testa in preventing cellular rupture during imbibition of legume seeds. Plant Physiol 67:449–456

Duke SO (1978) Interactions of seed water contents with phytochrome-mediated germination of *Rumex crispus* (L) seeds. Plant Cell Physiol 19:1043–1049

Duke SO, Egley GA, Reger BJ (1977) Model for variable light sensitivity in imbibed dark-dormant seeds. Plant Physiol 59:244–249

Esashi Y, Leopold AC (1968) Physical forces in dormancy and germination of *Xanthium* seeds. Plant Physiol 43:871–876

Evans M, Black M, Chapman J (1975) Induction of hormone sensitivity by dehydration is one positive role for drying in cereals. Nature (London) 258:144–145

Evenari M, Koller D, Gutterman Y (1966) Effects of the environment of the mother plant on germination by control of seed coat impermeability in *Ononis sicula* Guss. Aust J Biol Sci 19:1007–1016

Fayemi AA (1957) Effect of temperature on the rate of seed swelling and germination of legume seeds. Agron J 49:75–76

Fussell LK, Dwarte DM (1980) Structural changes of the grain associated with black region formation in *Pennisetum americanum*. J Exp Bot 31:645–654

Galli MG, Levi M, Sparvoli E (1981) Lack of correlation between overall protein synthesis and the onset of cell elongation in germinating embryos of *Haplopappus gracilis*. Physiol Plant 51:321–325

Gorski T (1975) Germination of seeds in the shadow of plants. Physiol Plant 34:342–346

Grubert M (1974) Studies on the distribution of myxospermy among seeds and fruits of Angiospermae and its ecological importance. Acta Biol Venez 8:315–551

Gutterman Y, Evenari M (1972) The influence of day length on seed coat colour, an index of water permeability of the desert annual *Ononis sicula* Guss. J Ecol 60:713–719

Gutterman Y, Witztum A, Evenari M (1967) Seed dispersal and germination in *Blepharis persica* (Burm) Kuntze. Isr J Bot 16:213–234

Haber AH, Luippold HJ (1960a) Separation of mechanisms initiating cell division and cell expansion in lettuce seed germination. Plant Physiol 35:168–173

Haber AH, Luippold HJ (1960b) Effects of gibberellin, kinetin, thiourea and photomorphogenic radiation on mitotic activity in dormant lettuce seed. Plant Physiol 35:486–494

Hadas A (1968) Evaporation and drying processes in layered soils. Ph D Thesis, Hebrew Univ, Jerusalem (English summary)

Hadas A (1969) Effects of soil moisture stress on seed germination. Agron J 61:325–327

Hadas A (1970) Factors affecting seed germination under soil moisture stress. Isr J Agric Res 20:3–14

Hadas A (1974) Water transfer to germinating seeds as affected by soil hydraulic properties and seed soil water contact impedance. In: Vries de DA, Afgan G (eds) Heat and mass transfer in biosphere. Scripta, New York, pp 501–572

Hadas A (1976) Water uptake and germination of leguminous seeds under changing external water potential conditions in osmotic solutions. J Exp Bot 27:480–489

Hadas A (1977a) Water uptake and germination of leguminous seeds in soils of changing matric and osmotic water potential. J Exp Bot 28:977–985

Hadas A (1977b) A simple laboratory approach to test and estimate seed germination performance under field conditions. Agron J 69:582–585

Hadas A, Russo D (1974a) Water uptake by seeds as affected by water stress, capillary conductivity and seed soil contact. I. Experimental study. Agron J 66:643–647

Hadas A, Russo D (1974b) Water uptake by seeds as affected by water stress, capillary conductivity and seed soil contact. II. Analysis of experimental data. Agron J 66:647–652

Hadas A, Stibbe E (1973) Analysis of water uptake and growth patterns of seedlings of four species prior to emergence. In: Hadas A, Swartzendruber D, Rijtema PE, Fuchs M, Yaron B (eds) Physical aspects of soil water and salts in ecosystems. Ecol Stud Vol IV. Springer, Berlin Heidelberg New York, pp 97–106

Hagon MW, Ballard LAT (1970) Reversibility of strophiolar impermeability to water in seeds of subterranean clover (*Trifolium subterraneum* L). Aust J Biol Sci 23:519–528

Halloran GM, Collins WJ (1974) Physiological predetermination of the order of hardseededness breakdown in subterranean clover (*Trifolium subterraneum* L). Ann Bot (London) 38:1039–1044

Hamly DH (1932) Softening the seeds of *Melilotus alba*. Bot Gaz 93:345–375

Hanson AD (1973) The effects of imbibition drying treatments on wheat seeds. New Phytol 72:1063–1073

Harper JL, Benton RA (1966) The behaviour of seeds in soil. II. The germination of seeds on the surface of a water supplying substrate. J Ecol 54:151–166

Hegarty TW (1977) Seed activation and seed germination under moisture stress. New Phytol 78:349–359

Hegarty TW (1978) The physiology of seed hydration and dehydration and the relation between water stress and the control of germination: a review. Plant Cell Environ 1:101–119
Hegarty TW, Ross HA (1980/81) Investigations of control mechanisms of germination under water stress. Isr J Bot 29:83–92
Hendricks SB, Taylorson RB (1976) Variation in the germination and amino acid leakage with temperature related to membrane phase change. Plant Physiol 58:7–11
Heydecker W, Orphanos PI (1968) The effect of excess moisture on the germination of *Spinacia oleracea* L. Planta 83:237–247
Hillel D (1971) Soil and water: Physical principles and processes, Kozlowski TT (ed). Academic Press, London New York
Hsiao AI, Vidaver W (1971) Seed water content in relation to phytochrome-mediated germination of lettuce seeds (*Lactuca sativa* L. Var Grand Rapids). Can J Bot 49:111–115
Hunter JR, Erickson AE (1952) Relation of seed germination to soil moisture tension. Agron J 44:107–109
Hyde EOC (1954) The function of the hilum in some Papilionaceae in relation to the ripening of the seed and the permeability of the testa. Ann Bot (London) 18:241–256
Hyde EOC, McLeavy AM, Harris GS (1959) Seed development in ryegrass, and in red and white clover. N Z J Agric Res 2:947–952
Kahn A (1960) An analysis of 'dark-osmotic inhibition' of lettuce seeds. Plant Physiol 35:1–7
Karssen CM (1970) The light promoted germination of the seeds of *Chenopodium album* L. V. Dark reactions regulating quantity and rate of response to red light. Acta Bot Neerl 19:187–196
Karssen CM (1980/81) Environmental conditions and endogenous mechanisms involved in secondary dormancy of seeds. Isr J Bot 29:45–64
Katznelson J, Morley FHW (1965) A taxonomic revision of Sect *Calycomorphum* of the genus *Trifolium*. I. The geocarpic species. Isr J Bot 14:112–134
Kaufmann MR (1969) Effects of water potential on germination of lettuce, sunflower and citrus seeds. Can J Bot 47:1761–1764
Khan AA (1975) Primary, preventive and permissive roles of hormones in plant systems. Bot Rev 4:391–420
Khan AA (ed) (1977) The physiology and biochemistry of seed dormancy and germination. Elsevier, Amsterdam
Khan AA, Peck NH, Samimy S (1980/81) Seed osmoconditioning: Physiological and biochemical changes. Isr J Bot 29:133–144
Khan RA, Laude HM (1969) Influence of heat stress during seed maturation on germinability of barley seed at harvest. Crop Sci 9:55–58
King TJ (1975) Inhibition of seed germination under leaf canopies in *Arenaria serpyllifolia, Veronica arvensis* and *Cerastrum holosteoides*. New Phytol 75:87–90
Klein S, Pollock BM (1968) Cell fine structure of developing Lima bean seeds related to seed desiccation. Am J Bot 55:658–672
Koller D (1955) The regulation of germination in seeds. Bull Res Counc Isr Sect D 5:85–108
Koller D (1969) The physiology of dormancy and survival of plants in desert environments. Symp Soc Exp Biol 23:449–469
Koller D (1970) Analysis of the dual action of white light on germination of *Atriplex dimorphostegia* (Chenopodiaceae). Isr J Bot 19:499–516
Koller D (1972) Environmental control of seed germination. In: Kozlowski TT (ed) Seed biology Vol II. Academic Press, London New York, pp 1–101
Koller D, Cohen D (1959) Germination-regulating mechanisms in some desert seeds. VI. *Convolvulus lanatus* Vahl, *Convolvulus negevensis* Zoh and *Convolvulus secundus* Desr. Bull Res Counc Isr Sect D7:175–180
Koller D, Roth N (1964) Studies on the ecological and physiological significance of amphicarpy in *Gymnarrhena micrantha* (Compositae). Am J Bot 51:26–35
Koller D, Poljakoff-Mayber A, Berg A, Diskin T (1963) Germination-regulating mechanisms in *Citrullus colocynthis*. Am J Bot 50:597–603

Koller D, Sachs M, Negbi M (1964) Germination-regulating mechanisms in some desert seeds. VIII. *Artemisia monosperma*. Plant Cell Physiol 5:79–84

Kunze RJ, Kirkham D (1962) Simplified accounting for membrane impedance in capillary conductivity measurement. Soil Sci Soc Am Proc 26:421–426

Labouriau LG (1978) On the physiology of seed germination in *Vicia graminea* Sm. II. An analysis of the temperature dependence of the seed germination rate. Anais Acad Bras Cienc 44:477–534

Levitt J, Hamm PC (1943) A method for increasing the rate of seed germination of *Taraxacum kok-saghyz*. Plant Physiol 18:288–293

Loercher L (1974) Persistence of red light induction in lettuce seed of varying hydration. Plant Physiol 53:503–506

Maguire JD (1977) Seed quality and germination. In: Khan AA (ed) The physiology and biochemistry of seed dormancy and germination. Elsevier, Amsterdam, pp 219–235

Manohar MS, Heydecker W (1964) Effect of water potential on germination of pea seeds. Nature (London) 202:22–24

Matthews S (1973) Changes in developing pea (*Pisum sativum* L) seeds in relation to their ability to withstand desiccation. Ann Appl Biol 75:93–105

Matthews S (1976) Seed in relation to ecology. Adv Res Technol Seeds 2:86–106

Mayer AM, Poljakoff-Mayber A (1975) The germination of seeds. 2nd edn, Pergamon, New York London

Mayer AM, Shain Y (1974) Control of seed germination. Annu Rev Plant Physiol 25:167–193

McArthur JA (1978) Light effects upon dry lettuce seeds. Planta 144:1–5

McDonough WT (1975) Water potential of germinating seeds. Bot Gaz 136:106–108

McGinnies WJ (1960) Effects of moisture stress and temperature on germination of six range grasses. Agron J 52:159–163

McWilliam JR, Dowling PM (1970) Factors influencing the germination and establishment of pasture seeds on the soil surface. Proc 11th Grassl Conf. Queensl Univ Press, Brisbane, pp 578–583

McWilliam JR, Phillips PJ (1971) Effect of osmotic and matric potentials on the availability of water for seed germination. Aust J Biol Sci 24:423–431

Morinaga TI (1926) Germination of seeds under water. Am J Bot 13:126–140

Morris IJ, Campbell WF, Wiebe HH (1968) A refractory method for testing seed coat impermeability. Agron J 60:79–80

Mott JJ (1974) Factors affecting seed germination in three annual species from an arid region of Western Australia. J Ecol 62:699–709

Nabors MW, Lang A (1971) The growth physics and water relations of red-light-induced germination in lettuce seeds. I. Embryos germinating in osmoticum. Planta 101:1–25

Negbi M, Rushkin E, Koller D (1966) Dynamic aspects of water relations in germination of *Hirschfeldia incana* seeds. Plant Cell Physiol 7:363–376

Nicholls PB (1979) Induction of sensitivity to gibberellic acid in developing wheat caryopses: effect of rate of desiccation. Aust J Plant Physiol 6:229–240

Onckelen van H, Caubergs R, Horemans S, De Greef JA (1980) Metabolism of abscisic acid in developing seeds of *Phaseolus vulgaris* L and its correlation to germination and α-amylase activity. J Exp Bot 31:913–920

Orphanos PI, Heydecker W (1968) On the nature of the soaking injury of *Phaseolus vulgaris* seeds. J Exp Bot 19:770–784

Phillips RE (1968) Water diffusivity of germinating soybean, corn and cotton seeds. Agron J 60:568–571

Pollock BM, Toole VK (1966) Imbibition period as the critical temperature sensitive stage in germination of Lima bean seeds. Plant Physiol 41:221–229

Porter RH (1949) Recent developments in seed technology. Bot Rev 15:221–344

Pradet A, Narayanan A, Vermeersch J (1968) Etude des adénosine-5′-mono, di- et triphosphate dans les tissus végétaux. III. Métabolisme énérgetique au cours des premières stades de la germination des semences de laitue. Bull Soc Fr Physiol Veg 14:107–114

Quinlivan BJ (1961) The effect of constant and fluctuating temperatures on the permeability of hard seeds of some legume species. Aust J Agric Res 12:1009–1022

Quinlivan BJ (1965) The influence of the growing season and the following dry season on the hard seededness of subterranean clover in different environments. Aust J Agric Res 16:277–291
Quinlivan BJ (1966) The relation between temperature fluctuations and the softening of hard seeds of some legume species. Aust J Agric Res 17:625–631
Quinlivan BJ (1968) The softening of hard seeds of sand plain lupin (*Lupinus varius* L). Aust J Agric Res 19:507–515
Quinlivan BJ (1971) Seed coat impermeability in legumes. J Aust Inst Agric Sci 37:283–295
Reynolds T (1975) Characterization of osmotic restraints on lettuce fruit germination. Ann Bot (London) 39:791–796
Richards LA, Wadleigh CH (1952) Soil water and plant growth. In: Shaw BT (ed) Soil physical conditions and plant growth. Academic Press, London New York, pp 74–253
Rijtema PE (1959) Calculation of capillary conductivity from pressure plate outflow data with a non-negligible membrane impedance. Neth J Agric Sci 7:209–215
Roberts EH (1972) Dormancy: a factor affecting seed survival in the soil. In: Roberts EH (ed) Viability of seeds. Chapman and Hall, London, pp 321–359
Roberts EH, Smith RD (1977) Dormancy and the pentose phosphate pathway. In: Khan AA (ed) The physiology and biochemistry of seed dormancy and germination. Elsevier, Amsterdam, pp 385–411
Rogerson NE, Matthews S (1977) Respiratory and carbohydrate changes in developing pea (*Pisum sativum* L) seeds in relation to their ability to withstand desiccation. J Exp Bot 28:304–313
Scheibe J, Lang A (1965) Lettuce seed germination: evidence for a reversible light-induced increase in growth potential and for phytochrome mediation of the low temperature effect. Plant Physiol 40:485–492
Scheibe J, Lang A (1967) Lettuce seed germination: a phytochrome-mediated increase in the growth rate of lettuce seed radicles. Planta 72:348–354
Sedgeley RH (1963) The importance of liquid seed contact during the germination of *Medicago tribuloides*. Aust J Agric Res 14:646–654
Shaykewich CF (1973) Proposed method for measuring swelling pressure of seeds prior to germination. J Exp Bot 24:1056–1061
Shaykewich CF, Williams J (1971) Resistance to water absorption in germinating rape seed (*Brassica napus* L). J Exp Bot 22:19–24
Simon EW (1974) Phospholipids and plant membrane permeability. New Phytol 73:377–420
Simon EW, Wiebe HH (1975) Leakage during imbibition, resistance to damage at low temperature and the water content of peas. New Phytol 74:407–411
Smith BW (1956) *Arachis hypogaea*. Embryology and the effect of pod elongation upon embryo and endosperm growth. Am J Bot 43:233–240
Smith H (1973) Light quality and germination: ecological implications. In: Heydecker W (ed) Seed ecology. Butterworth, London, pp 219–231
Soriano A (1953) Estudios sobre germinacion. I. Rev Invest Agric 7:315–340
Spurny M (1973) The imbibition process. In: Heydecker W (ed) Seed ecology. Butterworth, London, pp 367–389
Spyropoulos CG, Lambiris MP (1980) Effect of water stress on germination and reserve carbohydrate metabolism in germinating seeds of *Ceratonia siliqua* L. J Exp Bot 31:851–857
Stiles IE (1948) Relation of water to the germination of corn and cotton seeds. Plant Physiol 23:201–222
Stone EC, Juhren G (1951) The effect of fire on the germination of the seed of *Rhus ovata* Wats. Am J Bot 38:368–372
Taylorson RB, Hendricks SB (1972) Rehydration of phytochrome in imbibing seeds of *Amaranthus retroflexus*. Plant Physiol 49:663–665
Taylorson RB, Hendricks SB (1977) Dormancy in seeds. Annu Rev Plant Physiol 28:331–354
Thomas H (1972) Control mechanisms in the resting seed. In: Roberts EH (ed) Viability of seeds. Chapman and Hall, London, pp 360–396

Tobin EM, Briggs WR (1969) Phytochrome in embryos of *Pinus palustris*. Plant Physiol 44:148–150

Uhvits R (1946) Effect of osmotic pressure on water absorption and germination of alfalfa seeds. Am J Bot 33:278–285

Ungar IA (1962) Influence of salinity on seed germination in succulent halophytes. Ecology 43:763–764

Vincent EM, Cavers PB (1978) The effects of wetting and drying on the subsequent germination of *Rumex crispus*. Can J Bot 56:2207–2217

Ward J, Shaykewich CF (1972) Water absorption by wheat seeds as influenced by hydraulic properties of the soil. Can J Soil Sci 52:99–105

Webster BD, Leopold AC (1977) The ultrastructure of dry and imbibed cotyledons of soybean. Am J Bot 64:1268–1293

Wellington PS (1964) Studies on the germination of cereals. 5. The dormancy of barley grains during ripening. Ann Bot (London) 28:113–126

Wellington PS, Durham VM (1961) Studies on the germination of cereals. 3. The effect of the covering layers on the uptake of water by the embryo of the wheat grains. Ann Bot (London) 25:185–196

Werker E, Marbach I, Mayer AM (1979) Relation between the anatomy of the testa, water permeability and the presence of phenolics in the genus *Pisum*. Ann Bot (London) 43:765–771

Wesson G, Wareing PF (1969) The role of light in the germination of naturally occurring populations of buried weed seeds. J Exp Bot 20:402–413

Wiggans SC, Gardner EB (1959) Effectiveness of various solutions for simulating drought conditions as measured by germination and seedling growth. Agron J 51:315–318

Williams AW, Elliott JR (1960) Ecological significance of seed coat impermeability to moisture in crimson, subterranean and rose clovers in a Mediterranean type climate. Ecology 41:785–790

Williams J, Shaykewich CF (1971) Influence of soil matric potential and hydraulic conductivity on the germination of rape seed (*Brassica napus* L.). J Exp Bot 22:586–597

Witztum A, Gutterman Y, Evenari M (1969) Integumentary mucilage as an oxygen barrier during germination of *Blepharis persica* (Burm) Kuntze. Bot Gaz 130:238–241

Woodstock LW, Pollock BM (1965) Physiological predetermination: imbibition, respiration and growth in Lima bean seeds. Science 150:1031–1032

Wooley JT, Stoller EW (1978) Light penetration and light-induced seed germination in soil. Plant Physiol 61:597–600

Young JA, Evans RA (1973) Mucilaginous seed coats. Weed Sci 21:52–54

Young JA, Evans RA, Kay BL (1975) Dispersal and germination dynamics of Broadleaf Filaree, *Erodium botrys* (Cav) Bertol. Agron J 67:54–57

Zohary M (1937) Die verbreitungsökologischen Verhältnisse der Pflanzen Palästinas. I. Die antitelechorischen Erscheinungen. Beih Bot Zbl 56:1–155

13 Environmental Aspects of the Germination of Spores

A.S. SUSSMAN and H.A. DOUTHIT, Jr.

CONTENTS

13.1 Introduction: Ecology of Microbial Spores

Two constraints were imposed in this review: the lack of well-developed theoretical models of ecological phenomena as applied to microorganisms, and the need to sort out those aspects of ecological interactions that relate specifically to spores. The first constraint, which is due in large part to difficulties in basic methodology, such as reliable methods for counting cells of diverse types which may have wide variability in generation times, means that reliable data on natural situations often are not available. As for spores and their special role in nature, considerable information exists but, again, much of it is limited to in vitro situations. Therefore, drawing analogies between microbial ecology and that developed for higher organisms, while tempting, may be treacherous.

The outline of this chapter is based upon the functions of spores we feel may be important to the ecology of spores, and include the enhancement of survivability and dissemination and their use as timing devices. This arrangement of the subject has been used before (SUSSMAN and HALVORSON 1966 and SUSSMAN 1968). Emphasis will be placed upon the literature since the earlier reviews were written but, where appropriate, the older literature will be discussed.

The definitions upon which this chapter is based are those of SUSSMAN (1968) for fungi, and VARY (1979) for bacteria. Further details of dormancy and germination in microbes can be found in the work of SUSSMAN and HALVORSON (1966), SUSSMAN and DOUTHIT (1973) and in the continuing reports of the International Spore Conference and the Spore Newsletter.

13.2 Enhancement of Survivability

Earlier bacteriologists like KNAYSI (1948) and LAMANNA (1952) have expressed doubt about this role for bacterial spores because of the frequent observation that the majority of bacteria found in soil are not known to form spores. Indeed, some fungi survive mainly as mycelium (WARCUP 1957; HANSEN 1976; WEARING and BURGESS 1977; KATO et al. 1977), although, as was noted above, the experimental techniques may be misleading. Nevertheless, VAN NIEL (1955) has suggested that the resistance of spores is indeed a factor in survival. He argued that since not a single spore-former is known that can thrive under nutritive conditions that are unsuited for bacteria that do not form spores, spore-formers must have gained survival value through increased resistance to adverse environmental conditions instead of through diversified metabolic abilities.

13.2.1 Longevity of Microorganisms

That spores are very resistant to environmental factors of various kinds has been discussed before (SUSSMAN and HALVORSON 1966, and SUSSMAN 1968) and extensive tabular data are provided for documentation, so the reader is referred to this earlier work.

13.2.2 Factors Involved in Survivability

As GARRETT (1956) has proposed, survivability in the mixed populations found in nature must be the result of several interacting variables. Those that he has suggested, modified by SUSSMAN (1968), include resistance to deleterious agents, competitive saprophytic ability, disseminability, responsiveness (timing), and mutational capacity.

Resistance to deleterious agents like extreme temperature and pressure, desiccation, radiation of various kinds, and chemicals often characterize spores, in contrast to vegetative stages. For example, BORUT (1960) has shown that most fungi of desert soils sporulate intensely, or form resistant stages of other kinds and that the optimum for sporulation is 4–10 °C higher than that for growth, suggesting adaptation to the temperatures found in that habitat. In general, spores of both bacteria and fungi are more resistant to extreme temperatures, chemicals, and radiation than are vegetative stages, although exceptions can be found, such as arthrospores of the human pathogen, *Trichophyton mentagrophytes,* which are more sensitive to a variety of antibiotics than is the mycelium (HASHIMOTO and BLUMENTHAL 1978).

Comparative saprophytic ability subsumes the following: rapid spore germination and high growth rate, effective metabolic potential, capacity to produce antibiotics, and resistance to antibiotics produced by competitors. Spores are passive in their contribution to this ability, in that often they are resistant to antibiotics and, unless they germinate rapidly, they do not contribute to comparative saprophytic ability otherwise.

Disseminability and responsiveness are important characteristics of spores that will be considered in a later section. As for mutational capacity, it is likely that spores are not more resistant than vegetative stages of the same organism so they do not confer selective advantage in this way.

13.2.3 Water in Relation to Germination and Resistance

That there is an inverse correlation in many cases between the resistance of microbial propagules and their water content has been documented by SUSSMAN (1965) and MURRELL (1967). On the other hand, their respiratory capacity often is directly proportional to water content (SUSSMAN 1965; DESSER and BRODA 1965). Moreover, germination is strongly affected by relative humidity, and most microbial propagules germinate best at high humidities although a few, like those of the powdery mildews, require lower ones (SUSSMAN 1965).

Even transitory contact with water can change the state of some microbial propagules, as the results of NICKERSON et al. (1981) illustrate. When spores of *Rhizopus stolonifer* are harvested under dry conditions, as compared with wet ones, they incorporate much less $^{14}CO_2$, possess no polyribosomes, do not survive even short storage at room temperature, and have more nutritional requirements for germination. Therefore, the germinative condition of spores of this fungus, and probably other dormant propagules as well, is a consequence of the physiological state of the cells that give rise to them and the environment in which they find themselves.

The central role of water in dormancy is made clear by a consideration of the "anhydrous core" hypothesis as applied to bacterial spores. This assumption, based on the relatively dry state of spore protoplasts, has fascinated microbiologists for almost a hundred years, although good data have proven difficult to obtain. Recent evidence in support of lowered water content of spore protoplasts (GERHARDT and MURRELL 1978) includes a large reduction in protoplast volume during sporulation prior to encasement within cortical and coat layers. Mechanisms that bring about and maintain this constriction and the exact relationship to the anhydrous state of the protoplast are not yet understood, and are currently under intense investigation.

13.3 Disseminability and Dispersal

Adaptive significance can derive from the liberation of spores, their transport, arrival, and settling on a substratum, as well as from their germination. Thus, the size, shape, color, contents, and arrangement of propagules may influence selective advantage in relation to spore dispersal. Although many fungi have developed a large number of ways to assist in the release and take-off of propagules, bacteria and yeasts have not evolved such mechanisms (INGOLD 1965; GREGORY 1966). But even this generalization may have to be qualified since the finding by LACHANCE et al. (1976) that strains of *Metschnikowia australis,* pathogens of brine shrimp, forcibly expel their needle-shaped spores from asci. Apparently, the propulsive force originates from a mucilage which, upon swelling, exerts the requisite force.

However, discharge adaptations may be difficult to evaluate because of their complexity. For example, SUSSMAN (1968) discusses the case of *Sphaerobolus stellatus,* which appears to discharge spores only during daylight hours under some circumstances, but will do so in the dark under others. Moreover, a discharge rhythm with a period of 10–12 days can override the other discharge mechanisms when the organism is kept under constant light and temperature, so this rhythm's effect upon spore release in nature must be evaluated in addition to the other factors.

Of the media for dissemination, air probably has been studied most intensively. According to the calculation of SUSSMAN and HALVORSON (1966), the theoretical terminal velocity of bacterial spores (0.00037 cm per second) probably is small enough to ensure that settling is more difficult than it is for larger spores, such as those of fungi, and for pollen (GREGORY 1961). In any case, given the presence of vegetative bacterial cells in the atmosphere along with spores, what seems to be unique about the latter is their survivability, not the fact that they can be dispersed in air. Nevertheless, viable cells of non-spore-formers as well as spore-forming bacteria are spread widely in the air and have been found in Antarctica, high elevations in temperate zones and over the Atlantic. Although spore-formers predominate in some of these habitats, vegetative cells also survive. Certain technical difficulties make the data less reliable than would be desired, such as the difficulty of distinguishing between colonies that arise from spores and vegetative cells, clumping of cells and the use of selective media.

Many data attest to the wide distribution of fungi through the air because of the intense study of plant pathogens over many years. A case in point is *Puccinia chondrilla,* the uredospores of which germinate at temperatures from 0° to 36 °C. This organism occurs as a pathogen in the cold continental climates of southern Siberia to the hot, mediterranean ones of Portugal and North Africa (HASAN and WAPSHERE 1973). That many other factors influence distribution in any medium may be seen in the data of SIMPSON et al. (1973) who have found that the spore's response to moisture is the determinant of geographical distribution of the parasites of cotton with which they worked.

Light is another factor that strongly affects the distribution of microorganisms in air. Thus, DOWDING (1969) and others have observed that nonpigmented conidia separated from one another lose viability rapidly when exposed to sunlight. In fact, the frequency of occurrence of viable spores in the air (ALI et al. 1976) over desert areas or in desert soils (NICOT 1960; RANZONI 1968) can be correlated with the resistance to insolation imparted by melanized walls, although resistance to desiccation must also play a part in survival in such regions. That resistance to ozone, an attribute aiding survival in the atmosphere, also is correlated with pigmentation of spores, has been shown by HIBBEN and STOTZKY (1969). Moreover, the spores of the fungi that populate the surface of leaves, and those that are carried by birds, are black and appear to be adapted to withstand insolation as well (PUGH 1971, 1972). Organisms in similar environments, such as epiphytic parasites, very often produce mycelium, spores, and other structures that are melanized, whereas endophytic parasites form similar pigments only in structures that are produced outside the host (PARBERY and EMMETT 1977).

Dispersal by animals has been studied but much remains to be learned in this important field. For example, MASER et al. (1978) have pointed out that the role of small mammals as primary vectors in the dissemination of hypogeous mycorrhizal fungi makes the poisoning of forest rodents in timber management risky if healthy coniferous forests are to be maintained. Coprophilous ascomycetes, among other fungi, sometimes are dependent upon passage through the gut of mammals for the activation of their spores, as well as for food after germination. This is true as well for *Phytophthora megasperma;* the germination of its spores is increased from 5% to 93% after ingestion by the land snail, *Helix aspersa* (SALVATORE et al. 1973). Similar results were obtained by using enzymes of the animal. Birds are a means of transport (PUGH 1972) and are vectors of diseases as well (HUBALEK and RUSH-MUNRO 1973).

As for soil as a habitat in which spores appear to play a role, dry ones show a predominance of spore-formers and actinomycetes and, in turn, probably give rise to much of the dust from which airborne bacteria arise (DOETSCH and COOK 1973). According to MISHUSTIN (1956), the type of soil determines the proportion of spores and spore-formers found in this habitat. It would be interesting to know the distribution of spores and spore-formers in the different zones of soil, including the rhizosphere, etc., but the experiments are difficult because they require observation at the microscopic level so that spores can be distinguished from vegetative stages.

In the sea, about 80% of the bacteria are gram-negative, implying that only a minority can be spore-formers, although the details remain to be studied. Adaptation to the marine environment has taken place in some fungi which can sporulate and germinate in seawater (GARETH JONES et al. 1971; BYRNE and GARETH JONES 1975), although terrestrial fungi are less tolerant of this environment. The spores of a number of fungi have evolved the "tetraradiate" habit, which appears to be useful in anchoring these organisms under the turbulent conditions of streams (INGOLD 1975; WEBSTER et al. 1978).

13.4 Responsiveness

Microorganisms living in environments in which moisture, food, temperature, light, and other factors fluctuate gain selective advantage from mechanisms that ensure development when conditions are most favorable. Dormant spores and other propagules may serve this function by sensing the environment and germinating under appropriate conditions. Among microorganisms, such timing devices can operate at two stages of development, sporulation and spore germination, but only the latter will be discussed herein.

Our approach will be that of SUSSMAN and HALVORSON (1966), who consider dormant spores to represent a simple timing system in which an arresting device must be overcome by an "activator". Activators usually are likened to triggers, or signals, which may be physical or chemical factors in the environment. In what follows, we will discuss cases in which dormant spores appear to act as timing devices by virtue of their responses to environmental signals.

13.4.1 Responses to Physical Signals

When physical factors that influence spore germination differ significantly in amount, intensity, or quality from the requirements for vegetative growth they may be defined as triggers or activators. Thus, SUSSMAN (1965) has shown that the temperatures at which dormant fungal spores are activated usually are beyond the range at which growth takes place. Therefore, when words like "high" or "low" temperatures are used in this connection, they are meant to be compared with the requirements for vegetative growth and development. For example, PLOTNIKOVA et al. (1977) have shown that uredospores of *Puccinia graminis tritici* need a "heat shock" of 30 °C in order to germinate. Such a moderate temperature is a trigger in this case because the mycelium of this parasite develops only at lower temperatures.

That heat in excess of that appropriate for optimal growth of endospore-forming bacteria activates their spores for rapid germination has long been known and has been discussed in some detail by KEYNAN and EVENCHIK (1969). Recently, SOPER et al. (1976) have shown that treatment of spores of *Bacillus megaterium* with vacuum dehydration at high temperatures activates by a mechanism they feel is unlike that thought to be involved in the usual heat activation. Moreover, the opposite situation has been reported by KIRILLOVA et al. (1975) for spores of *Thermoactinomyces vulgaris,* which respond to treatment with low temperature by increased spore germination. Thus, temperature-induced spore activation is a complicated phenomenon, the basic biochemistry of which is not known in any detail, and the extent of which depends upon species, composition of growth media, age of spore suspension, and a number of quite unknown variables (KEYNAN and EVENCHIK 1969).

As for the fungi, temperatures between 37 °C and 40 °C have been used to activate conidia of ascomycetes, slime mold spores and those of some rusts (SUSSMAN 1965). Temperatures between 50 °C and 70 °C are necessary for many ascospores and for some other types of spores, like those of *Phycomyces blakesleeanus.* Many of the spores that require such high temperatures are from coprophiles, and it has been suggested that the requirement for heat is associated with passage through animals and subsequent incubation in habitats in which such temperatures are generated by microbial action.

In contrast to fungi requiring high temperatures for activation, most of which are saprophytes, those that require low temperatures for activation are pathogens or mycorrhizal fungi. For example, the rust, *Tilletia caries* (GASSNER and NIEMANN 1954), germinates at 3 °C and the mite parasite, *Hirsutella thompsonii,* requires treatment at 5 °C before it will germinate (KENNETH et al. 1979). A distinction has to be made between the temperature that is optimal for after-ripening and that needed for germination, as noted previously by SUSSMAN (1965).

In order to assess the ecological significance of these observations, the fact that the temperatures at which particular fungal spores are activated usually are beyond the range at which growth takes place must be considered. This means that incubation of such dormant spores at temperatures optimal for vegetative development will delay or suppress their germination. The example of basidiospores of *Flammula alnicola,* a fungus causing decay of conifers, helps to suggest the selective advantage that accrues to temperature as an activating force (SUSSMAN and HALVORSON 1966). Thus, these spores require storage at −7 °C, followed by incubation at higher temperatures, before they develop (DENYER 1960), thereby restricting infection to the spring season. In addition, such infections usually affect only the roots because the spores cannot survive the temperatures usually found on the surface of the soil in winter. Therefore, it is likely that the development of this parasite is timed to coincide with the beginning of the growth of its host, as well as with the start of the short growing season in Canada. A different kind of ecological role for heat activation has been postulated for ascospores of carbonicolous ascomycetes by WICKLOW and ZAK (1979), who argue that a fire not only serves to activate these spores but to reverse the widespread soil fungistasis that will be mentioned in a later section.

Light often is a requirement for the germination of basidiomycetous pathogens although spores of other kinds respond to light as well, such as oospores of *Phytophthora cactorum* (CARDOSO and SCHMITTHENNER 1975) and spores of *Botrytis cinerea* and *Penicillium roqueforti* (ZEMLYANUKHIN and CHEBOTAREV 1973). Blue light usually is most effective, as with spores of *Tilletia controversa* (SCHAUZ 1977). In the case of oospores of *P. cactorum,* the action spectrum shows two maxima, one at 400–450 nm, on the basis of which was suggested the participation of a phytochrome system (CARDOSO and SCHMITTHENNER 1975). In the case of spores of *B. cinerea* and *P. roqueforti,* wavelengths in the 500–550 nm range are effective, as are those below 470 nm (ZEMLYANUKHIN and CHEBOTAREV 1973), so there is variation in the action spectra of different spores. That light and temperature can interact in breaking dormancy has been shown by GASSNER and NIEMANN (1954), and others, in that the proper combination of these two factors is required for the optimal germination of several rusts.

Evidence of the variability in the response of spores to light can be found in the uredospores of *Puccinia graminis tritici,* which are inhibited by red light (720 nm), whereas simultaneous irradiation with ineffective wavelengths (653 nm) nullifies the effect (LUCAS et al. 1975). Once again, the possible presence of a phytochrome system is invoked by the authors but there is, as yet, insufficient evidence to implicate these pigments in the effects noted.

Other physical factors that influence germination include alternate wetting and drying, which has been suggested to be the equivalent of freezing (ARTHUR

1929), and freeze-thaw cycles. In fact, GOLD and LITTLEFIELD (1979) have demonstrated that teliospores of *Melampsora lini* can be induced to germinate by both wet-dry and freeze-thaw treatments. Perhaps these kinds of fluctuating regimes mimic the treatment received by many spores when they overwinter in nature.

Another physical treatment that reflects a situation in nature is the abrasion or removal of parts of the surface of spores. Thus, passage through animals may result in activation, as in the case of oospores of *Phytophthora megasperma* ingested by the land snail, *Helix aspersa* (SALVATORE et al. 1973).

Changes in osmotic potential can activate spores of *Dictyostelium discoideum* (COTTER 1977) and ascospores of *Neurospora crassa* (SUSSMAN et al. 1959) or reversibly deactivate them (COTTER 1977). Moreover, liquid water may be detrimental to spore germination as in the case of uredospores of *Puccinia graminis tritici* (MAHESHWARI and SUSSMAN 1970) and basidiospores of the mycorrhizal fungus, *Pisolithus tinctorius* (MARX 1976).

A reasonable supposition, based upon correlations between data like these and phenological observations in natural situations, is that the physical factors discussed are a means through which the activities of host and pathogen are synchronized, or through which saprophytes respond to nutritional needs, competition, etc.

13.4.2 Responses to Chemical Signals

Such well-known ecological phenomena as exclusion and succession may be explained on the basis of competition for, and availability of, nutrients or by the presence of toxic and stimulatory factors produced by previous or past inhabitants of a niche. Distinguishing among these explanations may not be easy but there are cases for which it is possible. When stimulatory factors in smaller amounts than those that can be used for food have been invoked, they have been described as "ecological ectocrines" (LUCAS 1947). Such substances have been defined more recently (SCHNEIDER 1967) as "chemical substances biosynthesized by one species and exerting an effect on the function of another via the external medium". In nature, it is likely that toxic and stimulatory ecological ectocrines interact on occasion to affect the timing of the development of microbial systems. Thus toxic substances serve to arrest growth and thereby induce exogenous dormancy (SUSSMAN and HALVORSON 1966) which, in turn, could be disrupted by appropriate stimulants of spore germination.

In situations in which lack of food serves to induce exogenous dormancy, provision of the limiting nutrient(s) is the signal that triggers development. Admittedly, making the distinction between foods and other substances as chemical signals may be difficult, but occasionally this can be made on the basis of the amount required and the metabolic role of the substances involved. The data summarized below are only rarely adequate to allow us to make this distinction with precision.

13.4.2.1 Chemical Inhibitors

The "Widespread Soil Fungistasis"

An extensive literature, most recently reviewed by LOCKWOOD (1977), documents the poor germination of fungal propagules placed in soils from many parts of the world. These studies, and extensions of them, may provide insights into the survival benefits that accrue to organisms when the germination of their dormant propagules is prevented under certain conditions. Generalizations from these studies probably will aid in understanding the biology not only of fungal propagules found in soils, but of spores of many organisms found in a variety of habitats.

The phenomenon of soil fungistasis includes the findings that many soils are fungistatic, that sterilization annuls soil fungistasis, that most fungal propagules are susceptible to soil fungistasis, and that addition of energy sources to soil reverses the fungistasis. Possible explanations of the available data include substances known to inhibit fungal spore germination which are found frequently in soils. Another type of explanation suggests that spores react to the absence or low concentration of soluble nutrients in soil by remaining ungerminated. It is likely that both of these hypotheses will be shown to be correct under some circumstances, as well, perhaps, as others. Because soil is a habitat generally impoverished for soluble nutrients, and enriched for both soluble and insoluble substances that are poorly metabolized by microbes, it is similar to many natural habitats, including aqueous ones. It is also likely that mechanisms that prevent the germination of dormant propagules in the presence of toxic substances, or in the absence of nutrients, have developed frequently throughout the microbial world. Thus, a careful study of the widespread soil fungistasis may provide considerable insight into microbial ecology.

Other Inhibitory Effects of Microorganisms

As in the case of the widespread soil fungistasis, other interactions among microorganisms must be considered in the light of the difference between the formation of inhibitors or nutrient deprivation. For example HSU and LOCKWOOD (1969) have shown that, of 20 isolates of *Streptomyces* tested, all were inhibitory to *Mucor ramannianus* and 18 to *Glomerella cingulata* but fewer than half formed inhibitors, the rest presumably being inhibitory by their ability to deprive the fungi of sufficient food.

That bacteria can have inhibitory effects on the germination of phylloplane fungi, presumably through nutrient deprivation, is known (BRIAN et al. 1977). Many other observations, especially in vitro, have established that antibiotics are produced by bacteria, some of which prevent the germination of fungal spores. Toxic bacterial metabolites produced by *Bacillus brevis* and a number of taxa of *Streptomyces* appear to reduce the germination of chlamydospores of the bunt fungus, *Tilletia caries*. KOLLMORGEN and JONES (1975) and WOOD and FRENCH (1965) suggest a similar effect by unidentified bacteria on ascospores of a parasite of quaking aspen, *Hypoxylon pruinatum*. In the latter case, at least, nutrient deprivation has not been ruled out as an explanation of the effect. The spread of *Aspergillus niger* is inhibited by *Serratia marcescens* in all the types of soils used by ROSENZWEIG and STOTZKY (1979) and conidia are more sensitive than hyphae in this instance. In controlled environments, *Bacillus cereus* prevents the germination of uredospores of *Puccinia allii* and the development of the rust disease of leek (DOHERTY and PREECE 1978).

Some fungal spores form toxins that inhibit higher plants, as in the case of *Helminthosporium maydis* on maize leaves (COMSTOCK and MARTINSON 1975), germinating uredospores of *Phakospora anachyrhizi* on soybean pods (DEVERALL et al. 1977) and *Alternaria kikuchiana* on pear leaves (OTANI et al. 1975). The latter authors speculate that the reason that spores of *Penicillium citrinum* cannot penetrate pear leaves is that citrinin, a toxin exhibiting similar effects to those of *A. kikuchiana,* is produced only by the hyphae of the former fungus.

While bacteria frequently have been shown to have deleterious effects upon fungal spores, the reverse has been reported much less often.

Inhibitors from Higher Organisms

Most of the data available relate to higher plants although some evidence exists for animals as well. The literature on this subject is large and has been reviewed before so we shall discuss this field only briefly.

Volatile inhibitors from swelling seeds of a number of cultivated plants, including peas, beans, wheat, maize, and others severely inhibit the germination of spores of *Botrytis cinerea, Mucor racemosus* and *Trichoderma viride* (CATSKA et al. 1975). Experiments with these and other fungi suggest that the effectiveness of the toxins is dependent mainly upon the species of fungi tested, rather than upon the source of the toxin. Clonal and seasonal variation in substances inhibitory to spores of *Hypoxylon mammatum* were found in assays of the bark and wood of the quaking aspen (FRENCH and MANNION 1976). Spores of *Arthroderma curreri* are inhibited or killed by feather fats from certain birds, so its distribution is influenced by this means (PUGH and EVANS 1970).

Phytoalexins, those substances arising in response to infection, often act to suppress spore germination as in the case of methyl-3,4-dimethoxycinnamate, which is produced by *Phaseolus vulgaris* and is inhibitory to uredospores of *Uromyces phaseoli* (ELNAGHY and HEITEFUSS 1976). An interesting aspect of this interaction is that this phytoalexin also has been identified as the self-inhibitor of the germination of these spores.

The relationship between pathogenicity and sensitivity of spores to phytoalexins is an obvious ecological derivative of these data. Additionally, AFIFI (1978) suggests that the effect of root exudates and volatile substances released from leaves upon fungal spores may influence the colonization of the rhizosphere and phyllosphere, respectively, especially in certain plant associations.

Self-Inhibitors of Germination

In contrast to the inhibitors discussed above which, when their effects are reversible, may impose a form of exogenous dormancy in spores, self-inhibitors are a means of inducing "constitutive" dormancy (SUSSMAN and HALVORSON 1966). These substances have been reviewed by the latter authors and by SUSSMAN and DOUTHIT (1973) and this review will be cursory.

Although many propagules can germinate even within sporocarps, such as the ascospores of the genus *Tympanis,* which form mycelium within asci (OUEL-

LETTE and PIROZYNSKI 1974), others are restrained from developing by the presence of self-inhibitors. Evidence that self-inhibitors are present includes the reduced capacity of spores to germinate in high concentrations, as in the case of the spores of *Smittium culisetae* (EL-BUNI and LICHTWARDT 1976). Other evidence is the increase in germination upon leaching or washing, as reported for oospores of *Albugo candida* by VERMA and PETRIE (1975). Nevertheless, the precautions discussed by SUSSMAN and DOUTHIT (1973) must be considered before conclusions are drawn on such evidence alone; once again, competition for food is an alternative that must be tested.

The self-inhibitor has been identified and in some cases its mode of action is known. The self-inhibitor in spores of *Agaricus bisporus* has been identified as carbon dioxide, according to RAST and her coworkers (GREUTER and RAST 1975; RAST and HOLLENSTEIN 1977), who have done detailed studies that suggest its locus of action to be succinic dehydrogenase, after the formation of oxaloacetate by carbon dioxide fixation (RAST et al. 1979). Other cases in which self-inhibitors have been identified include the rusts which, according to the work of MACKO and coworkers, produce a family of cinnamic acid inhibitors. Thus, the self-inhibitors from wheat stem rust, bean rust and stripe rust have been identified (MACKO et al. 1970), along with those of the snapdragon and sunflower rusts (MACKO et al. 1972), and that for the peanut rust (FOUDIN and MACKO 1974).

A controversy has developed over the identification of the self-inhibitor of spores of *Dictyostelium discoideum*, which has been identified as N^2-dimethylguanosine (2-dimethylamino-6-oxypurineriboside) by BACON et al. (1973). However, the toxicity of this compound to these spores has been challenged (TANAKA et al. 1974) and chromatography of an authentic sample and a crude spore extract showed clear differences between the two substances. Therefore, it is likely that N^2-dimethylguanosine is not the self-inhibitor from this organism. Recently, UCHIYAMA and ABE (1977) have accomplished the synthesis of the self-inhibitor, called discadenine, worked on by Tanaka and coworkers, establishing its structure to be 3-(3-amino-3-carboxypropyl)-6-(3-methyl-2-butenylamino) purine. An interesting attribute of this substance, in addition to its strong inhibitory effect on spores of *D. discoideum*, is activity on higher plants resembling that of cytokinin (NOMURA et al. 1977).

Prevention of the germination of spores in the immediate vicinity of their site of formation may help to ensure their dissemination. In other cases, germination may be suppressed until food becomes available, as seems to happen with spores of *Fusarium oxysporum*, the effect of the self-inhibitor of which is reversed by ethanol, a metabolite of glucose (OKAZAKI 1976).

13.4.2.2 Chemical Stimulators

Stimulators Produced by Other Organisms

It has been observed that the percentage germination of some spores is considerably greater in natural situations than in synthetic media, as in the case of basidiospores of the mycorrhizal fungus, *Rhizopogon luteolus*, which is part of the rhizosphere (THEODOROV and BOWEN 1973). Such observations are widespread and a number of factors that stimulate spore germination have been described that originate from a broad range of organisms. For the most part, these substances have not been identified, nor is their mechanism of action known, except in a few cases. Bacteria produce substances that activate spores of *Dictyostelium discoideum* (HASHIMOTO et al. 1976; DAHLBERG and COTTER 1978). A difference in the results of these two sets of workers lies in the lack of effect of *Escherichia coli* in the former case, which may be explained by differences in the strains of slime mold used.

Spores of *Bacillus subtilis* fail to germinate in acid forest soils unless fungal growth occurs nearby (SIALA and GRAY 1974). Volatile substances from *Ceratocystis fagacearum* stimulate the germination of spores of *Lactarius* spp. (OORT 1974), as do substances produced by the yeast *Rhodotorula glutinus* (FRIES 1977), so even unrelated organisms can act identically upon the same spore, as in the case of the bacteria that affect *D. discoideum*. Animals also may activate spores during their passage through the gut of insect larvae (LIM 1977) and of rabbits (HARPER and WEBSTER 1964).

By far the largest number of stimulatory substances has been reported from higher plants: exudates and volatiles from seeds, roots and leaves, guttation drops, and extracts of pollen yield substances that stimulate spore germination. That these effects may have important consequences in nature is shown by the large increase in the infection of maize by *Fusarium graminearum*, the germination of the macroconidia of which is stimulated by pollen from the same plant (NAIK and BUSCH 1978).

The chemical nature of germination stimulators is known in some cases and has been reviewed for fungal spores (SUSSMAN 1965; SUSSMAN and HALVORSON 1966; SUSSMAN 1976) and bacterial spores (KEYNAN and EVENCHIK 1969).

Morphogenic effects upon fungal spores also are induced by substances from higher plants. An important example of such effects is the formation of infection structures by some plant pathogens, a subject reviewed recently by STAPLES and MACKO (1980). Substances that induce such effects include acrolein, which induces differentiation of infection structures in the wheat stem rust fungus (MACKO et al. 1978). Factors of this kind may be part of the environment of stomata through which the germinating spore gains access to its host, according to GRAMBOW and RIEDEL (1977) and others. Another effect on the differentiation of germinating spores is the induction of indirect germination, as in the case of sporangia of *Phytophthora infestans*, in response to factors in potato tubers (WALMSLEY-WOODWARD et al. 1975). Inducers of sporulation also occur that are produced by higher plants, but these will not be covered in this review.

The same organism can produce stimulators and inhibitors of spore germination. Volatiles from the leaves of *Ocimum basilicum* inhibit the germination of spores of several fungi, whereas the germination of spores of *Alternaria tenuis* and *Penicillium chermesinum* is enhanced (AFIFI 1975a). Even within a single habitat, the rhizosphere, it has been shown that 11 species of bacteria antagonize the growth of the butt-rot fungus, *Fomitopsis annosa*, while 10 others are stimulatory (OREKHOV 1974).

In an environment like soil, in which the available nutrient supply usually is low, there would be selective advantage for spores that germinate in response to chemical stimulators. For example, chemicals released by host plant roots in the rhizosphere appear to induce the development of pathogens, perhaps at times of the year when metabolic activity is high and environmental conditions favorable.

On the other hand, premature commitment to germinate, or a response induced by nonspecific signals from "decoy plants", may be deleterious. BLACK and DIX (1976) have shown that ferulic acid can stimulate the germination of several fungi even though they may not be able to utilize this substance as a substrate. In fact, the survival of sclerotia of *Sclerotium rolfsii* in soil is reduced when they are stimulated to germinate by volatile substances released by alfalfa hay (LINDERMAN and GILBERT 1969). Also, while cowpea exudates induce the germination of several species of *Fusarium,* the germ tubes and some conidia are lysed by the extracts within 48 h (ODUNFA 1978). A similar effect may help explain the difference in response of cultivars of pea to pathogenic fusaria (WHALLEY and TAYLOR 1976).

Self-Stimulators of Germination

When large inocula support germination better than small, the presence of self-stimulators is suggested, as with the spores of *Aspergillus* spp. studied by TOTH (1973) and those of *Ustilago cynodontis* studied by SHARMA and AGRAWAL (1978). An extreme instance of this kind is that of the arthrospores of *Geotrichum candidum,* the germination of which is stimulated when two or more spores are in contact with one another (ROBINSON 1973). A similar conclusion is suggested by the decrease in germination caused by washing spores of *Botrytis cinerea, Cephalosporium coccorum, Fusarium solani, Mucor racemosus, Trichoderma viride,* and *Verticillium dahliae* (AFIFI 1975b). The latter situation probably illustrates Lockwood's thesis that leaching of nutrients can inhibit germination of many spores (see Sect. 13.4.2.1.1). Reciprocally, the provision of the substance(s) washed away can stimulate the germination of spores previously deprived of nutrients.

Several reports exist of the enhancement of pathogenesis through the use of increased spore density, as in the case of chlamydospores of *Phytophthora palmivora* infecting papaya (RAMIREZ and MITCHELL 1975) and basidiospores of *Cronartium ribicola* infecting pine needles (HANSEN and PATTON 1977). However, it is not known whether this effect is due to self-stimulators of germination or to other phases in the process of infection.

A special case of self-stimulation involves the interaction of two different stages of an organism, such as the increased germination of spores of *Paxillus involutus* and *Leccinum scabrum* caused by factors from their respective mycelia (FRIES 1978). Spores of *Agaricus bisporus* represent a classic case of this kind and iso-valeric and iso-amyl alcohols may be the activators involved (LOSEL 1964). However, it remains to be established whether the parent mycelium is the sole producer of the activating substance or whether, as in the case of *Agaricus bisporus,* other microorganisms can produce the active factor as well.

A germination stimulator produced by spores themselves has been identified as pelargonaldehyde, which activates spores of the wheat stem rust (FRENCH and WEINTRAUB 1958).

References

Afifi AF (1975a) Effect of volatile substances from species of Labiatae on rhizospheric and phyllospheric fungi of *Phaseolus vulgaris*. Phytopathol Z 83:296–302

Afifi AF (1975b) Effect of colchicine on ribose and galactose absorption in relation to spore germination of some fungi. Indian Phytopathol 28:476–478

Afifi AF (1978) Effect of volatile substances released from *Origanum majorana* and *Ocimum basilicum* on the rhizosphere and phyllosphere fungi of *Phaseolus vulgaris*. Folia Microbiol 23:399–405

Ali MI, Salama AM, Tahani MA (1976) possible role of solar function on the viability of some air fungi in Egypt. Zentralbl Bakteriol Parasitenkd Infektionskr Hyg Abt 2 131:529–534

Arthur JC (1929) The plant rusts. Wiley and Sons, New York

Bacon CW, Sussman AS, Paul AG (1973) Identification of a self-inhibitor from spores of *Dictyostelium discoideum*. J Bacteriol 113:1061–1063

Black RLB, Dix NJ (1976) Utilization of ferulic acid by microfungi from litter and soil. Trans Br Mycol Soc 66:313–317

Borut ST (1960) An ecological and physiological study of soil fungi of the northern Negev (Israel). Bull Res Counc Isr Sect D8:65–80

Brian A, Dickinson CH, Goodfellow M (1977) Antagonistic interactions of phylloplane bacteria with *Drechslera dictyoides* (Drechsler) Shoemaker. Can J Microbiol 23:710–715

Byrne P, Gareth Jones EB (1975) Effect of salinity on spore germination of terrestrial and marine fungi. Trans Br Mycol Soc 64:497–504

Cardoso EJBN, Schmitthenner AF (1975) Influence of light on the germination of *Phytophthora cactorum* oospores. Summa Phytopathol 1:23–30

Catska V, Affifi AF, Vancura V (1975) The effect of volatile and gaseous metabolites of swelling seeds on germination of fungal spores. Folia Microbiol 20:152–156

Cole ALJ (1976) Fungal spore germination inhibitors from wheat leaves. Mauri Ora 4:3–7

Comstock JC, Martinson CA (1975) Involvement of *Helminthosporium maydis* race t toxin during colonization of maize leaves. Phytopathology 65:616–61

Cotter DA (1977) The effects of osmotic pressure changes on the germination of *Dictyostelium discoideum* spores. Can J Microbiol 23:1170–1177

Dahlberg KR, Cotter DA (1978) Activators of *Dictyostelium discoideum* spore germination released by bacteria. Microbios Lett 9:139–146

Denyer WBG (1960) Cultural studies of *Flammula alnicola* (Fr.) and *Flammula conissans* (Fr.) Gillet. Can J Bot 38:909–920

Desser H, Broda E (1965) Radiochemical determination of the endogenous and exogenous respiration of bacterial spores. Nature (London) 206:1270–1271

Deverall BJ, Keogh RC, McCleod S (1977) Responses of soybean to infection by, and to germination fluids from, urediniospores of *Phakospora pachyrhizi*. Trans Br Mycol Soc 69:410–416

Doetsch RN, Cook TM (1973) Introduction to bacteria and their ecology. Univ Park Press, Baltimore

Doherty MA, Preece TF (1978) *B. cereus* prevents germination of uredospores of *Puccinia allii* and development of rust disease of leek *Allium porrum* in controlled environments. Physiol Plant Pathol 12:123

Dowding P (1969) The dispersal and survival of spores of fungi causing bluestain in pine. Trans Br Mycol Soc 52:125–137

El-Buni AM, Lichtwardt RW (1976) Spore germination in axenic cultures of *Smittium* spp. (Trichomycetes). Mycologia 68:573–582

Elnaghy MA, Heitefuss R (1976) Permeability changes and production of antifungal compounds in *Phaseolus vulgaris* infected with *Uromyces phaseoli*. I. Role of the spore germination inhibitor. Physiol Plant Pathol 8:253–267
Foudin AS, Macko V (1974) Identification of the self-inhibitor and some germination characteristics of peanut rust uredospores. Phytopathology 64:990–993
French JR, Mannion PD (1976) *Hypoxylon mammatum* ascospore germination and mycelial growth on bark and wood media from young branches of trembling aspen. Can J Bot 54:1438–1442
French RC, Weintraub RL (1958) Pelargonaldehyde as an endogenous germination stimulator of wheat stem rust spores. Arch Biochem Biophys 72:235–237
Fries N (1977) Germination of *Laccaria laccata* spores in vitro. Mycologia 69:848–850
Fries N (1978) Basidiospore germination in some mycorrhiza forming hymenomycetes. Trans Br Mycol Soc 70:319–324
Gareth-Jones EB, Byrne P, Alderman DJ (1971) The response of fungi to salinity. In: Vie et milieu. 3rd Eur Symp Mar Biol Suppl 22:265–280
Garrett SD (1956) Biology of root-infecting fungi. Cambridge Univ Press, London New York
Gassner G, Niemann E (1954) Untersuchungen über die Temperatur- und Lichtabhängigkeit der Sporenkeimung verschiedener *Tilletia*-Arten. Phytopathol Z 21:367–394
Gerhardt P, Murrell WG (1978) Basis and mechanism of spore resistance: a brief preview. In: Chambliss G, Vary JC (eds) Spores Vol VII. Am Soc Microbiol Press, Washington, pp 18–20
Gold RE, Littlefield LJ (1979) Light and scanning electron microscopy of the telial, pycnial and aecial stages of *Melampsora lini*. Can J Bot 57:629–638
Grambow HJ, Riedel S (1977) The effect of morphogenically active factors from host and nonhost plants on the in vitro differentiation of infection structures of *Puccinia graminis* f. sp. *tritici*. Physiol Plant Pathol. 11:213–224
Gregory PH (1961) The microbiology of the atmosphere. Wiley-Interscience, New York
Gregory PH (1966) Dispersal. In: Ainsworth GC, Sussman AS (eds) The fungi Vol II. Academic Press, London New York, pp 709–732
Greuter B, Rast D (1975) Ultrastructure of the dormant *Agaricus bisporus* spore. Can J Bot 53:2096–2101
Gupta SK, Banerjee AB, Achari B (1976) Isolation of ethyl p-methoxycinnamate, the major antifungal principle of *Curcuma zedoaria*. Lloydia 39:218–222
Hansen EM (1976) Twenty-year survival of *Phellinus* (*Poria*) *weirii* in Douglas-fir stumps. Can J For Res 6:123–128
Hansen EM, Patton RF (1977) Factors important in artificial inoculation of *Pinus strobus* with *Cronartium ribicola*. Phytopathology 67:1108–1112
Harper JE, Webster J (1964) An experimental analysis of the coprophilous fungus succession. Trans Br Mycol Soc 47:511–530
Hasan S, Wapshere AJ (1973) The biology of *Puccinia chondrillina*, a potential biological control agent of skeleton weed. Ann Appl Biol 74:325–332
Hashimoto T, Blumenthal HJ (1978) Survival and resistance of *Trichophyton mentagrophytes* arthrospores. Appl Environ Microbiol 35:274–277
Hashimoto Y, Tanaka Y, Yamada T (1976) Spore germination promoter of *Dictyostelium discoideum* excreted by *Aerobacter aerogenes*. J Cell Sci 21:261–271
Hibben CR, Stotzky G (1969) Effects of ozone on the germination of fungus spores. Can J Microbiol 15:1187–1196
Hora FK, Baker R (1974) Abiotic generation of a volatile fungistatic factor in soil by liming. Phytopathology 74:624–629
Hsu SC, Lockwood JL (1969) Mechanisms of inhibition of fungi in agar by Streptomycetes. J Gen Microbiol 57:149–158
Hubalek Z, Rush-Munro FM (1973) A dermatophyte from birds, *Microsporum ripariae* sp. nov. Sabouraudia 11:287–292
Ingold CT (1965) Spore liberation. Oxford Univ Press, London, New York
Ingold CT (1975) Hooker lecture: convergent evolution in aquatic fungi: the tetraradiate spore. Biol J Linn Soc 7:1–25

Kato H, Yamaguchi T, Nishihara N (1977) Seed transmission, pathogenicity and control of ragi blast fungus and susceptibility of ragi to *Pyricularia* spp. from grasses, cereals and mioga. Ann Phytopathol Soc Jpn 43:392–401

Kenneth R, Muttath TI, Gerson U (1979) *Hirsutella thompsonii,* a fungal pathogen of mites. I. Biology of the fungus in vitro. Appl Biol 91:21–28

Keynan A, Evenchik Z (1969) Activation. In: Gould GW, Hurst A (eds) The bacterial spore. Academic Press, London New York, pp 359–396

Kirillova IP, Agre NS, Kalakutskii LV (1975) Temperature regimes of the formation and germination of *Thermoactinomyces vulgaris* endospores. Jzv Acad Nauk SSSR Ser Biol 6:910–913

Knaysi G (1948) The endospore of Bacteria. Bacteriol Rev 12:19–77

Kollmorgen JF, Jones LC (1975) The effects of soil-borne micro-organisms on the germination of chlamydospores of *Tilletia caries* and *T. foetida.* Soil Biol Biochem 7:407–410

LaChance MA, Miranda M, Miller MW, Phaff HJ (1976) Dehiscence and active spore release in pathogenic strains of the yeast *Metschnikowia bicuspidata* var. *australis:* possible predatory implications. Can J Microbiol 22:1756–1761

Lamanna C (1952) Biological role of spores. In: Symposium on the biology of bacterial spores. Bacteriol Rev 16:90–93

Lim TM (1977) Production, germination and dispersal of basidiospores of *Ganoderma pseudoferreum* on *Hevea.* J Rubber Res Inst Malays 25:93–100

Linderman RG, Gilbert RG (1969) Stimulation of *Sclerotium rolfsii* in soil by volatile components of alfalfa hay. Phytopathology 59:1366–1372

Lockwood JL (1977) Fungistasis in soils. Biol Rev Cambridge Philos Soc 52:1–43

Losel DM (1964) Stimulation of spore germination in organisms by living mycelium. Ann Bot (London) 28:541–554

Lucas CE (1947) The ecological effects of external metabolites. Biol Rev Cambridge Philos Soc 22:270–295

Lucas JA, Kendrick RE, Givan C (1975) Photocontrol of fungal spore germination. Plant Physiol 56:847–849

Macko V, Staples RC, Gershon H, Renwick JAA (1970) Self-inhibitor of bean rust uredospores: methyl 3,4-dimethoxycinnamate. Science 170:539–540

Macko V, Staples RC, Allen PJ, Renwick JAA (1971) Identification of the germination self-inhibitor from wheat stem rust uredospores. Science 173:835–836

Macko V, Staples RC, Renwick JAA, Pirone J (1972) Germination self-inhibitors of rust uredospores. Physiol Plant Pathol 2:347–355

Macko V, Staples RC, Yaniv Z, Granados RR (1976) Self inhibitors of fungal spore germination. In: Weber DJ, Hess WM (eds) The fungal spore: form and function. Wiley-Interscience, New York, pp 73–98

Macko V, Trione EJ, Young SA (1977) Identification of the germination self-inhibitor from uredospores of *Puccinia striiformis.* Phytopathology 67:1473–1474

Macko V, Renwick JAA, Rissler JF (1978) Acrolein induces differentiation of infection structures in the wheat stem rust fungus. Science 199:442–443

Maheshwari R, Sussman AS (1970) The nature of cold-induced dormancy in uredospores of *Puccinia graminis tritici.* Plant Physiol 47:289–295

Marx DH (1976) Synthesis of ectomycorrhizae on loblolly pine seedlings with basidiospores of *Pisolithus tinctorius.* For Sci 22:13–20

Maser C, Trappe JM, Nussbaum RA (1978) Fungal-small mammal interrelationships with emphasis on Oregon coniferous forests. Ecology 59:799–809

Mineo K, Uritani I (1976) Possible involvement of furanoterpenoid phytoalexins in establishing host-parasite specificity between sweet potato and various strains of *Ceratocystis fimbriata.* Physiol Plant Pathol 8:97–111

Mineo K, Uritani I (1978) Isolation and characterization of factors in sweet potato root which agglutinate germinated spores of *Ceratocystis fimbriata,* black rot fungus. Plant Physiol 62:751–753

Mishra RR, Pandey KK (1978) Studies on soil fungistasis: effect of certain physical and biological factors. Plant Soil 49:355–366

Mishustin EN (1956) The law of zonality and the study of microbial associations of the soil. Soils Fert 19:385–390

Mitchell DJ (1975) Density of *Pythium myriotylum* oospores in soil in relation to infection of rye. Phytopathology 65:570–575

Murrell WG (1967) Biochemistry of the bacterial endospore. In: Rose AH, Wilkinson JF (eds) Advances in microbial physiology Vol I. Academic Press, London New York, pp 133–251

Naik DM, Busch LV (1978) Stimulation of *Fusarium graminearum* by maize pollen. Can J Bot 56:1113–1117

Nickerson KW, Freer SN, van Etten JL (1981) *Rhizopus stolonifer* sporangiospores: a wet-harvested spore is not a native spore (unpublished)

Nicot J (1960) Some characteristics of the microflora of desert sands. In: Parkinson D, Waid JS (eds) The ecology of soil fungi. Univ Press, Liverpool, pp 94–99

Niel van CB (1975) Natural selection in the microbial world. The second Marjory Stephenson memorial lecture. J Gen Microbiol 13:201–217

Nomura T, Tanaka Y, Abe H, Uchiyama M (1977) Cytokinin activity of discadenine: a spore germination inhibitor of *Dictyostelium discoideum*. Phytochemistry 16:1819–1820

Odunfa VSA (1978) Root exudation in cowpea and sorghum and the effect on spore germination and growth of some soil fusaria. New Phytol 80:607–612

Okazaki H (1976) Stimulation of chlamydospore germination of *Fusarium oxysporum* f. sp. Raphani by volatile metabolites produced by mycelium. Ann Phytopathol Soc Jpn 42:442–449

Oort AJP (1974) Activation of spore germination in *Lactarius* species by volatile compounds of *Ceratocystis fagacearum*. Proc K Ned Akad Wet Ser C 77:301–307

Orekhov DA (1974) Interrelation between butt-rot fungus *Fomitopsis annosa* (Fr.) Karst and bacteria of the rhizosphere of trees. Mikol Fitopatol 8:245–249

Otani H, Nishimura S, Kohmoto K, Yano K, Seno T (1975) Nature of specific susceptibility to *Alternaria kikuchiana* in Nijisseiki cultivar among Japanese pears: V. Ann Phytopathol Soc Jpn 4:467–476

Ouellette GB, Pirozynski KA (1974) Reassessment of *Tympanis* based on types of ascospore germination within asci. Can J Bot 52:1889–1912

Parbery DG, Emmett RW (1977) Hypotheses regarding appressoria, spores, survival and phylogeny in parasitic fungi. Rev Mycol 41:429–448

Plotnikova Yu M, Andreev LN, Serez GV, Zaitseva LG (1977) Conditions of the formation of infectious structures of *Puccinia graminis tritici* in vitro. Izv Akad Nauk SSSR Ser Biol 1:90–94 (in Russian)

Pugh GJF (1971) Factors which influence the early colonization of organic matter by fungi. In: Organismes du sol et production primaire. IVth Int Colloq Soil Zool Paris, pp 319–327

Pugh GJF (1972) The contamination of bird's feathers by fungi. Ibis 114:172–177

Pugh GJF, Evans MD (1970) Keratinophilic fungi associated with birds. II Physiological studies. Trans Br Mycol Soc 54:241–250

Ramirez BN, Mitchell DJ (1975) Relationship of density of chlamydospores and zoospores of *Phytophthora palmivora* in soil to infection of papaya. Phytopathology 65:780–785

Ranzoni FV (1968) Fungi isolated in culture from soils of the Sonoran desert. Mycologia 60:356–371

Rast D, Hollenstein GO (1977) Architecture of the *Agaricus bisporus* spore wall. Can J Bot 55:2251–1162

Rast D, Stussi H, Zobrist P (1979) Self-inhibition of the *Agaricus bisporus* spore by CO_2 and/or γ-glutaminyl-4-hydroxybenzene and γ-glutaminyl-3,4-benzoquinone: a biochemical analysis. Physiol Plant 46:227–234

Robinson PM (1973) Chemotropism in fungi. Trans Br Mycol Soc 61:303–313

Rodrigues CJ Jr, Meideiros EF, Lewis BG (1975) Relationship between a phytoalexin-like response in coffee leaves (*Coffea arabica*) and compatibility with *Hemileia vastatrix*. Berk et Br Physiol Plant Pathol 6:35–42

Romine M, Baker R (1973) Soil fungistasis: evidence for an inhibitory factor. Phytopathology 63:756–759

Rosenzweig WD, Stotzky G (1979) Influence of environmental factors on antagonism of fungi by bacteria in soil: clay minerals and pH. Appl Environ Microbiol 38:1120–1126
Salvatore MA, Gray FA, Hine RB (1973) Enzymatically induced germination of oospores of *Phytophthora megasperma*. Phytopathology 63:1083–1084
Schauz K (1977) The role of light quantity and quality in spore germination and spordial production of the dwarf bunt of wheat, *Tilletia controversa* Kuehn. Phytopathol Z 89:134–145 (in German)
Schneider HA (1967) Ecological ectocrines in experimental epidemiology. Science 158:597–603
Sharma KD, Agrawal DK (1978) Cultural studies on *Ustilago cynodontis* (Pers) P. Henn. Indian J Microbiol 18:66
Shirata A (1978) Production of phytoalexin in cortex tissue of mulberry shoot. Ann Phytopathol Soc Jpn 44:485–492
Shirata A, Takahashi K, Takasugi M, Nagao S, Masamune R (1978) Antifungal and antibacterial activities of phytoalexins, moracin A and B, isolated from mulberry. Bull Seric Exp Stn (Tokyo) 27:647–660
Siala A, Gray TRG (1974) Growth of *Bacillus subtilis* and spore germination in soil observed by a fluorescent antibody technique. J Gen Microbiol 81:191–198
Simpson ME, Marsh PB, Filsinger EC (1973) Geographical localization of infections of cotton bolls with *Colletotrichum gossypii*. Plant Dis Rep 57:828–832
Smith Shirley N (1977) Comparison of germination of pathogenic *Fusarium oxysporum* chlamydospores in host rhizosphere soils conducive and suppressive to wilts. Phytopathology 67:502–510
Soper CJ, Whistler JM, Davies DJG (1976) The response of bacterial spores to vacuum treatments: II. Germination and viability studies. Cryobiology 13:71–79
Staples RC, Macko V (1980) Formation of infection structures as a recognition response in fungi. Exp Mycol 4:2–16
Sussman AS (1965) Physiology of dormancy and germination in the propagules of cryptogamic plants. In: Ruhland W (ed) Encyclopedia of plant physiology Vol XV/2, Springer, Berlin Göttingen Heidelberg, pp 933–1025
Sussman AS (1968) Longevity and survivability of fungi. In: Ainsworth GC, Sussman AS (eds) The fungi Vol III. Academic Press, London New York, pp 447–486
Sussman AS (1976) Activators of fungal spore germination. In: Weber DJ, Hess WM (eds) The fungal spore: form and function. Wiley-Interscience New York, pp 101–137
Sussman AS, Douthit HA (1973) Dormancy in microbial spores Annu Rev Plant Physiol 24:311–352
Sussman AS, Halvorson HO (1966) Spores: their dormancy and germination. Harper and Row, New York
Sussman AS, Lowry RJ, Tyrrell E (1959) Activation of *Neurospora* ascospores by organic solvents and furans. Mycologia 51:237–247
Tanaka Y, Yanagisawa K, Hashimoto Y, Yamaguchi M (1974) True spore germination inhibitor of a cellular slime mold *Dictyostelium discoideum*. Agric Biol Chem 38:689–690
Theodorov C, Bowen GD (1973) Inoculation of seeds and soil with basidiospores of mycorrhizal fungi. Soil Biol Biochem 5:765–771
Toth JA (1973) The influence of spore number per surface unit on the course of germination in some *Aspergillus* species on solid medium. Acta Bot Acad Sci Hung 18:385–389
Uchiyama M, Abe H (1977) A synthesis of (±)-discadenine. Agric Biol Chem 41:1549–1552
Vary JC (1979) Bacterial spore germination. Spore Newsl 6:191
Verma PR, Petrie GA (1975) Germination of oospores of *Albugo candida*. Can J Bot 53:836–842
Walmsley-Woodward D, Lewis BG, Akerman AM (1975) Behaviour of *Phytophthora infestans* (Mont.) de Bary on potato tubers in relation to lenticel resistance. Physiol Plant Pathol 7:293–302
Warcup JH (1957) Chemical and biological aspects of soil sterilization. Soils Fert 20:1–5
Wearing AH, Burgess LW (1977) Distribution of *Fusarium roseum* "Graminearum" group 1, and its mode of survival in eastern Australian wheat belt soils. Trans Br Mycol Soc 69(3):429–442

Webster J, Sanders PF, Descals E (1978) Tetraradiate aquatic propagules in two species of *Entomophthora*. Trans Br Mycol Soc 70:472–479

Whalley WM, Taylor GS (1976) Germination of chlamydospores of physiologic races of *Fusarium oxysporum* f. sp. *Pisi* in soil adjacent to susceptible and resistant pea cultivars. Trans Br Mycol Soc 66:7–13

Wicklow DT, Zak JC (1979) Ascospore germination of carbonicolous ascomycetes in fungistatic soils: an ecological interpretation. Mycologia 71:238–242

Wood FA, French DW (1965) Microorganisms associated with *Hypoxylon pruinatum*. Mycologia 57:766–775

Zemlyanukhin AA, Chebotarev LN (1973) Effect of visible light on the germination of mold spores. Biol Nauki 16:57–60 (in Russian)

14 Physiological Responses to Flooding

R.M.M. CRAWFORD

CONTENTS

14.1 Introduction

Flooding denotes an environmental transformation which is of limited but uncertain duration. Plants that are absent from flood-prone sites will include species that are directly injured by inundation of their roots, as well as species which remain physiologically viable but become inefficient competitors under these conditions. On the other hand, plants that inhabit flood-prone sites will include species that can grow actively in flooded soils, as well as other species that survive flooding in a quiescent or dormant condition. The adverse effects of flooding on unadapted species will therefore be discussed separately from the various reactions of flood-tolerant plants, as tolerance and intolerance of flooding does not necessarily depend on positive and negative aspects of the same phenomena.

Despite the extensive research that has been carried out on flooding injury on agricultural crops (SCOTT RUSSELL 1977; CANNELL 1977), it is still difficult to make any generalizations on the initial causes of flooding damage to unadapted plants. In flood-tolerant species a certain degree of rationalization is possible (HOOK and CRAWFORD 1978), provided the type of flooding is defined. Whether or not the flood-water is moving, stationary, nutrient-rich, nutrient-poor of short or long duration has important implications for the survival strategy of the adapted plants.

The adaptations found in flood-tolerant plants do not include those of the truly aquatic species. Aquatic species do not suffer the stress of a sudden removal of oxygen from the soil atmosphere. In aquatic plants the greater part of the plant is in contact with a large body of free water and the problems of oxygen deficiency and toxin accumulation are largely avoided. By contrast, in the terrestrial plant adapted to withstand flooding, the amphibious nature of its existence creates problems both for the supply of oxygen and the removal of toxins produced either internally in the plant or externally due to the reducing conditions in the soil.

14.2 Flooding Injury

14.2.1 Symptoms

When examining the symptoms of flooding damage to plants it is often difficult to distinguish between the immediate and original causes of injury. Paradoxically, the commonest sign of the roots having an excess of water is the development of a water deficit in the leaves (KRAMER 1951; KRAMER and JACKSON 1954). The wilting and defoliation that is found on flooding can be traced to an increased resistance to water flux in the roots (MEES and WEATHERLEY 1957). Roots of intact tobacco plants that were sealed in a chamber where they received their nutrients in a spray began to reduce their water uptake within 6 min of starting to flush the root chamber with nitrogen (WILLEY 1970). The reduction in leaf water potential on flooding is not confined to flood-intolerant plants. Flooding of *Alnus glutinosa* also reduces leaf water potential, which is not lessened by the possession of adventitious roots (GILL 1975). Reduction in root aeration can reduce or cause imbalance in ion absorption (LEE 1977). Symptoms of mineral deficiency, in particular nitrogen, are a common symptom of flooding injury (KRAMER 1969; LEE 1978). In barley nitrate level in the shoot is reduced to one third of that in the controls within 48 h of flooding. The shoot, however, still continues to grow and chlorosis rapidly develops in the lower leaves as nitrogen is translocated to the growing regions (DREW and SISWORO 1979). Other symptoms of flooding damage include epinasty (JACKSON 1955) and alterations in tropic responses irrespective of whether or not the plants are injured by flooding (KORDAN 1975; BENDIXEN and PETERSON 1962).

In flood-sensitive species, root and shoot growth are usually rapidly reduced on flooding (JACKSON 1979) and in particular the root tip may be damaged so that when aeration is restored, growth can be renewed only from the regions proximal to the stem (MORISSET 1978). However, growth alone is not a reliable indicator of the ability of a plant to survive flooding. Many well-adapted species reduce their growth when flooded and resume normal growth rates only when the water table has subsided below the main rooting zone (CRAWFORD 1972). Some wetland species even show improved growth when the water table is lowered. The growth of *Narthecium ossifragum* is increased when the drainage is improved (MILES 1976), as is also that of *Empetrum nigrum* (BELL and TALLIS 1974). In the swamp trees *Taxodium distichum* and *Nyssa sylvatica* growth is better in saturated soils that dry out between successive waterings (DICKSON and BROYER 1972).

14.2.2 Survival Time

No higher plant can endure prolonged anoxia. Even rice is unable to extend more than the coleoptile and the first seminal roots in the absence of oxygen (VARTAPETIAN et al. 1978). Much of the observed variation that is found in the relative endurance of higher plants to flooding can be accounted for in the varying time that it takes to kill different species by flooding. Sudden death due to flooding is normally observed only in species of well-drained habitats where inundation is a rare event, e.g., *Centaurium erythrea, Geranium molle* and *Luzula campestris* (CRAWFORD 1966). The most intolerant of all plants to flooding are those that contain cyanogenic glycosides. The stem and root bark of apricot, plum and peach trees is rich in cyanogenic glycosides and these species can be killed by as little as 24 h flooding due to anaerobic hydrolysis yielding free hydrogen cyanide (ROWE and BEARDSELL 1973). In warm greenhouse conditions reduction in yield in peas and tomatoes of up to 43% and 50% respectively can be caused by as little as 24 h flooding (ERICKSON and VAN DOREN 1960).

The most reliable data for the length of time that plants can survive flooding is found with trees. When water reservoirs have been completed without the trees being felled it is possible to collect accurate data on their growth and survival (GILL 1970). Table 14.1 lists the flooding survival time for a number of flood-tolerant trees. For most species the length of the flooding period is only critical if it occurs during the growing season. The tulip tree (*Liriodendron tulipifera*) can withstand prolonged flooding in the dormant season but will die after 4 days flooding in June (MCALPINE 1961). This seasonal aspect of flooding tolerance has been documented for over 2,000 years. CATO (234–149 B.C.) in his work *De Re Rustica* stressed the need to avoid spring and autumn flooding of winter cereal crops.

If the water table is lowered for short periods, the length of time in which flooding can be endured is much lengthened. All-year-round flooding can be tolerated by some trees provided it happens only in isolated years (GILL 1970). In general even the most flood-tolerant of trees need to be unflooded for at least 55%–60% of the growing season. Once a habitat is flooded for more than 40% of the growing season it cannot be colonized by woody species although established trees may survive for a whole season. The upper limit of flooding tolerance in woody species appears to be related to the need to

Table 14.1. Survival time under inundation of some flood-tolerant trees

Species	Flooding survival years	Species	Flooding survival years
Quercus lyrata [1]	3	*Carya aquatica* [3]	2
Q. nuttali [1]	3	*Salix interior* [3]	2
Q. phellos [1]	2	*Cephalanthus occidentalis* [3]	2
Q. nigra [3]	2	*Nyssa aquatica* [3]	2
Q. palustris [2]	2	*Taxodium distichum* [3]	2
Q. macrocarpa [2]	2	*Celtis laevigata* [2]	2
Acer saccharinum [3]	2	*Quercus falcata* [4]	1
A. rubrum [3]	2	*Acer negundo* [4]	0.5
Diospyros virginiana [3]	2	*Crataegus mollis* [4]	0.5
Fraxinus pennsylvanica [3]	2	*Platanus occidentalis* [4]	0.5
Gleditsia triacanthos [3]	2	*Pinus contorta* [5]	0.3
Populus deltoides [3]	2		

References: 1 BROADFOOT and WILLISTON 1973; 2 YAEGER 1949; 3 HALL et al. 1946; 4 BELL and JOHNSON 1974; 5 VESTER and CRAWFORD 1978

regenerate the younger absorbing roots every two years. As this cannot take place under anaerobic conditions, most tree species succumb by the third year of inundation and only in exceptional circumstances do they manage to persist for 4 years (Table 14.1).

14.2.3 Causes of Injury

14.2.3.1 Metabolic Damage Due to Hypoxia

The problem of identifying the causes of flooding injury is illustrated by the contrasting effects obtained when peas are subjected to flooding in soil or kept in nitrogen bubbled culture solution without aeration (JACKSON 1979). Cessation of aeration to water-cultured pea seedlings causes no immediate harm, whereas flooding in soil induces rapid injury. Experiments such as this show that in many instances flooding injury involves factors other than just a lack of oxygen. When roots are in water culture they will not be affected by the reduced toxins that will be formed in flooded soils. However, the solution bathing the roots will allow a free diffusion of toxins out of the root and it is possible therefore that there is more than one explanation of why unaerated water culture is less injurious to peas than flooded soils.

The simple question of how much oxygen plant roots need to sustain aerobic respiration has not yet received a satisfactory answer. Many authors have argued that due to the low apparent K_m values for cytochrome oxidase (4×10^{-6} to 2×10^{-8} M; LONGMUIR 1954; HAYAISHI 1962) the concentration of oxygen in the plant tissue must be less than 1% (v/v) before availability alone proves a limit to its utilization (GREENWOOD 1968). Nevertheless, reduction in aeration

Table 14.2. The critical oxygen pressure (COP) necessary for sustaining aerobic unrestricted respiratory activity in roots of flood tolerant plants measured in vitro and in vivo

Species	COP (Atmospheres of oxygen)	References
Oryza sativa (in vitro)	0.21	LUXMOORE et al. 1970
O. sativa (in vivo)	0.024	ARMSTRONG and GAYNARD 1976
Eriophorum angustifolium (in vitro)	0.14	ARMSTRONG and GAYNARD 1976
E. angustifolium (in vivo)	0.026	ARMSTRONG and GAYNARD 1976
Senecio aquaticus (in vitro)	0.14	LAMBERS 1976

by flooding the soil or by the cessation of aeration in water culture experiments can give rise to an immediate increase in the ethanol concentration in plant roots (CRAWFORD and BAINES 1977). These results suggest that the oxygen supply to the roots was not sufficient and the glycolytic rate exceeded that of aerobic respiration. Even in a normal atmosphere of air, ethanol can always be detected in root tips (BETZ 1957; RUHLAND and RAMSHORN 1938), suggesting that for some regions of the root any reduction of the oxygen concentration below that in normal air will cause glycolysis to predominate over aerobic respiration.

Numerous experiments have been carried out to determine the effect of decreasing oxygen supply on its rate of consumption in plant roots. The point at which the rate of oxygen consumption falls as concentration is reduced is referred to as the critical oxygen pressure (COP – BERRY and NORRIS 1949). Criticism has been made of this type of experiment as it is usually performed with detached roots where the normally gas-filled intercellular spaces will become flooded during the period of measurement. Experiments of this type may overestimate, by at least an order of magnitude, the COP values of roots due to the greater resistance to oxygen diffusion that exists when the normally gas-filled spaces are flooded with water (ARMSTRONG and GAYNARD 1976). Table 14.2 summarizes the COP values recorded in vivo and in vitro for the roots of a number of flood-tolerant species.

Despite these optimistic predictions that only very low concentrations of oxygen are necessary for roots to maintain fully operative aerobic metabolism, there are a number of biochemical indications that oxygen diffusion is not able to support fully aerobic respiration in all regions of the flooded root. The accumulation of ethanol on flooding was mentioned above as an indication of an oxygen deficiency (hypoxia) in roots. An immediate effect of removing the oxygen from around roots is to reduce the concentration of ATP. This change can take place in as little as 15 min even in a flood-tolerant plant such as rice. When the roots of rice plants are enclosed in anaerobic chambers with their shoots exposed to normal air, there is a significant fall in energy charge as soon as the atmosphere around the root becomes anoxic (RAYMOND et al. 1978).

It is possible to recognize at least three different patterns in the accumulation of ethanol that invariably takes place when roots are flooded.

1. Homeostatic Flood-Tolerant Plants. In flood-tolerant plants of this type ethanol-production rates are not markedly increased by flooding (see Fig. 14.1, and CRAWFORD and BAINES 1977). This behaviour is found in flood-tolerant plants which have root tissues where ethanol is not readily removed by either diffusion or translocation.

2. Accelerated Glycolysis with Ethanol Removal. This strategy is found in flood-tolerant plants where the root system is either adventitious or the roots are in contact with moving water or there is the development of special ventilation mechanisms so that, either by direct diffusion from the roots or else by transport to other parts of the plant, ethanol can be readily removed. In these cases an increase in ethanol production does not lead to a proportionately higher level of ethanol in the root. Thus in rice, which relies on ethanol production as the main product of an active glycolytic metabolism during flooding (JOHN et al. 1974) there is only a 2.5-fold increase in the ethanol content of the roots on flooding (CHIRKOVA 1978). Similarly, in the swamp tupelo tree (*Nyssa sylvatica*), a shallow rooting system and the close proximity of well-developed lenticel tissue at the base of the trunk results in only a twofold increase in ethanol content during the first month of flooding (KEELEY 1979).

3. Accelerated Glycolysis with Ethanol Accumulation. Where there is no means of ethanol removal or oxidation, flooding can cause large increases in root-ethanol-content in flood-sensitive species. This happens in the flood-intolerant coniferous tree *Picea sitchensis* with a 16-fold increase in root ethanol content within 24 h of flooding (CRAWFORD and BAINES 1977). The problem of ethanol removal is particularly acute in deep anchoring roots, storage tissues, over-wintering plants and in woody species in general. Table 14.3 lists some of the anatomical modifications that are found in trees which are able to withstand flooding and which may contribute to root aeration and ethanol removal.

Although in the third class of intolerant plants there are numerous correlations between flooding sensitivity and increased ethanol production, there is no clear demonstration that these concentrations of ethanol are directly toxic to the roots. External application of ethanol in such physiological concentrations as occur naturally in roots are seldom toxic to plants (CHIRKOVA 1978). The same situation is found in rapidly fermenting yeast where high ethanol concentrations are thought to be a cause of cell death. Application of these concentrations will not kill the yeast cells and it is surmised that ethanol produced internally is considerably more toxic than that given externally to plant tissue (NAGODAWITHANA and STEINKRAUSS 1976). It is also important to note that the sensitivity of plants to ethanol under anoxia cannot be assessed under aerobic conditions as, when oxygen is available, plants have a very high capacity for oxidizing ethanol (COSSINS 1978). In germinating seeds, where before the rupture of the testa lack of aeration induces periods of severe hypoxia, there is further evidence that high anaerobic respiration rates are associated with poor survival after prolonged soaking (CRAWFORD 1977). Seeds have proved also more suitable material than roots for testing the effects of externally administered ethanol. Concentrations of as little as 0.1% ethanol administered in the soak-water

Table 14.3. Examples of morphological adaptations to flooding in woody plants

Adaptation	Species	References
Buttress and knee roots	*Taxodium distichum*	HOOK and SCHOLTENS (1978)
Pneumatophores	*Rhizophora* spp.	SCHOLANDER et al. (1955)
	Avicennia spp.	
Adventitious roots and Root branching	*Salix* spp.	GILL (1970)
	Alnus glutinosa	GILL (1970)
	Cephalanthus occidentalis	GILL (1970)
	Pinus contorta	GILL (1970)
	Thuja plicata	GILL (1970)
	Tsuga heterophylla	GILL (1970)
	Ulmus americana	GILL (1970)
Lenticell proliferation and hypertrophy	*Nyssa aquatica*	HOOK and SCHOLTENS (1978)
	Salix spp.	CHIRKOVA and GUTMAN (1972)
Reduction in suberization of epidermis and casparian strip	*Nyssa aquatica*	HOOK and SCHOLTENS (1978)
Stelar embolism with oxygen transport	*Pinus contorta*	COUTTS and PHILIPSON (1978)

can drastically reduce germination (THIESS and LICHTENTHALER 1973; BELATSKAYA 1977). By contrast, lactate produced under anaerobic conditions in seeds will inhibit germination but not cause death (MISSEN and WILSON 1970).

Indirect damage due to ethanol production on flooding has been reported in lupin (ALLEN 1974), where leaking ethanol has been reported as a specific stimulant for the chemotactic behaviour of *Phytophthora cinnamomi* zoospores in the infection of lupin radicles.

14.2.3.2 Hormonal Imbalance

The root apex is the major site for the synthesis of gibberellins (JONES and PHILLIPS 1966) and cytokinins (ITAI and VAADIA 1965) and when this region of the root is subjected to anaerobic conditions their synthesis and translocation will be reduced. Many of the symptoms of flooding injury seen in shoots are growth effects which are characteristic of hormone imbalance. To remove these symptoms it is sufficient in some cases merely to remove the plant roots before standing the shoots in water (JACKSON et al. 1978). In sunflower, synthesis and translocation of cytokinin is reduced by flooding (BURROWS and CARR 1969). In some cases the injury can be reduced by spraying the leaf with gibberellin (REID and CROZIER 1971) or with gibberellin plus cytokinin (JACKSON and CAMPBELL 1979). In tomato (RAILTON and REID 1973; SELMAN and SANDAMAN 1972) and in barley (DREW et al. 1979b) spraying the leaves with 6-benzylaminopurine will reduce the effects of flooding on leaf senescence and also help to restore stem growth. Simultaneous application of gibberellin and cytokinin can alleviate growth retardation as a result of flooding in tomato (JACKSON and CAMPBELL 1979; SELMAN and SANDAMAN 1972).

Abscisic acid increases within hours of flooding in a number of crop plants, and as this leads to stomatal closure it may be a contributory cause for reduced growth in flooded conditions (MILBORROW 1974). Why flooding should cause an increase in abscisic acid levels in roots is not clear. It has been suggested that desiccation stress caused to plants by flooding may be responsible (WRIGHT 1977), but it is doubtful if this level of water deficit is sufficiently high to induce abscisic acid synthesis.

Ethylene has received much attention in relation to the varying ability of plants to withstand flooding. Increased concentrations of ethylene have been reported in a number of crop species susceptible to flooding; these include dwarf bean, broad bean, chrysanthemum, maize, radish, sunflower, tomato and wheat (KAWASE 1974; EL-BELTAGY and HALL 1974; SMITH and JACKSON 1974; JACKSON and CAMPBELL 1975a, b), whereas in rice and barley no increase in ethylene was found in the shoots (JACKSON and CAMPBELL 1975c). In most soils ethylene concentrations rise on flooding, but it is not this ethylene that has been considered to be of importance in relation to the flooding tolerance of plants. Instead, it is the ability of the plant tissues themselves to produce ethylene that is associated with the flooding injury symptoms. In this connection it is of interest that large changes in ethylene production can be found in the shoot of flooded tomato plants (JACKSON et al. 1978). Although there are also increases in the roots this appears to be caused by the prevention of ethylene escape by water, rather than its increased synthesis. The synthesis of ethylene produced in roots ceases under total anoxia and its production appears to be dependent on hypoxic or normoxic conditions. The ethylene that is produced in the aerial parts of waterlogged tomato plants is in response to a substance translocated to the shoot from the flooded root (JACKSON et al. 1978). This substance has recently been identified as 1-amino cyclopropane-1-carboxylic acid (BRADFORD and YOUNG 1980).

Any attempt to relate flooding tolerance to ethylene production must also take into account the fact that ethylene is not fatal to plant tissues even in high concentrations. At concentrations above about 1 ppm it can act as a powerful growth-retardent in roots, but the roots remain viable and will resume growth when the ethylene concentration has fallen. The ability to halt growth and resume it again is a part of the survival strategy that is found in flood-tolerant plants (CRAWFORD 1972). Ethylene may also act in adapting plants as a morphological controller in that its production under hypoxic conditions has been shown to stimulate the production of aerenchyma in maize (DREW et al. 1979a). It would appear, therefore, that while ethylene is intimately connected with a variety of growth effects on flooding in many species it is not possible to distinguish flood-tolerant and non-flood-tolerant species from each other either on the level of production of ethylene or its effects.

14.2.3.3 Soil Toxins

Internal metabolic oxygen deficits alone are not sufficient to account for the injury that many plants sustain on flooding. In most regions of the world there is a sharp distinction between the floras of eutrophic and oligotrophic mires. Although this may reflect the different mineral needs of distinct vegetation types, it will also be associated with varying concentration of potentially toxic reduced ions such as iron and manganese as well as hydrogen sulphide. When the oxygen concentration falls to zero it is followed by the sequential appearance of reduced ions in the soil (GEMBRELL and PATRICK 1978). The reduction of organic matter can also produce short-chain organic acids which have phytotoxic effects in their undissociated form (LYNCH 1977) at pH 4 or below (LEE 1977).

Manganous Ions. Many crop plants that are sensitive to waterlogging are adversely affected by reduced manganous ions in water culture experiments (DIONNE and PESANT 1976; BENAC 1976; ROBSON and LONERAGAN 1970). Manganese poisoning has been attributed to two progressive conditions. In the first, high manganese concentrations induce iron chlorosis. This condition can be relieved by giving additional supplies of iron. The second condition is marked by the appearance of marginal chlorosis often with brown necrotic spots and cannot be relieved by additional supplies of iron. The balance of other ions also affects the degree to which manganese can produce toxic effects (OSAWA and IKEDA 1976). An increase in calcium decreases manganese uptake (ROBSON and LONERAGAN 1970) and its toxicity is reduced when phosphorus supplies are adequate (BORTNER 1935). The genus *Maytenus* (Celastraceae), which also includes species characteristic of wet lakeside soils in Patagonia, has been reported capable of accumulating very high concentrations of manganese. *Maytenus bureaviana* can have concentrations of up to 32,000 ppm in its leaves (JAFFRE and HEIM 1977). Manganese is highly mobile within the plant and the level at which it causes toxicity can be greatly modified, depending on the interactions with other ions in the plant.

Iron Toxicity. Higher plants able to withstand flooding usually have a greater ability to withstand iron poisoning than non-flood-tolerant plants (BARTLETT 1961). The quantities of iron that are sometimes present in flood waters can be sufficient to overcome the detoxifying capacity of roots even of such flood-tolerant plants as rice (HOWELER 1973). In a study of the effects of flooding on some woodland plants MARTIN (1968) was able to show that the possession of aerenchyma is correlated not only with tolerance of wet soils but in particular with resistance to iron toxicity. The ranges of iron concentration affecting the survival of these woodland plants is shown in Table 14.4. It is generally considered (GREEN and ETHERINGTON 1977; ARMSTRONG and BOATMAN 1967) that the escape from the root of oxygen that has diffused downwards from the shoot protects the root against the toxic effects of soluble ferrous ions by oxidation to the insoluble ferric form. The deposition of a core of ferric iron oxide or hydroxide round the roots is so characteristic of plants in wetland soils that like the calcium concretions in arid soils they have been given the name of pedotubules (BREWER 1964). The ability of some species to tolerate concentrations that are toxic to others, as well as being due to the presence

Table 14.4. Tolerance of certain woodland species to ferrous iron in water culture (MARTIN 1968)

	ppm Fe^{2+} (observations on root system)	
	Survival	Death
Mercurialis perennis	2	4
Endymion non-scripta	5	10
Brachypodium sylvaticum	–	15[a]
Geum urbanum	10	10–20
Circaea lutetiana	10	15
Primula vulgaris	10–20	20+
P. elatior	20	30
Carex sylvatica	30	30–40
Deschampsia caespitosa	50	80–100

[a] Not tested below 15 ppm Fe^{2+}

of aerenchyma, may be related to the varying mechanisms that plants adopt to obtain adequate supplies of iron in different soil conditions. Plants that live in soils where the available iron concentration is low and yet succeed in obtaining adequate supplies are termed iron-efficient (AMBLER et al. 1971). Such plants are able to excrete reductants (CLARK and BROWN 1974; BROWN 1978) which reduce the insoluble ferric iron to the soluble ferrous form. It is inevitable, therefore, that plants which produce a reducing environment at their root surface for the reduction of ferric iron will be totally unadapted to prevent the uptake of ferrous iron by oxidation when it occurs in toxic quantities in flooded soils.

In comparing the varying efficiency of rice plants to withstand iron poisoning TADANO (1975) lists three properties which assist resistance, namely:

1. a capacity to reduce the iron concentration in the medium around the root, either by oxidation or by changes in pH,
2. the prevention of iron that has reached the root surface from penetrating any further,
3. the prevention of that iron that has entered the root from being translocated to the shoot.

This last property is particularly important, as it is when iron is translocated from root to shoot that the effects of iron toxicity begin to affect plant growth (BARTLETT 1961). It is of particular interest in rice that detached segments of roots have the ability to oxidize ferrous iron. This appears to be an enzymatic property as it is lost on boiling (YAMADA and OTA 1958). The possibility of a metabolically assisted oxidation of ferrous iron may exist in other species. ARMSTRONG (1967) has observed that the amount of methylene blue oxidized by roots of *Menyanthes trifoliata* and *Molinia caerulea* was greater than that which might be expected from estimations based merely on the diffusion rate of oxygen down to the root.

14.3 Survival Mechanisms

14.3.1 Maintenance of Oxygen Supplies to Flooded Roots

In most flood-tolerant species, flooding induces morphological changes in the roots. These modifications usually involve root thickening with an increase in porosity. This increase in porosity effectively reduces the resistance of the roots to gaseous diffusion and facilitates the continued aerobic functioning of the inundated roots. However, this downward diffusion of oxygen can also be readily detected in the roots of a number of non-flood-tolerant plants (LUXMOORE and STOLZY 1972; HEALY and ARMSTRONG 1972). In most flood-intolerant species the resistance to oxygen diffusion increases with root length. In barley, the oxygen status of the root tissue falls by 75% after only 10 cm growth and this effectively limits the ability of the root to grow below the water surface (ARMSTRONG 1978). In flood-tolerant species, such as lodgepole pine (*Pinus contorta*) and rice, not only is the resistance to oxygen diffusion less, but there is some impediment to oxygen escaping in the upper regions of the root. Thus in lodgepole pine, oxygen can be channeled through the root to a depth of 30 cm below the level of the water table. The radial oxygen loss from roots has frequently been cited as an indication that there is an adequate supply to the roots for all their metabolic needs (ARMSTRONG 1978). An alternative approach to determining whether or not the oxygen requirements of the root have been met is based on the detection of anaerobic products of metabolism. When these products increase in concentration in roots, then it can be presumed that the oxygen supply is not adequate for the needs of the oxidative processes of the root. When the roots of *Pinus contorta* are flooded there is a steady increase in ethanol concentration in spite of the oxygen diffusion capacity noted above. Although this increase is not as large as in the flood-intolerant *Picea sitchensis,* it does indicate that root ventilation does not suppress the accumulation of ethanol (CRAWFORD and BAINES 1977). Similarly, as already discussed (p. 457) when roots are gassed with nitrogen, the immediate drop in energy charge within 15 min indicates that even in rice the downward diffusion of oxygen is not sufficient to maintain the root adenylate nucleotide pool at its normal state of phosphorylation (RAYMOND et al. 1978). In addition, the fact that roots alter their growth form when flooded must mean that certain regions of the root must suffer an insufficiency of oxygen or else inundation has changed some other physical parameter such as the amount of ethylene or carbon dioxide to which the root tissues are exposed. Concentrations of carbon dioxide have to be in the region of 20% to reduce growth by the same amount as a fall in the oxygen concentration to 1% (WILLIAMSON 1970). Due to the greater solubility of carbon dioxide in water than oxygen, CO_2 gradients in the rhizosphere are never so marked as those for oxygen and it is unlikely that roots are affected directly by changes in the carbon dioxide content of the soil atmosphere, although internal concentrations may have a metabolic regulatory function.

Although some degree of root ventilation clearly takes place in many species the aerenchyma normally associated with it is not present in all wet-land vegeta-

Table 14.5. Percent air space in roots of various wet-land species of higher plants determined by weighing intact and disintegrated roots in pycnometer bottles. (Unpublished data of CRAWFORD and SMIRNOFF). Mean values based on observations on whole root system unless otherwise stated

Species	Percentage air space ± S.E.
Nardus stricta (primary root)	45 ± 1.2
Eriophorum angustifolium	42 ± 1.9
Glyceria maxima (primary root)	35 ± 0.3
Eriophorum vaginatum	31 ± 1.8
Trichophorum caespitosum	26 ± 3.1
Juncus effusus	25 ± 3.5
Spartina townsendii	22 ± 1.4
Narthecium ossifragum (primary root)	19 ± 1.5
Deschampsia caespitosa	19 ± 2.7
Ranunculus lingua (primary root)	19 ± 0.4
Ranunculus lingua (lateral root)	12 ± 1.7
Lythrum salicaria	12 ± 1.8
Glyceria maxima (lateral root)	11 ± 0.8
Potentilla palustris	10
Nardus stricta	8 ± 2.0
Mentha aquatica	7 ± 0.6
Phalaris arundinacea	6 ± 0.6
Myosotis scorpioides	4.9 ± 0.7
Narthecium ossifragum (lateral root)	3.3 ± 1.4
Filipendula ulmaria	2.0 ± 0.5

tion (Table 14.5). Table 14.6 records the root porosity of a number of species from wet-land plants together with the oxygen reservoir capacity of the root air space. Two facts emerge. The first is that the oxygen reservoir capacity of root air spaces is extremely low. The high figure for *Eriophorum angustifolium* agrees with the findings of ARMSTRONG (1975) in that the internal oxygen content will not support aerobic respiration for more than 2 h unless renewed. The second fact is that wet-land species differ greatly in the air space content of their roots. In *Filipendula ulmaria* the root air space is only 2%. There would appear, therefore, to be two classes of roots in wet-land species;

1. those which possess a degree of ventilation which allows oxygen to diffuse a certain distance below the water table; and
2. those roots where there is no evidence of any specialized ventilation system.

In the first group, the plants with specialized ventilation mechanisms, the monocots are predominant. The possession of an adventitious root system also aids ventilation. Furthermore, as adventitious roots are renewed annually it is not necessary for them to survive the prolonged period of anaerobiosis that inevitably arises during winter when the shoots die down and the water table rises. Adventitious root systems are also found in trees tolerant of flooding such as willow, where lenticel development is also well marked in the lower

Table 14.6. The capacity of root air space to serve as an oxygen reservoir calculated on the length of time that the air volume of the root could maintain full activity of aerobic dark respiration at 18 °C without replenishment (CRAWFORD and SMIRNOFF unpublished data) assuming that the oxygen content of the root atmosphere is 20%

Species	Oxygen reservoir capacity (min)	Oxygen consumption $\mu l O_2$ mg^{-1} dry wt. h^{-1}	Air space in ml g^{-1} dry wt.
Eriophorum vaginatum	126	0.362	3.81
Eriophorum angustifolium	102	0.781	6.62
Trichophorum caespitosum	67	0.168	0.94
Glyceria maxima	37	0.520	1.61
Ranunculus lingua	32	0.670	1.76
Juncus effusus	34	0.645	1.84
Carex curta	22	0.633	1.15
Lythrum salicaria	22	0.515	0.94
Narthecium ossifragum	14	0.388	0.46
Mentha aquatica	13	0.650	0.71
Lycopus europaeus	12	0.755	0.76
Potentilla palustris	9	0.520	0.39
Filipendula ulmaria	2	0.375	0.07

regions of the trunk and assists both the downward diffusion of oxygen and the upward movement of toxic substances such as ethanol and acetaldehyde (CHIRKOVA and GUTMAN 1972).

Apart from the cases where anatomical modifications assist root ventilation there are many cases where no such provision is found and the plant tissues are able to survive prolonged periods in an anaerobic environment. It would appear that here metabolic adaptation rather than ventilation is necessary for survival. The deeper anchoring roots of trees as well as the roots and perennating organs of most helophytes have frequently to endure hypoxia in winter and early spring, although the hazards of anaerobiosis will be reduced during dormant periods. In addition, there are striking examples from the flood-plain forests of the tropics where roots are able to endure prolonged periods of flooding and where there is no evidence of root ventilation. The flood-plain forest of the upper Amazon has to endure flooding to a depth of 8–10 m for periods that can be as long as 6 months (GESSNER 1968), and there is no evidence that in the still waters of the lagoons (Igapo) away from river currents there is any ventilation of the tree roots.

14.3.2 Adaptive Metabolic Responses to Anoxia

14.3.2.1 Aerobic Metabolic Rate

The demands on the root ventilation system can be reduced by lowering the metabolic rate after the roots have been flooded. When the roots of *Senecio aquaticus* are grown in a solution of low oxygen tension their anaerobic respiration rate is reduced by over 50% (LAMBERS et al. 1978). It must be noted, however, that these values are obtained using excised roots and the entry of water into the air spaces of the roots will increase the resistance to oxygen diffusion (ARMSTRONG and GAYNARD 1976). In intact organs a surface film of water has been shown to have an inhibitory effect on oxygen uptake especially when this rate is high (RUHLAND and RAMSHORN 1938; KANDLER 1950). In a detailed study of this phenomenon OHMURA and HOWELL (1960) showed that the reduction in oxygen uptake was due to an increased resistance to the diffusion of oxygen. It is therefore possible that the physical effect of flooding will reduce the rate of oxygen uptake in intact root systems as well as in excised sections.

Oxygen uptake may also be reduced for metabolic reasons when roots are flooded. When the roots of *Senecio aquaticus* are transferred to a culture solution with a low oxygen concentration there is a decrease in the activity of the alternate respiration pathway (LAMBERS and SMAKMAN 1978; LAMBERS et al. 1978). This selective decrease in activity of the CN-resistant chain is probably related to a lower affinity for oxygen of the alternative oxidase than that of cytochrome oxidase (SOLOMOS 1977).

14.3.2.2 Anaerobic Metabolic Rate

A number of technical and physiological problems make it difficult to determine if under flooded conditions in the field, flood-tolerant plants differ from intolerant plants in their anaerobic respiration rates. Most studies are based on extrapolations from excised roots incubated in a suitable medium under total anoxia (CRAWFORD 1966, 1967; JOHN and GREENWAY 1976; SMITH and AP REES 1979a, b). Total anoxia is fatal to all higher plant tissues and rapidly halts most physiological processes. Translocation ceases (WILKINS and MARTIN 1967), nutrient uptake is inhibited (LEE 1978; JOHN et al. 1974) and even carbon dioxide production declines to less than 10% of its initial rate within 4 h (SMITH and AP REES 1979a).

From oxygen diffusion studies it appears unlikely that total anoxia ever occurs in the flooded root. It is more realistic, therefore, to try to investigate the differential effects of hypoxia on the roots of flood-tolerant and non-flood-tolerant plants. To this end some experiments have used intact plants with flooded root systems, but in these studies it is impossible to control the exact degree of hypoxia or to isolate the effects of translocation and leakage on the fate of root metabolites. In experiments that have used excised roots, an increase in anaerobic respiration rate determined manometrically from carbon dioxide evolution and enzymatically from ethanol production was shown to be associated with sensitivity to flooding (CRAWFORD 1966, 1967). A careful study confined to the apical 2 cm of roots of similar species, however, could detect no differences between flood-tolerant and non-flood-tolerant plants (SMITH and AP REES 1979b). However, in seeds the ability to withstand hypoxia during germination appears to be linked to low metabolic rates. In germinating seeds, species able to withstand prolonged soaking (96 h) showed lower rates of anaerobic carbon dioxide and ethanol production than those species which failed to survive this soaking treatment (CRAWFORD 1977).

14.3.2.3 Enzyme Polymorphism

Attempts to relate habitat fitness to metabolic adaptations have been investigated in relation to enzyme polymorphism in alcohol dehydrogenase. In most species so far studied there is a high degree of polymorphism in the genetic control of alcohol dehydrogenase isoenzyme production (MARSHALL et al. 1973; FRANCIS et al. 1974). In maize a fast and a slow migrating form of ADH can be detected by gel electrophoresis. These two isoenzymes are specified by two alleles, Adh_1^F (fast) and Adh_1^S (slow). The two enzyme forms have different physical and chemical properties. The fast variant has greater activity and is more heat-stable than its slow counterpart. The heterodimeric hybrid molecule in Adh_1^F/Adh_1^S heterozygotes is intermediate between the homodimers in activity and is as heat-stable as the fast homodimer (MARSHALL et al. 1973). Previous studies have shown that in various tissues susceptible to anaerobic injury there is a marked induction of ADH activity with little effect being produced on the flood-tolerant species (CRAWFORD and MCMANMON 1968; CHIRKOVA 1978). Others claim, however, that if the assay conditions are optimized carefully for each species this difference will disappear (SMITH and AP REES 1979b). By using homozygotes for Adh_1^F and Adh_1^S, MARSHALL and co-workers were able to test whether the varying isoenzyme properties conferred any degree of fitness on the roots when flooded. Flooding reduced the growth of both genotypes but the reduction was greater for the Adh_1^F homozygotes, which produced the catalytically more active enzyme. The Adh_1^S homozygotes were virtually unaffected by flooding. Repetition of these experiments with *Bromus mollis* (BROWN et al. 1974) and *Trifolium subterraneum* (FRANCIS et al. 1974) suggests that the maintenance of polymorphism in isoenzymes of ADH confers improved fitness in populations that come from sites with variable soil moisture. It is even suggested (FRANCIS et al. 1974) that ADH activity can be used as a predictive test for flooding tolerance provided the experimental conditions are carefully controlled. For two competing alleles to maintain their presence in any population implies that each is advantageous to its possessor under certain conditions. Thus, it is possible that the low-activity ADH isoenzymes confer an advantage when the roots are flooded that is of great enough importance to maintain their presence against the more active form, which is associated with higher growth rates in unflooded conditions. Conversely, the more active form must carry a disadvantage to plants growing in sites where flooding is frequent. The deleterious effect of the higher activity of ADH associated with the fast migrating isoenzyme would fit the metabolic theory of flooding tolerance suggested by MACMANMON and CRAWFORD (1971). However, it must be noted that this theory still requires more direct evidence that greater ADH activity is associated with ethanol concentrations that are toxic to flooded roots.

14.3.2.4 Metabolic Products

A number of substances have been found to accumulate in plant tissues on flooding and it has been suggested by analogy with their role in animal metabolism that they thus serve to accumulate the oxygen debt (Table 14.7). As with

Table 14.7. Substances reported to accumulate under anaerobic conditions in higher plants and animals. (References are intended only as a source of further information and do not indicate priority of discovery or complete range of occurrence)

Substance	Animal occurrence	Plant occurrence
Lactic acid	Vertebrate skeletal muscle[1]	Germinating seeds[2], tubers[3]
Pyruvic acid	Vertebrate skeletal muscle[1]	Willow roots[4]
Formic acid	Parasitic helminths[5]	–
Acetic acid	Bivalve molluscs, cestodes[6]	–
Acetoin	Nematodes[5]	–
Propionic acid	Molluscs, cestodes[5]	–
Butyric acid	Parasitic protozoa[5]	–
Succinic acid	Bivalve molluscs[6]	Seeds[2]
Malic acid	–	Roots of marsh plants[8,9]
Shikimic acid	–	Iris and water lily roots[10]
Glycolic acid	–	Willow roots[4]
Ethanol	Parasitic protozoa[5], helminths[5]	Flood-intolerant roots and seeds[10]
Sorbitol	Insects[16]	–
Glycerol	Insects[16]	Alder roots[10]
Alanine	Sea turtles[7], molluscs[19]	Flood-tolerant roots[4]
Aspartic acid	Marine annelids[11]	Flood-tolerant roots[4]
Glutamic acid	Marine annelids[11]	Flood-tolerant roots[4]
Serine	–	Flood-tolerant roots[4]
Proline	–	Flood-tolerant roots[4]
Octopine	Cephalopods[12]	Crown-gall tissues[13]
γ-amino butyric acid	–	Tomato roots[14] and radish leaves[17]
Methyl butyrate	Parasitic nematodes[15]	–
Methyl valerate	Parasitic nematodes[15]	–
Glycerophosphate	Insects[16]	–
Hydrogen	Parasitic protozoa[5]	–
Ethylene	–	Roots and fruits[18]

References: 1 HOCHACHKA and STOREY 1975; 2 WAGER 1961; 3 DAVIES et al. 1974; 4 DUBININA 1961; 5 VON BRAND 1966; 6 GÄDE et al. 1975; 7 HOCHACHKA et al. 1975; 8 CRAWFORD and TYLER 1969; 9 LINHART and BAKER 1973; 10 CRAWFORD 1972; 11 ZEBE 1975; 12 HOCHACHKA et al. 1977; 13 MENAGE and MOREL 1964; 14 FULTON et al. 1964; 15 BRYANT 1971; 16 GILMOUR 1965; 17 STREETER and THOMPSON 1972; 18 PRATT and GOESCHL 1969; 19 DE ZWAAN and WIJSMAN 1976

the studies on ethanol accumulation (see Sect. 14.3.2.2) quantification of the importance of these substances is lacking in many of these studies due to the problems of movement up or out of the hypoxic root. One of the best-studied end-products of glycolysis is lactate, which serves to accumulate the oxygen debt in potato tubers and *Equisetum* rhizomes (JAMES 1953), as well as for a brief period during the germination of seeds (SHERWIN and SIMON 1969). The extent to which pyruvate is able to serve as a proton acceptor during anaerobiosis (forming lactate) appears to be strictly limited. DAVIES (1973) has suggested that there is a switch to ethanol production from lactate as the pH of the cytoplasm drops. This is questioned by SMITH and AP REES (1979b) who found that lactate production ceased very quickly under total anoxia before any pH change was likely to have taken place.

MAZELIS and VENNESLAND (1957) suggested, on grounds of enzymatic equilibria that malic acid should serve as an end-product of glycolysis in plants. Under totally anoxic conditions and following the path of C^{14} from radioactive sucrose (U^{14}) SMITH and AP REES (1979b) were unable to detect any production of malate by the excised root apices of a number of flood-tolerant species. The only evidence that malic acid may play a role comes from an examination of the pool-size of this metabolite in the roots of plants that have been grown under flooded and unflooded conditions for periods varying from a few days to 1 month. In a study of a number of herbaceous species it was found that malate increased in concentration in the flood-tolerant species and fell in the intolerant (CRAWFORD and TYLER 1969). This interspecific difference was also shown to operate at a provenance level in a study of 20 populations of *Veronica peregrina*. Populations that came from a pond centre where they grew under flooded conditions increased their malic acid content when flooded. No regularity of behaviour could be found in plants taken from sites only a short distance away in the drier ground at the edge of the pond (LINHART and BAKER 1973). These studies, however, give no quantitative evidence as to how much of the presumed oxygen debt is accumulated by malic acid. KEELEY (1978) in a critical examination of the respective roles of ethanol and malate in accumulating a possible oxygen debt in the submerged roots of the swamp tupelo tree (*Nyssa sylvatica* var. *biflora*) showed that in the first month of flooding ethanol fermentation rates accelerate rapidly then fall to a low level after a year of inundation. Malic acid, on the other hand, reaches a peak later than ethanol and remains at this level to the end of the year's flooding observations. It is, therefore, suggested that as malic acid has a different time course in its accumulation, it does not serve as an alternative to ethanol in servicing the oxygen debt. It must be noted, however, that after 1 month's flooding, both ethanol and malate were higher than at the beginning of the experiment. As malic acid is a strong acid (stronger than lactic) its continued production requires a certain degree of translocation as well as electrical balancing with cation uptake and that it only accumulates when the degree of hypoxia is lessened by anatomical modification of the root.

In chick pea seeds (*Cicer arietinum*) malic and lactic acid and ethanol all increase during the typical period of hypoxia that occurs in the early stages of germination (ALDASORO and NICOLAS 1980). The maximum concentrations were reached for all three metabolites simultaneously 24 h after the beginning of the germination period. The concentration of malic acid increased one half as much that of ethanol. The peak in malic acid accumulation was matched by maximal activities for dark fixation of carbon dioxide and PEP carboxylase.

Thus in investigations in which intact plants are used, as when whole root systems still attached to the shoot are flooded, or in the course of germination of seeds, there are cases where malic acid accumulates and appears to match the role of ethanol in representing the anaerobic oxygen debt. It must be noted that malic acid accumulation under anoxia has not been observed in detached roots (SMITH and AP REES 1979b) and some doubt therefore remains in assessing its role in anaerobiosis.

Other metabolites which have been shown to accumulate under anaerobic conditions include shikimic acid (BUTT and BEEVERS 1960; BOULTER et al. 1963; TYLER and CRAWFORD 1970), glycerol (CRAWFORD 1972), proline and glutamic acid (LAMBERS 1976) and alanine (SMITH and AP REES 1979b). Further experimentation is, however, required to establish the quantitative importance of most of these metabolites in the hypoxic root.

14.3.2.5 Nitrate Respiration

The observation that resistance to flooding in certain crop species is enhanced by applications of nitrate and adversely affected by ammonium has prompted several investigators to determine if nitrate ions can function as oxidizing agents in root respiration when supplies of molecular oxygen are limiting. In many species of facultatively anaerobic micro-organisms nitrate respiration or dissimilatory nitrate reduction can permit growth in the complete absence of oxygen (FEWSON and NICHOLAS 1961). The possibility that this may take place in higher plants has been suggested by ARNON (1937) and MALAVOLTA (1954) who, working with barley and rice respectively, showed that these species withstood flooding better if the nitrogen supply was as nitrate rather than ammonium. The oxygen demand of tree roots can be reduced by substituting nitrate for ammonium in the root nutrients (REUTHER et al. 1958). In addition, flooding has been shown to increase nitrate reductase activity in the roots of a wide range of species (GARCIA NOVO and CRAWFORD 1973; CHIRKOVA 1971; LAMBERS 1976). The logical expectation of these studies would be that nitrate absorbed by the roots would reduce the oxygen debt and should in consequence lower the amount of ethanol produced under anaerobic conditions.

When barley roots are excised and incubated under total anoxia there is no absorption of nitrate and ethanol production is independent of the amount of nitrate accumulated by the roots before the period of anoxia (LEE 1978). The production of ethanol is related solely to the amount of carbohydrate stored in the roots before the experiment. It is possible that nitrite reduction and glycolytic activity do not compete for the same pool of NADH as glycolysis uses cytoplasmic NADH while nitrite reductase in cereal roots is located in the plastids (DALLING et al. 1972; MIFLIN 1974) together with at least part of the activity of glutamine synthetase and glutamate synthase (MIFLIN 1974; MIFLIN and LEA 1975).

There therefore remains the problem of resolving the observed effects of nitrate on respiration rates and improving flooding tolerance, with the absence of any observable action on the metabolism of excised roots kept under total anoxia. It is possible, as in several other instances noted above (e.g., SMITH and AP REES 1979b), that total anoxia and root excision is too severe a stress to investigate a phenomenon which requires some degree of nutrient uptake and translocation for its operation. Under conditions of natural flooding the normal limitation of oxygen stress to hypoxic rather than anoxic conditions would allow both nutrient uptake and nitrate reductase induction (GARCIA NOVO and CRAWFORD 1973) and in these conditions nitrate would then be able to alleviate the shortage of oxygen caused by flooding, particularly in roots near the soil surface where the relief may be sufficient to restore hormone balance (DREW et al. 1979a).

14.3.3 The Evolution of Flooding Tolerance

Evolution of flooding tolerance can be observed on a microscale both in the natural ecotypes that are found in species such as *Veronica peregrina* (LINHART and BAKER 1973), *Trifolium subterraneum* (FRANCIS et al. 1974) and *Bromus mollis* (BROWN et al. 1974) and in a number of varieties of crop plants including maize (MARSHALL et al. 1973) and rice (INTERNATIONAL RICE RESEARCH INSTITUTE 1973). At the species level there are also many examples of pairs of species which are similar in many of their attributes but differ sharply in their tolerance to flooding (CRAWFORD 1966; LAMBERS 1976). At the family level there are whole groups such as the Juncaginaceae, Typhaceae, Juncaceae, Mayaceae and Cyperaceae where flooding tolerance is almost a uniform characteristic. It can even be argued that the primary division of plants into monocots and dicots is due to a differential response to flooding. It is now over 150 years since DE CANDOLLE proposed an aquatic origin for the monocots. Although taxonomic opinion differs in detail there is general agreement that the modern monocots have evolved from a primitive dicotyledenous group not unlike the modern Nymphaeales (TAKTAJAN 1969). The simplified form of the monocots, like the aquatic angiosperms, aids the physiological fitness that is necessary for survival in flood-prone sites. Hollow stems, adventitious roots and the absence of woody tissue, as well as aiding the downward diffusion of oxygen to the submerged root, assist the removal of the more volatile products of anaerobiosis, such as ethanol and acetaldehyde. In the monocots it appears that a one-gene change at single locus can be sufficient to favour flooding survival (MARSHALL et al. 1973). How this is implemented at the molecular level of metabolism under hypoxia is still doubtful. Depending on the type of flooding, different plants have developed distinctive metabolic strategies. Where ethanol cannot be removed, there is evidence that homeostatic mechanisms enable plants to withstand periods of flooding. The role of certain organic and amino acids has been suggested, but not proven. It is probable that many of the organic acid transformations of the TCA cycle originally evolved under anaerobic conditions where they served to produce an electron transfer system that liberated phosphorylations from being bound to substrate linkage (HORVARTH 1974). No plant tissue can sustain metabolic processes under total anoxia and it may be that investigations under the more experimentally difficult regime of hypoxia will reveal the existence of various metabolic solutions to flooding tolerance in higher plants that are not too dissimilar from those found in many facultative anaerobes, diving reptiles and mammals (CRAWFORD 1978). External observations of growth, as well as internal measurements of respiratory activity and ATP levels (BRÄNDLE 1980), clearly show that ambient oxygen concentration has to fall only from 21% to 15% to have a severely limiting effect on growth and metabolism even in such a flood-tolerant species as the bulrush (*Schoenoplectus lacustris*). The consequences of hypoxia, its severity and duration are as yet poorly understood in relation to flooding tolerance.

Acknowledgement. I am much indebted to Dr. M.C. Drew and Dr. M.B. Jackson for many helpful suggestions in improving this review.

References

Aldasoro J, Nicolas G (1980) Fermentative products and dark CO_2 fixation during germination of seeds of *Cicer arietinum*. Phytochemistry 19:3–5

Allen RN (1974) Ethanol in lupin radicles. N Z J Bot 12:179–183

Ambler JE, Brown JC, Gauch HG (1971) Sites of iron reduction in soybean plants. Agron J 63:95–97

Armstrong W (1967) The oxidising activity of roots in waterlogged soils. Physiol Plant 20:920–926

Armstrong W (1975) Waterlogged soils. In: Etherington JR (ed) Environment and plant ecology. Wiley and Sons, New York, pp 181–218

Armstrong W (1978) Root aeration in the wetland condition. In: Hook DD, Crawford RMM (eds) Plant life in anaerobic environments. Ann Arbor, Michigan, pp 269–297

Armstrong W, Boatman DJ (1967) Some field observations relating the growth of bog plants to conditions of soil aeration. J Ecol 55:101–110

Armstrong W, Gaynard TJ (1976) The critical oxygen pressures for respiration in intact plants. Physiol Plant 37:200–206

Arnon DI (1937) Ammonium and nitrate nitrogen nutrition of barley at different seasons in relation to hydrogen ion concentration, manganese, copper and oxygen supply. Soil Sci 44:5590–5753

Bartlett RJ (1961) Iron oxidation proximate to plant roots. Soil Sci 92:372–379

Baskin JM, Baskin CC (1976) Evidence for metabolic adaptation to flooding in *Leavenworthia uniflora*. J Chem Ecol 2:441–447

Belatskaya EK (1977) Changes in metabolism of winter crops during their adaptation to flooding. Sov Plant Physiol 24:750–756

Bell DT, Johnson FL (1974) Flood-caused tree mortality around Illinois reservoirs. Trans Ill State Acad Sci 67:28–37

Bell JNB, Tallis JH (1974) The response of *Empetrum nigrum* L. to different mire water regimes, with special reference Wynburry Moss, Cheshire and Featherbed Moss, Derbyshire. J Ecol 62:75–95

Benac R (1976) Effect of manganese concentration in the nutrient solution on groundnuts (*Arachis hypogea* L.). Oleagineaux 31:539–543

Bendixen L, Peterson ML (1962) Tropism as a basis for flooding tolerance of strawberry clover to flooding conditions. Crop Sci 2:223–228

Berry LJ, Norris Jr WE (1949) Studies on onion root respiration. I. Velocity of oxygen consumption in different segments of root at different temperatures as a function of partial pressure of oxygen. Biochem Biophys Acta 3:593–606

Betz A (1957) Zur Atmung wachsender Wurzelspitzen. Planta 50:122–143

Bortner C (1935) Manganese toxicity in tobacco. Soil Sci 39:15–33

Boulter D, Coult DA, Henshaw GG (1963) Some effects of gas concentration on metabolism of the rhizome of *Iris pseudacorus* L. Physiol Plant 16:541–548

Bradford KJ, Young SF (1980) Xylem transport of 1-amino cyclopropane-1-carboxylic acid, an ethylene precursor in waterlogged tomato plants. Plant Physiol 65:322–326

Brändle R (1980) Die Überflutungstoleranz der gemeinen Teichbinse *Schoenoplectus lacustris* (L.) Palla: Abhängigkeit des ATP-Spiegels und des Sauerstoffverbrauchs in Wurzel- und Rhizomgewebe von der Sauerstoffkonzentration und Temperatur in der Umgebung. Flora 171:20–27

Brand von T (1966) Biochemistry of parasites. Academic Press, London New York

Brewer R (1964) Fabric and mineral analysis of soils. Wiley and Sons, New York
Broadfoot WM, Williston HL (1973) Flooding effects on southern forests. USJ For 71:584–587
Brown ADH, Marshall DR, Albrecht L (1974) The maintenance of alcohol dehydrogenase polymorphism in *Bromus mollis* L. Aust J Sci 27:545–559
Brown JC (1978) Mechanism of iron uptake by plants. Plant Cell Environ 1:249–257
Bryant C (1971) The biology of respiration. Arnold, London
Burrows WJ, Carr DT (1969) Effect of flooding the root system of sunflower plants on the cytokinin content in the xylem sap. Physiol Plant 22:1105–1112
Butt VS, Beevers H (1960) Hexose metabolism in maize roots. Biochem J 76:51
Cannell RQ (1977) Soil aeration and compaction in relation to shoot growth and soil management. In: Coaker TH (ed) Applied biology Vol II. Academic Press, London New York
Chirkova TV (1971) The role of nitrate respiration of roots on the viability of certain trees under temporary anaerobiosis. Vestn Leningr Univ 21:118–124 (in Russian)
Chirkova TV (1978) Some regulatory mechanisms of plant adaptation to temporal anaerobiosis. In: Hook DD, Crawford RMM (eds) (1978) Plant life in anaerobic environments. Ann Arbor, Michigan, pp 137–154
Chirkova TV, Gutman TC (1972) The physiological role of branch lenticels of willow and poplar under conditions of root anaerobiosis. Fiziol Rast 19:352–359 (in Russian)
Clark RB, Brown JC (1974) Internal root control of iron uptake and utilization in maize genotypes. Plant Soil 40:669–677
Cossins EA (1978) Ethanol metabolism in plants. In: Hook DD, Crawford RMM (eds) Plant life in anaerobic environments. Ann Arbor, Michigan, pp 169–202
Coutts MP, Philipson JJ (1978) The tolerance of tree roots to waterlogging. II. Adaptation of sitka spruce and lodgepole pine to waterlogged soil. New Phytol 80:71–77
Crawford RMM (1966) The control of anaerobic respiration as a determining factor in the distribution of the genus *Senecio*. J Ecol 54:403–413
Crawford RMM (1967) Alcohol dehydrogenase activity in relation to flooding tolerance in roots. J Exp Bot 18:458–464
Crawford RMM (1972) Some metabolic aspects of ecology. Trans Bot Soc Edinburgh 41:309–322
Crawford RMM (1977) Tolerance of anoxia and ethanol metabolism in germinating seeds. New Phytol 79:511–517
Crawford RMM (1978) Biochemical and ecological similarities in marsh plants and diving animals. Naturwissenschaften 65:194–201
Crawford RMM, Baines M (1977) Tolerance of anoxia and ethanol metabolism in tree roots. New Phytol 79:519–526
Crawford RMM, McManmon M (1968) Inductive responses of alcohol and malic dehydrogenases in relation to flooding tolerance in roots. J Exp Bot 19:435–441
Crawford RMM, Tyler PD (1969) Organic acid metabolism in relation to flooding tolerance in roots. J Ecol 57:235–244
Dalling MJ, Tolbert NE, Hageman RH (1972) Intracellular location of nitrate reductase and nitrite reductase. II. Wheat roots. Biochem Biophys Acta 283:513–519
Davies DD (1973) Control of and by pH. Symp Soc Exp Biol 27:513–529
Davies DD, Grego S, Kenworthy P (1974) The control of production of lactate and ethanol by higher plants. Planta 118:297–310
Dickson RE, Broyer TC (1972) Effect of aeration, water supply and nitrogen source on growth and development of tupelo gum and bald cypress. Ecology 53:626–634
Dionne JL, Pesant AR (1976) Effects of pH and water regime on the yield and mineral content of lucerne and birdsfoot trefoil grown in the glasshouse. Can J Plant Sci 56:919–928
Drew MC, Jackson MB, Giffard S (1979a) Ethylene promoted adventitious rooting and development of cortical air spaces (Aerenchyma) in roots may be adaptive responses to flooding in *Zea mays* L. Planta 147:83–88
Drew MC, Sisworo EJ, Sakar LR (1979b) Alleviation of waterlogging damage to young barley plants by application of nitrate and a synthetic cytokinin, and comparison between

the effects of waterlogging, nitrogen deficiency and root excision. New Phytol 82:315–329
Dubinina IM (1961) Metabolism of roots under various levels of aeration. Fiziol Rast 8:395–406
El-Beltagy AS, Hall MA (1974) Effect of water stress upon endogenous ethylene levels in *Vicia faba*. New Phytol 73:47–60
Erickson AE, Doren van DM (1960) The relation between plant growth and yield to oxygen availability. Proc 7th Int Congr Wis IV 54:429–434
Fewson CA, Nicholas DJD (1961) Utilization of nitrate by microorganisms. Nature (London) 190:2–7
Francis CM, Devitt AC, Steele P (1974) Influence of flooding on the alcohol dehydrogenase activity of roots of *Trifolium subterraneum* L. Aust J Plant Physiol 1:9–13
Fulton JM, Erickson AE, Tolbert NE (1964) Distribution of C^{14} among metabolites of flooded aerobically grown tomato plants. Agron J 56:527–529
Gäde G, Wilps H, Kluytmans JH, Zwaan DE A (1975) Glycogen degradation and end products of anaerobic metabolism in the fresh water bivalve *Anodonta cygnea*. J Comp Physiol 104:79–85
Garcia Nova F, Crawford RMM (1973) Soil aeration, nitrate reduction and flooding tolerance in higher plants. New Phytol 72:1031–1039
Gembrell RP, Patrick Jr WH (1978) Chemical and microbiological properties of anaerobic soils and sediments. In: Hook DD, Crawford RMM (eds) Plant life in anaerobic environments. Ann Arbor, Michigan, pp 375–423
Gessner F (1968) Zur ökologischen Problematik der Überschwemmungswälder des Amazonas. Int Rev Gesamten Hydrobiol 53:525–547
Gill CJ (1970) The flooding tolerance of woody species – a review. For Abstr 31:671–688
Gill CJ (1975) The ecological significance of adventitious rooting as a response to flooding in woody species with special reference to *Alnus glutinosa* (L.) Gaertn. Flora 164:85–97
Gilmour D (1965) The metabolism of insects. Oliver and Boyd, Edinburgh
Girton RE (1979) Effects of oxygen concentration on the respiration of excised root tip segments of maize and rice and of germinating grains of rice and buckwheat. Physiol Plant 46:58–62
Green MS, Etherington JR (1977) Oxidation of ferrous iron by rice (*Oryza sativa* L.) roots: a mechanism for waterlogging tolerance. J Exp Bot 28:678–690
Greenwood DJ (1968) Effect of oxygen distribution in the soil on plant growth. Proc 15th Easter Sch Agric Sci Univ Nottingham, pp 202–221
Hall TF, Penfound WT, Hess AD (1946) Water level relationships of plants in the Tennessee valley with particular reference to malaria control. J Tenn Acad Sci 21:18–59
Hayaishi O (1962) Biological oxidations. Annu Rev Biochem 31:25–46
Healy MT, Armstrong W (1972) The effectiveness of internal oxygen transport in a mesophyte (*Pisum sativum* L.). Planta 103:302–309
Hochachka PW, Storey KB (1975) Metabolic consequence of diving in animals and man. Science 613–621
Hochachka PW, Owen TG, Allen JF, Whittow GG (1975) Multiple end-products of anaerobic metabolism in diving vertebrates. Comp Biochem Physiol 50B:17
Hochachka PW, Hartline PH, Fields HHA (1977) Octopine as an end-product in the chambered Nautilus. Science 195:72–74
Hook DD, Crawford RMM (eds) (1978) Plant life in anaerobic environments. Ann Arbor, Michigan
Hook DD, Scholtens JR (1978) Adaptation and flood tolerance in trees. In: Hook DD, Crawford RMM (eds) Plant life in anaerobic environments. Ann Arbor, Michigan, pp 299–331
Horvarth RS (1974) Evolution of anaerobic energy-yielding pathways of the procaryote. J Theor Biol 47:361–371
Howeler RH (1973) Iron induced oranging disease of rice in relation to physicochemical changes in a flooded oxisol. Soil Sci Soc Am Proc 37:898–903
Int Rice Inst Annu Rep (1973) Los Banos, Laguna, Philippines, pp 100–104

Itai C, Vaadia Y (1965) Kinetin-like activity in root exudates of water-stressed sunflower plants. Physiol Plant 18:941–944
Jackson MB (1979) Rapid injury to peas by soil waterlogging. J Sci Food Agric 30:143–152
Jackson MB, Campbell DJ (1975a) Movement of ethylene from roots to shoots, a factor in the reponses of tomato plants to waterlogged soil conditions. New Phytol 74:397–406
Jackson MB, Campbell DJ (1975b) Ethylene and waterlogging effects in tomato. Ann Appl Biol 81:102–105
Jackson MB, Campbell DJ (1975c) Hormones and the responses of plants to waterlogged soil. Rep Agric Res Counc Letcombe Lab 1974:45–48
Jackson MB, Campbell DJ (1979) Effects of benzyladenine and gibberellic acid on the responses of tomato plants to anaerobic root environments and to ethylene. New Phytol 82:331–340
Jackson MB, Gales K, Campbell JD (1978) Effects of waterlogged soil conditions on the production of ethylene and on water relationships in tomato plants. J Exp Bot 29:183–193
Jackson WT (1955) The role of adventitious roots in recovery of shoots following flooding of the original root system. Am J Bot 42:816–819
Jaffre T, Heim R (1977) Accumulation du manganese par des éspèces associées aux terrain ultrabasiques de Nouvelle Caledonie. CR Acad Sci 284:1573–1575
James WO (1973) Plant respiration. Clarendon Press, Oxford
John CD, Greenway H (1976) Alcoholic fermentation and activity of some enzymes in rice roots under anaerobiosis. Aust J Plant Physiol 3:167–178
John CD, Limpinuntana V, Greenway H (1974) Adaptation of rice to anaerobiosis. Aust J Plant Physiol 1:513–520
Jones RL, Phillips IDJ (1966) Organs of gibberellin synthesis in light grown sunflower plants. Plant Physiol 41:1381–1386
Kandler O (1950) Untersuchungen über den Zusammenhang zwischen Atmungsstoffwechsel und Wachstumsvorgängen bei in vitro kultivierten Maiswurzeln. Z Naturforsch 5:203–211
Kawase M (1974) The role of ethylene in induction of flooding damage in sunflower. Physiol Plant 31:29–38
Keeley JE (1978) Malic acid accumulation in roots in response to flooding: evidence contrary to its role as an alternative to ethanol. J Exp Bot 29:1345–1349
Keeley JE (1979) Population differentiation along a flood frequency gradient: physiological adaptation to flooding in *Nyssa sylvatica*. Ecol Monogr 49:89–108
Kordan HA (1975) Oxygen and the establishment of normal shoot geotropism in germinating rice seedlings. Phytomorphology 25:201–205
Kramer PJ (1951) Causes of injury to plants from flooding of the soil. Plant Physiol 26:722–736
Kramer PJ (1969) Plant and soil water relationships. McGraw-Hill, New York
Kramer PJ, Jackson WT (1954) Causes of injury to flooded tobacco plants. Plant Physiol 29:241–245
Lambers H (1976) Respiration and NADH oxidation of the roots of flood-tolerant and flood-intolerant *Senecio* species as affected by anaerobiosis. Physiol Plant 37:117–122
Lambers H, Smakman G (1978) Respiration of the roots of flood-tolerant and flood-intolerant *Senecio* species: affinity for oxygen and resistance to cyanide. Physiol Plant 42:163–166
Lambers H, Steingrover E, Smakman G (1978) The significance of oxygen transport and of metabolic adaptation in flood-tolerance of Senecio species. Physiol Plant 43:277–281
Lee RB (1977) Effect of organic acids in the loss of ions from barley roots. J Exp Bot 28:578–587
Lee RB (1978) Inorganic nitrogen metabolism in barley roots under poorly aerated conditions. J Exp Bot 29:693–708
Linhart YB, Baker G (1973) Intra-population differentiation of physiological responses to flooding in a population of *Veronica peregrina* L. Nature (London) 242:275
Longmuir IR (1954) Respiration of bacteria as a function of oxygen concentration. Biochem J 57:81–87

Luxmore RJ, Stolzy LH (1972) Oxygen diffusion in the soil plant system. VI. A synopsis with commentary. Agron J 64:725–729
Luxmoore RJ, Stolzy LH, Letey J (1970) Oxygen diffusion in the soil plant system. I. A model. Agron J 62:317–332
Lynch JM (1977) Phytoxicity of acetic acid produced in the anaerobic decomposition of wheat straw. J Appl Bacteriol 42:81–87
Malavolta E (1954) Studies on the nitrogen nutrition of rice. Plant Physiol 29:98–99
McAlpine RG (1961) Yellow poplar seedlings intolerant to flooding. J For 59:566–568
McManmon M, Crawford RMM (1971) A metabolic theory of flooding tolerance: the significance of enzyme distribution and behaviour. New Phytol 70:299–306
Marshall DR, Broué P, Pryor AJ (1973) Adaptive significance of alcohol dehydrogenase isoenzymes in maize. Nature (London) New Biol 244:16–17
Martin MH (1968) Conditions affecting the distribution of *Mercurialis perennis* in certain Cambridgeshire woodlands. J Ecol 56:777–793
Mazelis M, Vennesland B (1957) Carbon dioxide fixation into oxalacetate in higher plants. Plant Physiol 32:591–600
Mees GC, Weatherley PE (1957) The mechanism of water absorption by roots. II. The role of hydrostatic pressure gradients. Proc R Soc London Ser B 147:381–391
Menage A, Morel G (1964) Sur la présence d'octopine dans les tissus de crown gall. CR Acad Sci 259:4795–4796
Miflin BJ (1974) The location of nitrate reductase and other enzymes related to amino-acid biosynthesis. Plant Physiol 54:550–555
Miflin BJ, Lea PJ (1975) Glutamine and asparagine as nitrogen donators for reductase dependent synthesis in pea roots. Biochem J 149:403–409
Milborrow BV (1974) The chemistry and physiology of abscisic acid. Annu Rev Plant Physiol 25:259–307
Miles J (1976) The growth of *Narthecium ossifragum* in some southern English mires. J Ecol 64:849–858
Missen AW, Wilson AT (1970) The metabolism of *Sinapis alba* seeds in water under anaerobic conditions. Phytochemistry 9:1473–1478
Morisset C (1978) Structural and cytoenzymological aspects of the mitochondria in excised roots of oxygen-deprived *Lycopersicon* cultivated in vitro. In: Hook DD, Crawford RMM (eds) Plant life in anaerobic environments. Ann Arbor, Michigan, pp 497–537
Nagodawithana TW, Steinkrauss KH (1976) Influence of rate of ethanol production and accumulation on the viability of *Saccharomyces cerevisiae* in "Rapid Fermentation". Appl Environ Microbiol 31:158–162
Ohmura T, Howell RW (1960) Inhibitory effect of water on oxygen consumption by plant materials. Plant Physiol 35:184–188
Osawa T, Ikeda H (1976) Heavy metal toxicities in vegetable crops. I. The effect of iron concentration in the nutrient solution on manganese toxicities in vegetable crops. J Jpn Soc Hortic Sci 45:50–58
Pratt HK, Goeschi JD (1969) Physiological role of ethylene in plants. Annu Rev Plant Physiol 20:541–584
Railton ID, Reid DM (1973) Effects of benzyladenine on the growth of water-logged tomato plants. Planta 111:261–266
Raymond P, Bruzau F, Pradet A (1978) Etude du transport d'oxygène des parties aeriennes aux racines à l'aide d'un paramètre du metabolisme: la charge energétique. CR Acad Sci 286:1061–1063
Reid DM, Crozier A (1971) Effect of waterlogging on gibberellin content and growth of tomato plants. J Exp Bot 22:39–48
Reuther W, Embleton TW, Jones WW (1958) Mineral nutrition of tree crops. Annu Rev Plant Physiol 9:175–206
Robson AD, Loneragan JF (1970) Sensitivity of annual *Medicago* species to manganese toxicity as affected by calcium and pH. Aust J Agric Res 21:223–232
Rowe RN, Beardsell DV (1973) Waterlogging of fruit trees. Hortic Abstr 43:534–548
Ruhland W, Ramshorn K (1938) Aerobe Gärung in aktiven pflanzlichen Meristemen. Planta 28:471–514

Scholander PF, Dam VAN L, Scholander SI (1955) Gas exchange in the roots of mangroves. Am J Bot 42:92–98
Scott Russell R (1977) Plant root systems. McGraw-Hill, New York
Selman IW, Sandaman S (1972) Growth responses of tomato plants in non-aerated water culture to foliar sprays of gibberellic acid and benzyladenine. Ann Bot 36:837–848
Sherwin T, Simon EW (1969) The appearance of lactic acid in *Phaseolus* seeds germinating under wet conditions. J Exp Bot 20:776–785
Smith AM, Rees ap T (1979a) Effects of anaerobiosis on carbohydrate oxidation by roots of *Pisum sativum*. Phytochemistry 18:1453–1458
Smith AM, Rees ap T (1979b) Pathways of carbohydrate fermentation in the roots of marsh plants. Planta 146:327–334
Smith KA, Jackson MB (1974) Ethylene, waterlogging and plant growth. Rep Agric Res Counc Letcombe Lab 1973:60–75
Solomos T (1977) Cyanide resistant respiration in higher plants. Annu Rev Plant Physiol 28:279–297
Streeter JG, Thompson JF (1972) Anaerobic accumulation of γ aminobutyric acid in radish leaves (*Raphanus sativum* L.) Plant Physiol 49:572–578
Tadano T (1975) Devices of rice roots to tolerate high iron concentration in growth media. Jpn Agric Res Q 9:34–39
Taktajan A (1969) Flowering plants: origin and dispersal. Oliver and Boyd, Edinburgh
Thiess DE, Lichtenthaler HK (1973) Hemmung der Samenkeimung durch niedere einwertige Alkohole. Naturwissenschaften 60:302
Tyler PD, Crawford RMM (1970) The role of shikimic acid in waterlogged roots and rhizomes of *Iris pseudacorus* L. J Exp Bot 21:677–682
Vartapetian BB, Andreeva IN, Nuritdinov N (1978) Plant cells under oxygen stress. In: Hook DD, Crawford RMM (eds) Plant life in anaerobic environments. Ann Arbor, Michigan, pp 13–88
Vester G, Crawford RMM (1978) Verschiedene Provenienzen von *Pinus contorta* Loudon und Überflutungsstress: Klassifikation auf Grund morphologischer und metabolischer Kriterien. Flora 167:433–444
Wager HG (1961) The effect of anaerobiosis on acids of the tricarboxylic cycle in peas. J Exp Bot 12:34–46
Wilkins MB, Martin M (1967) Dependence of basipetal polar transport of auxin upon aerobic metabolism. New Phytol 42:831–839
Willey CR (1970) Effect of short periods of anaerobiosis and near anaerobiosis on water uptake by tobacco roots. Agron J 62:224–229
Williamson RE (1970) Effect of soil gas composition and flooding on growth of *Nicotiana tabacum* L. Agron J 62:80–83
Wright STC (1977) The relationship between leaf water potential (γleaf) and the levels of abscisic acid and ethylene in excised wheat leaves. Planta 134:183–189
Yaeger LE (1949) Effect of flooding on a river-bottom timber area. Bull Ill Nat Hist Surv 25:33–65
Yamada N, Ota Y (1958) Study on the respiration of crops 7. Enzymatic oxidation of ferrous iron by rice roots. Proc Crop Sci Soc Jpn 26:201–210
Zebe E (1975) In vivo Untersuchungen über den Glucose-Abbau bei *Arenicola marina* (Annelida, Polychaeta). J Comp Physiol 101:133–145
Zwann de A, Wijsman TCM (1976) Anaerobic metabolism in Bivalva (Mollusca); characteristics of anaerobic metabolism. Comp Biochem Physiol 54B:313–324

15 Functional Significance of Different Pathways of CO_2 Fixation in Photosynthesis

C.B. OSMOND, K. WINTER, and H. ZIEGLER

CONTENTS

15.1 Introduction

The functional significance of different photosynthetic CO_2 fixation pathways is a question which can be answered in many ways, each being appropriate to certain scales of enquiry. In physiological ecology our purpose should be to integrate these different scales of enquiry as comprehensively as possible, and to show how photosynthesis contributes, directly or indirectly, to performance and survival of plants in diverse habitats. Studies of photosynthetic CO_2 fixation were afflicted with a post-Calvin cycle chauvinism in the 1950's, which may have been responsible for the slow and tentative revelation of the C_4 pathway of photosynthetic carbon assimilation in the USA and USSR (BURR et al. 1957; KARPILOV 1960; KORTSCHAK et al. 1965). Elucidation of this pathway undoubtedly stimulated new interest in the carbon metabolism of photosynthesis in the next decade (HATCH and SLACK 1966, 1970; CC BLACK 1973) and led to an upsurge in comparative studies of higher plant photosynthesis (BLACK 1971; BJÖRKMAN 1973). Largely as a result of this stimulus, the peculiar dark CO_2 fixation processes of succulent plants, known as crassulacean acid metabolism (CAM), were also recognized as a distinctive photosynthetic process (KLUGE and TING 1978; OSMOND 1978). It also led to the present revival of interest in the photosynthesis of aquatic plants.

Discussions of the functional significance of different photosynthetic CO_2 fixation pathways usually emphasize plant productivity. In the physiological context, superior productivity may be evidence for superior adaptation, suggesting that component physiological systems are better fitted together, giving optimum performance in a particular habitat. Many ecological studies are based on records of plant occurrence (survival) in different habitats and it is not easy to anticipate the "manifold consequences in ecological relationships which may arise from two- to threefold differences in primary productivity" (BLACK 1971).

Because the biochemical processes relevant to this chapter have been reviewed at length by TREBST and AVRON (1978) and GIBBS and LATZKO (1979), and because many of the physiological processes are reviewed in companion chapters, our treatment of them is about inversely proportional to the publications available. In spite of the stimulus to ecophysiological studies of photosynthetic process in the last decades, relatively few experiments have given satisfying insights into the functional significance of different photosynthetic pathways in an ecological context. The techniques of reciprocal transplant and competition experiments, which might uncover evidence in support of many speculations, appear to hold little fascination for modern physiologists or ecologists. Few such experiments are available for review in this chapter, and we therefore hope to place the question of the functional significance of different pathways of CO_2 fixation into more appropriate contexts which might cause such experiments to become more commonplace in the future. Our main purpose is to attempt a critical evaluation of the large-scale experiments of nature, presented to us as complex patterns of photosynthetic response to seasonal and geographic climates, and displayed as patterns of distribution of plants with different photosynthetic pathways.

15.2 The Biochemical Context: Comparative Biochemistry of Photosynthetic CO_2 Fixation Pathways

In the whole of evolution the enzymic chemistry of photosynthetic CO_2 assimilation, "the lone reaction that counterbalances the vast expenditures of respiration, that reverses decomposition and death" (UPDIKE 1968), appears to have been successfully resolved in only one way. All three photosynthetic CO_2 fixation pathways ultimately depend on the photosynthetic carbon reduction cycle (PCRC) for net CO_2 fixation. The PCRC is a unique association of ribulose bisphosphate (RuP_2) carboxylase-oxygenase, a reduction step, and the enzymes of the reductive pentose-phosphate pathway which catalyze the fixation and reduction of CO_2 and permit the simultaneous regeneration of the carboxylation substrate. Although there are differences in the primary and secondary structure of RuP_2 carboxylase-oxygenase from different autotrophic organisms (AKAZAWA 1979), and some differences in kinetic properties of the enzyme (BADGER 1980; HOCH HIN et al. 1980, 1981), the chemistry of this "lone reaction" is essentially similar in all plants so far examined, with the possible exception of some bacteria (QUANDT et al. 1977). The reaction mechanism, as far as it is understood (LORIMER and ANDREWS 1973, 1981), also predisposes the carboxylase to also react as an oxygenase (Fig. 15.1), allowing O_2 and CO_2 to react with RuP_2 at the same active site on the enzyme, and thus to behave as competitive inhibitors. The products of oxygenation of RuP_2 are then consumed in reactions of the photorespiratory carbon oxidation pathway (PCOP) which is integrated with the PCRC (Fig. 15.1).

15.2.1 Biochemistry of a Compromise: the Integration of Reduction and Oxidation Pathways in C_3 Photosynthesis

The functional significance of the oxygenase activity of the fundamental carboxylase in photoautotrophic organisms is obscure. The reaction itself is inevitable when the carboxylase functions in photosynthetic systems which are in equilibrium with the present atmosphere of 210 mbar O_2 and 330 μbar CO_2. Presumably the oxygenase activity is a phylogenetic relic of the evolution of autotrophic CO_2 fixation in photosynthetic organisms which derived energy from processes other than the photolysis of water and which existed in a primitive reducing atmosphere rich in CO_2 (LORIMER and ANDREWS 1981). In the present atmosphere RuP_2 carboxylase in leaves of C_3 plants functions at something less than its K_m (CO_2) and in the vicinity of the $K_i(O_2)$ (see Chap. 16, this Vol.). If the PCRC in chloroplasts of C_3 plants was to function in isolation, a large part of the assimilated carbon would be excreted as P-glycolate. When isolated intact chloroplasts of C_3 plants are provided with limiting $^{14}CO_2$ in the presence of 210 mbar O_2, most of the ^{14}C is converted to glycolic acid and excreted to the suspending medium. Ultimately, the chloroplast is depleted of PCRC metabolites (KIRK and HEBER 1976).

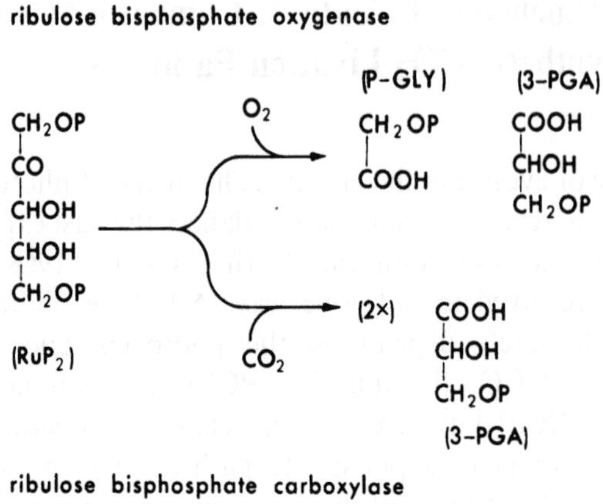

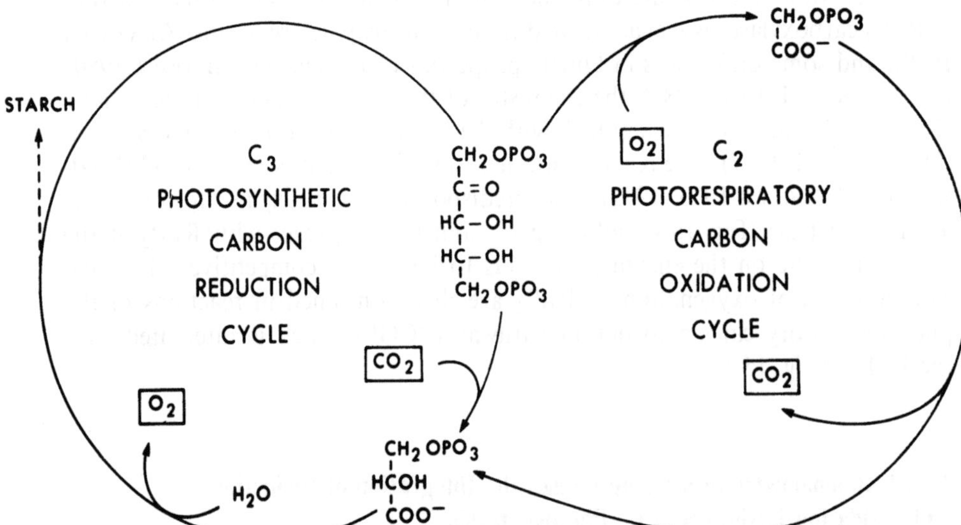

Fig. 15.1. Linkage of the photosynthetic carbon reduction cycle and the photorespiratory carbon oxidation cycle by RuP_2 carboxylase-oxygenase and its reaction products

The reactions of the PCOP in peroxisomes and mitochondria in cells of C_3 plants, reviewed by SCHNARRENBERGER and FOCK (1976), BERRY et al. (1978), LORIMER et al. (1978) and shown schematically in Fig. 15.2, conserve carbon which is lost from the chloroplast as a result of oxygenation. The integration of PCRC and PCOP, unlike RuP_2 oxygenase itself, can be viewed as functionally significant in several ways.

1. Integration of the reactions of the PCOP with those of the PCRC in C_3 photosynthesis permits the recovery of 75% of the carbon excreted from the PCRC as P-glycolate. Figure 15.2 shows that for each 2 mol of P-glycolate (C_2) formed as an inevitable consequence of oxygenation, 1 mol of 3-PGA (C_3) is returned to the chloroplast. The remaining 25% is released as the CO_2 of photorespiration and it is this CO_2 production which results in the relatively high CO_2 compensation point in leaves of C_3 plants. The flow of carbon through

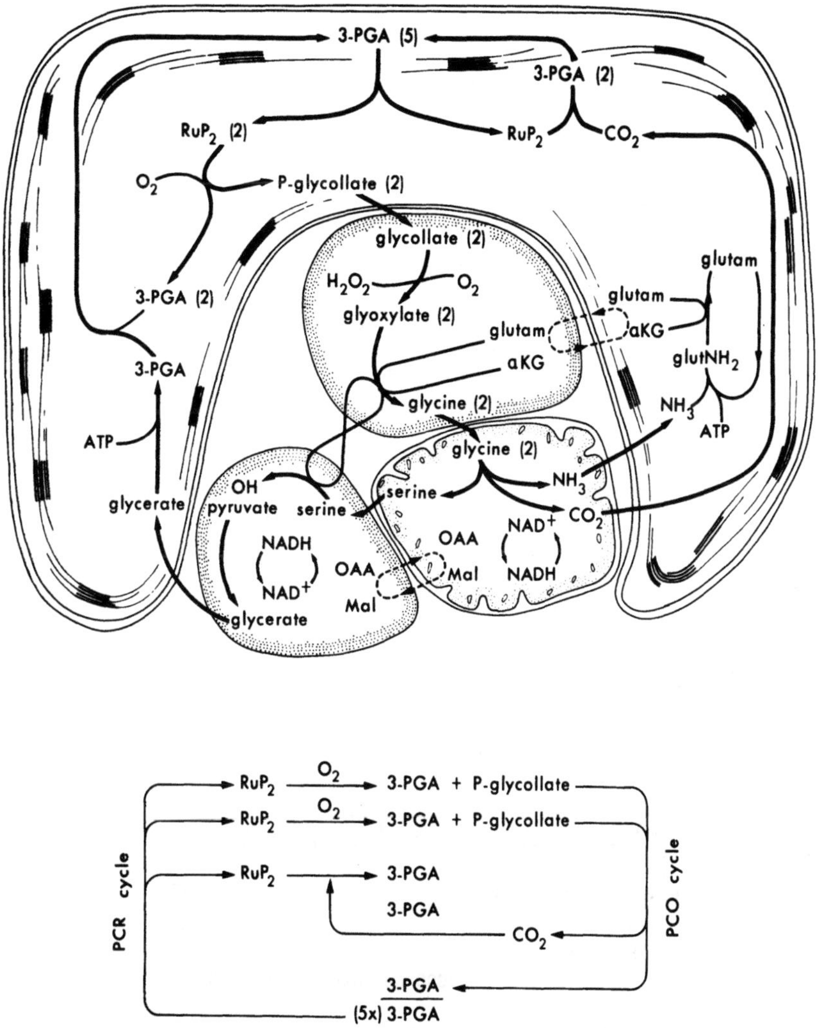

Fig. 15.2. Schematic outline showing the integration of the reduction and oxidation cycles at the CO_2 compensation point in leaves of C_3 plants, and the location of steps of the oxidation cycle in leaf chloroplasts, peroxisomes, and mitochondria

the PCOP following the oxygenation of RuP_2 also consumes ATP and reducing potential, and accounts for depression of the quantum yield in C_3 plants by O_2 (EHLERINGER and BJÖRKMAN 1977). The stoichiometry of carbon and energy metabolism in these integrated cycles is detailed in Chapter 16, this Volume.

2. Although only 75% of the carbon diverted from the PCRC as P-glycolate is recovered in the PCOP, it seems likely that all of the NH_3, released stoichiometrically with CO_2, must be recovered by the integrated PCRC and PCOP (LEA and MIFFLIN 1979; FARQUHAR et al. 1980). The fluxes of N through reactions of the PCOP are an order of magnitude greater than the capacity of leaves to assimilate NO_3^-, so the conservation of N in these reactions is essential.

3. Recent experiments suggest that by providing an internal source of CO_2, photorespiration and integration of PCRC and PCOP may have another impor-

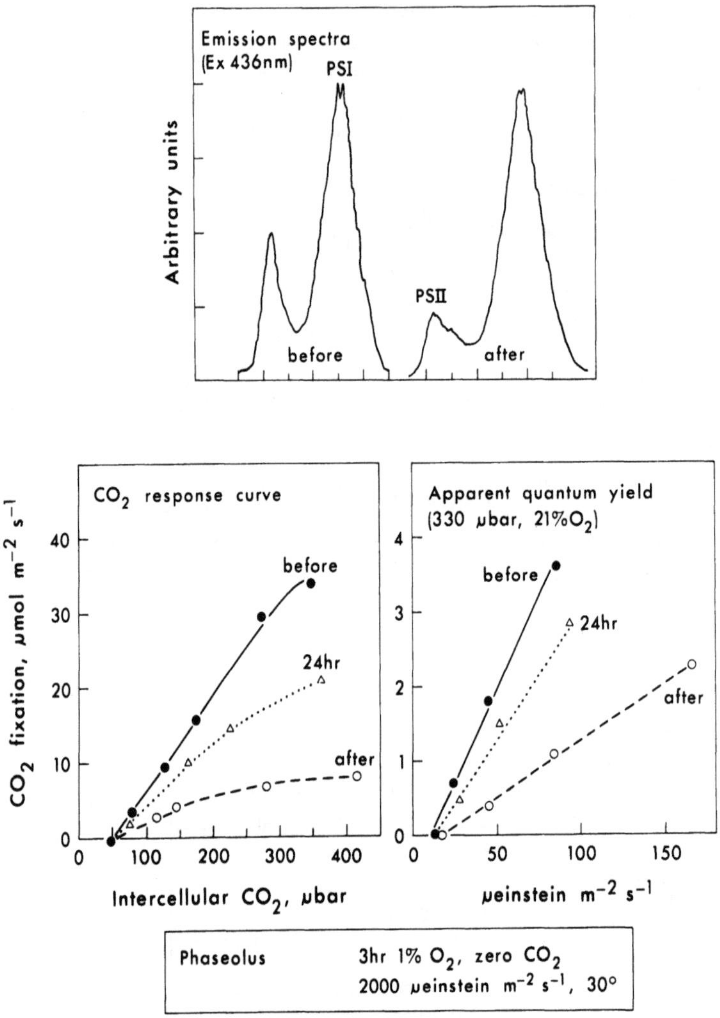

Fig. 15.3. Photoinhibition in leaves of *Phaseolus vulgaris* (C_3) after 3-h illumination in the absence of CO_2 and at low O_2 (no photorespiration), as shown by reduction in the capacity (CO_2 response curve), and efficiency (quantum yield) of photosynthesis. The fluorescence emission spectra indicate photosystem II has been selectively damaged. Partial recovery from photoinhibition is observed after 24 h in normal atmosphere. No photoinhibition is observed if leaves are kept at the CO_2 compensation point in 21% O_2. (After OSMOND et al. 1980a)

tant function in terrestrial plants. If leaves of C_3 plants are exposed to normal light intensities in the absence of CO_2 and at low O_2 concentrations which prevent photorespiration, the orderly dissipation of photochemical energy does not take place. Under these conditions the excess excitation energy causes the destruction of photochemical reaction centers and photoinhibition results (CORNIC 1976; POWLES and OSMOND 1978). Photoinhibition reduces both the capacity and efficiency of photosynthesis (Fig. 15.3), with substantial damage to photosystem II reactions in chloroplast thylakoids (OSMOND et al. 1980a).

Conquest of land by higher plants presumably depended upon the evolution of stomatal mechanisms to control water loss in the water-limited habitat, and indeed stomata are found in the shoot of the first land plant, the leafless *Rhynia*. However, if stomata close tightly in the light and deprive the leaf of access to external CO_2, the limited capacity to use the products of photosynthetic light reactions may lead to photoinhibition. Internal sources of CO_2, such as photorespiratory CO_2 production in the PCOP might mitigate photoinhibition. In some species the production of large amounts of highly reduced secondary plant substances (terpenes etc.) under stress conditions might also serve as a sink for photochemical energy under conditions of limiting CO_2. However, proof of the significance of these hypotheses depends on the demonstration that, in arid habitats, stomata close sufficiently tightly in the light to allow the internal leaf atmosphere to decline to the CO_2 compensation point (see Chap. 7, this Vol.). Of course, the presence of 210 mbar O_2 in the present atmosphere ensures that C_3 plants are unlikely to suffer photoinhibition, even if stomata close completely. However, these speculations imply that it may be undesirable to attempt to eliminate photorespiration in land plants.

Not surprisingly, with the emphasis on photosynthetic rate as the determinant of plant productivity (BLACK 1971), substantial programs have been directed to the control of photorespiration. The wisdom of these is questionable because, as will be evident later, photosynthetic rate itself can rarely be shown to limit plant productivity. Indeed, there is good evidence that increased harvestable yield of C_3 crops such as wheat has been achieved by selecting varieties which have lower rates of photosynthesis per unit leaf area, but improved display of photosynthetic tissues (DUNSTONE et al. 1973). Moreover it seems improbable that the complex biochemistry of the PCOP would have evolved under natural selection if it was totally without function. In broad terms, photorespiration and the integrated PCRC and PCOP represent an adaptive compromise which compensates, at some cost, for imperfectations of the carboxylation mechanisms of plants in the terrestrial environment.

15.2.2 CO_2-Concentrating Mechanisms in Photosynthesis

The high affinity of RuP_2 carboxylase for CO_2 [$K_m(CO_2) \approx 20$ μM] compared to the affinity for O_2 as an inhibitor [$K_i(O_2) \approx 300$ mM] means that the oxygenation of RuP_2 can be largely prevented by a two-to threefold increase in CO_2 concentration. It is not surprising that, rather than devise a new chemistry of carboxylation, natural selection has favored plants which have evolved different mechanisms for effecting an increase in CO_2 concentration in the vicinity of RuP_2 carboxylase-oxygenase. The different photosynthetic pathways effectively serve this function and in each of these the diversion of carbon from the PCRC to the PCOP as a result of oxygenation during CO_2 fixation is largely prevented.

Three main types of CO_2-concentrating mechanism are known and these are based on the following physiological processes:

1. Intercellular transport of C_4 dicarboxylic acids in the symplast of adjacent cells in C_4 plants (BJÖRKMAN 1971; HATCH and OSMOND 1976).

2. Intracellular transport of malic acid across the tonoplast, into and out of the vacuole, in CAM plants (KLUGE and TING 1978; OSMOND 1978; LÜTTGE and BALL 1979).

3. HCO_3^- transport systems in the plasma membrane of aquatic organisms (BADGER et al. 1978; RAVEN 1981).

15.2.2.1 C_4 Photosynthesis

On the basis of anatomical observations, VOLKENS (1887), HABERLANDT (1884) and RHOADES and CORVALHO (1944) speculated on the possibility of a "division of labor" during photosynthesis in plants which displayed the Kranz arrangement of photosynthetic tissues. The C_4 pathway of photosynthesis is always associated with Kranz anatomy in the leaves in one or more of its complex forms (CAROLIN et al. 1975; HATTERSLEY et al. 1977) and KARPILOV (1960) proposed the term "cooperative photosynthesis" which aptly describes the coordinated functioning of component reactions in adjacent cells. During C_4 photosynthesis CO_2 is first assimilated into C_4 carboxylic acids in outer mesophyll cells and these acids are transported through the symplast to bundle-sheath cells. Here the acids are decarboxylated by one of three different pathways (HATCH and OSMOND 1976; EDWARDS and HUBER 1981), to yield CO_2 and a C_3 compound which is returned in the symplast to mesophyll cells where it serves again as substrate for the primary carboxylation reaction (Fig. 15.4). The CO_2 thus released is refixed by reactions of the PCRC in bundle-sheath chloroplasts. The concentration gradients of C_4 acids and the activity of the decarboxylase enzymes are such that CO_2 concentration in the bundle-sheath cells is about ten times greater than that in the atmosphere (HATCH 1971). The CO_2 concentrating mechanism is based on a spatial separation of carboxylation and decarboxylation events.

Under these conditions, O_2 inhibition of the carboxylase is largely overcome (BJÖRKMAN 1971), oxygenation activity is inhibited by CO_2, and proportionally less carbon flows through the PCOP (OSMOND and BJÖRKMAN 1972; SCHNARRENBERGER and FOCK 1976). Because PEP-carboxylase is not sensitive to O_2, and as a result of the CO_2 concentrating mechanism, C_4 photosynthesis does not respond to O_2 partial pressure between 10 and 210 mbar. In this sense, C_4 photosynthesis is an elaborate accessory to the conventional C_3 pathway which is functionally significant in that it overcomes the inhibitory effects of 210 mbar O_2 on net CO_2 fixation in the present atmosphere. As noted by BLACK (1971) C_4 plants are thus potentially capable of substantially higher rates of CO_2 fixation. We shall later examine circumstances in which this potential may be realized.

Because the activity of PEP-carboxylase in mesophyll cells is so high, practically all CO_2 leaking from the PCRC in the bundle sheath can be refixed and returned as C_4 acids in the symplast. Thus C_4 plants display a CO_2 compensation point near to zero. In CO_2-free air BERRY and FARQUHAR (1978) believe the CO_2 concentration in the bundle sheath cells is similar to that in C_3 meso-

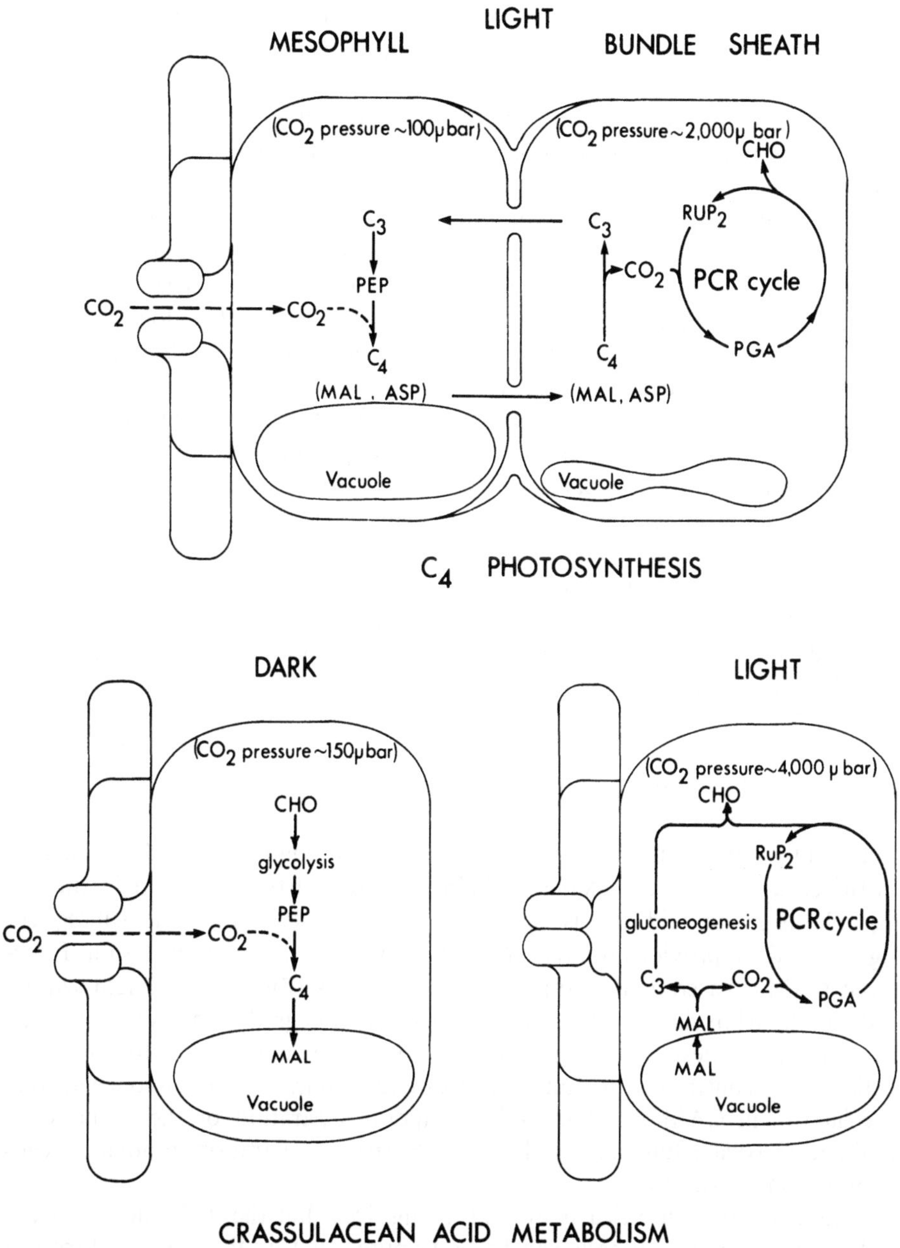

Fig. 15.4. Schematic outline of C_4 photosynthesis and CAM showing the biochemical analogies between the two pathways, as well as the differences in spatial (C_4) and temporal (CAM) organization of the component processes. (After BJÖRKMAN 1973)

phyll cells, but no methods have yet been devised to verify these assumptions. It is known that many C_4 plants have only low activities of PCOP enzymes in the bundle-sheath cells (HATCH and OSMOND 1976) and their capacity for photorespiration is not easily assessed (MOROT-GAUDRY et al. 1980). What is clear is that by concentrating CO_2 in the bundle-sheath cells, O_2 inhibition of RuP_2 carboxylase is prevented and the carbon recovery mechanisms of the PCOP are presumably not required to the same extent (RATHNAM and CHOLLET 1980).

This leaves the question of whether C_4 plants might be more vulnerable to photoinhibition because they lack measurable photorespiration. Moreover, the mesophyll cell chloroplasts of C_4 plants lack RuP_2 carboxylase altogether and have no prospect for CO_2 production via the PCOP. When illuminated in the absence of CO_2 at 10 mbar O_2, C_4 plants show photoinhibition which in many respects is indistinguishable from that in C_3 plants (POWLES et al. 1980). Photoinhibition in C_4 plants cannot be prevented by O_2 alone, testifying to the absence of effective internal CO_2 production via the PCOP and, incidentally, to a photochemical O_2-reducing system (Mehler reaction) capable of significant energy dissipation (cf. RADMER and KOK 1976). However, photoinhibition can be largely prevented by very low CO_2 partial pressure (10 μbar), presumably because high affinity C_4 metabolism serves as a sink for dissipation of photochemical energy under these conditions. This in itself enables C_4 plants to retain normal function at very low stomatal conductance and intercellular CO_2 concentration. Proof of the function of these processes under natural conditions depends on evidence of the extent of stomatal closure in response to water stress. We cannot yet discern whether C_4 plants are more susceptible to photoinhibition because they lack photorespiration.

15.2.2.2 CAM as a Photosynthetic Process

Succulent plants with CAM normally carry out the bulk of their CO_2 fixation in the dark via PEP carboxylase. The assimilated carbon is stored in the cell vacuole as malic acid and is released and decarboxylated in the subsequent light period, to provide an internal source of CO_2 at high concentration. Effectively CAM is based on the same carboxylase (responsible for malic acid synthesis) and decarboxylases (responsible for malic acid degradation) as are involved in C_4 photosynthesis. These disparate functions are contained within the same cell, but are constrained to function at different times. The CO_2-concentrating mechanism in CAM is achieved by temporal separation of the carboxylase and decarboxylase functions, rather than by spatial separation in adjacent cells as in C_4 photosynthesis.

These analogies are clearly evident in Fig. 15.4, but the differences between CAM and C_4 photosynthesis can be best appreciated by the size and turnover rates of the intermediate C_4 acid pools. The capacity of CAM plants to assimilate CO_2 in the dark depends on the supply of carboxylation substrate and the capacity of the vacuole to store malic acid. Generally this is limited at about 200 $\mu mol\ g^{-1}$ fr. wt. and the pool turns over with a half-time of about 10,000 s. In C_4 plants the turnover pool is very small (about 2 $\mu mol\ g^{-1}$ fr. wt.) but turns over very rapidly with a half-time of about 10 s. In a 10-h day C_4 plants have a potential to assimilate about 20 times as much CO_2 as CAM plants in a 10-h night. Obviously, the existence of CAM in many organisms is clear evidence that high rates of CO_2 fixation are not the only property upon which natural selection has acted in the evolution of higher plant photosynthetic metabolism.

During the decarboxylation phase of CAM, the release of CO_2 from stored malic acid generates extraordinarily high intercellular CO_2 concentrations, in the vicinity of 4,000–10,000 μbar CO_2 (SPALDING et al. 1979; COCKBURN et al. 1979). Under these conditions the diversion of carbon through the reactions of the PCOP is likely to be much reduced (OSMOND 1978; SPALDING et al. 1979). Because the stomatal conductance of CAM plants is unusually low during the early light period, as a consequence of the high internal CO_2 concentration, little CO_2 is lost either from decarboxylation of malate or from photorespiration.

In arid conditions it is not uncommon that stomata of CAM plants remain tightly closed in the light for about 100 days (SZAREK et al. 1973) and frequently they remain closed in the dark as well, so that CAM is reduced to the refixation of respiration CO_2 in the dark. The internal generation of CO_2 in the subsequent light period, first by deacidification and later by photorespiration, presumably prevents photoinhibition in the high light intensity arid environments. Succulent plants with CAM are thus able to retain a functional photosynthetic apparatus at high plant water potential, in the absence of net CO_2 exchange. The functional significance of CAM in these terms has yet to be explored experimentally, but it may be more important than the interpretations in terms of water economy, developed below.

15.2.2.3 Aquatic Plants

The Aqueous Environment. Submerged aquatic plants experience a physical and chemical environment which is in certain respects fundamentally different from that of land plants (SPENCE 1976, 1981; see Chap. 8, this Vol.). The following environmental attributes are important in the present context:

1. In terms of photosynthetic metabolism, carbon may be available for uptake by leaves or other green tissue either as CO_2 or as HCO_3^- ions. Bicarbonate use is defined as the uptake of HCO_3^- ions, probably through active transport, followed by their dehydroxylation within the cytoplasm, the assimilation of CO_2 and the efflux of OH^- ions.

The equilibrium distribution of solutes in aqueous carbon solutions may be calculated according to methods set out in the first series of the Encyclopedia of Plant Physiology (BUCH 1960a, b; RUTTNER 1960; KERN 1960) or in STUMM and MORGAN (1970). Using the terminology of the latter authors:

a) free carbon dioxide, $H_2CO_3^* = [CO_2(\text{aqueous})] + [H_2CO_3^*]$ (mol m^{-3})

b) total carbon, $C_T = [H_2CO_3^*] + [HCO_3^-] + CO_3^{2-}]$ (mol m^{-3})

c) total alkalinity, $Alk = [HCO_3^-] + [2\ CO_3^{2-}] + [OH^-] - [H^+]$ (Eq m^{-3}).

In what follows, "CO_2" is used instead of "$H_2CO_3^*$".

The proportion of different carbon species in any volume of water is strongly pH-dependent. With a pH range of 7.4 to 8.3 in the ocean the proportional range of the components of total carbon are 4% to 1% CO_2, 96% to 89% HCO_3^- and 0.2% to 11% CO_3^{2-} (Fig. 15.5).

2. In freshwater the proportional range of different carbon components is much wider than in the sea because of the greater variation in alkalinity and

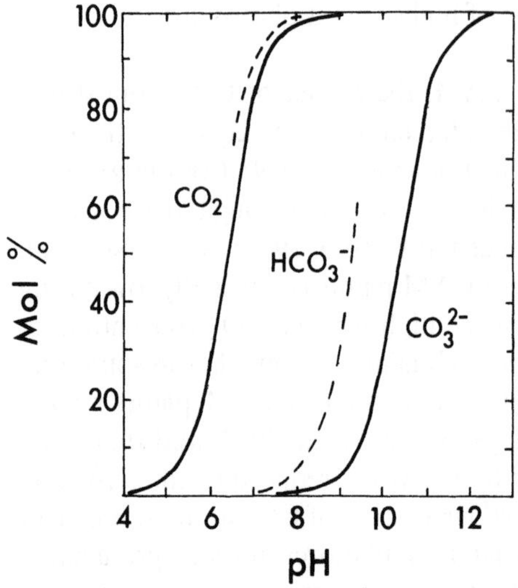

Fig. 15.5. Components of inorganic carbon in distilled water at 20 °C (–) and in sea water at 4 °C (––) at different pH values. With decreasing temperature the two curves show a parallel shift to the right with values corresponding to the temperature coefficients of the relevant pK values. (After BUCH 1960a, b)

in pH which also depends, particularly at low alkalinity, on photosynthetic activity (see below). Not only do different freshwater bodies have different alkalinities, ranging from less than 0.01 to about 4.4 Eq. m^{-3} (ALLEN and SPENCE 1981); a single water body can vary considerably in alkalinity, C_T, and pH over one growing season: e.g., Esthwaite Water (TALLING 1976).

In the open water of nutrient-rich but poorly buffered, low-alkalinity lakes, photosynthetic demand by phytoplankton can reduce CO_2 concentration below its air-equilibrium value during the day. If demand continues to outstrip supply this deficit increases through the remainder of the growing season and in turn further reduces $[C_T]$. Bicarbonate use enhances C_T deficit and in poorly buffered water of this type, like Esthwaite Water, can lead to pH values approaching the theoretical maximum where all aqueous inorganic carbon is absent, HCO_3^- and CO_3^{2-} being replaced by OH^- (viz. pH 11.01 at an alkalinity of 1 Eq. m^{-3}). Similar trends are also well known in stands of submerged macrophytes. For instance, CO_2 concentration in such stands can fall to less than 1 μM a few hours after sunrise and $[HCO_3^-]$ can approach zero later in the day (e.g., BROWN et al. 1974; VAN et al. 1976).

3. The oxygen concentration in the surface waters of lakes may by contrast become considerably over-saturated during photosynthesis of aquatic plants (30 kPa or 300 mbar, corresponding to 30%, in the above-mentioned cases). In certain submerged macrophytes with well-developed lacunae or intercellular air spaces, $[O_2]$ can reach relatively high values during photosynthesis (HARTMAN and BROWN 1976).

4. Carbon is supplied to submerged plants along an entirely aqueous diffusion pathway; for CO_2 at 20 °C aqueous diffusive resistance is 10^4 times slower than in air. This means that boundary layer resistance around aquatic plants is likely to be high, even in well-stirred conditions (SMITH and WALKER 1981). Since the diffusive coefficients of CO_2 and HCO_3^- in water are similar, boundary layer resistance for both compounds will be almost the same (JARVIS 1971; BROWSE et al. 1979b), although for most submerged freshwater macrophytes

Table 15.1. Submerged aquatic angiosperms in which photosynthesis evidently proceeds via the C_3 pathway

Species	Reference
Elodea canadensis	BROWN et al. (1974)
Lagarosiphon major	—″—
Egeria densa	BROWN et al. (1974); BROWSE et al. (1979a)
Vallisneria spiralis	PRINS and WOLFF (1974)
Hydrilla verticillata	PRINS and WOLFF (1974); VAN et al. (1976); BOWES et al. (1977)
Thalassia hemprichii	ANDREWS and ABEL (1979)
Halophila spinosa	ANDREW and ABEL (1979)
Myriophyllum spicatum	VAN et al. (1976); BOWES et al. (1977)
Ceratophyllum demersum	VAN et al. (1976); BOWES et al. (1977)
Potamogeton pectinatus	WINTER (1978)
Sagittaria sp.	BJÖRKMAN (1966)
Najas flexilis	HOUGH (1974)
Scirpus subterminalis	HOUGH (1974)
Ranunculus pseudofluitans	WESTLAKE (1967)

in nature the external concentration gradient for CO_2 is likely, because of greater demand, to be steeper than that for HCO_3^- (ALLEN and SPENCE 1981).

Some Functional Attributes of Aquatic Plants. Thus submerged freshwater plants may under certain circumstances be confronted with low CO_2 and high O_2 concentrations during daylight. The inhibitory effects of O_2 on photosynthesis and the losses of CO_2 through photorespiration could be reduced by increasing the partial pressure of CO_2 inside the photosynthesising cells. High diffusive resistance to CO_2 loss by the plant may be related to the existence of gas-filled lacunae adjacent to the chloroplasts. As discussed below, cells of many microalgae and macrophytes may achieve an increase in the internal partial pressure of CO_2 by means of HCO_3^- uptake. This is probably important in all waters, except those of low alkalinity, and could account for some of the very low CO_2 compensation points reported in the literature.

Several aquatic macrophytes have been checked to determine if they have a CO_2-concentrating mechanism analogous to C_4 photosynthesis (Table 15.1). Although there are many conflicting interpretations of incomplete data (e.g., BENEDICT and SCOTT 1976), there are no reports of an aquatic plant which meets the structural and functional requirements of C_4 photosynthesis. Of 27 submerged angiosperms so far examined only one, *Potamogeton preaelongus,* has any structural similarity with Kranz anatomy. In this species, enhancement of photorespiration by high O_2 pressure, and low PEP carboxylase levels, confirm the absence of C_4 photosynthesis (HOUGH and WETZEL 1977). Table 15.1 lists studies which suggest that normal C_3 photosynthesis proceeds in a wide range of submerged macrophytes.

In view of the general evidence that submerged aquatic macrophytes are C_3 plants, the apparently low CO_2 evolution in the light has been attributed to efficient refixation of CO_2 from gas-filled lacunae, coupled with high aqueous diffusive resistance to CO_2 loss from the plant surfaces (see, for example, HOUGH and WETZEL 1972, 1978; LLOYD et al. 1977). SØNDERGAARD (1979) investigated apparent respiration and refixation of CO_2 in four macrophytes with air spaces ranging from 55% of leaf volume in *Lobelia,* 48% in *Litorella,* 14% in *Elodea* to zero in *Fontinalis.* Comparing intact leaves and those with cut lacunae, SØNDERGAARD found that the efficiency of refixation of CO_2 evolved in the light had a maximum of 50%–70% and decreased with a decrease in the relative volume of air space in leaves.

The most complete evidence for normal C_3 photosynthesis in an aquatic angiosperm is found in the comprehensive pulse-chase labeling studies of ANDREWS and ABEL (1979). Yet there are many interesting reports which suggest that alternative or mixed pathways of CO_2 assimilation are possible in some plants. HOLADAY and BOWES (1980) reported changes in the enzyme complement and physiological properties of *Hydrilla verticillata* in response to summer and winter growth conditions which suggest that CO_2 fixation in this plant does not fit the three major pathways found in terrestrial plants. On the other hand KEELEY (1981) has demonstrated nocturnal malic acid synthesis and deacidification similar to CAM in submerged *Isoetes.*

In the algae, in which after all, the C_3 pathway of photosynthesis was first described, there is no evidence for a significant contribution of alternate pathways of CO_2 fixation. Even the most highly developed morphological types in the Phaeophyceae and Rhodophyceae display conventional C_3 metabolism during photosynthetic CO_2 fixation (YAMAGUCHI et al. 1969; MORRIS and FARRELL 1971; KREMER and WILLENBRINK 1972; KREMER and KÜPPERS 1977; WILLENBRINK et al. 1975). Nevertheless, in some Phaeophyceae a high rate of CO_2 fixation via PEP-carboxykinase is observed which is independent of light, and presumably depends on mannitol as a carbon source. The rate of this carboxylation reaction is particularly high in young, actively growing cells or tissues, e.g., in the growing zone of the thallus (WEIDNER and KÜPPERS 1973; WILLENBRINK et al. 1975; KÜPPERS and KREMER 1978) or in gametophytes, gametes, and zygotes (KREMER and MARKHAM 1979). This is presumably an anaplerotic reaction and should not be confused with the C_4 pathway of photosynthesis in terrestrial plants.

Evidence that different aquatic macrophytes may differ in their ability to utilize CO_2 and HCO_3^- in photosynthesis was reviewed by STEEMANN NIELSEN (1960) and RAVEN (1970). Since then, either by indirect methods like comparison of rates of net O_2 evolution by plants in CO_2 solutions and in HCO_3^- solutions at the same pH, or by indirect methods involving the uptake of $^{14}CO_2$ and $H^{14}CO_3^-$, BLACK MA (1973), KEENAN (1975), LUCAS (1975, 1976), FINDENEGG (1976), LUCAS et al. (1978) and BADGER et al. (1978) established that HCO_3^- is used by various cyanophyte, chlorophyte, and charophyte species, defining these groups cytologically (PICKETT-HEAPS 1975; STEWART and MATTOX 1975) and biochemically (e.g., RAVEN and GLIDEWELL 1978). Accepting that there can be degrees of adaptation to HCO_3^- use, as in some microalgae (BRIGGS and WHITTINGHAM 1952; ÖSTERLIND 1951, 1952; GRAHAM and REED 1971; REED and GRAHAM 1977; BADGER et al. 1978) and macrophytes (HOLADAY and BOWES 1980), the prevalent view has tended to be that there are two physiological groups, HCO_3^- users and nonusers, represented among macrophytes by *Potamogeton crispus* and *Fontinalis antipyretica* respectively.

Recently ALLEN and SPENCE (1981) studied this differential ability in 15 microalgae and macrophytes (Angiospermae, Bryophyta, and Charophyta) occurring in freshwaters where the range in concentrations and proportions of CO_2 and HCO_3^- is far wider than in the sea. They found that the microalgae had considerably greater apparent affinities for HCO_3^- and slightly greater apparent affinities for CO_2 than the macrophytes, including several previously established HCO_3^- users, while the macrophytes had larger apparent affinities for CO_2 than for HCO_3^- and larger diffusive resistances to CO_2. For species with different affinities for CO_2 and HCO_3^-, specific CO_2 compensation points were determined, in carbon-depleted solutions having a range of constant alkalinities, such as those in C_3 plants. At still lower total carbon values, any assimilation was held to be due to HCO_3^- uptake, and "HCO_3^- compensation points" for some species were also determined.

Rather than "users" and "nonusers" of HCO_3^- it was concluded that a gradation exists, use depending on $[HCO_3^-]$, as affected by alkalinity and pH of the bathing solution, and HCO_3^- compensation point. It was further concluded that rates of photosynthesis of macrophytes and some microalgae in nature are usually functions of exogenous $[CO_2]$, while those of other microalgae are functions of $[CO_2]$ and $[HCO_3^-]$, and that a knowledge of CO_2 and HCO_3^- compensation points is essential in characterizing the response of aquatic plants to inorganic carbon supply. This knowledge would avoid, inter alia, mistaking an apparent C_4 attribute like a very low total carbon compensation point for the CO_2 compensation point, where the true CO_2 compensation point has a higher, typically C_3, value.

So land plants that have become adapted to living under freshwater conditions have to differing extents acquired the ability to take up HCO_3^- through the surface of their leaves. Also to differing extents they have developed gas-filled lacunae. Both these adaptations act as carbon-concentrating mechanisms and would tend to preclude selection for C_4 metabolism. The same argument should apply to marine angiosperms.

Since all carbon in photosynthesis of submerged marine and freshwater plants is fixed finally by RuP_2 carboxylase, and since this enzyme uses free CO_2, HCO_3^- has to be dehydrated inside the cell before the CO_2 can be used by RuP_2 carboxylase (MILLER and COLEMAN 1980). This reaction is catalyzed by carbonic anhydrase (cf. RAVEN 1970). The activity of this enzyme in the submerged aquatic macrophytes analyzed by BOWES et al. (1977) was sufficient to support HCO_3^- use in photosynthesis, and in algae the activity increases when cells adapt to low CO_2 pressure (GRAHAM et al. 1971). The capacity of these cells to use HCO_3^- and the affinity for CO_2 increase simultaneously with the increase in carbonic anhydrase activity (FINDENEGG 1976; BERRY et al. 1976; BADGER et al. 1978).

The increase in affinity for CO_2 is a consequence of a CO_2-concentrating mechanism based on HCO_3^- transport, and BADGER et al. (1977) showed that adapted cells are able to achieve intracellular CO_2 pressures ten times higher than those in the external solution. A similar CO_2-concentrating mechanism

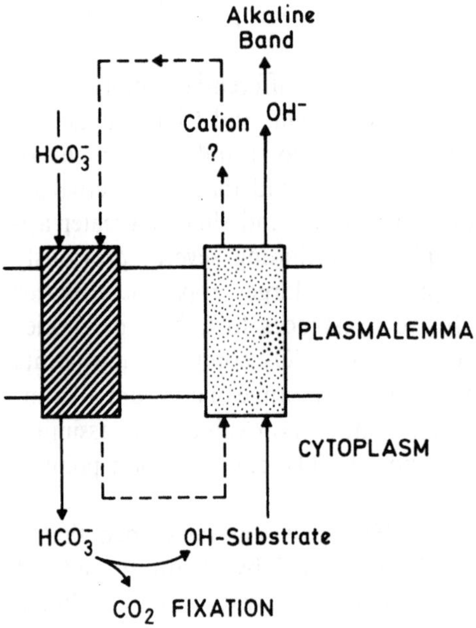

Fig. 15.6. Schematic representation of a model for the formation of alkaline bands formed on the surface of *Chara corallina* cells. HCO_3^- and OH^- transporters exist at spatially different sites within the plasmalemma (LUCAS 1979)

based on HCO_3^- transport is known from blue-green algae (MILLER and COLEMAN 1980) and in *Chara*. In *Chara* spp. the long-established alkaline banding phenomenon (ARENS 1939) is due to spatial separation of HCO_3^- uptake and OH^- extrusion functions during light-dependent HCO_3^- uptake (LUCAS 1979, 1980; see Fig. 15.6).

Considering the preponderance of HCO_3^- in seawater it is likely that marine plants may also have a CO_2-concentrating mechanism based on HCO_3^- transport, but experimental evidence is lacking. Some marine algae use HCO_3^- in photosynthesis (BLINKS 1963; THOMAS and TREGUNNA 1968; JOLIFFE and TREGUNNA 1970), but the brown alga *Carpophyllum* evidently fixes CO_2 (DROMGOOLE 1978), as do the sea-grasses (BENEDICT et al. 1980; cf. BEER et al. 1977).

15.2.3 Stable Isotope Discrimination in Relation to Photosynthetic Pathway and Environment

One of the more significant developments for physiological ecology, with respect to the different pathways of photosynthetic CO_2 fixation, was the observation by BENDER (1968) that plants with the C_4 pathway discriminated less against the heavy isotope ^{13}C during carboxylation than did C_3 plants. The carbon isotope discrimination properties of plants are readily evaluated by measurement of the ratio of ^{13}C to ^{12}C in CO_2 produced by the combustion of dried plant material. The ratio is expressed as a $\delta^{13}C$ value, in parts per thousand, after reference to a limestone standard, i.e.,

$$\delta^{13}C\ (‰) = \left[\frac{^{13}C/^{12}C\ \text{sample}}{^{13}C/^{12}C\ \text{standard}} - 1\right] \times 10^3$$

Table 15.2. Decrease of the $\delta^{13}C$ value in the organic dry substance of *Chara ceratophylla* with the water depth. This is in accordance with the report of RAU (1978) that the $\delta^{13}C$ value of soluble inorganic carbon compounds decreases with increasing water depth. (ZIEGLER 1979)

Water depth	$\delta^{13}C$ [$^0/_{00}$]
1 m	−20.80
2 m	−20.75
5 m	−23.40
7 m	−24.72
9 m	−26.79
11 m	−29.21

With access to an appropriate mass spectrometer, the method provides a rapid and reliable method for distinguishing between terrestrial C_3 and C_4 plants; it is more difficult to interpret results for CAM plants and for aquatic plants (BENDER 1971; SMITH and EPSTEIN 1971; LERMAN 1975; OSMOND and ZIEGLER 1975; BENEDICT 1978; ZIEGLER 1979; TROUGHTON 1979; VOGEL 1980). Later in this chapter we shall make extensive use of this technique in surveys of photosynthetic pathway among plant populations, but it is appropriate here to discuss the biochemical and physiological bases of the discrimination process.

When CO_2 fixation in terrestrial plants takes place via RuP_2 carboxylase in mesophyll cells of C_3 plants under conditions which allow large stomatal conductances to prevail (adequate water supply and low leaf-air vapor pressure difference), the $\delta^{13}C$ value is largely determined by the biochemistry of the carboxylation step itself (PARK and EPSTEIN 1960), but to a certain extent also by diffusional processes. In vitro, the carboxylation of RuP_2 leads to an isotopic fractionation of $-27^0/_{00}$ (WONG WW et al. 1979). Since atmospheric CO_2 has an average $\delta^{13}C$ of −7%, $\delta^{13}C$ values of C_3 plants should be around $-34^0/_{00}$. Under optimal growth conditions nonenzymic isotope fractionations are small (BENDER and BERGE 1979), but if CO_2 is fixed under stress conditions with low stomatal conductance, then isotope fractionation during CO_2 diffusion to the leaf may yield $\delta^{13}C$ values $-2^0/_{00}$ to $-10^0/_{00}$ less negative than controls (GUY et al. 1980; G.D. FARQUHAR, M.H. O'LEARY, K. WINTER, unpublished). In fact, the bulk of the C_3 species shows less negative $\delta^{13}C$ values ($-25^0/_{00}$ to $-28^0/_{00}$). This difference suggests an incomplete equilibrium between the atmospheric and intercellular CO_2 pool (SCHMIDT and WINKLER 1979), and diffusional fractionation is implicated (VOGEL 1980). If the CO_2 fixed by C_3 plants is enriched with respiratory carbon from decaying vegetable matter – plants growing on the floor of a dense forest (MEDINA and MINCHIN 1980) or at the bottom of a lake (Table 15.2) – or animal respiration – plants in growth chambers in the basement of a laboratory building (OSMOND 1975) – or the burning of fossil fuels (heated or CO_2 enriched commercial greenhouses) then the $\delta^{13}C$ value of the C_3 plants may be significantly more negative than $-27^0/_{00}$ (LERMAN 1975).

As pointed out above, the few species of aquatic plants carefully researched thus far assimilate CO_2 mainly via RuP_2 carboxylase and the PCRC. However, the $\delta^{13}C$ values of aquatic plants are extremely variable (even within one species), ranging from $-5.6^0/_{00}$ to $34.3^0/_{00}$ for marine plants (BLACK and BENDER 1976; ANDREWS and ABEL 1979; H. ZIEGLER, W. STICHLER, W. HUBER and A. EDER

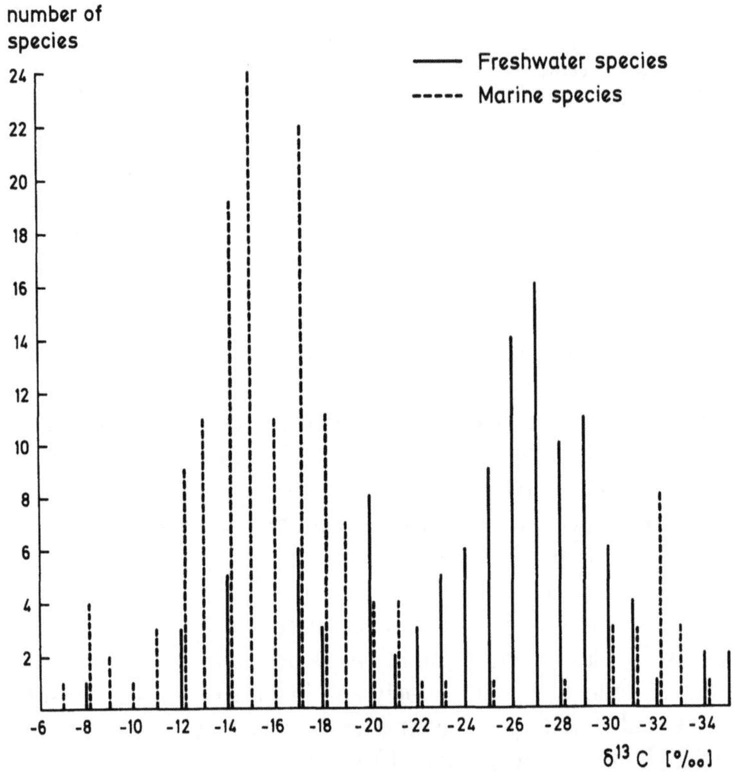

Fig. 15.7. $\delta^{13}C$ values for some species of freshwater and marine submerged aquatic plants. Most of the marine species show values in the order of the terrestrial C_4 plants (the possible reasons for this are discussed in the text), while the freshwater plants mostly have C_3 values. (Unpublished data of H. ZIEGLER, W. STICHLER, W. HUBER and A. EDER)

unpublished; cf. Fig. 15.7) and from $-8.7^0/_{00}$ to $-42.8^0/_{00}$ for aquatic plants (PARK and EPSTEIN 1960). Three factors contributing to these variable values are the $\delta^{13}C$ value of the source carbon (seawater HCO_3^- has a $\delta^{13}C$ value about $+1^0/_{00}$ to $-2^0/_{00}$, WEBER and WOODHEAD 1970; BENEDICT et al. 1980), the species of carbon fixed (dissolved CO_2 will have a $\delta^{13}C$ value more nearly that of air, $-7^0/_{00}$ to $-10^0/_{00}$) and factors associated with diffusion and uptake of the carbon species by the submerged plant. ANDREWS and ABEL (1979) point out that in *Thalassia* there appears to be a variable intermediate pool of CO_2/HCO_3^- between the CO_2/HCO_3^- of the plant and the medium, possibly due to a CO_2/HCO_3^- uptake mechanism. The kinetic properties of this intermediate pool can produce a substantial change in the $\delta^{13}C$ value of the carbon it contains, and this in turn will be preserved in the metabolically fixed carbon irrespective of subsequent biochemical fractionations. BENEDICT et al. (1980) found a small difference between the $\delta^{13}C$ value of CO_2 in seawater $(-10.3^0/_{00})$ and the carbohydrates of the seagrass *Thalassia testudinum* $(-11.5^0/_{00})$; they also explained the lack of discrimination during CO_2 fixation by RuP_2 carboxylase in seagrasses in terms of the complete fixation of a separate internal CO_2 pool. Whether the kinetics of physical uptake processes result in a discrete internal pool, or whether they simply establish an unstirred boundary layer (ABELSON and HOERING 1961; BENEDICT 1978) remains to be assessed in different species and environments. A very lucid account of the different factors contributing to the $\delta^{13}C$ value of submerged plants is given by SMITH and WALKER (1980), with supporting data by OSMOND et al. (1981).

Since the discrimination against ^{13}C by RuP_2-carboxylase is a kinetic isotope effect (cf. SIMON 1982), if all the carbon entering the organism is immediately fixed, there is no possibility for further discrimination. The $\delta^{13}C$ value of the incorporated carbon will be identical to that of the carbon source. A very clear indication that the $\delta^{13}C$ value of submerged water plants is strongly influenced by the $\delta^{13}C$ of the source carbon, and not by the photosynthetic machinery is found in the variation of $\delta^{13}C$ value of *Chara* with depth (Table 15.2), and the different $\delta^{13}C$ value of submerged and emerged leaves of some water plants. These factors require careful evaluation in aquatic plants and it is incorrect to conclude that $\delta^{13}C$ values in the vicinity of $-10‰$ indicate CO_2 fixation via PEP carboxylase in these plants OSMOND et al. (1981).

The reported $\delta^{13}C$ values of C_4 plants range between $-6‰$ and $-17‰$ (SMITH and EPSTEIN 1971). It was thought that this value is primarily determined by the kinetic isotope discrimination properties of PEP carboxylase, because in vitro this enzyme shows a small discrimination of about $-2‰$ (BENEDICT 1978). However, PEP carboxylase functions with HCO_3^-, rather than CO_2, and HCO_3^- in equilibrium with atmospheric CO_2 will have a $\delta^{13}C$ value of about $0‰$, due to thermodynamic isotope fractionation during hydration (DEUSER and DEGENS 1967). In addition, as a consequence of the greater mass of HCO_3^- the kinetic isotope effect is smaller than with CO_2 (SIMON and PALM 1966). If PEP carboxylase alone is responsible for the isotope fractionation in C_4 plants, $\delta^{13}C$ values in the vicinity of $-2‰$ should be found. It is also widely accepted that the sum of these isotope fractionations, which determines the $\delta^{13}C$ value of the C_4 acids formed in C_4 plants, is then passed on to all products of carbon reduction in the bundle-sheath cell. Further significant fractionation due to RuP_2 carboxylase is unlikely, because the decarboxylations and refixation occur in a relatively closed compartment. Essentially all the carbon entering as the [4-C] carboxyl of a C_4 acid is fixed via the PCRC. This interpretation is consistent with the concept of CO_2-concentrating mechanism in C_4 photosynthesis, and with the observation that C_4 plants do not release CO_2 to CO_2-free air. However, diffusional leakage of CO_2 between mesophyll and bundle-sheath cells, and diffusional fractionation from air to mesophyll cells, are now considered to be the principal determinants of the observed $\delta^{13}C$ value of C_4 plants (O'LEARY 1981).

In CAM plants $\delta^{13}C$ values range from about $-9‰$ to $-25‰$ and by analogy with C_3 and C_4 plants, it was early recognized that this range reflects the proportion of total CO_2-fixation which takes place via PEP carboxylase in dark CO_2 fixation, and RuP_2 carboxylase during CO_2 fixation in the light (OSMOND et al. 1973; BENDER et al. 1973; LERMAN et al. 1974). Malic acid formed in the dark is decarboxylated in the light and the refixation of CO_2 by RuP_2 carboxylase takes place when stomata are tightly closed. As in C_4 plants, there is little opportunity for further isotope fractionation during this process, so that the $\delta^{13}C$ value imparted by PEP carboxylase is transferred to the carbohydrate products of deacidification. The uncertainties surrounding enzymic and diffusional components of isotope fractionation in C_3 and C_4 plants also apply to CAM plants (O'LEARY and OSMOND 1980; O'LEARY 1981).

The proportion of carbon from each source, and hence the $\delta^{13}C$ value of CAM plants differs between species (stem succulents engage in less net CO_2 fixation in the light than leaf succulents), with leaf age (young leaves engage in less dark CO_2 fixation than older leaves) and with many environmental factors (water-stressed CAM plants do not normally fix external CO_2 in the light). Changes in the $\delta^{13}C$ value of CAM plants can be correlated with environmental factors (OSMOND 1978) and this is particularly useful in ecological studies. However, because the $\delta^{13}C$ values of CAM plants range between those of C_3 and C_4 plants, additional criteria are necessary for the classification of a species as a CAM plant. The least equivocal evidence for CAM, if the $\delta^{13}C$ value is in the range found for C_4 plants, is a nocturnal increase in the free titratable acid content, a decrease in this acid the following day, and the absence of Kranz leaf anatomy. The leaf succulent *Zygophyllum simplex* was thought to be a CAM plant on the basis of $\delta^{13}C$ value (MOONEY et al. 1977), but evidence of Kranz anatomy (CROOKSTON and MOSS 1972) suggests it is a C_4 plant.

A further development of significance for physiological ecology is the analysis and interpretation of hydrogen isotope discrimination in terrestrial plants (ZIEGLER et al. 1976). The factors which affect the δD value in the dry organic matter of plants are complex. Since HDO has a lower vapor pressure than H_2O, HDO enrichment in the liquid phase is observed following the preferential evaporation of H_2O. Although the reference standard seawater contains 320 molecules of HDO in each 1,000,000 molecules of H_2O (CRAIG and GORDON 1965) and remains relatively constant, evaporation is responsible for large variations in the δD value of precipitation and soil water. Variation in the δD value of source water can account for large variation in the δD value of water in plant tissues (SCHIEGL 1970; SCHIEGL and VOGEL 1970). However, the δD value of tissue water is also determined by the rate of water turnover in the plant. The δD value of plant dry matter is then a product of biochemical fractionations during metabolism in this water, as well as exchange reactions subsequent to metabolic processes. This metabolic hydrogen, that is the hydrogen released as water on combustion of dry plant material, shows a discrimination against deuterium which is larger than, but in the same direction as, the carbon isotope discrimination.

In C_3 plants the δD value is about $-120‰$ and in C_4 plants is about $-85‰$ when grown side by side with the same water supply (δD value $-78‰$). CAM plants grown in the same study were slightly enriched in deuterium (δD value $-75‰$) and if exposed to water stress conditions, enrichment in deuterium in CAM plants may be substantial (δD value $+50‰$). The tissue water of CAM plants has a much higher deuterium content, because succulents absorb a large amount of water when it is available, store it and transpire it slowly over a long period without further water uptake (ZIEGLER 1979). These plants remain metabolically active during this time so that the δD value of dry matter

Table 15.3. $\delta^{13}C$ and δD values of the organic dry substance in CAM, C_4, and C_3 species respectively each from the same site. Notice the high deuterium contents of the CAM plants. (ZIEGLER 1979)

Type	Species, location	$\delta^{13}C$[⁰/oo]	δD[⁰/oo]
CAM	*Mesembryanthemum crystallinum* (Aizoaceae), Tenerife, Bajamar	−21.40	−35.3
C_3	*Beta patellaris* (Chenopodiaceae) as before	−26.28	−68.6
CAM	*Sempervivum wulfenii* (Crassulaceae), S-Tyrol; Kellerbauerweg	−17.94	−69.7
C_3	*Hyperzia selago* (Lycopodiaceae) as before	−24.61	−98.0
CAM	*Portulaca karmesina* (Portulacaceae), SW-Africa; 30 km N of Okahandja	−12.01	− 5.1
CAM	*Cyphostemma currorii* (Vitaceae) as before	−14.22	−28.8
CAM	*Aloe litoralis* (Liliaceae) as before	−13.02	−27.6
C_4	*Schmidtia pappophoroides* (Poaceae) as before	−12.84	−62.0
C_3	*Leucas glabrata* (Lamiaceae) as before	−26.53	−73.0
CAM	*Cissus quadrangularis* (Vitaceae), Saudi-Arabia	−15.14	+49.6
C_3	*Citrullus colocynthis* (Cucurbitaceae) as before	−28.47	−72.0

equilibrates with that of the water (Table 15.3). In CAM plants, less negative $\delta^{13}C$ values in the range −15⁰/oo to −10⁰/oo, which are correlated with predominance of dark CO_2 fixation under drought conditions, are also correlated with δD values in the range of −25⁰/oo to +50⁰/oo. On the other hand, C_4 plants with the same $\delta^{13}C$ value show δD values in the vicinity of −75⁰/oo to −50⁰/oo. Together, $\delta^{13}C$ value and δD value provide strong evidence for CAM.

15.2.4 Summary

In the biochemical context the different pathways of photosynthetic CO_2 fixation can be viewed as mechanisms which compensate for, or largely overcome, the deficiencies of RuP_2 carboxylase-oxygenase in the present atmosphere. In a presumably long evolutionary history the carbon recovery mechanism of the integrated PCRC and PCOP in C_3 photosynthesis was selected in the majority of terrestrial plants. In several species of aquatic plants the problem of O_2 inhibition was resolved by selection of a rapid HCO_3^- transport system which concentrates CO_2 in the vicinity of the carboxylase. Functionally similar processes, based on C_4 acid transport mechanisms, evolved in C_4 plants and CAM plants. It is fascinating to reflect that the presumed selective pressure for these different pathways over hundreds of millions of years (the low CO_2 partial pressure of the present atmosphere), may well be removed in the space of a century. The increase in atmospheric CO_2 partial pressure as a result of the burning of fossil fuels (cf. Chapt. 15, Vol. 12D) may abolish the O_2 inhibition of CO_2 fixation in C_3 plants, and abolish any marginal benefit enjoyed by

C_4 plants on this account. It is naive, however, to interpret the functional significance of these pathways in such simple terms. There are "manifold consequences" of these biochemical processes which give rise to very different physiological responses to light, temperature, and water. These physiological parameters may be more important in determining plant responses to the rapidly increasing high CO_2 atmosphere, which will also almost certainly result in a hotter and more arid environment.

15.3 The Physiological Context: Comparative Responses of Plants with Different Photosynthetic Pathways to Environmental Factors

If our interpretation of the functional significance of the C_4 pathway in the biochemical context is correct, namely that it is a mechanism to overcome the deficiencies of RuP_2 carboxylase in the present atmosphere, then, as BJÖRKMAN (1975, 1976) has pointed out, we can expect that the C_4 pathway might confer an advantage in terms of photosynthetic production under any conditions which tend to deplete intercellular CO_2 partial pressure in leaves. Other things being equal, we have no reason to suppose that possession of C_4 photosynthesis should be a liability in conditions in which CO_2 partial pressure was not limiting. By the same reasoning, if internal CO_2 generation in the light in photorespiration and CAM are important in protection against photoinhibition, we would expect these processes to confer an advantage in terms of stability of the photosynthetic apparatus under conditions which lead to CO_2 deprivation (OSMOND et al. 1980a). Investigations of a wide range of physiological responses of photosynthesis under laboratory and field conditions tend to confirm these hypotheses.

15.3.1 Comparative Water Economy of Different Photosynthetic Pathways

Since water vapor shares the same stomatal diffusion pathway as CO_2, water is inevitably lost when stomata open and CO_2 is absorbed. As discussed in Chapter 7, this Volume, stomatal conductance appears to adjust in response to the assimilatory capacity of the mesophyll tissue. That is, other things being equal, stomata open to the extent required to provide CO_2 at rates sufficient to meet the CO_2 fixation requirements of the metabolic pathway (WONG SC et al. 1979).

The CO_2-concentrating function of C_4 photosynthesis leads to steady-state CO_2 concentrations of about 100 μbar in mesophyll cells (Fig. 15.8) and about 2,000–3,000 μbar in bundle-sheath cells (HATCH and OSMOND 1976; WONG SC et al. 1979). As a consequence, CO_2 fixation in C_4 plants is practically CO_2-saturated in the normal atmosphere. On the other hand, C_3 plants in the normal atmosphere function with intercellular CO_2 concentrations in the vicinity of 220 μbar (WONG SC et al. 1979) and under these conditions CO_2 fixation rate is somewhat short of saturation with respect to CO_2, especially in high-N plants

Fig. 15.8. Photosynthetic CO_2 response curves of a C_3 plant (*Gossypium*) and a C_4 plant (*Zea*) grown at low and high levels of nitrogen nutrition. (Unpublished data, courtesy of S.C. WONG)

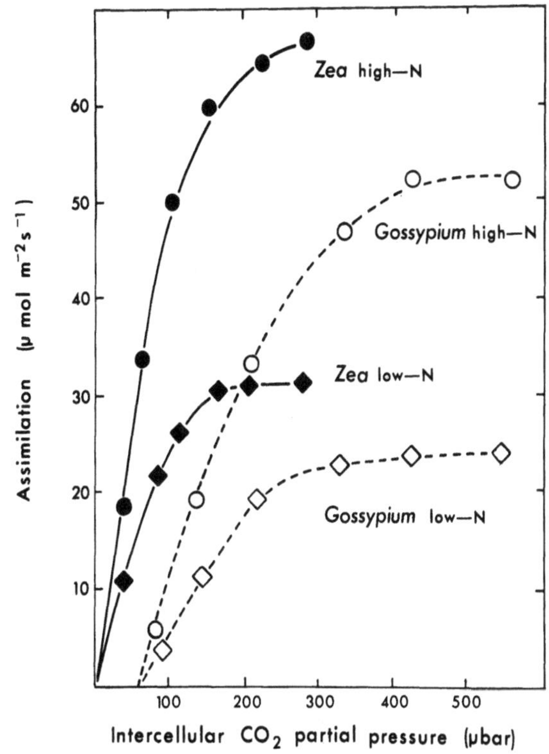

(Fig. 15.8). These marked differences in the operational intercellular CO_2 concentrations in C_3 and C_4 plants are achieved by maintenance of somewhat lower stomatal conductance in C_4 plants. Under exactly similar conditions of leaf temperature and leaf-air vapor pressure difference, the lower stomatal conductance of C_4 plants is responsible for substantially lower transpiration (E) from leaves of C_4 plants than from leaves of C_3 plants. Because the CO_2-concentrating mechanism of C_4 plants overcomes both O_2 inhibition of carboxylase and depresses oxygenase, assimilation (A) of C_4 plants in air is often greater than in C_3 plants. Consequently the transpiration ratio E/A is greater in C_3 plants than in C_4 plants: C_4 plants have a high water use efficiency (see Chaps. 7 and 17, this Vol.).

When C_3 and C_4 plants are compared under exactly similar conditions this distinction between the water economy is maintained at different levels of organization (LUDLOW and WILSON 1972), and it also persists over long growth periods (Table 15.4). However, these observations should not be taken to imply that the lower transpiration ratios of C_4 photosynthesis necessarily result in more water-efficient carbon acquisition under natural conditions. The ratio E/A is markedly dependent on the water vapor difference and hence leaf and air temperature during growth period. A C_3 plant growing in the cooler part of the year (a warm desert annual for example) could have a lower transpiration ratio than a C_4 plant which grows in the hot season.

The importance of taking seasonal activity patterns into account in comparative studies of the water economy of C_3 and C_4 plants is well illustrated in

Table 15.4. Transpiration ratio of C_3 and C_4 plants, data calculated from LUDLOW (1976), assuming 0.65 g dry wt = 1 g CO_2

Tissue	Transpiration ratio ($mol\ H_2O\ mol^{-1}\ CO_2$)	
	C_4 Pasture grasses	C_3 Legumes
Leaf[a]	83	191
Whole plant	338	662
Sward	536	1165
Long term whole plant[b]	512	1045

[a] CAM plants, dark (VPD 10 mbar) 50 mol H_2O mol^{-1} CO_2; light (VPD 10 mbar) 100 mol H_2O mol^{-1} CO_2; light (VPD 25 mbar) 250 mol H_2O mol^{-1} CO_2

[b] calculated as above from SCHANTZ and PIEMEISEL (1927)

experiments with the Chenopodiaceae *Atriplex confertifolia* (C_4) and *Ceratoides lanata* (C_3) in the cool Great Basin Desert of Utah. CALDWELL et al. (1977b) found that the daily transpiration ratios of these two perennial plants of comparable life form were not markedly different during the period of maximum photosynthesis, in the cool season, but that during the warm dry summer months the C_4 plant showed a considerably superior transpiration ratio. Integrating over the whole spring and summer seasons, and taking 0.65 g dry weight g^{-1} CO_2, the average annual transpiration ratio for the C_4 species was 882 mol H_2O mol^{-1} CO_2, and for the C_3 species was 1306 mol H_2O mol^{-1} CO_2. Even though the C_3 plant completed most of its growth in the cool season with minimum water vapor pressure difference, it continued to lose water without fixing carbon in the hot season. The C_4 plant continued to acquire carbon at an only slightly increased water cost, presumably because it was able to maintain better cellular water relations as a result of salt uptake and osmotic adjustment.

The water economy of CAM plants is superior to that of C_3 and C_4 plants, principally because CAM plants assimilate the bulk of their carbon at night when leaf temperature and water vapor pressure difference is low. Effectively, the temporal considerations, important in long-term comparisons of the water economy of C_3 and C_4 plants, apply to CAM plants on a day/night basis. CO_2 fixation in the dark by means of high affinity PEP carboxylase has an intrinsically high water use efficiency because CO_2 fixation proceeds at maximal rates at lower CO_2 partial pressure, just as it does in C_4 plants (cf. Figs. 15.8, 15.9). At high plant water potentials CAM plants may also fix external CO_2 via RuP_2 carboxylase and the PCRC after malic acid has been consumed. Assimilation in the light saturates at higher intercellular CO_2 partial pressures which can only be sustained by high stomatal conductances. However, the difference between the transpiration ratio in the dark and light under comparable conditions is not as great as the difference between C_3 and C_4 plants. In *Kalanchoe pinnata*, for example, at 20 °C and 9 mbar leaf-air vapor pressure difference,

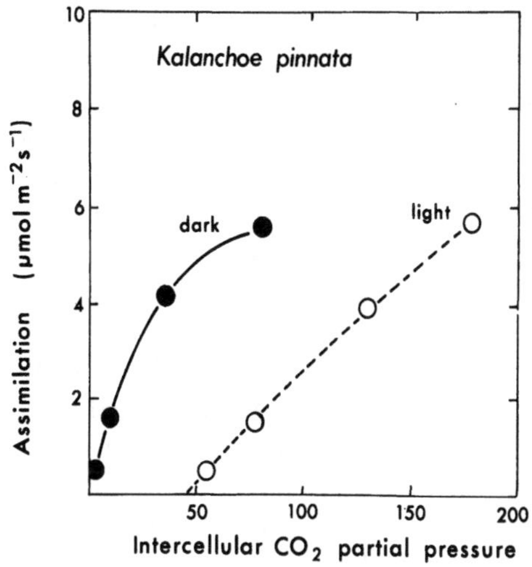

Fig. 15.9. Photosynthetic CO_2 response curves in the dark and light for the CAM plant *Kalanchoe pinnata*. (After WINTER 1980)

the transpiration ratio in the dark is 90 mol H_2O mol^{-1} CO_2, and in the light it is 130 mol H_2O mol^{-1} CO_2 (WINTER 1980). However, at the water pressure differences commonly encountered in the dark (10 mbar) the transpiration ratio of *Opuntia* is about 49 mol H_2O mol^{-1} CO_2 and at the highest water vapor pressure differences prevailing in the light (20–25 mbar) this ratio is four to six times larger (OSMOND et al. 1979b).

The stomatal frequency of CAM plants (≈2,500 stomata cm^{-2}) is about ten times lower than in C_3 and C_4 plants (≈20,000 stomata cm^{-2}) (BLACK et al. 1976; KLUGE and TING 1978) and this presumably contributes to the extraordinarily low stomatal conductance of CAM plants. Maximum rates of dark CO_2 fixation in *K. pinnata* are attained with stomatal conductance of <0.1 mol m^{-2} s^{-1}, conductances which are as low as those of many C_3 plants in the dark with so-called "closed" stomata.

In C_3 and C_4 plants which are capable of maintaining turgor at low tissue water potentials, the differences in transpiration ratio are preserved during water stress. Studies of *Atriplex hymenelytra* (C_4) and *Larrea divaricata* (C_3) in Death Valley by BJÖRKMAN and his colleagues (reviewed by OSMOND et al. 1980b) show that the intrinsic water use efficiency (assimilation/stomatal conductance) actually increases at low tissue water potentials in hot dry seasons, thereby compensating for the increased water vapor difference during the dry season. As a result, the daily transpiration ratios in both C_3 and C_4 plants remained relatively unchanged as plant water potential declined. Another consequence of this response is that intercellular CO_2 partial pressure decreases. In plants of *Cicer arietinum* which were repeatedly water-stressed during growth, the intercellular CO_2 partial pressure declined from approximately 230 μbar to 100–150 μbar (K. WINTER unpublished).

There is no a priori reason to assume that the less expensive water economy of C_4 plants makes them any more tolerant of low tissue water potentials (LUDLOW 1976). In the absence of competing C_3 plants, the less expensive water economy of C_4 plants at most may enable them to conserve limited water resources. However, long-term studies of CO_2 fixation by plants in the

Negev Desert show that with advancing water stress, the C_4 plant *Hammada scoparia* maintained higher maximum photosynthetic rates and higher daily carbon balance, relative to an irrigated control, than did C_3 plants (SCHULZE et al. 1980a, b). These plants never showed their full photosynthetic potential under arid conditions, even when irrigated, and maximum photosynthetic rate was more sensitive than daily carbon gain. Although the lower water use efficiency of C_4 photosynthesis might contribute to the superior performance of *Hammada,* factors such as osmotic adjustment, morphology, and life form remain to be evaluated.

The potential benefits of assimilating carbon at low water cost in CAM plants are curtailed by the limited capacity for CO_2 fixation in the dark. Thus it is unlikely that the low water cost of CO_2 fixation in the dark can confer any advantage in terms of productivity in a competitive situation. Rather, the low water cost of dark CO_2 fixation enables these plants to survive as functional photosynthetic systems in arid habitats. A potential advantage of CAM is that it may allow some additional carbon assimilation in the night at low water cost (but high energy cost) after soil water potential drops below that of the plant and hence ceases to be available. The roots of cacti become nonfunctional at low soil water potential and water loss from the plant to the soil is presumably prevented by suberization. When soil water becomes available again, new roots are formed (KAUSCH 1965) and water uptake commences.

The large water content, low surface-to-volume ratio, and the low stomatal frequency per unit surface area of succulent plants prolong the period in which nocturnal stomatal opening, water loss, and CO_2 uptake is possible. NOBEL (1976, 1977) estimated that this buffering effect, which is the sum of several properties of succulent CAM plants, allowed an extra 8 days nocturnal uptake in *Agave* after soil water potential dropped below that of the plant and an extra 40 days in *Ferocactus.* Thus the functional significance of this combination of high tissue water reservoir and low water cost of carbon assimilation must be assessed against the frequency and duration of available water and total net carbon gain in these habitats. Clearly, it may be the most important factor in carbon gain in the desert habitats of *Agave* and *Ferocactus,* but in the regions of Australia invaded by *Opuntia,* the frequency of summer rainfall is such that prolonged periods of CO_2 fixation in the light and of maximum CO_2 fixation in the dark are observed (OSMOND et al. 1979a) and the plants may be very productive. In this case, the buffering effect of succulence is probably unimportant in most seasons.

In most CAM plants stomata remain closed if tissue water potentials decline much below -10 bar. Although recycling of respiratory CO_2 continues, questions of water economy do not then arise. There are exceptions, however, and in a number of extreme habitats stomatal closure in the light period is not completely effective in preventing water loss (PATTEN and DINGER 1969; DESPAIN et al. 1970; LANGE et al. 1975). MEINZER and RUNDEL (1973) observed a similar net loss of CO_2 in the light in *Echeveria pumila,* and presumably for this reason found no correlation between growth and maximum dark CO_2 fixation rate. There was an inverse correlation between growth and transpiration ratio during dark CO_2 fixation (based on short-term measurements). However, the slow-

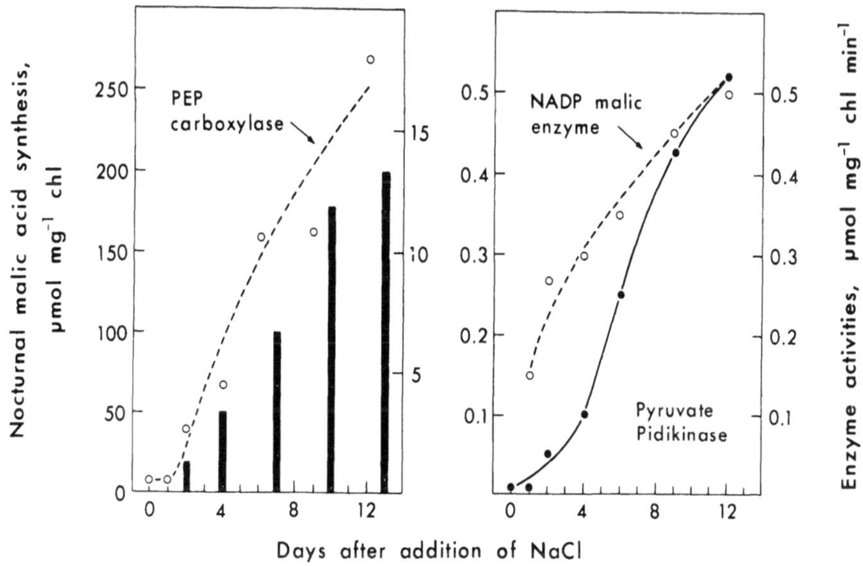

Fig. 15.10. Induction of CAM in *M. crystallinum* in response to salinity treatment in solution culture (commencing day 1), showing changes in nocturnal malic acid synthesis and in the activity of three enzymes required for acid synthesis (PEP carboxylase), deacidification (NADP malic enzyme) and gluconeogenesis (pyruvate, Pi dikinase). (Unpublished data, J.A.M. HOLTUM, K. WINTER)

growing plants with high transpiration ratio in the dark were those which showed a high proportion of total transpiration in the light period and were presumably those with the greatest losses of dark-fixed carbon in the light period. It should also be noted that nonphotosynthetic tissues were also included in some of these experiments and that respiratory CO_2 from such tissues might complicate these interpretations.

The identification of an important new class of plants which only engage in CAM in response to water (salinity) stress (WINTER and VON WILLERT 1972; WINTER and LÜTTGE 1976) holds much promise for establishing functional relationships. These "inducible" CAM plants, predominantly succulent halophytes such as *Mesembryanthemum crystallinum*, show conventional C_3 photosynthesis until subject to water stress, whereupon the biochemical machinery of CAM is induced (Fig. 15.10). The total daily carbon gain of these plants declines as CAM is induced, but the overall transpiration ratio increases because a proportion of carbon gain is transferred to the dark (WINTER 1979b). Obviously, the relationships between water economy and growth of CAM plants are complex and we are not yet in a position to speculate further on the functional significance of the potentially low transpiration ratio associated with dark CO_2 fixation.

15.3.2 Comparative Light Requirements of Different Photosynthetic Pathways

The difference between the light response curves for CO_2 fixation in normal air was one of the most obvious features of early physiological studies with

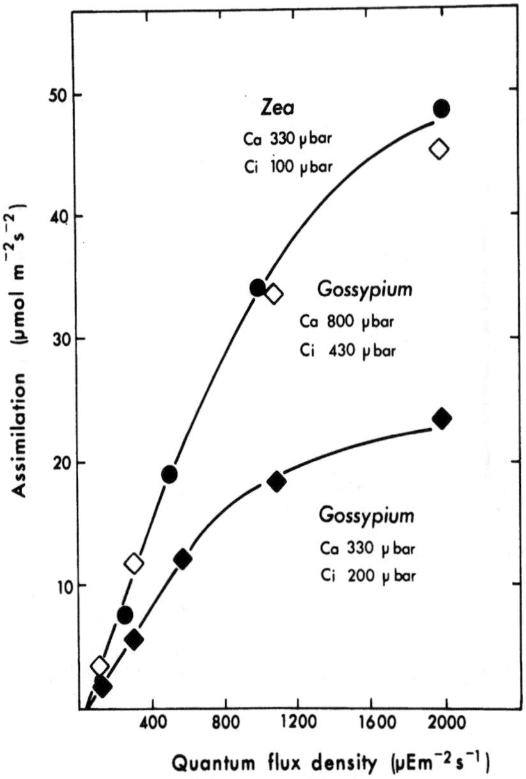

Fig. 15.11. Light response curve for photosynthetic CO_2 fixation in a C_4 plant (*Zea*) in normal air (●) and for a C_3 plant (*Gossypium*) in normal air (◆) or at high CO_2 partial pressure (◇). (Unpublished data, courtesy of S.C. WONG)

C_3 and C_4 plants (BLACK 1971; see Chap. 3, Vol. 12 A). Under otherwise identical conditions, this difference arises because the leaves of C_4 plants assimilate carbon in bundle-sheath cells at near saturating intercellular CO_2 partial pressure when the leaf is in normal air. On the other hand, C_3 plants function at only about half CO_2 saturation, and if the light responses of C_3 and C_4 photosynthesis are examined under CO_2 saturation or at low O_2 partial pressure, then the difference in light saturation properties is abolished (Fig. 15.11). The potential advantage of C_4 plants in photosynthetic productivity at high light intensities is thus a result of the CO_2 concentrating mechanism (BJÖRKMAN 1975, 1976). These differences are also modified by N status, and C_3 plants with elevated RuP_2 carboxylase activity as a result of high N nutrition do not saturate until full sunlight.

The energy costs of biochemical reactions in C_4 plants (5 ATP and 2 NADPH mol^{-1} CO_2 fixed) are greater than those in C_3 plants in the absence of photorespiration (3 ATP and 2 NADPH mol^{-1} CO_2 fixed). However, in 330 μbar CO_2 and 210 mbar O_2 the diversion of carbon by the oxygenation of RuP_2 and its recovery by the PCOP raises the energy cost of C_3 photosynthesis (about 5 ATP and 3 NADPH mol^{-1} CO_2 fixed). These energy costs are reflected in the quantum yields of leaf photosynthesis measured under light-limiting conditions. The intrinsic quantum yield of C_4 plants is 0.056 mol CO_2 $einstein^{-1}$ in air, and is independent of temperature and O_2 concentration. In C_3 plants the intrinsic quantum yield is 0.080 in low O_2. Because of the variable cost of carbon flux through the PCOP quantum yield is a function of temperature and O_2 concentration (EHLERINGER and BJÖRKMAN 1977). In normal air, between 25° and 30 °C, the quantum yield of C_3 and C_4 plants is almost identical, as expected on the basis of the carbon biochemistry (see Chap. 3, Vol. 12 A).

The light responses of CO_2 assimilation in CAM plants are complex and have yet to be studied in detail. Light intensity in the preceding light period determines the extent of malic acid synthesis (KAPLAN et al. 1977; NOBEL and HARTSOCK 1978), presumably by controlling the level of glucan which can be mobilized to provide PEP for CO_2 fixation (OSMOND and HOLTUM 1981). Light intensity also controls the rate of deacidification during the early light period in CAM plants and in *Opuntia inermis,* the rate of deacidification continued to increase until full sunlight intensity (OSMOND et al. 1979b). This response is reminiscent of that in C_4 plants and presumably has a similar explanation in that photosynthesis is CO_2-saturated during deacidification.In the late afternoon net CO_2 fixation appears to be saturated at low light intensities (ALLAWAY et al. 1974; OSMOND et al. 1979b) as one would expect of C_3 plants, but with high N status, light saturation shifts to near full sunlight (K. WINTER unpublished).

There are few data on quantum yields of CAM plants, and these measurements are prone to complications. If the light period is extended to facilitate steady-state measurements, CAM plants switch to dark CO_2 fixation during experiments when light level is lowered to the quantum yield range (OSMOND et al. 1979b). SPALDING et al. (1980) estimated the quantum requirement during deacidification in *Sedum* to be 0.065 mol CO_2 einstein^{-1}, which is intermediate between that of C_4 plants and C_3 plants at CO_2 saturation. During CO_2 fixation in the light in air or in air with 2% O_2 the quantum yield was 0.024 and 0.053 mol CO_2 einstein^{-1} respectively. These low quantum yields may imply that reactions in addition to the carboxylation of RuP_2 and the PCOP are involved. For example, a "futile cycle" of CO_2 fixation via PEP-carboxylase and subsequent decarboxylation of malic acid in the light might be involved. Alternatively, the optical properties of the thick leaves of CAM plants may differ from those of C_3 and C_4 plants and cause a greater proportion of incident energy to be dissipated by nonphotosynthetic processes. NOBEL's (1980) analysis of light interception by stem succulents allows no generalization about the significance of differing morphologies. The overall quantum yields of CAM plants are determined by environmental conditions which regulate the extent of CO_2 fixation in the light and dark. NOBEL and HARTSOCK (1978) studied the effect of previous light period on dark malic acid synthesis in *Agave* which did not engage in net CO_2 fixation in the light. They calculated an overall quantum yield for net CO_2 fixation in this CAM plant of only 0.015 mol CO_2 einstein^{-1}.

There have been few attempts to determine the functional significance of the differences in the light responses of C_3 and C_4 plants. Light intensity and temperature commonly co-vary in open habitats, so interpretation of experiments under field conditions is complex. In a comparison of two cool-tolerant *Atriplex* species, *A. sabulosa* (C_4) and *A. glabriuscula* (C_3), grown at 16 °C the quantum yield of the C_3 plant was greater than that of the C_4 plant, but then even at this low temperature, the C_4 plant had a higher light-saturated rate of photosynthesis (reviewed in OSMOND et al. 1980b). However, the measured growth and calculated total carbon gain of the two species was about the same, because the greater quantum yield of *A. glabriuscula* compensated for its lower rate of CO_2 fixation at light saturation. A similar comparison was made between *Tidestromia oblongifolia* (C_4) and *Larrea divaricata* (C_3), two heat-tolerant species from desert habitats, when grown at 40 °C. The light response curves at this temperature showed that the C_3 species was inferior to the C_4 species both in terms of lower quantum yield and lower light saturated CO_2 fixation rate. Both the superior light-saturated CO_2 fixation rate and quantum yield in the C_4 plant at high temperature can be ascribed to the CO_2 concentrating function of the biochemical reactions.

15.3.3 Comparative Temperature Responses of Different Photosynthetic Pathways

The thermal stability of the photosynthetic apparatus, both at high and low temperatures, is determined by a variety of properties, none of which is unique to C_3, C_4, or CAM pathways of photosynthetic CO_2 fixation. These properties have been discussed in detail by BERRY and BJÖRKMAN (1980) and in Chaps. 10, 12, and 14, Volume 12 A, and for our purposes it is sufficient to note that they are primarily concerned with temperature responses of the photochemical reactions of photosynthesis, which in turn control the supply of energy to, and the activation of, different steps in the carbon assimilation pathways. In high light environments CO_2 fixation by C_4 pathway plants at higher temperatures is superior to that of C_3 plants, because the C_4 plants operate near to CO_2 saturation. In C_3 plants, oxygenase activity increases relative to carboxylase at high temperatures and photorespiration is greater. If the species being compared have similar intrinsic thermal stabilities, then this difference in response of C_3 and C_4 plants at high temperatures can be largely abolished by increasing the CO_2 partial pressure to levels which are saturating for the C_3 plant, or by reducing O_2 concentration. At lower temperatures the differences between C_3 and C_4 plants in rates of CO_2 fixation largely disappear. In cold, low-light environments the lower quantum requirement of C_3 plants may sometimes allow the daily carbon gain to exceed that of C_4 plants. Under such conditions the higher energy costs of C_4 photosynthesis may become a liability (OSMOND et al. 1980b).

Although a large number of C_4 plants, particularly those of the NADP malic enzyme type, are unable to function effectively at low temperatures, CO_2 fixation in many other C_4 plants is not impaired by low temperature (BJÖRKMAN et al. 1976; LONG et al. 1975). CALDWELL et al. (1977a) showed that low growth temperature had little effect on the pulse-chase kinetics of $^{14}CO_2$ fixation in cool and warm desert C_4 *Atriplex* spp. Substantial changes in leaf morphology were noted in plants grown at low temperature, and substantial deposits of starch accumulated in both mesophyll and bundle-sheath cells, but by all criteria the carbon assimilation pathway was unaltered.

Impaired photosynthesis in C_4 plants which are sensitive to low temperature, such as the tropical grasses, is manifest in chlorosis, abnormal chloroplast structure, impaired photochemistry and subsequent severe disruption of the CO_2 fixation pathway (LUDLOW and WILSON 1971; TAYLOR and ROWLEY 1971; SLACK et al. 1974). It is conceivable that some of these changes are light-dependent following the chilling injury and the consequences of photoinhibition which follows whenever the capacity of illuminated chloroplasts to dispose of photochemical energy in an orderly fashion is impaired (OSMOND et al. 1980a; POWLES and CRITCHLEY 1980). Chilling injury impairs the capacity of chloroplasts to generate ATP and $NADPH_2$ and the excess excitation energy absorbed by the choroplasts may be dissipated in destructive oxidative processes, analogous to those responsible for photoinhibition following illumination of leaves in CO_2 free air under conditions which prevent photorespiration. Interestingly, when leaves of freezing-tolerant evergreens are exposed to high light intensity at low temperature, they too suffer impairment of photochemical reactions (MARTIN et al. 1978; ÖQUIST et al. 1978), presumably because at low temperature, slow carbon turnover limits the orderly dissipation of photochemical products (OSMOND 1981).

The temperature responses of CO_2 fixation in CAM plants have been widely discussed, but in most studies stomatal regulation of CO_2 supply has not been taken into account. Stomata of CAM plants are responsive to vapor pressure difference (OSMOND et al. 1979b; LANGE and MEDINA 1979) and this changes rapidly with leaf and air temperature. NOBEL (1976, 1977) and NOBEL and HARTSOCK (1978) showed the declining dark CO_2 fixation at high temperature was accompanied by stomatal closure in three hot desert CAM plants. In *Opuntia inermis* night temperature between 7° and 27 °C had little effect on malic acid synthesis, or on rate of CO_2 fixation in short-term measurements, until stomata began to close at higher temperature (OSMOND et al. 1979a, b). When leaf-air water vapor pressure difference was kept constant at 15° and 25 °C, stomata of *Kalanchoe pinnata* were slightly more open at 25 °C, but net carbon gain during the night was reduced by only 20% (K. WINTER unpublished). Furthermore, in many studies the capacity for CO_2 assimilation at high night temperature may be under-estimated by the inclusion of nonphotosynthetic, respiring tissues in the gas exchange curvette. It is not clear whether stomatal closure contributed to the sharp optimum (10 °C) for malic acid synthesis in *Echeveria columbiana* (MEDINA and DELGADO 1976). MEDINA and OSMOND (1981) established that the 15 °C optimum for dark CO_2 fixation in *K. daigremontiana* is not due to stomatal control of CO_2 uptake. Moreover, they showed that the reduction in dark CO_2 uptake at high temperatures was relatively lower in plants grown at high night temperatures and proposed that these temperature responses may be mediated by effects on transport of malate to and from the vacuole.

Comprehensive analyses of the effects of day and night temperature on the gas exchange patterns of CAM plants have been published by NEALES (1973a, b) and by LANGE and ZUBER (1980). The latter authors confirmed the observations of WAGNER and LARCHER (1981) who showed that *Sempervivum montanum* does not deacidify at low temperatures in the light and has instead a net CO_2 exchange pattern similar to that of non-CAM plants. At higher temperatures some species such as *Caralluma negevensis* show substantial loss of CO_2 in the light (LANGE et al. 1979). There are several ways in which the temperature in the light period influences net CO_2 fixation of CAM plants. High temperature, at high light intensity, accelerates the rate of deacidification and may therefore increase the possibility of additional CO_2 fixation in the light. In some experiments, higher day temperatures increase the capacity for subsequent dark CO_2 fixation (GERWICK and WILLIAMS 1978; cf. NOBEL and HARTSOCK 1978).

15.3.4 Nutrition and Salinity in Relation to Different Photosynthetic Pathways

A wide range of photosynthetic responses to nutrition has been recorded but few of these have been investigated in such a way that the functionally significant relationships can be discerned. Some, such as the early stages of K^+ deficiency in maize (PEASLEE and MOSS 1968) can be attributed almost entirely to declining stomatal conductance, which is not surprising in view of the importance of K^+ fluxes in stomatal opening in many species (cf. RASCHKE 1975). TERRY (1979) reported that limiting Fe^{2+} supply controls the number of photosynthetic units per unit leaf area, and hence photosynthetic capacity.

The best-documented responses by far are those to nitrogen nutrition, in which declining photosynthesis in response to declining nitrogen is closely associated with the level of the major leaf protein, RuP_2 carboxylase-oxygenase (RYLE and HESKETH 1969; MEDINA 1970, 1971; NEVINS and LOOMIS 1970). As pointed out by BJÖRKMAN et al. (1976) and BROWN (1978), C_4 plants are likely to be very much more efficient in the use of nitrogen than C_3 plants. Because the CO_2 concentrating mechanism allows RuP_2 carboxylase in bundle-sheath cells of C_4 plants to function near CO_2 saturation, higher rates of photosynthesis can be sustained by lower concentrations of this enzyme in C_4 plants. In C_3 *Atriplex* spp. grown with adequate nutrition this enzyme comprises 41%–56% of soluble leaf protein whereas in a C_4 *Atriplex* spp. grown under the same conditions it comprises only 19%–21%. In *Tidestromia*, a C_4 plant which grows rapidly at 45 °C, this enzyme comprises only 4%–8% of soluble leaf protein. The high nitrogen use efficiency of C_4 plants may be of particular significance in hot arid regions with soils of low N status.

Similar arguments have been advanced by K. WINTER and S.C. WONG (unpublished) for CAM plants which may also occupy low nitrogen arid habitats or ephiphytic habitats in rainforests. Dark CO_2 fixation via PEP carboxylase, followed by the reassimilation of CO_2 at high internal concentrations, should require lower investment in RuP_2 carboxylase than conventional C_3 photosynthesis. *Kalanchoë pinnata* was grown at three levels of nitrogen nutrition in the expectation that low nitrogen plants might engage in a greater proportion of total CO_2 fixation in the dark than high N plants. Although this expectation was realized in gas exchange studies and practically no net CO_2 fixation in the late light period was observed in low-N plants, δ^{13} values, which should be an integrated measure of the total carbon acquisition in light and dark, did not confirm this trend during long-term growth.

Photosynthetic responses to salinity have not been extensively studied. It was recognized that some halophytes show a requirement for Na^+ as a micronutrient ion (BROWNELL and WOOD 1957) and later it was shown that the requirement for this micronutrient appears to be restricted to C_4 plants (BROWNELL and CROSSLAND 1972). HUBER and SANKHLA (1976) reviewed many studies which have claimed to show that the level of NaCl supply may modify the pathway of photosynthetic CO_2 assimilation. Although salinity may often influence labeling patterns during photosynthetic assimilation of $^{14}CO_2$ in C_3 or C_4 plants, or cause shifts in the $\delta^{13}C$ value of C_3 halophytes (GUY et al. 1980), there is no evidence that the carbon assimilation pathway is essentially changed. Sodium-deficient C_4 plants remain functional C_4 plants in all respects (BOAG and BROWNELL 1979). The concept of balance between C_3 and C_4 pathways of carbon assimilation is itself probably erroneous. Even in so-called intermediate C_3/C_4 plants there is no convincing evidence, from $\delta^{13}C$ value for example, that a significant proportion of total carbon is assimilated by C_4 photosynthesis. RATHNAM and CHOLLET (1980) conclude that limited C_4 pathway activity in some so-called intermediate C_3/C_4 plants may effect some increase in CO_2 concentration in bundle-sheath cells and thereby account for the different O_2 sensitivity of CO_2 fixation, and of CO_2 compensation point, in these plants.

By and large, successful halophytes are those which are able to control NaCl accumulation to the shoot, use it for osmotic adjustment in the cell vacuole and prevent the interference of electrolytes in the normal functioning of leaf cells (OSMOND et al. 1980b). Salt-sensitive plants are by and large unable to achieve one or all of these functions. There is no reason to suspect that C_3, C_4, or CAM pathways of photosynthetic CO_2 fixation should facilitate the functionally important relationships. It could be argued that if NaCl transport to the shoot is largely dependent on transpiration, then shoots of C_4 plants with lower transpiration ratios may be better able to retain shoot salt load within manageable levels. However, to our knowledge there have been no comprehensive comparative analyses of the salt, water, and carbon economies of halophytes which bear on this speculation. Certainly the studies of WILLIAMS (reviewed in OSMOND et al. 1980b) show that salinity and low water potential in arid zone soils had similar effects on transpiration, growth, and transpiration ratios in C_3 and C_4 halophytes.

In most respects the effects of salinity on photosynthesis are mediated via water relations of the leaves and closely associated processes. It is not therefore surprising that the induction to CAM in succulent annuals such as *Mesembryanthemum crystallinum* in response to NaCl can also be brought about by any treatment which reduces water uptake, such as growth in nutrient solutions at low temperature or low O_2 partial pressure (WINTER 1979b). Although these plants are most commonly found in arid saline habitats which are seasonally arid, the change in CO_2 fixation pathway is a response to water stress, not to salinity per se. This change to CAM is functionally significant in that it allows net carbon gain and normal reproductive development to continue.

Some CAM plants show Na^+ deficiency symptoms when grown under conditions which maximize their dependence on dark CO_2 fixation (BROWNELL and CROSSLAND 1974). *Bryophyllum tubiflorum* grown without Na^+ may suffer strongly reduced growth, but still exhibits all features of CAM and thus the implication that Na^+ is a "prerequisite for the onset of CAM" (BLOOM 1979a) is misleading. It has also been suggested (LAETSCH 1974) that salinity-induced CAM is analogous to malate synthesis as a balance of cation surplus (OSMOND 1976). It has been further proposed that, in *M. crystallinum,* inorganic cations move between epidermis and mesophyll to balance the changing level of malate in the mesophyll during the day/night cycle (BLOOM 1979b). These analogies seem to be irrelevant, because dark CO_2 fixation leads to the accumulation of equivalent amounts of malate anions and protons (hydrogen malate = malic acid) in mesophyll cells of *M. crystallinum,* as it does in all other CAM plants. There are no suitable analogies for the bidirectional fluxes of hydrogen malate in CAM (LÜTTGE and BALL 1979); the process is evidently unique.

15.3.5 Summary

In the physiological context the different pathways of photosynthetic CO_2 fixation can be viewed as the foundation of diverse responses of C_3, C_4, and CAM plants to environment. The biochemical properties of the pathways are such that, in leaves of C_3 plants in the present atmosphere, photosynthesis proceeds at an intercellular CO_2 partial pressure of about 200 μbar. In C_4 plants the biochemical properties permit the same or higher rates of photosynthesis with an intercellular CO_2 partial pressure of about 100 μbar so that, at given leaf temperature and water vapor pressure difference between leaf and air, C_4 plants can fix CO_2 at much lower water cost than C_3 plants because they can function with lower stomatal conductance.

If adequate water is available for transpiration, and the thermal stability of the photosynthetic apparatus is suited to high temperature, then this potential advantage of C_4 photosynthesis is likely to be better expressed at high temperature. The biochemical properties of CAM plants, which permit C_4-like CO_2 fixation in the dark by PEP carboxylase, enable carbon assimilation to proceed at very low water cost, at very low stomatal conductance. However, the physiological constraints arising from the storage of malic acid in the vacuole limit the capacity for nocturnal carbon assimilation in these plants. Additional carbon assimilation in the light by CAM plants is closely controlled by plant water relations. The relative contributions of CO_2 fixation in the dark and the light are discussed in the next section. Some CAM plants display a distinct temperature optimum for nocturnal CO_2 fixation; others do not. The physiological bases for these responses are not clear.

The biochemical processes of C_4 and CAM plants have higher intrinsic energy costs and hence lower quantum yields than C_3 plants. However, because these processes allow C_4 plants to function near saturating intercellular CO_2 partial pressure, leaves of these plants tend to display nonsaturating light response curves for photosynthesis. The dependence of CAM on light intensity is a complex function of component processes separated in time, and the overall efficiency of light energy conversion is markedly dependent on stomatal responses to internal and external factors.

The functional relationship between nutrition and photosynthetic CO_2 fixation pathways is not yet well understood. Because the biochemical processes of C_4 and CAM pathways permit RuP_2 carboxylase to function at elevated intercellular CO_2 partial pressures, lower concentrations of this enzyme are needed to sustain high photosynthetic rates. Because this enzyme comprises the major leaf protein, both pathways may be considered to be nitrogen-efficient. In both C_4 and CAM pathways sodium has a specific but as yet undefined role.

15.4 The Ecological Context: Productivity and Survival in Different Habitats

The functional significance of different CO_2 fixation pathways in the ecological context will be displayed by natural selection to the extent that photosynthetic properties contribute to performance and survival of the whole plant in different habitats. The translation of carbon-assimilating processes into productivity of the whole plant is primarily determined by carbon allocation processes which are poorly understood. In the same environment, greater CO_2 assimilation rate on a leaf area basis of a C_4 plant may result in growth rates lower than those of a C_3 plant, if a greater proportion of the assimilated carbon is allocated to nonphotosynthetic structures in the C_4 plant. SLATYER (1970) showed that the initially higher net assimilation rate in C_4 *Atriplex spongiosa* declined to less than that of the C_3 *Atriplex hastata* after 3 weeks growth because the C_3 plant developed greater leaf area. On the other hand, survival of the organism may be determined by the allocation of assimilate between photosynthetic and reproductive structures, rather than on vegetative production.

Carbon allocation patterns are reflected in plant life forms, as discussed in Chapter 18, this Volume. With the exception of tree form *Euphorbia* spp. of Hawaii (PEARCY and TROUGHTON 1975), as well as Chenopodiaceae and Polygonaceae of Middle Asia (WINTER 1981), C_4 plants are shrubs and grasses. In some cool desert species, such as *Atriplex confertifolia*, root/shoot ratios of seven have been measured (CALDWELL and CAMP 1974). Most CAM plants are shrub-like with little biomass invested in roots and nonphotosynthetic tissues. Although some Cactaceae and Euphorbiaceae develop tree form, the stems and branches remain photosynthetic. In most Didieraceae small succulent leaves capable of CAM are borne on woody stems with only a thin chloroplast containing cortex (WINTER 1979a), and in some species, leaves with normal C_3 photosynthesis are borne on green stems which carry out CAM (*Frerea indica*, Asclepiadaceae: LANGE and ZUBER 1977; *Euphorbia trigona*, Euphorbiaceae: EDER et al. 1981). These C_3 leaves are dropped in response to water stress.

Although allocation pattern is unlikely to be related to rates of translocation, it is important to note that C_4 grasses generally have translocation rates about twice those of C_3 grasses, and translocation in C_4 plants is two to three times faster on a phloem area basis (LUSH and EVANS 1974). Moreover, C_4 plants generally translocate a greater proportion of recent photosynthate from leaves than do C_3 plants (HOFSTRA and NELSON 1969; LUSH and EVANS 1974; GALŁAHER et al. 1975). Photosynthate which is not exported in the light is stored as starch and mobilized in the dark. In the C_4 plants *Paspalum dilatatum* and *Panicum maximum,* but not in C_3 *Lolium temulentum,* the mobilization of stored starch is abruptly slowed by night temperatures below 10 °C (LUSH and EVANS 1974) and this may be responsible for the accumulation of starch in leaves of many C_4 plants grown at low temperatures (HILLIARD and WEST 1970; CALDWELL et al. 1977a).

15.4.1 Environmental Effects on Allocation of Assimilates

There have been few comparative studies of the effects of environmental factors on assimilate partitioning in plants differing in photosynthetic CO_2 fixation pathway. However, interactions in other experiments serve to emphasize the importance of assimilate partitioning in modulating carbon-assimilating capacity in terms of productivity and survival.

It is commonly observed that decreasing light intensity causes an increase in the proportion of assimilate retained for leaf growth. The extent to which this increase in photosynthetic surface compensates for the reduction in photosynthetic capacity at lower light intensities varies between species. In sun plants such as *Amaranthus spinosus* (C_4) and *Datura stramonium* (C_3) there was no compensation at all, and relative growth rate declined more than proportionally with light intensity (SINGH and GOPAL 1973). In shade plants, the compensation is more effective, so that maximum relative growth rate is achieved at a fraction of full sunlight. It is quite impossible to generalize these relationships with respect to photosynthetic pathway, and indeed SINGH and GOPAL describe both "sun" and "shade" response types in two varieties of *Portulaca oleracea* (C_4) which grow intermixed in open habitats. The broad leaf variety maintained higher dry weight and seed production at decreasing light intensity, whereas the narrow leaf variety declined in dry weight and showed a more than proportional decline in seed production over the same range.

When *Avena sativa* (C_3) and *Amaranthus edulis* (C_4), or when *Panicum maximum* (C_4) and *Atriplex hortensis* (C_3) were grown alone or in mixed cultures, root weight of all species increased with decreased water supply (Fig. 15.12). However, in the C_4 plants the ratio root/shoot decreased, whereas in the C_3 plants this ratio increased. When the same experiments were run under shade and full sunlight, growth of the C_4 plants was stimulated by high light and low water supply, whereas growth of C_3 plants was inhibited (Fig. 15.13). The increased growth of C_4 plants was associated with a decrease in the shoot/root ratio. Evidently the higher water use efficiency of C_4 plants enables them to maintain higher leaf area under drier conditions and hence to attain higher rates of photosynthesis per plant and higher growth. Further experiments with C_3 and C_4 plants of different life forms are needed to determine whether, in competitive situations, water spared by C_4 plants is expended by C_3 plants.

Fig. 15.12. Dry weight of the aboveground and belowground parts of 12-week-old *Avena sativa* (C_3) and *Panicum miliaceum* (C_4) plants cultivated in pure culture or under competitive conditions on garden soil with different levels of ground water. *Avena*, pure culture ●---●, mixed culture with *Panicum* o---o; *Panicum*, pure culture ●———●, mixed culture with *Avena* o———o. Aboveground parts reach the highest value under competitive conditions on the driest soil in the C_4 plant, *Panicum*, under competitive conditions on the wettest soil in the C_3 plant *Avena*. The yield of *Panicum* is generally higher in mixed culture with *Avena*. The yield of *Avena* on the other hand is higher in pure culture. There is a tendency for higher dry weight in the belowground parts in the driest soil. (After Öztürk et al. 1981)

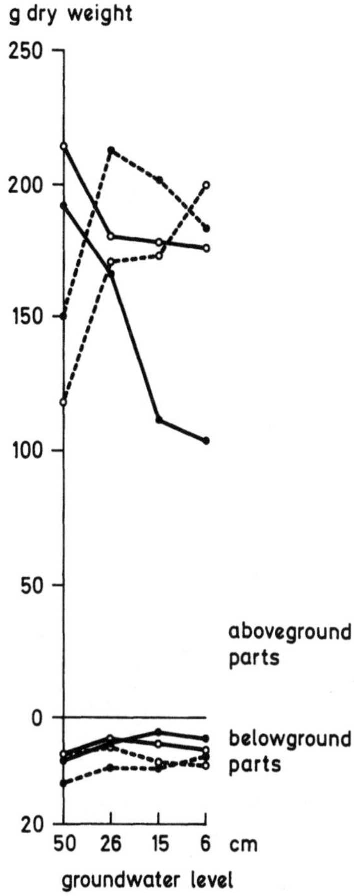

Pearcy et al. (1981) found limitation of water supply did not influence competitive interactions between C_4 *Amaranthus* and C_3 *Chenopodium*.

The effects of temperature on assimilate partitioning are likely to be very complex because leaf temperature is an important determinant of transpiration and hence of water acquisition by the root system. The vast difference between the root/shoot ratio of warm desert *Atriplex* spp. (C_4) in south eastern Australia (0.3) and of cool desert *A. confertifolia* (C_4) in Utah (7.0) may be related to the utilization of unpredictable superficial rainfall in Australia, and to the predictable winter recharge of the whole soil profile in Utah. In experiments with warm desert *Atriplex* spp. (C_4) in which only the air temperature was controlled, higher temperatures resulted in reduced allocation of carbon to root (Nicholls 1972) and increased total growth. In similar experiments with cool coastal *Atriplex* spp. (C_3) the same response was observed up to moderate temperatures, but at higher temperatures reduced growth was accompanied by an increase in the proportion of assimilates allocated to roots (Fig. 15.14). Osmond et al. (1980b) suggested that these responses to temperature may be generalized. The response of cool coastal *A. hortensis* (C_3) matches that of C_3 winter annuals studied by Davidson (1969a) and that of warm desert *A. argentea* (C_4) matches Davidson's observations with subtropical C_4 grasses.

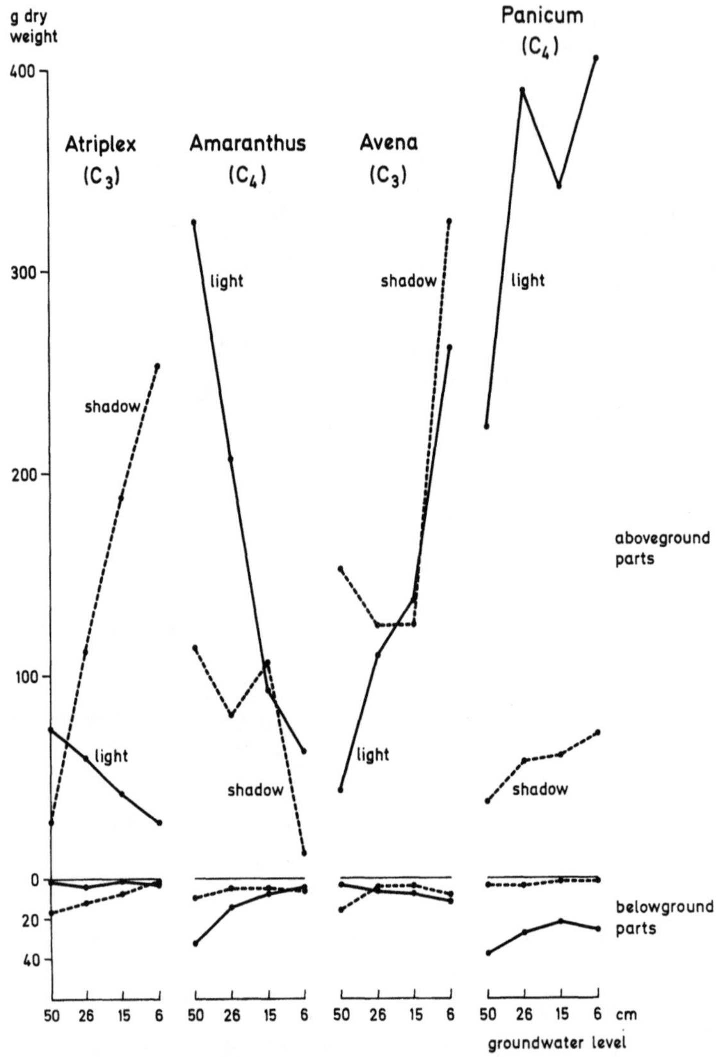

Fig. 15.13. Dry weight of the aboveground and belowground parts of 12-week-old plants of the grasses *Avena sativa* (C_3) and *Panicum miliaceum* (C_4) and of the herbaceous plants *Atriplex hortensis* (C_3) and *Amaranthus retroflexus* (C_4), cultivated in mixed culture of all four species on garden soil with different levels of ground water. One plot was grown in full light (——) and the other was shadowed to about 50% of the light intensity by a screen (---). The dry weight of the aboveground parts of C_4 plants was strongly promoted under full light conditions and – at least relatively to the C_3 plants – under dry soil conditions. The absolute dry weight of the belowground parts generally increases with decreasing soil moisture, while the shoot/root ratio shows variable values. (After ÖZTÜRK et al. 1981)

Clearly, these responses in well-irrigated plants indicate that assimilate partitioning in plants which are capable of CO_2 fixation at high temperatures is such as to reinforce their photosynthetic potential by maintaining the minimal allocation of photosynthate to roots commensurate with the water requirements of

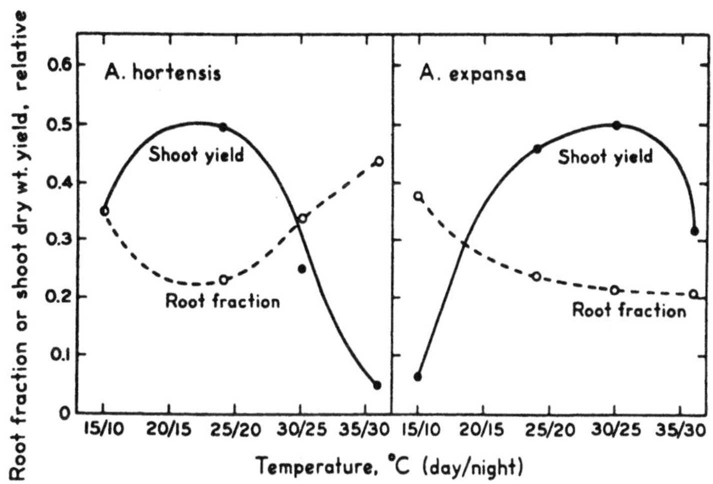

Fig. 15.14. The effect of day/night temperature regime on assimilate partitioning in a cool temperate C_3 plant (*Atriplex hortensis*) and a warm desert C_4 plant (*Atriplex expansa*) (reproduced from OSMOND et al. 1980b)

the plant. Even if adequate leaf water relations and CO_2 fixation capacity are maintained, the allocation of assimilate to roots rather than shoots will decrease productivity and growth. Whether the higher water use efficiency of C_4 plants enables them to maintain lower root/shoot ratios and hence higher productivity in water-limited environments remains to be determined.

DAVIDSON (1969b) developed a similar hypothesis with respect to the effects of nutrients on assimilate partitioning. He observed that the growth of shoots decreased, relative to the growth of roots during nutrient deficiency, and proposed that photosynthate was partitioned to roots in increase proportion to their functional activity, whether it be the absorption of water or nutrients. Because C_4 plants can achieve high rates of CO_2 fixation at lower leaf nitrogen concentrations, we might speculate that they would show differences in assimilate partitioning with respect to nutrient supply. To our knowledge no comparisons between C_3 and C_4 plants have been made.

The partitioning of assimilates between roots and shoots also changes in response to salinity, and the response is similar to the response to water stress. HOFFMAN et al. (1971) noted that depressed growth of cotton in response to salinity or water deficit was associated with an increase in the root/shoot ratio. In comparisons of halophytic C_3 and C_4 *Atriplex* spp. stimulation of growth in response to low levels of salinity was associated with a decrease in the root/shoot ratio as photosynthate was used to develop further photosynthetic tissue. At higher levels of salinity growth was depressed and the root/shoot ratio increased (OSMOND et al. 1980b). Salinity had little effect on the rate of CO_2 fixation of these halophytes over the range of salt concentrations tested, so that the growth response was controlled principally by the allocation of assimilates.

15.4.2 Performance (Productivity) of Plants with Different Photosynthetic Pathways in Relation to Environment

The range of maximum rates of net CO_2 fixation in air by C_3 and C_4 plants overlaps because plants differ, among other things, in the temperature optimum for photosynthesis. Table 15.5 shows that some C_3 plants adapted to high temperatures are capable of CO_2 fixation at rates comparable with those of high temperature-adapted C_4 plants. Some C_4 plants, which are adapted to low temperatures, show maximum CO_2 fixation rates comparable with other low temperature adapted C_3 plants. Clearly most of the C_4 plants listed in Table 15.5 are adapted to higher temperatures than the C_3 plants. However, GIFFORD (1974) pointed out that although at the highest natural light intensities a crop of *Zea* (C_4) at 29 °C showed 1.5 times higher CO_2 fixation than *Triticum* (C_3) at 14°–17 °C (equal leaf area indices), this advantage disappeared below about 60% full sunlight. Attenuation of the higher potential productivity shown by individual leaves of C_4 plants at high light intensity arises because in dense stands light levels are usually well below the noon maximum for most of the day. Moreover, night time respiratory losses of carbon are likely to be greater in high temperature habitats occupied by C_4 plants.

There is abundant evidence that differences in photosynthetic CO_2 fixation as a function of temperature in individual leaves are translated into differences in growth of individual plants irrespective of whether they are C_3 or C_4 plants. The dry matter yields of *Atriplex* spp. from cool coastal and warm interior habitats were compared when plants were grown under full sunlight, with adequate water at 15 °C day/10 °C night and 36 °C day/31 °C night. All the plants of cool coastal origin, whether C_3 or C_4, grew only half to one third as rapidly at 36°/31 °C compared to 15 °/10 °C (OSMOND et al. 1980b). All the C_4 plants of warm interior habitats grew three to five times more rapidly at 36°/31 °C compared to 15°/10 °C (OSMOND et al. 1980b).

GIFFORD (1974) compared maximum crop growth rates of C_3 and C_4 plants and concluded that "there is no apparent difference between the best examples of the two groups when grown in their own preferred environments". However, reappraisal of these and other data sets led other authors (LOOMIS and GERAKIS 1975; MONTEITH 1978) to conclude that C_4 crop plants do indeed have higher maximum short-term crop growth rates (51–54 g m^{-2} day^{-1}) than C_3 crop plants (27–39 g m^{-2} day^{-1}). The higher crop growth rates which have been reported for some C_3 plants, which were responsible for indistinct separation of C_3 and C_4 groups in Gifford's data, seem to have been obtained from small plots in which substantial lateral interception of light evidently permitted higher growth rates. This is, of course, an academic distinction, which is relevant to the efficiency of crop production and of little concern in natural communities where lateral interception of light is likely to be significant, but in which other factors such as water and nutrition are likely to be limiting. More extensive analyses of productivity in different ecosystems are given in Volume 12 D of this series.

The higher vegetative productivity (dry matter production) of C_4 plants relative to C_3 plants is sometimes a product of a longer growing season. Esti-

Table 15.5. Maximum values of net CO_2 fixation in C_3 and C_4 plants. (Data from GIFFORD 1974; BJÖRKMAN et al. 1975, and MOONEY et al. 1976)

Pathway and species	Leaf temp. (°C)	Net CO_2 fixation (nmol cm^{-2} s^{-1})
C_3 plants		
Camissonia claviformis	30	5.9
Typha latifolia	25	4.3
Triticum boeoticum	21	4.5
Medicago sativa	–	5.4
Glycine max	29	2.7
Helianthus annuus	–	2.8
C_4 plants		
Tidestromia oblongifolia	40	5.0
Saccharum hybrid	25	6.4
Pennisetum typhoides	35	6.4
Zea mays	32	5.5
Sorghum sudanense	28	5.7
Atriplex sabulosa	16	2.8

mates of long-term productivity of C_3 and C_4 plants are summarized in Tables 15.6, 15.7. Although the range of values for annual productivity overlaps between the two groups, full-year tropical C_4 plants, such as *Pennisetum* and *Saccharum*, are more productive than full-year tropical C_3 plants such as *Manihot*. The average growth rate over the whole season is only about 40% of the maximum crop growth rates disputed above, but these longer-term studies tend to confirm that the higher productivity of C_4 plants is based on higher growth rates. The maintained superior productivity of vegetative C_4 crop plants at low latitudes (*Saccharum, Pennisetum*) is about twice that of *Zea* (Fig. 15.15), presumably because selection for seed production has been at the cost of vegetative production. In fact, the productivity of C_3 root crops such as *Manihot* and *Beta* is comparable with that of C_4 plants such as *Zea* and *Cynodon* over a wide range of latitudes in which these plants are grown.

CAM plants are not noted for their high productivity because CO_2 fixation in the dark is limited by availability of substrate carbon and the capacity to store malic acid. However, in some communities CAM plants can attain relatively high biomass, nearly all of which is photosynthetically active surface. Estimates of maximum biomass of 100–300 t ha^{-1} for *Opuntia* in parts of eastern Australia (DODD 1940) indicate the potential productivity of this persistent organism. OSMOND et al. (1979a) estimate that at this biomass, which corresponds to a leaf area index of about 12–15, productivity of 30–40 g m^{-2} day^{-1} could be attained. The pineapple (*Ananas sativus*) is frequently cultivated at leaf area indices of 10–12 and sustains crop growth rates of 20 g m^{-2} day^{-1} and a productivity over two years of 34 t ha^{-1} yr^{-1} (BARTHOLOMEW and KADZIMIN 1975). Cultivated *Opuntia* spp. have yielded approximately 40 t ha^{-1} yr^{-1} over 2 years and harvests of 15–30 t ha^{-1} yr^{-1} of nopales have been recorded

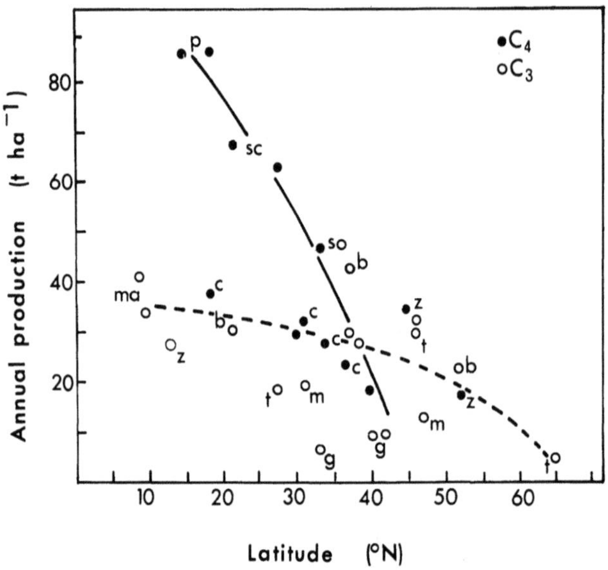

Fig. 15.15. Maximum short-term crop growth rates as a function of latitude. C_4 plants are shown by *closed symbols* and C_3 plants by *open symbols*. The crop species are identified as follows: C_4 plants, *p Pennisetum purpureum; sc Saccharum* sp.; *so Sorghum* sp.; C_3 plants, *ma Manihot esculenta; b Beta vulgaris; m Medicago* sp.; *g Glycine* max; *t Triticum vulgare.* (Redrawn from LOOMIS and GERAKIS 1975)

(cited by KLUGE and TING 1978; field weights corrected to 15% dry weight). Under natural rainfall conditions in the Sonoran Desert, productivity of *Agave* and *Ferocactus* is about only 8% of these values for *Ananas* and *Opuntia,* because in these habitats CAM plants engaged in net CO_2 fixation only for brief periods following rain.

CALDWELL (1975) observed that "although many agronomic species possessing C_4 metabolism are known for high yields ... data do not support this hypothesis for natural grazing ecosystems. Even when C_4 and C_3 species are cultured together or are found in similar microhabitats, the C_4 species is not necessarily more productive" (Table 15.7). The annual production and average growth rate of wild C_3 plants in habitats in which water is not likely to be limiting (emergent hydrophytes, old field in New Jersey) may equal or exceed that of C_3 crop plants (Table 15.6). However, in water-limited habitats such as the grasslands of USA and USSR, the long-term productivity of C_3 and C_4 plants appears to be comparable, and only 10%–25% of crop growth rates.

In North American grasslands, SIMS et al. (1978) and SIMS and SINGH (1978a) found that maximum short-term growth rates of cool season C_3 grasses at the Cottonwood site (3.7 to 6.5 g m^{-2} day^{-1}) were comparable with those of warm season C_4 grasses at the Osage site (4.8 to 6.2 g m^{-2} day^{-1}) and only twice as productive as a relatively arid site at Pawnee (2.4 to 4.4 g m^{-2} day^{-1}), the latter site was dominated by C_4 grasses and CAM plants. The annual productivity of these different sites is plotted as a function of annual precipitation in Fig. 15.16. There is no clear distinction between the productivity of C_3-

Table 15.6. Long-term productivity of crop plants. (Data from BOARDMAN 1978)

Species and location		Growth period (days)	Production ($t\ ha^{-1}$)	Growth rate ($g\ m^{-2}\ day^{-1}$)
C_3 plants				
Manihot esculenta	(Java)	365	41.0	11.2
Medicago sativa	(California, USA)	365	29.7	8.1
Beta vulgaris	(California, USA)	240	33.8	14
Beta vulgaris	(The Netherlands)	160	22.0	12.5
Glycine max	(Japan)	130	8.9	6.8
C_4 plants				
Pennisetum purpureum	(El Salvador)	365	85.3	23.3
Saccharum officinarum	(Hawaii, USA)	365	67.3	18.4
Cynodon dactylon	(Puerto Rico)	365	37.3	10.2
Zea mays	(Colorado, USA)	117	26.6	22.7
Sorghum spp.	(California, USA)	120	27.6	23.0
Pennisetum typhoides	(Northern Territory, Australia)	112	21.3	19.0

Table 15.7. Long-term productivity of grass and shrub ecosystems. (Data from CALDWELL 1975; CALDWELL et al. 1977b; JONES et al. 1970)

Ecosystem and principal CO_2 fixation pathway	Growth period (days)	Production ($t\ ha^{-1}$)	Growth rate ($g\ m^{-2}\ day^{-1}$)
Dominated by C_3 plants			
Typha, Phragmites, Scirpus (Czechoslovakia)	130	29.6–40.4	28.6
Old field (New Jersey, USA)	120	23.4	19.5
Festuca thurberi (USA)	95	8.5	9.0
Grassland (short grass prairie northern USA)[a b]	139	6.3	4.5 (5.6)
Shortgrass savannahs (Tajikistan, USSR)	153	5.5	3.6
Ceratoides lanata (Utah, USA)	127	4.2	3.3
Dominated by C_4 plants			
Grass savannah (Ivory Coast)	365	22.0	6.0
Tall-grass prairie			
(Missouri, USA)	180	10.3	5.7
(Colorado, USA)	140	11.2	6.2
Short-grass prairie (Southern USA)[b]	218	4.5	2.1 (4.2)
Atriplex nummularia[a] (south eastern Australia)	500	30.3	6.1
Atriplex confertifolia (Utah, USA)	154	5.4	3.5

[a] root/shoot=0.3, irrigated plantations and early seedling growth, 1 plant m^{-2}
[b] Maximum short term growth rate (SIMS and SINGH 1978a)

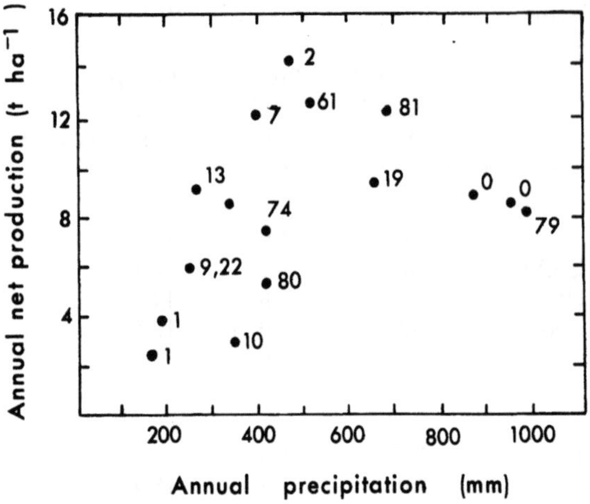

Fig. 15.16. The relationship between total annual productivity and precipitation for ungrazed grassland ecosystems in Western United States. The *numbers adjacent to each data point* refer to the percentage of C_3 plants in the harvested biomass at that site during the growth season. (Reconstructed from SIMS and SINGH 1978b; and SIMS et al. 1978)

or C_4-dominated grassland at either end of the rainfall spectrum. Moreover, these authors found that water use efficiency of C_3-dominated mountain grasslands was not necessarily inferior to C_4-dominated desert grasslands, presumably because growth season vapor pressure difference counterbalanced intrinsic photosynthetic properties.

There are few studies of photosynthetic productivity of C_3 and C_4 plants as a basis of plant competition (HOFSTRA and STIENSTRA 1977; BASKIN and BASKIN 1978). The difference in photosynthetic pathway in the two competing dicotyledons *Amaranthus* (C_4) and *Chenopodium* (C_3) did not rate a comment (KROH and STEPHENSON 1980). *Amaranthus* showed the highest productivity and relative competitive ability in diversity/yield experiments with *Chenopodium* and two C_4 grasses, *Setaria* and *Panicum*. However, in pair competition *Chenopodium* showed higher production and competitive ability than all three C_4 plants. A great deal more research is required to evaluate the significance of photosynthetic pathway and production in plant competition (PEARCY et al. 1981; ÖZTÜRK et al. 1981).

There is more evidence that in many mixed communities plants with different photosynthetic pathways do not compete but tend to share resources in time. The photosynthetic temperature responses of short grass prairie dominant *Agropyron smithii* (C_3) and *Bouteloua gracilis* (C_4) are consistent with the early spring growth of the C_3 plants at low temperature and the later spring and summer growth of the C_4 plants (WILLIAMS 1974). *Opuntia polyacantha*, a CAM plant common in the short grass prairie, is also summer-active, responding to rainfall only at warm temperatures and engaging in substantial CO_2 fixation in the light under these conditions (GERWICK and WILLIAMS 1978).

In one attempt to assess the significance of the potentially superior water economy of C_4 plants beyond the question of dry matter production NOBS et al. (1972) compared *A. triangularis* (C_3) and *A. rosea* (C_4) in transplant experiments. The plants were grown with and without irrigation in a cool coastal habitat (Half Moon Bay) and drier warm habitat (Stanford). In the cool moist habitat, there was no growth response to irrigation and both species produced copious seed. In the warm dry habitat, the C_3 plant declined rapidly in the absence of irrigation, and plants died before ripe seed were produced. There is no evidence that these plants are markedly different in their tolerance of low tissue water potential or in the depth of penetration of their rooting systems. It is likely that, because the C_4 plant used less water per unit time than the C_3 plant, it was able to postpone the onset of low soil water potentials until it completed its life cycle. More detailed data are required to substantiate these conclusions. Yet it is only in such experiments that we can assess the functional ecological significance of the more economic water use of C_4 plants.

15.4.3 Distribution (Survival) of Plants with Different Photosynthetic Pathways in Relation to Environment

There have been attempts to interpret the distribution of plants with C_3, C_4, and CAM pathways of photosynthesis in terms of the physiological responses of photosynthesis in these plants to physical factors such as temperature, light, and water, and the environmental factors of the habitats in which the plants are found. Although it is true that the distribution of plants presupposes their ability to fix CO_2 in photosynthesis, to grow and reproduce in the habitat in which they are found, there is no reason to suspect that survival in that habitat is limited by photosynthetic response. Distribution is determined by the tolerance limits of the most sensitive process in the life cycle of the organism.

Only if the photosynthetic CO_2 fixation pathway is the most sensitive process would we expect to find a close correlation with environmental factors. It is unlikely that a process so fundamental to the survival of an autotrophic organism would be the primary factor limiting its distribution. Consequently, one would not anticipate a simple relationship between CO_2 fixation pathway and habitat factors.

Interpretation of relationships is likely to be further complicated because of the limited data bases available for comparison. The bulk of photosynthetic physiology is concerned with environmental response functions well within the tolerance limits of the organism, rather than with the tolerance limits themselves. Photosynthetic physiology may give insights into productivity relationships and productivity may be related to relative abundance in a particular environment. However, plant distribution data frequently indicate little more than that the organism has survived in a particular habitat only to be collected by a botanist who is likely to notice the exception rather than the rule. Given these qualifications it is surprising perhaps that any broad generalizations are possible at all.

Ehleringer (1978) attempted to project plant distribution on the basis of predicted photosynthetic productivity. His premises were that decreased quantum yield is the only disadvantage of the C_4 pathway and, with the temperature dependence of light-saturated photosynthesis, it could be used to simulate productivity. He further assumed that the C_3 and C_4 plants were identical in all other respects and that the plant type with greater productivity in a particular habitat would replace the less productive type. He chose to make this comparison with grasses because the C_3 and C_4 pathways are well represented in this family, because these plants form canopies of high leaf area index in which light might be limiting, and because grasses are found over wide geographical regions. These premises and assumptions predict superior productivity of C_4 grasses from April to October in Tucson, Arizona. This agrees with the observation that all 16 common summer grasses of the Sonoran Desert are C_4 plants (JOHNSON 1975; MULROY and RUNDEL 1977). The simulation of shaded habitats confirms the observation of HOFSTRA et al. (1972) in several parts of Java. They examined all herbaceous plants in 1 m^2 quadrats and in the sun, 38% were C_4 plants whereas in the shade more than 95% were C_3 plants. In a shaded habitat (10% sunlight) C_4 plants are predicted to have a higher productivity than C_3 plants only at high temperatures (above 33 °C). Such temperatures are uncommon in shaded habitats, except in the tropics, where several shade-tolerant C_4 plants are found, including the tree-form *Euphorbia* spp. in the understorey of Hawaiian forests (PEARCY and TROUGHTON 1975). Shade-tolerant C_4 grasses are not common in Hawaii (RUNDEL 1980) but are represented in the genera *Microstegium* (Southern Asia) and *Ptychophyllum* (mainly South America). However, some *Mühlenbergia* species (C_4) are also found in southern Canada, where they grow in the shade of moist woods and thickets, that is, under conditions of low light intensity, low temperature, and high soil moisture (BROWN 1977).

Perhaps the best illustration of the insecurity of such generalizations and their underlying complexities is to be found by comparison of EHLERINGER (1978) with TEERI and STOWE (1976), who estimated the presence or absence of C_3 and C_4 taxa grass floras from 32 regions in North America (Fig. 15.17). They then used a linear stepwise multiple regression procedure to identify which of 20 climatic variables yielded the most accurate prediction of the percent C_4 taxa in each region. Ehleringer's simulation of daily carbon gain in identical canopies of C_3 and C_4 plants suggests that the latter will have an advantage below about latitude 45 °N. TEERI and STOWE found that the proportion of grass taxa with C_4 photosynthesis was greater than 50% below about 37 °N (Fig. 15.18). There is thus substantial agreement between the prediction of distribution based on productivity and the observations on the basis of survivals.

Closer inspection, however, gives little cause for confidence in this coincidence. The productivity estimates of Ehleringer are about an order of magnitude greater than the measured productivity of grasslands in these habitats (Table 15.7; Fig. 15.16). One would expect that if the two approaches were equally meaningful, there would be a good correlation between distribution and irradiance factors, as well as with temperature during the light period. TEERI and STOWE found a poor correlation between percent C_4 grass taxa and irradiance factors ($r=0.228$ to 0.560) and with the linked variables irradiance, low water availability, and maximum temperature. Rather the distribution of percent C_4 taxa in grass floras was best correlated with night temperature in the growing season, a climatic variable which has no direct relationship to the processes of CO_2 fixation.

A number of other analyses of the distribution of C_3 and C_4 plants have confirmed the close correlation with night temperature. DOLINER and JOLLIFFE

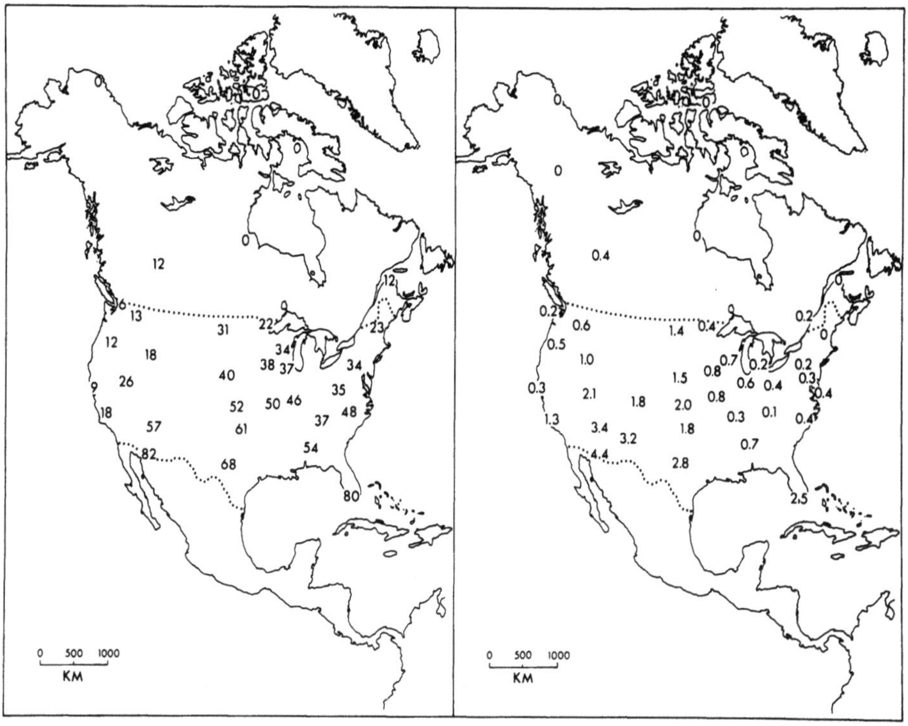

Fig. 15.17. (*left*) Percentage of grass taxa which are C_4 plants and (*right*) percentage of dicotyledon taxa which are C_4 plants in regional floras throughout North America. (Redrawn from TEERI and STOWE 1976; STOWE and TEERI 1978)

(1979) applied nonparametric analysis of variance, and multiple regression analysis techniques to vegetation and climatic data from California and Central Europe. The photosynthetic pathways of 155 Californian and 55 European species were known and no significant difference between the distribution of C_3 and C_4 species could be ascribed to the light factor, but distribution was highly correlated with temperature. Analysis of more than a thousand grass species in South Africa (VOGEL et al. 1978) also established that night temperature during the growing season was the principal determinant of the distribution of C_4 plants. It will be recalled that in many C_4 grasses low night temperature impairs the normal development of the photosynthetic apparatus and also impairs the translocation of assimilate from the leaf. These indirect effects of night temperature presumably reflect the chilling sensitivity of cell and organelle membranes in C_4 plants of tropical origin which, not surprisingly, set the tolerance limits of these grasses, independently of CO_2 fixation processes.

Problems of evaluating relative biological significance of differences in correlation coefficients, and of major discrepancies between models and measurement can be avoided by more comprehensive field measurements. When productivity related parameters, such as biomass, are measured along with presence or absence of photosynthetic pathway types, we have a better picture of the relationship between photosynthetic performance and survival. For example, TIESZEN et al. (1979) studied the distribution of C_3 and C_4 grasses along a 20×350 km

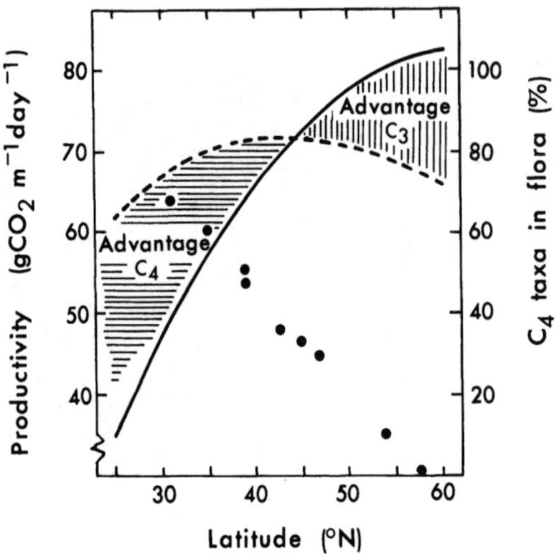

Fig. 15.18. Simulation of the daily carbon gain of C_3 and C_4 grasslands (leaf area index 4) at different latitudes within the Great Plains of North America, on the basis of temperature responses of quantum yield. Over this is plotted the percentage of grass taxa which are C_4 plants at different latitudes in this region. (Redrawn from EHLERINGER 1978, and TEERI and STOWE 1976)

transect in Kenya, rising from 610 to 4,000 m. The leaf anatomy of a 530 species was examined and used to identify 102 C_3 and 428 C_4 taxa. The percentage composition of C_3 and C_4 plants in floras at 11 sites were calculated (Fig. 15.19). These authors also measured the relative abundance of C_3 and C_4 plants by clipping small quadrats and measuring the $\delta^{13}C$ value of the aboveground biomass.

In agreement with TEERI and STOWE (1976), C_4 plants were rarely found where July minimum temperatures are below 8 °C and none were found where mean annual minimum temperature was below 8 °C. However, the percentage of C_4 plants was also closely correlated with an aridity index (available soil moisture index). There was a sharp transition from 100% C_4 grasses to 100% C_3 grasses between 2,000 m and 3,000 m, which is matched by changes in their contribution to biomass. On Hawaii a similar transition was found between 1,000 and 1,400 m (RUNDEL 1980). TIESZEN et al. (1979) found some minor exceptions to this distribution. Some C_3 grasses were found at low altitudes, but these were found mainly in the understorey of closed forests and were from tribes composed principally of C_4 species (cf. HOFSTRA et al. 1972). A few C_4 species were found at high altitudes including the tussock grass *Andropogon amethystinus* at tree line and these species are evidently chilling-tolerant.

A similar comprehensive analysis of species distribution, photosynthetic pathway, and biomass has been made by BOUTTON et al. (1980) along a grassland transect from 1,400 m to 2,650 m elevation in Wyoming. Figure 15.19 shows that C_4 grasses comprised 26% of species, and 88%–85% of the biomass at low elevation, and declined progressively to zero at the highest elevation. Although CAM plants were present at all sites, no trends in biomass were detected with elevation in this study. The proportion of biomass due to C_4 plants was significantly higher in August (high summer) than in June (late spring) at all sites, although only at the lowest site was biomass itself higher in August. Although the contribution of C_4 plants to biomass declined with altitude, these plants retained higher biomass throughout summer, indicating perhaps a contri-

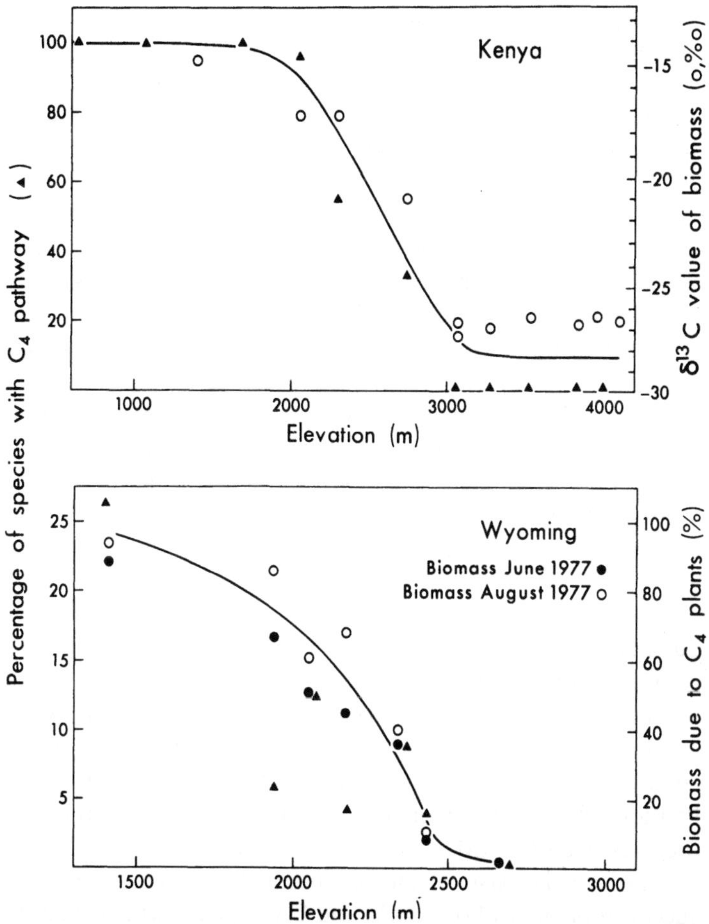

Fig. 15.19. Correlations between the percentage of taxa which are C_4 plants and the biomass due to C_4 plants for two grassland ecosystems. Contribution of C_4 plants to biomass was estimated by $\delta^{13}C$ value (Kenya) or by taxonomic evaluation (Wyoming). (Redrawn from data of TIESZEN et al. 1979; BOUTTON et al. 1980)

bution of water use efficiency to production. Regression analysis of relative abundance as a function of climate indicated that mean annual temperature and annual precipitation were equally reliable predictors of C_3 and C_4 biomass along this transect.

The percent C_4 taxa among North American dicotyledonous floras is much smaller than in the grass floras. STOWE and TEERI (1978) showed that the distribution of these is quite clearly different to that of the C_4 grasses (Fig. 15.17). Two closely related climatic variables, annual evaporation and dryness ratio, were both highly correlated with the percent C_4 taxa among dicots ($r = 0.934$, 0.931). July minimum temperature was poorly correlated. The higher efficiency of water use during CO_2 fixation by the C_4 pathway immediately suggests itself as the basis for such a good correlation with water relations parameters. However, STOWE and TEERI (1978) noted that in families of dicots which contain C_4 species, the C_3 species are also well correlated with summer pan evaporation and dryness ratio. They concluded that "there appear to exist adaptive properties of these families, in addition to the type of photosynthetic pathway, which favor their presence in arid regions". These observations thus suggest that the efficiency of water use during CO_2 fixation may be of less significance to survival in arid habitats than other properties shared by C_3 and C_4 plants in the Chenopodiaceae, Amaranthaceae, and Euphorbiaceae.

TIEZEN et al. (1979) confirmed suggestions by BROWN (1977) that those C_4 grasses occupying the most arid regions and growing during the hot dry season belong to the NAD-malic enzyme subtype of C_4 photosynthesis. They showed, for instance, that members of the Eragrostoideae and Aristideae occupied environments with the lowest available soil moisture, whereas members of the Paniceae and Andropogoneae (all C_4) were common to more mesic sites. This trend was also recognized by VOGEL et al. (1978) who reported that 81% of the grasses in the Southern Kalahari were "aspartate formers" which presumably belong in this same biochemical sub-group of C_4 plants. It is difficult to imagine any biochemical or physiological basis for this correlation and presumably factors other than C_4 pathway sub-group are responsible for this distribution.

Statistical analysis of the distribution of members of the succulent families Cactaceae and Crassulaceae in North America (TEERI et al. 1978) showed that in both families distribution is highly correlated with climatic indices of dry habitats. However, the authors noted that Cactaceae were best correlated with habitats characterized by high evaporation, whereas Crassulaceae were best correlated with low precipitation regions. Unlike C_3 and C_4 pathways of CO_2 fixation, which can be assigned with some certainly on the basis of leaf anatomy, the presence of CAM and its contribution to total CO_2 fixation is not easily ascribed on the basis of morphological features such as succulence. There are succulent plants without CAM and CAM plants which are not succulent. However, succulents with CAM provide the most direct and readily evaluated evidence for the functional significance of different photosynthetic CO_2 fixation pathways.

15.4.4 The Compromise Between Productivity and Survival in CAM Plants

Ideally, to test the ecological significance of different photosynthetic pathways one should have access to the same or closely related genotypes which differ in that property alone. For this reason the few comparisons that have been made with C_3 and C_4 species of *Atriplex* have been especially valuable, and for this reason the prospect of functionally intermediate hybrids was particularly exciting (OSMOND et al. 1980b). However, as far as is known there are no wild plants, or artificial hybrids which are functionally intermediate and which have been the subject of ecophysiological evaluation. Comparisons of the so-called intermediate C_3/C_4 species of *Panicum* would be rewarding (BROWN and BROWN 1975; KU and EDWARDS 1978; RATHNAM and CHOLLET 1980). There is no single genotype in which the potential advantages of a C_3 property can be evaluated against a C_4 property in response to the selective pressures of the environment. Yet in constitutive CAM plants (OSMOND 1975) we may separately evaluate the response and functional significance of CO_2 fixation in the dark (C_4-like) and the light (C_3-like) in relation to performance and survival in different habitats. Moreover, in inducible CAM plants, such as *M. crystallinum*, we are able to evaluate the induction of nocturnal CO_2 fixation and the decline in conventional C_3 photosynthesis in response to environmental factors.

These engaging possibilities have yet to attract much experimental proof. Because the $\delta^{13}C$ value in leaf succulents is a useful integrator of the contributions of CO_2 fixation in the dark and light, it has been used in pilot laboratory studies (OSMOND et al. 1976) and widely applied to infer patterns of carbon assimilation among succulent plants in the field. By and large, our enthusiasm in application of this technique in the field has exceeded our understanding of the processes involved. One notable example of this lapse concerns the contribution of CO_2 fixation in the light to total carbon gain in *Opuntia* under mesic conditions. More negative $\delta^{13}C$ values for specimens collected in more mesic sites (OSMOND 1975) suggested a substantial amount of net CO_2 fixation in the light in this stem succulent. Although this response in the patterns of CO_2 fixation was later confirmed in field experiments, it was found that nocturnal CO_2 fixation also increased markedly with improved plant water status, and this obscured any significant trend in the $\delta^{13}C$ value (OSMOND et al. 1979a). Other studies with stem succulents, under natural conditions and under experimental irrigation, confirm that although net CO_2 fixation in the light may sometimes be observed, it evidently contributes little to overall carbon gain (SUTTON et al. 1976; EICKMEIER and BENDER 1976; HANSCOMB and TING 1978; RUNDEL et al. 1979). Surveys of the $\delta^{13}C$ value of stem succulents in southern Africa, California, and Chile (MOONEY et al. 1974, 1977), confirm that most lie between $-10‰$ and $-17‰$ (Fig. 15.20), suggesting that most carbon is acquired by dark CO_2 fixation. Similar results were obtained with the endemic stem succulents of Madagascar (WINTER 1979a).

In contrast to the stem succulents, leaf succulents of the Crassulaceae known to be capable of CAM display much more variable $\delta^{13}C$ values which, in laboratory studies, are readily correlated with the proportions of CO_2 fixation in the light and dark (OSMOND 1975, 1978). Studies of samples from the field indicate three categories of CAM in leaf succulents:

1. Exclusive dark CO_2 fixation in species from very dry habitats, such as the Crassulaceae from Mexico (RUNDEL et al. 1979).
2. Sensitive responses to water stress with an increase in dependence on dark CO_2 fixation.
3. No dark CO_2 fixation in some succulents from natural habitats, but CAM can be induced under extreme conditions as was found in Central-European *Sedum* spp. (OSMOND et al. 1975; KLUGE 1977).

Sempervivum species from the driest habitats in the European Alps showed slightly less negative $\delta^{13}C$ values than those from moist habitats (OSMOND et al. 1975). Similar variations were found in *Dudleya* spp. from different habitats in the Santa Monica Mountains (TROUGHTON et al. 1977). More convincing evidence for this response along an aridity gradient was found in *Welwitschia mirabilis* by SCHULZE et al. (1976). In the most arid lichen desert regions of the Namib the $\delta^{13}C$ value averaged $-19.6\% \pm 1.0‰$, whereas in the more

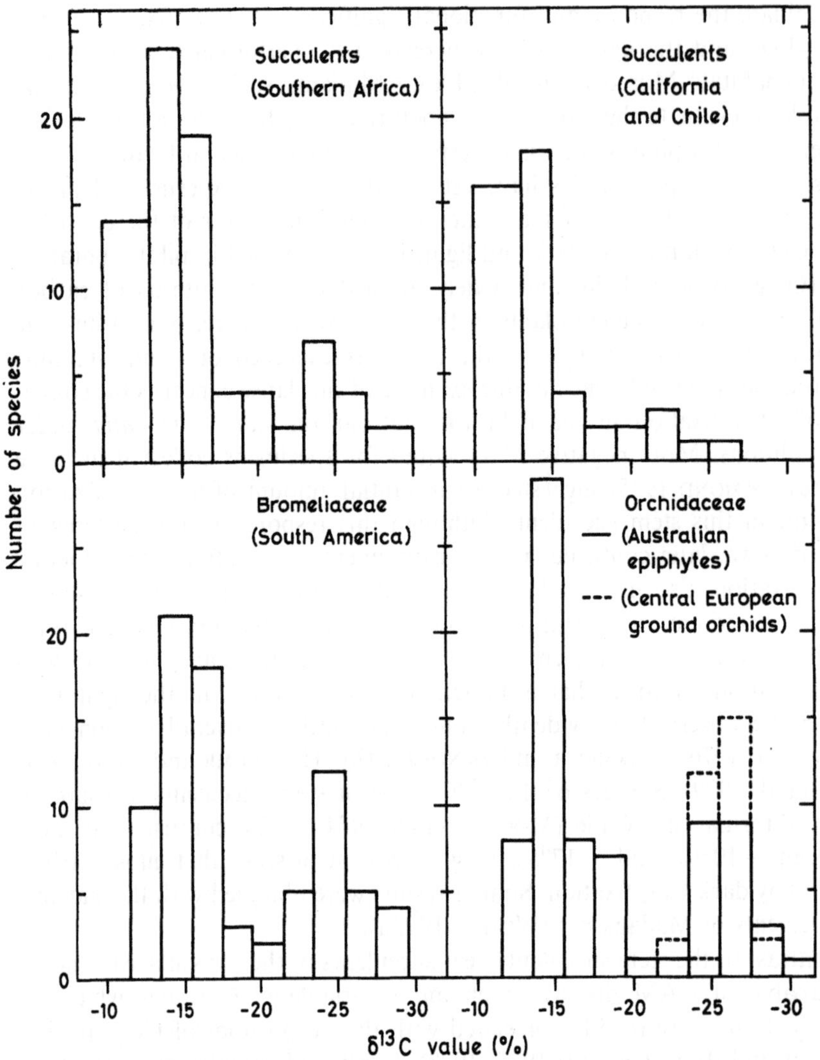

Fig. 15.20. Indications of the frequency of CAM among terrestrial succulents from Southern Africa and Chile, among epiphytic and terrestrial Bromeliads of South America and among epiphytic orchids in Australian rainforests. The preponderance of less negative $\delta^{13}C$ values indicates that the majority of succulents are dependent on nocturnal CO_2 fixation via CAM in these habitats. $\delta^{13}C$ values of all European ground orchids are indicative of normal C_3 photosynthesis. (Redrawn from data of MOONEY et al. 1974, 1977; MEDINA et al. 1977; unpublished data of K. WINTER; and unpublished data of H. ZIEGLER and W. STICHLER)

mesic savanna grasslands it was $-21.6\% \pm 0.6‰$. The $\delta^{13}C$ values of C_3 and C_4 plants remained distinct and were unchanged in these two communities. The most convincing correlation between $\delta^{13}C$ value and the change in CO_2 fixation patterns in leaf succulents is that found on induction of CAM in *M. crystallinum* (Fig. 15.21).

Because the $\delta^{13}C$ value of leaf succulents is so responsive to environmental and biological factors, it is often difficult to exclude the possibility of CAM

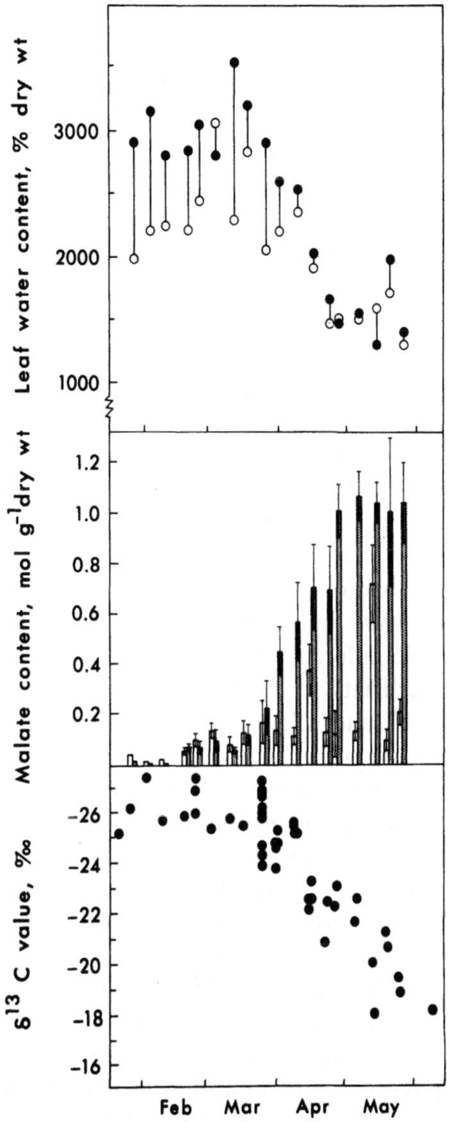

Fig. 15.21. Induction of CAM in *Mesembryanthemum crystallinum* growing in its natural habitat on rocky coastal cliffs of the Mediterranean Sea, Israel. (Redrawn from WINTER et al. 1978)

in these plants. Nevertheless, surveys of the $\delta^{13}C$ value in Bromeliaceae in South America (MEDINA et al. 1977) and ephiphytic Orchidaeae from tropical rainforests in northeastern Australia (K. WINTER, unpublished) suggest that the majority of the species examined fix most of their CO_2 in the dark (Fig. 15.20). European ground orchids are, on the other hand, exclusively C_3 plants (H. ZIEGLER and W. STICHLER, unpublished). If further studies extend this conclusion to the majority of epiphytic orchids and bromeliads (COUTINHO 1969; BENZING and RENFROW 1971), these groups are probably the most numerous of CAM plants, in terms of species diversity.

Although a great deal remains to be done before the flexible CO_2 fixation pathways of CAM plants can be fully evaluated in terms of ecological processes, some principles are already evident. The stem succulents appear to rely on nocturnal CO_2 fixation for carbon acquisition throughout their life cycle. This process can "idle", recycling respiratory CO_2 during periods of excessive water stress (SZAREK et al. 1973; KLUGE and TING 1978), or assimilate external CO_2 to the limit of the vacuolar capacity when water is readily available. Whether the functional significance of CAM in these plants really lies in the low water costs of dark CO_2 assimilation, or in terms of the maintenance of a functional photosynthetic apparatus in brightly illuminated tissues by internal generation of CO_2 in the absence of net CO_2 exchange, remains to be evaluated. Clearly, however, in the compromise between productivity and survival, CAM has immense potential for maintaining viable photosynthetic tissues throughout stress periods which other species survive only in the form of seeds or dormant reproductive structures. In some communities, such as the C_3 sclerophyll forests invaded by *Opuntia* in Eastern Australia, CAM plants may achieve high biomass and become highly productive following summer rainfall.

On the other hand, there is less evidence that leaf succulents employ the CO_2 fixation processes of CAM as a means of survival via preservation of the photosynthetic apparatus. The behavior of massive leaf succulents such as *Agave* and other Liliaceae and Crassulaceae (RUNDEL et al. 1979) is probably little different to that of stem succulents. However, many of the smaller leaf succulents in the Crassulaceae shed their leaves in response to severe water stress and seldom persist long by means of nocturnal recycling of CO_2 via CAM. Some, such as *Bryophyllum* and *Kalanchoe* have very effective systems of vegetative reproduction by means of plantlets born on the leaf margins or tips. In others, such as *M. crystallinum,* survival is dependent on seed set. *M. crystallinum* is most productive in mesic seasons and this productivity, which is largely achieved by C_3 photosynthesis, allows occupation of the site to the exclusion of other species. Nevertheless, by extending the period of net CO_2 fixation into the arid season, CAM permits the organism a longer life cycle. The declining productivity which is associated with the switch to CAM is of little consequence, provided it remains sufficient for normal reproductive development in the face of increasing aridity. That this pattern of biochemical, physiological and ecological events might also be common to other succulents is indicated by the observation of MOONEY et al. (1977) that the $\delta^{13}C$ value of fruits from sixteen succulent species is less negative than leaf material. This suggests that CO_2 fixed in the dark, late in the life cycle, makes a proportionally greater contribution to the completion of reproductive development than CO_2 fixed in the light.

15.5 Conclusions

There is little doubt that our understanding of the biochemical bases of the different photosynthetic pathways enables us to describe their functional significance in a physiological context. By and large the processes of photorespiration compensate for, and those of C_4 photosynthesis overcome, the deficiencies of RuP_2 carboxylase function in contemporary atmospheric conditions. Moreover, it can be clearly demonstrated that the biochemistry of C_4 photosynthesis permits higher potential rates of CO_2 fixation in high light, high temperature environments at lower costs in terms of water lost, and for relatively low investment of N in leaf protein. It is equally clear that the biochemistry of CAM permits a low rate of net CO_2 assimilation at low water cost which can function to conserve respiratory carbon and preserve the photosynthetic apparatus at high water potential in high light, high temperature environments. These conclusions, based in the main on ecophysiological studies of wild plants from contrasting habitats, provide satisfactory accounts of photosynthetic adaptation in the physiological sense, of the way component processes are fitted together to give superior performance in a particular environment.

However, as yet we cannot derive much satisfaction from our understanding of photosynthetic physiology in the ecological context. Here, adaptation may depend upon any attribute of an organism which confers a reproductive advantage in relation to its competitors (HARPER 1977). We know very little about the relationships between photosynthesis and survival, beyond that a certain amount of net carbon assimilation must occur for production of seed or organs of vegetative reproduction.

The fact that the special photosynthetic pathways have arisen in relatively few families of terrestrial plants suggests it is only one of many features which contribute to ecological success in different environments. As far as is known to date, from the 538 families of the flowering plants and ferns (WILLIS 1973) only 17 contain C_4 taxa (Cyperaceae, Poaceae, Acanthaceae, Aizoaceae, Amaranthaceae, Asteraceae, Boraginaceae, Caryophyllaceae, Chenopodiaceae, Cleomaceae, Euphorbiaceae, Molluginaceae, Nyctaginaceae, Portulacaceae, Polygonaceae, Scrophulariaceae, Zygophyllaceae; cf. ZIEGLER 1979; WINTER et al. 1977) and only 20 contain CAM-taxa (Polypodiaceae; Welwitschiaceae; Agavaceae, Bromeliaceae, Liliaceae, Orchidaceae, Aizoceae, Asclepiadaceae, Asteraceae, Cactaceae, Crassulaceae, Cucurbitaceae, Didieraceae, Euphorbiaceae, Geraniaceae, Lamiaceae, Oxalidaceae, Piperaceae, Portulacaceae, Vitaceae; cf. SZAREK and TING 1977).

Although it is certain that additional families with C_4 or CAM species will be discovered, the above suggests that genetic limitations may primarily determine the expression of these metabolic processes in many habitats. These genetic limitations are most evident, for example, in the virtual absence of perennial, stem succulent CAM plants in the Australian deserts, although recent studies show that this metabolic pathway may be well expressed on this continent among rainforest epiphytes (Fig. 15.20). In addition to the limitations of genetic resources, distribution of plants with different photosynthetic pathways may also reflect recent biological history. The distribution of C_4 plants in North America (Fig. 15.17) could reflect migration of tropical C_4 grasses from Florida, and arid region C_4 dicotyledons from south-western deserts, since the last glaciation.

Perhaps it was the replacement of C_3 by C_4 species within a given family along climatic gradients which has led to the widespread generalization that the presence of C_4 plants is favored in hot, arid environments, often with the implication that C_3 species are less well represented in these environments. This is particularly evident in the Gramineae. Although 98% of the grasses in the Indian deserts are C_4 plants (ZIEGLER et al. 1981), considering the whole flora, the majority of species in many desert environments show C_3 photosynthesis (PHILPOTT and TROUGHTON 1974; JOHNSON 1975). Loessial plains of the Negev highlands which receive run-off water are dominated by C_4 plants, whereas the more arid slopes and hilltops are dominated by C_3 plants (WINTER and TROUGHTON 1978). In the Namib desert in south-west Africa among the nonsucculent species analyzed for $\delta^{13}C$ value, 120 were C_3 plants and 35 were C_4 plants, of which 30 were grasses (E-D SCHULZE, H. ZIEGLER and W. STICHLER, unpublished). Perhaps the most comprehensively researched evidence that factors other than photosynthetic pathway contribute to success in the ecological context is provided by comparison of C_3 and C_4 perennial chenopods in the Great Basin Desert (CALDWELL et al. 1977b).

If any generalizations are possible they would be that C_4 plants enjoy an ecological advantage among herbaceous life forms in hot, moist, high light (tropical) regions and, when water is available, in hot arid regions. Cool temperatures during the wet season may reduce the potential advantage in productivity of C_4 plants in many deserts, such as the Negev (MOONEY et al. 1974). There is little evidence that the high water use efficiency of C_4 plants confers an advantage in tolerance of water stress or survival mechanisms. The latter depend principally on the shedding of photosynthetic (and transpiring) tissues in the dry season (EVENARI et al. 1971), except in the case of CAM plants. Even here, only the stem succulents retain photosynthetic biomass, in arid habitats, and CAM plants are excluded from the most arid habitats like the floor of Death Valley. Rather, the preferred habitats of terrestrial CAM plants are semi-arid regions with seasonal rainfall, coastal fog deserts such as the Namib and Atacama deserts and, if the trends in the Orchidaceae and Bromeliaceae are confirmed, the epiphytic habitats in tropical rainforests.

The extent to which the physiological efficiencies of different photosynthetic CO_2 fixation pathways may contribute to competitive advantage, through superior productivity or superior exploitation of available resources, remains relatively obscure. The integration of photosynthetic physiology into the longer term goals, as they are perceived by the plant, is a major challenge in ecophysiology. Sometimes these larger questions, such as the patterns of resource (nutrient and water) utilization and of assimilate partitioning in respect of reproductive behavior, may be difficult to express in other than teleological language. Implausible though some of these poorly articulated hypotheses may seem, they provide a legitimate framework for contriving further experiments which better enable us to assess the functional, if at times marginal, significance of different photosynthetic CO_2 fixation pathways.

Acknowledgments. We are very grateful to many colleagues for their critical advice, but particularly to Professor D.H.N. Spence for his advice and contributions to our comments on aquatic plants.

References

Abelson PH, Hoering TC (1961) Carbon isotope fractionation in formation of amino acids by photosynthetic organisms. Proc Natl Acad Sci USA 47:623–632

Akazawa T (1979) Ribulose-1,5-bisphosphate carboxylase. In: Gibbs M, Latzko E (eds) Photosynthesis II. Carbon metabolism and related processes. Encyclopedia of plant physiology new ser Vol VI. Springer, Berlin Heidelberg New York, pp 208–229

Allaway WG, Austin B, Slatyer RO (1974) Carbon dioxide and water vapour exchange parameters of photosynthesis in a Crassulacean plant *Kalanchoe daigremontiana*. Aust J Plant Physiol 1:397–405

Allen ED, Spence DHN (1981) The differential ability of aquatic plants to utilize the inorganic carbon supply in fresh waters. New Phytol 87:269–283

Andrews TJ, Abel KM (1979) Photosynthetic metabolism in seagrasses: ^{14}C-labeling evidence for the C_3 pathway. Plant Physiol 63:650–656

Arens K (1939) Physiologische Multipolarität der Zelle von *Nitella* während der Photosynthese. Protoplasma 33:295–300

Badger MR (1980) Kinetic properties of ribulose-1,5-bisphosphate carboxylase-oxygenase from *Anabaena variabilis*. Arch Biochem Biophys 201:247–254

Badger MR, Kaplan A, Berry JA (1977) The internal CO_2 pool of *Chlamydomonas reinhardtii*: response to external CO_2. Carnegie Inst Washington Yearb 76:362–366

Badger MR, Kaplan A, Berry JA (1978) A mechanism for concentrating CO_2 in *Chlamydomonas reinhardtii* and *Anabaena variabilis* and its role in photosynthetic CO_2 fixation. Carnegie Inst Washington Yearb 77:251–261

Bartholomew DP, Kadzimin SB (1975) The ecophysiology of pineapple. In: Alvim PT (ed) The ecophysiology of tropical crops Vol 1. CEPLAC, Bahia, pp 1–58

Baskin JM, Baskin CC (1978) A discussion of the growth and competitive ability of C_3 and C_4 plants. Castanea 43:71–76

Beer S, Eshel A, Waisel Y (1977) Carbon metabolism in seagrasses. I. The utilization of exogenous inorganic carbon species in photosynthesis. J Exp Bot 28:1180–1189

Bender MM (1968) Mass spectrometric studies of carbon-13 variations in corn and other grasses. Radiocarbon 10:468–472

Bender MM (1971) Variation in the $^{13}C/^{12}C$ ratios of plants in relation to the pathway of photosynthetic carbon dioxide fixation. Phytochemistry 10:1239–1244

Bender MM, Berge AJ (1979) Influence of N and K fertilization and growth temperature on $^{13}C/^{12}C$ ratios in Thimothy (*Phleum pratense* L.). Oecologia 44:117–118

Bender MM, Rouhani I, Vines HM, Black CC (1973) $^{13}C/^{12}C$ ratio changes in crassulacean acid metabolism plants. Plant Physiol 52:427–430

Benedict CR (1978) Nature of obligate photoautotrophy. Annu Rev Plant Physiol *29*:67–93

Benedict CR, Scott JR (1976) Photosynthetic carbon metabolism of a marine grass. Plant Physiol 57:876–880

Benedict CR, Wong WWL, Wong JHH (1980) Fractionation of the stable isotopes of inorganic carbon by seagrasses. Plant Physiol 65:512–517

Benzing DH, Renfrow A (1971) Significance of the patterns of CO_2 exchange to the ecology and phylogeny of the Tillandsioideae (Bromeliaceae). Bull Torrey Bot Club 98:322–327

Berry JA, Björkman O (1980) Photosynthetic response and adaptation to temperature in higher plants. Annu Rev Plant Physiol 31:491–543

Berry JA, Farquhar GD (1978) The CO_2 concentrating function of C_4 photosynthesis. A biochemical model. In: Hall DO, Coombs J, Goodwin TW (eds) Photosynthesis 77. Biochem Soc, London, pp 119–131

Berry JA, Boynton J, Kaplan A, Badger MR (1976) Growth and photosynthesis of *Chlamydomonas reinhardtii* as a function of CO_2 concentration. Carnegie Inst Washington Yearb 75:432–433

Berry JA, Osmond CB, Lorimer GH (1978) Fixation of $^{18}O_2$ during photorespiration: kinetic and steady state studies of the photorespiratory carbon oxidation cycle with intact leaves and isolated chloroplasts of C_3 plants. Plant Physiol 62:954–967

Björkman O (1966) The effect of oxygen on photosynthesis in higher plants. Physiol Plant 19:618–633

Björkman O (1971) Comparative CO_2 exchange in higher plants. In: Hatch MD, Osmond CB, Slatyer RO (eds) Photosynthesis and photorespiration, Wiley-Interscience, New York, pp 18–32

Björkman O (1973) Comparative studies on photosynthesis in higher plants. In: Giese A (ed) Current topics in photobiology, photochemistry, and photophysiology Vol VIII. Academic Press, London New York, pp 1–63

Björkman O (1975) Environmental and biological control of photosynthesis: inaugural address. In: Marcelle R (ed) Environmental and biological control of photosynthesis. Junk, The Hague, pp 1–16

Björkman O (1976) Adaptive and genetic aspects of C_4 photosynthesis. In: Burris RH, Black CC (eds) CO_2 metabolism and plant productivity. Univ Park Press, Baltimore, pp 287–309

Björkman O, Mooney HA, Ehleringer J (1975) Photosynthetic responses of plants from habitats with contrasting thermal environments. Carnegie Inst Washington Yearb 74:743–748

Björkman O, Boynton J, Berry J (1976) Comparison of the heat stability of photosynthesis, chloroplast membrane reactions, photosynthetic enzymes, and soluble protein in leaves of heat-adapted and cool-adapted C_4 species. Carnegie Inst Washington Yearb 75:400–407

Black CC (1971) Ecological implications of dividing plants into groups with distinct photosynthetic production capacities. Adv Ecol Res 7:87–114

Black CC (1973) Photosynthetic carbon fixation in relation to net CO_2 uptake. Annu Rev Plant Physiol 24:253–286

Black CC, Bender MM (1976) $\delta^{13}C$ values in marine organisms from the Great Barrier Reef. Aust J Plant Physiol 3:25–32

Black CC, Goldstein LD, Ray TB, Kestler DP, Mayne BC (1976) The relationship of plant metabolism to internal leaf and cell morphology and to efficiency of CO_2 assimilation. In: Burris RH, Black CC (eds) CO_2 metabolism and plant productivity. Univ Park Press, Baltimore, pp 113–139

Black MA (1973) Exogenous inorganic carbon sources in the photosynthesis of aquatic plants. M Sc Thesis, Univ St Andrews

Blinks LR (1963) The effect of pH upon the photosynthesis of littoral marine algae. Protoplasma 57:126–136

Bloom AJ (1979a) Salt requirement for Crassulacean acid metabolism in the annual succulent, *Mesembryanthemum crystallinum*. Plant Physiol 63:748–753

Bloom AJ (1979b) Diurnal ion fluctuations in the mesophyll tissue of the Crassulacean acid metabolism plant *Mesembryanthemum crystallinum*. Plant Physiol 64:919–923

Boag S, Brownell PF (1979) C_4 photosynthesis in sodium-deficient plants. Aust J Plant Physiol 6:431–434

Boardman NK (1978) Solar energy conversion in photosynthesis: potential contribution to world demand for liquid and gaseous fuels. In: Hall DO, Coombs J, Goodwin TW (eds) Photosynthesis 77. Biochem Soc, London, pp 636–644

Boutton TW, Harrison AT, Smith BN (1980) Distribution of biomass of species differing in photosynthetic pathway along an altitudinal transect in southeastern Wyoming grassland. Oecologia 45:287–298

Bowes G, Holaday AS, Van TK, Haller WT (1977) Photosynthetic and photorespiratory carbon metabolism in aquatic plants. In: Hall DO, Coombs J, Goodwin TW (eds) Photosynthesis 77. Biochem Soc, London, pp 289–298

Briggs GE, Whittingham CP (1952) Factors affecting the rate of photosynthesis of *Chlorella* at low concentrations of carbon dioxide and in high illumination. New Phytol 51:236–249

Brown JMA, Dromgoole FI, Towse MW, Browse JA (1974) Photosynthesis and photorespiration in aquatic macrophytes. Roy Soc NZ Bull 12:243–249

Brown RH (1978) A difference in the N use efficiency in C_3 and C_4 plants and its implications in adaptation and evolution. Crop Sci 18:93–98

Brown RH, Brown WV (1975) Photosynthetic characteristics of *Panicum milioides*, a species with reduced photorespiration. Crop Sci 15:681–685

Brown WV (1977) The Kranz syndrome and its subtypes in grass systematics. Mem Torrey Bot Club 23:1–97

Brownell PF, Crossland CL (1972) The requirement for sodium as a micronutrient by species having the C_4 dicarboxylic acid photosynthetic pathway. Plant Physiol 49:794–797

Brownell PF, Crossland CJ (1974) Growth responses to sodium by *Bryophyllum tubiflorum* under conditions inducing Crassulacean acid metabolism. Plant Physiol 54:416–417

Brownell PF, Wood JG (1957) Sodium as an essential micronutrient element for *Atriplex vesicaria* Heward. Nature (London) 179:635–636

Browse JA, Brown JMA, Dromgoole FI (1979a) Photosynthesis in the aquatic macrophyte *Egeria densa*. II. Effects of inorganic carbon conditions on ^{14}C fixation. Aust J Plant Physiol 6:1–9

Browse JA, Dromgoole FI, Brown JMA (1979b) Photosynthesis in the aquatic macrophyte *Egeria densa*. III. Gas exchange studies. Aust J Plant Physiol 6:499–512

Buch K (1960a) Dissoziation der Kohlensäure, Gleichgewichte und Puffersysteme. In: Ruhland W (ed) Handbuch der Pflanzenphysiologie Vol I. Springer, Berlin Göttingen Heidelberg, pp 1–11

Buch K (1960b) Kohlendioxyd im Meerwasser. In: Ruhland W (ed) Handbuch der Pflanzenphysiologie Vol I. Springer, Berlin Göttingen Heidelberg, pp 47–61

Burr GO, Hartt CE, Brodie HW, Tanimoto T, Kortschak HP, Takahashi D, Ashton FM, Coleman RE (1957) The sugar cane plant. Annu Rev Plant Physiol 8:275–298

Caldwell MM (1975) Primary production of grazing lands. In: Copper JP (ed) Photosynthesis and productivity in different environments. IBP Vol 3. Univ Press, Cambridge, pp 41–73

Caldwell MM, Camp LB (1974) Below ground productivity of two cool desert communities. Oecologia 17:123–130

Caldwell MM, Osmond CB, Nott DL (1977a) C_4 pathway photosynthesis at low temperature in cold-tolerant *Atriplex* species. Plant Physiol 60:157–164

Caldwell MM, White RS, Moore RT, Camp LB (1977b) Carbon balance, productivity, and water use of cold-winter desert shrub communities dominated by C_3 and C_4 species. Oecologia 29:275–300

Carolin RC, Jacobs SWL, Vesk M (1975) Leaf structure in Chenopodiaceae. Bot Jahrb Syst 95:226–255

Cockburn W, Ting IP, Sternberg LO (1979) Relationship between stomatal behavior and internal carbon dioxide concentration in Crassulacean acid metabolism plants. Plant Physiol 63:1029–1032

Cornic G (1976) Effet exercé sur l'activité photosynthetique du *Sinapis alba* L. par une inhibition temporaire de la photorespiration se déroulant dans un air sans CO_2. CR Acad Sci Ser D 282:1955–1958

Coutinho LM (1969) Novas observacoes sobre a ocorrencia do "efeito de de Saussure" e suas relacoes com a suculencia, a temperatura folhear e os movimentos estomaticos. Botanica 24:44–102

Craig H, Gordon LI (1965) Deuterium and oxygen-18 variations in the ocean and the marine atmosphere. Proc Spoleto Conf Stable Isotopes in Oceanographic studies and paleotemperatures, Pisa

Crookston RK, Moss DN (1972) C-3 and C-4 carboxylation characteristics in the genus *Zygophyllum* (Zygophyllaceae). Ann Mo Bot Gard 59:465–470

Davidson RL (1969a) Effect of root-leaf temperature differentials on root/shoot ratios in some pasture grasses and clover. Ann Bot (London) 33:561–569

Davidson RL (1969b) Effects of soil nutrients and moisture on root/shoot ratio in *Lolium perenne* L. and *Trifolium repens*. Ann Bot (London) 33:571–577

Despain DG, Bliss LC, Boyer JS (1970) Carbon dioxide exchange in suguaro seedlings. Ecology 51:912–914

Deuser WG, Degens ET (1967) Carbon isotope fractionation in the system CO_2 (gas) – CO_2 (aqueous) – HCO_3^- (aqueous). Nature (London) 215:1033–1035

Dodd AP (1940) The biological campaign against prickly pear. Queensl Gov Printer, Brisbane
Doliner LH, Joliffe PA (1979) Ecological evidence concerning the adaptive significance of the C_4 dicarboxylic acid pathway of photosynthesis. Oecologia 38:23–34
Dromgoole FI (1978) The effects of pH and inorganic carbon on photosynthesis and dark respiration of *Carpophyllum* (Fucales, Phaeophyceae). Aquat Bot 4:11–22
Dunstone RL, Gifford RM, Evans LT (1973) Photosynthetic characteristics of modern and primitive wheat species in relation to ontogeny and adaptation to light. Aust J Biol Sci 26:295–307
Eder A, Stichler W, Ziegler H (1981) Mechanismen der CO_2-Fixierung bei *Euphorbia trigona* Haw. und einigen *Pachypodium*-Arten. Biol Biochem Pflanz 176:1–12
Edwards GE, Huber SC (1981) C_4 pathway. In: Hatch MD, Boardman NK (eds) The biochemistry of plants: a comprehensive treatise Vol VIII. Academic Press, London New York pp 237–282
Ehleringer JR (1978) Implications of quantum yield differences on the distributions of C_3 and C_4 grasses. Oecologia 31:255–267
Ehleringer J, Björkman O (1977) Quantum yields for CO_2 uptake in C_3 and C_4 plants. Plant Physiol 59:86–90
Eickmeier WG, Bender MM (1976) Carbon isotope ratios of Crassulacean acid metabolism species in relation to climate and phytosociology. Oecologia 25:341–347
Evenari M, Shanan L, Tadmor N (1971) The Negev. The challenge of a desert. Harvard Univ Press, Cambridge, Massachusetts
Farquhar GD, Firth PM, Wetselaar R, Weir B (1980) On the gaseous exchange of ammonia between leaves and the environment: Determination of the ammonia compensation point. Plant Physiol 66:710–714
Findenegg GR (1976) Correlations between accessibility of carbonic anhydrase for external substrate and regulation of photosynthetic use of CO_2 and HCO_3^- by *Scenedesmus obliquus*. Z Pflanzenphysiol 79:428–447
Gallaher RN, Ashley DA, Brown RH (1975) ^{14}C-photosynthate translocation in C_3 and C_4 plants as related to leaf anatomy. Crop Sci 15:55–59
Gerwick BC, Williams GJ III (1978) Temperature and water regulation of gas exchange of *Opuntia polyacantha*. Oecologia 35:149–159
Gibbs M, Latzko E (1979) (eds) Photosynthesis II: Photosynthetic carbon metabolism and related processes. Encyclopedia of plant physiology new ser Vol VI. Springer, Berlin Heidelberg New York
Gifford RM (1974) A comparison of potential photosynthesis, productivity and yield of plant species with differing photosynthetic metabolism. Aust J Plant Physiol 1:107–117
Graham D, Reed ML (1971) Carbonic anhydrase and the regulation of photosynthesis. Nature (London) 231:81–83
Graham DC, Atkins CA, Reed ML, Patterson BD, Smillie RM (1971) Carbonic anhydrase, photosynthesis, and light-induced pH changes. In: Hatch MD, Osmond CB, Slatyer RO (eds) Photosynthesis and photorespiration. Wiley-Interscience, New York, pp 267–274
Guy RD, Reid DM, Krause HR (1980) Shifts in carbon isotope ratios of two C_3 halophytes under natural and artifical conditions. Oecologia 44:241–247
Haberlandt G (1884) Physiologische Pflanzenanatomie. Engelmann, Leipzig
Hanscomb Z III, Ting IP (1978) Irrigation magnifies CAM-photosynthesis in *Opuntia basilaris* (Cactaceae). Oecologia 33:1–15
Harper JL (1977) Population biology of plants. Academic Press, London New York
Hartman RT, Brown DL (1967) Changes in internal atmosphere of submersed vascular hydrophytes in relation to photosynthesis. Ecology 48:252–258
Hatch MD (1971) The C_4-pathway of photosynthesis. Evidence for an intermediate pool of carbon dioxide and the identity of the donor C_4-dicarboxylic acid. Biochem J 125:425–432
Hatch MD, Osmond CB (1976) Compartmentation and transport in C_4 photosynthesis. In: Heber U, Stocking CR (eds) Transport in plants III. Encyclopedia of plant physiology new ser Vol III. Springer, Berlin Heidelberg New York, pp 144–184

Hatch MD, Slack CR (1966) Photosynthesis by sugarcane leaves. A new carboxylation reaction and the pathway of sugar formation. Biochem J 101:103–111
Hatch MD, Slack CR (1970) The C_4 carboxylic acid pathway of photosynthesis. In: Reinhold L, Liwschitz Y (eds) Progress in phytochemistry. Wiley-Interscience, New York, pp 35–106
Hattersley PW, Watson L, Osmond CB (1977) In-situ immunofluorescent labelling of ribulose-1,5-bisphosphate carboxylase in leaves of C_3 and C_4 plants. Aust J Plant Physiol 4:523–539
Hilliard JH, West SH (1970) Starch accumulation associated with growth reduction at low temperatures in a tropical plant. Science 168:494–496
Hoch Hin Y, Badger MR, Watson L (1980) Variation in $K_m(CO_2)$ of ribulose-1,5-bisphosphate carboxylase among grasses. Plant Physiol 66:1110–1112
Hoch Hin Y, Badger MR, Watson L (1981) Variation in kinetic properties of ribulose-1,5-bisphosphate carboxylase among plants. Plant Physiol 67:1151–1155
Hoffman GJ, Rawlins ST, Garber MJ, Cullen EM (1971) Water relations and growth of cotton as influenced by salinity and relative humidity. Agron J 63:822–826
Hofstra G, Nelson CD (1969) A comparative study of translocation of assimilated ^{14}C from leaves of different species. Planta 88:103–112
Hofstra JJ, Stienstra AW (1977) Growth and photosynthesis of closely related C_3 and C_4 grasses, as influenced by light intensity and water supply. Acta Bot Neerl 26:63–72
Hofstra JJ, Aksornkoae S, Atmowidjojo S, Banaag Santosa JF, Sastrohoetoma RA, Thu LTN (1972) A study on the occurrence of plants with a low CO_2 compensation point in different habitats in the tropics. Ann Bogor 5:143–157
Holaday AS, Bowes G (1980) C_4 acid metabolism and dark CO_2 fixation in a submersed aquatic macrophyte (*Hydrilla verticillata*). Plant Physiol 65:331–335
Hough RA (1974) Photorespiration and productivity in submersed aquatic vascular plants. Limnol Oceanogr 19:912–927
Hough RA, Wetzel RG (1972) A ^{14}C-assay for photorespiration in aquatic plants. Plant Physiol 49:987–990
Hough RA, Wetzel RG (1977) Photosynthetic pathway of some aquatic plants. Aquat Bot 3:297–313
Hough RA, Wetzel RG (1978) Photorespiration and CO_2 compensation point in *Najas flexilis*. Limnol Oceanogr 23:719–724
Huber W, Sankhla N (1976) C_4 pathway and regulation of the balance between C_4 and C_3 metabolism. In: Lange OL, Kappen L, Schulze E-D (eds) Water and plant life: problems and modern approaches. Ecol Stud Vol 19. Springer, Berlin Heidelberg New York, pp 335–363
Jarvis PG (1971) The estimation of resistances to carbon dioxide transfer. In: Šesták K, Čatský J, Jarvis PG (eds) Plant photosynthetic production: manual and methods. Junk, The Hague, pp 49–110
Johnson HB (1975) Gas exchange strategies in desert plants. In: Gates DM, Schmerl M (eds) Perspectives of biophysical ecology. Ecol Stud Vol 12. Springer, Berlin Heidelberg New York, pp 105–120
Joliffe EA, Tregunna EB (1970) Studies on HCO_3^- ion uptake during photosynthesis in benthic marine algae. Phycologia 9:293–303
Jones R, Hodgkinson KC, Rixon AJ (1970) Growth and productivity in rangeland species of *Atriplex*. In: Jones R (ed) The biology of *Atriplex*. CSIRO, Canberra, pp 31–42
Kaplan A, Gale J, Poljakoff-Mayber A (1977) Effect of oxygen and carbon dioxide concentration on gross dark CO_2 fixation and dark respiration in *Bryophyllum daigremontianum*. Aust J Plant Physiol 4:745–752
Karpilov YS (1960) The distribution of radioactivity of ^{14}C among the products of photosynthesis in maize. Trans Kazan Agric Inst 41:15–24
Kausch W (1965) Beziehungen zwischen Wurzelwachstum, Transpiration und CO_2-Gaswechsel bei einigen Kakteen. Planta 66:229–238
Keeley J (1981) *Isoetes howellii*: A submerged aquatic CAM plant? Am J Bot 68:420–424
Keenan JD (1975) Bicarbonate utilization in *Anabaena*. Physiologia Plant 34:157–161
Kern DM (1960) The hydration of carbon dioxide. J Chem Educ 37:14–23

Kirk M, Heber U (1976) Rates of synthesis and source of glycolate in intact chloroplasts. Planta 132:131–141

Kluge M (1977) Is *Sedum acre* a CAM-plant? Oecologia 29:77–83

Kluge M, Ting IP (1978) Crassulacean acid metabolism: analysis of an ecological adaptation. Ecol Stud Vol 30. Springer, Berlin Heidelberg New York

Kortschak HP, Hartt CE, Burr GO (1965) Carbon dioxide fixation in sugar cane leaves. Plant Physiol 40:209–213

Kremer BP, Küppers U (1977) Carboxylating enzymes and pathway of photosynthetic carbon assimilation in different marine algae – evidence for the C_4-pathway? Planta 133:191–196

Kremer BP, Markham JW (1979) Carbon assimilation by different developmental stages of *Laminaria saccharina*. Planta 144:497–501

Kremer BP, Willenbrink J (1972) CO_2-Fixierung und Stofftransport in benthischen marinen Algen. I. Zur Kinetik der $^{14}CO_2$-Assimilation bei *Laminaria saccharina*. Planta 103:55–64

Kroh GC, Stephenson SN (1980) Effect of diversity and pattern on relative yields of four Michigan first year fallow field plant species. Oecologia 45:366–371

Ku SB, Edwards GE (1978) Photosynthetic efficiency of *Panicum hians* and *Panicum milioides* in relation to C_3 and C_4 plants. Plant Cell Physiol 19:665–675

Küppers U, Kremer BP (1978) Longitudinal profiles of CO_2 fixation capacities in marine macroalgae. Plant Physiol 62:49–53

Laetsch WM (1974) The C_4 syndrome: a structural analysis. Annu Rev Plant Physiol 25:27–52

Lange OL, Medina E (1979) Stomata of the CAM plant *Tillandsia recurvata* respond directly to humidity. Oecologia 40:357–363

Lange OL, Zuber M (1977) *Frerea indica*, a stem succulent CAM plant with deciduous C_3 leaves. Oecologia 31:67–72

Lange OL, Zuber M (1980) Temperaturabhängigkeit des CO_2-Gaswechsels stammsukkulenter Asclepiadaceen mit Säurestoffwechsel. Flora 170:529–553

Lange OL, Schulze E-D, Kappen L, Evenari M, Buschbom U (1975) CO_2 exchange pattern under natural conditions of *Caralluma negevensis*, a CAM plant of the Negev desert. Photosynthetica 9:318–326

Lea PJ, Mifflin J (1979) Photosynthetic ammonia assimilation. In: Gibbs M, Latzko E (eds) Photosynthesis II: Photosynthetic carbon metabolism and related processes. Encyclopedia of plant physiology new ser. Vol VI. Springer, Berlin Heidelberg New York, pp 445–456

Lerman JC (1975) How to interpret variations in the carbon isotope ratio of plants: biologic and environmental effects. In: Marcelle R (ed) Environmental and biological control of photosynthesis. Junk, The Hague, pp 323–335

Lerman JC, Deleens E, Nato A, Moyse A (974) Variation in the carbon isotope composition of a plant with Crassulacean acid metabolism. Plant Physiol 53:581–584

Lloyd NDH, Canvin DT, Bristow JM (1977) Photosynthesis and photorespiration in submerged aquatic vascular plants. Can J Bot 55:3001–3005

Long SP, Incoll LD, Woolhouse HW (1975) C_4 photosynthesis in plants from cool temperate regions with particular reference to *Spartina townsendii*. Nature (London) 257:622–624

Loomis RS, Gerakis PA (1975) Productivity of agricultural ecosystems. In: Cooper JD (ed) Photosynthesis and productivity in different environments. IBP Vol III. Univ Press, Cambridge, pp 145–172

Lorimer GH, Andrews TJ (1973) Plant photorespiration – an inevitable consequence of the existence of atmospheric oxygen. Nature (London) 243:359

Lorimer GH, Andrews TJ (1981) The C_2 photo- and chemorespiratory carbon oxidation cycle. In: Hatch MD, Boardman NK (eds) The biochemistry of plants: a comprehensive treatise Vol VIII. Academic Press, London New York, pp 329–374

Lorimer GH, Woo KC, Berry JA, Osmond CB (1978) The C_2 photorespiratory carbon oxidation cycle in leaves of higher plants: pathway and consequences. In: Hall DO, Coombs J, Goodwin TW (eds) Photosynthesis 77. Biochem Soc, London, pp 311–322

Lucas WJ (1975) Photosynthetic fixation of 14 carbon by internodal cells of *Chara corallina*. J Exp Bot 26:331–346

Lucas WJ (1976) The influence of Ca^{2+} and K^+ on $H^{14}CO_3^-$ influx in internodal cells of *Chara corallina*. J Exp Bot 27:32–42

Lucas WJ (1979) Alkaline band formation in *Chara corallina*. Plant Physiol 63:248–254

Lucas WJ (1980) Control and synchronization of HCO_3^- and OH^- transport during photosynthetic assimilation of exogenous HCO_3^-. In: Spanswick RM, Lucas WJ, Dainty J (eds) Plant membrane transport: current conceptual issues. Elsevier/North-Holland, Biomedica Press, Amsterdam New York, pp 317–327

Lucas WJ, Tyree MT, Petrov A (1978) Characterization of photosynthetic ^{14}carbon assimilation by *Potamogeton lucens* L. J Exp Bot 29:1409–1421

Ludlow MM (1976) Ecophysiology of C_4 grasses. In: Lange OL, Kappen L, Schulze E-D (eds) Water and plant life: problems and modern approaches. Ecol Stud Vol 19. Springer, Berlin Heidelberg New York, pp 364–386

Ludlow MM, Wilson GL (1971) Photosynthesis of tropical pasture plants. I. Illuminance, carbon dioxide concentration, leaf temperature and leaf-air vapour pressure difference. Aust J Biol Sci 24:449–470

Ludlow MM, Wilson GL (1972) Photosynthesis of tropical pasture plants. IV. Basis and consequences of differences between grasses and legumes. Aust J Biol Sci 25:1133–1145

Lüttge U, Ball E (1979) Electrochemical investigation of active malic acid transport at the tonoplast into the vacuoles of the CAM plant *Kalanchoe daigremontiana*. J Membr Biol 47:401–422

Lush WM, Evans LT (1974) Translocation of photosynthetic assimilate from grass leaves, as influenced by environment and species. Aust J Plant Physiol 1:417–431

Martin B, Martensson O, Öquist G (1978) Effects of frost hardening and dehardening on photosynthetic electron transport and fluorescence properties in isolated chloroplasts of *Pinus silvestris*. Physiol Plant 43:297–305

Medina E (1970) Relationships between nitrogen level, photosynthetic capacity and carboxydismutase activity in *Atriplex patula* leaves. Carnegie Inst Washington Yearb 69:655–662

Medina E (1971) Effect of nitrogen supply and light intensity during growth on the photosynthetic capacity and carboxydismutase activity of leaves of *Atriplex patula* ssp. *hastata*. Carnegie Inst Washington Yearb 70:551–559

Medina E, Delgado M (1976) Photosynthesis and night CO_2 fixation in *Echeveria columbiana* v. Poellr. Photosynthetica 10:155–163

Medina E, Minchin P (1980) Stratification of $\delta^{13}C$ values of leaves in Amazonian rain forests. Oecologia 45:377–378

Medina E, Osmond CB (1981) Temperature dependence of dark CO_2 fixation and acid accumulation in *Kalanchoe daigremontiana*. Aust J Plant Physiol 8:641–649

Medina E, Delgado M, Troughton JH, Medina JD (1977) Physiological ecology of CO_2 fixation in Bromeliaceae. Flora 166:137–152

Meinzer FC, Rundel PW (1973) Crassulacean acid metabolism and water use efficiency in *Echeveria pumila*. Photosynthetica 7:358–364

Miller AG, Coleman B (1980) Evidence for HCO_3^- transport by the blue-green alga (Cyanobacterium) *Coccochloris peniocystis*. Plant Physiol 65:397–402

Monteith LJ (1978) Reassessment of maximum growth rates for C_3 and C_4 crops. Exp Agric 14:1–5

Mooney HA, Troughton JH, Berry JA (1974) Arid climates and photosynthetic systems. Carnegie Inst Washington Yearb 73:793–805

Mooney HA, Björkman O, Ehleringer J, Berry JA (1976) Photosynthetic capacity of in situ Death Valley plants. Carnegie Inst Washington Yearb 75:410–413

Mooney HA, Throughton JH, Berry JA (1977) Carbon isotope measurements of succulent plants in southern Africa. Oecologia 30:295–306

Morot-Gaudry JF, Farineau JP, Huer JC (1980) Oxygen effect on photosynthetic and glycolate pathways in young maize leaves. Plant Physiol. 66:1079–1084

Morris I, Farrell K (1971) Photosynthetic rates, gross patterns of carbon dioxide fixation and activities of ribulose diphosphate carboxylase in marine algae grown at different temperatures. Physiol Plant 25:372–377

Mulroy TW, Rundel PW (1977) Annual plants; adaptations to desert environments. Bioscience 27:109–114
Neales TF (1973a) The effect of night temperature on CO_2 assimilation, transpiration and water use efficiency in *Agave americana* L. Aust J Biol Sci 26:705–714
Neales TF (1973b) Effect of night temperature on the assimilation of carbon dioxide by mature pineapple plants *Ananas comosus* (L.) Merr. Aust J Biol Sci 26:539–546
Nevins DJ, Loomis RS (1970) Nitrogen nutrition and photosynthesis in sugar beet. Crop Sci 10:21–25
Nicholls AO (1972) An analysis of the growth of seedlings of four *Atriplex* L. species in controlled environments. Ph D Thesis, Univ Melbourne
Nobel PS (1976) Water relations and photosynthesis of a desert CAM plant, *Agave deserti*. Plant Physiol 58:576–582
Nobel PS (1977) Water relations and photosynthesis of a barrel cactus, *Ferocactus acanthodes*, in the Colorado desert. Oecologia 27:117–133
Nobel PS (1980) Interception of photosynthetically active radiation by cacti of different morphology. Oecologia 45:160–166
Nobel PS, Hartsock TL (1978) Resistance analysis of nocturnal carbon dioxide uptake by a Crassulacean acid metabolism succulent, *Agave deserti*. Plant Physiol 61:510–514
Nobs MA, Pearcy RW, Berry JA, Nicholson F (1972) Reciprocal transplant responses of C_3 and C_4 Atriplexes. Carnegie Inst Washington Yearb 71:164–169
Öquist G, Martensson O, Martin B, Malmberg G (1978) Seasonal effects on chlorophyll-protein complexes isolated from *Pinus silvestris*. Physiol Plant 44:187–192
Österlind S (1951) Inorganic carbon sources of green algae. III. Measurements of photosynthesis in *Scenedesmus quadricauda* and *Chlorella pyrenoidosa*. Physiol Plant 4:242–254
Österlind S (1952) Inorganic carbon sources of green algae. IV. Further experiments concerning photoactivation of bicarbonate assimilation. Physiol Plant 5:403–408
Öztürk M, Rehder H, Ziegler H (1981) Biomass production of C_3- and C_4-plant species in pure and mixed culture with different water supply. Oecologia 50:73–81
O'Leary MH (1981) Carbon isotope fractionation in plants. Phytochemistry 20:553–567
O'Leary MH, Osmond CB (1980) Diffusional contribution to carbon isotope fractionation during dark CO_2 fixation in CAM plants. Plant Physiol 66:931–934
Osmond CB (1975) Environmental control of photosynthetic options in Crassulacean plants. In: Marcelle R (ed) Environmental and biological control of photosynthesis. Junk, The Hague, pp 311–321
Osmond CB (1976) Ion absorption and carbon metabolism in cells of higher plants. In: Lüttge U, Pitman MG (eds) Transport in plants II. Encyclopedia of plant physiology, new ser Vol IIA. Springer, Berlin Heidelberg New York, pp 347–372
Osmond CB (1978) Crassulacean acid metabolism: a curiosity in context. Annu Rev Plant Physiol 29:379–414
Osmond CB (1981) Photorespiration and photoinhibition: some implications for the energetics of photosynthesis. Biochim Biophys Acta 639:77–98
Osmond CB, Björkman O (1972) Simultaneous measurements of oxygen effects on net photosynthesis and glycolate metabolism in C_3 and C_4 species of *Atriplex*. Carnegie Inst Washington Yearb 71:141–148
Osmond CB, Holtum JAM (1981) Crassulacean acid metabolism. In: Hatch MD, Boardman NK (eds) The biochemistry of plants: a comprehensive treatise Vol VIII. Academic Press, London New York, pp 283–328
Osmond CB, Ziegler H (1975) Schwere und leichte Pflanzen: Stabile Isotope im Photosynthesestoffwechsel und in der biochemischen Ökologie. Naturwiss Rundsch 28:323–328
Osmond CB, Allaway WG, Sutton BG, Troughton JH, Queiroz O, Lüttge U, Winter K (1973) Carbon isotope discrimination in photosynthesis of CAM plants. Nature (London) 246:41–42
Osmond CB, Ziegler H, Stichler W, Trimborn P (1975) Carbon isotope discrimination in alpine succulent plants supposed to be capable of Crassulacean acid metabolism (CAM). Oecologia 18:209–217

Osmond CB, Bender MM, Burris RH (1976) Pathways of CO_2 fixation in the CAM plant *Kalanchoe daigremontiana* III. Correlation with $\delta^{13}C$ value during growth and water stress. Aust J Plant Physiol 3:787–799
Osmond CB, Nott DL, Firth PM (1979a) Carbon assimilation patterns and growth of the introduced CAM plant *Opuntia inermis* in Eastern Australia. Oecologia 40:331–350
Osmond CB, Ludlow MM, Davis R, Cowan IR, Powles SB, Winter K (1979b) Stomatal responses to humidity in *Opuntia inermis* in relation to control of CO_2 and H_2O exchange patterns. Oecologia 41:65–76
Osmond CB, Winter K, Powles SB (1980a) Adaptive significance of carbon dioxide recycling during photosynthesis in water stressed plants. In: Turner NC, Kramer PJ (eds) Plant responses to water and high temperature stress. Wiley-Interscience, New York, pp 139–154
Osmond CB, Björkman O, Anderson DJ (1980b) Physiological processes in plant ecology: toward a synthesis with *Atriplex*. Ecol Stud Vol 36. Springer, Berlin Heidelberg New York
Osmond CB, Valaane N, Haslam SM, Uotila P, Roksandic Z (1981) Comparisons of $\delta^{13}C$ values in leaves of aquatic macrophytes from different habitats in Britain and Finland: some implications for photosynthetic processes in aquatic plants. Oecologia 50:117–124.
Park RB, Epstein S (1960) Carbon isotope fractionation during photosynthesis. Plant Physiol 36:133–138
Patten DT, Dinger BE (1969) Carbon dioxide exchange patterns of cacti from different environments. Ecology 50:686–688
Pearcy RW, Troughton JH (1975) C_4 photosynthesis in tree form *Euphorbia* species from Hawaiian rainforest sites. Plant Physiol 55:1054–1056
Pearcy RW, Tumosa N, Williams K (1981) Relationships between growth, photosynthesis and competitive interactions for a C_3 and C_4 plant. Oecologia 48:371–376
Peaslee DE, Moss DN (1968) Stomatal conductivities in K-deficient leaves of maize (*Zea mays* L.) Crop Sci 8:427–430
Philpott J, Troughton JH (1974) Photosynthetic mechanisms and leaf anatomy of hot desert plants. Carnegie Inst Washington Yearb 73:790–793
Picket-Heaps JD (1975) Green algae: structure, reproduction and evolution in selected genera. Sinauer Sunderland
Powles SB, Critchley C (1980) Effect of light intensity during growth on photoinhibition of intact attached bean leaflets. Plant Physiol 65:1181–1187
Powles SB, Osmond CB (1978) Inhibition of the capacity and efficiency of photosynthesis in bean leaflets illuminated in a CO_2-free atmosphere at low oxygen: a possible role for photorespiration. Austr J Plant Physiol 5:619–629
Powles SB, Chapman KSR, Osmond CB (1980) Photoinhibition in intact attached leaves of C_4 plants: dependence on CO_2 and O_2 partial pressures. Aust J Plant Physiol 7:737–747
Prins HBA, Wolff RW (1974) Photorespiration in leaves of *Vallisneria spiralis*: the effect of oxygen on the carbon dioxide compensation point. Proc Kon Ned Akad Wet Ser C 77:239–245
Quandt T, Gottschalk G, Ziegler H, Stichler W (1977) Isotope discrimination by photosynthetic bacteria. FEMS Microbiol Lett 1:125–128
Radmer RJ, Kok B (1976) Photoreduction of O_2 primes and replaces CO_2 assimilation. Plant Physiol 39:336–340
Raschke K (1975) Stomatal action. Annu Rev Plant Physiol 26:309–340
Rathnam CKM, Chollet R (1980) Photosynthetic carbon metabolism in C_4 plants and C_3-C_4 intermediate species. Prog Phytochem 6:1–48
Rau G (1978) Carbon-13 depletion in a subalpine lake: carbon flow implications. Science 201:901–902
Raven JA (1970) Exogenous inorganic carbon sources in plant photosynthesis. Biol Rev 45:167–221
Raven JA (1981) Nutritional strategies of submerged benthic plants: the acquisition of C, N, and P by rhizophytes and halophytes. New Phytol 88:1–30

Raven JA, Glidewell SM (1978) C_4 characteristics of photosynthesis in the C_3 alga *Hydrodictyon africanum*. Plant Cell Environ 1:185–197
Reed ML, Graham D (1977) Carbon dioxide and the regulation of photosynthesis: activities of photosynthetic enzymes and carbonate dehydratase (carbonic anhydrase) in *Chlorella* after growth or adaptation to different carbon dioxide concentrations. Aust J Plant Physiol 4:87–98
Rhoades MM, Carvalho A (1944) The function and structure of the parenchyma sheath plastids of the maize leaf. Bull Torrey Bot Club 71:335–346
Rundel PW (1980) The ecological distribution of C_4 and C_3 grasses in the Hawaiian Islands. Oecologia 45:354–359
Rundel PW, Rundel JA, Ziegler H, Stichler W (1979) Carbon isotope ratios of central Mexican Crassulaceae in natural and greenhouse environments. Oecologia 38:45–50
Ruttner F (1960) Von Kohlendioxyd und Kohlensäure im Süßwasser. In: Ruhland W (ed) Handbuch der Pflanzenphysiologie Vol I. Springer, Berlin Göttingen Heidelberg, pp 62–69
Ryle GJA, Hesketh JD (1969) Carbon dioxide uptake in nitrogen deficient plants. Crop Sci 9:451–454
Schantz HL, Piemeisel LN (1927) The water requirement of plants at Akron, Colorado. J Agr Res (Washington DC) 34:1093–1189
Schiegl WE (1970) Natural deuterium in biogenic materials. Influence of environment and geophysical applications. Ph D Thesis, Univ South Africa, Pretoria
Schiegl WE, Vogel JC (1970) Deuterium content of organic matter. Earth Planet Sci Lett 7:307–313
Schmidt H-L, Winkler FJ (1979) Einige Ursachen der Variationsbreite von $\delta^{13}C$-Werten bei C_3- und C_4-Pflanzen. Ber Dtsch Bot Ges 92:185–191
Schnarrenberger C, Fock H (1976) Interactions among organelles involved in photorespiration. In: Heber U, Stocking CR (eds) Transport in plants III. Encyclopedia of plant physiology new ser Vol III. Springer, Berlin Heidelberg New York, pp 185–234
Schulze E-D, Ziegler H, Stichler W (1976) Environmental control of Crassulacean acid metabolism in *Welwitschia mirabilis* Hook. fil. in its natural range of distribution in the Namib desert. Oecologia 24:323–334
Schulze E-D, Hall AE, Lange OL, Evenari M, Kappen L, Buschbom U (1980a) Long-term effects of drought on wild and cultivated plants in the Negev Desert. I. Maximal rates of net photosynthesis. Oecologia 45:11–18
Schulze E-D, Lange OL, Evenari M, Kappen L, Buschbom U (1980b) Long-term effects of drought on wild and cultivated plants in the Negev Desert II. Diurnal patterns of net photosynthesis and daily carbon gain. Oecologia 45:19–25
Simon H (1982) Tracer-Methoden in der Biologie. In: Hoppe W, Lohmann W, Markl H, Ziegler H (eds) Biophysik, 2nd ed. Springer, Berlin Heidelberg New York
Simon H, Palm D (1966) Isotope effects in organic chemistry and biochemistry. Angew Chem 5:920–933
Sims PL, Singh JS (1978a) The structure and function of ten Western North American grasslands. II. Intraseasonal dynamics in primary producer compartments. J Ecol 66:547–572
Sims PL, Singh GH (1978b) The structure and function of ten Western North American grasslands. III. Net primary production, turnover and efficiencies of energy capture and water use. J Ecol 66:573–579
Sims PL, Singh JS, Lauenroth WK (1978) The structure and function of ten Western North American grasslands. I. Abiotic and vegetational characteristics. J Ecol 66:251–285
Singh KD, Gopal B (1973) The effects of photoperiod and light intensity on the growth of some weeds of crop fields. In: Slatyer RO (ed) Plant response to climatic factors. UNESCO, Paris, pp 73–75
Slack CR, Roughan RG, Bassett HCM (1974) Selective inhibition of mesophyll chloroplast development in some C_4 pathway species by low night temperature. Planta 118:67–73
Slatyer RO (1970) Comparative photosynthesis, growth and transpiration of two species of *Atriplex*. Planta 93:175–189

Smith BN, Epstein S (1971) Two categories of $^{13}C/^{12}C$ ratios for higher plants. Plant Physiol 47:380–384

Smith FA, Walker NA (1980) Photosynthesis by aquatic plants: effects of unstirred layers in relation to assimilation of CO_2 and HCO_3^- and to carbon isotopic discrimination. New Phytol: 86:245–259

Søndergaard M (1979) Light and dark respiration and the effect of the lacunal system on refixation of CO_2 in submerged aquatic plants. Aquat Bot 6:269–283

Spalding MH, Stumpf DK, Ku MSB, Burris RH, Edwards GE (1979) Crassulacean acid metabolism and diurnal variations of internal CO_2 and O_2 concentrations in *Sedum praealtum* D.C. Aust J Plant Physiol 6:557–567

Spalding MH, Edwards GE, Ku MSB (1980) Quantum requirement for photosynthesis in *Sedum praealtum* during two phases of Crassulacean acid metabolism. Plant Physiol 66:463–465

Spence DHN (1976) Light and plant response in fresh water. In: Evans GC, Bainbridge R, Rackham O (eds) Light as an ecological factor Vol II. 16th Symp Br Ecol Soc. Blackwell, Oxford, pp 93–133

Spence DHN (1981) Zonation of plants in freshwater lakes. Adv Ecol Res

Steemann Nielsen E (1960) Uptake of carbon dioxide by the plant. In: Ruhland W (ed) Handbuch der Pflanzenphysiologie. Vol I. Springer, Berlin Göttingen Heidelberg, pp 70–84

Stewart KD, Mattox KR (1975) Comparative cytology, evolution and classification of the green algae with some consideration of the origin of other organisms with chlorophylls a and b. Bot Rev 41:104–135

Stowe LG, Teeri JA (1978) The geographic distribution of C_4 species of the dicotyledonae in relation to climate. Am Nat 112:609–623

Stumm W, Morgan JJ (1970) Aquatic chemistry. Wiley-Interscience, New York

Sutton BG, Ting IP, Troughton JH (1976) Seasonal effects on carbon isotope composition of cactus in a desert environment. Nature (London) 261:42–43

Szarek SR, Ting IP (1977) The occurrence of Crassulacean acid metabolism among plants. Photosynthetica 11:330–342

Szarek SR, Johnson HB, Ting IP (1973) Drought adaptation in *Opuntia basilaris*. Significance of recycling carbon through Crassulacean acid metabolism. Plant Physiol 52:539–541

Talling JF (1976) The depletion of carbon dioxide from lake water by phytoplankton. J Ecol 64:79–121

Taylor AO, Rowley JA (1971) Plants under climatic stress. I. Low temperature, high light effects on photosynthesis. Plant Physiol 47:713–718

Teeri JA, Stowe LG (1976) Climatic patterns and the distribution of C_4 grasses in North America. Oecologia 23:1–12

Teeri JA, Stowe LG, Murawski DA (1978) The climatology of two succulent plant families, Cactaceae and Crassulaceae. Can J Bot 56:1750–1758

Terry N (1979) The use of mineral nutrient stress in the study of limiting factors in photosynthesis. In: Marcelle R, Clijsters H, Pucke M van (eds) Photosynthesis and plant development. Junk, The Hague, pp 151–160

Thomas EA, Tregunna EB (1968) Bicarbonate ion assimilation in photosynthesis by *Sargassum muticum*. Can J Bot 46:411–415

Tieszen LL, Senyimba MM, Imbamba SK, Troughton JH (1979) The distribution of C_3 and C_4 grasses and carbon isotope discrimination along an altitudinal and moisture gradient in Kenya. Oecologia 37:337–350

Trebst A, Avron M (1978) (eds) Photosynthesis. I. Photosynthetic electron transport and photophosphorylation. Encyclopedia of plant physiology new ser Vol V. Springer, Berlin Heidelberg New York

Troughton JH (1979) $\delta^{13}C$ as an indicator of carboxylation reactions. In: Gibbs M, Latzko E (eds) Photosynthesis II. Encyclopedia of plant physiology new ser Vol 6. Springer, Berlin Heidelberg New York, pp 140–147

Troughton JH, Mooney HA, Berry JA, Verity D (1977) Variable carbon isotope ratios of *Dudleya* species growing in natural habitats. Oecologia 30:307–312

Updike J (1968) Couples. Fawcett, Greenwich
Van TK, Haller WT, Bowes G (1976) Comparison of the photosynthetic characteristics of three submerged aquatic plants. Plant Physiol 58:761–768
Vogel JC (1980) Fractionation of the carbon isotopes during photosynthesis. Springer, Berlin Heidelberg New York
Vogel JC, Fuls A, Ellis RP (1978) The geographical distribution of Kranz grasses in South Africa. S Afr J Sci 74:209–215
Volkens G (1887) Die Flora der ägyptisch-arabischen Wüste. Sitzungsber Preuss Acad Wiss Berl Phys Math Kl
Wagner J, Larcher W (1981) Dependence of CO_2 gas exchange and acid metabolism of the alpine CAM plant *Sempervivum montanum* on temperature and light. Oecologia 50:88–93
Weber JN, Woodhead P (1970) Carbon and oxygen isotope fractionation in the skeletal carbonate of reef building corals. Chem Geol 6:93–123
Weidner M, Küppers U (1973) Phosphoenolpyruvat-Carboxykinase und Ribulose-1,5-Diphosphat-Carboxylase von *Laminaria hyperborea* (Gunn.) Fosl: Das Verteilungsmuster der Enzymaktivitäten im Thallus. Planta 114:365–372
Westlake DF (1967) Some effects of low velocity currents on the metabolism of aquatic macrophytes. J Exp Bot 18:187–205
Willenbrink J, Rangoni-Kübbeler M, Tersky B (1975) Frond development and CO_2 fixation in *Laminaria hyperborea*. Planta 125:161–170
Williams GJ III (1974) Photosynthetic adaptation to temperature in C_3 and C_4 grasses. A possible ecological role in the short-grass prairie. Plant Physiol 54:709–711
Willis JC (1973) A dictionary of the flowering plants and ferns, 8th edn. Univ Press, Cambridge
Winter K (1978) Short-term fixation of 14-Carbon by the submerged aquatic angiosperm *Potamogeton pectinatus*. J Exp Bot 29:1169–1172
Winter K (1979a) $\delta^{13}C$ values of some succulent plants from Madagascar. Oecologia 40:104–112
Winter K (1979b) Photosynthetic and water relationships of higher plants in saline environments. In: Jefferies RL, Davy AJ (eds) Ecological processes in coastal environments. Blackwell, Oxford, pp 297–320
Winter K (1980) Carbon dioxide and water vapor exchange in the Crassulacean acid metabolism plant *Kalanchoe pinnata* during a prolonged light period. Plant Physiol 66:917–921
Winter K (1981) C_4 plants of high biomass in arid regions of Asia – occurrence of C_4 photosynthesis in Chenopodiaceae and Polygonaceae from the Middle East and USSR. Oecologia 48:100–106
Winter K, Lüttge U (1976) Balance between C_3 and CAM pathway of photosynthesis. In: Lange OL, Kappen L, Schulze E-D (eds) Water and plant life: problems and modern approaches. Ecol Stud Vol 19. Springer, Berlin Heidelberg New York, pp 323–334
Winter K, Troughton JH (1978) Photosynthetic pathways in plants of coastal and inland habitats of Israel and the Sinai. Flora 167:1–34
Winter K, Willert von DJ (1972) NaCl-induzierter CAM bei *Mesembryanthemum crystallinum*. Z Pflanzenphysiol 67:166–170
Winter K, Kramer D, Throughton JH, Card KA, Fischer K (1977) C_4 pathway of photosynthesis in a member of the Polygonaceae: *Calligonum persicum* (Boiss. & Buhse) Boiss. Z Pflanzenphysiol 81:341–346
Winter K, Lüttge U, Winter E, Troughton JH (1978) Seasonal shift from C_3 photosynthesis to Crassulacean acid metabolism in *Mesembryanthemum crystallinum* growing in its natural environment. Oecologia 34:225–237
Wong SC, Cowan IR, Farquhar GD (1979) Stomatal conductance correlates with photosynthetic capacity. Nature (London) 282:424–426
Wong WW, Benedict CR, Kohel RJ (1979) Enzymic fractionation of the stable carbon isotopes of carbon dioxide by ribulose-1,5-bisphosphate carboxylase. Plant Physiol 63:852–856

Yamaguchi T, Ikawa T, Nisizawa K (1969) Pathway of mannitol formation during photosynthesis in brown algae. Plant Cell Physiol 10:425–440

Ziegler H (1979) Diskriminierung von Kohlenstoff- und Wasserstoffisotopen: Zusammenhänge mit dem Photosynthesemechanismus und den Standortbedingungen. Ber Dtsch Bot Ges 92:169–184

Ziegler H, Osmond CB, Stichler W, Trimborn P (1976) Hydrogen isotope discrimination in higher plants: correlation with photosynthetic pathway and environment. Planta 128:85–92

Ziegler H, Batanouny KH, Sankhla N, Vyas OP, Stichler W (1981) The photosynthetic pathway types of some desert plants from India, Saudi Arabia, Egypt and Iraq. Oecologia 48:93–99

16 Modelling of Photosynthetic Response to Environmental Conditions

G.D. FARQUHAR and S. VON CAEMMERER

CONTENTS

16.1 Introduction

Photosynthesis is the incorporation of carbon, nitrogen, sulphur and other substances into plant tissue using light energy from the sun. Most of this energy is used for the reduction of carbon dioxide and, consequently, there is a large body of biochemical and biophysical information about photosynthetic carbon assimilation. In an ecophysiological context, we believe that most of today's biochemical knowledge can be summarized in a few simple equations. These equations represent the rate of ribulose bisphosphate (RuP_2)-saturated carboxylation, the ratio of photorespiration to carboxylation, and the rates of electron transport/photophosphorylation and of "dark" respiration in the light. There are many other processes that could potentially limit CO_2 assimilation, but probably do so rarely in practice. Fundamentally this may be due to the expense, in terms of invested nitrogen, of the carboxylase and of thylakoid functioning. To reach our final simple equations we must first discuss the biochemical and biophysical structures – as they are understood at present – that finally reduce the vast number of potentially rate-limiting processes to the four or five listed above. A diagrammatic representation of these processes is given in Fig. 16.1.

We then discuss integrated functioning in individual leaves of C_3 species, and later briefly discuss canopy photosynthesis. Models of C_4 photosynthesis are less well developed than those of C_3 photosynthesis; we discuss how they may be extended. (For description of these pathways of photosynthesis see also Chap. 15, this Vol.)

Models of leaf photosynthesis have been reviewed by THORNLEY (1976), JASSBY and PLATT (1976), and TENHUNEN et al. (1980a). In this article we emphasize recent progress made in mechanistic models, but recognize the validity of other types (see Chap. 8, this Vol.) and give brief references to successful empirical models. Stomatal and boundary layer conductances affect the intercellular partial pressure of CO_2, C, and the leaf temperature, T. We examine responses of the rate of assimilation to C and T but do not consider submodels

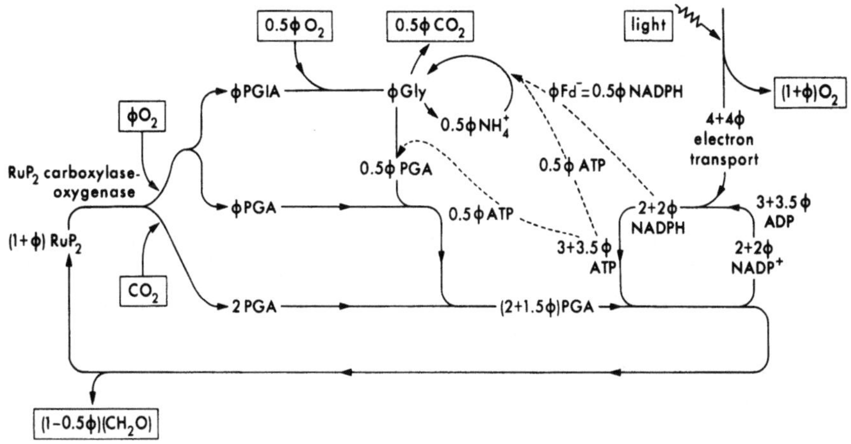

Fig. 16.1. Simplified photosynthetic carbon reduction (PCR) and photorespiratory carbon oxidation (PCO) cycles, with the cycle for regeneration of NADPH and ATP linked to light-driven electron transport. For each carboxylation, ϕ oxygenations occur. *Gly* glycine; *Fd⁻* reduced ferredoxin (assumed equivalent to $^1/_2$ NADPH); *PGA* 3-phosphoglycerate; *PGlA* phosphoglycolate. At the compensation point $\phi=2$

of the response of stomatal conductance to environmental perturbations. These are dealt with in Chapter 8, this Volume (see also Chaps. 7 and 17, this Vol.).

16.2 Stromal and Extrachloroplastic Reactions

16.2.1 Kinetics of Ribulose Bisphosphate Carboxylase-Oxygenase

16.2.1.1 RuP_2 Saturated Rates

The currently accepted equation to describe the rate, V_c, of carboxylation of ribulose bisphosphate (RuP_2), in the presence of competitive inhibition by oxygen, and of saturating concentrations of RuP_2 is

$$V_c = W_c = \frac{V_{c\,max}\,C}{C + K_c(1 + O/K_o)}. \tag{16.1}$$

We denote RuP_2 carboxylase rates in general as V_c, and RuP_2 saturated rates as W_c. $V_{c\,max}$ is the maximum velocity, C and O are partial pressures of CO_2 and O_2, $p(CO_2)$ and $p(O_2)$ respectively, in equilibrium with their dissolved concentrations in the chloroplast stroma, K_c is the Michaelis-Menten constant for CO_2, and K_O is the Michaelis-Menten constant for O_2. This equation is used in the models of LAING et al. (1974), KU and EDWARDS (1977b), TENHUNEN et al. (1977), BERRY and FARQUHAR (1978), FARQUHAR et al. (1980), COOKE and RAND (1980), BAUWE et al. (1980) and RAVEN and GLIDEWELL (1981). Earlier attempts to include competitive inhibition by oxygen in models of leaf photosynthesis were made by LAISK (1970), HALL (1971), PEISKER (1974), HALL and BJÖRKMAN (1975), and CHARTIER and PRIOUL (1976).

The analogous equation for the oxygenation of RuP_2 is

$$V_o = W_o = \frac{V_{o\,max}\, O}{O + K_o(1 + C/K_c)}, \tag{16.2}$$

where $V_{o\,max}$ is the maximum rate. Equation (16.2) is used in the models of LAING et al. (1974), TENHUNEN et al. (1977), BERRY and FARQUHAR (1978), SINCLAIR and RAND (1979), FARQUHAR et al. (1980), COOKE and RAND (1980), BAUWE et al. (1980), and RAVEN and GLIDEWELL (1981). The ratio of the activities of oxygenase and carboxylase, denoted ϕ by FARQUHAR et al. (1980), is (LAING et al. 1974)

$$\phi = \frac{V_o}{V_c} = \frac{V_{o\,max}}{V_{c\,max}} \cdot \frac{O K_c}{K_o C}. \tag{16.3}$$

16.2.1.2 RuP_2 Limited Rates

Equations (16.1) and (16.2) apply when the RuP_2 is present in saturating amounts. Equation (16.3) applies regardless, provided the mechanisms of the reactions are ordered with RuP_2 binding first (FARQUHAR 1979). When RuP_2 is not saturating, (16.1) becomes, in the absence of oxygen (BADGER and COLLATZ 1977),

$$\frac{V_{c\,max}}{V_c} = 1 + \frac{K_{rc}}{R} + \frac{K_c}{C} + \frac{K_{ir}}{R} \cdot \frac{K_c}{C} \tag{16.4}$$

and (16.2) becomes, in the absence of CO_2,

$$\frac{V_{o\,max}}{V_o} = 1 + \frac{K_{ro}}{R} + \frac{K_o}{O} + \frac{K_{ir}}{R} \cdot \frac{K_o}{O}, \tag{16.5}$$

where R is the concentration of free RuP_2, K_{rc} and K_{ro} are Michaelis-Menten constants for RuP_2 in the presence of saturating CO_2 and O_2 respectively, and K_{ir} is the dissociation constant for the enzyme-RuP_2 complex.

The mechanism of the reactions is uncertain, being either random, or ordered with RuP_2 binding to the enzyme first (BADGER and COLLATZ 1977). FARQUHAR (1979) showed that the latter case modified (16.1) and (16.2)

$$V_c = W_c \cdot \frac{R}{R + K_r'} \tag{16.6}$$

$$V_o = W_o \cdot \frac{R}{R + K_r'}, \tag{16.7}$$

where W_c and W_o are the RuP_2 saturated rates given by Eqs. (16.1) and (16.2), and K_r' is the effective Michaelis-Menten constant for RuP_2 given by

$$K_r' = \frac{K_{ro} O/K_o + K_{rc} C/K_c + K_{ir}}{O/K_o + C/K_c + 1}. \tag{16.8}$$

An implicit form of these equations was used by HALL (1979).

16.2.1.3 On the High Concentration of Enzyme Sites

PEISKER (1974), aware of the high concentration of enzyme sites in the chloroplast, E_t, presented the equations which would apply if this concentration exceeded that of total RuP_2, R_t. These equations, written in terms of the rate constants of the individual processes, were complex, and FARQUHAR (1979) showed that they could be written more simply as

$$V_c = W_c \cdot \frac{R_t}{E_t + K_r'} \tag{16.9}$$

$$V_o = W_o \cdot \frac{R_t}{E_t + K_r'}. \tag{16.10}$$

The analysis suggested that although the K_m of RuP_2 in vitro (where the concentration of enzyme sites is extremely small, typically 1 μM) is only about 20 μM, half maximal rates in vivo, when the enzyme site concentration is about 4 mM (JENSEN and BAHR 1977), would only be seen when the RuP_2 concentration exceeded 2 mM. FARQUHAR (1979) predicted a non-rectangular hyperbolic response of carboxylase activity to total RuP_2 concentration (as distinct from the rectangular hyperbolic response to free CO_2 concentration)

$$\left(\frac{V_c}{W_c}\right)^2 - \frac{V_c}{W_c}\left(1 + \frac{R_t + K_r'}{E_t}\right) + \frac{R_t}{E_t} = 0 \tag{16.11}$$

$$\left(\frac{V_o}{W_o}\right)^2 - \frac{V_o}{W_o}\left(1 + \frac{R_t + K_r'}{E_t}\right) + \frac{R_t}{E_t} = 0. \tag{16.12}$$

He showed that with the high concentration of enzyme sites in vivo this could be closely approximated by

$$V_c = W_c \cdot \min\{1, R_t/E_t\}, \tag{16.13}$$

where min{ } denotes "minimum of" and E_t is the total concentration of enzyme sites, and R_t is the total concentration (free plus bound) of RuP_2. Equation (16.3) remains valid. Equations (16.4)–(16.7) also remain valid provided R refers to the concentrations of free (unbound) RuP_2. However, the usual assays for RuP_2 measure the total concentration, R_t.

Data supporting the use of Eq. (16.13) have been obtained by COLLATZ (1978) in *Chlamydomonas* and spinach cells. COLLATZ measured CO_2 assimilation rate, A, and RuP_2 concentrations, R_t, at various irradiances. When A is plotted

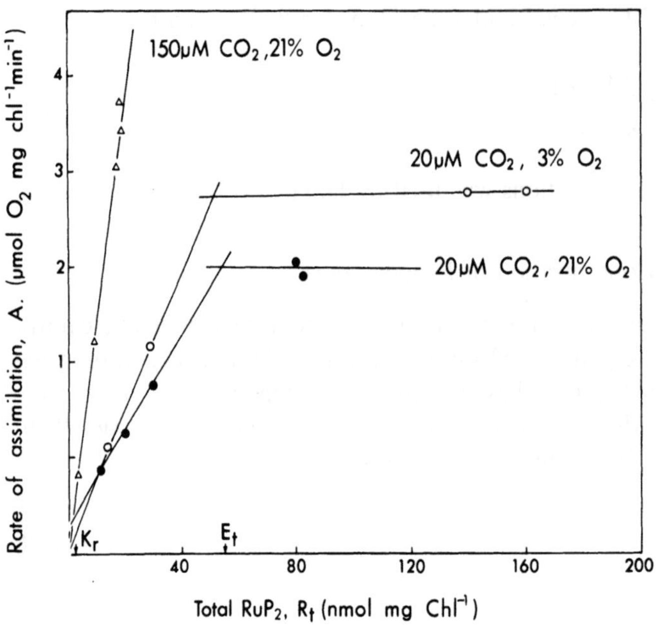

Fig. 16.2. Net rate of photosynthesis, A, vs. total RuP_2 concentration, R_t, in *Chlamydomonas reinhardtii* cells as irradiance varies at various concentrations of CO_2 and O_2. (Replotted from the data of COLLATZ 1978). Total concentration of enzyme sites, E_t, is approximately 55 nmol mg Chl^{-1} in this species. A K_m for free RuP_2, K_r, of approximately 20 μM would correspond to 0.6 nmol RuP_2 mg $Chl^{-°}$, since 1 mg Chl is approximately 30 μl stroma

against R_t (Fig. 16.2), it is apparent that saturation does not occur until concentrations vastly in excess of the in vitro K_m for (free) RuP_2 are reached.

16.2.1.4 Activation of RuP_2 Carboxylase-Oxygenase

The degree of activation of the carboxylase-oxygenase enzyme depends on the concentrations of CO_2 and of Mg^{2+}; this has been modelled by LORIMER et al. (1976) and LAING and CHRISTELLER (1976). In a steady, partially activated state, $V_{c\,max}$ and $V_{o\,max}$ are modified so that W_c is given by (FARQUHAR 1979)

$$W_c = \frac{C \cdot [Mg^{2+}]}{C[Mg^{2+}] + K_1 \cdot K_2} \cdot \frac{V_{c\,max} C}{C + K_c(1 + O/K_o)}. \qquad (16.14)$$

W_o is modified in an analogous manner. K_1 and K_2 are kinetic parameters. $[Mg^{2+}]$ refers to the concentration of free (unbound) Mg^{2+} and for complete activation, with a Mg^{2+} ion bound to each site, the total concentration must exceed E_t, which is typically 4 mM. The effective K_m for RuP_2 should also be modified when the enzyme is not fully activated (FARQUHAR 1979). Data of MEIDNER (1970) and VON CAEMMERER and FARQUHAR (1981) may be interpreted in the present context as suggesting that the effective CO_2 Michaelis constant of activation for intact leaves (given by $K_1 K_2/[Mg^{2+}]$) is typically about 10 μbar.

Thus under normal conditions ($C \approx 170$–270 μbar) the enzyme is likely to be fully activated with W_c given by Eq. (16.1). Thus while the effects of enzyme inactivation on the properties of the integrated photosynthetic system are of physiological and biochemical interest, they appear at present to be unimportant in an ecophysiological context. This view could change if it were shown that stomatal closure causes low intercellular $p(CO_2)$ in natural conditions.

16.2.1.5 "Enzyme Associated" Rate of CO_2 Assimilation

In an attempt to gradually develop full equations for CO_2 assimilation rate, we anticipate later discussion and note that oxygenation of 1 mol of RuP_2 causes the release of 0.5 mol of CO_2 in the photorespiratory carbon oxidation (PCO) cycle. The rate of CO_2 assimilation, A, associated with RuP_2 carboxylase-oxygenase is then

$$A = V_c - 0.5\ V_o. \tag{16.15}$$

[LAING et al. (1974), SINCLAIR et al. (1977) and COOKE and RAND (1980) have incorrectly assumed 0.25 mol of CO_2 per mol of oxygenation, confusing this with the fraction of glycolate carbon released.] At a whole leaf level, the above rate of assimilation needs to be offset by CO_2 evolution due to other processes.

16.2.2 "Day" Respiration

The extent to which "dark" respiration continues in the light has been discussed elsewhere (GRAHAM and CHAPMAN 1979). A proportion of the cells in a leaf is heterotrophic, and respiration may be expected to continue in these cells. MANGAT et al. (1974) showed that mitochondrial respiration in autotrophic cells was almost completely eliminated by high levels of ATP in the cytoplasm in experiments carried out in CO_2-free air. Data of PEISKER et al. (1981) suggest that at the CO_2 compensation point, CO_2 evolution by leaf respiration was inhibited by 50%. As we discuss later, there is likely to be less ATP available at the compensation point than in CO_2-free air, and less still at normal partial pressures of CO_2. Mitochondrial respiration in the light may, under normal ambient conditions, differ little from dark respiration at the same temperature; more experiments are required. Whether or not mitochondrial oxygen uptake and electron transport associated with normal dark respiration are partially inhibited by illumination, the tricarboxylic acid cycle should continue to supply α-ketoglutarate for amino acid biosynthesis. This results in CO_2 evolution. Some glycine decarboxylation may result from glycolate produced by reactions other than those associated with the rate of oxygenation of RuP_2, V_o, although recent evidence suggests this is minimal (SOMMERVILLE and OGREN 1979). Indeed glycine decarboxylation in photorespiration is probably the major potential source of electrons for mitochondria in the light. We denote as R_d ("day respiration") all CO_2 evolution from the cells inside the leaf, in addition to that due (via PCO) to V_o. The net rate of CO_2 assimilation A, is then given by

$$A = V_c - 0.5\ V_o - R_d. \tag{16.16}$$

16.2.3 CO_2 Compensation Point, Γ

16.2.3.1 Introduction

If stomata were to close completely, so that CO_2 assimilation rate went to zero, then the intercellular $p(CO_2)$, C, should be at the CO_2 compensation point, Γ. There is little evidence to suggest that stomata behave this way in practice. Nevertheless, Γ is a widely studied and modelled parameter. The fact that, in ordinary partial pressures of CO_2 (338 μbar in 1980), there is an irradiance, the "light compensation point", at which net CO_2 assimilation is zero, means that strictly the compensation point should be called the light and CO_2 compensation point, since it will depend on both irradiance and CO_2 concentration.

In our opinion, and consistent with the data of COLLATZ (1978), *most* studies of the CO_2 compensation point are carried out in conditions where RuP_2 is not limiting. In such conditions, Γ is independent of irradiance. We discuss this situation first.

In what follows, we ignore the consequences of the sites of CO_2 evolution, leaf mitochondria, being seperated from those of carboxylation, the chloroplast stroma. These effects have been discussed by CHARLES-EDWARDS (1978) and PEISKER (1978a), and are most likely of secondary importance in the present context.

16.2.3.2 $R_t > E_t$, $R_d = 0$

In the absence of R_d, we see from (16.16) that assimilation rate is zero when $\phi = V_o/V_c = 2$. Thus Γ_*, the CO_2 compensation point in the absence of day respiration, called the CO_2-photocompensation point by LAISK (1977), is given by

$$\Gamma_* = \frac{0.5\, V_{o\,max} K_c O}{V_{c\,max} K_o}. \qquad (16.17)$$

We note from this equation, together with Eq. (16.3), that the ratio of RuP_2 oxygenation to carboxylation is given by

$$\phi = 2\Gamma_*/C. \qquad (16.18)$$

Equation (16.17) may be rewritten as

$$\Gamma_* = \gamma_* \cdot O, \qquad (16.19)$$

where

$$\gamma_* = 0.5 \frac{V_{o\,max} K_c}{V_{c\,max} K_o}. \qquad (16.20)$$

Γ_* increases linearly with $p(O_2)$, as discussed by LAING et al. (1974).

16.2.3.3 $R_t > E_t$, $R_d > O$

From earlier equations it may be shown that the CO_2 compensation point, Γ, in the presence of day respiration, R_d, is given by (TENHUNEN et al. 1980b; FARQUHAR et al. 1980)

$$\Gamma = \frac{\Gamma_* + K_c(1 + O/K_o)\, R_d/V_{c\,max}}{1 - R_d/V_{c\,max}}. \tag{16.21}$$

Again Γ depends linearly on $p(O_2)$ and may be written as

$$\Gamma = \gamma \cdot O + \Gamma_o, \tag{16.22}$$

where

$$\gamma = \frac{K_c}{K_o} \cdot \frac{0.5 V_{o\,max}/V_{c\,max} + R_d/V_{c\,max}}{1 - R_d/V_{c\,max}}, \tag{16.23}$$

$$\gamma = \frac{\gamma_* + \dfrac{K_c R_d}{K_o V_{c\,max}}}{1 - R_d/V_{c\,max}}, \tag{16.23a}$$

and

$$\gamma > \gamma_*,$$

and

$$\Gamma_o = \frac{K_c R_d/V_{c\,max}}{1 - R_d/V_{c\,max}} \tag{16.24}$$

(TENHUNEN et al. 1980b; AZCON-BIETO et al. 1981).

It has been observed experimentally that $V_{o\,max}/V_{c\,max}$ is virtually constant over a range of temperatures and is approximately 0.21 (BADGER and ANDREWS 1974; BADGER and COLLATZ 1977; JORDAN and OGREN 1981). [This evidence has been ignored by TENHUNEN in his papers (TENHUNEN et al. 1976, and subsequent papers) and he has assumed a ratio of 2.0.] Consequently the temperature dependence of Γ is due to the temperature dependencies of K_c/K_o and of $R_d/V_{c\,max}$. From Eqs. (16.21) to (16.24) it may be seen that whenever R_d occurs it is divided by $V_{c\,max}$. CHARLES-EDWARDS (1978) looked for a dependence of Γ on the rate of respiration in the dark, R_n. The appropriate dependence is on $R_d/V_{c\,max}$. PEISKER et al. (1981) observed a relationship between Γ and $R_n/V_{c\,max}$. They concluded that $R_d \approx 0.5\, R_n$. Since Γ_o depends on $R_d/V_{c\,max}$ it is easy to see why Γ_o has been nearly zero in some cases and larger in others, as discussed by AZCON-BIETO et al. (1981).

16.2.3.4 $R_t < E_t$, $R_d = 0$

At small irradiances, the RuP_2 carboxylase-oxygenase is RuP_2 limited and $R_t < E_t$, i.e., only a proportion of the sites have RuP_2 available. PEISKER (1974) derived a linear dependence of Γ_* on $p(O_2)$ for this situation with $R_d = 0$. From Eq. (16.6) and (16.7) it is clear that carboxylation and oxygenation are affected equally, and Eqs. (16.17)–(16.20) are valid.

16.2.3.5 $R_t < E_t$, $R_d > 0$

The situation where day respiration is occurring is more complex. From Eq. (16.13) it emerges that in Eqs. (16.21)–(16.24) $V_{c\,max}$ must be multiplied by R_t/E_t, the proportion of enzyme sites with RuP_2 attached. As R_t decreases, as occurs at low irradiances, Γ increases, especially at higher temperatures where R_d becomes large. Relevant recent data on compensation points have been presented by ČATSKÝ and TICHÁ (1979). Unlike earlier circumstances, we do not expect a linear dependence of Γ on O_2 here. Changing $p(O_2)$ at the compensation point causes changes in patterns of RuP_2 consumption and regeneration, and hence of R_t/E_t.

The estimation of these patterns is the subject of the next section, since the rate of carboxylation depends on R_t/E_t [Eq. (16.13)].

16.2.4 Regeneration of RuP_2 – Stromal and Extrachloroplastic Reactions

16.2.4.1 Introduction

The carboxylation of RuP_2 results in the production of two molecules of phosphoglyceric acid (PGA). The regeneration of the acceptor RuP_2 is achieved in the Calvin, or reductive pentose phosphate or photosynthetic carbon reduction (PCR) cycle, reviewed by BASSHAM (1979). The photorespiratory carbon oxidation (PCO) cycle functions to regenerate RuP_2 after it has been oxygenated, and was described by BERRY and FARQUHAR (1978) following studies by LORIMER et al. (1978). In Chap. 15, this Volume, it is discussed as a pathway (PCOP; Figs. 15.1 and 15.2) rather than as a cycle, although this distinction is debatable.

The integrated PCR and PCO cycles consume ATP and NADPH. We believe that, in cases where the supply of RuP_2 is limiting the rate of RuP_2 carboxylation in natural conditions, it is the supply of ATP and NADPH (discussed in Sect. 16.2.5) that is usually limiting rather than the activities of enzymes, or pool sizes of intermediates, in the PCR and PCO cycles. A possible exception is the supply of inorganic phosphate (see Sect. 16.5.1).

16.2.4.2 Photorespiratory Carbon Oxidation Cycle

Oxygenation of 1 mol of RuP_2 produces 1 mol of PGA and 1 mol of phosphoglycolate (PGlA) (see Fig. 16.3). Following this 1 mol PGlA gives rise to 0.5 mol PGA, requiring 0.5 mol ATP in the final step, and releases 0.5 mol of CO_2. As well, 0.5 mol ammonia is released and refixed. The refixation of

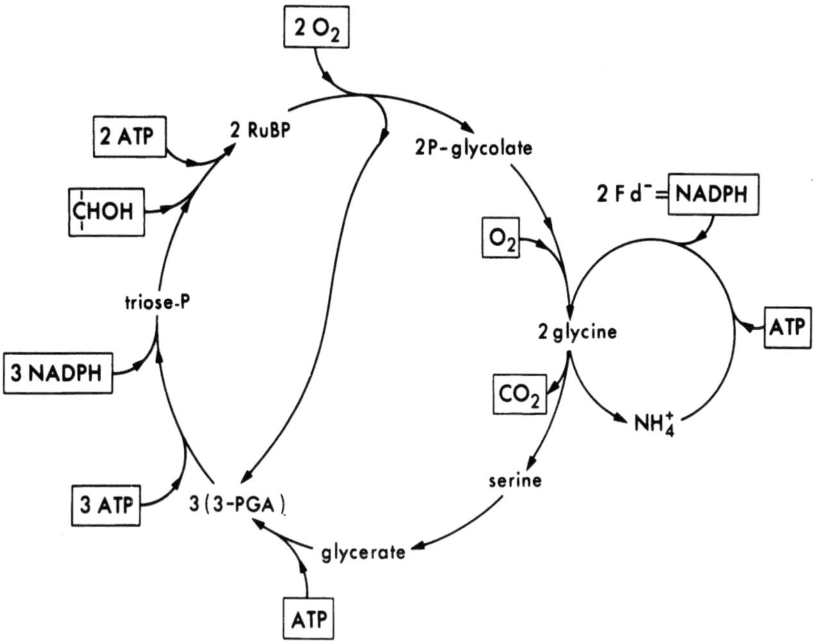

Fig. 16.3. The photorespiratory carbon oxidation cycle. (After BERRY and FARQUHAR 1978)

0.5 mol NH_4^+ requires 1 mol F_d^- (reduced ferredoxin) (which, in terms of electron transport, is the equivalent to 0.5 mol NADPH) and 0.5 mol ATP. The PGA produced is then converted to RuP_2 as in the PCR cycle.

16.2.4.3 PGA Production, ATP and NADPH Consumption in the Integrated PCR and PCO Cycles.

In the PCR cycle, 2 mol of PGA are consumed to produce 1 mol RuP_2 and 1 mol (CH_2 O) (see Fig. 16.1). In the process, 2 mol NADPH and 3 mol ATP are consumed. From the considerations of BERRY and FARQUHAR (1978) and FARQUHAR et al. (1980)

$$\text{rate of PGA production} = 2\,V_c + 1.5\,V_o = (2 + 1.5\,\phi)\,V_c = (2 + 3\,\Gamma_*/C)\,V_c, \quad (16.25)$$

$$\text{rate of NADPH consumption} = 2\,V_c + 2\,V_o = (2 + 2\,\phi)\,V_c = (2 + 4\,\Gamma_*/C)\,V_c, \quad (16.26)$$

$$\text{rate of ATP consumption} = 3\,V_c + 3.5\,V_o = (3 + 3.5\,\phi)\,V_c = (3 + 7\,\Gamma_*/C)\,V_c. \quad (16.27)$$

We next examine the steps in which PGA, NADPH and ATP are consumed, and RuP_2 is produced.

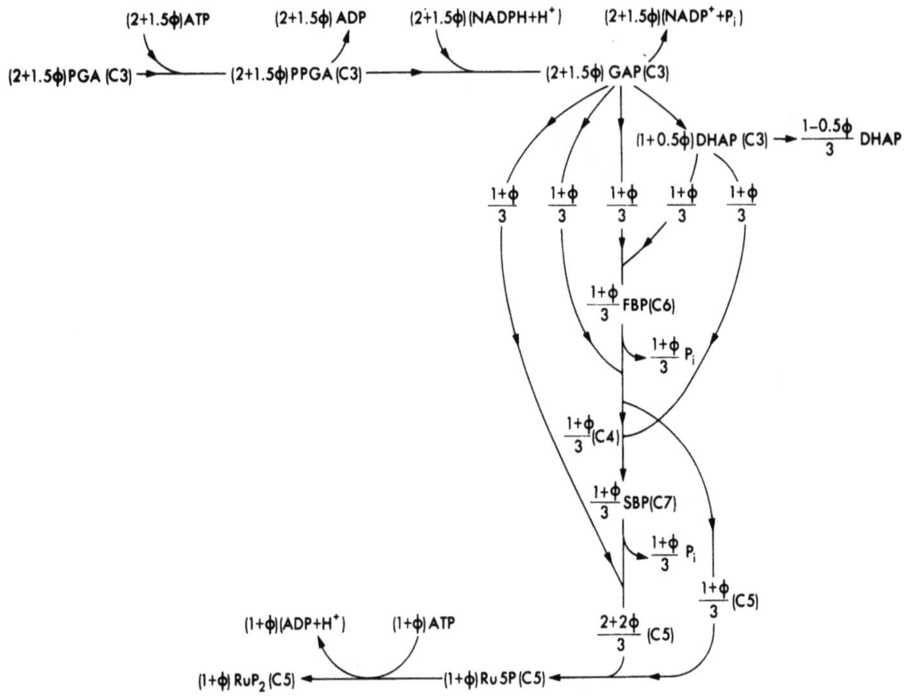

Fig. 16.4. The carbon pathway from PGA to RuP_2 (where ϕ is the number of oxygenations per carboxylation; see Fig. 16.1)

16.2.4.4 Enzymatic Steps from PGA to RuP_2

There are many enzymatic steps involved in the pathway from PGA to RuP_2 (BASSHAM 1979), and these are outlined in Fig. 16.4. The reactions are reversible apart from those catalyzed by fructose bisphosphatase (FBPase) and by sedoheptulose bisphosphatase (SBPase). SBPase activity also increases with SBP levels (WOODROW and WALKER 1980). The activities of both FBPase and SBPase are modulated by pH and the concentration of Mg^{2+} and of reduced thioredoxin. The latter substance is reduced by ferredoxin from the electron transport chain and this ensures that futile cycles do not waste energy in the dark (BASSHAM 1979). Furthermore, since these enzymes have only a limited capacity in the light the concentration of DHAP is maintained at levels appropriate for export to the cytoplasm. However, while the maximum rates of FBPase and SBPase are regulating factors, we do not believe they are primary limiting factors in CO_2 assimilation by leaves. From Fig. 16.4 it may be seen that they only need to run at $^1/_3$ the rate of RuP_2 regeneration. LEEGOOD and WALKER (1980) measured FBPase rates of 120 μmol g Chl^{-1} h^{-1} in wheat chloroplasts, which were more than sufficient (by a factor of 4 in fact) to account for an assimilation rate of 100 μmol g Chl^{-1} h^{-1}. Similarly high rates have been measured by ROBINSON and WALKER (1980) in spinach chloroplasts. Rates of SBPase sufficient to account for CO_2 assimilation have been measured by WOODROW and WALKER (1980) in wheat chloroplasts. Since FBPase and SBPase activities reflect the amount of reduced ferridoxin, it is possible that they may co-limit RuP_2 regeneration with electron transport, the latter being the more fundamental limitation.

The reactants involved in the steps between PGA and DHAP are nearly in equilibrium (LILLEY et al. 1977) and have been modelled by GIERSCH et al. (1980b). Detailed modelling of the RuP_2 regeneration cycle could usefully be extended. Progress might be made by assuming that other reactions, apart from those catalyzed by RuP_2 carboxylase-oxygenase, FBPase and SBPase, are also at equilibrium. A detailed analysis of the enzyme system involved in RuP_2 regeneration will probably give information about pool sizes of Calvin cycle intermediates, but not affect the estimation of CO_2 assimilation by leaves. To reach this conclusion we must examine the existing, crude representations of RuP_2 regeneration, before integrating these with the thylakoid reactions in Section 16.4.

HALL (1971) simplified the regeneration steps to a single one, first order in [PGA] and [NADPH]. Thus HALL wrote, using different symbols,

$$\text{rate of reduction of PGA} = \frac{[\text{PGA}]}{R_p} \cdot \frac{[\text{NADPH}]}{N_t} \cdot M, \tag{16.28}$$

where M is the maximum rate, a fictional composite of many reactions, R_p is the potential pool size of RuP_2 which would occur in the absence of carboxylation and oxygenation, and N_t is the total concentration of NADPH and $NADP^+$, i.e.,

$$N_t = [\text{NADPH}] + [\text{NADP}^+]. \tag{16.29}$$

This has formed the basis of models of RuP_2 regeneration in many subsequent models (HALL and BJÖRKMAN 1975; PEISKER 1976; HALL 1979; FARQUHAR et al. 1980). In the models of HALL (1971), HALL and BJÖRKMAN (1975) and PEISKER (1976) an increased pool of PGA requires a decreased pool of RuP_2 on an equimolar basis. However, LILLEY et al. (1977) have shown that total phosphate is conserved in the chloroplast; FARQUHAR et al. (1980) and FARQUHAR and VON CAEMMERER (1981) optimistically interpreted this as conservation of esterified phosphates and assumed that 1 mol extra PGA resulted in 0.5 mol less RuP_2, since RuP_2 contains twice as many phosphates.
They wrote

$$0.5[\text{PGA}] + R_t = R_p. \tag{16.30}$$

The potential PGA pool was then 2 R_p and they used this to replace R_p in Eq. (16.28). However, their equation is inadequate because it does not take into account changing concentrations of inorganic phosphate, P_i (STITT et al. 1980). Both it, and the analogous equation used by HALL (1971)

$$[\text{RuP}_2] + [\text{PGA}] = R_p$$

ignore the autocatalytic nature of the Calvin cycle, whereby the total concentration of esterified phosphates may change.

An important distinction between Eq. (16.30) and an analogous one used by HALL (1979) is that Eq. (16.30) refers to total RuP_2 (free plus bound).

In HALL's model, Eqs. (16.6)–(16.8) were used in an implicit form and all RuP_2 was effectively assumed to be free. This is unlikely to be the case (COLLATZ 1978) and means that HALL's model has no distinct change from RuP_2 carboxylation limitation to regeneration limitation as $p(CO_2)$ increases, an important feature discussed in Section 16.5.

FARQUHAR and VON CAEMMERER (1981) considered that ATP may be more limiting than NADPH (see Sect. 16.4) and wrote

$$\text{rate of phosphorylation of PGA} = \frac{[\text{PGA}]}{2R_p} \cdot \frac{[\text{ATP}]}{[\text{ATP}]+[\text{ADP}]} \cdot M. \qquad (16.31)$$

This change was unimportant given that M, the maximum rate of PGA "reduction or phosphorylation to RuP_2", is still a fictional composite. Nevertheless, for the estimation of CO_2 assimilation rate in conditions where RuP_2 supply is limiting, we can afford to use crude representations of the enzyme steps if it is shown that their capacity is in excess of the capacity for ATP and NADPH production.

16.3 Thylakoid Reactions

16.3.1 Introduction

NADPH and (indirectly) ATP are produced by electron transport in the chloroplast thylakoids. Despite the vast amount of research that has been carried out in this field (ARNON 1977), it was only recently that models of leaf photosynthesis incorporated electron transport explicitly. FARQUHAR et al. (1980) introduced a "potential rate of whole-chain electron transport", J, depending on temperature and absorbed irradiance, I, that would occur if the acceptor $NADP^+$ were at its maximum concentration, and which they related to uncoupled rates of electron transport in vitro. The irradiance dependence was empirical and was arranged to give appropriate quantum yields, and to saturate at high irradiances.

SIGGEL (1976) modelled the function of plastoquinone as electron and proton carrier. CRAMER et al. (1981) derived a model of electron transport from plastoquinone to P-700 in photosystem I (PSI). FARQUHAR and VON CAEMMERER (1981) presented a mechanistic model of whole-chain electron transport, which included diffusion limitation on the plastoquinone shuttle, and the tendency for reverse electron flow from plastoquinone. They simplified the result to

$$J = \frac{J_{max}\, I}{I + 2.1\, J_{max}} \qquad (16.32)$$

and presented supporting data. The form preserves the maximum rate and initial slope of the full equation, but may introduce a non-linearity of response to irradiance too early.

In this section we first examine the temperature dependence of J_{max}. We then derive equations for the rate of production of NADPH, including the effects of insufficient amounts of $NADP^+$. In an analogous manner we derive equations for the rate of ATP production. We examine various mechanisms by which the conflict in requirements for NADPH and ATP may be resolved.

16.3.2 Temperature Dependence of Potential Electron Transport Rate

The light-saturated potential rate of electron transport, J_{max}, depends on temperature (ARMOND et al. 1978; BJÖRKMAN et al. 1980). Several authors have suggested that the temperature dependence of whole-chain electron transport is closely linked to properties of the thylakoid membranes (see discussion in Chap. 10, Vol. 12 A). They suggest that membrane viscosity and lipid composition are important in determining temperature adaptation to high and low temperatures (NOLAN and SMILLIE 1976; RAISON and BERRY 1979; PIKE and BERRY 1979). The reactions associated with photosystem II (PS II) are the most temperature-sensitive. ARMOND et al. (1978) found that electron transport through PS I increased with temperature up to 55 °C, whereas the whole-chain transport peaked at 45 °C and then declined.

Acclimation occurs when plants are grown at different temperatures. The threshold temperature at which electron transport bccomes irreversibly inhibited appears to be a function of growth temperature, high growth temperatures resulting in higher thermostability at high temperatures (ARMOND et al. 1978; BJÖRKMAN 1981). It also appears from the data of ARMOND et al. (1978) that low growth temperatures result in improved performance at temperatures below the optima. The temperature response of whole-chain electron transport may therefore be an important parameter in modelling gas exchange. FARQUHAR et al. (1980) used data of NOLAN and SMILLIE (1976) to determine the coefficient B, and parameters E, S, and H for use in the following expression

$$J_{max} = B \cdot \exp[-E/RT]/\{1+\exp[(ST-H)/RT]\}, \tag{16.33}$$

where R is the universal gas constant.

Equation (16.33) is a simplified version of equations developed by JOHNSON et al. (1942) and SHARPE and DE MICHELLE (1977) to describe the effects of temperature on enzyme inactivation. This equation was used by TENHUNEN et al. (1976) and HALL (1979) to model related parameters. S and H are parameters affecting the rate at high and low temperatures, respectively. The temperature optimum, T_{opt}, of J_{max} may be different for different species and the following expression was derived by differentiation of Eq. (16.33) with respect to T

$$T_{opt} = H[S + R \ln(H/E - 1)]. \tag{16.34}$$

16.3.3 NADPH Production

The reduction of $NADP^+$ to NADPH requires two electrons and the rate of electron transport required to satisfy the NADPH requirement of the integrated PCR and PCO cycles is, from Eq. (16.26)

$$\text{rate of whole chain electron transport} = 4\,V_c + 4\,V_o = (4+4\,\phi)\,V_c = (4+8\,\Gamma_*/C)\,V_c. \quad (16.35)$$

The enzyme which catalyzes the production of NADPH is $NADP^+$-ferredoxin reductase, which occurs after (reducing side of) PS I in the electron transport chain (ARNON 1977). There are approximately 500 chlorophyll (Chl) molecules per PS I in leaves (MELIS and BROWN 1980) and since 1 mol Chl is associated with 25 l stroma (HELDT and SAUER 1971), there are $1/(500 \times 25)$ mol of PS I sites per l stroma, i.e., the effective concentration of sites is approximately 80 μM (2 mmol/mol Chl). SHIN and OSHINO (1978) have purified the equivalent of 10.6 mg purified reductase from 2 kg fresh weight of spinach. A comparison of its diaphorase activity with that of the original crude homogenate suggests that 2 kg contains 424 mg reductase. Since the molecular weight of the dimer is 80,000 (SHIN et al. 1981) and assuming a chlorophyll content of 1 mmol/kg fr. wt. (VON CAEMMERER unpublished), this corresponds to one mol reductase per 204 mol Chl, or approximately 196 μM (4.9 mmol/mol Chl). The K_m for $NADP^+$ is only about 8 μM (SHIN 1971). Thus we have a situation analogous to the high concentration of RuP_2 carboxylase-oxygenase relative to the K_m for RuP_2. The rate of $NADP^+$ reduction should have a non-rectangular hyperbolic dependence on $[NADP^+]$, similar to that in Eq. (16.11) and (16.12). By analogy with Eq. (16.13) we can approximate the situation by

$$\text{rate} = 0.5\,J \cdot \min\{1, [NADP^+]/F\}, \quad (16.36)$$

where again min { } means "minimum of", and F is the concentration of $NADP^+$-ferredoxin reductase sites. Thus for maximum rates of electron transport, $[NADP^+]$ should be slightly greater than F. NADPH will be required in the stroma, and so the total pool $N_t (= [NADP^+] + [NADPH]$, should also be greater than F, which we estimated above as 5 mmol/mol Chl. LENDZIAN and BASSHAM (1976) report a value for N_t of 36 mmol/mol Chl.

The above considerations help to explain the discrepancy (LILLEY and WALKER 1979) between the in vivo concentration of ferredoxin (19–38 mmol/mol Chl or ~1 mM in the stroma) and that required in in vitro studies (~10 μM). To the extent that ferredoxin is soluble, it will need to be at a concentration in excess of the sum of PS I and reductase sites, for maximum rates of electron transport. Measurements are needed of the levels of pyridine nucleotides, ferredoxin, and reductase sites in the same chloroplasts.

16.3.4 ATP Production

16.3.4.1 Introduction

A situation analogous to the high concentration of reductase sites may exist for photophosphorylation, the conversion of ADP to ATP on the thylakoid membranes. STROTMANN et al. (1973) have measured 0.42 g of coupling factor (CF_1) per g Chl. CF_1 is the ATP synthetase without hydrophobic subunits which anchor in the membrane. YOUNIS et al. (1977) found 0.4 to 0.5 g CF_1/g Chl. Assuming CF_1 has a molecular weight of 325,000 (FARRON 1970), this corresponds to 1.2 mmol/mol Chl, giving an effective concentration of 46 μM. BERZBORN et al. (1981) have measured 1.5 mmol/mol Chl, or 62 μM. BERZBORN and MÜLLER (1977) showed that the ratio depends on the irradiance during growth. There are multiple binding sites for ADP, although usually only two show tight binding, giving half-maximal binding at about 2 μM [ADP] (MCCARTY 1979). The concentration of tight binding sites, approximately 0.1 mM, is again much larger than the appropriate half-saturation concentration. There is evidence that these tight binding sites are not necessarily the sites of catalysis and various models of ATP synthesis have been suggested (MCCARTY 1979). Some schemes and experiments (e.g., HOCHMAN and CARMELI 1981) still favour two sites of phosphorylation, with as yet unknown affinities for ADP. In our opinion, when structural and substrate forms of bound ADP are taken into account, the dependence of the rate of photophosphorylation on the total (bound and free) ADP concentration will probably be approximated by analogy with Eq. (16.36) as

$$\text{rate} = \text{maximum rate} \cdot \min\{1, [\text{ADP}]/\mathscr{A}\}, \tag{16.37}$$

where $\mathscr{A}$ is the concentration of sites of ADP phosphorylation. Accurate measurements of $\mathscr{A}$ are needed.

At first sight it would appear from Eq. (16.37) that A_t (=[ADP]+[ATP]) would only need to be slightly in excess of $\mathscr{A}$ (≈2 mmol/mol Chl). However, this is not the case. A_t has variously been estimated as 70–100 (HEBER and SANTARIUS 1970), 9.2 (LILLEY et al. 1977), 27 (WIRTZ et al. 1980) and 44 mmol/mol Chl (GIERSCH et al. 1980a). This may be because a large concentration of ATP is required to keep the reversible reaction between Ru5P and RuP_2 favouring production of the latter. Expressed differently, it is known that ribulose-5-P kinase requires a large energy charge to support RuP_2 regeneration (PREISS and KOSUGE 1976).

16.3.4.2 Non-Cyclic Photophosphorylation via Whole-Chain Electron Transport

It is now widely accepted that three protons move across the thylakoid into the stroma every time an ATP molecule is produced (SHAVIT 1980). From Eq. (16.27) we see that in the steady state, the required

$$\begin{aligned} \text{rate of proton production} &= 9\,V_c + 10.5\,V_o = (9 + 10.5\,\phi)\,V_c \\ &= (9 + 21\,\Gamma_*/C)\,V_c. \end{aligned} \tag{16.38}$$

The movement of one electron through the whole electron transport chain results in the accumulation of two protons in the thylakoid spaces, one from the splitting of water in PS II and one from the shuttle of reduced plastoquinone across the membrane (JUNGE 1977). Thus if the proton production is by whole-chain electron transport alone, the required

$$\text{rate of whole chain electron transport} = 4.5\ V_c + 5.25\ V_o = (4.5 + 5.25\ \phi)\ V_c = (4.5 + 10.5\ \Gamma_*/C)\ V_c. \quad (16.39)$$

16.3.4.3 Pseudo-Cyclic Electron Transport (Mehler Reaction)

Comparing Eq. (16.39) with Eq. (16.35) we see that there is a disparity of $(0.5+1.25\ \phi)\ V_c$. For this reason FARQUHAR and VON CAEMMERER (1981) developed equations to model the situation where the extra whole-chain electron transport was to an acceptor other than $NADP^+$, such as O_2, as occurs in the Mehler reaction (MEHLER 1951). This reaction was reported to be half saturated at a $p(O_2)$ of 80 mbar (8%) (RADMER et al. 1978). An increased requirement for O_2 at high rates of electron transport may explain the observation by VIIL et al. (1977) that at high $p(CO_2)$, photosynthetic rate is lowered when $p(O_2)$ is lowered from 210 to 5 mbar.

The required

$$\text{rate of electron transport to } O_2 = (0.5 + 2.5\ \Gamma_*/C)\ V_c, \quad (16.40)$$

which, with $\Gamma_* = 31$ and $C = 230$ μbar, respectively, is about 16% of the rate given by Eq. (16.35). This is comparable with the measured capacities of the Mehler reaction in higher plants (HEBER et al. 1978; MARSHO et al. 1979). Nevertheless, there are other mechanisms by which chloroplasts may meet the conflicting requirements for ATP and NADPH. The incorporation of nitrate into amino acids reduces the disparity. Other mechanisms are discussed in the following two subsections.

16.3.4.4 Cyclic Photophosphorylation

Whole-chain electron transport to $NADP^+$ alone leaves a deficiency in proton production of $(1+5\ \Gamma_*/C)\ V_c$. This could be met by a portion of the electron transport being cyclical around PS I. Since this cycle will probably only contribute one proton per electron involved, the required

$$\text{cyclic-electron transport} = (1 + 5\ \Gamma_*/C)\ V_c. \quad (16.41)$$

FARQUHAR and VON CAEMMERER (1981) considered this possibility and wrote an expression for whole-chain electron transport

$$J = \frac{J_{max}\, I}{I + K_I}, \quad (16.42)$$

where

$$K_I = \frac{9.4 + 10.9\phi}{4 + 4\phi} J_{max} \simeq 2.4 J_{max}. \quad (16.43)$$

In this case, J_{max} is the maximum rate of whole chain transport, while cyclic flow is occurring simultaneously. If the limitations to electron transport, other than irradiance, are common to both cyclic and non-cyclic flow, then J_{max} is $(4.5+5.25\,\phi)/(5+6.5\,\phi)$ times the rate which would occur in the absence of cyclic flow (FARQUHAR and VON CAEMMERER 1981).

16.3.4.5 DHAP/PGA Shuttle

A third mechanism which may contribute to the balancing of conflicting requirements is the DHAP/PGA shuttle. In this shuttle (WALKER 1976), which does not actually involve the thylakoid, DHAP is exported from the chloroplasts, and oxidized in the cytoplasm to yield ATP, NADH and 3-PGA, the latter returning to the stroma together with a proton. If we consider that x mol PGA return in this manner per mol of carboxylations, Eqs. (16.26) and (16.27) are replaced by

$$\text{rate of NADPH consumption} = (2+2\,\phi+x)\,V_c \tag{16.44}$$

$$\text{rate of ATP consumption} = (3+3.5\,\phi+x)\,V_c \tag{16.45}$$

and the required rates of whole-chain electron transport become, respectively,

$$\text{rate} = (4+4\,\phi+2x)\,V_c \tag{16.46}$$

and

$$\text{rate} = (4.5+5.25\,\phi+1.5x)\,V_c. \tag{16.47}$$

If this were the *sole* mechanism for balancing requirements then (16.46) and (16.47) would have to be equated, yielding

$$x = 1+2.5\,\phi. \tag{16.48}$$

This, as we see later, would be energetically expensive for the chloroplast.

16.3.4.6 Photosynthetic Control of Whole-Chain Electron Transport

It is known that electron transport is reduced when a proton motive force (p.m.f.) occurs across the thylakoid membrane. This has been called photosynthetic control (WEST and WISKICH 1968). Thus each time a mole of ATP is produced on the stromal side, and 3 mol of protons move across the membrane, reducing the p.m.f., 1.5 mol of electrons (1.5 equivalents) should be able to move along the whole chain, or 3 mol cyclically around PS I, perhaps by removing limitations on the diffusion of reduced PQ (SIGGEL 1976). Thus

$$\begin{aligned}\text{rate of ATP production} = {} & \tfrac{2}{3}\cdot\text{rate of whole chain electron transport} \\ & +\tfrac{1}{3}\cdot\text{rate of cyclic electron transport.}\end{aligned} \tag{16.49}$$

An appropriate equation describing photosynthetic control in detail has not been incorporated into a model of leaf photosynthesis. Such an equation, when combined with others describing the rates of consumption and production of protons in the stroma would be useful.

Photosynthetic control by p.m.f. operates in the same direction as the ADP dependence of photophosphorylation rate [Eq. (16.37)], when $[ADP] < \mathscr{A}$. As the RuP_2 regeneration reactions release 1 mol ADP, which binds to the coupling factor, 3 mol protons move across the thylakoid, reducing the p.m.f. and allowing 1.5 equivalents to move along the chain. Thus we write

$$\text{rate of whole-chain electron transport} = J \cdot \min\{1, [ADP]/\mathscr{A}\}. \tag{16.50}$$

It appears from Eq. (16.36) that electron transport could be controlled by $[NADP^+]$. However, when $[NADP^+]$ is low, electrons will normally be able to move from Fd^- to O_2 in the Mehler reaction. Equation (16.50) emerges as our link between thylakoid functioning and the stromal regeneration of RuP_2.

16.4 Integration of Factors Limiting RuP_2 Regeneration

In the following developments we assume that the shortfall in ATP production associated with normal whole-chain electron transport to $NADP^+$ via Fd is overcome by whole-chain electron flow to O_2 via Fd. When equations representing cyclic phosphorylation and the DHAP/PGA shuttle are included, the resulting models are not substantially different from the one developed here.

Equating two expressions for the rate of electron transport [Eqs. (16.39) and (16.50)], we obtain

$$V_c = J' \cdot \min\{1, [ADP]/\mathscr{A}\}, \tag{16.51}$$

where

$$J' = J/(4.5 + 10.5\,\Gamma_*/C). \tag{16.52}$$

J' is the maximum carboxylation velocity allowed by the potential electron transport rate (FARQUHAR and VON CAEMMERER 1981). When $R_t > E_t$, the carboxylation rate is given by W_c, and the electron transport is forced to go at the rate $(4.5 + 10.5\,\Gamma_*/C)\,W_c$. The important question here is whether, when carboxylation is RuP_2 limited ($R_t < E_t$), V_c is less than J', i.e., is $[ADP] < \mathscr{A}$? Put another way, with $[ADP] \geqq \mathscr{A}$, so that $V_c = J'$, is there sufficient [ATP] ($\leqq A_t - \mathscr{A}$) to ensure that the regeneration of RuP_2 is sufficiently rapid? We are now forced to use the crude model of RuP_2 regeneration presented earlier. Combining Eqs. (16.25) and (16.31) we obtain

$$(2 + 3\Gamma_*/C)\,V_c = \frac{[PGA]}{2R_p} \cdot \frac{[ATP]\,M}{A_t}. \tag{16.53}$$

Provided M is sufficiently large, the answer is yes. More precisely, by combining Eqs. (16.51) and (16.53) for $R_t < E_t$, then $V_c = J'$, provided

$$M' > \frac{2R_p}{[PGA]} \cdot \frac{J'}{1 - A/A_t}, \tag{16.54}$$

where

$$M' = M/(2 + 3\Gamma_*/C), \tag{16.55}$$

the maximum rate of carboxylation allowed by the enzymes regenerating RuP_2, given that $2\Gamma_*/C$ oxygenations occur per carboxylation.

We see now why the pool size of adenylates ($A_t \approx 0.45$ mM) is approximately five times larger than the concentration of ADP binding sites ($\mathscr{A} \approx 92$ μM). This ensures that at low irradiance, when carboxylation is RuP_2 limited, there is sufficient ATP, after all the binding sites are filled by ADP, for the RuP_2 regeneration enzymes (e.g., Ru-5 P kinase and PGA kinase) to maintain a slight excess capacity.

We note that at the transition from carboxylation limitation to regeneration limitation, $R_t = E_t$, $V_c = W_c = J'$ and the "phosphorylation potential" should be given by

$$\text{for } R_t = E_t \quad \frac{[ATP]}{[ADP]} = \frac{A_t}{\mathscr{A}} - 1 \approx 3.9 \tag{16.56}$$

The model predicts that the potential should be lower when the irradiance is low, or even at saturating irradiance if the $p(CO_2)$ is high. In vivo estimates range between 2 and 5 (Krause and Heber 1976).

When $R_t > E_t$, electron transport rates should increase with increasing $p(CO_2)$ and, from Eq. (16.50), [ADP] should also increase, cf. Sect. 16.2.2.

It must be emphasized that our understanding of the integrated control of RuP_2 regeneration is meagre. One of our greatest difficulties is to explain why ATP is present in chloroplasts in the dark. More realistic models of the enzyme system will modify the expression for the required capacity for regeneration [Eq. (16.54)], but hopefully not the conclusion that when RuP_2 is limiting, V_c can be usefully approximated by J'.

16.5 Integrated C_3 Metabolism

16.5.1 Rate of CO_2 Assimilation, A

We are now in a position to examine the integrated metabolism. The net rate of assimilation of CO_2 is given by Eq. (16.16):

$$A = V_c - 0.5\, V_o - R_d$$

which may be rewritten using Eq. (16.18) as

$$A = V_c(1 - \Gamma_*/C) - R_d. \tag{16.57}$$

We note that Γ_*/C represents the ratio of the rates of photorespiration and of carboxylation.

In the previous sections we suggested that

$$V_c = \min\{W_c, J'\}, \tag{16.58}$$

where W_c is the RuP_2-saturated rate of carboxylation given by

$$W_c = V_{c\,max} \cdot \frac{C}{C + K_c(1 + O/K_0)} \tag{16.59}$$

(assuming the enzyme is fully activated) and J′ is the electron transport/photophosphorylation limited rate of RuP_2 regeneration given variously as

$$J' = \frac{JC}{4C + 8\Gamma_*} \tag{16.60a, b}$$

$$J' = \frac{JC}{4.5C + 10.5\Gamma_*} \tag{16.60c}$$

$$J' = \frac{JC}{6C + 18\Gamma_*} \tag{16.60d}$$

depending on whether the additional ATP to balance the NADPH production is (a) not required, or generated by (b) cyclic electron transport, (c) pseudocyclic electron transport (to O_2 in a Mehler reaction), or (d) via a DHAP/PGA shuttle. The form of J′ is not necessarily fixed, but we favour the use of Eq. (16.60c). J is the potential rate of whole chain electron transport, which depends on irradiance as given approximately by Eq. (16.32)

$$J = \frac{J_{max}\,I}{I + 2.1\,J_{max}}.$$

Eq. (16.32) probably underestimates J at intermediate irradiances. For case (b) (extra ATP produced by cyclic electron transport) 2.1 in the denominator is replaced by

$$(9.4\,C + 21.8\,\Gamma_*)/(4\,C + 8\,\Gamma_*) \simeq 2.4.$$

From Eqs. (16.57) to (16.59) we obtain the RuP_2-saturated rate of CO_2 assimilation, given by (FARQUHAR et al. 1980)

$$A = V_{c\,max} \cdot \frac{C - \Gamma_*}{C + K_c(1 + O/K_o)} - R_d. \tag{16.61}$$

Comparing this with Eq. (16.1), C in the numerator is offset by Γ_* but, contrary to KU and EDWARDS (1977b), not in the denominator.

From Eqs. (16.57) and (16.60) we obtain the RuP_2 limited rates, e.g.,

$$A = J \cdot \frac{C - \Gamma_*}{4.5C + 10.5\Gamma_*} - R_d. \tag{16.62}$$

For brevity, we now omit analogous equations referring to cases a, b and d, wherever possible.

As an aside we note that when DHAP is exported from the chloroplast to the cytoplasm, orthophosphate moves in to maintain the supply of P_i (WALKER and ROBINSON 1978). If the latter process is limiting, A becomes independent of C and O. However, in what follows, we assume that in cases where RuP_2 is limiting, its regeneration is limited by the potential rates of electron transport/photophosphorylation.

16.5.2 Quantum Yield

At low irradiances, $V_c = J'$ and Eq. (16.62) are appropriate. In cases where cyclic electron transport is not involved (a, c and d above) 2.1 mol quanta are absorbed to move 2 mol electrons (equivalents) through the whole chain and at low irradiances (I), J is replaced by I/2.1. When cyclic transport occurs (case b) approximately 2.4 mol quanta are required

$$\underset{I \to 0}{\text{limit}} \frac{dA}{dI} = \frac{C - \Gamma_*}{4C + 8\Gamma_*} \cdot \frac{I}{2.1} \tag{16.63a}$$

$$\text{or} = \frac{C - \Gamma_*}{9.4C + 21.8\Gamma_*} \cdot I \tag{16.63b}$$

$$\text{or} = \frac{C - \Gamma_*}{4.5C + 10.5\Gamma_*} \cdot \frac{I}{2.1} \tag{16.63c}$$

$$\text{or} = \frac{C - \Gamma_*}{6C + 18\Gamma_*} \cdot \frac{I}{2.1}. \tag{16.63d}$$

The effects of CO_2 and oxygen concentrations, and of temperature, are summarized by Eq. (16.63) together with Eq. (16.19), see also discussion in Chapter 3, Volume 12 A.

Equations giving similar dependencies on concentrations and temperature to those of (16.63a) have been derived by PEISKER (1978b), HALL (1979), and FARQUHAR et al. (1980), although PEISKER neglected energetic costs of the reassimilation of NH_4^+ in the photorespiratory cycle (WOO et al. 1978; BERRY and FARQUHAR 1978; KEYS et al. 1978). Earlier predictions were made by HALL (1971). All these modelling approaches reflect competition between the PCR and PCO cycles for limiting NADPH and ATP. FARQUHAR et al. (1980) derived an equation similar to (16.63 c) and numerical solutions of this equation were graphed by BERRY and FARQUHAR (1978). All of these models are able, with differing degrees of precision, to match the CO_2, O_2 and temperature dependencies observed by EHLERINGER and BJÖRKMAN (1977). (It is interesting to note here that KU and EDWARDS (1978) found that the quantum yield (mol CO_2 fixed per mol quanta absorbed) of wheat leaves was independent of temperature when the solubility ratio of O_2/CO_2 was kept constant. They recognized that this may, in part, be fortuitous.)

All of the published models cited above, while successful in predicting effects of environmental changes, were empirical in the sense that parameters were chosen in order to make the quantum yield close to observed values at high $[CO_2]$ or low $[O_2]$ (i.e., $C \gg \Gamma_*$). HALL (1979) chose a value for a fundamental parameter he called the "photochemical efficiency" and FARQUHAR et al. (1980) chose a value for the "fraction of light not effectively absorbed by chloroplasts". Eq. (16.63) suffer less from this deficiency. Under conditions where $C \gg \Gamma_*$, the quantum requirements (mol quanta absorbed per mol CO_2 fixed) become (a) 8.4, (b) 9.4, (c) 9.5, and (d) 12.6. Using white light, EHLERINGER and BJÖRKMAN (1977) observed requirements of 12.3 in one set of experiments and 13.7 in another, while KU and EDWARDS (1978) found requirements of 12.0. TERRY (1980), using red light, found values ranging between 8 and 11, but averaging 10.

Eq. (16.63) have other deficiencies. Firstly, there are chloroplastic requirements for ATP and NADPH outside the PCR and PCO cycles (RAVEN 1972) including additional amino acid biosynthesis, lipid metabolism and, in some cases, nitrate reduction. These will raise the apparent quantum requirement. Secondly, the quantum yields are measured at finite irradiances, often between 50 and 150 $\mu E\, m^{-2}\, s^{-1}$, where J may not necessarily increase linearly with I. From Eq. (16.32)

$$\frac{dJ}{dI} = \frac{2.1\, J_{max}^2}{(I + 2.1\, J_{max})^2} \tag{16.64}$$

and for case (b) (cyclic), from Eq. (16.43)

$$\frac{dJ}{dI} = \frac{K_I\, J_{max}}{(I + K_I)^2}. \tag{16.65}$$

For $J_{max} = 200\ \mu Eq\, m^{-2}\, s^{-1}$, at $I = 100\ \mu E\, m^{-2}\, s^{-1}$, dA/dI becomes, for $C \gg \Gamma_*$, (a) 0.078, (b) 0.072, (c) 0.069, (d) 0.052, giving apparent quantum requirements of (a) 12.9, (b) 13.8, (c) 14.6, (d) 19.2. We see that case (d) would be energetically expensive. Inclusion of (b) in a strict comparison with the others may be misleading since with the same electron transport capacity J_{max} should be smaller in this case (cf. Sect. 16.3.4.4) giving a greater quantum requirement than 13.8. As was mentioned before, Eq. (16.32) may lead to an underestimate of dJ/dI.

In 21% oxygen, at normal CO_2 concentrations, and 25 °C, the apparent quantum requirements are increased, theoretically and in practice (EHLERINGER and BJÖRKMAN 1977) by about 40%; i.e., quantum yield is inhibited by about 30%.

16.5.3 Carboxylation Efficiency

16.5.3.1 Introduction

FARQUHAR et al. (1980) differentiated Eq. (16.61) to obtain

$$\frac{dA}{dC} = V_{c\,max} \cdot \frac{\Gamma_* + K_c(1+O/K_o)}{[C+K_c(1+O/K_o)]^2}, \tag{16.66}$$

i.e., the slope, sometimes called the "mesophyll conductance" and, more appropriately, the "carboxylation efficiency" (KU and EDWARDS 1977b), should have a slight dependence only on CO_2 concentration because K_c is relatively large. Inclusion of activation effects [Eq. (16.14)] may lead to an even more linear response above the compensation point. At $C = \Gamma_*$

$$\frac{dA}{dC} = \frac{V_{c\,max}}{\Gamma_* + K_c(1+O/K_o)}. \tag{16.67}$$

VON CAEMMERER and FARQUHAR (1981) have measured dA/dC using gas exchange techniques and $V_{c\,max}$ using biochemical techniques and found good agreement between these results and Eq. (16.67). Leaves of *Phaseolus vulgaris* were used and the sources of variation were age, nitrogen nutrition, iron nutrition, previous irradiance, and CO_2 concentration during growth and defoliation.

16.5.3.2 Oxygen Dependence of Carboxylation Resistance

The inverse of dA/dC, the "carboxylation resistance", is

$$\frac{1}{dA/dC} = \frac{\Gamma_* + K_c(1+O/K_o)}{V_{c\,max}} \tag{16.68}$$

$$= \frac{K_c + (\gamma_* + K_c/K_o)\,O}{V_{c\,max}} \tag{16.69}$$

which is linearly dependent on oxygen concentration, as observed by KU and EDWARDS (1977b), with a slope of $(\gamma_* + K_c/K_o)/V_{c\,max}$.

Anomalous results have been reported by PEISKER and APEL (1971), who observed that the oxygen dependence of carboxylation resistance increased with oxygen concentration. PEISKER et al. (1979) also reported anomalous changes in carboxylation resistance at high temperatures and high O_2 concentrations. However, KU and EDWARDS (1977b) found no such anomalies.

16.5.3.3 Temperature Dependence of dA/dC

The temperature dependence of dA/dC is determined by the kinetic constants of RuP_2 carboxylase-oxygenase and their respective temperature dependencies.

FARQUHAR et al. (1980) used the temperature dependencies of K_c, K_o, $V_{c\,max}$ and $V_{o\,max}$, determined by BADGER and COLLATZ (1977). This model predicts that dA/dC should increase slowly with temperature (2% per degree at 25 °C) at normal $p(O_2)$, but be independent of temperature at zero $p(O_2)$. HALL (1979) used the same activation energies for K_c and K_o, but assumed an in vivo temperature optimum for $V_{c\,max}$ and $V_{o\,max}$ of about 30 °C. Recent data of WEIS (1981) appear to support this assumption. There is, as yet, no in vitro biochemical support. KU and EDWARDS (1977b) observed increasing "carboxylation efficiency" with increasing temperature, when expressed in terms of concentration in solution. When expressed in terms of partial pressures of CO_2, the increase was almost zero. PEISKER et al. (1979) observed that at normal $p(O_2)$, dA/dC were similar at 23 °C and 33 °C in wheat.

16.5.4 Transition from Limitation Due to RuP_2 Carboxylation Capacity to One Due to RuP_2 Regeneration Capacity

The above equations [(16.66)–(16.69)] apply only when RuP_2 is saturating. FARQUHAR et al. (1980) suggested that this would be so at low CO_2 concentrations, but that at higher $p(CO_2)$ a point is reached where RuP_2 regeneration is limiting, and the dependence on CO_2 changes in nature. Their prediction of a change in the response of A to C has been confirmed by VON CAEMMERER and FARQUHAR (1981) (see Fig. 16.5). To the extent that RuP_2 regeneration is limited by photophosphorylation, the dependence on C is given by Eq. (16.62c). Note that A continues to increase with C in this region as limiting ATP is diverted from the PCO cycle to the PCR cycle. Analyses of CO_2 exchange under high CO_2 concentration which ignore the RuP_2 limitation (SINCLAIR and RAND 1979) are unreasonable.

From Eq. (16.59) and (16.60c), we can predict that the transition from RuP_2 carboxylation limitation to RuP_2 regeneration limitation occurs at

$$C=\frac{K_c(1+O/K_o)\,J/(4.5V_{c\,max})-7/3\,\Gamma_*}{1-J/(4.5V_{c\,max})}. \qquad (16.70)$$

Note that the transition depends on the ratio $J/V_{c\,max}$ and since this ratio is somewhat conservative at the growth irradiance (VON CAEMMERER and FARQUHAR 1981), the transition C also tends to be conservative.

The change in CO_2 dependence moving from RuP_2 carboxylation limitation to RuP_2 regeneration limitation causes faster saturation than would occur due to the kinetics of RuP_2 carboxylase-oxygenase alone. Failure to recognize this has led TENHUNEN et al. (1977) to fit sunflower data at 25 °C with the physiologically unreasonable values of 0.45 μbar and 1 mbar for K_c and K_o respectively, and wheat data at 25 °C (TENHUNEN et al. 1980a) with $K_c = 12$ μbar and $K_o = 82$ mbar. TENHUNEN et al. (1980a) emphasized the linear relationship between A and C as indicating support for mesophyll resistance to diffusion. However, all that is needed for such a response is to use biochemically reasonable values for the kinetic constants. FARQUHAR et al. (1980) used 460 μbar and 210 mbar,

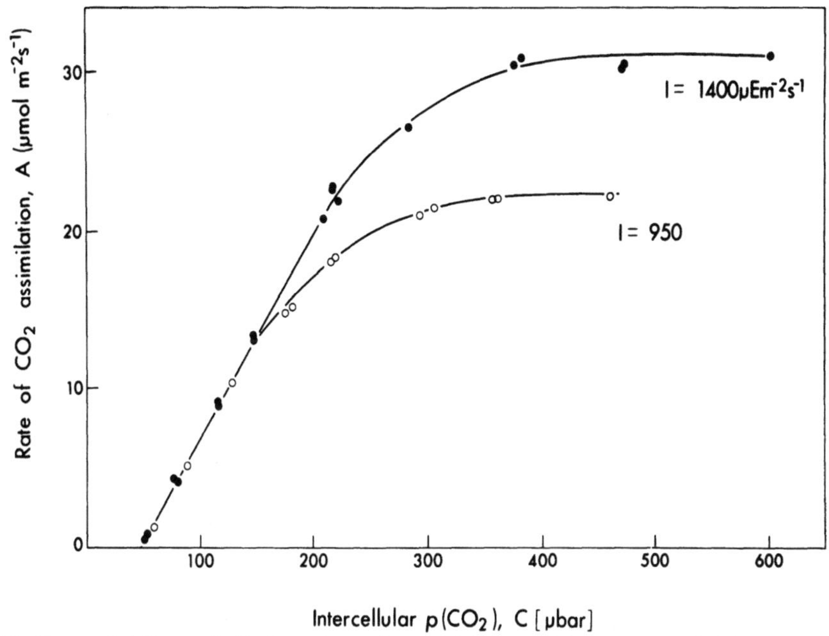

Fig. 16.5. Rate of CO_2 assimilation *A*, vs. intercellular partial pressure of CO_2, *C*, in *Phaseolus vulgaris* at an irradiance, *I*, of 1,400 μE $m^{-2} s^{-1}$ (●), and 950 μE $m^{-2} s^{-1}$ (○) at 28°C (after VON CAEMMERER and FARQUHAR 1981)

for K_c and K_o at 25 °C. There is still some uncertainty about exact values which, although reasonably consistent among higher C_3 land plants, may show a twofold variation (YEOH et al. 1981). All measurements indicate values much higher than those used by TENHUNEN.

At low CO_2 concentrations the response of A to C is not linear, as demonstrated by MEIDNER (1970). It is possible that this curvature may be due to carboxylase-oxygenase inactivation [cf. Eq. (16.14)].

16.5.5 Temperature Optimum

The dependence on temperature of the CO_2 assimilation rate, A, depends on that of its subprocesses, but the analysis is nevertheless complex (see also Chap. 10, Vol. 12 A). At normal $p(CO_2)$ and $p(O_2)$ the RuP_2 saturated rate, given by Eq. (16.61), increases with temperature. This dependence is greater at high $p(CO_2)$. At sufficiently high temperatures there is a decline due to increased day respiration, R_d. The latter effect is more pronounced at low $p(CO_2)$ since R_d then has a greater fractional impact on A. The temperature dependence of the RuP_2-limited rate, given by Eq. (16.62c) is affected more by that of J_{max}, and the optimum is closer to that of J_{max}, which occurs at lower temperatures. We show in Fig. 16.6 that by increasing the ratio $J_{max}/V_{c\,max}$, which tends to release the RuP_2 limitation, an increase in the temperature optimum is obtained.

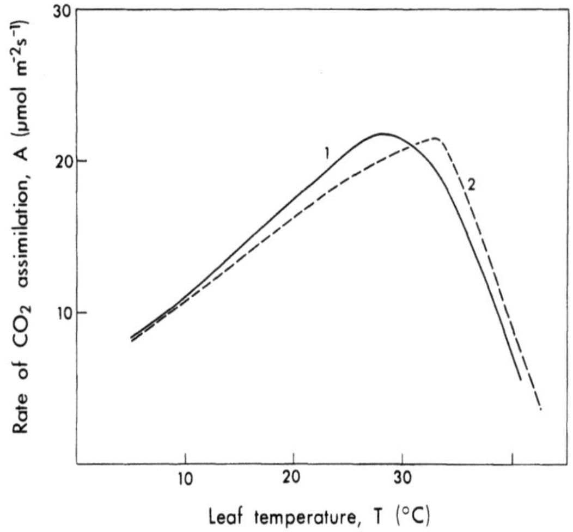

Fig. 16.6. Modelled rate of CO_2 assimilation, *A*, vs. leaf temperature, *T*, for two ratios of $J_{max}/V_{c\,max}$:

$$1 \quad \frac{J_{max}}{V_{c\,max}} = \frac{190\,\mu\text{Eq m}^{-2}\,\text{s}^{-1}}{115\,\mu\text{mol m}^{-2}\,\text{s}^{-1}} = 1.65, \qquad 2 \quad \frac{J_{max}}{V_{c\,max}} = \frac{210\,\mu\text{Eq m}^{-2}\,\text{s}^{-1}}{98\,\mu\text{mol m}^{-2}\,\text{s}^{-1}} = 2.14.$$

The irradiance is 2,000 µE $m^{-2}\,s^{-1}$, intercellular $p(CO_2)$ is 230 µbar

Increasing $p(CO_2)$ from a low level causes an increase in the optimum temperature, as does increasing the irradiance from a low level. Both also cause the temperature response to become sharper, as discussed by HALL (1979). All these phenomena are observed in practice (ENOCH and SACKS 1978; BERRY and BJÖRKMAN 1980). Decreasing the $p(O_2)$ from an initially high value causes a similar response as observed by JOLLIFFE and TREGUNNA (1968).

16.5.6 Resistance of the Intercellular Spaces

YOCUM and LOMMEN (1975) estimated the drop in the concentration of CO_2 at cells some distance from the substomatal cavities. They used equations developed to describe a system where enzymes are uniformly distributed in a diffusing medium without carriers for facilitated transport. They regarded the distributed cells as being analogous to the distributed enzymes, and estimated that the CO_2 concentration at the farthest distance from the substomatal cavities was 0.91 of that in the cavities, C_o. However, their dimensionless Michaelis-Menten constant, K_m/C_o, was probably too low by at least an order of magnitude, as they chose a value for K_m of only 25 µbar. The latter value was determined, presumably, using the erroneous method of assessing K_m from A vs C response curves, without considering electron transport and photophosphorylation limitations. If a more reasonable estimate of $K_m[=K_c(1+0/K_o)]$ is used in their equations, with the parameters they determined for an ivy leaf, the estimated drawdown in CO_2 concentration in remote regions of the intercellular spaces is negligible.

RAND (1977) estimated the maximum drawdown to be 50% of the substomatal concentration. His sink strengths were unreasonable and he subsequently (RAND 1978) revised this estimate to 2%–14%. The substomatal concentration is typically 220 $\mu l\ l^{-1}$ in C_3 species (WONG et al. 1979). COOKE and RAND (1980) estimated the intercellular air space (ias) resistance to be typically 0.25 $m^2\ s\ mol^{-1}$ for CO_2. PARKHURST (1977) has derived a three-dimensional model for CO_2 uptake in which he estimated an ias resistance of 6.3 $m^2\ s\ mol^{-1}$. The reasons for the disparity are not obvious to us.

We now give independent estimates. The maximum drawdown, at cells most remote from the stomata, in a hypostomatous leaf assimilating uniformly throughout its volume is approximately $A_1\, r_i/2$, where A_1 ($\mu mol\ m^{-2}\ s^{-1}$) is the flux of CO_2 into the lower surface, and r_i is the resistance to diffusion of CO_2 across the whole leaf through the intercellular spaces. In an amphistomatous leaf the maximum drawdown $\simeq \frac{A_1\, A_2}{A_1 + A_2} \cdot \frac{r_i}{2}$, where A_2 is the CO_2 flux into the upper surface. FARQUHAR and RASCHKE (1978) measured the resistance to the diffusion of helium across leaves of cotton and *Xanthium strumarium*. After allowing for stomatal and boundary layer resistances on each side of the leaf, and for the ratio of the diffusivities of CO_2 and helium in air, the total intercellular resistances were approximately 8 and 3.2 $m^2\ s\ mol^{-1}$ for cotton and *X. strumarium*, respectively. With $A_1 = A_2 = 10$, and $A_1 = 8$, $A_2 = 5$, respectively, the maximum drawdowns are then 20 and 5 $\mu l\ l^{-1}$. The average drawdowns in the leaves are approximately half these values.

At a whole leaf level these predictions for average drawdown may be mimicked by resistances to the diffusion of CO_2 in the intercellular air spaces of 0.5 and 0.2 $m^2\ s\ mol^{-1}$, respectively, in agreement with the estimates of COOKE and RAND (1980). These resistances are relatively unimportant, but may underestimate the effects of close packing in, for example, sclerophyllous leaves.

16.5.7 Liquid-Phase Resistance

YOCUM and LOMMEN (1975) pointed out that most of the liquid diffusion path is within the chloroplasts themselves. They used the equations relevant to distributed sinks with Michaelis-Menten kinetics to examine CO_2 diffusion across a chloroplast. They estimated that the CO_2 concentration at the surface away from the wall would be 0.5 of that nearest it if the diffusing species were CO_2 alone. Unfortunately the K_m chosen for CO_2 was only 25 μbar and with a more appropriate value the drawdown is again very small. YOCUM and LOMMEN (1975) estimated that carbonic anhydrase activity would reduce the depletion to almost zero.

SINCLAIR et al. (1977) have developed equations to describe the net characteristics associated with both diffusion through cytoplasm and kinetics of RuP_2 carboxylase-oxygenase. When geometric characteristics of C_3 plants are used, certain simplifications are possible, but the equations still *appear* to depend on geometry. In fact the resulting equations may be further reduced to a description based on the amount of carboxylase per unit area of leaf, with no dependence on liquid-phase resistance, or on cell geometry. COOKE and RAND (1980) treated the results of SINCLAIR et al. (1977) as an expression of "liquid-phase resistance", but the terminology in this case is inappropriate.

HALL (1971) estimated that the liquid-phase resistance contributed an effective resistance, in terms of gaseous diffusion of only 0.4 m^2 s mol^{-1}. With a net assimilation rate of 20 μmol m^{-2} s^{-1}, the depletion in partial pressure would then be only 8 μbar. Considering carbonic anhydrase activity and the aqueous and lipid portions of the pathway, RAVEN and GLIDEWELL (1981) estimated the liquid-phase resistance as 25 to 38 m^2 s mol^{-1} on a cell area basis. NOBEL et al. (1975) observed a correlation between mesophyll cell surface area per unit leaf area and assimilation rate, irradiance during growth being the independent variable. In *Plectranthus parviflora* when the surface area ratio was 50, the assimilation rate was 7.3 μmol m^{-2} s^{-1}. On this basis the above estimate of liquid-phase resistance would correspond to a depletion of 3.6 to 5.5 μbar. However, ratios are often smaller, and assimilation rates are often higher in other species (RAVEN and GLIDEWELL 1981), and if these changes do occur simultaneously the drawdown could be greater. LONGSTRETH and NOBEL (1980) found that with varying nutrition, cotton gave varying rates, but that the area ratio remained relatively constant ($\approx$25). At the highest assimilation rate, 30 μmol m^{-2} s^{-1}, RAVEN and GLIDEWELL's estimate of liquid-phase resistance would correspond to a drawdown of 30 to 45 μbar.

FARQUHAR et al. (1980) and BJÖRKMAN (1981) have pointed out that an increase in mesophyll cell area is usually associated with increases in photosynthetic enzymes and electron carriers, which are the prime cause of increased assimilation rates. An excess of photosynthetic machinery per cell will cause inefficiency (RAVEN and GLIDEWELL 1981), but there is little evidence that this problem occurs. A reduced ratio of surface areas, because it is generally accompanied by decreased photosynthetic capacity, does not necessarily cause a significant CO_2 drawdown.

16.5.8 On the Appropriate Measure of CO_2 Concentration

Summarizing the previous two sections, it appears that, unless there is bicarbonate pumping in cells, as occurs in some algae (BADGER et al. 1980) and submerged macrophytes (see Chap. 15, this Vol.), the CO_2 concentration, C, at the sites of carboxylation is usually only marginally less than in the substomatal cavities. FARQUHAR et al. (1982) came to the same conclusion from considerations of observed carbon isotope fractionations. Controversy exists whether the most appropriate measure of C is the molar concentration in solution (KU and EDWARDS 1977a, b, 1978) or the equilibrium partial pressure (BADGER and COLLATZ 1977). The question will be resolved when it is known whether or not the CO_2 in solution is in equilibrium with that bound to the carboxylase sites. KU and EDWARDS (1977a, b) have shown that much of the temperature dependence of various processes is reduced when the solubilities of CO_2 and O_2 are taken into account. However, it is unlikely that the kinetic constants are completely independent of temperature. If the molar concentration is the appropriate measure, the temperature dependence of the kinetic constants can still be treated in terms of partial pressures by absorbing the temperature dependence of solubility (HALL 1979).

16.6 Long-Term Effects of Environment on Leaf Photosynthesis

Photosynthetic CO_2-assimilation by leaves is affected by plant nutrition, light and temperature regimes, leaf age and other physiological factors. If analytical, mechanistic models are to be used to estimate the influence of individual components on the integrated performance, an understanding of which parameters may change and which are invariant under environmental changes is important. As discussed previously, there is evidence that the ratio $V_{o\,max}/V_{c\,max}$ is constant among species and under different growth conditions. Similarly the Michaelis-Menten constants for the RuP_2 carboxylase-oxygenase, K_c and K_o, are not affected by different growth conditions such as high temperature (BERRY and BJÖRKMAN 1980) and high $[CO_2]$ (YEOH et al. 1981). However, the amounts of carboxylase and of electron transport components may vary greatly. This has been shown by several authors. For example, BJÖRKMAN et al. (1972) and POWLES and CRITCHLEY (1980) found that RuP_2 carboxylase activity, $V_{c\,max}$, and maximum electron transport rate, J_{max}, were higher, on a leaf area basis,

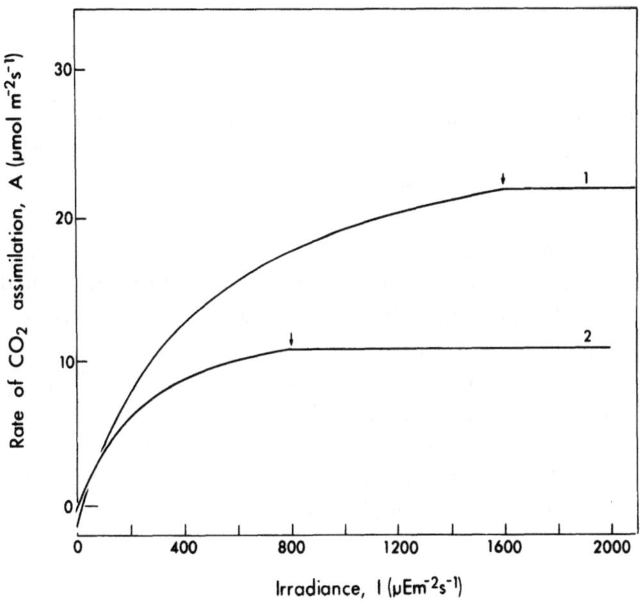

Fig. 16.7. Modelled rate of CO_2 assimilation, *A*, vs. irradiance, *I*: *1* represents high light grown plant with $V_{c\,max}=115\ \mu mol\ m^{-2}\ s^{-1}$ and $J_{max}=210\ \mu Eq\ m^{-2}\ s^{-1}$; *2* represents low light grown plant with $V_{c\,max}=57\ \mu mol\ m^{-2}\ s^{-1}$ and $J_{max}=105\ \mu Eq\ m^{-2}\ s^{-1}$. Values of J_{max} and $V_{c\,max}$ chosen are typical of those determined experimentally by the authors. Rates of day respiration have been scaled accordingly. *Arrows* indicate transition from RuP_2 regeneration limitation to RuP_2 carboxylation limitation

in plants grown at high light than in plants grown at low light (see Chap. 3, Vol. 12 A). The modelled effects are shown in Fig. 16.7. Plants grown at differing levels of nitrogen nutrition also have differing values of $V_{c\,max}$ and J_{max} (MEDINA 1969; VON CAEMMERER and FARQUHAR 1981). The decrease in CO_2-assimilation rate observed during leaf ageing is associated with a decline in $V_{c\,max}$ and J_{max}. Similarly the increase in CO_2-assimilation rate after defoliation of other leaves (WAREING et al. 1968) has been correlated with an increase in $V_{c\,max}$. WONG (1979) has shown that the lowered assimilation rate at normal $p(CO_2)$ of cotton plants grown at enhanced $p(CO_2)$, is associated with a lowered carboxylase activity.

Temperature adaptations may also to some extent be explained by changes in the amount and ratio of J_{max} and $V_{c\,max}$. BERRY and BJÖRKMAN (1980) reported that high temperature grown plants have less carboxylase than low temperature grown plants. This will cause an increase in the temperature optimum of CO_2-assimilation rate as discussed earlier. The temperature optimum of electron-transport rate itself did not change significantly in the example cited, although there was greater thermal stability at temperatures greater than the optimum.

Thus it appears that many longer-term environmental effects may be modelled by suitable changes in two key parameters, $V_{c\,max}$ and J_{max}.

16.7 C_4-Photosynthesis

The mathematical modelling of CO_2 assimilation by C_4-species is in its infancy. Nevertheless, many of the gas exchange characteristics observed with intact leaves have been predicted using models based on the known biochemical and anatomical characteristics. PEISKER (1978a) and BERRY and FARQUHAR (1978) predicted the low compensation point, insensitive to $p(O_2)$. BERRY and FARQUHAR predicted the insensitivity of quantum yield to $p(CO_2)$, $p(O_2)$ and temperature, and the lack of O_2 effects on assimilation rate. They predicted that photorespiration should nevertheless continue at a low rate in the bundle-sheath cells, as observed in several studies and most recently by MOROT-GAUDRY et al. (1980). They also predicted the greater nitrogen use efficiency of C_4 plants, especially at higher temperatures.

Factors needing consideration in future models are the mechanism of action of phosphoenol pyruvate (PEP) carboxylase, the malate and aspartate inhibitions of PEP carboxylation, the feedback inhibition by free CO_2 of C_4 acid decarboxylation (RATHNAM and CHOLLET 1980), and the higher K_c for RuP_2 carboxylase in C_4 species (YEOH et al. 1981).

16.8 Canopy Photosynthesis

Photosynthesis of canopies of C_3-species has been the subject of a great deal of experimentation and modelling (HESKETH 1980; see also Chaps. 4–10, Vol.

12D). It is useful to examine how the equations for leaf photosynthesis may be utilized in this context.

Earlier we suggested that for an individual leaf the rate of assimilation should be given by

$$A=(1-\Gamma_{*}/C)\cdot\min\{W_c, J'\}-R_d. \qquad (16.71)$$

Since W_c and J' are usually correlated (VON CAEMMERER and FARQUHAR 1981) and since irradiance is often limiting in the canopy, we can expect that to a good approximation $W_c \leqq J'$ for most leaves. Substituting for J' using Eq. (16.60c) we obtain the RuP_2 limited rate of assimilation for a single leaf (16.62c) and using Eq. (16.32) this yields

$$A=\frac{J_{max} I}{I+2.1 J_{max}}\cdot\frac{C-\Gamma_{*}}{4.5C+10.5\Gamma_{*}}-R_d. \qquad (16.72)$$

The problem of predicting canopy photosynthesis, the rate of assimilation per unit ground surface, reduces to that of determining how J_{max}, R_d, I, T and C vary through the canopy, and integrating appropriately. It is possible that treatments could be developed which exploit the correlation that can occur between J_{max} and growth irradiance to predict how J_{max} is reduced lower in the canopy. Further, R_d may be correlated with A (ANGUS and WILSON 1976). Intercellular $p(CO_2)$ depends on stomatal conductance, but at a particular temperature is often insensitive to I (GOUDRIAAN and VAN LAAR 1978; WONG et al. 1979), except at low irradiances where contributions to photosynthesis are, in any case, minimal. The main problem, therefore, relates to the penetration, propagation and absorption of radiation (DE WIT 1965), which in turn depends on canopy structure (MONSI et al. 1973).

Regardless of the details of light absorption in the canopy, the irradiance dependence of canopy photosynthetic rate (per unit ground area) will be similar to that given by Eq. (16.72), but with J_{max} much greater than that for a single leaf, and with R_d summed over the whole canopy. This result may be seen in the classical papers of BOYSEN JENSEN (1932, 1949).

16.9 Empirical Models

At present, even in the models of short-term responses discussed earlier, the dependence of J_{max} on temperature is fairly empirical. So too are the effects of water stress and salinity. Models concerned with time scales in which $V_{c\,max}$, J_{max} and R_d change are also, of necessity, empirical at present. Models which successfully deal with these longer times include that of LANGE et al. (1977), in which the productivity of lichens is predicted. Even over the short-time scales discussed earlier, some authors have found it easier to make statistical analyses of gas exchange characteristics such as the response of A to appropriately spaced changes in C, T and I than to assume a particular underlying biochemical structure (ENOCH and SACKS 1978). Nevertheless it is probably easier for most ecophysiologists interested in C_3-photosynthesis to assume a structure such as the one described earlier, measure CO_2-assimilation rate under certain conditions, and fit values to $V_{c\,max}$, and to R_d and J_{max} and their temperature dependencies, using techniques similar to those described by HALL (1979). Hopefully further research and modelling of the mechanistic kind will continue to decrease the areas where empiricism is still required.

References

Angus JF, Wilson JH (1976) Photosynthesis of barley and wheat leaves in relation to canopy models. Photosynthetica 10:367–377

Armond PA, Schreiber U, Björkmann O (1978) Photosynthetic acclimation to temperature in the desert shrub, *Larrea divaricata*. II. Light harvesting efficiency and electron transport. Plant Physiol 61:411–415

Arnon DI (1977) Photosynthesis 1950–75: Changing concepts and perspectives. In: Trebst A, Avron M (eds) Photosynthesis I. Photosynthetic electron transport and photophosphorylation. Encyclopedia of plant physiology New Ser Vol V. Springer, Berlin Heidelberg New York, pp 7–56

Azcon-Bieto J, Farquhar GD, Caballero A (1981) Effects of temperature, oxygen concentration, leaf age and seasonal variations on the CO_2 compensation point of *Lolium perenne* L: Comparison with a mathematical model including non photorespiratory CO_2 production in the light. Planta 152:497–504

Badger MR, Andrews TJ (1974) Effects of CO_2, O_2 and temperature on a higg-affinity form of ribulose diphosphate carboxylase-oxygenase from spinach. Biochem Biophys Res Commun 60:204–210

Badger MR, Collatz GJ (1977) Studies on the kinetic mechanism of ribulose-1,5-bisphosphate carboxylase and oxygenase reactions, with particular reference to the effect of temperature on kinetic parameters. Carnegie Inst Washington Yearb 76:355–361

Badger MR, Kaplan A, Berry JA (1980) Internal inorganic carbon pool of *Chlamydomonas reinhardtii*. Evidence for a carbon dioxide-concentrating mechanism. Plant Physiol 66:407–413

Bassham JA (1979) The reductive pentose phosphate cycle. In: Gibbs M, Latzko E (eds) Photosynthesis II. Photosynthetic carbon metabolism and related processes. Encyclopedia of plant physiology New Ser Vol VI. Springer Berlin Heidelberg New York, pp 9–30

Bauwe H, Apel P, Peisker M (1980) Ribulose 1,5-bisphosphate carboxylase/oxygenase and CO_2 exchange characteristics in C_3 and C_3–C_4 intermediate species checking mathematical models of carbon metabolism. Photosynthetica 14:550–556

Berry JA, Björkman O (1980) Photosynthetic response and adaptation to temperature in higher plants. Annu Rev Plant Physiol 31:491–543

Berry JA, Farquhar GD (1978) The CO_2 concentrating function of C_4 photosynthesis. A biochemical model. In: Hall D, Coombs J, Goodwin T (eds) Proc 4th Int Congr Photosynthes. Biochem Soc London, pp 119–131

Berzborn RJ, Müller D (1977) Correlation of grana in chloroplasts with the variability in the size of 'photophosphorylation unit'. In: Coombs J (ed) Read Abstr, pp 30–31

Berzborn RJ, Müller D, Roos P, Andersson B (1981) Significance of different quantitative determinations of photosynthetic ATP-synthase CF_1 for heterogeneous CF_1 distribution and grana formation. In: Akoyunoglou G (ed) Proc Fifth Int Congr Photosynthesis Vol. III. Balaban Philadelphia, pp 107–120

Björkman O (1981) Ecological adaptation of the photosynthetic apparatus. In: Akoyunoglou G (ed) Proc Fifth Int Congr Photosynthesis Vol. VI. Balaban Philadelphia, pp 191–202

Björkman O, Badger MR, Armond PA (1980) Response and adaptation of photosynthesis to high temperatures. In: Turner NC, Kramer PJ (eds) Adaptation of plants to water and high temperature stress. Wiley and Sons, New York, pp 233–249

Björkman O, Boardman NK, Anderson JM, Thorne SW, Goodchild DJ, Pyliotis NA (1972) Effect of light intensity during growth of *Atriplex patula* on the capacity of photosynthetic reactions, chloroplast components and structure. Carnegie Inst Washington Yearb 71:115–135

Boysen Jensen P (1932) Die Stoffproduktion der Pflanzen. Fischer, Jena 108 pp

Boysen Jensen P (1949) The production of matter in agricultural plants and its limitation. Biol Med 21:1–28

Caemmerer von S, Farquhar GD (1981) Some relationships between the biochemistry of photosynthesis and the gas exchange of leaves. Planta 153:376–387

Čatský J, Tichá I (1979) CO_2 compensation concentration in bean leaves: Effect of photon flux density and leaf age. Biol Plant 21:361–364

Charles-Edwards DA (1978) Leaf carbon dioxide compensation points at high light flux densities. Ann Bot (London) 42:733–739

Chartier P, Prioul JL (1976) The effects of irradiance, carbon dioxide and oxygen on the net photosynthetic rate of the leaf: A mechanistic model. Photosynthetica 10:20–24

Collatz GJ (1978) The interaction between photosynthesis and ribulose-P_2 concentration – effects of light, CO_2 and O_2. Carnegie Inst Washington Yearb 77:248–251

Cooke JR, Rand RH (1980) Diffusion resistance models. In: Hesketh JD, Jones JW (eds) Predicting photosynthesis for ecosystem models Vol I. CRC Press, Boca Raton, pp 93–122

Cramer WA, Whitmarsh J, Widger W (1981) On the properties and function of cytochromes *b*-559 and *f* in chloroplast electron transport. In: Akoyunoglou G (ed) Proc Fifth Int Congr Photosynthesis Vol. II. Balaban Philadelphia, pp 509–522

Ehleringer J, Björkman O (1977) Quantum yields for CO_2 uptake in C_3 and C_4 plants. Dependence on temperature, CO_2 and O_2 concentrations. Plant Physiol 59:86–90

Enoch HZ, Sacks JM (1978) An empirical model of CO_2 exchange of a C_3 plant in relation to light, CO_2 concentration and temperature. Photosynthetica 12:150–157

Farquhar GD (1979) Models describing the kinetics of ribulose bisphosphate carboxylase-oxygenase. Arch Biochem Biophys 193:456–468

Farquhar GD, Caemmerer von S (1981) Electron transport limitations on the CO_2 assimilation rate of leaves: a model and some observations in *Phaseolus vulgaris* L. In: Akoyunoglou G (ed) Proc Fifth Int Congr Photosynthesis Vol. IV. Balaban Philadelphia, pp 163–175

Farquhar GD, Raschke K (1978) On the resistance to transpiration of the sites of evaporation within the leaf. Plant Physiol 61:1000–1005

Farquhar GD, Caemmerer von S, Berry JA (1980) A biochemical model of photosynthetic CO_2 assimilation in leaves of C_3 species. Planta 149:78–90

Farquhar GD, O'Leary MH, Berry JA (1982) On the relationship between carbon isotope discrimination and the intercellular carbon dioxide concentration in leaves. Aust J Plant Physiol 9:121–137

Farron F (1970) Isolation and properties of a chloroplast coupling factor and heat-activated adenosine triphosphatase. Biochemistry 9:3823–3828

Giersch Ch, Heber U, Kobayashi Y, Inoue Y, Shibata K, Heldt HW (1980a) Energy charge, photophosphorylation potential and proton motive force in chloroplasts. Biochim Biophys Acta 590:59–73

Giersch Ch, Heber U, Krause GH (1980b) ATP transfer from chloroplasts to the cytosol of leaf cells during photosynthesis and its effect on leaf metabolism. In: Spanswick RM, Lucas WJ, Dainty J (eds) Plant membrane transport: current conceptual issues. Elsevier/North-Holland Biomedical Press, Amsterdam New York, pp 65–79

Goudriaan J, Laar van HH (1978) Relations between leaf resistance CO_2-concentration and CO_2-assimilation in maize, bean, lalang grass and sunflower. Photosynthetica 12:241–249

Graham D, Chapman EA (1979) Interactions between photosynthesis and respiration in higher plants. In: Gibbs M, Latzko E (eds) Photosynthesis II: Photosynthetic carbon metabolism and related processes. Encyclopedia of plant physiology New Ser Vol VI. Springer, Berlin Heidelberg New York, pp 150–162

Hall A (1971) A model of leaf photosynthesis and respiration. Carnegie Inst Washington Yearb 70:530–540
Hall A (1979) A model of leaf photosynthesis and respiration for predicting carbon dioxide assimilation in different environments. Oecologia 143:299–316
Hall A, Björkman O (1975) A model of leaf photosynthesis and respiration. In: Gates D, Schmerl R (eds) Perspectives of biophysical ecology. Ecol Stud Vol 12. Springer, Berlin Heidelberg New York, pp 55–72
Heber Uh, Hank EU, Jensen M, Koster S (1978) Regulation of photosynthetic electron transport in intact chloroplasts and leaves of *Spinacia oleraceae* L. Planta 143:41–49
Heber U, Santarius KA (1970) Direct and indirect transfer of ATP and ADP across the chloroplast envelope. Z Naturforsch 25b:718–728
Heldt HW, Sauer F (1971) The inner membrane of the chloroplast envelope as the site of specific metabolite transport. Biochim Biophys Acta 234:83–91
Hesketh JD (1980) Predicting canopy photosynthesis from gas exchange studies in controlled environments. In: Hesketh JD, Jones JW (eds) Predicting photosynthesis for ecosystem model Vol I. CRC Press, Boca Raton, pp 37–50
Hochman Y, Carmeli C (1981) Binding of manganese ions to CF_1 and their effect on the kinetics of ATPase activity. In: Akoyunoglou G (ed) Proc Fifth Int Congr Photosynthesis Vol. II. Balaban Philadelphia, pp 821–827
Jassby AD, Platt T (1976) Mathematical formulation of the relationship between photosynthesis and light for phytoplankton. Limnol Oceanogr 21:540–547
Jensen RG, Bahr JT (1977) Ribulose 1,5-bisphosphate carboxylase-oxygenase. Annu Rev Plant Physiol 28:379–400
Johnson F, Eyring H, Williams R (1942) The nature of enzyme inhibitions in bacterial luminescence: Sulfanilamide, urethane, temperature, and pressure. J Cell Comp Physiol 20:247–268
Jolliffe PA, Tregunna EB (1968) Effect of temperature, CO_2 concentration, and light intensity on oxygen inhibition of photosynthesis in wheat leaves. Plant Physiol 43:902 906
Jordan DB, Ogren WL (1981) A sensitive assay procedure for simultaneous determinations of ribulose-1-5, bisphosphate carboxylase and oxygenase activities. Plant Physiol 67:237–245
Junge W (1977) Physical aspects of light harvesting, electron transport and electrochemical potential generation in photosynthesis of green plants. In: Trebst A, Avron M (eds) Photosynthesis I. Photosynthetic electron transport and photophosphorylation. Encyclopedia of plant physiology new Ser Vol V. Springer, Berlin Heidelberg New York, pp 59–93
Keys AJ, Bird IF, Cornelius MJ, Lea PJ, Wallsgrove RM, Miflin BJ (1978) Photorespiratory nitrogen cycle. Nature (London) 275:741–743
Kirk JTO, Tilney-Bassett RAE (1978) The plastids, 2nd edn. Elsevier/North-Holland Biomedical Press, Amsterdam New York 960 pp
Krause GH, Heber U (1976) Energetics of intact chloroplasts. In: Barber J (ed) The intact chloroplast. Academic Press, London New York, pp 171–214
Ku S, Edwards G (1977a) Oxygen inhibition of photosynthesis. I. Temperature dependence and relation to O_2/CO_2 solubility ratio. Plant Physiol 59:986–990
Ku S, Edwards G (1977b) Oxygen inhibition of photosynthesis. II. Kinetic characteristics as affected by temperature. Plant Physiol 59:991–999
Ku S, Edwards G (1978) Oxygen inhibition of photosynthesis. III. Temperature dependence of quantum yield and its relation to O_2/CO_2 solubility ratio. Planta 140:1–6
Laing WA, Christeller JT (1976) A model for the kinetics of activation and catalysis of ribulose 1,5-bisphosphate carboxylase. Biochem J 159:563–570
Laing WA, Ogren W, Hageman R (1974) Regulation of soybean net photosynthetic CO_2 fixation by the interaction of CO_2, O_2 and ribulose-1,5-diphosphate carboxylase. Plant Physiol 54:678–685
Laisk A (1970) A model of leaf photosynthesis and photorespiration. In: Prediction and measurement of photosynthetic productivity. Proc IBP/PP Tech Meet Třeboň 1969. PUDOC, Wageningen, pp 295–306

Laisk A (1977) Modelling of the closed Calvin cycle. In: Unger K (ed) Biophysikalische Analyse pflanzlicher Systeme. Fischer, Jena, pp 175–182
Lange OL, Geiger IL, Schulze E-D (1977) Ecophysiological investigations on lichens of the Negev Desert. V. A model to simulate net photosynthesis and respiration of *Ramalina maciformis*. Oecologia 28:247–259
Leegood RC, Walker DA (1980) Autocatalysis and light activation of enzymes in relation to photosynthetic induction in wheat chloroplasts. Arch Biochem Biophys 200:575–582
Lendzian K, Bassham JA (1976) NADPH/NADP$^+$ ratios in photosynthesising reconstituted chloroplasts. Biochim Biophys Acta 430:478–489
Lilley RMcC, Chon CJ, Mosbach A, Heldt HW (1977) The distribution of metabolites between spinach chloroplasts and medium during photosynthesis in vitro. Biochim Biophys Acta 460:259–272
Lilley RMcC, Walker DA (1979) Studies with the reconstituted chloroplast system. In: Gibbs M, Latzko E (eds) Photosynthesis II: Photosynthetic carbon metabolism and related processes. Encyclopedia of plant physiology New Ser vol VI. Springer, Berlin Heidelberg New York, pp 41–53
Longstreth DJ, Nobel PS (1980) Nutrient influences on leaf photosynthesis. Effects of nitrogen, phosphorus and potassium for *Gossypium hirsutum* L. Plant Physiol 65:541–543
Lorimer GH, Badger MR, Andrews TJ (1976) The activation of ribulose-1,5-bisphosphate carboxylase by carbon dioxide and magnesium ions. Equilibria, kinetics, a suggested mechanism, and physiological implications. Biochemistry 15:529–536
Lorimer GH, Woo KC, Berry JA, Osmond CB (1978) The C2 photorespiratory carbon oxidation cycle in leaves of higher plants: Paths and consequences. In: Hall DO, Coombs J, Goodwin TW (eds) Photosynthesis 77. Biochem Soc London, pp 311–322
Mangat BS, Levin WB, Bidwell RGS (1974) The extent of dark respiration in illuminated leaves and its control by ATP levels. Can J Bot 52:673–681
Marsho TV, Behrens PW, Radmer RJ (1979) Photosynthetic oxygen reduction in isolated intact chloroplasts and cells from spinach. Plant Physiol 64:656–659
McCarty RE (1979) Roles of a coupling factor for photo-phosphorylation in chloroplasts. Annu Rev Plant Physiol 30:79–104
Medina E (1971) Relationships between nitrogen level, photosynthetic capacity and carboxydismutase activity in *Atriplex patula* leaves. Carnegie Inst Washington Yearb 69:655–662
Mehler AH (1951) Studies on reactions of illuminated chloroplasts. I. Mechanism of the reduction of oxygen and other Hill reagents. Arch Biochem Biophys 33:65–77
Meidner H (1970) Precise measurements of carbon dioxide exchange by illuminated leaves near the compensation point. J Exp Bot 21:1067–1075
Melis A, Brown JS (1980) Stoichiometry of System I and System II reaction centres and of plastoquinone in different photosynthetic membranes. Proc Natl Acad Sci USA 77:4712–4716
Monsi M, Uchijima Z, Oikawa T (1973) Structure of foliage canopies and photosynthesis. Annu Rev Ecol Syst 4:301–327
Morot-Gaudry JF, Farineau J, Huet JC (1980) Oxygen effect on photosynthetic and glycolate pathways in young maize leaves. Plant Physiol 66:1079–1084
Nobel PS, Zaragoza LJ, Smith WK (1975) Relation between mesophyll surface area, photosynthetic rate and illumination level during development for leaves of *Plectranthus parviflorus* Henckel. Plant Physiol 55:1067–1070
Nolan WG, Smillie RM (1976) Multi-temperature effects on Hill reaction activity of barley chloroplasts. Biochim Biophys Acta 440:461–475
Parkhurst DF (1977) A three-dimensional model for CO_2 uptake by continuously distributed mesophyll in leaves. J Theor Biol 67:471–488
Peisker M (1974) A model describing the influence of oxygen on photosynthetic carboxylation. Photosynthetica 8:47–50
Peisker M (1976) Ein Modell der Sauerstoffabhängigkeit des photosynthetischen CO_2-Gaswechsels von C_3-Pflanzen. Kulturpflanze 24:221–235
Peisker M (1978a) Der Einfluß von Sauerstoff auf die CO_2-Kompensationskonzentration von C_3- und C_4-Pflanzen und von Intermediärformen. Kulturpflanze 26:81–98

Peisker M (1978b) A comment on the effects of carbon dioxide, oxygen and temperature on photosynthetic quantum yield in C_3 plants. Acta Physiol Plant 1:23–26
Peisker M, Apel P (1971) Untersuchungen zum Einfluß von Sauerstoff auf den CO_2-Gaswechsel assimilierender Blätter. Biochem Physiol Pflanz 162:165–176
Peisker M, Tichá I, Apel P (1979) Variations in the effect of temperature on oxygen dependence of CO_2 gas exchange in wheat leaves. Biochem Physiol Pflanz 174:391–397
Peisker M, Tichá I, Cǎtský J (1981) Ontogenetic changes in the internal limitations to bean-leaf photosynthesis. 7. Interpretations of the linear correlation between CO_2 compensation concentration and CO_2 evolution in darkness. Photosynthetica 15:161–168
Pike CS, Berry JA (1979) Phase separation temperatures of phospholipids from warm and cool climate plants. Carnegie Inst Washington Yearb 78:163–168
Powles SB, Critchley C (1980) Effect of light intensity during growth on photoinhibition of intact attached bean leaflets. Plant Physiol 65:1181–1187
Preiss J, Kosuge T (1976) Regulation of enzyme activity in metabolic pathways. In: Bonner J, Varner JE (eds) Plant biochemistry, 3rd edn. Academic Press, London New York, pp 277–336
Radmer R, Kok B, Ollinger O (1978) Kinetics and apparent K_M of oxygen cycle under conditions of limiting carbon dioxide fixation. Plant Physiol 61:915–917
Raison JK, Berry JA (1979) Viscotropic denaturation of chloroplast membranes and acclimation to temperature by adjustment of lipid viscosity. Carnegie Inst Washington Yearb 78:149–152
Rand RH (1977) Gaseous diffusion in the leaf interior. Trans Am Soc Agric Eng 20:701–704
Rand RH (1978) A theoretical analysis of CO_2 absorption in sun versus shade leaves. J Biomech Eng 100:20–24
Rathnam CKM, Chollet R (1980) Photosynthetic carbon metabolism in C4 plants and C_3–C_4 intermediate species. Prog Phytochem 6:1–48
Raven JA (1972) Endogenous inorganic carbon sources in plant photosynthesis. I. Occurrence of the dark respiratory pathway in illuminated green cells. New Phytol 71:227–247
Raven JA, Glidewell SM (1981) Processes limiting carboxylation efficiency. In: Johnson CB (ed) Processes limiting plant productivity. Butterworth, London, pp 109–136
Robinson RP, Walker DA (1980) The significance of light activation of enzymes during the induction phase of photosynthesis in isolated chloroplasts. Arch Biochem Biophys 202:617–623
Sharpe PSH, De Michelle DW (1977) Reaction kinetics of poikilothermic development. J Theor Biol 64:649–670
Shavit N (1980) Energy transduction in chloroplasts: Structure and function of the ATPase complex. Annu Rev Biochem 49:111–138
Shin M (1971) Ferredoxin-NADP reductase from spinach. In: San Pietro A (ed) Methods in enzymology Vol 23. Academic Press, London New York, pp 440–447
Shin M, Oshino R (1978) Ferredoxin-sepharose 4B as a tool for the purification of ferredoxin-$NADP^+$ reductase. J Biochem (Tokyo) 83:357–361
Shin M, Wakita R, Yamasaki Y, Oshino R (1981) Interrelation of two forms of ferredoxin-$NADP^+$ reductase with different molecular weights. Plant Cell Physiol 22:461–464
Siggel U (1976) The function of plastoquinone as electron and proton carrier in photosynthesis. Bioelectrochem Bioenerg 3:302–318
Sinclair TR, Goudriaan J, Wit de CT (1977) Mesophyll resistance and CO_2 compensation concentration in leaf photosynthesis models. Photosynthetica 11:56–65
Sinclair TR, Rand RH (1979) Mathematical analysis of cell CO_2 exchange under high CO_2 concentrations. Photosynthetica 13:279–244
Sommerville CR, Ogren WL (1979) A phosphoglycolate phosphatase-deficient mutant of *Arabidopsis*. Nature (London) 280:833–835
Stitt M, Wirtz W, Heldt HW (1980) Metabolite levels during induction in the chloroplast and extra chloroplast compartments of spinach protoplasts. Biochim Biophys Acta 593:85–102
Strotmann H, Hesse H, Edelmann K (1973) Quantitative determination of coupling factor CF_1 of chloroplasts. Biochim Biophys Acta 314:202–210

Tenhunen JD, Weber J, Filipek L, Gates D (1977) Development of a photosynthesis model with an emphasis on ecological applications. III. Carbon dioxide and oxygen dependencies. Oecologia 30:189–207
Tenhunen JD, Hesketh JD, Gates DM (1980a) Leaf photosynthesis models. In: Hesketh JD, Jones JW (eds) Predicting photosynthesis for ecosystem models Vol I. CRC Press, Boca Raton, pp 123–182
Tenhunen JD, Hesketh JD, Harley PC (1980b) Modelling C_3 respiration in the light. In: Hesketh JD, Jones JW (eds) Predicting photosynthesis for ecosystem models Vol II. CRC Press, Boca Raton, pp 17–47
Tenhunen JD, Yocum CS, Gates DM (1976) Development of a photosynthesis model with an emphasis on ecological applications. I. Theory. Oecologia 26:89–100
Terry N (1980) Limiting factors in photosynthesis. I. Use of iron stress to control photochemical capacity in vivo. Plant Physiol 65:114–120
Thornley J (1976) Mathematical models in plant physiology – a quantitative approach to problems in plant and crop physiology. Academic Press, London New York 318 pp
Viil J, Laisk A, Oja V, Pärnik T (1977) Enhancement of photosynthesis caused by oxygen under saturating irradiance and high CO_2 concentrations. Photosynthetica 11:251–259
Walker DA (1976) CO_2 fixation by intact chloroplasts: Photosynthetic induction and its relation to transport phenomena and control mechanisms. In: Barber J (ed) The intact chloroplast. Elsevier/North-Holland Biomedical Press, Amsterdam New York, pp 235–278
Walker DA, Robinson SP (1978) Regulation of photosynthetic carbon carbon assimilation. In: Siegelman HW, Hind G (eds) Photosynthetic carbon assimilation. Plenum, New York London, pp 43–59
Wareing PF, Khalifa MM, Treharne KJ (1968) Rate-limiting processes in photosynthesis at saturating light intensities. Nature (London) 220:453–457
Weis E (1981) Reversible heat inactivation of the Calvin cycle: A possible mechanism of the temperature regulation of photosynthesis. Planta 151:33–39
West KR, Wiskich JT (1968) Photosynthetic control by isolated pea chloroplasts. Biochem J 109:527–532
Wirtz W, Stitt M, Heldt HW (1980) Enzymic determination of metabolites in the subcellular compartments of spinach chloroplasts. Plant Physiol 66:187–193
Wit de CT (1965) Photosynthesis of leaf canopies. Agric Res Rep 663:1–57
Wong SC (1979) Elevated atmospheric partial pressure of CO_2 and plant growth. I. Interactions of nitrogen and photosynthetic capacity in C_3 and C_4 plants. Oecologia 44:68–74
Wong SC, Cowan IR, Farquhar GD (1979) Stomatal conductance correlates with photosynthetic capacity. Nature (London) 282:424–426
Woo KC, Berry JA, Turner GL (1978) Release and refixation of ammonia during photorespiration. Carnegie Inst Washington Yearb 77:240–245
Woodrow IE, Walker DA (1980) Light-mediated activation of stromal sedoheptulose-1,7-bisphosphate. Biochem J 191:845–849
Yeoh H-H, Badger MR, Watson L (1981) Variations in kinetic properties of ribulose-1,5-bisphosphate carboxylases among plants. Plant Physiol 67:1151–1155
Yocum CS, Lommen PW (1975) Mesophyll resistances. In: Gates DM, Schmerl RD (eds) Perspectives of biophysical ecology. Ecol Stud Vol 12. Springer, Berlin Heidelberg New York, pp 45–54
Younis HM, Winget GD, Racker E (1977) Requirement of the δ subunit of chloroplast coupling factor 1 for photophosphorylation. J Biol Chem 252:1814–1818

17 Regulation of Water Use in Relation to Carbon Gain in Higher Plants

I.R. Cowan

CONTENTS

17.1 Introduction

Growth in terrestrial plants is hazardous. An essential substrate for growth, carbon, is acquired only at the expense of losing water and therefore depleting the reserves of solvent in which growth processes take place. Because replenishment of the reserves by rainfall is irregular and unpredictable, the course plants follow in growing and using water cannot be invariably successful. Plants sometimes fail to survive, due to drought exacerbated by their own activity. There tends, therefore, to be an inverse relationship between growth rate and the probability of survival. We may presume that selective adaptation has tended to "optimize" the relationship. That is to say, plants have acquired characteristics which enable them to grow more quickly without diminishing the probability of survival, and to increase the probability of survival without diminishing their growth rate. The aim of this chapter is to identify some of these characteristics, particularly those relating to temporal regulation of carbon gain and water loss.

17.2 A Theory of Short-Term Regulation

17.2.1 Optimal Stomatal Variation

Stomatal movements are the most rapid means that plants have to adjust to changes in the environment of their photosynthesizing organs. The movements are reversible, and are inexpensive in terms of energy. Because stomata tend to open with changes in environment which promote carbon fixation and close with changes which promote water loss, their operation has the effect of reducing the average rate of water loss relative to the average rate of carbon fixation (see also Chap. 7, this Vol.). It is this empirical observation that gave rise to the theory of optimal variation proposed by COWAN (1977), and COWAN and FARQUHAR (1977).

A particular leaf at a particular instant of time has a certain rate of assimilation of CO_2 per unit area, A, and a certain rate of transpiration, E. These fluxes are in part determined by the aperture of the stomata. If the stomata were more widely open, A and E would in general be greater; if they were less widely open A and E would in general be smaller. We can conceive, then, of a relationship $E = E(A)$ in which the aperture of the stomata is an implicit parameter. It represents all the possible compromises that could have been made between taking up CO_2 and losing water. This concept can be extended to a whole plant during a finite period of time. As the external physical and leaf-internal physiological factors that influence gas exchange differ from leaf to leaf and vary with time, the transpiration function is written as

$$E = E(A, s, t) \tag{17.1}$$

where s is a coordinate representing a particular element of leaf surface and t is time.

The average rates of assimilation $\bar{A}$ and transpiration $\bar{E}$ over the whole surface, area S, of the foliage of a plant during a period of time T are

$$\bar{A} = \frac{\int_0^T \int_0^S A \, ds \, dt}{TS}; \quad \bar{E} = \frac{\int_0^T \int_0^S E(A, s, t) \, ds \, dt}{TS}. \tag{17.2}$$

Clearly A will vary with s and t in a particular way in a particular plant during a particular period of time, and this will result in particular magnitudes not only of $\bar{A}$, but of $\bar{E}$ also. Nevertheless there will exist an infinite number of other, conceivable variations in A (each implicitly associated with a different variation of stomatal aperture), and, provided the function E(A, s, t) were known, the magnitudes of $\bar{A}$ and $\bar{E}$ corresponding to each could, in principle, be found. The criterion of optimality may be defined as follows.

A variation in A and E with s and t is optimal if $\bar{A}$ could not conceivably be increased without increasing $\bar{E}$, or, what amounts to the same thing, $\bar{E}$ could not conceivably be decreased without decreasing $\bar{A}$. A derivation, from first principles, of the implications of this criterion is provided by COWAN

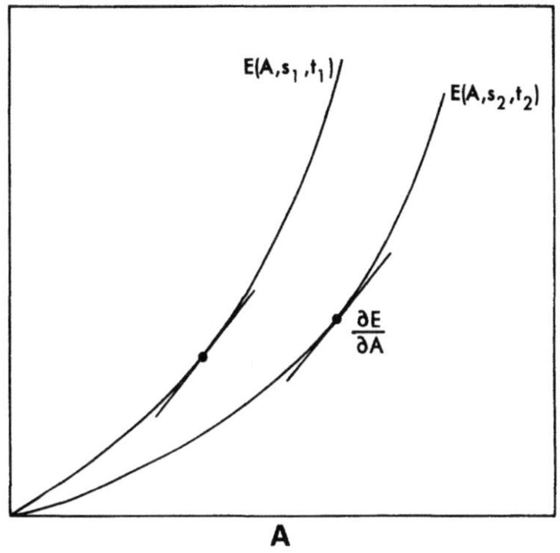

Fig. 17.1. Examples of the transpiration function. A is rate of CO_2 assimilation, E is rate of transpiration

(1977) and COWAN and FARQUHAR (1977), but the result is readily found using standard methods of variational calculus (see, for example, COURANT and HILBERT 1953). The criterion requires that

$$\int_0^T \int_0^S [E(A, s, t) - \lambda A]\, ds\, dt = \text{minimum} \tag{17.3}$$

where λ, a Lagrange multiplier, is a constant. The condition that the integral should be an extremum is given by Euler's equation which, in this particularly simple application, reduces to

$$\left(\frac{\partial E}{\partial A}\right)_{s,t} = \lambda. \tag{17.4}$$

That the integral be a minimum requires also that

$$\left(\frac{\partial^2 E}{\partial A^2}\right)_{s,t} > 0. \tag{17.5}$$

The result may be illustrated with the aid of Fig. 17.1 in which two imaginary examples of the relationship $E = E(A, s, t)$ are drawn. They might be associated with two different leaves (s_1 and s_2) on the same plant, or with the same leaf at two different instants of time (t_1 and t_2). Both relationships are curved in the sense defined by Eq. (17.5). If A is regarded as benefit to the plant and E as cost, the curves are ones of increasing cost. To use economic jargon, the slope, $\partial E/\partial A$, of such a curve at any point is the unit marginal cost. The analysis leading to Eq. (17.4) shows that optimisation, as it has been defined, requires that a plant maintain the unit marginal cost uniform in all its leaves, and constant in time. The theory, so far, has had nothing to say about the absolute magnitude of the unit marginal cost, λ, sustained by a plant. However, it may be helpful to anticipate the conclusion of later discussion by interpreting λ as a physiological parameter that depends on the amount of water available to a plant. Therefore the period, T, during which the theory, as it has so far been developed, may apply must certainly be short in terms of relative changes in the amount of water in the soil explored by the roots of the plant.

17.2.2 The Transpiration Function

The movement of water vapour and CO_2 across the epidermis and boundary layer of a leaf may be expressed as

$$E = g\frac{e_i - e_a}{1 - \bar{e}/P} \tag{17.6}$$

$$A = 0.63\,g \cdot (p_a - p_i) - \frac{\bar{p}}{P}E \tag{17.7}$$

where g is the conductance to vapour transfer, e_i and p_i are partial pressures of water vapour and CO_2 in the leaf internal air space, e_a and p_a are partial pressures of water vapour and CO_2 in the ambient atmosphere, $\bar{e} = (e_a + e_i)/2$, $\bar{p} = (p_a + p_i)/2$, P is atmospheric pressure, and the coefficient 0.63 is the ratio of the diffusion coefficients of CO_2 and water vapour in air. No allowance is made here for the fact that the ratio of the transfer coefficients for CO_2 and water vapour in the boundary layer is somewhat greater than 0.63 (see COWAN 1972), because the influence of the boundary layer on g is often relatively small compared with that of the stomata and in any event cannot be measured with great accuracy. Nor is cuticular transfer of water vapour considered (if it is appreciable, it may be approximately taken into account by replacing g in Eq. (17.6) by $g + g_c$, where g_c is cuticular conductance). Allowance is made in Eq. (17.6) for the influence of net mass transfer on vapour diffusion, and in Eq. (17.7) for the influence of vapour diffusion on CO_2 diffusion (see JARMAN 1974; VON CAEMMERER and FARQUHAR 1981). The second effect can reduce the effective transfer coefficient for CO_2 by as much as 20% in hot dry conditions, but has generally been neglected in treatments of gas exchange in leaves, including that of COWAN and FARQUHAR (1977).

Eqs. (17.6) and (17.7) do not by themselves define the interrelationship between E and A, because the partial pressures e_i and p_i are partly dependent on the fluxes. It may be assumed that the leaf internal air space is saturated with water vapour (see Chap. 1, Vol. 12 A). Therefore

$$e_i = e'(T_l) \tag{17.8}$$

where $e'(T_l)$ is saturation vapour pressure at leaf temperature T_l. From a consideration of the energy balance of a leaf, it may be shown (PEISKER 1977) that, to a very good approximation,

$$T_l = T_a + \frac{\Phi_0 - LE}{C_p g_b + 8\sigma T_a^3} \tag{17.9}$$

in which T_a is ambient air temperature, ϕ_o is the flux of solar and terrestrial radiation absorbed per unit area of leaf minus the flux of radiation that would be emitted per unit area of leaf if leaf temperature were the same as air temperature, L is the latent heat of vaporisation of water (44 kJ mol^{-1}), C_p is the molar heat capacity of air ($\approx$29.2 J mol^{-1} K^{-1}), g_b is the conductance of the leaf boundary layer to transfer of sensible heat, and σ is the Stefan-Boltzmann constant (5.9×10^{-8} W m^{-2} K^{-4}).

The final (and far more complicated) process to be considered is that of net carbon fixation in leaf mesophyll tissue. In the short term, the variables which primarily determine the net rate of fixation in a given leaf are internal partial pressure of CO_2, temperature, and irradiance. Thus

$$A = A(p_i, T_l, I). \tag{17.10}$$

Attempts to relate empirical measurements of net CO_2 fixation in leaves with the biochemistry of photosynthesis and respiration are progressively more successful, and have yielded models which go a long way towards defining the function above in terms of a small number of measurable metabolic parameters (see Chap. 16, this Vol.). However it is not yet clear whether rate of fixation is also affected in the short term by previous accumulation of photosynthate in the leaf, or by changes in leaf water potential as affected by rate of transpiration (see Chaps. 7 and 8, this Vol.). If it should prove necessary to extend the function to allow for such affects, it will greatly complicate the application of the optimization hypothesis.

The foregoing equations together constitute a relationship between rate of transpiration and rate of carbon assimilation, for the other variables that are in part dependent on the physiological functioning of a leaf, g, e_i, p_i, and T_l may in principle be eliminated. Of the remaining variables e_a, p_a and T_a are by definition independent of leaf function. Provided the optical properties and orientation of the leaf are not under physiological control (at least during the time-span considered) then variations in ϕ_o, I, and g_b stem entirely from variations in the radiation and wind environments. Variation in all of these environmental factors, and differences from leaf to leaf in the parameters of internal carbon metabolism may notionally be subsumed in the general spatial and temporal variables s and t, as in Eq. (17.1).

In practice the relationship between E and A must be found numerically, unless drastic simplifications to the system of equations are made (see COWAN 1977). The method is first to use Eqs. (17.6), (17.8) and (17.9) to determine E and T_l for various assumed magnitudes of g. Equation (17.7) and the model underlying Eq. (17.10) are then employed to find the corresponding magnitudes of A. Providing the differences in g are sufficiently small, the slope $\partial E/\partial A$ may be approximated as the ratio of finite differences.

17.2.3 Illustration

The surface in Fig. 17.2 depicts the transpiration function for a single leaf, E(A, t), during the course of a day. It is a hypothetical example, being derived

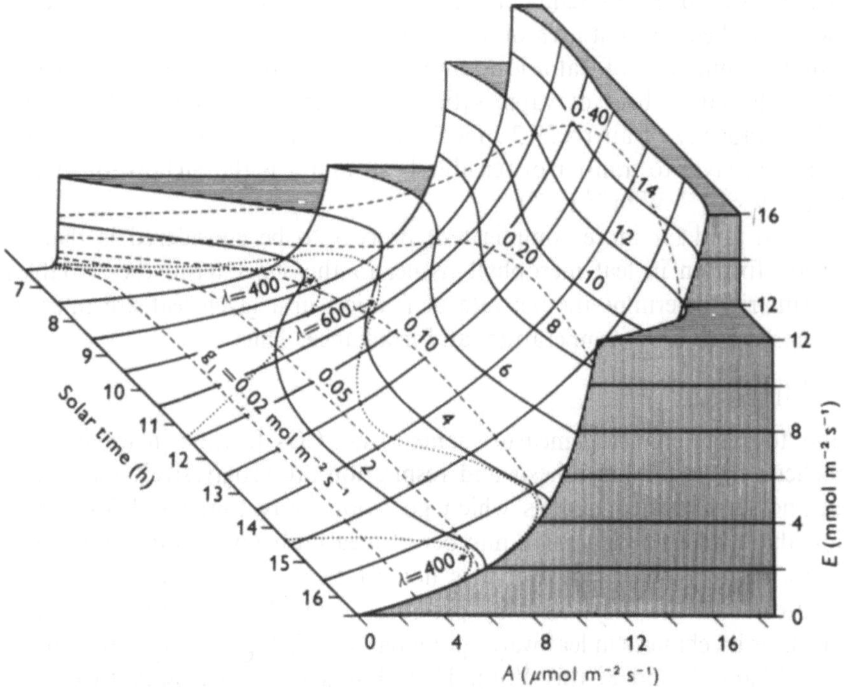

Fig. 17.2. A transpiration function for a single leaf during the course of a day. The *dotted lines* are optimal relationships between rate of transpiration, E, and rate of assimilation, A, corresponding to two constant magnitudes, $\lambda=400$ and $\lambda=600$, of $\partial E/\partial A$. The *broken lines* represent constant magnitudes of leaf conductance to vapour transfer, g_l

for a horizontal exposed leaf having assumed characteristics of C_3 photosynthetic carbon fixation and subject to an assumed diurnal variation in environment typical of a hot, dry, clear day in an arid sub-tropical environment (for details, see COWAN and FARQUHAR 1977). The diurnal behaviour of a leaf is represented by a particular trajectory on such a surface. The broken trajectories correspond to constant magnitudes of leaf conductance to gas exchange, g_l (the component of conductance g relating to the leaf epidermis, as distinct from that relating to the leaf boundary layer); the contours correspond to constant rates of transpiration. The dotted trajectories traverse the surface in such a way that the slope, $\partial E/\partial A$ is constant. As the surface is curved in the sence $\partial^2 E/\partial A^2 > 0$, they are optimal trajectories, two of an infinite number of optimal trajectories each one of which corresponds to a particular optimal variation in stomatal aperture and a particular magnitude of λ. Figure 17.3 shows a number of trajectories, represented as temporal variations in E and A. When λ is large, that is to say unit marginal cost is large, the optimal variations in gas exchange do not deviate greatly from those which would occur if leaf conductance were constant. With decreasing λ, the deviations become relatively greater. Gas exchange is increasingly confined to those periods, early in the morning and late in the afternoon, when irradiance and temperature are appropriate for rapid photosyn-

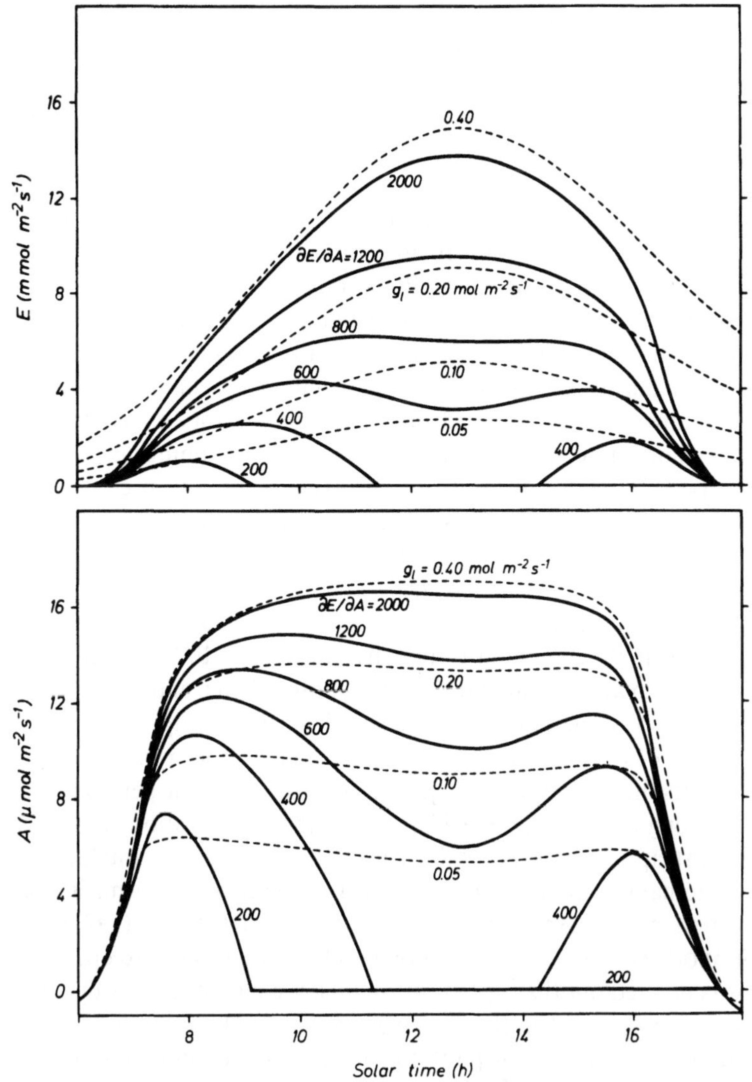

Fig. 17.3. Optimal time courses of rate of assimilation, A, and rate of transpiration, E, for various magnitudes of λ (*solid lines*). Also shown are A and E for various magnitudes of leaf conductance, g_l

thesis but, because the ambient vapour pressure deficit is relatively small, potential rate of transpiration is relatively small. COWAN and FARQUHAR (1977) pointed out that curves of the type shown in Fig. 17.3 resemble observations of gas exchange in plants out of doors, particularly those of LANGE et al. (1975) with wild plants in the Negev desert at various times of the year. However, the shapes of optimal variations in rate of transpiration and rate of assimilation depend not only on the magnitude of an undetermined parameter, λ, but also on characteristics of the physical environment, and of leaf orientation and exposure which are inevitably somewhat disturbed when leaves are placed in a gas exchange cuvette, and on characteristics of leaf internal metabolism which are exceedingly difficult to determine. For these reasons it will be very difficult to test the optimality hypothesis with plants in their natural condition.

17.2.4 The Evidence

If diurnal variation in natural physical environment were regular and predictable, then optimization would require only that there be an appropriate circadian rhythm in stomatal aperture. But variation in environment is, in part, stochastic, and therefore optimization would require that the plant respond appropriately to stochastic changes. This is the justification for examining the variance of $\partial E/\partial A$ during a sequence of single-factor step changes in humidity, irradiance or temperature of the kind imposed in laboratory experiments. Of course one does not know what peculiarities of preconditioning may take place in glasshouse or growth cabinet. Only one experiment (FIELD et al. 1981) to determine degree of optimization has been conducted with a plant growing in the field, *Lepechinia calycinia*. The results of that experiment, and also of laboratory experiments with *Nicotiana glauca* and *Corylus avellana* (FARQUHAR et al. 1980a), *Vigna unguiculata* (HALL and SCHULZE 1980), and *Pseudotsuga menziesii* (MEINZER 1982), demonstrate conservation in $\partial E/\partial A$. In all four investigations, changes in ambient humidity were imposed. HALL and SCHULZE changed temperature also; some of their results are illustrated in Fig. 17.4. No specific examination of $\partial E/\partial A$ with varying irradiance has yet been made.

Despite the paucity of quantitative information pertaining to $\partial E/\partial A$, there is strong circumstantial evidence that the optimization theory provides at least a first approximation to the truth. It relates to the qualitative characteristics of stomatal responses. Discussions of stomatal mechanisms in relation to optimization of gas exchange are to be found in COWAN (1977) and COWAN and FARQUHAR (1977) and will not be elaborated here, except in two respects. It was pointed out that optimization in some circumstances demands that stomata should respond directly to changes in external environmental conditions which influence rate of transpiration and rate of assimilation; that is to say, they should be capable of controlling gas exchange by a process of feed-forward, for only then is it possible for actual rate of transpiration to decrease when environmental changes tend to enhance rate of transpiration, or for intercellular partial pressure of CO_2 to increase when environmental changes tend to enhance rate of assimilation. That stomata are directly sensitive to difference in the partial pressure of water vapour across the epidermis had already been established (LANGE et al. 1971; SCHULZE et al. 1972; HALL et al. 1976). It has now been shown that stomatal movements due to changes in irradiance are predominantly of the feedforward kind (WONG et al. 1978) and that intercellular partial pressure sometimes increases with irradiance (SHARKEY and RASCHKE 1981; VON CAEMMERER unpublished). In short, the nature of the stomatal responses to the predominant environmental determinants of E and A is of the kind required to bring about optimization. Consider now spatial, rather than temporal, optimization, associated with differences in metabolic characteristics amongst leaves. There is much evidence, summarized by WONG et al. (1979)

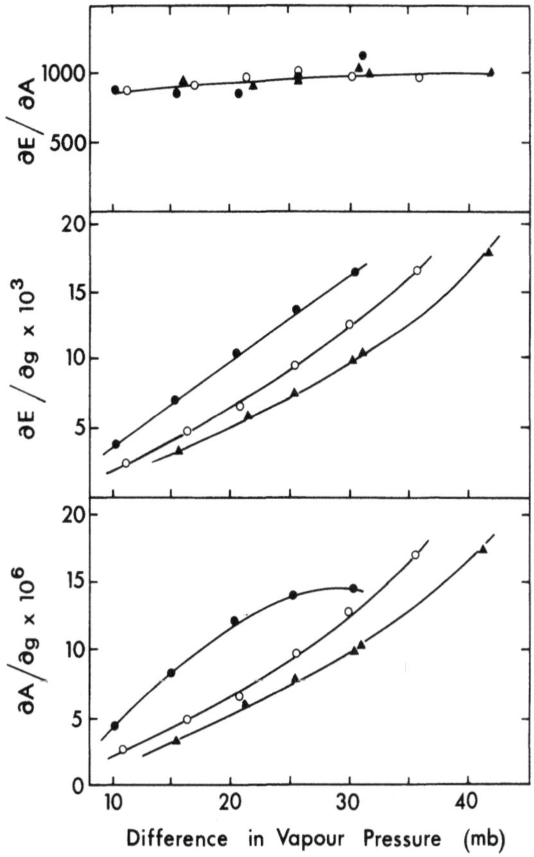

Fig. 17.4. Components of $\partial E/\partial A = (\partial E/\partial g)/(\partial A/\partial g)$ as functions of difference in leaf internal vapour pressure and external vapour pressure in *Vigna unguiculata*. Leaf temperatures: (●) 26 °C, (○) 30 °C, (▲) 34 °C. (Redrawn from HALL and SCHULZE 1980)

and in Chapter 7, this Volume, and recently confirmed by VON CAEMMERER (1981) that stomatal conductance is positively correlated with leaf mesophyll capacity for photosynthetic carbon fixation, whether differences in the latter are due to differences in age, in exposure to light, or in nutritional status. Assuming this correlation persists amongst the different leaves within one plant, it is clearly a phenomenon conforming qualitatively with the requirements of optimization.

Perhaps it is not too extravagant to summarize the argument in this way. Suppose that the hypothesis had been proposed, and its implications worked out, without knowing much of the responses of stomata; knowing only that the apertures of stomata may change in response to a variety of external physical, and internal physiological stimuli. Then, when it had been discovered that stomata indeed generally open in the light (to some extent, even in plants which are short of water), close in the dark (even in well-watered plants), close with increase in vapour pressure deficit (even to the extent that rate of water loss may be less than that at a lower deficit), and tend to be opened or closed as the intrinsic capacity of the leaf to fix carbon is large or small, the theory would surely have been thought a great success. That it has been developed with hindsight does not detract from its value in summarizing so much of what plants do in one neat relationship, $\partial E/\partial A = \lambda$. Doubtless the behaviour is not in exact quantitative accord with the relationship, but certainly it tends to reduce water loss in relation to carbon gain. It presumably has been adapted towards doing so. F. DARWIN (1898) put the matter thus: "When we remember the innumerable adaptations which serve to economise water, it is inconceivable that the closure of the stomata should not cooperate. Those who deny that the stomata close adaptively as a precaution against too great evaporation must assume, I imagine, that the closure on withering is a chance outcome or bye-product of the machine. They cannot deny, however, that this closure does effect an economy, and this makes it even more difficult to understand their position."

17.2.5 Some Problems

Whatever the worth of the empirical evidence that plants tend to optimize gas exchange, the theory is of limited conceptual value if it is without a plausible explanation as to why plants are so adapted. An explanation will have to contend with several problems. Firstly, the economy in water use that may be achieved appears generally to be small. Figure 17.5 shows results of some computations of daytime averages of rate of assimilation and rate of transpiration. They relate to three different radiation regimes, associated with three differing leaf orientations. The climatic circumstances, similar to those assumed in constructing Fig. 17.2, are shown in Fig. 17.6; they are chosen as typical of a hot, clear, dry day in an arid sub-tropical region. The characteristics

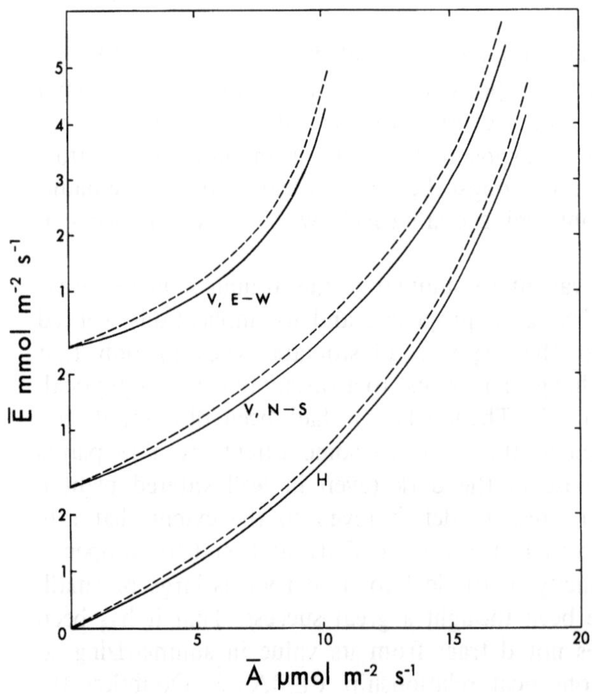

Fig. 17.5. Relationships between average daily rates of transpiration, $\bar{E}$, and assimilation, $\bar{A}$, in leaves of differing orientations: *H* horizontal; *V* vertical and *E-W* east-west; *V* vertical and *N-S* North-South. The *full curve* in each case is the optimal relationship, each *point* corresponding to a particular constant magnitude of λ. Each point on the *broken curve* corresponds to a constant magnitude of leaf conductance

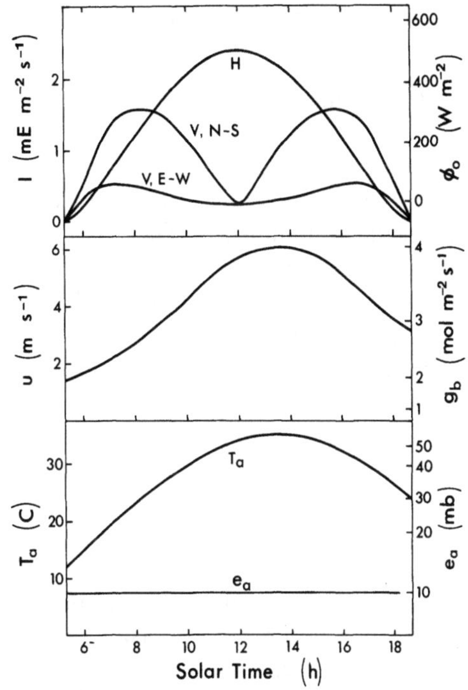

Fig. 17.6. Climatic variables used in the computations shown in Fig. 17.5. T_a, e_a, u, g_b, I, and ϕ_o are respectively ambient air temperature, ambient vapour pressure, windspeed, boundary layer conductance, irradiance, and net radiation

of leaf carbon metabolism are represented by the model of FARQUHAR et al. (1980b, see also Chap. 16, this Vol.). The lower of each pair of curves is the relationship between $\bar{E}$ and $\bar{A}$ that obtains if gas exchange is regulated optimally; the upper curve is that which obtains if stomatal conductance is constant. It is worth mentioning that much greater differences are found between the optimal and "constant-conductance" relationships when calculations are made for higher temperatures than those in Fig. 17.6. Nevertheless, if the theory is to have general validity, it must relate to conditions which occur frequently in the natural plant habitat.

A second problem that must be addressed relates to competition. Whatever the economy in water use that may potentially be achieved by optimization, and whatever the value of that economy to a plant growing in isolation, what can be the advantage of economy in a plant if the water conserved is available for use by competitors?

As suggested in Chapter 1, this Volume, competition might demand that λ is very large, and this leads one to the central question that cannot be answered by studying short-term regulation only: what is the optimal magnitude of λ? If optimization is in some way to be related to fitness, then the roles of carbon fixation and water use must be considered over periods which are at least long enough to encompass significant relative plant growth and development. How, on such a time scale, does the function of short-term reversible regulatory mechanisms relate to longer-term irreversible adjustments, associated for example with partitioning of carbon to leaves and roots?

These various problems will be touched on in the remainder of this chapter.

17.3 Longer-Term Regulation

17.3.1 Premises

As time progresses, the cumulative amount of carbon fixed by a plant is increased and, assuming there has been no rainfall, the amount of water available to the plant is diminished. There is no reason to suppose that the disadvantage of further water loss relative to the advantage of further carbon gain remains constant. COWAN and FARQUHAR (1977) remarked that it is only in short periods, perhaps a single day, that "the grand strategy associated with water relations and carbon metabolism is likely to degenerate to a tendency for $\partial E/\partial A$ to be constant".

The principal topic of this section is variation of λ with soil water content. Arguments will be based on a number of simplifying assumptions. Most of them will not cause concern, for it is not difficult to imagine how the conclusions that derive from them would be modified if they were relaxed and made more realistic.

We shall restrict discussion to plants which require that soil water be replenished on a number of occasions during the period between their emergence as seedlings and their production of seed. Falls of rain are assumed to be sufficient to saturate all the soil within the zone explored by the roots of a plant. If the probability of rain falling during an infinitesimal period of time is dt/τ, it follows that the probability of there being no rain for a period of at least t is

$$P(t) = \exp(-t/\tau). \quad (17.11)$$

The interval τ is the average interval between successive falls, and is taken as short compared with the life span of a plant. It is important to note that the expectation of rain does not increase with increase in time since the last rainfall. It is supposed that there is a certain amount of "available" water in the plant root zone immediately after rain has fallen. Let this be q (conveniently expressed as mole water per unit area of plant foliage). It is assumed that the plant dies if all of this water is extracted from the soil. It is extracted by the roots of the plant itself in association with loss of water through the stomata, and by other processes. The latter include water loss from the plant itself of a kind that is not under stomatal control and therefore not associated with carbon fixation, evaporation from the soil surface, drainage, and extraction by the encroaching root systems of other plants. Let the total of these other processes amount to an equivalent additional loss, F, per unit area of the plant foliage.

If ϕ is the amount of water extracted from the soil, as a proportion of q, after time t following a fall of rain, then

$$\frac{d\phi}{dt} = \frac{\bar{E} + F(\phi)}{q} \quad (17.12)$$

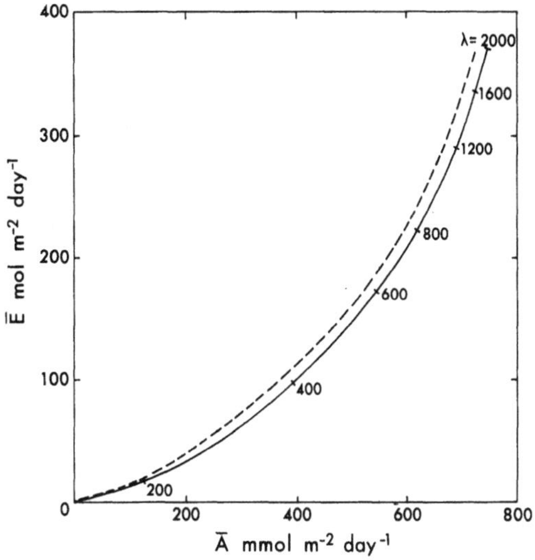

Fig. 17.7. Optimal (*full curve*) and constant conductance (*broken curve*) relationships between mean daily rates of transpiration, $\bar{E}$, and assimilation, $\bar{A}$. The relationships are the averages of those shown in Fig. 5. They are used in numerical examples pertaining to long-term regulation

where $\phi=0$ at $t=0$, and plant death occurs at $\phi=1$. In writing $F=F(\phi)$, we suppose that F may be, in part, a function of soil water content.

The characteristics of leaf carbon metabolism, i.e. of the function defined in Eq. (17.10), are taken to be independent of soil water content within the range $0<\phi<1$. It is assumed that the plants each day "optimize" the relationship between carbon fixation, $\bar{A}$, and water loss, $\bar{E}$, in the way that has been described. In reality the relation must change from day to day with fluctuations in diurnal climate. This source of variation is ignored; the function $\bar{E}=\bar{E}(\bar{A})$ is taken as constant (one might care to think of it as deriving from a statistical average for a particular species in a particular habitat). The function illustrated in Fig. 17.7 will be used in numerical examples; it is a mean of those functions for leaves of three different orientations shown in Fig. 17.5. Because we are about to deal with periods of many days the average daily fluxes $\bar{A}$ and $\bar{E}$ will be treated as though they were continuous functions of time.

17.3.2 The Optimal Variation

Let the plant vary its rate of water loss, E, following a fall of rain in such a way that the available water in the soil would be exhausted in time t_1 if no further rain occurred. It follows that only those plants that are fortunate enough not to encounter a rain-free period greater than t_l in their normal life span will survive and propagate. The critical time, t_l, defines the chance that a plant will die within a given period of time, once the probability function in Eq. (17.11) is fixed.

However, t_l does not define the time course of water use by the plant. There is an infinity of possible ways in which the plant may vary $\bar{E}$ with time following rainfall all of which lead to $\phi=1$ at $t=t_1$. The problem is to find the one which yields the greatest probable average growth rate given that the plant will not, in fact, die (the growth rate of plants that will die before reaching maturity is of no interest). Relative growth rate is proportional to rate of assimilation per unit area of foliage, other things, such as carbon partitioning for leaf growth, being equal. Therefore the problem is solved by finding the condition that the probable average rate of assimilation in a surviving plant is maximum. Let the probability of a plant surviving a period T, where T is long compared with t_1, be $\exp(-MT)$. The probability of the last rain-free interval within this period being between t and t+dt is, from Eq. (17.11), $(1/\tau)\exp(-t/\tau)dt$. Only if $t<t_1$ will the plant have survived. As the probability of the plant having survived the sum, T−t, of all the previous intervals is $\exp[-M(T-t)]$ it follows that

$$\exp(-MT)=\int_0^{t_1}\frac{1}{\tau}\exp(-t/\tau)\cdot\exp[-M(T-t)]\,dt$$

leading to the equation

$$M\tau=\exp\left[\frac{-t_1}{\tau}(1-M\tau)\right] \tag{17.13}$$

This has two real roots: M is to be taken as the root $\neq 1/\tau$. Let the probable average rate of assimilation in a surviving plant be $\hat{A}$. Then $\hat{A}$ must satisfy the integral equation

$$\hat{A}T\exp(-MT)=\int_0^{t_1}\left[\int_0^{t}\bar{A}(t')\,dt'+\hat{A}\cdot(T-t)\right]\cdot\frac{1}{\tau}\exp(-t/\tau)\cdot\exp[-M\cdot(T-t)]\,dt$$

from which it may be shown that

$$\hat{A}=\frac{1}{\tau(1-Mt_1)}\int_0^{t_1}\bar{A}(t)\cdot[(M\tau)^{t/t_1}-M\tau]\,dt \tag{17.14}$$

Eqs. (17.13) and (17.14) were first derived, as particular results of a more general analysis, by Dr. H.N. Comins, to whom I am greatly indebted. The presupposition made here, that probable relative mortality, M, and probable average rate of assimilation remain constant with time, requires comment. It is valid of the period T is independent of any particular rainfall event. But it is not strictly valid, for example, if the beginning of the period is chosen to coincide with a rainfall (as it would implicitly tend to be if taken at the time of seed germination). In that instance, mortality would be zero during the first t_1 days of the period. However, the analysis by Dr. Comins shows, *inter alia*, that, for all the numerical examples presented in this chapter ($t_1/\tau>2$), the probable relative mortality, and the probable average rate of assimilation, within the following t_1 days would differ from the asymptotic estimates provided by Eqs. (17.13) and (17.14) by no more than a few percent. The convergence is greater the greater the magnitude of t_1/τ.

The next step is to find the condition that the integral in Eq. (17.14) is an extremum. As $\bar{A}$ is uniquely related to $\bar{E}$, and $E = q\dot{\phi} - F(\phi)$ from Eq. (17.12), the integrand has the form $Q = Q(t, \phi, \dot{\phi})$. Therefore the Euler equation is

$$\frac{d}{dt}\left(\frac{\partial Q}{\partial \dot{\phi}}\right) - \frac{\partial Q}{\partial \phi} = 0$$

or

$$\frac{d}{dt}\left\{\frac{q}{\lambda}\left[(M\tau)^{t/t_1} - M\tau\right]\right\} + \frac{1}{\lambda}\frac{dF}{d\phi}\left[(M\tau)^{t/t_1} - M\tau\right] = 0$$

with $\lambda = \partial\bar{E}/\partial\bar{A}$ as before. Integration yields

$$\lambda = \lambda_0 \frac{(M\tau)^{t/t_1} - M\tau}{1 - M\tau} \exp\left(\frac{1}{q}\int_0^t \frac{dF}{d\phi} dt'\right) \qquad (17.15)$$

where $\lambda_o = \lambda$ at $t = t_o$. The magnitude of λ_o is not arbitrary. It must be chosen so $\phi = 1$ at $t = t_1$.

Equation (17.15) determines the time course of $\bar{E}$ and $\bar{A}$ following rain. As it determines the time course of $\bar{E}$ and $\bar{A}$ it also determines the probable average growth rate of plants which survive, $\hat{A}$, as defined by Eq. (17.14), and the corresponding probable average rate of water use, $\hat{E}$, which is defined by an equation analogous to (17.14).

17.3.3 Interpretation

If rainfall occurred at regular intervals, and the soil water available to a plant was not subject to depletion by physical processes or the action of competing plants, then plants would tend to assimilate and transpire water at constant rates from day to day so that the available water was almost, but never quite, exhausted at the end of each rain-free period. The total water loss corresponding to a given total of carbon uptake is a minimum when the rates of loss and uptake are constant, provided the sense of the curvature of the $\bar{E} = \bar{E}(\bar{A})$ relationship is that shown in Fig. 17.6 (a fact first pointed out by COWAN and TROUGHTON 1972).

Opposed to the influence of curvature in the $\bar{E} = \bar{E}(\bar{A})$ relation are influences associated with randomness in the occurrence of rainfall and depletion of the available soil water by other processes. A plant should fix carbon and use water rapidly immediately after rain, partly because a second fall of rain may occur shortly, and partly because the rate at which some of the other competing processes extract soil water is dependent on soil water content, and is greatest when the content of water is greatest. The rates should then decline with time because the risks of further water use increase; the reserve of available water is diminished, while the expectation of replenishment by rain is not increased. Of course, there is no risk unless competing processes continue to remove water as soil water content approaches the critical level. In fact there is always

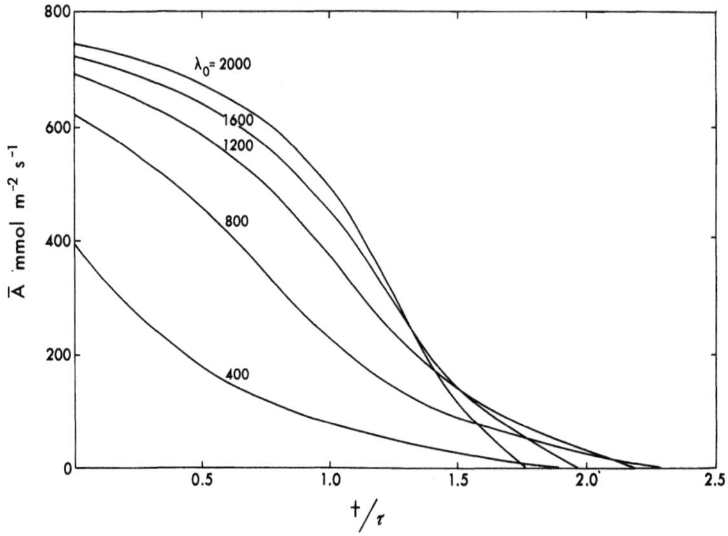

Fig. 17.8. Optimal variations of mean daily rate of assimilation, $\bar{A}$, with time, t, following rain. τ is the average interval between rainfalls. λ_o is the magnitude of $\partial\bar{E}/\partial\bar{A}$ at $t=0$

some loss of water from soils even when they have become too dry to support higher plant life.

All of these considerations are incorporated in Eq. (17.15). The equation shows how λ decreases with time; to put it graphically, how the plant slides down the curve in Fig. 17.7. The rate of descent depends on the statistics of rainfall as represented by the terms involving τ. It also depends on the way in which the processes which compete for soil water vary with soil water content. The term $dF/d\phi$ will invariably be negative; if $|dF/d\phi|$ is large – that is to say, F is very much diminished with decrease in water content – then λ will decline rapidly. But it may also be shown that λ_o is increased. Thus the plant uses the available water early in order that it should not be lost.

Some illustrations are shown in Figs. 17.8, 17.9 and 17.10. It has been assumed that $q/\tau = 500$ mol m^{-2} day^{-1}, and $F = 100$ mol m^{-2} day^{-1} (a case with F varying with water content is discussed later). Thus a constant rate of transpiration $E = 400$ mol m^{-2} day^{-1} would lead to the exhaustion of available water within the average period, τ, between successive periods of rain. If the plant did not transpire at all, the available water would be lost in a period five times as long.

Figure 17.8 shows the decline of $\bar{A}$ with time for various magnitudes of λ_o. Three points need to be emphasized. The graph does not imply that $\bar{A}$ runs the full time course shown after each fall of rain; the length of time actually run varies stochastically in accordance with the function P(t). Secondly, the times at which $\bar{A}$ becomes zero are not the critical times, t_1, corresponding to $\phi = 1$. Optimization requires that gas exchange cease before the available water is exhausted. Whether or not a plant, having reached this stage, will die depends on whether the next rain occurs before the processes contributing

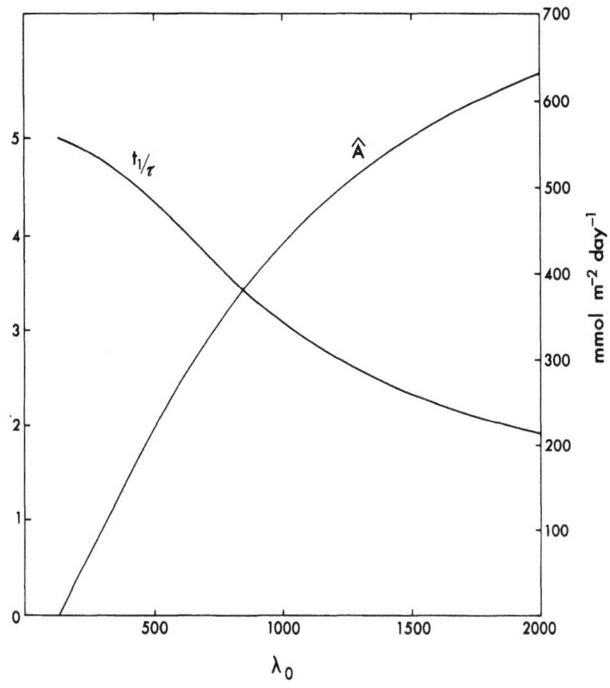

Fig. 17.9. Probable average assimilation rate (*right ordinate*) of surviving plants, $\hat{A}$, and the relative critical time t_1 (*left ordinate*), as functions of λ_0, the magnitude of $\partial\bar{E}/\partial\bar{A}$ immediately following rain. τ is the average interval between rainfalls

to F remove the remaining available water. Thirdly, the critical times are different for the different time courses in Fig. 17.8.

The significance of the parameter distinguishing the different curves in Fig. 17.8 becomes clear in Fig. 17.9, which shows the average probable assimilation rate $\hat{A}$, computed from Eq. (17.13), and the critical time relative to the average period between falls of rain, t_1/τ, as functions of λ_0. When λ_0 is small, $\hat{A}$ is small. However t_1 is large, implying that plants operating in this way are relatively unlikely to die from drought. When λ_0 is large, $\hat{A}$ is large and t_1 is small. Plants operating in this way grow rapidly but are relatively likely to die from drought.

In the next section the quantities $\hat{A}$ and t_1 will be translated in terms of growth and mortality. But this may be an appropriate juncture at which to address the inevitable question: what physiological mechanism might be associated with optimization of this kind? A possible answer is made obvious by plotting the optimal daily rate of evaporation, $\hat{E}$, against the deficit of soil water, ϕ, as in Fig. 17.10. The decline in rate of evaporation is, of course, associated with a decline in leaf conductance; and increase in soil water deficit would lead to a decrease in the potential of water not only in the soil, but in the plant itself. We may suppose that the actual responses of stomata to changes in plant internal water relations have become adapted towards optimization of the kind described. These responses, it is becoming clear, are associated with a complex hormonal regulation system (see Chap. 9, this Vol.). They are feedback responses, as distinct from the feedforward responses required to optimize stomatal behaviour in relation to variation in the environment of the foliage.

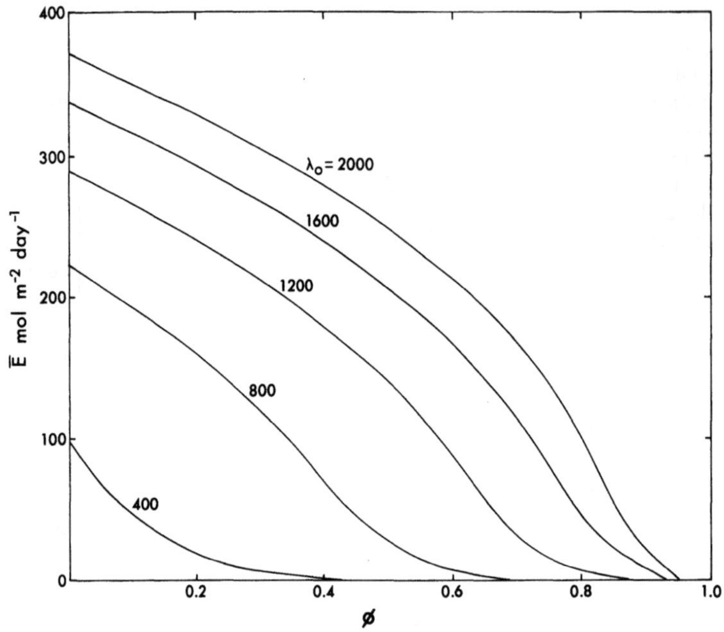

Fig. 17.10. Optimal variations of mean daily rate of transpiration, $\bar{E}$, with soil water deficit, ϕ

17.3.4 Growth and Mortality

An equation for plant relative growth rate that is in common use may be expressed in the form

$$G=\left(\frac{1}{W}\frac{dW}{dt}\right)=\alpha\left[\frac{S}{W}\hat{A}-\beta\right] \tag{17.16}$$

where W is the amount of carbon in the plant in moles, S is leaf area, $\hat{A}$ is the rate of assimilation in moles per unit area, α is the "growth efficiency" and β is the proportion of the plant carbon lost in unit time in association with "maintenance respiration". Typical magnitudes of α and β are 0.75 and 0.015 day^{-1} (HESKETH et al. 1980; HUNT and LOOMIS 1979).

The curves in Fig. 17.11 correspond to those in Fig. 17.9. The probable average relative growth rate of a surviving plant, G, has been calculated from Eq. (17.16) with S/W being taken as 0.1 $m^2\ mol^{-1}$. The probable relative mortality, M, derives from Eq. (17.13) with τ being taken as 10 days. It is seen that M is an accelerating function, and G is a decelerating function, of λ_o. We may take G as a rough indication of the reproductive capacity of an individual mature plant, but the significance of the relation between G and M in terms of population dynamics depends on many factors that are beyond the compass of this article. Nevertheless, in certain circumstances the quantity G-M is related to the relative increase in a plant population. It is seen that it has a maximum. That it happens to fall within the range of λ_o considered is, of course, a result

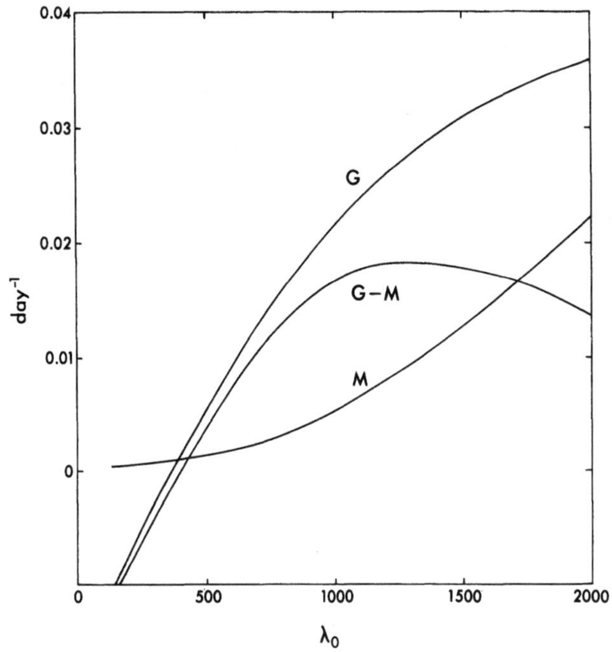

Fig. 17.11. Probable average relative growth rate of surviving plants, G, and probable average mortality, M, as functions of λ_o, the magnitude of $\partial\bar{E}/\partial\bar{A}$ immediately following rain

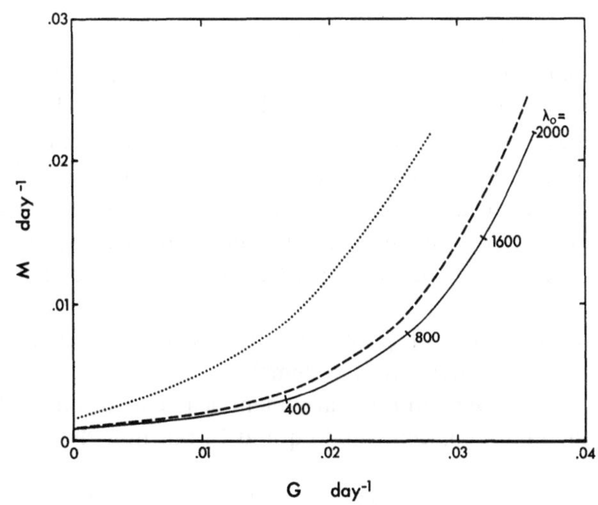

Fig. 17.12. Probable average mortality, M, as a function of probable average relative growth rate, G. The *full curve* is the optimal relationship; the *broken curve* represents long term regulation, with an absence of short-term regulation (i.e. constant leaf conductance during each day). The *dotted curve* represents the result of maintaining constant conductance at all times

of the particular numbers that have been used as a basis for computation, but, insofar as water resources are of major importance in limiting the performance of terrestrial plants, it seems likely that the relationship between growth, as it affects soil and plant hydrology, and mortality, as it is affected by soil and plant hydrology, should exhibit an optimum within the domain of real physiological and environmental conditions. But rather than attempting to discuss its location, it is perhaps better simply to graph M as a function of G, as in Fig. 17.12, and remark, as of the first figure in this article, that if the abscissa is a measure of benefit and the ordinate of cost, then the curve is one of increasing costs.

Two other curves are shown in Fig. 17.12. One is the relationship between G and M corresponding to a constant rate of assimilation, $\bar{A}$, and transpiration, $\bar{E}$. Plainly the decrease in mortality for a given growth rate (or the increase in growth rate for a given mortality) associated with optimal variation as compared with no variation is very large indeed. The other relationship is based on the assumption that $\bar{E}$ and $\bar{A}$ are constant within each day (see Fig. 17.7), but are varied optimally from day to day. The cost of such an imperfection in short-term regulation is appreciable when expressed in terms of M.

17.3.5 Competition

It is unrealistic to assume that the independent loss of water F is constant with time, as has been done in the previous numerical examples. The physical processes of evaporation and drainage in the soil will surely decrease as soil water content decreases. It has been noted that if F decreases with ϕ, then λ and therefore $\bar{E}$ and $\bar{A}$ must be initially larger and then decrease more rapidly with time. But what happens when the loss of water F is partially due to the uptake by the root systems of competing plants leads to more subtle considerations.

Let us compare two populations of plants, identical in all respects except that in one the root systems do not overlap, and in the other each plant shares its water resources with n others (n need not be integral, as the root systems may only partially overlap). Then, if we adapt Eq. (17.12) to describe the overlapping case we must replace q by $(n+1)q$ and F by $n\bar{E}+(n+1)F$. Clearly the net result is that the equation remains unchanged. As Eq. (17.14) is also unaffected, the optimal solution is still described by Eq. (17.15). This might be expected, because there is nothing to be gained or lost on the average by individual plants in having overlapping root systems (provided we neglect the diversion of photosynthate involved in creating more extensive, but no more productive root systems).

However, another argument leads to a very different result. If we make the same replacements to q and F, but now directly in Eq. (17.15), then

$$\lambda=\lambda_0\frac{(M\tau)^{t/t_1}-M\tau}{1-M\tau}\exp\left[\frac{1}{q}\int_0^t\frac{dF}{d\phi}dt'\right]\cdot\exp\left[\frac{n}{(n+1)q}\int_0^t\frac{d\bar{E}}{d\phi}dt'\right] \qquad (17.17)$$

This equation represents an optimisation that cannot be bettered by any *individual* plant as a result of its choosing a slightly different behaviour. It is akin to an "evolutionary stable strategy" (see MAYNARD-SMITH and PRICE 1973). It is different from the strategy represented by Eq. (17.15). That is the strategy which is optimal for a community of plants as an entity. When it was assumed that depletion of available water was in part due to the activity of competing plants *before* evaluating the optimal strategy, it was implicitly assumed, also, that the characteristics of an advantageous mutation would automatically and

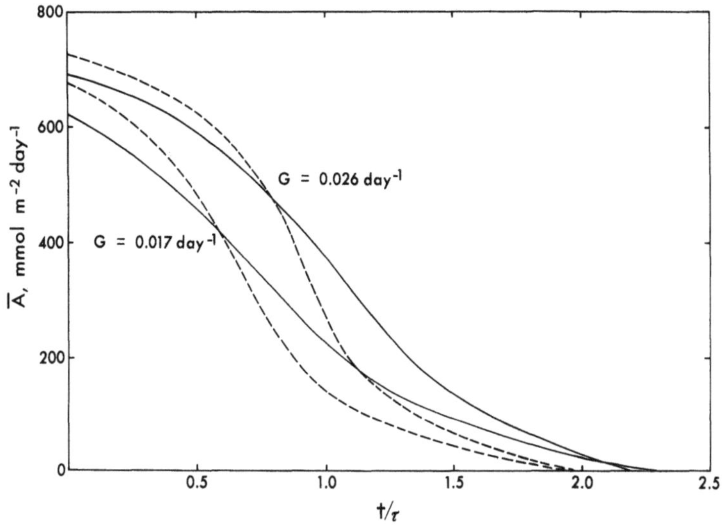

Fig. 17.13. Optimal variations, without competition (–) and with competition (---), in the mean daily rates of assimilation, $\bar{A}$, corresponding to two magnitudes of the probable average growth rate, G

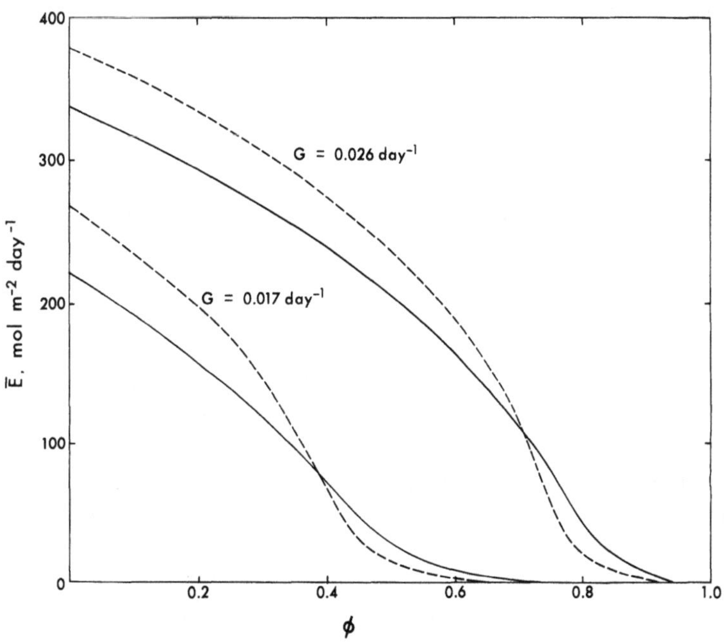

Fig. 17.14. Optimal variations of mean daily rates of transpiration, $\bar{E}$, with soil water deficit, ϕ, corresponding to the variations in Fig. 17.13

immediately be transferred to all members of the plant community. By so doing competition in any real sense of the word was excluded. Figure 17.13 shows two optimal time courses of $\bar{A}$ extracted from Fig. 17.7 ($\lambda_0 = 1{,}200$ and 800). The probable growth rates, G, are 0.026 and 0.017 day^{-1}. The broken curves are also optimal time courses, yielding the same two probable growth rates, but computed using Eq. (17.17) with $n=1$. The difference in the shapes of the curves is clear and needs no comment. What is not obvious is that the competitive plants use more water in achieving the same probable growth rate. As a result, their mortality, M, is increased by about 3%. The variations in daily rate of transpiration with soil water deficit are shown in Fig. 17.14.

17.4 Discussion and Conclusions

It is a gross simplification to assume that the only reversible effect of water depletion on plant functioning is stomatal closure, and that all irreversible effects are concentrated in a single catastrophe at a particular, critical soil water deficit. Nevertheless, the separation of the reversible and irreversible does provide an insight to certain aspects of the regulation of gas exchange in plants. Because of the unpredictability of rainfall, there is an adaptive advantage in conserving water by closure of stomata, quite apart from any immediate influence of water "stress" on plant metabolism. The conservation is not, of course, conservation in an absolute sense. It is conservation with respect to carbon fixation. That is why it has been possible to identify it with reduction in the probability of death due to drought corresponding to any particular probable rate of growth. Other things being equal, we may associate probable rate of growth with probable rate of reproduction. Therefore the variables G and M come close to defining the Malthusian fitness of a species, insofar as water deficiency may be the dominant limitation on fitness. However, there is no reason to suppose that the hypothesis of optimization is relevant only with plants in habitats that are generally arid. Severe droughts occur in mesic regions also. Even were they to do so at intervals much longer, on the average, than the life span of any individual plant, it is possible that they exert a considerable selection pressure for optimal regulation of water use. Nor are the principles of optimization changed when competition is taken into account. It might, perhaps, have been supposed that individual plants would waste water in a community in which there is strong competition for water. The truth is that they do tend to take up and lose water more rapidly when it is available, but do not neglect to fix as much carbon as possible in relation to the water used. As the theory of evolutionary stable strategies indicates, there is a sense in which the community as a whole is wasteful, but it is one which has no relevance to the individual.

None of these general conclusions is likely to be altered if we allow that plant carbon metabolism is affected by the state of plant water. Optimization would still require that $\partial E/\partial A$ be maintained nearly constant in the short term, and that E and A should decline with increase in soil water deficit. Of course, the determination of $\partial E/\partial A$ would be greatly complicated if it were necessary to take into account the influence of short-term changes in plant water relations due to changes in rate of transpiration, Eq. (17.10) being replaced by $A=A\ (p_i,$

T_1, I, E). And it would no longer be possible to describe the decline in $\bar{A}$ and $\bar{E}$ in terms of decrease in the parameter λ, as in Eq. (17.15), if the relationship between $\bar{E}$ and $\bar{A}$ were to change with increase in soil water deficit. Indeed it may be shown that $\partial\bar{E}/\partial\bar{A}$ would increase if the intrinsic decline in photosynthetic metabolism with increase in ϕ were sufficiently large, as was in fact observed by FARQUHAR et al. (1980a).

To introduce an empirical description of the direct effects of diminishing water supply on plant metabolism into the mathematical treatment of the implications of the optimization would tend to confound an important question: to what extent may the effects themselves be also a manifestation of optimization? If conductance and rates of gas exchange should decline with increase of soil water deficit as a precautionary measure, so to speak, it follows that other physiological disturbances might be explicable as economic consequences of this. A particularly dramatic example is that of leaf abscission. Leaves become inefficient when conductance becomes small, because they continue to lose some water through the cuticle without taking up much CO_2 through the stomata. This topic could have been treated by including cuticular transpiration in E, the transpiration function then making a positive intercept with the E-axis (see JONES 1980. Parenthetically JONES correctly pointed out that there is then a region, corresponding to small conductances, in which $\partial^2E/\partial A^2 < 0$. Optimization demands that the plant avoid operating in such a region. For an account of other circumstances in which $\partial^2E/\partial A^2 < 0$ and their implications see COWAN and FARQUHAR 1977). Cuticular transpiration has not been treated in this way here, because it is not under stomatal control. However it may be controlled in another, rather drastic way. If the risk of further water loss exceeds the potential worth of the established capacity for carbon fixation, then the leaf should be excised. The circumstances in which the risk exceeds the potential worth depends on the magnitude of $\partial G/\partial M$ at which the plant operates. In other words, the excision strategy should be tuned in a way which relates to the stomatal optimization of conductance, as indeed should all other strategies relating to the compromise between carbon gain and water loss. An example of this principle, relating to short-term regulation, is given by COWAN (1978). In those plants that make rapid adjustments of leaf angle in a way depending both on atmospheric environment and plant internal water relations, optimization suggests that $\partial E/\partial A = \lambda$ shuld be the same whether conductance or angle is taken as the physiological parameter linking E and A. Likewise long-term changes in leaf angle (or reflectance, or size) should be related to $\partial G/\partial M$.

It will have been noticed that no explicit allowance for the role of growth per se in optimization has been made. If the area of foliage increases at the same relative rate as the available water capacity of the extending root zone, then the parameter q remains constant. If the roots are extending into soil in which the water is no more and no less subject to depletion by competing processes than is that in the established root zone, then the function $F(\phi)$ is also invariant with time. Then the mathematics remains unchanged. In point of fact, however, it seems probable that increase in q, or decrease in $F(\phi)$, or both, do take place with seedling development. SILCOCK (1977) found that seedlings in semi-arid Queensland, Australia were most likely to die in the first 14 days after emergence. After 30 days deaths were relatively few (11%–17%) of the original population, and after 60 days the average relative mortality was only 3% per week. Changes in root:shoot ratio also seem to take place in response to decrease in the availability of water. However, the adaptive advantage, if one exists, of investing a greater proportion of photosynthate in water-gathering organs as distinct from photosynthesizing organs is not clear. As leaf conductance and rates of gas exchange decline, there would seem to be a diminishing need for water-gathering tissue. Perhaps the need for a disproportionately increasing root system stems from a decrease in the efficiency of roots in taking up water rather than an increased need for water (see Chaps. 1 and 3, this Vol.). For an elementary treatment of the relationship between optimization theory and carbon partitioning to roots and leaves see COWAN (1978).

The principle of optimization that has been expounded in this chapter is simply one of balancing risk of water use against benefit of carbon fixation in a changing environment. It has been illustrated with respect to diurnal variation in leaf microenvironment, and variation in soil water supply on a time scale characterized by the average interval between falls of rain. It presumably has relevance also to plant functioning on time scales appropriate to the life histories of plants, and to seasonal variations in environment. A premonition of its development in this wider context is provided in the next chapter of this volume.

References

Caemmerer von S (1981) On the relationship between chloroplast biochemistry and gas exchange of leaves. PhD Thesis, Aust Natl Univ, Canberra

Caemmerer von S, Farquhar GD (1981) Some relationships between the biochemistry of photosynthesis and the gas exchange of leaves. Planta 153:376–387

Courant R, Hilbert D (1953) Methods of mathematical physics Vol I. Wiley-Interscience, New York

Cowan IR (1972) Mass and heat transfer in laminar boundary layers with particular reference to assimilation and transpiration in leaves. Agric Meteorol 10:311–319

Cowan IR (1977) Stomatal behaviour and environment. Adv Bot Res 4:117–228

Cowan IR (1978) Water use in higher plants. In: McIntyre AK (ed) Water, planets, plants and people. Aust Acad Sci, Canberra, pp 71–107

Cowan IR (1981) Coping with water stress. In: Pate JS, McComb AS (eds) The biology of Australian plants. Univ West Australia Press, Nedlands, pp 1–32

Cowan IR, Farquhar GD (1977) Stomatal function in relation to leaf metabolism and environment. In: Jennings DH (ed) Integration of activity in the higher plant. Univ Press, Cambridge, pp 471–505

Cowan IR, Troughton JH (1972) The relative role of stomata in transpiration and assimilation. Planta 106:185–189

Darwin F (1898) Observations on stomata. Philos Trans R Soc London Ser B 190:531–621

Farquhar GD, Schulze E-D, Küppers M (1980a) Responses to humidity by stomata of *Nicotiana glauca* L. and *Corylus avellana* L. are consistent with the optimisation of carbon dioxide uptake with respect to water loss. Aust J Plant Physiol 7:315–327

Farquhar GD, Caemmerer von S, Berry JA (1980b) A biochemical model of photosynthetic CO_2 fixation in leaves of C_3 species. Planta 149:78–90

Field C, Berry JA, Mooney HA (1982) A portable system for measuring carbon dioxide and water vapour exchange of leaves. Plant Cell Environ 5:179

Hall AE, Björkman O (1975) Model of leaf photosynthesis and respiration. In: Gates DM, Schmerl RB (eds) Perspectives of biophysical ecology. Ecol Stud 12. Springer, Berlin Heidelberg New York, pp 187–202

Hall AE, Schulze E-D (1980) Stomatal response to environment and a possible interrelation between stomatal effects on transpiration and CO_2 assimilation. Plant Cell Environ 3:467–474

Hall AE, Schulze E-D, Lange OL (1976) Current perspectives of steady state stomatal responses to environment. In: Lange OL, Kappen L, Schulze E-D (eds) Water and plant life, problems and modern approaches. Ecol Stud 19. Springer, Berlin Heidelberg New York, pp 169–188

Hesketh JD, Alberte RS, Jones JW (1980) Predicting dark respiration in the soil-plant system. In: Hesketh JD, Jones JW (eds) Predicting photosynthesis for ecosystem models, vol I. CRC Press, Boca Raton, pp 69–84

Hunt WF, Loomis RS (1979) Respiration modelling and hypothesis testing with a dynamic model of sugar beet growth. Ann Bot 44:5–17

Jarman PD (1974) The diffusion of carbon dioxide and water vapour through stomata. J Exp Bot 25:927–936

Jones HG (1976) Crop characteristics and the ratio between assimilation and transpiration. J Appl Ecol 13:605–622

Jones HG (1980) Interaction and integration of adaptive responses to water stress: the implications of an unpredictable environment. In: Turner NC, Kramer PJ (eds) Adaption of plants to water and high temperature stress. Wiley-Interscience, New York, pp 353–365

Lange OL, Lösch R, Schulze E-D, Kappen L (1971) Responses of stomata to changes in humidity. Planta 100:76–86

Lange OL, Schulze E-D, Kappen L, Buschbom U, Evenari M (1975) Photosynthesis of desert plants as affected by internal and external factors. In: Gates DM, Schmerl RB (eds) Perspectives of biophysical ecology. Ecol Stud 12. Springer, Berlin Heidelberg New York, pp 121–143

Maynard-Smith J, Price GR (1973) The logic of animal conflicts. Nature (London) 246:15–18

Meinzer (1982) The effect of vapour pressure on stomatal control of gas exchange in Douglas fir – *Pseudotsuga menziesii*. Oecologia (in press)

Peisker M (1977) Transpiration and CO_2 uptake at varying stomatal aperture. In: Unger K (ed) Biophysikalische Analyse pflanzlicher Systeme. Fischer, Jena, pp 151–153

Schulze E-D, Lange OL, Buschbom U, Kappen L, Evenari M (1972) Stomatal responses to change in humidity in plants growing in the desert. Planta 108:259–270

Sharkey TD, Raschke K (1981) Separation and measurement of direct and indirect effects of light on stomata. Plant Physiol 68:33–40

Silcock RG (1977) A study of the fate of seedlings growing on sandy red earths in the Charleville district, Queensland. Aust J Bot 25:337–346

Wong SC, Cowan IR, Farquhar GD (1978) Leaf conductance in relation to assimilation in *Eucalyptus pauciflora* Sieb. ex Spreng. Plant Physiol 62:670–674

Wong SC, Cowan IR, Farquhar GD (1979) Stomatal conductance correlates with photosynthetic capacity. Nature (London) 282:424–426

18 Plant Life Forms and Their Carbon, Water and Nutrient Relations

E.-D. SCHULZE

CONTENTS

18.1 Introduction: Approaches to Defining Structural Types of Plants

The vegetation of the earth is characterized by distinct zones which have a rather uniform physiognomical appearance or structure even though they may be composed of many different plant species, genera, and families and may have a very different evolutionary history. The classic example of such "convergence" of morphological structure in a plant community is the evergreen sclerophyllous vegetation of the mediterranean-type climates (GRISEBACH 1872), but also the rain forest, the tropical and temperate grasslands, the savannahs or boreal forests and other formation types (WHITTAKER 1973) may be considered.

Early attempts to describe this physiognomic aspect of the earth's vegetation and to determine its ecological basis date back to the beginning of botanical science (review by DU RIETZ 1931; SZUJKÓ-LACZA and FEKETE 1969; FEKETE and SZUJKÓ-LACZA 1971). Two different approaches emerged: The first was inductive and descriptive. In the tradition of HUMBOLDT (1806) the aim was to illustrate exotic plant communities for readers quite unfamiliar with the flora concerned. This phytogeographical aspect of community description will not be discussed here since this has been comprehensively covered by GOOD 1974; SCHMITHÜSEN 1968; TROLL 1956 and others. Rather, in this chapter consideration will be given to the second or deductive approach, through which an attempt is made to explain morphological and structural features of plants and their pattern of occurrence under specific habitat conditions, on a functional basis.

In order to serve the descriptive as well as the analytical approach the term "Lebensform" was introduced by WARMING (1895) in contrast to the earlier term "Vegetationsform" of GRISEBACH (1872). In his first English edition WARMING (1909) translated "Lebensform" into "growth-form". It was CLEMENTS (1920) who redefined "growth form" as the direct and quantitative response made by a plant to different habitats and conditions, whereas "life-form" or "vegetation-form" is a morphological feature of a species which is insensitive to environmental changes. Since the separation of genetic and environmental influences is often difficult or impossible, MEUSEL (1978) used only the term "Wuchsform = growth-form" to investigate evolutionary trends of speciation in different taxa. The present paper will primarily try to explain the main structural types of plants and their ecology. In this context it seems appropriate to follow the terms "life-form" and "growth-form" according to the foregoing definition of CLEMENTS (1920) with certain reservations since it is quite clear that all our observations are made on growth-form of specific life-forms. The growth-form determines the range within which a specific life-form operates. For plants it appears not to be appropriate to use the zoological terminology of k- and r-strategy since by far the major proportion of plant species is perennial and would fall into one category only. Also the system of GRIME (1979) of r-, s-, and c-strategies does not lead any further, since it is not possible to make clear distinctions among groups at the species level and many species change between categories, depending on environment and ontogeny. GRIME

(1979) was only able to discuss his categories by using RAUNKIAER life-forms as sub-divisions.

As early as 1818 DE CANDOLLE suggested a classification system of plant life forms which was aiming at a functional interpretation rather than a description of plants (Table 18.1). It was based on the longevity of the individual and its organs and on characteristics of flower formation. Later, RAUNKIAER (1904) suggested what has become the best-known functional plant life form system. His principle aim was to "describe ... only those structural characters that reflect the essential dependence of the plants upon climate, i.e., their adaptation to survive the unfavourable season". For this purpose the classification was based on a single morphological character only, namely the height of the renewal bud above soil surface. It is a merit of the RAUNKIAER system that it has only a few distinct categories, although, the functional interpretation of bud height as being a principle adaptation to unfavourable seasons attracted criticism (DU RIETZ 1931; SZUJKÓ-LACZA and FEKETE 1969).

BOYSEN-JENSEN (1932) and MONSI (1960) first evaluated cormophytic plant life forms (plants with root, stem and leaves) in terms of carbon gains and losses, and considered marginal costs for the achievement of a certain product as the driving force of adaptation during evolution. MONSI (1960) and WALTER (1973) distinguished only two major types of plants on the basis of matter production, namely herbaceous and woody plants. The former were further divided into annuals, biennials and perennial herbs with deciduous and evergreen leaves. The woody plants were not grouped into trees and bushes according to height but were separated into deciduous and evergreen forms according to leaf longevity. It is quite clear that in nature these categories are expressed by a vast number of types, and that on the species level the definitions are far more difficult. Since only a treatment of the simplest case will be possible in this chapter, with some modification these categories are adopted in the subsequent analysis of growth and functioning of plant life forms. With respect to plant carbon relations, the distinction of annual and perennial plants, the latter grouped into herbaceous and woody species, seems to be appropriate. Because of their significance in terms of competition and total carbon turnover, a distinction is made between shrubs and trees.

The purpose of this chapter is to describe the allocation of carbon in different plant life forms and its relation to the water and nutrient balance. Consideration will be given to cormophytic plants only. Competitive advantages and limitations of their growth under different environmental conditions will be discussed, although these functional characteristics are still poorly understood. One reason for this is that ecological research has devoted considerable effort to understanding certain plant processes in single plant species, while comparative studies on different plant life forms have been neglected. A strong bias in ecological research is obvious. Due to human demands, a large amount of information is available on growth and performance of annual crop plants. In contrast, the information available on wild plants is rather limited and has often been restricted to only a few commonly studied types. The comparisons presented in this chapter are based upon limited experimental evidence and we are a

Table 18.1. Systems of plant life forms (The distinction between "phytogeographically" and "ecologically" oriented systems of plant life forms is not in all cases strict and contains overlaps
a) More phytogeographically oriented life-form systems

HUMBOLDT (1806)	GRISEBACH (1872)	DRUDE (1887)	DRUDE (1928)	DU RIETZ (1931)	SCHMITHÜSEN (1968)
Palmen	Holzgewächse	Holzpflanzen mit Laub	Aerophyten	*Main life forms:*	Kronenbäume
Bananenform	Sukkulente	Bäume	Holzgewächse	Woody plants	Schopfbäume
Malvenform	Schlinggewächse	Sträucher	Winder, Klimmer	Trees	Baumgräser
Mimosenform	Epiphyten	Lianen	Epiphyten	Shrubs	Baumwürger
Heidekräuter	Kräuter	Mangroven	Sukkulente	Dwarf shrubs	Lianen
Cactusform	Gräser	Holzparasiten	Grasartige	Cushion plants	Sträucher
Orchideen	Zellenpflanzen	Blattlose Holzpflanzen	Kräuter	Lianes	Zwergbäume
Casuarinen		Stamm-Sukkulente	Therophyten	Half shrubs	Stammsukkulente
Nadelhölzer	(60 subdivisions)	Blattlose Sträucher	Bryophyten	Herbaceous plants	Krautstammpflanzen
Arumform		Halbsträucher	Lichenen	Chtonophytic	Epiphyten
Lianen		perenne Kräuter	Myceten	Epiphytoidic	Zwergsträucher
Aloegewächse		Rosettenpflanzen	Hydrophyten	Parasitic	Halbsträucher
Grasform		Blatt-Sukkulente		Lianes	Zwerg-Sukkulenten
Farne		Epiphyten		(24 subdivisions)	Chamaephytische Stauden
Liliengewächse		hapaxanthe Pflanzen			Hemikryptophytische
Weidenform		Landpflanzen		*Growth forms:*	Holzpflanzen
Myrtenform		Wasserpflanzen		Periodicity life forms	Hemikryptophytische
Melastomeenform		Flechten		Bud height life forms	Stauden
Lorbeerform		Saprophyten/Parasiten		Bud type life forms	Winterannuelle
				Leaf life forms	Geophyten-Stauden
					Therophyten-Kräuter
					Schwimmblattpflanzen
					Submerse Kräuter

b) More ecologically oriented life-form systems

De Candolle (1818)	Warming (1884)	Raunkiaer (1904)	Clements (1920)	Warming (1923)	Monsi (1960) Walter (1973)
monocarpia	Hapaxanthe	Phanerophytes	Annuals	Autotrophs	Herbaceous plants
annua	monocyclisch	Chamaephytes	Biannuals	Hydrophytes (7 classes)	Annual
biennis	bicyclisch	Hemicryptophytes	Herbaceous	Aerophytes	Biennial
perennis	polycyclisch	Cryptophytes	perennials	Autonomous	Deciduous
polycarpia	Perenne	Therophytes	Sod grass	Epiphytoid (2 classes)	perennial
rhizocarpia	kein Wanderungsvermögen		Bunchgrass	Terrestrial	Evergreen
caulocarpia	Primärwurzel bleibend	subdivisions:	Bush grass	10. Chylophytes	perennial
suffrutex	verholzt/krautig	leaf size	Cushion herb	11. Halophytes	Woody plants
frutex	Primärwurzel absterbend	leaf duration	Mat herb	12. Agrophytes (herb)	Deciduous
arbuscula	mit Wanderungsvermögen	vegetative	Rosette herb	13. Poioides	woody
arbor	oberirdisch/unterirdisch	propagation	Carpet herb	14. Xyloides (ligneous)	Evergreen
	Wasserpflanzen		Succulents	Non-autonomous	woody
	(14 main groups)		Woody perennials	15. Klinophytes	
			Half shrub	Allotrophs	
			Bush	16. Saprophytes	
			Succulent	17. Parasites	
			Shrub		
			Tree		

long way from understanding the diversity of plant life forms on a functional basis.

18.2 Dry Matter Partitioning and Longevity in Plant Life Forms

Understanding of plant life forms on a functional basis requires at least a general understanding of the plant carbon, water and nutrient relations. In the following only those characteristics will be explained which are most pertinent to a more general discussion of the ecology of plant life forms.

The photosynthetic carbon gain (P) of a plant is a function of the rate of CO_2 assimilation per unit of time and leaf tissue (A) and the amount of leaf tissue (F):

$$P=f(A \times F) \tag{18.1}$$

To describe plant growth it is necessary to distinguish between growth of green, photosynthesizing biomass ($\Delta F/\Delta t$) and of supplementary, non-green biomass ($\Delta S/\Delta t$, e.g., roots, stems and conducting tissue) (MONSI 1960; DE WIT 1978). The sum of both is the total growth rate over time ($\Delta G/\Delta t$). Its integral over time is also referred to as primary production (P_N)

$$P_N=\int_0^T \frac{\Delta G}{\Delta t}=\int_0^T \left(\frac{\Delta F}{\Delta t}+\frac{\Delta S}{\Delta t}\right). \tag{18.2}$$

It is thought that the regulation of the proportion of carbon which is allocated into green (m) and non-green (1−m) tissue has special significance for the growth and success of plant life forms under particular ecological conditions. The growth of new foliage and of new non-photosynthesizing biomass may be expressed as:

$$\frac{\Delta F}{\Delta t}=m \cdot P \cdot (1-R_g^F)-F_{(t_0)} \cdot R_m^F \tag{18.3}$$

$$\frac{\Delta S}{\Delta t}=(1-m) \cdot P \cdot (1-R_g^S)-S_{(t_0)} \cdot R_m^S \tag{18.4}$$

where R_m represents maintenance respiration per unit time and per unit living tissue of existing biomass, and R_g represents growth respiration expressed as a fraction of carbon used in synthesis of new growth. The superscripts F and S indicate respiration rates in green and non-green biomass respectively, and $F_{(t_0)}$ and $S_{(t_0)}$ indicate the total green and non-green biomass at the time (t_0): Eqs. (18.3) and (18.4) apply in this form only to young, vegetatively growing plants, since they do not include reallocation of stored carbohydrates or transport phenomena, which occur, for example, during seed filling of plants. The equations explain the phenomenon of carbon allocation although the mechanism of control is uncertain.

The living biomass may still be different from the additive sums of growth rate and initial biomass because there occur permanent senescence, death and abscission of living tissue, which include predation by herbivorous animals. The turnover rate may be quite different in the roots or in leaves. The living biomass may be expressed by:

$$F_{(t)} = F_{(t_0)} + \frac{\Delta F}{\Delta t} - \frac{\Delta D^F}{\Delta t} \quad (18.5)$$

$$S_{(t)} = S_{(t_0)} + \frac{\Delta S}{\Delta t} - \frac{\Delta D^S}{\Delta t} \quad (18.6)$$

where $\Delta D^F/\Delta t$ and $\Delta D^S/\Delta t$ are the rates of carbon loss from the living biomass due to death and abscission in green and non-green biomass respectively. The dead portion may either be retained if it is essential to the functioning of the individual, like wood, or may be shed, having completed its function, like deciduous leaves. If the rate of biomass loss is slow with respect to new growth, then the individual organ has a high longevity. The rate changes with season, age, and with environmental conditions. Therefore, it bears strong influence on the overall carbon balance of a plant. (For detailed discussion of modelling of plant growth and production see Chap. 4, Vol. 12 D).

The interaction of water and carbon relations has been investigated through the crop growth model of DE WIT (1978). It is known that plant water status influences the carbon relations at the gas exchange and at the growth level, but it is presently not clear which water relations parameters are the principle controlling factors in each case. The DE WIT model predicts the partitioning of carbohydrates into root and shoot according to Fig. 18.1A as a function of the relative water content of the plant. This is determined by the rate of water uptake of the root biomass and by the evaporation rate of the leaf biomass. In order to ensure a functional balance between root and shoot it is implicitly assumed that the plant maintains a higher relative water content in the root than in the shoot, since generally the root appears to be more sensitive to desiccation than the shoot. A higher relative water content of the absorbing part of the root would account for the observation that shoot growth is more sensitive to a decrease in whole plant water content than the root. The result is that root growth is promoted at low relative water contents of whole plants, and shoot growth at higher ones.

The existence of a functional balance between root and shoot growth during perturbations in the plant water status is suggested by Fig. 18.1B (BROUWER 1963), which shows a linear relationship between root and leaf fresh weight of an undisturbed plant. Following a sudden disturbance in the root/leaf ratio, after removal of parts of the leaves or parts of the roots, for example, a gradual and full recovery of the equilibrium which existed before the perturbation occurs. With a relative shortage of roots, the time for recovery is shorter than with a relative shortage of leaves, because in the latter case not only the distribution ratio, but also the primary carbon gain through assimilation is reduced.

For an understanding of plant life forms and their adaptation to specific environmental conditions, it is not only important to know the functional inter-

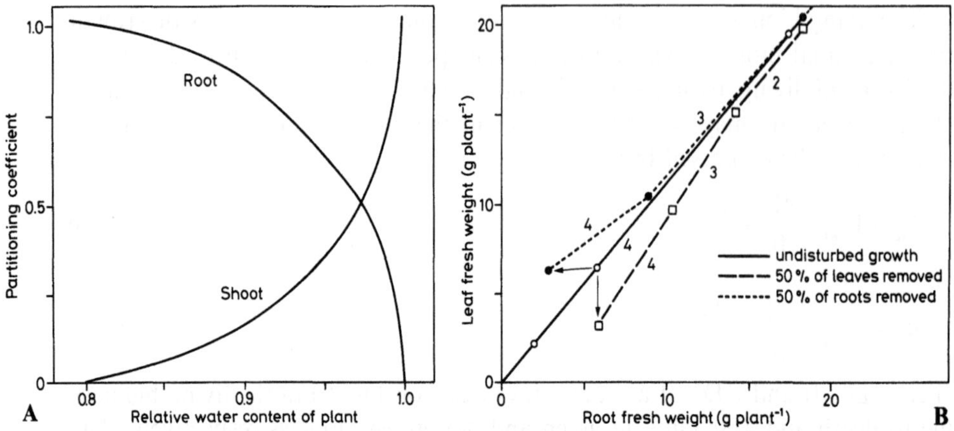

Fig. 18.1. A The partitioning of carbohydrates into root and shoot (y-axis: relative scale) with changing relative plant water content (x-axis). (After DE WIT 1978). **B** The relationship between leaf (x-axis) and root (y-axis) fresh weight for an undisturbed growing bean plant (*open circles*) and recovery after removal of 50% of the leaves (□) or 50% of the roots (●). *Numbers* indicate successive days of measurement. A sudden disturbance of the leaf/root ratio is followed by a recovery towards the original equilibrium. The period needed for recovery is 5–10 days and it is longer with a relative shortage of shoots. (After BROUWER 1963)

action of water and carbon relations, but also the degree to which a certain plant is able to maintain a certain plant water status along with a maximal rate of growth. At issue is, whether plants are at all able to optimize production in relation to plant water status and whether plant life forms represent different systems which are adapted in this respect to specific environments.

The problem of optimization of carbon and water relations may be investigated at the gas exchange and at the growth level. In Chapter 17 of this Volume there is given a theoretical basis for the hypothesis that stomatal behaviour is optimal if minimal loss of water is sustained for a given amount of carbon assimilated in a certain period of time (see also Chap. 1, this Vol.). This corresponds to minimizing $\int_0^T E\,dt$ for conditions in which $\int_0^T A\,dt$ is constant. Since both H_2O and CO_2 exchange underlie stomatal control, optimal water use during CO_2 assimilation could be approached if during a sequence of perturbations of the stomata which result in no net change in total assimilation, $\int_0^T \partial A\,dt = 0$, the resulting water loss becomes smaller, $\int_0^T \partial E\,dt < 0$. This condition is fulfilled if, for a given change in water loss, ∂E, during a change in stomatal conductance, ∂g, the marginal gain in assimilation, ∂A, with respect to water-loss remains constant under various conditions during the day. Mathematically optimal gas exchange would require that

$$\frac{\partial E}{\partial g} \Big/ \frac{\partial A}{\partial g} = \lambda \qquad (18.7)$$

where λ is a constant which is dependent on plant conditioning. HALL and SCHULZE (1980) gave experimental evidence for the existence of such regulation.

Applying the same principle to whole plant water status and growth, it is considered that the ratio of carbohydrate distribution into growth of photosynthesizing and non-photosynthesizing biomass, m, acts – at least phenomenologically — as the principle independent variable. Optimal regulation of growth with respect to whole plant water status would require for a certain growth over a period of time, $\int_0^T P_N\,dt = \text{constant}$, that the relative water content (W) is maximal, $\int_0^T W\,dt = \text{maximum}$. This would be the case if for certain perturbations, which result in no net change of total growth, $\int_0^T \partial P_N\,dt = 0$, the resulting water status of the plant is not improved, $\int_0^T \partial W\,dt \leqq 0$. For optimal regulation of growth in respect to plant water status in principle one could require as a hypothesis that $\partial W/\partial m$ and $\partial P_N/\partial m$ are balanced. But the conditions for growth are certainly more complicated than for gas exchange, since growth of photosynthetic and non-photosynthetic parts does not necessarily occur at the same time during the growth cycle and transfer of water from storage sites may be significant under fluctuating environments. Additionally a second variable exists, which determines the living biomass as an integral over time, namely the rate by which living biomass is lost or converted into non-living biomass [see Eq. (18.5) and (18.6)]. Plant life forms vary to a large degree by the longevity of different organs, e.g., deciduous and evergreen leaves. Therefore the hypothesis should be extended to include ontogenetic differences of growth responses and the rate of biomass loss. Consideration must be given to the standing functional biomass, which may include dead parts, like wood.

We are far from a mathematical or experimental solution of this problem, but a theoretical measure needs to be developed in future research by which we would be able to compare the degree of adaptation of specific plant life forms to certain environmental conditions. There are indications that such regulation may occur. Since plant water status for a given root biomass is largely determined by whole plant transpiration, optimal regulation would imply an approximately linear relationship between water use and growth. This has been shown for various crops (DE WIT 1958; see also Section 18.4).

In addition to the effects of plant water status, growth is strongly governed by the plant nutrient relations. However, the underlying functional bases are even more complicated and, with respect to whole plant processes, less understood than the interaction of water relations and growth. Since nutrient loss does not occur on the same order of magnitude as water loss during transpiration or carbon loss during respiration, the uptake of nutrients through roots serves primarily the incorporation of nutrients into newly grown biomass. The balance is complicated by the fact that only part of the total nutrient content in a plant is available for new growth and the availability in the plant may depend on other factors, e.g., nitrate reduction may be inhibited by water stress. Nutrient losses occur through root exudation and through loss of biomass (e.g., leaves), although the plant is able to recover some ions before natural abscission occurs.

For the regulation of growth in response to nutrient status, it may be hypothesized that the main independent variable is again (m), the portion of carbohydrates allocated for growth of new foliage. The factor (m) will become smaller with the decreasing availability of a certain nutrient element for growth of new biomass. The growth of roots should be less sensitive to the depletion of available nutrients than the shoot, in order to account for greater root than shoot growth with nutrient deficiency. In an analogous approach taken with respect to the plant water relations, one could envisage a control which operates via the carbohydrate allocation such that for a certain perturbation by the carbon allocation ratio, m, which results in no net change of total growth, $\int_0^T \partial P_N \, dt = 0$, the resulting status of available nutrients, N, in the plant, is not improved, $\int_0^T \partial N \, dt \leqq 0$. Again, $\partial N/\partial m$ and $\partial P_N/\partial m$ should be balanced. However, the reality of plant growth with respect to nutrient status appears to be even more complex, since nutrient uptake and growth may not occur at the same time in the growing season; plants frequently accumulate nutrients if they are abundant. Furthermore great differences exist in the mobility of various ions. Additionally, the living biomass as integral of growth over time is determined by the longevity of different plant organs, and the problem of evergreen and deciduous habit, as well as of xeromorphy and sclerophylly, is strongly dependent on the plant nutrient status. This is not yet included in the foregoing hypothesis. One may speculate to what extent plants are able to optimize both water and nutrient relations via the same control mechanisms, namely carbon allocation and longevity. In the case of optimal control of growth with respect to water and nutrients, it would be required that the marginal gain of both processes, $\partial W/\partial m$ and $\partial N/\partial m$, are balanced with respect to $\partial P_N/\partial m$. Numerous interactions between water relations and nutrient status are known (e.g., proline and nitrate accumulation with stress: TURNER and JONES 1980). Some of these interactions are discussed in Sections 18.3.3.2 and 18.3.3.3.

For deeper understanding of the interaction and existence of different life forms in plant communities, future plant ecology has to define the possible and alternative ways of adaptation in carbon turnover which served as "goals" of natural selection at environmental stress situations. In animal ecology different life cycle "strategies" have been defined (STEARNS 1976). This concept has been introduced into plant ecology by GRIME (1979) at a descriptive level. At present it is not clear with plants what "fitness" of the individual means in terms of the competitive interaction within plant communities – maximization of living or total biomass or careful expenditure of resources, a problem also discussed in insect ecology (STEARNS 1976). Some measure of "fitness" may emerge from the above hypothesis which proposes that carbon allocation is regulated on a whole-plant basis by the rules of equality of marginal returns (extra carbon return per unit carbon investment in the organs) which govern, for example, the input optimization in farms (e.g., amount of yield per unit investment of fertilizer). The functioning of whole-plant processes is tremendously complex, and one may doubt that carbon allocation could be considered in terms of equalizing the marginal returns for all the various possible organs, i.e., photosynthetic, support, transport, nutrient gathering, water gathering, productive or reproductive organs or others. In spite of these doubts and admitted difficulties, this chapter examines the hypothesis that the evolution of different plant life forms is the result of optimizing carbon returns for specific environmental conditions. This hypothesis underlies the following discussion of different plant life forms and their ecological characteristics. The nature and the existence of an optimization which takes the marginal gain of changes in water and nutrient status into account is not known, but principles of adaptation to various habitat conditions can already be discussed from the above equations.

18.2.1 The Photosynthetic Capacity of Plant Species

The light-saturated photosynthetic rate at ambient CO_2 concentration (300 to 330 ppm) and optimal environmental conditions, referred to in this context as photosynthetic capacity, A_{max}, may vary by a factor of 10 to 20 (KÖRNER et al. 1979). It is generally highest in C_4 plants, e.g., rates of 50 μmol m^{-2} s^{-1} were measured in the annual desert shrub *Tidestromia oblongifolia* (BJÖRKMAN et al. 1972). Even higher rates were reported from the C_3 desert annual *Camissonia claviformis* (60 μmol m^{-2} s^{-1}: MOONEY et al. 1976) even though C_3 plants on the average have rates which are 50% less than those of C_4 plants (see Chap. 15, this Vol.). The photosynthetic capacity of C_3 species is generally greatest among grasses and herbaceous dicots and progressively decreases in deciduous trees and evergreen species, with lowest rates found in succulents and conifers. According to KÖRNER et al. (1979) the photosynthetic capacity of plant species grouped on the basis of similar morphology or growth conditions is less a function of plant form as defined by MONSI (1960), than a function of the type of leaf, which in turn commonly reflects a certain set of ecological conditions and thus their growth form. Among herbaceous plants, for instance, photosynthetic capacity varies over the entire known range, with succulent leaves and shade leaves representing one extreme, and mesomorphic sun leaves the other.

It is presently not clear why plants do not invest a larger portion of their available nitrogen into photosynthetic enzymes in order to attain a higher rate of photosynthesis which would also imply an improved water use efficiency (cf. Chap. 7, this Vol.). Important for understanding plant life forms are considerations as to the marginal costs of constructing "better" leaves. Here, habitat conditions must be taken into account. A large discrepancy exists between the photosynthetic capacity of a species and the light climate of a habitat. Comparison with the actual light conditions of a habitat shows that even sun leaves operate at the level of light saturation for only about 20% of the total daytime (SCHULZE 1970; MOSER 1973). LANGE et al. (1974) found that maximal rates of CO_2 uptake at a certain plant water status were very rarely reached during the course of a day in an arid climate. Therefore, the opposite conclusion may hold, namely that plants do invest more nitrogen into leaf enzymes than they can actually use under average natural climatic conditions. However, experimental evidence as to the relationship between A_{max} and leaf nitrogen content and as to the actual level of operation for a given set of habitat conditions is lacking (see also Chap. 7, this Vol.).

In spite of this, it appears that photosynthetic capacity is more closely related to the plant environment than to plant life form, although the success of a species in its given environment and in relation to its competitors may well be related to its total carbon gain.

18.2.2 The Respiration Rate of Plant Species

It is apparent from Eqs. (18.3) and (18.4) that primary production is strongly dependent on carbon losses resulting from growth and maintenance respiration. Very few whole-plant carbon balances have been measured; most carbon balances contain extrapolations at the respiratory level. PENNING DE VRIES (1975) estimated that carbon loss from respiration may amount to about 28% of net photosynthesis in maize and wheat. This estimate may even be too low. The proportion of dry weight produced for a certain amount of CO_2 assimilation – a transformation factor which accounts for growth and maintenance respiration – was found to be 0.6 to 0.7 for various crop plants (WATANABE 1976). The environmental parameters, day and night temperature (LARCHER 1976) and day length, determine the respiratory portion and therefore growth. It appears that the difference in CO_2 release between various species and plant organs (leaves, roots, stems) is greater than the difference between plant life forms and environmental conditions. Maintenance respiration at a given temperature was found to increase with altitude (SMITH and HADLEY 1974), but it decreases with water stress (BROWN and TRLICA 1977) and with shade (SCHULZE 1970). For a detailed discussion of growth and maintenance respiration the reader is referred to Chapter 16, this Volume, and see Chapters 4 to 10, Volume 12D.

18.2.3 The Allocation of Carbohydrates into Photosynthetic and Non-Photosynthetic Tissues, and the Longevity of the Plant Individual

The partitioning of carbohydrates into plant organs and the longevity of the plant individual and its organs are major characteristics which determine plant life forms. In Eqs. (18.3) and (18.4) the portion (m) of carbohydrates allocated for growth of photosynthetic structures and the regulation of this factor strongly influences the plant carbon balance. The ratio of photosynthetic to non-photosynthetic tissue varies greatly in different plant life forms and within the same species this ratio is highly dependent on environmental conditions. The productivity of the plant increases as (m) approaches 1 since growth of new leaf area acts in a positive feed-back manner on the total productivity until full light interception is reached. If (m) is close to 1 the plant produces comparatively few roots or stems, a "strategy" approached through evolution by lianas or autotrophic parasites (mistletoe). The opposite is true for heterotrophic parasites, like *Lathraea*, which maximize the growth of non-photosynthetic organs. However, in this case the plant must rely on an external carbohydrate supply.

In the following section the principle types of carbon allocation in different plant life forms are discussed, although it is quite clear that in nature numerous exceptions and intermediate types exist at the species level. Because the natural variety of species overrides any classification which is based on a few distinct categories, the following discussion will primarily consider the simplest cases of carbon turnover. Further characteristic features, exceptions and transient types are discussed in Section 18.3.3.

18.2.3.1 Carbon Allocation and Ecological Features of Annuals

Annuals are generally herbaceous although their stems may lignify during senescence (e.g., *Anastatica hierochunta*). They have a short life cycle which may be completed in 2–3 weeks (*Linaria haelava*, EVENARI, personal communication), or within a year (*Triticum aestivum*) and the plant individual usually dies after seed formation. Generally annuals survive unfavourable cold or dry seasons as seeds. Since survival of annuals, more than in any other plant life form, depends on the ripening of new seeds, the formation of seeds within a short production period is characteristic of annuals. The harvest index (seed dry matter per aboveground dry matter) may reach 0.5 in annual crops and weeds (see Sect. 18.3.2.2). Although other herbaceous (orchids) or woody (*Salix*) species may also produce very large seed numbers the relative proportion of biomass allocated for seeds appears to be generally higher in annuals than in any other plant life form. Additionally annual species are characterized by rapid development after germination. WALTER (1981) illustrated this fact by taking the development of sunflower (*Helianthus annuus*) and the beech tree (*Fagus silvatica*) as an example. One year after germination sunflower had reached 600 g dry weight, whereas the beech seedling attained only 1.5 g. If growing sites are available and if the environmental conditions allow growth, the population size of annuals increases rapidly during consecutive seasons due to the annual increment in seed formation and the frequently repeated seedling and plant establishment. Therefore, annuals are most effective in colonizing bare land. However, if space is limited, intense intraspecific competition may develop within a few generations, such that the production of the individual will be restricted even before the critical leaf area index of the population is reached (see Sect. 18.2.3.2).

The structural forms of annuals range widely (SCHMITTHÜSEN 1968) including plants with strongly lignified stems (*Helianthus annuus*), small erect herbs (*Ranunculus arvense*), rosette plants (*Centaurea cyanus*), winding "liana types" (*Vicia hirsuta*) or parasites (*Cuscuta europaea*). A large number of crop species are annual (e.g., *Triticum, Hordeum, Zea* and others) or weakly perennial depending on the environment (e.g., *Oryza sativa, Sorghum vulgare*). Annuals occur in almost all climatic regions of the world (WALTER 1968, 1973) and are found in most open habitat communities as early successional species following disturbance. In animal ecology the equivalent "strategy" is performed by fugitive species which are specialized to survival in unpredictable habitats (r-strategy: STEARNS 1976; GRIME 1979). Annual plants are most successful in semi-arid and arid regions, where they may predominate over other plant forms following rainfall, even though during dry periods perennial woody plants appear to be the dominant plant form. However, because the whole life cycle must be completed in a single growing season, annuals are not very effective in arctic and alpine environments. In these vegetation zones they contribute <1% of the general flora (BILLINGS and MOONEY 1968). Although endemism and subspecies are common in annuals, some have a wide range of distribution, and one of the few cosmopolitan representatives is *Poa annua*.

18.2.3.2 Carbon Allocation and Ecological Features of Perennial Herbaceous Plants

Perennial herbaceous plants are similar to annuals in that they maintain living non-photosynthesizing parts. The significant difference between annual and perennial plants (herbaceous and woody) occurs with respect to dry matter production in successive years. Annuals rely solely on successive generations of new individuals for the continuation of dry matter production, whereas in perennials production is directed to the prolonged survival of the plant individual. (In animal ecology referred to as k-strategy: STEARNS 1976). In a seasonally cold or dry environment, for example, perennials accumulate and store carbohydrates or lipids to ensure survival and growth of the same individual in successive years. The stored reserves of buds or rhizomes enable the perennial plants to start growth more quickly at the onset of the growing season. Seed production may be delayed in a given growing season. Although the storage of reserves is discussed as being of ecological significance for survival of the individual in seasonal climates, it is important to realize that humid tropical environments are also dominated by perennial plants.

During occupation of bare land in the process of succession, a population of perennials cannot increase its dry matter as rapidly as annuals because of differences in carbon allocation after germination and because of the differences in seed yield following the initial years of colonization. But after several years and with the accumulation of reserves and deeper-established root systems, perennials are more successful in interspecific competition mainly through competition for light. Also production of seeds may then be quite high. As a result, the perennial plant community usually succeeds the annual pioneers. Yet since the critical leaf area index is constant in a closed plant community (see Sect. 18.2.4) for a given radiation climate and plant species, an upper limit to the increase in size of the individual exists. This may be a significant distinction between perennial herbaceous and woody species. If the living non-photosynthetic but supporting portion of the plant increases, the maintenance respiration also increases. Accordingly, for a given partitioning ratio [Eqs. (18.3) and (18.4)] the net production of the individual becomes smaller with increasing plant size. Eventually, one expects that a limit is reached where the primary production will be zero for a given leaf biomass and the plant individual will be maintained by respiration but no further growth will occur. Perennial herbaceous plants differ from woody species with respect to respiratory carbon loss in the non-photosynthetic plant portion. In herbaceous species, this remains to a large degree alive. Additionally herbaceous perennial plants lack secondary growth.

Perennial herbaceous plants may complete their life cycle after 2 years (biennials) while others require many years (perennials sensu strictu). They may flower only once, like all biennials or like *Agave deserti,* which directs all of its stored carbon into seed production at the end of a 10–30-year lifetime. Others flower repeatedly upon reaching maturity or under extreme conditions, perennial herbaceous plants may reproduce only vegetatively. Unfavourable dry or cold seasons may be endured in an evergreen state (many temperate zone grasses in winter) or as dormant buds (arid and temperate zone grasses at drought), rhizomes or bulbs. It is clear that within the category of perennial herbaceous plants a large variety of structural types occur (SCHMITHÜSEN 1968). Herbaceous perennials are most successful in environments where cold temperatures, shade, seasonal drought, wetness or mechanical disturbance create extreme but stable conditions. They lose efficiency when the probability of rainfall is low or the rainy season is of short duration, in which case annuals are more successful.

Perennial herbaceous plants occur further north than all other vascular plants (*Saxifraga oppositifolia:* 83°39′ northern latitude: HOLMEN 1957) and occupy the cold deserts (*Puccinellia* sp.: BILLINGS and MOONEY 1968). They are the only vascular plants on the Antarctic continent (*Deschampsia antarctica,* WALTER 1968). *Stellaria decumbens* reaches the highest altitude known for higher plants (6,136 m, Himalaya, SWAN 1967), and *Allocasia macrocarpa* shows the lowest known light saturation at 80 µmol photons $m^{-2}\ s^{-1}$ on the forest floor in a subtropical rain forest (BJÖRKMAN et al. 1972). Herbaceous perennials are predominant in vegetation types which cover about a fourth of the earth's land surface, namely the subtropical grasslands and the temperate and boreal steppes. Together with scattered trees they are dominant constituents of the subtropical savannah. Herbaceous perennials also dominate in regions of natural forest vegetation, as long as regular disturbance by grazing or mowing excludes competition from woody species. Within temperate and tropical forest regions perennial herbaceous plants contribute major portions of the forest floor vegetation.

18.2.3.3 Carbon Allocation and Ecological Features of Woody Plants

Woody plants are characterized by secondary growth and by the continuous conversion of structural tissue into non-living and therefore non-respiring biomass. This dry matter remains an integral part of the individual, essential to its functioning. It is part of the conducting system, it allows physical support of the large body of biomass, and it may also act as a storage reservoir (e.g., for water, nutrients or carbohydrates). The process of wood formation, therefore, improves the carbon balance with respect to respiratory CO_2 loss when compared with herbaceous plants on a relative scale. Trees and shrubs generally have a prolonged life cycle. The germination and seedling stage normally extends over an entire growing season and is followed by a sapling stage which may last for several years. Flowering and fruiting is further reduced in comparison to perennial herbaceous species. Many trees reach maturity only after several decades (*Picea excelsa* stands 60–70 years) and may even then produce fruits only periodically (every 3–8 years). In contrast, there are also tree species which produce abundant fruit (*Malus, Salix, Pinus*), but only after years of maturation.

However, in addition to other physiological and hydrological limitations on growth, there is also a limitation to growth in woody plants which is reached when the respiratory rates of the living portion of the stem and root approach the rate of carbon supply of the foliage. As with herbaceous plants, at this point an increase in non-photosynthetic biomass is possible only if less carbohydrate is allocated into growth of new foliage. This leads to a decrease in total leaf biomass and therefore to further deterioration of the carbon balance. This situation can be observed in seedling establishment on forest floors or with the oldest living tree, *Pinus aristata,* which is more than 4,000 years old. This age is reached partially because of the low rate of wood destruction in the semi-arid alpine environment of California, but also because of a unique dieback of the supporting stem and root which probably acts to maintain a balance between photosynthesizing and non-photosynthesizing living biomass (LA MARCHE 1969). Trees and shrubs differ mainly with respect to size and height, which are primarily determined by differences in branching pattern and by the extent of apical dominance.

Trees and shrubs are generally long-lived, with maximum ages ranging from about a hundred years (most northern temperate forest trees) to more than 1,000 years. The age of a single *Larrea tridentata* individual was estimated to be more than 11,000 years (VASEK 1980). Because of their longevity and dimensions, trees outcompete all other plant forms in a long-term succession provided environmental conditions allow establishment. The seedling is the most vulnerable stage in the life cycle of a woody species because of competition with other plant life forms. Many temperate-zone tree species establish most effectively in the shade of the parent tree where other plant species are suppressed (e.g., *Fagus*). This may be in contrast to tropical environments where seedlings have a greater chance of survival at larger distances from the parent plant because of highly specialized seed-predators (e.g., Bruchidae). JANZEN (1971) explained tree species diversity in tropical rainforest environment on the basis of this mechanism.

18.2.4 Leaf Area Index and Light Interception

For plant growth the carbon-partitioning between leaf area expansion and growth of non-photosynthesizing tissues becomes increasingly evident with the age of the individual. MONSI (1968) made a growth estimate for a hypothetical plant in which carbohydrates were distributed with the ratio of 0.7/0.3 for growth of new leaves and non-photosynthetic tissues respectively, at a constant rate of CO_2 assimilation and respiration. Neglecting competition for light between leaves it is predicted that within 3 years this plant would be 15 times larger than the opposite case with a ratio of 0.3/0.7 for leaf and non-photosynthetic biomass growth. As an example of this principle, MONSI and MURATA (1970) mentioned the growth and competition of lianas and rainforest trees. Lianas allocate relatively more carbohydrates into leaf area development than do trees, but use the latter as support in order to expose their leaves to full daylight above the canopy. Through this "light parasitism" the liana rapidly covers the host plant and eventually even causes its destruction. The dual effects of an increase in the allocation of carbohydrates for leaf growth, namely, increasing the photosynthetic area as well as decreasing the new growth of respiratory tissues, explains the frequent observation of a close correlation between growth rates and leaf area. Even within the same genus, EL-SHARKAWY et al. (1968) showed three-fold differences in net photosynthesis of 24 *Gossypium* species, but dry matter accumulation was associated with the rate of leaf area development and not with photosynthesis.

Theoretically, therefore, in order to obtain maximal growth, it may be most effective to maximize the portion of photosynthates allocated to build new leaves. This has, in fact, been done in wheat cultivars. EVANS and DUNSTONE (1970) have compared wheat from all stages of its evolution as a crop. They found that photosynthetic rate per unit flag leaf area has fallen by factor of 3 in highly productive modern cultivars compared with the diploid *Triticum boeoticum*, but flag leaf area and duration has increased considerably and also grain size. However, there is an upper limit to which leaf area can be increased, since there is an increased self-shading effect at high leaf density due to increasing amounts of supporting tissues and leaves. Figure 18.2A shows that growth rates are dependent on leaf area index, LAI (integral of projected leaf area per unit ground area), but as leaf area increases the growth rate reaches a maximum value. Thereafter, it may decline. The existence of an optimum LAI for herbaceous plants was debated by MCCREE and TROUGHTON (1966). At the turning point of the slope (critical LAI: BROUGHAM 1958) an increase in LAI and its associated CO_2 uptake will not counterbalance the reduction of CO_2 uptake in the existing leaf area because of self-shading. The community might still continue to gain in biomass, but at a lower rate.

If all leaves were horizontal and opaque and were arranged to cover the soil surface completely, then a LAI of 1 would be sufficient to use all the incident photosynthetically active radiation. Communities of *Petasites hybridus* which have a LAI of about 2 are so efficient in shading the undergrowth that they even build a stage of subclimax during succession. In most plant communities LAI ranges between 3 and 19 (Table 18.3), and may reach 22 in conifers (WARING et al. 1978: LAI of 53 taking total needle surface and not projected leaf area as a basis). Plant growth and life form strongly affects optimum LAI through leaf angle and clustering and vertical distribution of leaves. For example, leaf disposition in kale tends to be horizontal, whereas sugar beet has more erect foliage. At increasing LAI, fewer leaves within the

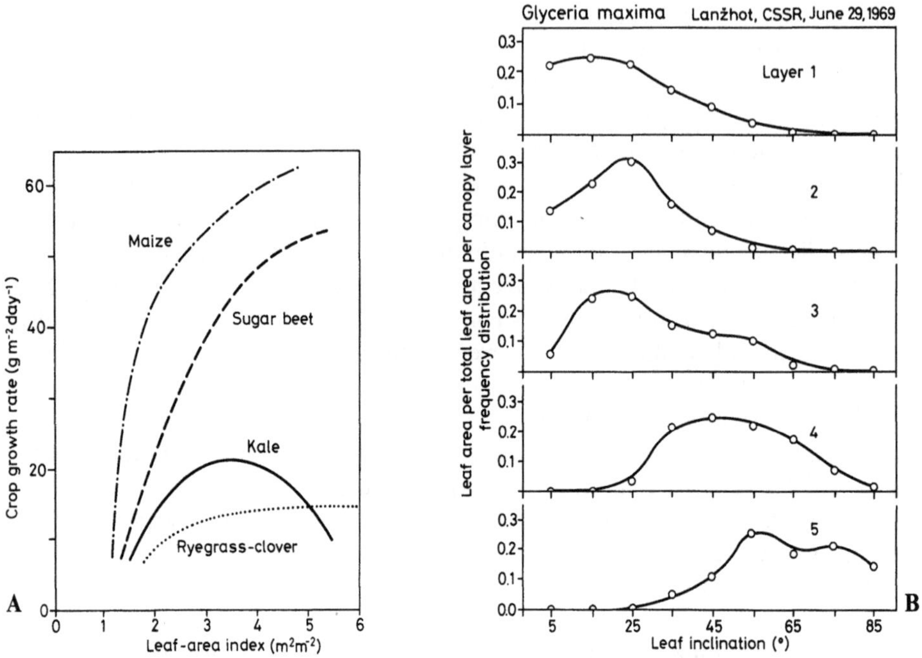

Fig. 18.2. A Crop growth rates of different crops as related to leaf area index. (After WATSON 1958, LOOMIS and GERAKIS 1975). WATSON proposed the existence of an optimum leaf area index for kale. This was questioned later. Most crops show a critical leaf area index at the point where the crop growth rate levels off. **B** The distribution of leaf inclination as relative fraction of the total leaf area in the respective layer. The stand was divided into five layers numbered from the top of the canopy to the botton. 0° is a vertical and 90° a horizontal leaf orientation. Leaves are predominantly vertically oriented in the upper layer of the canopy and horizontally oriented in the lower part of the canopy. (After GLOSER 1977)

kale canopy are illuminated adequately, whereas in sugar beet the opposite is true. Consequently, both species have not only a different critical LAI but also show strong differences in total crop growth at the critical LAI despite very small differences at a LAI of 1 (Fig. 18.2A). SAEKI (1960) showed that the critical LAI is 9–10 for vertical leaves but 3–4 for horizontal leaves. The maximal photosynthetic gain of a plant community is greater if leaves are erect at high light intensity, whereas at low light intensity the effect of leaf inclination is reversed. In the shade maximum productivity occurs in horizontal leaves. Figure 18.2B shows that such a change in leaf inclination can be observed even in a monospecific community of *Glyceria maxima* where the top leaves are erect and the leaves in the lower canopy are exposed horizontally. The high LAI of coniferous stands is based on the clustering effect of the needles, which allows deep penetration of light into the stand.

Although leaf area and disposition may vary greatly between plant forms, the effect in terms of total light extinction of a plant cover is quite similar. Radiation values of 7.5% of total natural radiation were measured at the soil surface in an alpine heath, 10 cm high, and in a forest stand 45 m in height. The discussion on leaf arrangement is directed in this chapter towards understanding of plant productivity. However, it should be realized that erect or horizontal exposure of leaves may be seen also as a means of avoidance of climate extremes, as is most obvious for plants showing leaf movement (SHACKEL and HALL 1979).

18.2.5 The Longevity of Leaves: Implications of the Evergreen and Deciduous Habits

In addition to the control of carbohydrate allocation into photosynthesizing and non-photosynthesizing biomass there is another important mechanism for adjustment of the carbon relations in plant life forms, namely, leaf longevity. Its ecological significance has been discussed since the beginnings of ecological and phytogeographical research (SCHIMPER 1898; GRISEBACH 1872), but the functional significance has not been assessed experimentally so far.

Deciduousness and evergreenness are the two most obvious leaf "strategies", but one should recognize that in nature there are many intermediate types (AXELROD 1966). Some "evergreen" broad-leaved trees shed their old leaves just after the new leaves have unfolded (*Eucalyptus*), others may exhibit only a very short deciduous period just before new leaf development (*Cyclobalanopsis*), and others may be deciduous in one part of the crown while the remainder of the canopy retains leaves (trees in tropical environments). Functionally with respect to the carbon balance the important distinction is, whether or not the leaves are perennial to such a degree that only a part of the foliage is renewed annually. In the case of evergreens the older leaves contribute to the photosynthetic gain so that the total carbon gain is maintained although the proportion of carbon allocation for new leaves may be very small. In *Picea excelsa*, for example, only 15% of the total needle biomass is renewed annually, but the total carbon gain is maintained at a high level almost irrespective of the annual needle growth (SCHULZE et al. 1977b).

It is held that the mesophytic deciduous- or herbaceous-type leaf generally has a high photosynthetic rate at a low carbon cost of construction (STOCKER 1976). A large portion of the total carbon gain is available for growth of non-photosynthesizing stems, roots, fruits or storage organs. When external nutrient supply is sufficient and when other environmental conditions are not limiting, this leaf type is most productive per unit area in all plant forms, if it does not need to be renewed too frequently. It is found in herbaceous species as well as in trees. In contrast, the evergreen leaf type generally occurs when carbon assimilation is restricted in a permanently unfavourable habitat.

All environmental conditions which tend to reduce the photosynthetic rate, e.g., low light or low temperature, reduce the photosynthetic carbon gain. Under these conditions the plant can maintain the absolute amount of carbon allocation for growth of non-photosynthetic water and nutrient absorbing tissues only by reducing the allocation of photosynthates for leaf growth [m in Eqs. (18.3) and (18.4)]. However, this would again reduce the carbon input. Therefore, the photosynthetic gain can then only be maintained if part of the foliage serves for a prolonged period of time or perennially. This will demand higher structural costs for protection against adverse climatic conditions (e.g., heat, cold or desiccation) and predation (ZWÖLFER 1978), which in turn, will favour an evergreen leaf biomass.

Low nutrient uptake reduces the amount of nutrients available for incorporation into new living biomass. In particular a shortage of phosphorus and nitrogen severely affects the photosynthetic capacity (see Chap. 7, this Vol.). In this case photosynthetic production will be reduced. In addition it is expected that the partitioning of carbohydrates will favour construction of a larger root biomass for nutrient uptake. Again, the functioning of the plant at low carbon input and high carbon costs for nutrient-uptake are expected to be achieved by the maintenance of leaf biomass (see Sect. 18.3.2.2).

Plant water stress is associated with an imbalance of water uptake and water loss. SCHULZE et al. (1980) showed that long-term water stress reduced the photosynthetic capacity of several contrasting species. This, in principle, should lead to perennial leaf forms if a recovery after stress is possible, an evolutionary trend followed by numerous desert xerophytes. But in contrast to carbohydrates and nutrients, in terms of water the conditions are more complicated, since the storage capacity of water is generally very low in relation to the turnover rates. Significant storage capacity is found only in special cases e.g., large trees (HINCKLEY et al. 1978), succulents, or bottle trees (WALTER 1973). Therefore the probability and seasonality of water availability strongly governs the success of deciduous or evergreen plant forms (see Sect. 18.3.3.3).

AXELROD (1966) suggested that the present geographical distribution patterns of evergreen and deciduous plant forms in the northern and southern hemisphere have evolved since the Cretaceous period, mainly in response to habitat conditions. But the competitive balance which may exist between plant forms of a specific habitat in the northern or southern hemisphere may be seriously disturbed by the introduction of an exotic species. For example, alpine and arid tree lines in the southern hemisphere, e.g., in New Zealand, can be changed dramatically following the introduction of northern hemisphere species such as *Pinus radiata* and *Pinus contorta*. Therefore the success of a certain plant life form in a specific habitat is always related to the ability of the neighbour species of the community which is strongly influenced by geological history.

18.3 The Influence of the Environment on the Growth Characteristics of Specific Plant Life Forms

In the following an attempt is made to describe those relations between structural forms, growth and environment which may be most significant for the main groups of plant life forms.

18.3.1 The Influence of Environmental Conditions on the Growth Characteristics of Annuals

The development of wheat is shown in Fig. 18.3 as an example of growth in annual plants (FISCHER 1975). In the early growth stages all photosynthates

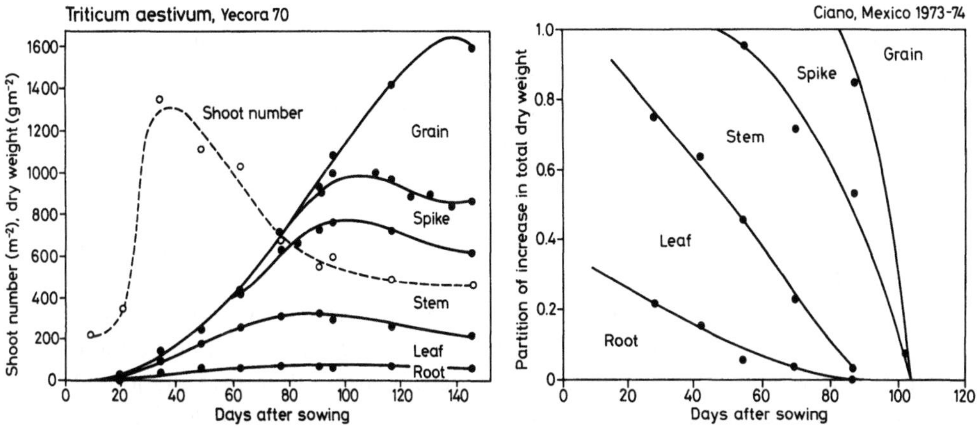

Fig. 18.3. A Distribution of total dry weight (including roots) and life shoot number during development in a crop of the dwarf spring wheat variety Yecora 70 grown under conditions of optimal agronomy. (After FISCHER 1975.) **B** Approximate distribution of various organs of a wheat plant during increase in total dry weight at different times in the life of the canopy shown in Fig. 18.7A. The figure does not consider transfer of dry matter from one organ to another after temporary storage in the former. Transfer from the stem, which decreases in dry weight, to grain, thereby augmenting its increase in dry weight, is quite important from day 110 onwards

are allocated for the formation of roots and leaves. Stem formation reaches its maximum just before anthesis. At that time the flow of carbohydrates is partitioned such that the assimilates of the flag leaf primarily serve the formation of the seed head, whereas the lower leaves mainly supply stem and root with assimilates for anabolic and catabolic processes. At the time of flower formation root growth slows down and leaf growth declines rapidly. At the end of the life cycle only seeds show a net increase in weight. The same principles as shown in Fig. 18.3 may apply for most annual species, although some may mobilize stored carbohydrates for seed formation (*Sorghum,* aubergines: DONALD and HAMBLIN 1976; CLAUSSEN and LENZ 1979) or redistribute carbon flow at stress (maize: JURGENS et al. 1978).

The following paragraphs discuss those growth characteristics of annuals which appear to be of greatest significance for their existence in natural environments.

18.3.1.1 The Phenological Control of Growth in Annuals

The phenological control of germination, growth, flowering and fruiting is probably quite similar in different plant life forms, but since completing the life cycle in one season is crucial for the reproductive success of annuals, some of the implications of this are discussed.

The environmental factors which control phenological development are different in plants, depending on their origin. Many tropical species are day-neutral and in these growth processes may be triggered by temperature, water status or other factors, whereas non-tropical plants are generally strongly dependent on photoperiod. With respect to annuals of seasonally dry and cold environments, the phenological control generally operates so that the life cycle is largely completed during a period of favourable growth conditions. Certainly, exceptions are found in nature, e.g., summer annuals in deserts may germinate after the first winter rains, but complete the main part of their life cycle, including flower and fruit formation, under extreme conditions of summer drought (*Salsola inermis*, EVENARI et al. 1971).

The control of germination is the first and major step of phenological control of annuals (see Chap. 11, this Vol.).

After germination the steps of development are triggered by various environmental factors depending on the day-neutrality of the species. Photoperiodic control of development explains hastened development after germination at the end of the wet season, in many non-tropical annuals, since calendar date rather than germination is a better predictor of the termination of the wet season (FISCHER and TURNER 1978). In *Anagallis arvensis* 24 days were required for floral initiation at an 11 h photoperiod, but only 9 days were necessary at a 15 h photoperiod (SINGH and GOPAL 1973). This pattern may be altered by water stress. Sudden or severe stress has been shown to delay floral initiation in annuals. More commonly, however, floral initiation was found to be stress-insensitive and flowering generally hastened slightly (TURNER and BEGG 1977). This may be different in principle in perennial herbaceous plants, where stress prolongs rather than hastens development, e.g., *Daucus carota* (DALE 1974). Temperature (heat sums) seems to be the principle environmental trigger for phenological development in many day-neutral species (HALL et al. 1979).

18.3.1.2 Seed Production

Seed production is the main event in the life cycle of an annual, and annuals maximize the commitment of growth to seed production (HARPER and OGDEN; see also Chap. 16, Vol. 12 C). Table 18.2 shows that seed production in wild annual species is comparable to that of cultivated crops, and the yield of wheat without fertilizer may be less than that of *Thlaspi arvense* (BEST and MCINTYRE 1975). Crop plants and wild annuals may allocate an average of 30%, in many cases even more than 50% of their above ground dry matter product (defined as harvest index: HI) to seed formation which is more than in a perennial plant community. DONALD and HAMBLIN (1976) provide a comprehensive review of harvest indices of cereals and their dependence on environmental factors. DE RIDDER et al. (1981) gave a complete review comparing wild and cultivated species.

The grain yield of annual cereals is affected by all environmental factors acting on floral initiation and on photosynthesis during grain filling. Therefore, seed yield is determined by a subtle balance of factors determining seed number and seed size (FISCHER and TURNER 1978). It is necessary to distinguish between determinate and indeterminate annuals. Many annuals including several grain crops from the temperate or cold arid region are determinate, i.e., flowering respresents an abrupt and irreversible end to further vegetative growth. In this

Table 18.2. Harvest indices (yield per aboveground biomass %) for some wild and cultivated species (HARPER and OGDEN 1970; DONALD and HAMBLIN 1976; DE RIDDER et al. 1981)

Helianthus annuus	14–40	*Trigonella arabica*	42
Carthamus tinctorius	16–25	*Zea mays*	38–49
Senecio vulgaris	25–27	*Oryza* var.	39–56
Chrysanthemum segetum	26–27	*Triticum* tall var.	25–36
Calendula officinalis	27	*Triticum* short var.	38–41
Matricaria matricarioides	35		

case environmental conditions during particular growth periods such as temperature during floral initiation, or drought during anthesis can irreversibly affect yield. In contrast, many species evolved from the semi-arid subtropical and tropical environments are indeterminate (*Arachis, Vigna*). In these species drought stress may prevent flowering and vegetative growth, but both may be resumed if favourable conditions return. Indeterminate species retain the capacity for high seed yield even after a period of severe stress (TURK and HALL 1980). In contrast, with certain indeterminate crops (tomato, cotton) factors stimulating vegetative growth (e.g., N-fertilization, irrigation) will negatively affect yield above a certain optimum (DONALD and HAMBLIN 1976).

The source/sink relationships of assimilate partitioning in annuals are certainly more complicated and still unclear (e.g., FARRAR 1980). It was suggested that the demand for assimilates determines the photosynthetic rate (KING et al. 1967), although this is still controversial and can be reproduced with pruned plants only. Ear photosynthesis may contribute 13% to grain filling and awns may increase that proportion to 34% (EVANS et al. 1972). Although the effect of awns is still under discussion, they appear to be advantageous with late drought (HALL et al. 1979). That additional factors besides source and sink determine yield is shown by the observation that insect attack may increase plant yield in some species because of early removal of apical dominance (HARRIS 1974).

Water stress severely affects seed yield in annuals not only under natural conditions but also in crops (HSIAO et al. 1976). In this respect it must be considered that the major portion of arable land for annual crops occurs in semi-arid environments. In the annual weed *Senecio vulgaris* the relative change in carbohydrate allocation during water stress (Fig. 18.4; HARPER and OGDEN 1970) indicated that total dry matter production under the high stress treatment was strongly suppressed (only 2% of the well-watered treatment) and dry matter was allocated mainly into the root leaving little or no surplus for seed formation. Similar responses are found in perennial species (HEIM et al. 1979; EVENARI et al. 1977). A close relationship was found between relative dry matter production or yield and water use (DE WIT 1958), which seems to be related not to leaf water potential but to the cumulative water stress experienced (TURK and HALL 1980). Despite the changes in absolute growth and yield numbers with stress, it is remarkable that annuals have the ability to maintain their HI for a considerable range of water stress provided stress develops early or is mild and evenly distributed over the life cycle (DAY et al. 1978).

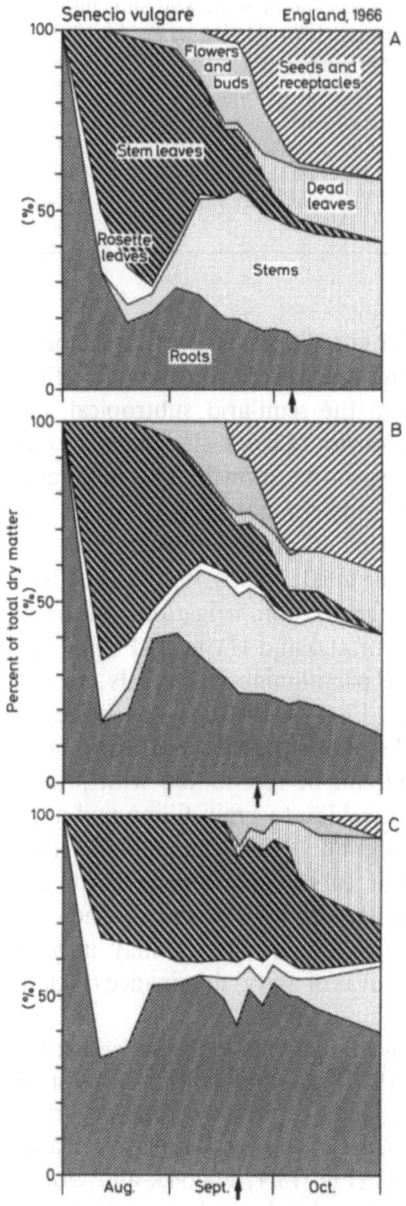

Fig. 18.4. A–C Percentage allocation of total calories of biomass throughout the life cycle of *Senecio vulgaris;* **A** "low stress" treatment with 1700 ml pot per plant grown in 50:50 coarse sand and loam mixture; **B** "medium" stress treatment with 300 ml pot per plant; **C** "high stress" treatment with 20 ml pot per plant. *Vertical arrows* mark dates at which maximum total calorific values were recorded. (After HARPER and OGDEN 1970)

In semi-arid environments higher yields than in those crops which are presently not successful could be expected if the vegetation period were shortened so that sufficient water was available throughout the seed-filling period. The success of plant breeders to adapt crops to drier regions has been linked closely to earlier flowering (FISCHER and TURNER 1978; HALL et al. 1979). Another approach is the breeding of dwarf cultivars, which have a higher yield potential (FISCHER and MAURER 1978). Although both tall and dwarf cultivars show the same linear relation between yield and drought intensity, it may be considered that dwarf crops are better adapted for cultivation in areas with a short wet season (LAING and FISCHER 1977). It is clear, however, that the dwarf stature would not be competitive under natural conditions where height determines the access to light during the time of active growth when water is still available.

Although the necessity for seed production in annuals is obvious, ecological research has devoted very little attention to the question of how few seeds are really necessary for survival of the species. Some information is available from investigations on biological control of weeds. For *Carduus nutans* it was found that a destruction of up to 88% of the viable seeds by the introduction of the seed predator *Rhinocyllus conicus* caused a decrease in the population of about 95% (ZWÖLFER 1980). In contrast, a 99% destruction of the viable seeds in the perennial *Ulex europaea* in New Zealand decreased the population negligibly over many years (HUFFAKER 1959).

18.3.1.3 Competition and Vegetative Growth

MONSI (1960) pointed out that competition between individuals is especially high in annuals following germination. In general water is not the main limiting factor in the early stages of growth since the soil is still moistened. If a large number of seeds per soil surface area germinate it is the competition for light and nutrients which causes selection between individuals. DONALD (1963) gave a very comprehensive review of our present understanding of growth competition among crop and pasture plants (see also Chap. 11, Vol. 12 C). Significant functional changes occur in plants during competition. The protein content of leaves decreased with increasing stand density of pure stands of *Anthemis arvensis* and *Sagina procumbens,* but *Anthemis* was more successful at nitrogen uptake in mixed stands than in pure stands (BORNKAMM 1970). Generally, increasing density reduces root growth and favours shoot elongation, thereby increasing the discrepancy between requirements for both extensive root systems and long stems in order to compete for nutrients and light respectively. As a result, the respiratory portions of the plant increase with respect to the amount of foliage, causing a decline in production and seed yield. The continuity of dry matter production at the community level is only achieved if the number of individuals remains unchanged, a principle of crop cultivation.

In the natural vegetation of humid environments annuals are only rarely able to occupy a habitat during successive years. Competition for light is avoided by winders (*Galium aparine*) and in epiphytic parasites (*Cuscuta*). Other annuals may start growth in winter and complete their life cycle before the main development occurs in the perennial vegetation (many *Cruciferae*). WHITTAKER and NIERING (1975) show that the portion of annuals increases as soon as competition for light decreases, as for instance with increasing aridity or due to disturbance.

18.3.2 The Influence of Environmental Conditions on the Growth Characteristics of Perennial Herbaceous Plants

Growth and development of perennial herbaceous plants is in many respects similar to annuals. But a principle difference exists due to the fact that leaves and roots may be shed continually (tropical environments) or seasonally (temperate and arid environments) without the death of the individual. Under seasonally arid or cold conditions dieback of the aboveground parts may leave only a dormant main bud and very few living roots containing reserves for new growth (e.g., *Stipagrostis obtusa*). Therefore the annual gain in living biomass of the individual may be very low although the seasonal primary production may be quite high. A typical example of this response type is found for the energy flow through grassland ecosystems (SIMS and SINGH 1978). Despite the great amount of total energy being captured, the seasonal energy balance shows only a minute increase of the perennially persisting energy of the system. From the numerous interactions, the following paragraphs discuss those which appear to be most significant for the success of perennial herbaceous plants in relation to other plant forms.

18.3.2.1 Growth Initiation and Carbon Allocation

It is obvious that perennial herbaceous plants have a major advantage over annuals due to the fact that the life cycles need not be completed in a single season. Furthermore, vegetative growth and photosynthesis may start from stored reserves at a time before annuals can germinate and can continue after annuals are occupied with seed production. Under identical environmental conditions, this leads to a higher total net primary production in perennial pasture grasses than in annual crops (DE WIT 1968). Because of the limited time for vegetative growth in annuals, some have evolved mechanisms to prolong the growing season. Winter annuals, such as winter wheat, achieve higher yields than spring wheat by having a greater interval between seedling establishment and flowering (DONALD 1963). The special feature of biennials in terms of dry matter production was emphasized in this context by WALTER (1973).

Perennial herbaceous plants have special adaptations for initiating rapid growth from buds and stored carbohydrates, especially in extreme environments such as arctic alpine areas (BILLINGS 1974). A large number of arctic alpine plants in the northern and southern hemisphere have shoot and flower buds, which are initiated 1 year, in some Greenland species even 2 years (SØRENSEN 1941), before flowering. Shoot elongation and flowering may then occur very quickly after snow melt (WIELGOLASKI and KÄRENLAMPI 1975). The reproductive period may also be extended by postponing fruit development. In *Braya (=Sisymbrium) humilis,* the fruiting inflorescence regularly overwinters and completes development during the following summer (BILLINGS and MOONEY 1968).

In perennial herbaceous species a relatively large proportion of carbohydrates is translocated to the belowground system (MARSHALL and SAGAR 1968 and others). Only a relatively small portion is used as reserves, but a large portion serves for growth of the underground root or rhizome system. There are very few studies on the root turnover rate of herbaceous and graminoid species. CHRISTIE (1978) found a root turnover rate of 1.1–2.9 g day^{-1} m^{-2}, which suggests that roots of an Australian *Astrebla* grassland may live on the average 2–3 years.

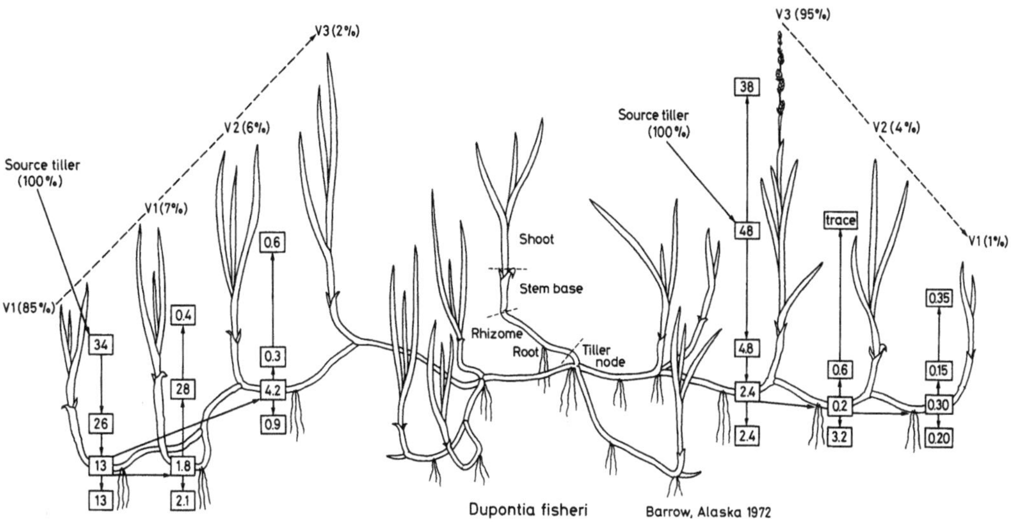

Fig. 18.5. Distribution of ^{14}C radioactivity in tiller compartments of single plants as influenced by tiller age. Values are percentage of the total radioactivity recorded from the tiller. *V1* is a tiller with excerpted leaves in its first season of shoot production; *V2* is a tiller in its second year of aboveground growth; *V3* and *V4* are tillers in their third and fourth year of growth aboveground. Flowering normally takes place in (*V3*) or (*V4*). The *dotted line* represents the total translocation per tiller. *Squares* and *arrows* represent translocation into the different organs. (After ALLESSIO and TIESZEN 1978)

In *Dupontia fisheri* (Fig. 18.5) only about $^1/_3$ of all newly photosynthesized carbohydrates remained in the leaf for expansion whereas $^2/_3$ were translocated from the leaf; 13% of the total carbon gain was allocated to the rhizome for vegetative spreading of the clone, but export to new tillers of the rhizome declined as soon as these newly grown parts were sustained by their own photosynthesis. Thereafter a young tiller might translocate 15% of its carbohydrate gain to older sprouts. There it is used mainly in the rhizome formation of 2-year-old sprouts or may be added to the reserves of neighbouring 1-year-old sprouts. In contrast, only 1% of the photosynthates were allocated from old tillers to young organs. After 3 years, sufficient carbohydrates were accumulated to support flowering. At that stage carbohydrates were redistributed. Only about 5% of the photosynthates were allocated to the rhizome and root and a large proportion was used for flower formation.

The pattern of carbon translocation is similar in many grasses, even those in habitats more favourable to growth, e.g., *Agrostis tenuis* and *Festuca pratensis* (HARRIS et al. 1979) and *Phleum pratense* (WILLIAMS 1964). However, generalizations are not possible in the case of dicotyledonous herbs since most work has been conducted with graminoid species. Under conditions of stress the carbon allocation pattern changes immediately. In *Medicago sativa* defoliation resulted in a rapid translocation of carbohydrates from the root to the new leaves and stems and a partial defoliation of *Lolium perenne* delayed the normal age-related decline of photosynthetic rates in the remaining main shoots (GIFFORD and MARSHALL 1973). The assimilate partitioning also changes under

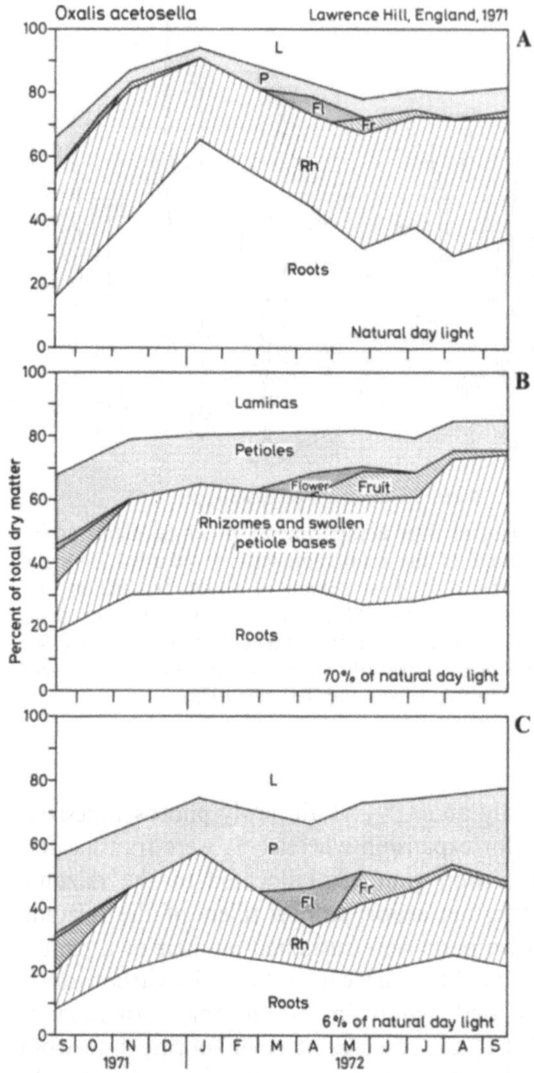

Fig. 18.6. A–C Dry matter allocation to different organs of *Oxalis acetosella* grown from cuttings under three light regimes of natural light; **A** 70%; **B** 27%; **C** 6% of natural midday light. (After PACKHAM and WILLIS 1977)

natural conditions of low carbon fixation, e.g., shade. In this case plants export a greater portion of their assimilates to growing leaves as shown in Fig. 18.6 for *Oxalis acetosella* (PACKHAM and WILLIS 1977). Thus, the root/shoot ratio, which is high for alpine and arctic plants, becomes very low in shaded habitats. Shade leaves are also broader and thinner than sun leaves (EAGLES 1973). A carbon allocation pattern similar to that of *Oxalis acetosella* is found in shade grown *Lolium temulentum,* but this may change to favour of translocation to roots following 2–3 days of exposure to high light intensity (RYLE and POWELL 1975).

The adaptations to shade and the competition between sun- and shade-tolerant plants may be more complicated than is obvious from growth analysis, since unrelated factors may be of significance. For example, *Cirsium palustre,* a shade-avoiding species, and *Geum urbanum,* a species avoiding exposed conditions, show similar growth under conditions of shade in the vegetative phase. However, only *Geum* continued to grow until fruit ripening. *Cirsium* developed a high leaf area in the shade but the stems were too weak to support themselves and collapsed, thus flowering was not achieved (PONS 1977). This shows that special environmental conditions may initiate a growth form which causes physical self-damage to an extent that the species is not reproductive in that habitat.

18.3.2.2 The Storage of Reserves

In a seasonal climate the storage of reserves, such as carbohydrates and lipids, is fundamental to the extension of the life cycle and initiation of growth in the following season as soon as conditions are favourable. Storage seems to be a principle ecological feature of herbaceous perennials and a basis for their success in arid and cold climates.

The development of bulbs and rhizomes as storage organs has been most commonly viewed as an adaptation to cold and dry seasons. Geophytes are typical elements of the mediterranean and the steppe vegetation. Despite this it should be noticed that many herbaceous plants of the humid tropics also develop rhizomes and in many cases are even planted as crops (e.g. *Allocasia, Zingiber*). One may say that although rhizomes and bulbs occur in both humid tropical and seasonal climates, as far as we know, they appear to have ecological significance only in the latter.

The seasonal change in the carbohydrate and lipid reserves reflects the growing activity of the plant (e.g., BILLINGS and MOONEY 1968). In *Polygonum bistortoides* carbohydrates were stored in large amounts in roots and rhizomes at the end of the growing season (Fig. 18.7A). But with breaking of dormancy in the following summer 50% of the rhizome reserves were used in a 1-week period of early and rapid shoot growth (MOONEY and BILLINGS 1965). Figure 18.7B shows that in *Primula glutinosa* the energy content of the whole plant drops significantly during the time of spring growth, indicating that photosynthesis does not balance growth and maintenance respiration.

FONDA and BLISS (1966) found in alpine regions that carbohydrate reserves may ensure growth even during adverse climatic periods within the growing season. There appears to be a tendency for carbohydrate contents to decrease with increasing altitude (MOONEY and BILLINGS 1965), although in some species (*Chinochloa rigida* and *C. macra*) the reserves in the root increased with altitude when growth was significantly retarded (PAYTON and BRASCH 1978). In alpine habitats, whenever plants are covered with snow during the summer period, carbohydrate reserves decrease and may not be restored during that season (BILLINGS and MOONEY 1968).

Although quite a large amount of information is available on the growth of herbaceous plants in alpine and arctic habitats, very little knowledge is available on carbohydrate storage in plants of arid regions.

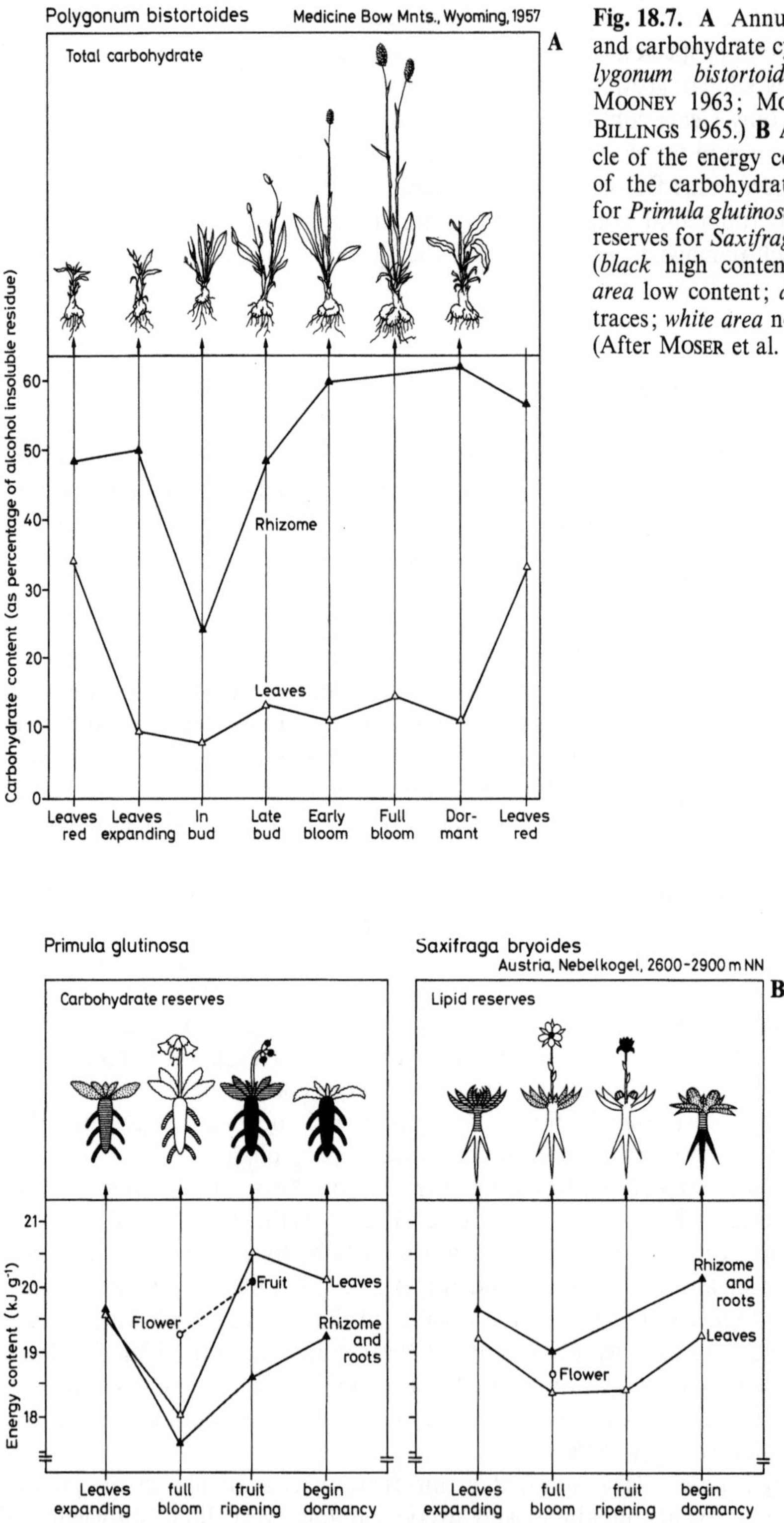

Fig. 18.7. A Annual growth and carbohydrate cycle of *Polygonum bistortoides*. (After MOONEY 1963; MOONEY and BILLINGS 1965.) **B** Annual cycle of the energy content and of the carbohydrate reserves for *Primula glutinosa* and lipid reserves for *Saxifraga bryoides* (*black* high content; *hatched area* low content; *dotted area* traces; *white area* no content). (After MOSER et al. 1977)

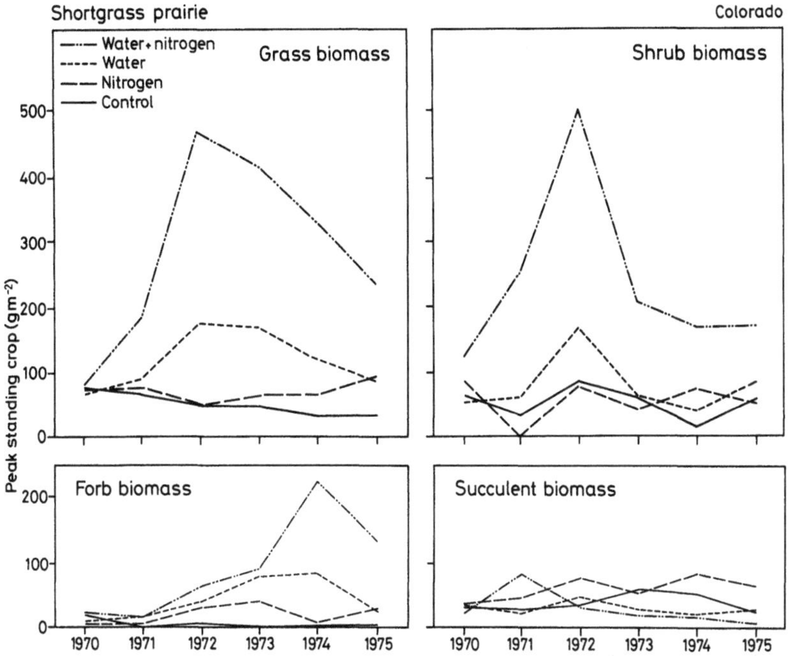

Fig. 18.8. Response of a grassland ecosystem to natural water stress, nitrogen fertilization (0.015 kg N m^{-2} in the first year and 0.005 kg N m^{-2} in subsequent years), and irrigation plus nitrogen fertilization (sprinkler irrigation to maintain soil matric potential between 0 and 0.8 bar at 10 cm soil depth plus 0.005 kg N m^{-2}). Average soil water content was 14% for the control and about 20% for the watered treatment. The *curves* show the peak standing crop of each year. (After DODD and LAUENROTH 1979)

18.3.2.3 Competition Between Graminoids and Dicotyledonous Herbaceous Species

The complex competitive interactions between graminoid and dicotyledonous herbaceous perennials have been investigated mainly in agricultural systems (DONALD 1963). Light penetration is one important factor in competition (see Sect. 18.2.4). Grasses tend to have erect leaves in the upper layers of the herbaceous vegetation, which allow sufficient light to penetrate to the horizontal lower leaves and to dicotyledonous herbaceous species at ground level.

Graminoid species have a basal meristem and a dense system of adventitious roots and seem more successful in extracting nutrients or water from the soil than other herbaceous plants. Dicotyledonous herbs may invade larger soil volumes but are not as efficient at obtaining water and nutrients for growth from a small soil volume or horizon. Under arid conditions (Fig. 18.8) the peak standing crop of grasses increased about six-fold in 2 years when the plants were fertilized and irrigated, but irrigation alone results in only a two-fold increase and fertilization without irrigation had no effect. The biomass of the half-shrub *Artemisia frigida* increased also, whereas succulents showed no response. The peak standing crop of forbs increased only after 3 to 4 years of water and fertilizer treatment. Similar findings were obtained in wet and boggy habitats (SUMMERS 1968) and under alpine conditions (WIELGOLASKI et al. 1975). The conclusion of VAN KEULEN (1975) that productivity of deserts is limited mainly by the supply of nitrogen is not correct in view of the above-mentioned grassland experiment.

Agriculturally, equilibrium in a man-made community of grasses and legumes can be maintained by withholding nitrogen and thus retarding grass growth as soon as it begins to dominate by shading the legume which grows close to the ground. In contrast grass growth is activated as soon as the nitrogen release from the legume nodules permits. A competitive equilibrium can be achieved in temperate *Lolium/Trifolium* pastures (HARRIS 1974; BROUGHAM 1958) or in subtropical *Panicum/Macroptilium* pastures (ROSS et al. 1972).

The strong response of the graminoid species in Fig. 18.8 was accompanied by a ten-fold increase in root biomass (DETLING 1979). The ability of grasses to rapidly initiate root growth from the adventitious root system may be one reason for the success of this plant form under arid conditions, since moisture always accelerates microorganism activity and thereby increases nitrogen availability. Grasses respond readily after fire for similar reasons (PAYTON and BRASCH 1978). Bare soil of burned areas may become dryer than unburned controls to such a degree that growth is retarded earlier in a dry season (REDMANN 1978). Thus, in areas subject to burning, only those species which are activated most quickly, namely the grasses, eventually succeed. Additionally, grasses, e.g., *Agrostis tenuis,* appear to be more effective in uptake of phosphorous than forbs, e.g., *Trifolium repens,* and this difference increased in favour of *Agrostis tenuis* during water stress (JACKMAN and MOUAT 1972). It remains unclear whether it is the availability of nutrients or the decline in uptake metabolism of the plant which first ceases with water stress. If grass growth in an arid environment is selectively harvested by grazing, a succession occurs leading not to forbs but to woody plants. The available water may at first be used by both, but woody species eventually suppress the other herbaceous plant forms through defense mechanisms against grazing and competition for light (WALTER 1973).

18.3.3 The Influence of Environmental Conditions on Growth Characteristics of Woody Shrubs and Bushes

It was pointed out in Section 18.2.3.3 that woody species are characterized by a larger potential growth of permanent functional biomass, whereas herbaceous plants generally may increase their plant volume each season but die back to relatively small, perennial storage organs. In woody plants the biomass of stems and roots therefore increases continually over time, although there is a superimposed seasonal variation in biomass because of shedding of plant parts. Although the secondary growth of wood is the major distinction between herbaceous and woody plant life forms, it appears justified with respect to the competitive importance of height growth to distinguish between shrubs or bushes, which have prostrate growth or little apical dominance, and trees, with a dominating vertical growth.

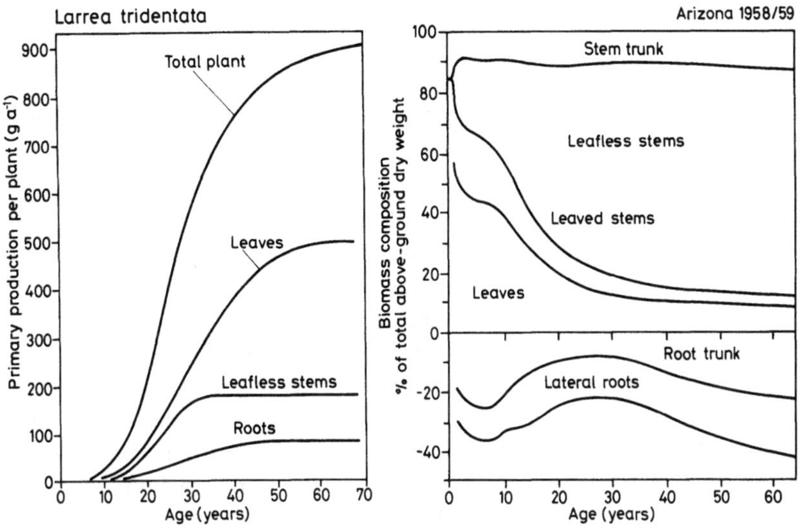

Fig. 18.9. Annual primary production of different organs and estimated relative composition of the shoot and root of *Larrea tridentata* of different ages. (After CHEW and CHEW 1965)

Carbon partitioning of shrubs is exemplarily shown in Fig. 18.9 for *Larrea tridentata*. It is quite obvious that the relative proportion of leaves decreases continually although annual leaf production increased. Partitioning of assimilates does not change significantly beyond age 30. With increasing age the carbon balance is more likely to become disturbed in unfavourable years. Growth patterns similar to those shown for *Larrea* are also found in temperate and arctic zone shrubs and bushes e.g., *Calluna vulgaris* (CHAPMAN and WEBB 1978). In many arctic and alpine shrubs, however, this ratio remains favourable by production of adventitious roots from a prostrate stem which acts comparably to a rhizome in herbaceous plants (e.g., *Salix arctica:* TIESZEN 1978).

The maintenance of a positive annual carbon balance in a large plant body over many years seems to be a functional problem which requires special adaptations in woody plants. Therefore, following the general discussion of Section 18.2.5 the consequences of the evergreen and deciduous habit in shrubs and bushes will be treated in the following paragraphs.

18.3.3.1 Evergreen and Deciduous Growth as Related to the Carbon Balance

Foliage longevity is significant for the survival in shaded habitats. Through a reduced leaf turnover at a comparable rate of total plant assimilation, a greater proportion for maintenance respiration and growth is achieved on a long-term basis. In fact low light intensity of shady habitats generally causes prolonged longevity of foliage in different plant forms. Leaf longevity increased in herbaceous *Oxalis acetosella* in shady habitats (PACKHAM and WILLIS 1977); the palm *Livistonia subglobosa* maintained leaves about 3 years in full daylight, but 12 years in the forest understory (MONSI 1960); 85-year-old spruce (*Picea abies*) had 12-year-old needles on similar age twigs only in the shade and not in the sun crown (SCHULZE et al. 1977a).

Low temperatures and a short vegetation period reduce the carbon input of plants especially in arctic/alpine environments (MOSER 1973). However, with respect to the evergreen or deciduous habit, general conclusions are not obvious. Both life forms occur in the arctic and alpine region. The carbon balance of plants is the result of a number of processes which favour either deciduous or evergreen plant life forms. A decrease in the length of the growing season would promote evergreen structures rather than deciduous leaves. The young leaves of deciduous species are in danger of suffering severe damage by late frost if bud break is early. Evergreen species avoid this critical period in the early season, since bud break is generally later in evergreen than in deciduous species. Therefore, evergreens are able to start CO_2 assimilation earlier in the season than deciduous species. But the length of the growing season is only one of several factors which determine the plant carbon balance. The rate of CO_2 assimilation was found to be three times higher in arctic deciduous *Betula nana* and *Salix arctica* than in evergreen *Ledum palustre* (JOHNSON and TIESZEN 1976). One may consider that the lower photosynthetic capacity of the evergreen leaf in comparison to the deciduous one is the structural cost of being cold- or drought-tolerant. The annual carbon gain per unit dry weight was higher in the deciduous *Rubus chamaemorus* and *Betula nana* than in the evergreen *Andromeda polifolia* and *Empetrum hermaphroditum* species (JOHANSSON and LINDER 1975). However, annual carbon investment into new foliage was found to be as much as three to four times higher in the deciduous than in the evergreen types. But damage by ice or wind and frost drought may cause the evergreen carbon balance to deteriorate. In summary JOHNSON and TIESZEN (1976) come to the conclusion that net carbon gain over the total life span of a leaf may be quite similar for deciduous and evergreen tundra species even though photosynthetic capacity and growth pattern may differ markedly.

Cushion plants are a special plant form typical of arctic and alpine environments. These are plant tufts of woody or herbaceous species which build a dense, cushion-like cover close to the ground. Because of a high boundary layer resistance, the microclimatic conditions within the cushion of *Loiseleuria procumbens* may exhibit a temperature increase of up to 15 °C above ambient (Fig. 18.10) which may even be close to the upper limit of heat tolerance in an alpine environment (LARCHER 1977). Since humidity is generally above 80% at reduced wind within the cushion, water stress was not found even with low humidities in the external air (KÖRNER and DE MORAES 1979). GRABHERR and CERNUSCA (1977) concluded that conditions within the cushion were more favourable than those outside during 83% of the 140 days of the growing season for this plant, and that this period was 49 days longer within the cushion than in an open shrub. Probably the carbon balance within the cushion is additionally improved by CO_2 supply from soil respiration, although this has not been investigated. Despite these favourable aspects of "cushion"-stature, photosynthesis of leaves in the lower strata of the cushion will be very negatively affected by self-shading from the upper strata.

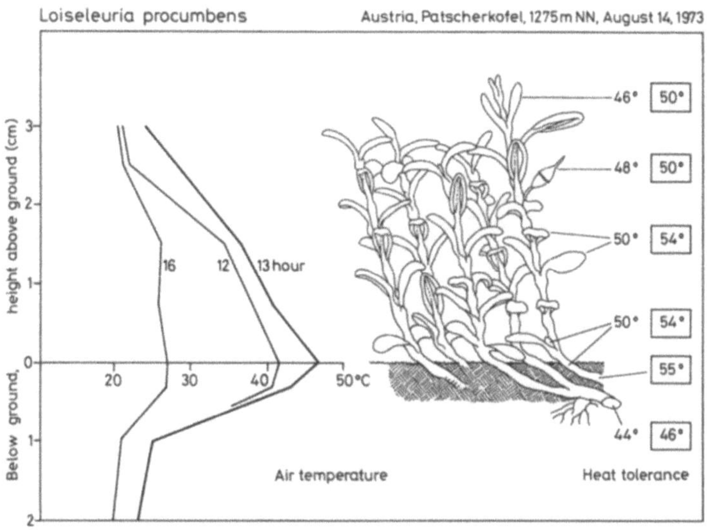

Fig. 18.10. Vertical temperature profiles in a cushion of *Loiseleuria procumbens* at 12 h, 13 h and 16 h of a clear summer day. Highest temperatures were 47.5° C at the soil surface, air temperature was 19.4° C at 10 cm above canopy and 16.6° C at 2 m above canopy. The heat tolerance is characterized by that temperature which was endured with 50% damage during cultivation following $^1/_2$ h treatment at that temperature. *Single numbers* indicate the actual heat tolerance at the time of climate measurements. *Numbers in squares* indicate the "potential" heat tolerance after a period of hardening. (After LARCHER 1977)

18.3.3.2 Evergreen and Deciduous Growth as Related to the Nutrient Balance: Xeromorphy and Sclerophylly

SCHIMPER (1898) recognized that xeromorphous plants occur on bogs with ample moisture supply. He interpreted this as "physiological drought" due to inhibition of water uptake by lack of oxygen. Since then it has been demonstrated that these plants may not be suffering water stress (SMALL 1972a), and that nutrient deficiencies, especially N, P and trace minerals (Cu) are responsible for the structural changes in the plant which are generally referred to as xeromorphic features (refer to the excellent review by SMALL 1973). These features include reduced stature, reduction of internode length, perenniality, and leaf modifications such as small size, small cells, thick cuticle and others (SHIELDS 1950). A sub-set of xeromorphic features is included in the term "sclerophylly", which refers to the "hardness" of leaves due to cutinization, sclerification, silicification and dehydration.

Sclerophylly appears to be associated with a relative decrease in protein rather than with an increase in fibre content per unit dry matter (Fig. 18.11). Additionally there is a correlation between protein and P content in sclerophyllous leaves (LOVELESS 1961). Plants growing in nutrient-deficient environments utilize their available nutrients very efficiently although high transpiration can help nutrient uptake. Bog species reabsorb significantly more N from their foliage preceding leaf-fall than non-bog species (SMALL 1972b). The potential

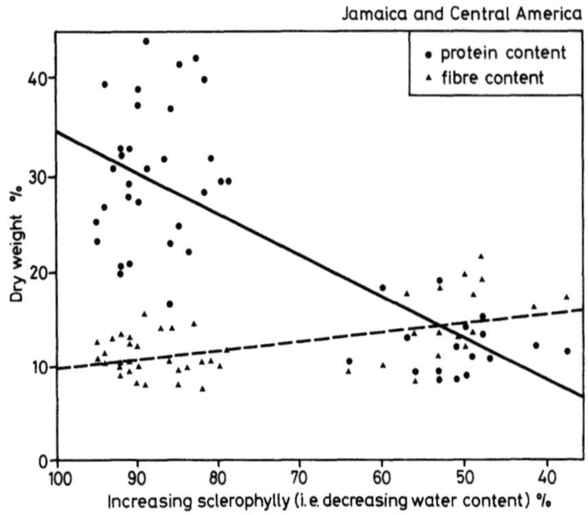

Fig. 18.11. Relative dry weight of protein content (*triangles*) and relative dry weight of fibre content (*circles*) as related to decreasing water content as measure of increasing sclerophylly of 19 evergreen Jamaican bushland species and 31 mesomorphic Central American species. (After LOVELESS 1961)

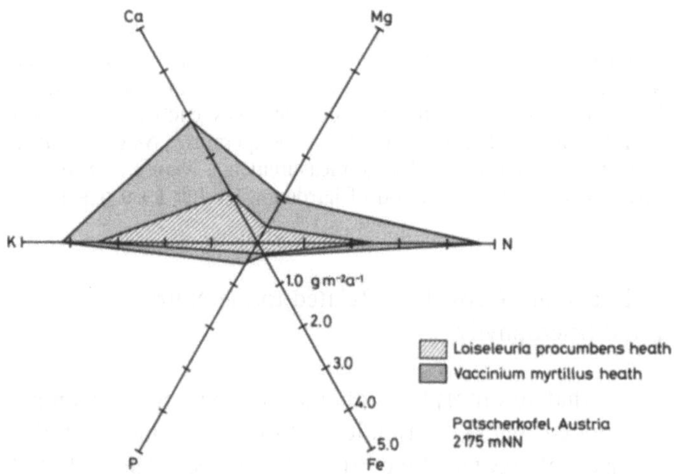

Fig. 18.12. Annual mineral uptake of a *Loiseleuria* heath and a *Vaccinium* heath. (After LARCHER 1977)

photosynthesis which a species could achieve during the time a unit nitrogen remained in the plant before being lost through leaf fall was found to be 235% higher for evergreen bog species than for bog deciduous species and even 275% higher than in non-bog deciduous species. One would therefore expect evergreen forms to be more widely distributed under nutrient-deficient conditions (i.e., humid tropical environments). The evergreen alpine *Loiseleuria procumbens* occupies habitats where leaves can incorporate only half as much N and P as the deciduous *Vaccinium myrtillus* incorporates in its habitat (Fig. 18.12).

It is not clear why, under conditions of limited nutrient supply, plants do not invest all extra carbon into a more extensive root system but produce leaves which require a great amount of structural carbohydrates. Apparently a limit is reached, at which the investment of additional roots together with their requirement for proteins exceeds the gain of additional nutrient uptake (e.g., in lateritic soils). In this case protection against environmental extremes and predation with secondary plant compounds (MORROW and MOONEY 1974) would appear to be more "productive" than increased rooting in terms of carbon return and nutrient conservation. However, this interpretation should be tested with additional experimental evidence. Hypothetically it may be considered that in nutrient-deficient environments carbon is produced in excess, since any further growth would require additional nutrients, and excess carbohydrates are deposited in cell walls or cutin and waxes, which do not require nutrients, leading to sclerophylly. Therefore both leaf longevity and sclerophylly appear to be adaptive mechanisms under nutrient-deficient conditions.

18.3.3.3 Evergreen and Deciduous Growth as Related to the Water Balance

Under arid conditions the evergreen habit is not necessarily more indicative of a limitation in water supply than the deciduous form (Sect. 18.2.5). It was even found that the deciduous form is more adapted to extremes of aridity than the evergreen form in mediterranean climates (MOONEY und DUNN 1970), in the Californian coastal sage vegetation (POOLE and MILLER 1975), and in the Sonoran (SZAREK and WOODHOUSE 1977) and Saharan desert (STOCKER 1971). The evergreen microphyllous tree *Olneya tesota* maintained a favourable water balance and a low rate of photosynthesis year-round. In contrast, in the deciduous *Ambrosia deltoidea* plant water potentials decreased from −10 bar at high rates of photosynthesis to almost −100 bar in the drought deciduous stage (Fig. 18.13). In the case of the evergreen life form, stress due to high evaporative demand in the atmosphere is avoided. They have a deep-reaching root system and are restricted to special habitats where at least some groundwater is available. In contrast drought deciduous plants have a shallow root system and water stress in both soil and atmosphere is tolerated albeit in a defoliated state. Thus they are able to inhabit very dry habitats of irregular or discontinuous water availability.

Under arid conditions evergreen and deciduous life forms are the extremes in a large variety of intermediate cases. One interesting form is the seasonal leaf dimorphism described for several Irano-Turanian and Saharan species (EVENARI et al. 1971). *Poterium spinosum,* an eastern mediterranean shrub, develops three sets of leaves as drought proceeds, each distinctly different. The ecological significance of this phenomenon is not clear, although it was proposed as a water-saving mechanism (ZOHARY 1961). Conservation of water could also be achieved through deciduousness, and the investment of carbohydrates into construction of new sclerophyllous summer leaves has not been shown to contribute with significance to the carbon balances of the plant in leaf dimorphous *Zygophyllum dumosum* and *Artemisia herba-alba* (SCHULZE et al. 1980). One is tempted to view seasonal dimorphism as a form of retarded growth occurring under stress and, perhaps, with unsufficient nutrient supply. SCHÄFER (1973) showed that the activity of microorganisms in a mediterranean climate is high only during winter (when moist) and that they remain in a dormant state

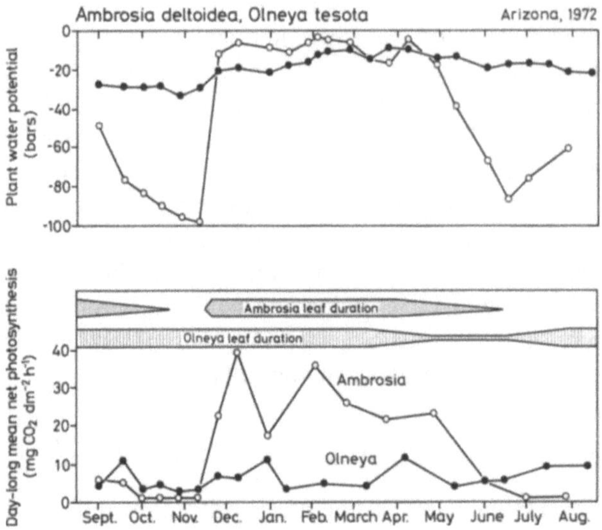

Fig. 18.13. The seasonal patterns of plant water potential and of day-long mean net photosynthesis in *Ambrosia deltoidea*, a drought deciduous shrub, and *Olneya tesota*, a non-riparian microphyllous tree. The duration of foliage is indicated by the *horizontal bars*. (After SZAREK and WOODHOUSE 1977)

during summer, which would result in a very limited nutrient supply in times of drought. In addition it is proposed that the activity of nitrate reductase is significantly reduced with water stress (MORILLA et al. 1973), and a two to three fold increase in nitrate occurs contributing to osmotic adjustment (TURNER and JONES 1980) at a time when protein content is reduced (HELLMUTH 1968). Once water stress has developed plants may enter a stage of "drought dormancy" in which watering or watering plus N fertilizer fail to stimulate growth (HODGKINSON et al. 1978). The tendency towards the more frequent occurrence of evergreen xeromorphic life forms in desert habitats may therefore be a complex problem of not only water but also nutrient supply.

18.3.4 The Influence of Environmental Conditions on Growth Characteristics of Trees

Principles similar to those described for shrubs apply to the production process of individual trees and tree stands. The primary production of forest stands reaches a high maximum at a very early age when leaf biomass is high in relation to the woody portion. Since total leaf biomass and thus photosynthetic carbon gain remains constant or even decreases with increasing age, primary production decreases as respiratory requirements increase.

Besides the constraints of the carbon balance there are other limitations to growth in tall trees with respect to construction and physical stability. Buttress formation in tropical trees has been discussed, although controversially, in terms of its stabilizing effect (RICHARDS 1952). The longitudinal tensile stress can reach 495 kg cm^{-2} in a 40 m high *Eucalyptus* tree (NICHOLSON 1973).

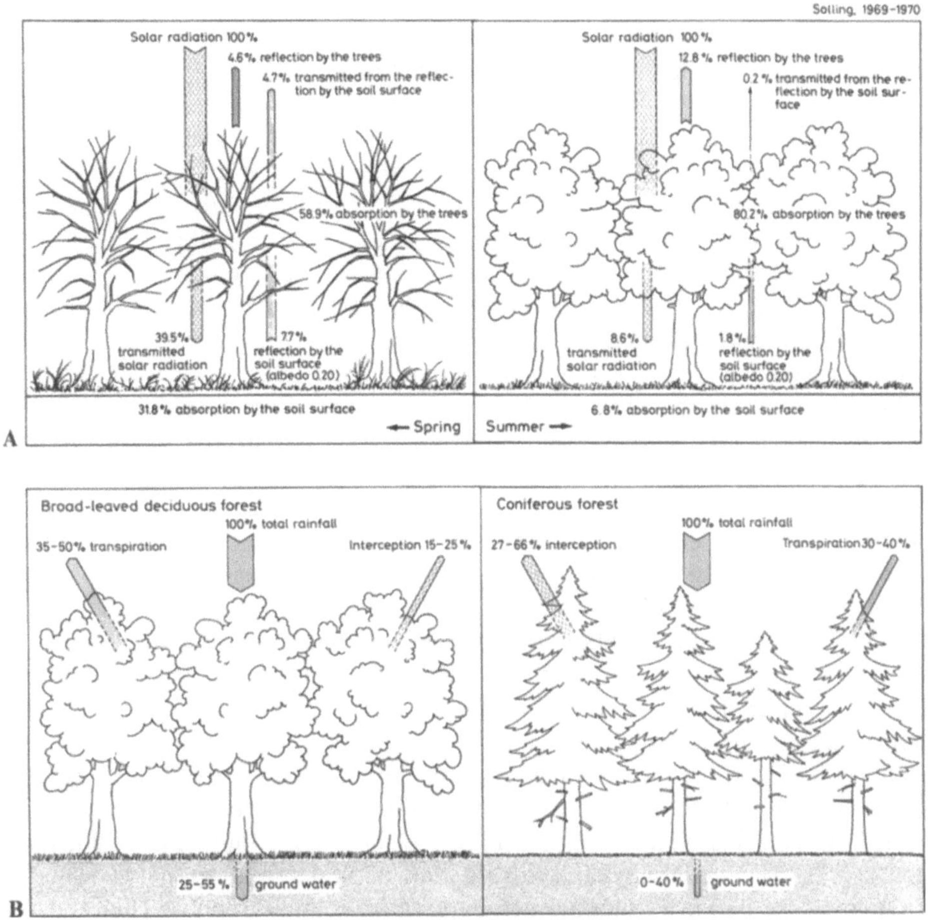

Fig. 18.14. A Total solar radiation (100%) and its reflectance and absorption in a broad-leaved forest of *Fagus silvatica* in its winter deciduous state in spring and its green state in summer. (After KIESE 1971.) **B** Total rainfall (100%) and its partitioning into transpiration, interception and groundwater recharge for an European broad-leaved *Fagus* forest and an evergreen *Picea* forest. The diagram shows average conditions as derived from literature data

18.3.4.1 The Effects of Evergreen and Deciduous Growth on Tree Stands

Since trees dominate all other plant life forms through competition for light, provided conditions are appropriate for tree growth, the deciduous and evergreen habits of trees have implications not only for the carbon balance of the trees as was discussed for shrubs, but also determine the supply of water and radiant energy available to other plant forms living on the forest floor. Therefore in forest vegetations not only the tree form, but also its effect on other forest floor species, needs to be considered, since significant changes in the habitat may follow changes in tree vegetation.

The energy balance of a stand of winter deciduous *Fagus silvatica* shows strong seasonal variation (Fig. 18.14A). About 40% of incoming radiation reaches the forest floor before and only 8.6% after spring leaf growth. This seasonal change in radiation climate does not exist in coniferous forests where permanently only 1.6% to 7.2% of incoming radiation reaches the ground (MITSCHERLICH et al. 1965). The supply of photosynthetically active light directly affects herbaceous plants. In deciduous forests, therefore, the life cycle of many species on the forest floor is restricted to the time before tree leaf growth in spring (GORYSHINA 1969). These species do not survive in the permanent shade of afforested conifers. The temperature climate affects herbaceous ground vegetation indirectly. If a comparison is made with the altitudinal adiabatic temperature decrease, the effect of an evergreen canopy on a warm summer day is comparable to a 1000-m altitudinal variation (MITSCHERLICH and KÜNSTLE 1970). The temperature effect is reversed in the deciduous forest in spring. Very high temperatures occur within the litter when temperatures are close to freezing in the open field. The high temperatures rapidly activate geophytes. The temperature climate also very likely affects soil microorganisms. The acidification of the soil in coniferous stands is related to the C/N ratio of the litter, but this effect may be enhanced by the micrometeorological differences in soil temperature.

Very few comparative studies of the water budgets of evergreen and deciduous forests have been reported and even then important information is lacking due to measurement and theoretical problems. In the temperate zone all studies report that on the average, total annual interception (evaporation of water from wet leaves) was 15%–20% higher for conifers than for broad-leaved trees (BENECKE 1976; MITSCHERLICH et al. 1966), (Fig. 18.14B). Loss of intercepted water from *Pseudotsuga* may even reach 36%–66% of gross rainfall (FRITSCHEN et al. 1977). The variability arises from the fact that interception depends strongly on the frequency and average duration of canopy wetness. The result is that the water budget and the water gain for groundwater may be 20% higher in broad leaved forests than in coniferous stands (STANHILL 1970). It may be concluded that a coniferous evergreen forest in Middle Europe is not only cooler and continually darker but also drier than a deciduous one.

The implications of evergreen and deciduous leaves for the carbon balance of trees are similar to those of shrubs (Sect. 18.3.3.1). Evergreen conifers have lower maximal rates of photosynthesis (4–6 mg CO_2 g^{-1} h^{-1}) than broad-leaved deciduous trees (10–14 mg CO_2 g^{-1} h^{-1}). In contrast, conifers have a higher wood production (0.8–1.6 kg m^{-2} a^{-1}) than broad-leaved trees (0.6–0.9 kg m^{-2} a^{-1}) (BONNEMANN and RÖHRIG 1972). Deciduous leaves of *Fagus silvatica* and *Betula verrucosa* have a higher photosynthetic gain per dry weight and year than conifers (Fig. 18.15). The low photosynthetic capacity of the coniferous needle is not counterbalanced sufficiently by a prolonged growing season, since a short photoperiod, low light and low temperatures restrict photosynthetic activity in autumn and spring in the European climate. Conditions are different in summer-dry and winter-mild climates, e.g., the Pacific Northwest of North America.

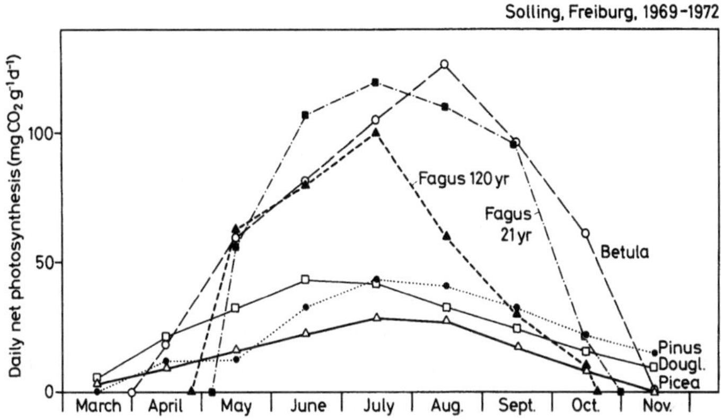

Fig. 18.15. Seasonal change of daily net photosynthesis for winter deciduous *Fagus silvatica* and *Betula verrucosa* and evergreen *Pinus silvestris, Pseudotsuga douglasii* and *Picea abies*. (After SCHULZE 1970; SCHULZE et al. 1977a; KÜNSTLE and MITSCHERLICH 1977)

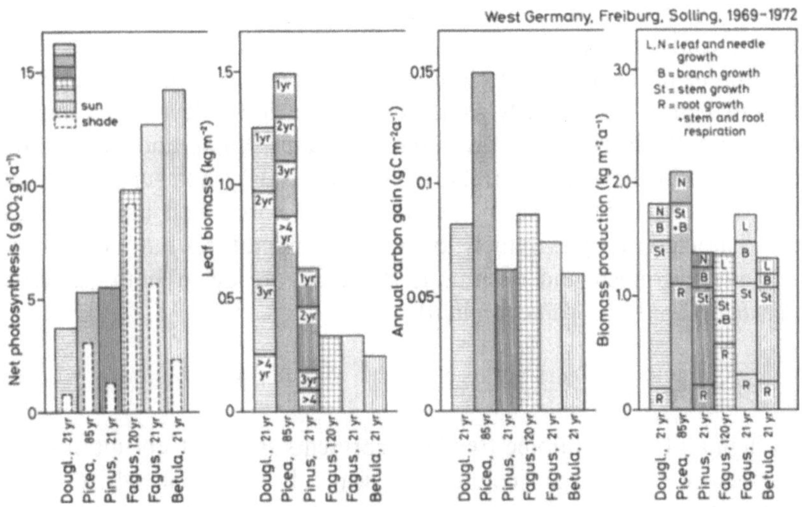

Fig. 18.16. Annual gain of net photosynthesis per gram leaf dry weight, leaf biomass of the canopy per m^2 stand, annual carbon gain of the stand and annual stand biomass production of evergreen *Pseudotsuga douglasii, Picea abies, Pinus silvestris* and winter deciduous *Fagus silvatica* and *Betula verrucosa.* (After SCHULZE 1970; SCHULZE et al. 1977a; KÜNSTLE and MITSCHERLICH 1977)

The differences in growth between the deciduous and evergreen form are a consequence of leaf longevity. The annual growth of new foliage is less in conifers than in deciduous trees (Fig. 18.16) but total leaf biomass is higher, because needles are maintained for up to 12 years in *Picea excelsa* (SCHULZE et al. 1977b). Therefore, despite a lower carbon gain per unit leaf weight, but because of a higher accumulation of total leaf biomass the annual carbon gain is greater for conifers than for deciduous trees. The significance of the old foliage for growth in conifers is shown in defoliation experiments. Removal of 1- to 2-year-old needles caused a 51% decrease in growth of *Pinus radiata*

(ROOK and WHYTE 1976). Needle retention time increases in *Pinus* with elevation and aridity and reaches a maximum of 45 years in *Pinus longaeva* (EVERS and SCHMID 1981).

The implications of these findings appear to be significant for the understanding of successional stages and geographical distributions of tree forms. Broadleaved trees grow more rapidly than evergreens during the sapling stage, as is known from successional *Populus, Betula,* and *Alnus,* because evergreen forms have to accumulate leaf biomass over many years before maximal productivity is achieved. With decreasing length of the growing season the annual costs of construction of deciduous leaves increase relative to the total carbon gain. Hence, evergreen trees are more efficient in the boreal climate and deciduous trees occur only on very nutritious soils at their northern and eastern border (KLÖTZLI 1976) or after disturbance (e.g., along water ways). However, evergreen trees are replaced by deciduous bushes, e.g., *Salix* and *Betula* in the transitional regions between the boreal forests and tundras. One may hypothesize that the metabolic costs of maintenance increase during dark periods lasting several months and also that the risk of snow and ice damage or eating away of the perennial needles becomes greater for the evergreen tree in the subarctic environment. Therefore, deciduous species are again more competitive in the tundra region, particularly as shrubs with a lower level of respiratory costs. However, the interaction of deciduous *Betula tortuosa* and the evergreen *Pinus* and *Picea* in the oceanic arctic region is certainly not sufficiently understood (see WALTER 1973). Also in the continental boreal and alpine climates which are characterized by extremely low winter temperatures but a distinct and warm summer evergreen and deciduous strategies are again both successful (*Larix decidua, L. sibirica, Pinus cembra*).

18.3.4.2 The Alpine Tree Line

At high altitude the alpine tree line represents the most abrupt change in plant dominance on a large geographical scale. It is a zone where the extension of forests is terminated by natural climatic barriers and competition. The major climatic constraints at high altitude are the decreasing length of the vegetation period, which is accompanied by decreasing temperatures and evaporative demand, but increasing radiation, wind and snow cover. The ecology of the alpine tree lines is discussed in Chapter 11, this Volume. Figure 18.17 gives a schematic presentation of the main functional processes.

At low temperatures the photosystem apparently becomes more susceptible to photoinhibition by high light (VAN HASSELT and STRICKWERD 1976). This effect, together with chlorophyll destruction due to increasing UV light causes a decrease in the rate of net photosynthesis at higher elevation (CALDWELL 1979). The annual primary production is also reduced due to the decreased length of the growing season. The main effect of the reduced carbon gain and the decreasing time of vegetative growth is the incomplete development of protective tissues in the evergreen needles. The effects of this become obvious mainly in late winter when low soil and air temperatures restrict water uptake and water transport. At high radiation needle temperature may rise above freez-

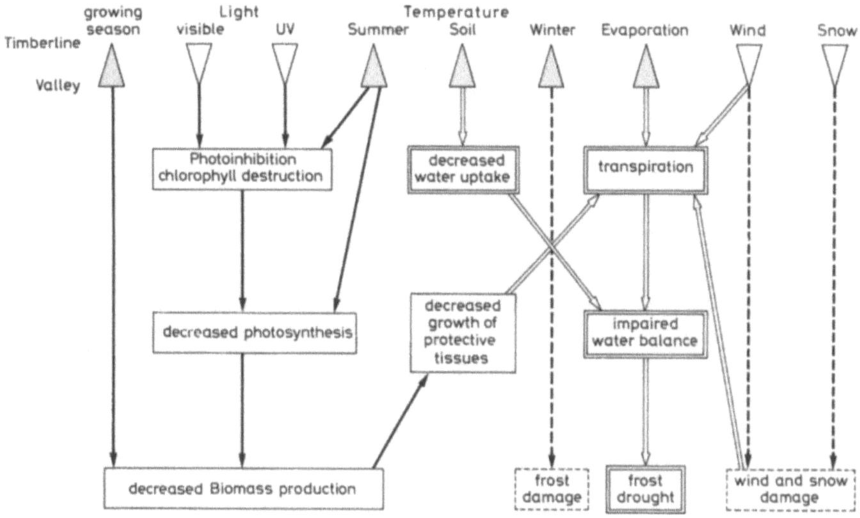

Fig. 18.17. Schematic presentation of environmental factors, which decrease in absolute value on a gradient from the valley to the timberline (*hatched triangles*) and which increase in absolute value (*open triangles*), influencing and restricting plant functions related to carbon relations (*solid arrows*) and to water relations (*open arrows*) and physical damage (*broken line*). The interaction of these factors and functions constrains tree growth at the alpine timberline

ing and despite closed stomata, the needle will lose water by cuticular transpiration. The result is a decrease of water content below a critical value resulting in "frost drought". Low temperatures are not decisive for the survival of conifers in alpine habitats. Frost tolerance is adequate for the minimum habitat temperatures (PISEK et al. 1967; see also Chap. 13, Vol. 12A) and is determined by photoperiod and hardening (SCHWARZ 1970). Wind may act as an additional factor by reducing boundary layer resistance although this effect is negligible when stomata are closed; see also Chap. 15, Vol. 12A. However, wind may cause physical damage to the xylem of the stem and of the needles, which again would enhance frost drought (MARCHANT and CHABOT 1978). It may very well be that the needle is partially separated from the stem by wind and that water stress develops in this half-damaged needle.

The environmental conditions are different for shrubs and needles which are buried beneath snow and which are thus protected against climatic extremes, but which may suffer from drought and frost as soon as they are exposed e.g., by wind (LARCHER 1963). Because of this and because of the decreasing density of the forest canopy and the concomitant increase in light in the understory, low shrubs and herbs are increasingly successful as ground vegetation within the upper alpine forest region. This competitive interaction of trees and other plant life forms is very pronounced at the tree line of *Eucalyptus pauciflora*, the sole tree line species on mainland Australia. It has been found to survive up to several hundred meters above timberline when transplanted into cleared plots (SLATYER personal communication). Seedling establishment does not occur in undisturbed shrub- and grass-dominated alpine vegetation.

In the New Zealand alpine region northern hemisphere conifers grow successfully up to 300 m higher in altitude than the native *Nothofagus*. Seedlings of *Nothofagus* are rather photo-sensitive and re-establishment occurs only under the protection of the crown of adult trees. The result is a very distinct and closed alpine forest line. This is in contrast to the sun-tolerant pine seedlings of the northern hemisphere which may re-establish independent of the main stand. Although the discrepancy between the tree- and forest line in Europe has been repeatedly attributed to the influence of man (ELLENBERG 1978), *Pinus aristata* shows a similar pattern of crippled growth forms and a separation of forest- and tree line in Nevada in areas where human disturbance is absent. It is alternatively suggested that whether or not the tree line and forest line coincide may strongly depend on the species and on the competitive situation during the stage of seedling establishment.

18.3.4.3 The Arid Tree Line

The ecological significance of the tree trunk is generally discussed in terms of light competition with other plant forms. Trees are also effective carbohydrate (GÄUMANN 1935) or nutrient (BURROWS 1976) storage bodies, although this may not be so different in principle from shrubs or herbaceous plants. However, trees are the only morphological types of plant life form other than highly specialized succulents, which can store considerable amounts of water and use this store during the course of a day. This does not only apply to specialized forms such as bottle trees (*Adansonia digitata, Brachychiton australe*) but also to temperate zone forest trees (HINCKLEY et al. 1978). Even in the humid climate of Scotland *Pinus silvestris* relies significantly on water stored in the sapwood over short periods (WARING et al. 1979).

Despite this storage capacity, tree growth is retarded with decreasing precipitation. WARING et al. (1978) found a decreasing leaf area with increasing drought. Additionally, a redistribution of carbohydrates favouring root growth occurs. A root/leaf dry weight ratio of 7.8 was found in the trees and shrubs of a tree savannah (LAMOTTE 1975). This increased to 9.4 and 11.4 with increasing drought in a shrub savannah and in an open savannah respectively.

The significance of drought for tree existence becomes obvious along an environmental gradient of a desert dry valley. The height of *Acacia constricta* (Fig. 18.18) decreased linearly with soil drought experience ($\Psi_{\text{soil x week}}$). At 400 "bar weeks" tree height was almost the same as for the desert shrub *Larrea tridentata*. At this point there was no competitive advantage of the tree structure. Both trees and shrubs exhibit further reductions in height as water stress increases and eventually reach the same height as the low half-shrub *Krameria glandulosa* (BALDING and CUNNINGHAM 1974).

WALTER (1973) has schematically described the competitive strategies of trees and grasses under semi-arid conditions. Trees and bushes have a large and deep root system whereas grasses are characterized by a dense adventitious root system in the upper horizon. With high rainfall sufficient water reaches

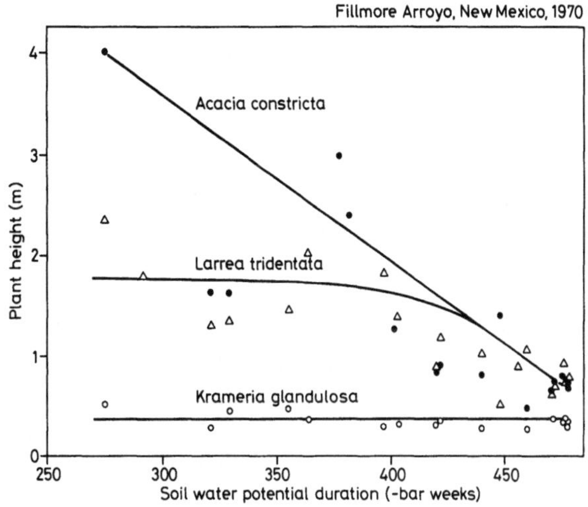

Fig. 18.18. Plant height of a tree (*dots: Acacia constricta*), a shrub (*triangles: Larrea tridentata*) and a half-shrub (*circles: Krameria glandulosa*) in a desert valley as related to cumulative soil water potential. (After BALDING and CUNNINGHAM 1974)

the low soil horizon to be used by trees, which grow rapidly and suppress grasses through competition for light. Australian *Acacia aneura* even exhibits a special growth form of erect branches which allow rain to be captured in a funnel fashion and channelled along the stem into the lower soil profile towards the roots. Furthermore the shading of the evergreen tree crown reduces soil evaporation close to the tree (SLATYER 1961; PRESSLAND 1976). With lower rainfall and retarded tree growth, the competitive advantage of trees in respect to light gathering decreases. At this stage grasses gain in dominance since they are more efficient at water uptake in the upper soil layers. Indirect evidence for these relationships can be seen in the dominance of trees and shrubs in stony habitats where grasses are unable to develop their dense root system. In contrast, sandy soils are grass-dominated as well as clay soils where water does not penetrate deeply.

On a phytogeographic scale the evergreen and deciduous tree habit is distributed in arid regions such that either one or the other form dominates vegetation. In the North Sahara the evergreen tree habit is dominant. The reverse holds for the South Sahara where deciduous *Acacia* ssp. are by far more common. STOCKER (1976) suggested that evergreen trees are adapted to the combination of cold winter and summer drought in the North Sahara, similarly to mediterranean shrubs, but that deciduous trees are favoured in the Sahel where summer rains occur, but where the mild winter is characterized by extreme and prolonged drought. In contrast, the arid regions of Australia are dominated by evergreen *Eucalyptus* and *Acacia*. WALTER (1968) pointed out a principle difference between the Australian and African semi-arid environments. Africa has a strict dry season of several months, whereas Australia has very irregular but recurring rainfalls throughout the year, although marked dry seasons do occur in North and Southwest Australia. Under conditions of uncertain but repeated rainfall evergreen sclerophyllous species appear to be more successful than deciduous trees, since the latter may have to replace foliage several times per year. Additionally, phytogeographical observations confirm that, similarly to shrubs, the evergreen structure in trees is better adapted to atmospheric drought, whereas the deciduous tree is more adapted to soil aridity.

18.4 The Effect of Plant Life Form on Biomass, Productivity and Water Use Efficiency in Different Vegetation Zones

The latitudinal zones of the earth's vegetation are dominated by different plant life forms (e.g., TROLL 1956; WALTER 1973, 1968). Tree vegetations generally occupy habitats of higher rainfall, with the exception of arctic and alpine climates, whereas shrubs and grasses generally succeed at the drier end of the scale. High rainfall may partially compensate for a long dry period. This section will investigate the extent to which plant form determines the accumulated biomass, primary production and water use efficiency of different vegetation types. Although numerous biomass studies have been conducted, it is very difficult to find data with complete and sufficient information for such a comparison (Table 18.3; see also Chaps. 4–9, Vol. 12D).

Certainly tree vegetation has the highest total biomass (Fig. 18.19), but when a comparison is made between different forest types it is remarkable that conifers of the temperate and boreal zone may reach and even exceed the total biomass of temperate and tropical broad-leafed forests. Despite the large variability in the primary production of forest stands, which is related to differences in temperature and drought between the various study sites, it can be seen that a 10- to 40-fold difference between species in A_{max} at the leaf level (Sect. 18.2.1) is reduced to only a factor of 2 to 4 at the primary production level. For forest vegetations this variation becomes even smaller if the differences in annual growing season are considered. Coniferous forests have the highest monthly primary production of all forest types. The large potential advantage of the deciduous tree at the leaf level is attenuated and even reversed at the stand level when compared with the coniferous tree (see Sects. 18.3.4.1 and 18.2.4).

In natural vegetation primary production of tropical and temperate grasslands may reach the same order of magnitude as forest vegetations, but a large variation within the data mainly due to drought is apparent. Extremely high biomass production was reported by WALTER (1981) for Siberian river flood plain vegetation of tall herbaceous plants. Leaf area index and leaf biomass are quite similar in grasslands and broad-leaved forests. As discussed in Section 18.2.4 it is the supply of radiant energy which eventually limits productivity in the temperate and tropical region independent of plant life form. Biomass and P_N decrease in the arctic and arid environment, but in these regions where vegetation types are not "light-limited", the diversity of plant life form increases (WHITTAKER and NIERING 1975).

The similarity of P_N per month of growing season (Fig. 18.19) implies that the water use efficiency (WUE) corrected for differences in evaporative demand may be rather similar in various vegetation types as long as water is available. It may be quite different if water availability is diminished, but usually then

Table 18.3. Biomass and primary production in different vegetation zones

Vegetation	Maximum biomass kg m^{-2}	Leaf biomass kg m^{-2}	Primary production kg m^{-2} a^{-1}	Leaf area index m^2 m^{-2}	Max. height m	Rain-fall mm	Growing season months	References
Tropical rainforest	41–65	1–3.5	6–17					LIETH 1975; RODIN et al. 1975
Manaus, Amazone	47.3	0.90	1.8	5.2	38	2000	12	KLINGE 1976a, b, JORDAN and UHL 1978
Thailand	41	0.76	2.7	6.7	56	1867	12	KIRA 1978
S. Thailand [a]	36.5	0.84	2.9	12.3 [b]	36	2696	12	OGAWA et al. 1965a, b; KIRA et al. 1967 [b]
Peninsular Malaysia	(34.6)	0.78	2.7	6.9	58	2054	12	KIRA 1978
Tropical seasonal forest	42		1.6–2.5	6–10				LIETH 1975
S.W. Cambodia	38.2	0.73	—	7.4	44	3700	9	HOZUMI et al. 1969a, b, c
Côte d'Ivoire	29.1	0.25	1.3	3.2	50	2000	10	MÜLLER and NIELSEN 1965
C. Thailand, dry evergreen forest	18.6 [b]	0.71	1.8	8.6	23	1290	9	OGINO et al. 1967
C. Thailand, dry *Dipterocarpus* forest	9.0 [b]	0.18	0.7	0.82	16	1290	9	OGINO et al. 1967
Evergreen temperate forest broadleaved								
Evergreen oak (*Castanopsis/ Cyclobalanopsis*)	46.7	0.67	1.8	6.7	24	2680	10	KIRA et al. 1978
Distylium racemosum [a]	(32.1)	1.14 [b]	2.1	8.8 [b]	25	5700	11	KIRUMA 1960
Cinnamomum camphora	(21.5)	0.47	(1.36)	4.9	17	2300	7	SATOO 1968
Castanopsis cuspidata	9.4	0.84	2.3	8.9	8	1800	11	TADAKI 1968
Evergreen temperate forest coniferous								
Tsuga sieboldii	73.2	1.02	0.9	7.4	27	2800	9	ANDO et al. 1977
Abies firma	64.7	1.78	1.1	9.2	31	2800	9	ANDO et al. 1977
Pinus radiata	31.6	1.0	2.5	7.5	31	1500	12	ROOK and BENECKE, personal communication
Summer deciduous forest broadleaved	18–58		0.4–2.5	3–12				LIETH 1975; DUVIGNEAUD et al. 1971, WHITTAKER 1975
Smoky Mts cove forest	(58.5)	0.60	(1.3)	6.2	34	700	—	WHITTAKER 1975
Fagus crenata	36.9	0.41	1.1	6.8	22	2500	6	OGINO et al. 1976
Fagus crenata	(36.4)	0.38	1.2	7.9	29	—	6	MARUYAMA 1977
Fagus silvatica	30.4	0.37	1.1	6.0	26	1065	5	ELLENBERG 1978
Querceto-Coryletum	31.6	0.35	1.5	5.7	24	800 [a]	5	DUVIGNEAUD et al. 1971

Table 18.3 (continued)

Vegetation	Maximum biomass kg m^{-2}	Leaf biomass kg m^{-2}	Primary production kg m^{-2} a^{-1}	Leaf area index m^2 m^{-2}	Max. height m	Rain-fall mm	Growing season months	References
Querceto-Crataegetum	18.6	0.31	1.2	4.4	20	800 [a]	5	DUVIGNEAUD et al. 1971
Querceto-Betuletum	20.4	0.33	1.3	4.8	21	800 [a]	5	DUVIGNEAUD et al. 1971
Betula maslimowitziana	(11.7)	0.22	(0.6)	4.1	20	1200	5	SATOO 1974a
Populus tremuloides	(20)	0.50	(1.2)	7.4	18	800	—	WHITTAKER and NIERING 1975
Populus davidiana	(11.5)	0.22	(0.9)	2.4	20	1200	5	SATOO 1956
Summer deciduous forest coniferous								
Larix leptolepis	(16.4)	0.36	(1.26)	4.2	19	1800	5	SATOO 1970a b
Metasequoia glyptostroboides	(8.9)	0.43	(1.62)	8.5	17	1800	7	SATOO 1974b
Boreal forest	20–92		0.2–1.5	7.19				LIETH 1975; WHITTAKER and NIERING 1975
Pseudotsuga menziesii	150	1.5–2.0	1.2	15.0	80	2400	5	WARING and FRANKLIN 1979
Pseudotsuga menziesii	(92)	2.5	(1.34)	19.0	45	800	—	WHITTAKER and NIERING 1975
Abies veitchii	(31.1)	1.88	(1.3)	10.6	20	2000	7	TADAKI et al. 1970
Abies lasiocarpa	(42)	1.60	(1.0)	14.0	35	800	—	WHITTAKER and NIERING 1975
Abies concolor	(32)	1.70	(1.3)	18.0	40	800	—	WHITTAKER and NIERING 1975
Pinus ponderosa	(30)	0.68	(0.7)	7.0	25	800	—	WHITTAKER and NIERING 1975
Thujopsis dolabrata	(16)	4.4	(1.92)	17.9	9	2300	6	SATOO 1974c
Picea abies	17	0.21	0.67	4	15	400	6	GORTINSKY 1975
Mediterranean and semi-arid scrub	26		0.25–1.5	4–12				LIETH 1975
Eucalyptus socialis, *E. dumosa*	6.7	0.24	(0.24)	0.67	7	407	4–6	BURROWS 1976
Acacia aneura, (100yrs)								
Eucalyptus populnea (55yrs)	7.0	0.70	(0.33)	1.96	10	467	2–3	BURROWS 1976
Eucalyptus ssp. (Malle) (13yrs)	3.9	0.30	(0.54)	0.85	3.3	407	4–6	BURROWS 1976
Calif. evergreen comm.	(2.8)	0.53	(0.67)	2.6	3–4	410	6–7	MILLER et al. 1977; MILLER and
Chile evergreen comm.	(1.2)	0.34	(0.25)	2.0	3–4	440	6–7	POOLE 1979; POOLE and MILLER 1978
Tropical grassland	< 5		0.02–3.4	1–7				
Eragrostis, India	4.45	2.73 [a]	4.55 [b]	8.5 [a]	2.00	1134	9–10	SINGH and JOSHI 1979
Panicum, India	2.92	1.74 [a]	3.53 [b]	8.5 [a]	3.00	790	10–11	SINGH and YADAVA 1974

Sehima, India	1.29	0.43	0.83	3.4	0.90	1257	9–10	Singh and Joshi 1979
Dichanthium, India	1.25	0.52	0.98	6.7	0.90	1030	9–10	Singh and Joshi 1979
Cenchrus, Australia	0.55	0.15	0.34	1.6[a]	0.40	650	3	Christie 1978
Thyridolepis, Australia	0.23	0.12	0.20	1.3[a]	0.40	660	3	Christie 1978
Temperate Grassland	<3		0.02–1.3	5–16				Lieth 1975
Phalaris arundinacea meadow CSSR	3.05	1.39	2.15	4.2	1.0	520	6–7	Rychnovská 1979
Festuca sulcata meadow, CSSR	2.51	0.96	1.09	1.3	0.5	520	6–7	Rychnovská 1979
Alopecurus pratensis meadow, CSSR	2.23	1.01	1.48	3.9	0.8	520	6–7	Rychnovská 1979
Mixed prairie, Hays, Cottonw., Dickinson, Matador	2.68	0.11	0.89	0.57[a]	–	425	6–7[c]	Sims and Coupland 1979
Shortgrass prairie, Pentax, Pawnee, USA	1.83	0.07	0.81	0.37[a]	–	422	7[c]	Sims and Coupland 1979 Brown 1974
Desert Grassland Jornada, USA	0.34	0.05	0.22	0.26[a]	–	228	10[c]	Sims and Coupland 1979
Bouteloua/Fouqueria	0.26	0.15	0.28	3.2	0.5/3	400	8[c]	Whittaker and Niering 1975
Desert	0.1–4		0.01–0.25	<1				Lieth 1975
Artemisia terrae-albae	4.29	0.14	0.76	–	0.7	120	6	Rodin 1979
Atriplex confertifolia	2.47	0.1	0.12	0.92	0.5	244	4–5	Caldwell et al. 1977
Ceratoides lanata	2.18	0.07	0.08	0.96	0.3	244	4–5	Caldwell et al. 1977
Haloxylon persicum	0.89	0.05	0.10	–	5	170–110	7	Rodin 1979
Larrea divaricata	0.43	0.04	0.14	0.93	1.5	300	8[c]	Whittaker and Niering 1975;
Haloxylon ammadendron	6.35	0.3	1.6		7	114	7	Rodin 1976
Tundra	0.1–3		0.1–0.4	0.5–1.3				Lieth 1975
Subarctic tundra	1.1–5.8	–	0.1	0.1–0.6	–	–	–	Wielgolaski 1975
Arctic tundra	0.6–1	–	0.1–0.5	0.6–3.3	–	–	–	Wielgolaski 1975
Alpine *Loiseleuria procumbens* heath	3.78	0.35	(0.29)	2.6	0.03	950	4–5	Schmidt 1977
Alpine *Vaccinium* heath	3.60	0.54	(0.49)	5.3	0.15	950	4–5	Schmidt 1977

() Above ground. [a] Estimate uncertain. [b] Probably overestimated. [c] Thermal potential growing season.
Personal communication with: T. Kira, T. Satoo, J.S. Singh, P.C. Miller, M.M. Caldwell, B. Burrows, E.K. Christie, M.J. Trlica, F.E. Wielgolaski, R.H. Whittaker, H. Klinge, E. Medina, U. Benecke, D. Rook, P.L. Sims, H.A.I. Madgwick.

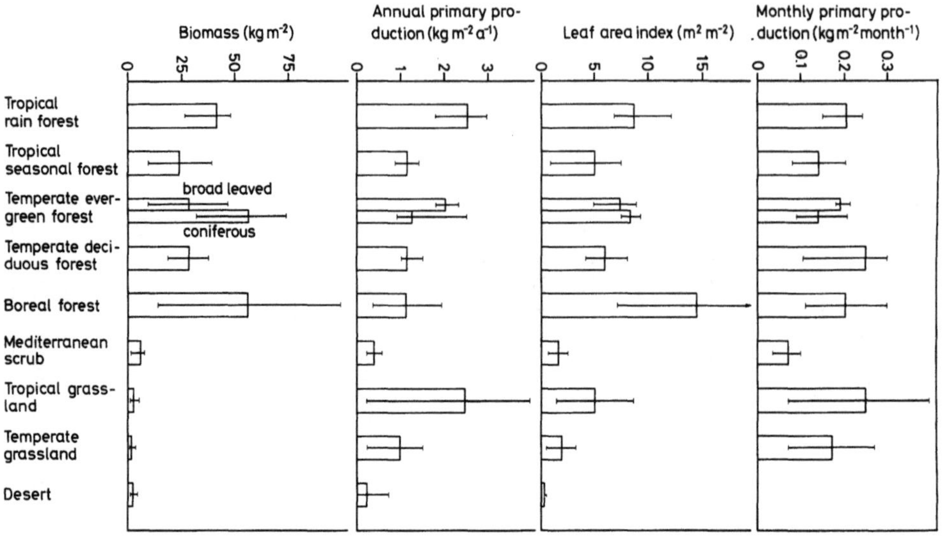

Fig. 18.19. Biomass, annual primary production, leaf area index and monthly primary production during the growing season of different vegetation formations according to Table 18.3. *Columns* averages; *line* range of conditions given in Table 18.3

growth is also retarded (FISCHER and TURNER 1978). DE WIT (1958) proposed that plant growth (P_N) is directly related to the ratio of the cumulative transpiration ($\sum E$) divided by the daily free evaporation ($\sum E_0$):

$$P_N = n \frac{\sum E}{\sum E_0}. \tag{18.9}$$

The data of DE WIT (1958) show that the "n" value varied between 10 and 14 g dry weight m^{-2} day^{-1} for many C_3 species at different locations in the Great Plains. It was 21–23 g dry weight m^{-2} day^{-1} for C_4 crops.

According to DE WIT (1958) the linear relationship between total dry matter production and transpirational water use, which was observed repeatedly, is due to the differences in evaporative demand. FISCHER and TURNER (1978) showed that, despite a large variation in WUE, the product of $WUE \cdot E_0$ is rather constant in field experiments with C_3 plants, including desert shrubs. The C_4 species *Zea mays* had a higher "n" value. FISCHER and TURNER (1978) maintain that although intensive consideration has been given to the many possible ways in which WUE may be affected by genotype and environment, the variation in WUE can be attributed largely to two factors, namely the evaporative demand in the atmosphere and the primary carboxylation enzyme in photosynthesis. These results imply that plants maximize WUE to such an extent that P_N is not very different between plant life forms in dry regions, provided allowance is made for differences in E_0 or that plants have not evolved means to control WUE at the canopy level. Apparently, there is a discrepancy between results based on a determination of WUE from total dry matter production and canopy transpiration and the results on T/P ratios at the leaf level which is not only determined by the vapour pressure deficit but also by stomatal response. Under natural conditions stomata are generally closed to such a degree that the plant assimilates CO_2 at the leaf level at less than 80% of its maximal rate (SCHULZE et al. 1980; LANGE et al. 1974). Therefore, although $WUE \cdot E_0$ may be constant for C_3 or for C_4 plants, it is not clear to what extent this WUE is also under stomatal control. Since $\sum E_0$ is related to the vapour pressure deficit of the air, the expression $\sum E/\sum E_0$ represents an expression of an average conductance of the canopy, $\bar{g}$. Furthermore, since P_N is related to the average CO_2 assimilation of the canopy, $\bar{A}$, one may rewrite Equ. 18.9 into $\bar{A}/\bar{g}=$ constant. Since the intercellular CO_2 concentration of the mesophyll is generally being expressed as "$c_i=c_a-A/g$" one could interpret the "n" factor of DE WIT to represent an average depletion of CO_2 in the mesophyll, which is different for C_3 and C_4 plants.

For ecological understanding of the primary production as related to the total water budget of the vegetation, interceptive water loss must be included in the consideration. WEBB et al. (1978) showed that P_N of semi-arid vegetation types was linearly related to average evapotranspiration (i.e., transpiration plus interception plus soil evaporation) above a minimum amount of water which may represent soil evaporation in open vegetation. However, this relation does not exist for forest systems, which were not water stressed mainly because of interceptive water loss. (For detailed discussions of productivity of different ecosystems see Vol. 12 D).

18.5 Conclusions

It is appropriate to enquire, first, what plant life forms do; second, how they do it. Only then is enough understanding present for the third question, if it may be asked at all, namely, "why they do it". Phytogeographical observations of the earth's vegetation have led to the first question and very practical systems for plant life form description were developed in order to classify and to compare the vegetation zones of the earth. The second problem, of how plant life forms operate, has been the subject of many decades of ecophysiological, physiological and biochemical work. It is obvious that we are presently able to understand some single plant processes even at the molecular basis. However, the integration of this knowledge into whole plant processes is still very limited. Therefore, it might not be appropriate at all to enquire into the third question, "why they do it", at this stage.

The present chapter has tried at its beginning to make assumptions about the "strategies" by which different plant life forms may operate. Strategy is used in the sense of animal ecology (ZWÖLFER 1978; STEARNS 1976) as an adaptive response which has been selected under ecological pressures. It is proposed that the performance of a specific plant life form in its natural habitat might be "measured" by the marginal returns of carbon relations, which means the extra carbon return per unit carbon, water or nutrient investment in an organ. It is hypothesized that the composition of plant life forms in the vegetation of the earth is governed by these rules. The way, in which plants could regulate carbon returns at limited carbon, water, or nutrient supply is seen in the regulation of carbon allocation into photosynthetic and non-photosynthetic biomass and in the regulation of leaf longevity.

A hypothesis as to "why plant life forms do it", is presently mathematically and experimentally not developed. There are almost no experimental studies known which investigate the immediate change of carbon returns due to a perturbation in the water and nutrient relations, but numerous descriptions of performance under certain conditions have been made. Therefore, extrapolations are not possible. In spite of this the implications of the foregoing assumptions were explored in terms of competitive features in different life forms and the result of this was compared with the behaviour and success of certain plant life forms in their natural vegetation. It appears that it is not the unfavourable season that determines the success of a specific plant life form, as was suggested by RAUNKIAER (1904), but the carbon relations during the favourable season. The environmental constraints during the growing season determine the plant life form of those species which can persist in a vegetation and compete with others. All species appear to have a large degree of adaptability to the climatic conditions of their habitat at the CO_2 assimilation level. Therefore, the adaptations to different environments occur with carbon allocation, which determines the structural forms of annuals, herbaceous perennials and of deciduous and evergreen woody shrubs and trees. These categories allow a functional insight into the phytogeographical composition of different vegetation types, although it is quite clear that the present study has taken only the simplest cases into account for each group; the phytogeographical diversity of species remains unexplained. The major reason for this is the complete lack of experimental evidence. The comparative study of plant life forms has been strongly neglected and we are still at the level of asking "what plant life forms do". Only for annual species we may be at a stage, mainly for crop species, to know, "how they do it". But solving this question, though it is essential for understanding the mechanisms of succession in natural communities, remains an important goal for further research. The ecological significance, which results from various growth strategies at the carbon allocation level, may be obvious whenever more is known about the functional bases of "success" and "fitness" and their role in community stability and organization.

Acknowledgement. This work was supported by the "Sonderforschungsbereich 137" of the Deutsche Forschungsgemeinschaft and by a grant from the Research School of Biological Sciences which enabled me to use the unique library facilities of the Australian National University. Some of the ideas expressed in this chapter developed during many discussions and in cooperation with Prof. Dr. I.R. COWAN, Prof. O.L. LANGE, Prof. Dr. A.E. HALL and Dr. G.D. FARQUHAR. I am grateful for this fruitful partnership. I thank Prof. E. BECK, Dr. U. BENECKE, Dr. R.A. FISCHER, Dr. H.G. JONES, and Prof. H. ZWÖLFER for reviewing the manuscript. Numerous personal communications were necessary for the compilation of Table 18.3 and I wish to thank these authors. Main appreciation, however, is given to Mrs. M. BALL and Mrs. L. TENHUNEN, who went to great pains to correct my English.

References

Allessio ML, Tieszen LL (1978) Translocation and allocation of ^{14}C-photoassimilate by *Dupontia fisheri*. In: Tieszen LL (ed) Vegetation and production ecology of an Alaskan arctic tundra. Ecol Stud Vol 29. Springer, Berlin Heidelberg New York, pp 393–413

Ando T, Chiba K, Nishimura T, Tanimoto T (1977) Temperate fir and hemlock forests in Shikoku. In: Shidei T, Kira T (eds) Primary productivity of Japanese forests. JIBP Synthesis Vol 16. Univ Press, Tokyo, pp 213–245

Axelrod DI (1966) Origin of deciduous and evergreen habits in temperate forests. Evolution 20:1–15

Balding FR, Cunningham GL (1974) The influence of soil water potential on the perennial vegetation of a desert arroyo. Southwest Nat 19:241–248

Benecke P (1976) Soil water relations and water exchange of forest ecosystems. In: Lange OL, Kappen L, Schulze E-D (eds) Water and plant life. Ecol Stud Vol. 19 Springer, Berlin Heidelberg New York, pp 101–132

Best KF, Macintyre GI (1975) The biology of Canadian weeds. 9. *Thlaspi arvense* L. Can J Plant Sci 55:279–292

Billings WD (1974) Adaptations and origins of alpine plants. Arct Alp Res 6:129–142

Billings WD, Mooney HA (1968) The ecology of arctic and alpine plants. Biol Rev 43:481–529

Björkman O, Ludlow MM, Morrow PA (1972) Photosynthetic performance of two rainforest species in their native habitat and analysis of their gas exchange. Carnegie Inst Yearb 72:94–102

Bonnemann A, Röhrig E (1972) Baumartenwahl, Bestandesgründung und Bestandespflege. Waldbau auf ökologischer Grundlage, Vol 2. Parey, Hamburg, Berlin

Bornkamm R (1970) Über den Einfluß der Konkurrenz auf die Substanzproduktion und den N-Gehalt der Wettbewerbspartner. Flora 159:84–104

Boysen-Jensen P (1932) Die Stoffproduktion der Pflanzen. Fischer, Jena

Brougham RW (1958) Interception of light by the foliage of pure and mixed stands of pasture plants. Aust J Agric Res 9:39–52

Brouwer R (1963) Some aspects of the equilibrium between overground and underground plant parts. Jaarb Inst Biol Scheikd Onderzoek (IBS) pp 31–39

Brown LF (1974) Photosynthesis of two important grasses of the shortgrass prairie as affected by several ecological variables. PhD Thesis, Colo State Univ, Fort Collins

Brown LF, Trlica MJ (1977) Interacting effects of soil water, temperature and irradiance on CO_2 exchange rates of two dominant grasses of the shortgrass prairie. J Appl Ecol 14:197–204

Burrows WH (1976) Aspects of nutrient cycling in semi-arid Mallee and Mulga communities. PhD Thesis, Aust Natl Univ, Canberra

Caldwell MM (1979) Plant life and ultraviolet radiation: some perspective in the history of the earth's climate. Biol Sci 29:520–525

Caldwell MM, White RS, Moore RT, Camp LM (1977) Carbon balance, productivity, and water use of cold-winter desert shrub communities dominated by C_3 and C_4 species. Oecologia 29:275–300

Candolle De AP (1818) Regni vegetabilis systema naturale Vol I. Treuttel et Würtz, Paris

Chapman SB, Webb NR (1978) The productivity of a *Calluna* heathland in Southern England. In: Heal OW, Perkins DF (eds) Production ecology of British moors and montane grasslands. Ecol Stud Vol 27. Springer, Berlin Heidelberg New York, pp 247–269

Chew RM, Chew AE (1965) The primary productivity of a desert-shrub (*Larrea tridentata*) community. Ecol Monogr 35:355–375

Christie EK (1978) Ecosystem processes in semi-arid grasslands. I. Primary production and water use of two communities possessing different photosynthetic pathways. Aust J Agric Res 29:773–787

Claussen W, Lenz F (1979) Die Bedeutung des Assimilatstaus in den Blättern für die Regulierung der Nettophotosyntheserate bei Auberginen (*Solanum melongena* L). Angew Bot 53:41–52

Clements FE (1920) Plant indicators. The relation of plant communities to process and practice. Carnegie Inst Washington Publ 290

Dale HM (1974) The biology of Canadian weeds. 5. *Daucus carota*. Can J Plant Sci 54:673–685

Day W, Legg BJ, French BK, Johnston AE, Lawlor DW, Jeffers WDeC (1978) A drought experiment using mobile shelters: The effect of drought on barley yield, water use and nutrient uptake. Agric Sci 91:599–623

Detling JK (1979) Processes controlling blue grama production on the short grass prairie. In: French NR (ed) Perspectives in grassland ecology, Ecol Stud Vol 32. Springer, Berlin Heidelberg New York, pp 25–42

DeWit CT (1978) Simulation of assimilation, respiration and transpiration of crops. Simulation Monographs. PUDOC, Wageningen

Dodd JL, Lauenroth WK (1979) Analysis of the response of a grassland ecosystem to stress. In: French NR (ed) Perspectives in grassland ecology. Ecol Stud Vol 32. Springer, Berlin Heidelberg New York, pp 43–58

Doanald CM (1963) Competition among crop and pasture plants. Adv Agron 15:1–118

Donald CM, Hamblin J (1976) The biological yield and harvest index of cereals as agronomic and plant breeding criteria. Adv Agron 28:361–405

Drude O (1887) Die systematische und geographische Anordnung der Phanerogamen. In: Schenk A (ed) Handbuch der Botanik Vol III/2. Trewendt, Breslau, pp 175–496

Drude O (1928) Pflanzengeographische Ökologie. In: Abderhalden's Handbuch der biologischen Arbeitsmethoden, Abt XI. Urban and Schwarzenberg, Berlin Wien, pp 1–56

Duvigneaud P, Kestemont P, Ambroes P (1971) Primary productivity of the temperate deciduous forests of Western Europe. In: Duvigneaud P (ed) Productivity of forest ecosystems. UNESCO, Paris, pp 259–270

Eagles CF (1973) Effect of light intensity on growth of natural populations of *Dactylis glomerata* L. Ann Bot (London) 37:253–262

Ellenberg H (1978) Vegetation Mitteleuropas mit den Alpen in ökologischer Sicht, 2nd edn. Ulmer, Stuttgart

El-Sharkawy M, Hesketh J, Muramoto H (1968) Leaf photosynthetic rates and other growth characteristics among 26 species of Gossypium. Crop Sci 8:670–674

Evans LT, Dunstone RJ (1970) Some physiological aspects of evolution in wheat. Aust J Biol Sci 23:725–738

Evans LT, Bingham J, Jackson P, Sutherland J (1972) Effects of awns and drought on the supply of photosynthate and its distribution within wheat ears. Ann Appl Biol 70:67–76

Evenari M, Shanan L, Tadmor N (1971) The Negev. The challenge of a desert. Harvard Univ Press, Cambridge, Mass

Evenari M, Lange OL, Schulze E-D, Kappen L, Buschbom U (1977) Net photosynthesis, dry matter production, and phenological development of apricot trees (*Prunus armeniaca* L.) cultivated in the Negev Highlands (Israel). Flora 166:383–414

Ewers FW, Schmid R (1981) Longevity of needle fascicles of *Pinus longaeva* (Bristlecone Pine) and other North American pines. Oecologia 51:107–115

Farrar JF (1980) Allocation of carbon to growth, storage and respiration in the vegetative barley plant. Plant Cell Environ 3:97–106

Fekete G, Szujkó-Lacza J (1971) A survey of the plant life-form systems and the respective research approaches. III. Raunkiaer's life-form concept. The application of life-forms in the characterization of phytoclimate and vegetation analysis. Ann Hist Nat Mus Nat Hung Botanica 63:37–50

Fischer RA (1975) Future role of physiology in wheat breeding. In: Johnson VA (ed) Proc 2nd international winter wheat congress. Univ Neb Misc Publ 32:178–196

Fischer RA, Maurer R (1978) Drought resistance in spring wheat cultivars. I. Grain yield responses. Aust J Agric Res 29:897–912

Fischer RA, Turner NC (1978) Plant productivity in the arid and semiarid zones. Ann Rev Plant Physiol 29:277–317

Fonda RW, Bliss LC (1966) Annual carbohydrate cycle of alpine plants on Mt Washington, New Hampshire. Bull Torrey Bot Club 93:268–277

Fritschen LJ, Hsia J, Doraiswamy P (1977) Evapotranspiration of a douglas fir determined with a weighing lysimeter. Water Resour Res 13:145–148

Gäumann E (1935) Der Stoffhaushalt der Buche im Laufe eines Jahres. Ber Schweiz Bot Ges 44:157–334

Gifford RM, Marshall C (1973) Photosynthesis and assimilate distribution in *Lolium multiflorum* Lam. following differential tiller defoliation. Aust J Biol Sci 16:517–526

Gloser J (1977) Photosynthesis and respiration of some alluvial meadow grasses: responses to soil water stress, diurnal and seasonal courses. Acta Sci Nat Acad Sci Bohemoslovacae Brno NS 11:1–36

Good R (1974) The geography of the flowering plants. 4th edn. Longman, London New York

Gortinsky GB (1975) Productivity of forests of the European part of the USSR. In: Resources of the biosphere (Synthesis of the Soviet studies for the International Biological Programme) Vol I. Leningrad, NAUKA, pp 34–42

Goryshina TK (1969) Prevernal photosynthesis in the leaves of nemoral herbaceous plants after overwintering. Bot J Acad Nauka CCCP 54:919–923

Grabherr G, Cernusca A (1977) Influence of radiation, wind, and temperature on the CO_2 gas exchange of the alpine dwarf shrub community *Leuseleurietum cetrariosum*. Photosynthetica 11:22–28

Grime JP (1979) Plant strategies and vegetation processes. Wiley and Sons, New York

Grisebach A (1872) Die Vegetation der Erde nach ihrer klimatischen Anordnung. Ein Abriß der vergleichenden Geographie der Pflanzen. Engelmann, Leipzig

Hall AE, Schulze E-D (1980) Stomatal response to environment and a possible interrelation between stomatal effects on transpiration and CO_2 assimilation. Plant Cell Environ 3:467–474

Hall AE, Foster KW, Waines JG (1979) Crop adaptation to semi-arid environments. In: Hall AE, Cannell GH, Lawton HW (eds) Agriculture in semi-arid environments, Ecol Stud Vol 34. Springer, Berlin Heidelberg New York, pp 148–179

Harper JL, Ogden J (1970) The reproductive strategy of higher plants. I. The concept of strategy with special reference to *Senecio vulgaris* L. J Ecol 58:681–698

Harris P (1974) A possible explanation of plant yield increases following insect damage. Agroecosystems 1:219–225

Harris W (1974) Competition among pasture plants. V. Effects of frequency and height of cutting on competition between *Agrostis tenuis* and *Trifolium repens*. NZ J Agric Res 17:251–256

Harris W, Pandey KK, Gray YS, Conchman PK (1979) Observations on the spread of perennial ryegrass by stolons in a lawn. NZ J Agric Res 22:61–68

Hasselt Van PR, Strickwerd JT (1976) Pigment degradation in discs of the thermophilic *Cucumis sativus* as affected by light, temperature, sugar application and inhibitors. Physiol Plant 37:253–257

Heim G, Landsberg JJ, Watson RL, Brian P (1979) Eco-physiology of apple trees: Dry matter production and partitioning by young golden delicious trees in France and England. J Appl Ecol 16:179–194

Hellmuth EO (1968) Eco-physiological studies on plants in arid and semi-arid regions in Western Australia. I. Autecology of *Rhagodia baccata* (Labill.) Moq. J Ecol 56:319–344

Hinckley TM, Lassoie JP, Running SW (1978) Temporal and spacial variations in the water status of forest trees. For Sci Monogr 20:1–72

Hodgkinson KC, Johnson PS, Norton BE (1978) Influence of summer rainfall on root and shoot growth of a cold-winter desert shrub *Atriplex confertifolia*. Oecologia 34:353–362

Holmen K (1957) The vascular plants of Peary Land, North Greenland. Medd Groenl 124:1–149

Hozumi K, Yoda K, Kokawa S, Kira T (1969a) Production ecology of tropical rain forests in southwestern Cambodia. I. Plant biomass. Ibid 6:1–51

Hozumi K, Yoda K, Kira T (1969b) Ibid. II. Photosynthetic production in an evergreen seasonal forest. Ibid 6:57–81

Hozumi K, Yoda K, Kira T (1969c) Production ecology of tropical rain forests in southwestern Cambodia. II. Photosynthetic production in an evergreen seasonal forest. In: Kira T, Iwata K (eds) Nature and life in Southeast Asia Vol 6. Jpn Soc Promotion Sci, Tokyo, pp 57–81
Hsiao TC, Ferreres E, Acevedo E, Henderson DW (1976) Water stress and dynamics of growth and yield of crop plants. In: Lange OL, Kappen L, Schulze E-D (eds) Water and plant life. Ecol Stud Vol 19. Springer, Berlin Heidelberg New York, pp 281–306
Huffaker CB (1959) Biological control of weeds with insects. Ann Rev Entomol 4:251–276
Humboldt von A (1806) Ideen zu einer Physiognomik der Gewächse. Cotta, Tübingen
Jackman RH, Mouat MCH (1972) Competition between grass and clover for phosphate. II. Effect of root activity, efficiency of response to phosphate, and soil moisture. NZ J Agric Res 15:667–675
Janzen DH (1971) Seed predation by animals. Annu Rev Ecol Syst 2:465–492
Johansson LG, Linder S (1975) The seasonal pattern of photosynthesis of some vascular plants on a subarctic mire. In: Wielgolaski FE (ed) Fennoscandian tundra ecosystems. I. Plants and microorganisms. Ecol Stud Vol 16. Springer, Berlin Heidelberg New York, pp 194–200
Johnson DA, Tieszen LL (1976) Above ground biomass allocation, leaf growth, and photosynthesis patterns in tundra plant forms in arctic Alaska. Oecologia 24:159–173
Jordan CF, Uhl C (1978) Biomass of a "terra firme" forest of the Amazon basin. Oecol Plant 13:287–400
Jurgens SK, Johnson RR, Boyer JS (1978) Dry matter production and translocation in maize subjected to drought during grain fill. Agron J 70:678–682
Keulen Van H (1975) Simulation of water use and herbage growth in arid regions. PUDOC Wageningen
Kiese O (1971) The measurement of climatic elements which determine production in various plant stands. In: Ellenberg H (ed) Integrated experimental ecology. Ecol Stud Vol 2. Springer, Berlin Heidelberg New York, pp 132–142
King RW, Wardlaw IF, Evans LT (1967) Effect of assimilate utilization on photosynthetic rate in wheat. Planta 77:261–276
Kira T (1978) Community architecture and organic matter dynamics in tropical lowland rain forests of southeast Asia with special reference to Pasoh Forest, West Malaysia. In: Tomlinson PB, Zimmermann MH (eds) Tropical trees as living systems. Cambridge Univ Press, London, pp 561–590
Kira T, Ogawa H, Yoda K, Ogimo K (1967a) Comparative ecological studies on three main types of forest vegetation in Thailand. IV. Dry matter production, with special reference to the Khao Chong rain forest. In: Kira T, Iwata K (eds) Nature and life in Southeast Asia Vol 6. Jpn Soc Promotion Sci, Tokyo, pp 149–174
Kira T, Ogawa H, Yoda K, Ogino K (1967b) Ibid. IV. Dry matter production, with special reference to the Khao Chong rain forest. Nature Life SE Asia (Kyoto) 5:149–174
Kira T, Ono Y, Hosokawa T (1978) Biological production in a warm-temperate evergreen oak forest of Japan. JIBP Synthesis Vol 18. Univ Press, Tokyo
Kiruma M (1960) Primary production of the warm-temperate laurel forest in the southern part of Osumi Peninsula, Kyushu, Japan. Misc Rep Res Inst Natl Recour (Tokyo) 52/53:36–47
Klinge H (1976a) Root mass estimation in lowland tropical rain forests of central Amazonia, Brazil. III. Nutrients in fine roots from giant humus podsols. Trop Ecol 16:28–38
Klinge H (1976b) Bilanzierung von Hauptnährstoffen im Ökosystem tropischen Regenwalds (Manaus) – Vorläufige Daten. Biogeographica 7:59–77
Klötzli F (1976) Grenzen von Laubwäldern in Europa. Ber Dtsch Bot Ges 89:371–380
Körner CH, Moraes De JAPV (1979) Water potential and diffusion resistance in alpine cushion plants on clear summerdays. Oecol Plant 14:109–120
Körner CH, Scheel JA, Bauer H (1979) Maximum leaf diffusive conductance in vascular plants. Photosynthetica 13:45–82
Künstle E, Mitscherlich G (1977) Photosynthese, Transpiration und Atmung in einem Mischbestand im Schwarzwald. IV. Bilanz. All Forst Jagdztg 148:227–238

Laing DR, Fischer RA (1977) Adaptation of semi-dwarf wheat cultivars to rainfed conditions. Euphytica 26:129–139
LaMarche VC (1969) Environment in relation to age of Bristlecone Pine. Ecology 50:53–59
Lamotte M (1975) The structure and function of a tropical savannah ecosystem. In: Golley EB, Medina E (eds) Tropical ecological systems. Ecol Stud Vol 11, Springer, Berlin Heidelberg New York, pp 179–222
Lange OL, Schulze E-D, Evenari M, Kappen L, Buschbom U (1974) The temperature related photosynthetic capacity of plants under desert conditions. I. Seasonal changes of the photosynthetic response to temperature. Oecologia 17:97–110
Larcher W (1963) Zur spätwinterlichen Erschwerung der Wasserbilanz von Holzpflanzen an der Waldgrenze. Ber Naturwiss Med Ver Innsbruck 53:125–137
Larcher W (1976) Ökologie der Pflanzen. UTB 232, 2nd edn. Ulmer, Stuttgart
Larcher W (1977) Ergebnisse des IBP-Projektes „Zwergstrauchheide Patscherkofel" Sitzungsber Oesterr Akad Wiss Math Naturwiss Kl Abt I 186:301–371
Lieth H (1975) Primary production of the major vegetation units of the world. In: Lieth H, Whittaker RH (eds) Primary productivity of the biosphere. Ecol Stud Vol 14. Springer, Berlin Heidelberg New York, pp 203–215
Loomis RS, Gerakis PA (1975) Productivity of agricultural ecosystems. In: Cooper JP (ed) Photosynthesis and productivity in different environments. Univ Press, Cambridge IBP 3:145–172
Loveless AR (1961) A nutritional interpretation of sclerophylly based on differences in the chemical composition of sclerophyllous and mesophytic leaves, Ann Bot (London) 25:168–184
Marchant PJ, Chabot BF (1978) Winter water relations of tree-line plant species on Mt Washington, New Hampshire. Arct Alp Res 10:105–116
Marshall C, Sagar GR (1968) The interdependence of tillers in *Lolium multiflorum* Lam – a quantitative assessment. J Exp Bot 19:785–794
Maruyama K (1977) Beech forests in the Naeba Mountains. Part I. Comparison of forest structure, biomass and net productivity between the upper and lower parts of beech forest zone. In: Shidei T, Kira T (eds) Primary productivity of Japanese forests. JIBP Synthesis Vol 16. Univ Press, Tokyo, pp 186–201
McCown B (1978) The interaction of organic nutrients, soil nitrogen, and soil temperature and plant growth and survival in the arctic environment. In: Tieszen LL (ed) Vegetation and production ecology of an Alaskan arctic tundra. Ecol Stud Vol 29. Springer, Berlin Heidelberg New York, pp 435–456
McCree KJ, Troughton JH (1966) Non-existence of an optimum leaf area index for the production rate of white clover grown under constant conditions. Plant Physiol 41:1615–1622
Meusel H (1978) Wuchsform und ökogeographisches Verhalten von *Bupleurum spinosum* Gouan im Vergleich mit einigen nahe verwandten Arten. Bot Jahrb Syst 99:222–248
Miller PC, Bradbury DE, Hajek E, La Marche V, Thrower NJ (1977) Past and present environment. In: Mooney HA (ed) Convergent evolution in Chile and California. Mediterranean climate ecosystems, US/IBP Synthesis 5. Dowden, Hutchinson and Ross, Stroudsburg, pp 27–72
Miller PC, Poole DK (1979) Patterns of water use by shrubs in Southern California. For Sci 25:84–98
Mitscherlich G, Künstle E (1970) Untersuchungen über die Bodentemperatur in einigen Nadel- und Laubholzbeständen in der Nähe von Freiburg/Br. Allg Forst Jagdztg 141:129–133
Mitscherlich G, Moll W, Künstle E, Maurer P (1965) Ertragskundlich-ökologische Untersuchungen im Rein- und Mischbestand. II. Wind, Globalstrahlung und Bestandeshelligkeit. Allg Forst Jagdztg 136:149–257
Mitscherlich G, Moll W, Künstle E, Maurer P (1966) Ertragskundlich-ökologische Untersuchungen im Rein- und Mischbestand. IV. Niederschlag, Stammablauf und Bodenfeuchtigkeit. Allg Forst Jagdztg 137:1–12
Monsi M (1960) Dry-matter reproduction in plants. I. Schemata of dry-matter reproduction. Bot Mag 73:81–90

Monsi M (1968) Mathematical models of plant communities. In: Eckardt FE (ed) Functioning of terrestrial ecosystems at the primary production level. UNESCO, Paris, pp 131–149

Monsi M, Murata Y (1970) Development of photosynthetic systems as influenced by distribution of matter. In: Prediction and measurement of photosynthetic productivity. PUDOC, Wageningen, pp 115–130

Mooney HA (1963) Physiological ecology of coastal, subalpine and alpine populations of *Polygonum bistortoides*. Ecology 44:813–816

Mooney HA, Billings WD (1965) Effects of altitude on carbohydrate content of mountain plants. Ecology 46:750–751

Mooney HA, Dunn EL (1970) Photosynthetic systems of mediterranean climate shrubs and trees of California and Chile. Am Nat 104:447–453

Mooney HA, Ehleringer JR (1978) The carbon gain benefits of solar tracking in a desert annual. Plant Cell Environ 1:307–312

Mooney HA, Ehleringer JR, Berry JA (1976) High photosynthetic capacity of a winter annual in Death Valley. Science 194:322–324

Mooney HA, Ferrar PJ, Slatyer RO (1978) Photosynthetic capacity and carbon allocation patterns in diverse growth forms of *Eucalyptus*. Oecologia 36:103–111

Morilla CA, Boyer JS, Hageman RH (1973) Nitrate reductase activity and polyribosomal content of corn (*Zea mays* L.) having low leaf water potentials. Plant Physiol 51:817–824

Morrow PA, Mooney HA (1974) Drought adaptations in two Californian evergreen sclerophylls. Oecologia 15:205–222

Moser W (1973) Licht, Temperatur und Photosynthese an der Station „Hoher Nebelkogel" (3184 m). In: Ellenberg H (ed) Ökosystemforschung. Springer, Berlin Heidelberg New York, pp 203–223

Moser W, Brzoska W, Zachhuber K, Larcher W (1977) Ergebnisse des IBP-Projektes „Hoher Nebelkogel 3184m". Sitzungsber Oesterr Akad Wiss Math Naturwiss Kl Abt I 186:387–419

Müller D, Nielsen J (1965) Production brute, pertes par respiration et production nette dans la forêt ombrophile tropicale. Forstl Forsoegsvaes Dan 29:69–160

Nicholson JE (1973) Growth stress differences in Eucalypts. For Sci 19:169–174

Ogawa H, Yoda K, Kira T, Ogino K, Shidei T, Ratanawongse D, Apasutaya C (1965a) Comparative ecological studies on three main types of forest vegetation in Thailand. I. Structure and floristic composition, Nature Life SE Asia (Kyoto) 4:13–48

Ogawa H, Yoda K, Ogino K, Kira T (1965b) Ibid. II. Plant biomass. Ibid 4:49–80

Ogino K (1977) A beech forest at Ashiu – biomass, its increment and net production. In: Shidei T, Kira T (eds) Primary productivity of Japanese forests. JIBP Synthesis, Vol 16. Univ Press, Tokyo, pp 172–186

Ogino K, Ratanawongse D, Tsutsumi T, Shidei T (1967) The primary production of tropical forest in Thailand. SE Asian Stud (Kyoto) 5:121–154

Orshan G (1973) Morphological and physiological plasticity in relation to drought. In: McKell CM, Blaisdell JP, Goodin JR (eds) Wildland shrubs – their biology and utilization, USDA forest service. Gen Tech Rep INT-1:245–259

Packham JR, Willis AJ (1977) The effects of shading on *Oxalis acetosella*. J Ecol 65:619–642

Payton IJ, Brasch DJ (1978) Growth and nonstructural carbohydrate reserves in *Chinochloa rigida* and *C. macra*, and their short-term response to fire. NZ J Bot 16:435–460

Penning de Vries FW (1975) Use of assimilates in higher plants. In: Cooper JP (ed) Photosynthesis and productivity in different environments. IBP 3. Cambridge London New York Melbourne, pp 459–480

Pisek A, Larcher W, Unterholzner R (1967) Kardinale Temperaturbereiche der Photosynthese und Grenztemperaturen des Lebens der Blätter verschiedener Spermatophyten. I. Temperaturminimum der Nettoassimilation, Gefrier- und Frostschadensbereiche der Blätter. Flora (Jena) Abt B 157:239–264

Pons TL (1977) An ecophysiological study in the field layer of ash coppice. III. Influence of diminishing light intensity during growth on *Geum urbanum* and *Cirsium palustre*. Acta Bot Neerl 26:251–263

Poole DK, Miller PC (1975) Water relations of selected species of chaparral and coastal sage communities. Ecology 56:1118–1128
Poole DK, Miller PC (1978) Water related characteristics of some evergreen sclerophyll shrubs in central Chile. Oecol Plant 13:289–299
Pressland AJ (1976) Soil moisture redistribution as affected by throughfall and stemflow in an arid zone shrub community. Aust J Bot 24:641–649
Raunkiaer C (1904) Biological types with reference to the adaptation of plants to survive the unfavourable season. In: Egerton FN (ed) History of ecology, life forms of plants and statistical plant geography. Arno Press, New York, Reprint 1977
Redmann RE (1978) Plant and soil water potentials following fire in a northern mixed grassland. J Range Manage 3:443–445
Richards PW (1952) The tropical rainforest. An ecological study. Univ Press, Cambridge
Ridder De N, Seligman NG, Keulen Van H (1981) Analysis of environmental and species effects on the magnitude of biomass investment in the reproductive effort of annual pasture plants. Oecologia (in press)
Rietz Du GE (1931) Life-forms of terrestrial flowering plants. Acta Phytogeogr Suec 3:1–95
Rodin LE (1976) Primary productivity of desert communities in North Africa and Asia. Probl Desert Dev 3–4:55–65 Ashkhabad
Rodin LE (1979) Productivity of desert communities in central Asia. In: Perry RA, Goodall DW (eds) Arid-land ecosystems: structure, functioning and management Vol I. Univ Press, Cambridge, pp 273–298
Rodin LE, Bazilivich NJ, Rozov HN (1975) Productivity of the world's main ecosystems. In: Reichle DE, Franklin JF, Goodall DW (eds) Productivity of world ecosystems. Natl Acad Sci, Washington, pp 13–26
Rook DA, Whyte AGD (1976) Partial defoliation and growth of 5-year-old *Radiata* pine. NZ J For Sci 6:40–56
Ross PJ, Henzell EF, Ross DR (1972) Effects of nitrogen and light in grass-legume pastures – a systems analysis approach. J Appl Ecol 9:535–556
Rychnovská M (1978) Energy flow and relevant processes in a meadow ecosystem. In: Sen DN (ed) Environmental physiology and ecology of plants. Bishen Singh Mahendra Pal Singh, Dehra Dun, pp 315–322
Rychnovská M (1979) Ecosystem synthesis of meadows. Energy flow. In: Coupland RT (ed) Grassland ecosystems of the world: analysis of grasslands and their uses. IBP 18. Univ Press, Cambridge, pp 165–169
Ryle GJA, Powell CE (1975) Defoliation and regrowth in the graminaceous plant: The role of current assimilate. Ann Bot (London) 39:297–310
Saeki T (1960) Interrelationship between leaf amount, light distrubition and total photosynthesis in a plant community. Bot Mag 73:55–63
Satoo T (1956) Materials for the study of growth in stands. III. Amount of leaves and production of wood in an aspen (*Populus davidiana*) second growth in Hokkaido. Bull Tokyo Univ For 51:33–51
Satoo T (1968) Materials for the studies of growth in stands. 7. Primary production and distribution of produced dry matter in a plantation of *Cinnamomum camphora*. Bull Tokyo Univ For 64:241–275
Satoo T (1970a) Primary production in a plantation of Japanese larch, *Larix leptolepis*: A summarized report of JPTE-66 KOIWAI. J Jpn For Soc 52:154–158
Satoo T (1970b) A synthesis of studies by the harvest method: Primary production relations in the temperate deciduous forests of Japan. In: Reichle DE (ed) Analysis of temperate forest ecosystems. Ecol Stud Vol I. Springer, Berlin Heidelberg New York, pp 55–72
Satoo T (1974a) Primary production relations in a natural forest of *Betula maximowicziana* in Hokkaido: materials for the studies of growth in forest stands. 1. Bull Tokyo Univ For 66:109–117
Satoo T (1974b) Primary production relations of a young stand of *Metasequoia glyptostroboides* planted in Tokyo: materials for the studies of growth in forest stands. 13. Bull Tokyo Univ For 66:153–164

Satoo T (1974c) Primary production relations in plantations of *Thujopsis dolabrata* in the Noto Peninsula: materials for the studies of growth in forest stands. 12. Bull Tokyo Univ For 66:139–151
Schäfer R (1973) Microbial activity under seasonal conditions of drought in mediterranean climates. In: DiCastri F, Mooney HA (eds) Mediterranean type ecosystems. Ecol Stud Vol 7. Springer, Berlin Heidelberg New York, pp 191–198
Schimper AFW (1898) Pflanzen-Geographie auf physiologischer Grundlage. Fischer, Jena
Schmidt L (1977) Phytomassevorrat und Nettoprimärproduktivität alpiner Zwergstrauchbestände. Oecol Plant 12:195–213
Schmithüsen J (1968) Allgemeine Vegetationsgeographie, 3rd edn. De Gruyter, Berlin
Schulze E-D (1970) Der CO_2-Gaswechsel der Buche (*Fagus silvatica* L.) in Abhängigkeit von den Klimafaktoren im Freiland. Flora 159:177–232
Schulze E-D, Fuchs MI, Fuchs M (1977a) Spacial distribution of photosynthetic capacity and performance in a montane spruce forest of Northern Germany. I. Biomass distribution and daily CO_2 uptake in different crown layers. Oecologia 29:43–61
Schulze E-D, Fuchs M, Fuchs MI (1977b) Spacial distribution of photosynthetic capacity and performance in a mountain spruce forest of Northern Germany. III. The significance of the evergreen habit. Oecologia 30:239–248
Schulze E-D, Hall AE, Lange OL, Evenari M, Kappen L, Buschbom U (1980) Long-term effects of drought on wild and cultivated plants in the Negev Desert. I. Maximal rates of net photosynthesis. Oecologia 45:11–18
Schwarz W (1970) Der Einfluß der Photoperiode auf das Austreiben, die Frosthärte und die Hitzeresistenz von Zirbe und Alpenrose. Flora 159:258–285
Shackel KA, Hall AE (1979) Reversible leaflet movements in relation to drought adaptation of cowpeas, *Vigna unguiculata* (L.) Walp. Aust J Plant Physiol 6:265–276
Shields LM (1950) Leaf xeromorphy as related to physiological and structural influences. Bot Rev 16:299–447
Sims PL, Coupland RT (1979) Producers. In: Coupland RT (ed) Grassland ecosystems of the world: analysis of grasslands and their uses. IBP 18. Univ Press, Cambridge, pp 49–72
Sims PL, Singh JS (1978) The structure and function of ten western North American grasslands. IV. Compartmental transfers and energy flow within the ecosystem. J Ecol 466:983–1010
Singh JS, Joshi MC (1979) Primary production. In: Coupland RT (ed) Grassland ecosystems of the world: analysis of grasslands and their use. IBP 18. Univ Press, Cambridge, pp 187–217
Singh KP, Gopal B (1973) The effects of photoperiod and light intensity on the growth of some weeds of crop fields. In: Slatyer RO (ed) Plant responses to climatic factors. Ecology and conservation 5. UNESCO, Paris, pp 77–85
Singh JS, Yadava PS (1974) Seasonal variation in composition, plant biomass and net primary productivity of a tropical grassland at Kwukshetra, India. Ecol Monogr 44:351–375
Slatyer RO (1961) Internal water balance of *Acacia aneura* F. Muell. in relation to environmental conditions. UNESCO Arid Zone Res 16:137–146
Small E (1972a) Water relations of plants in raised *Sphagnum* peat bogs. Ecology 53:726–728
Small E (1972b) Photosynthetic rates in relation to nitrogen recycling as an adaptation to nutrient deficiency in peat bog plants. Can J Bot 50:2227–2233
Small E (1973) Xeromorphy in plants as a possible basis for migration betweeen arid and nutritionally deficient environments. Bot Not 126:534–539
Smith EM, Hadley EB (1974) Photosynthetic and respiratory acclimation to temperature in *Ledum groenlandicum* populations. Arct Alp Res 6:13–27
Sørensen T (1941) Temperature relations and phenology of the northeast Greenland flowering plants. Medd Groenl 125:1–305
Stanhill G (1970) The water flux in temperate forests: Precipitation and evapotranspiration. In: Reichle DE (ed) Analysis of temperate forest ecosystems. Ecol Stud Vol I. Springer, Berlin Heidelberg New York, pp 247–256

Stearns SC (1976) Life-history tactics: A review of the ideas. Q Rev Biol 51:3–47
Stocker O (1971) Der Wasser- und Photosynthesehaushalt von Wüstenpflanzen der mauretanischen Sahara. II. Wechselgrüne, Rutenzweige und stammsukkulente Bäume. Flora 160:1–43
Stocker O (1976) The water-photosynthesis syndrome and the geographical plant distribution in the Saharan desert. In: Lange OL, Kappen L, Schulze E-D (eds) Water and plant life. Ecol Stud, Vol 19. Springer, Berlin Heidelberg New York, pp 506–522
Summers CF (1968) Production in montane dwarf shrub communities. In: Heal OW, Perkins DF (eds) Production ecology of British moors and montane grasslands. Ecol Stud Vol 27. Springer, Berlin Heidelberg New York, pp 263–276
Swan LW (1967) Alpine and aeolian regions of the world. In: Wright HE Jr, Osborn WH (eds) Arctic and alpine environments. Indiana Univ Press, Blommington, pp 29–54
Szarek SR, Woodhouse RM (1977) Ecophysiological studies of Sonoran Desert plants. II. Seasonal photosynthesis patterns and primary production of *Ambrosia deltoidea* and *Olneya tesota*. Oecologia 28:365–375
Szujkó-Lacza J, Fekete G (1969) A survey of the plant-form systems and the respective research approaches I. Ann Hist Nat Mus Natl Hung 62:129–139
Tadaki Y (1968) Studies on the production structure of forest (XIV). The third report on the primary production of a young stand of *Castanopsis cuspidata*. J Jpn For Soc 50:60–65
Tadaki Y, Hatiya K, Tochiaki K, Muyauchi H, Matsuda U (1970) Studies on the production structure of forest (XVI). Primary productivity of *Abies veitchii* forests in the subalpine zone of Mt Fuji. Bull Gov For Exp Stn 229:1–22
Tieszen LL (1978) Vegetation and production ecology of an Alaskan arctic tundra. Ecol Stud Vol 29. Springer, Berlin Heidelberg New York
Troll C (1956) Das Wasser als pflanzengeographischer Faktor. In: Ruhland W (ed) Handbuch der Pflanzenphysiol Vol 3, pp 750–786
Turk KJ, Hall AE (1980) Drought adaptations of cowpeas. II. Influence of soil water deficits and evaporative demand on plant water status and relation with seed yield. Agron J (in press)
Turner NC, Begg JE (1977) Response of pasture plants to water deficits. In: Wilson JR (ed) Plant relations in pastures. CSIRO, Melbourne, pp 50–66
Turner NC, Jones MM (1980) Turgor maintenance by osmotic adjustment: A review and evaluation. In: Turner NC, Kramer PJ (eds) Adaptation of plants to water and high temperature stress. Wiley-Interscience, New York London, pp 87–104
Vasek FC (1980) Creosote bush: long-lived clones in the Mojave desert. Am J Bot 67:246–255
Walter H (1968) Die Vegetation der Erde. II. Die gemäßigten und arktischen Zonen. Fischer, Stuttgart
Walter H (1973) Die Vegetation der Erde. I. Tropische und subtropische Zonen, 3rd edn. Fischer, Jena Stuttgart
Walter H (1979) Vegetation und Klimazonen, 4th edn. UTB 14. Ulmer, Stuttgart
Walter H (1981) Über Höchstwerte der Produktion von natürlichen Pflanzenbeständen in NO Asien. Vegetatio 44:37–41
Waring RH, Franklin JF (1979) Evergreen coniferous forests of the Pacific Northwest. Science 204:1380–1386
Waring RH, Emmingham WH, Gholz HL, Grier CC (1978) Variation in maximum leaf area of coniferous forests in Oregon and its ecological significance. For Sci 24:131–140
Waring RH, Whitehead D, Jarvis PG (1979) The contribution of stored water to transpiration in Scots Pine. Plant Cell Environ 2:309–318
Warming E (1884) Über perenne Gewächse. Bot Centralbl 18:184–188
Warming E (1895) Plantesamfund. Grundtrack af den økologiske plantegeografi. PG Philipsen, Kjøbenhavn
Warming E (1909) Ecology of plants. An introduction to the study of plant communities. Clarendon, Oxford
Warming E (1923) Økologiens Grundformer. Udkast til en systematisk ordning Rackke IV, 2 København. K Vidensk Selsk Skr Naturvidensk Math Afd 8

Watanabe I (1976) Transformation factor from CO_2 net assimilation to dry matter in crop plants. Jpn Agric Res Q 10:114–118
Watson DJ (1958) The dependence of net assimilation rate on leaf area index. Ann Bot NS 22:27–54
Webb W, Szarek S, Lauenroth W, Kinerson R, Smith M (1978) Primary productivity and water use in native forest, grassland, and desert ecosystems. Ecology 59:1239–1247
Whittaker RH (1973) Handbook of vegetation science. Ordination and classification of communities. Vol V. Junk, The Hague
Whittaker RH (1975) Communities and ecosystems, 2nd edn. MacMillan, New York
Whittaker RH, Niering WA (1975) Vegetation of the Santa Catalina Mountains, Arizona. V. Biomass, production, and diversity along the elevation gradient. Ecology 56:771–790
Wielgolaski FE (1975) Primary production of tundra. In Cooper JP (ed) Photosynthesis and productivity in different environments. IBP 3. Univ Press, Cambridge, pp 75–106
Wielgolaski FE, Kärenlampi L (1975) Plant phenology of Fennoscandian tundra areas. In: Wielgolaski FE (ed) Fennoscandian tundra ecosystems. I. Plants and microorganisms. Ecol Stud Vol. 16. Springer, Berlin Heidelberg New York, 94–102
Wielgolaski FE, Kjelvik S, Kallio P (1975) Mineral content of tundra and forest tundra plants in Fennoscandia. In: Wielgolaski FE (ed) Fennoscandian tundra ecosystems. I. Plant and microorganisms. Ecol Stud Vol 16. Springer, Berlin Heidelberg New York, pp 316–332
Williams RD (1964) Assimilation and translocation in perennial grasses. Ann Bot NS 28:419–425
Wit CT De (1958) Transpiration and crop yields. Versl Landbouwkd 2 onderz Agr Res Rep 64. 6. PUDOC, Wageningen
Wit CT De (1968) Plant production. Misc Pap Landbouwhogesch Wageningen 3:25–50
Zohary M (1961) On the hydro-ecological relations of the Near Eastern desert vegetation. In: Plant water relationships in arid and semi-arid conditions. UNESCO Arid Zone Res 16:199–212
Zwölfer H (1978) Mechanismen und Ergebnisse der Co-Evolution von phytophagen und entomophagen Insekten und höheren Pflanzen. Sonderbd Naturwiss Ver Hamburg 2:7–50
Zwölfer H (1980) Distelblütenköpfe als ökologische Kleinsysteme: Konkurrenz und Koexistenz in Phytophagenkomplexen. Mitt Dtsch Entoml Ges 2:21–37

Author Index

Page numbers in *italics* refer to the references

Taxonomic Index

Page numbers in **bold face** refer to tables or figures

Subject Index

Page numbers in **bold face** refer to tables or figures

MIX
Papier aus verantwortungsvollen Quellen
Paper from responsible sources
FSC® C105338

If you have any concerns about our products,
you can contact us on
ProductSafety@springernature.com

In case Publisher is established outside the EU,
the EU authorized representative is:
Springer Nature Customer Service Center GmbH
Europaplatz 3, 69115 Heidelberg, Germany

Printed by Libri Plureos GmbH
in Hamburg, Germany